Geography of the Physical Environment

The *Geography of the Physical Environment* book series provides a platform for scientific contributions in the field of Physical Geography and its subdisciplines. It publishes a broad portfolio of scientific books covering case studies, theoretical and applied approaches as well as novel developments and techniques in the field. The scope is not limited to a certain spatial scale and can cover local and regional to continental and global facets. Books with strong regional focus should be well illustrated including significant maps and meaningful figures to be potentially used as field guides and standard references for the respective area.

The series appeals to scientists and students in the field of geography as well as regional scientists, landscape planners, policy makers, and everyone interested in wide-ranging aspects of modern Physical Geography. Peer-reviewed research monographs, edited volumes, advance and undergraduate level textbooks, and conference proceedings covering the major topics in Physical Geography are included in the series. Submissions to the Book Series are also invited on the theme 'The Physical Geography of…', with a relevant subtitle of the author's/editor's choice. Please contact the Publisher for further information and to receive a Book Proposal Form.

More information about this series at http://www.springer.com/series/15117

Gouri Sankar Bhunia • Uday Chatterjee
K.C. Lalmalsawmzauva • Pravat Kumar Shit
Editors

Anthropogeomorphology

A Geospatial Technology Based Approach

Springer

Editors
Gouri Sankar Bhunia
Department of Geography
Visiting Faculty, Seacom Skills University
West Bengal, West Bengal, India

Uday Chatterjee
Department of Geography
Bhatter College, Dantan
Paschim Medinipore, India

K.C. Lalmalsawmzauva
Geography & Resource Management
Mizoram University
Aizawl, India

Pravat Kumar Shit
PG Department of Geography
Raja N. L. Khan Women's College
(Autonomous)
Midnapore, West Bengal, India

ISSN 2366-8865 ISSN 2366-8873 (electronic)
Geography of the Physical Environment
ISBN 978-3-030-77571-1 ISBN 978-3-030-77572-8 (eBook)
https://doi.org/10.1007/978-3-030-77572-8

This Springer imprint is published by the registered company Springer Nature Switzerland AG
The registered company address is: Gewerbestrasse 11, 6330 Cham, Switzerland

Dedicated to
Young Scholars in the Field of Applied Geomorphology and Geosciences

Foreword

Population and development have walked hand in hand since the dawn of civilization. Before the appearance of *Homo sapiens* on Earth, the purely natural system ruled our planet. Many geophysical events such as earthquakes, volcanic eruptions, landslides, and floods threatened only the prevailing flora and fauna. After the appearance of *Homo sapiens*, human society has evolved to such an extent that human beings, at present, have a great impact on the physical environment of Earth. Some scientists believe that the latest geological epoch, labelled Anthropocene or the Age of Man, began when man's activities set about impacting the planet's physical environment, which will endure in the geological record long after human civilization has perished.

To ease their existence, humans have also disturbed the spontaneous rhythm of nature, overexploiting resources to make life more comfortable. In the one-million-year-old history of man's arrival on this planet, humans have altered the Earth in just the past century or two, kicking off this all-new epoch, which has transformed geophysical events into natural disasters. Thus, the theme of this edited book is very much relevant to the present scenario.

There is constant confusion about the name of the subject, whether it should be anthropogeomorphology or anthropogenic geomorphology. Many discussions and deliberations have been going on, but geoscientists of the world are yet to draw a conclusive decision. Thus, use of the term anthropogeomorphology can be continued until scientists reach a unanimous decision.

The present book, entitled *Anthropogeomorphology: A Geospatial Technology-Based Approach*, edited by Gouri Sankar Bhunia, Uday Chatterjee, KC Lalmalsawmzauva, and Pravat Kumar Shit, is a worthy contribution to earth sciences because human activities change the Earth's surface continuously and are responsible for altering geomorphological processes and geomorphological hazards. Addressing such an important issue through geospatial technology has given an additional flavour to the subject.

The book is well designed, organized into 32 chapters and written by different authors covering almost all features of geomorphology, mapping, techniques, and management issues. I believe that the book will attract wide attention from the scientific community, including geomorphologists, environmentalists, environmental activists, and planners.

Vice President, International Association of Geomorphologists (IAG), Department of Geography, North-Eastern Hill University, Shillong, Meghalaya, India
26.01.2021

Sunil Kumar De

Preface

The intensity of involvement has been proportional to the size of the human population to its demands upon the environment and to the level of technological progress achieved to satisfy growing demands. Since the 1970s in the research of the physical environment regularly interweaving trends between natural environment and human economic intervention are protuberant. In the last two decades, concern with global environmental change has brought the role of anthropogeomorphology into sharper focus. Global warming has important implications for many geomorphological processes and phenomena as a result of the direct effects of warming, as a consequence of other related climatic changes. Anthropogeomorphology concerns about the number of landforms associations of extreme variability depending on the given way and goal of their formation, which have been completed by the man-induced activity. However, the scope of anthropogeomorphology does not only include the study of man-made landforms but also the investigation of man-induced surface changes, the prophecy of consequences of upset natural stabilities along with the preparation of proposals with the purpose of impede detrimental effects. Therefore, anthropogeomorphology can be regarded as a discipline that supports to explain both socio-economic as well as environmental and natural protection issues. This book provides an innovative approach and exercise to explore our physical environment with the growing demands from society against geography due to scientific-technical rebellion, to encourage competently the rational exploitation of natural resources and capacities, and to accomplish an environmental management sustaining social condition and prospects.

The organization of the books follows a general introduction of aims and scope of this discipline, and the individual chapters focus on the various sectors of human activity. Generally, the following fields of anthropogeomorphology have been considered, namely mountain, coast, riverine, forest, urban, agricultural, hydrometeorology, industry, mining, and tourism. Finally, qualitative and quantitative summary of the human impact on Earth's surface has been described. The fieldwork is combined with references and the comparison of the present-day situation with the achievement.

West Bengal, India — Gouri Sankar Bhunia
Paschim Medinipore, India — Uday Chatterjee
Aizawl, India — K.C. Lalmalsawmzauva
West Bengal, India — Pravat Kumar Shit

Acknowledgments

The preparation of this book has been guided by several geomorphologic pioneers. We are obliged to these experts for providing their time to evaluate the chapters published in this book. We thank the anonymous reviewers for their constructive comments that led to substantial improvement to the quality of this book. Because this book was a long time in the making, preparation of this book became a tedious and time-consuming job for the editors. We want to thank our family and friends for their continued support. This work would not have been possible without constant inspiration from our students, knowledge from our teachers, enthusiasm from our colleagues and collaborators, and support from our family. Finally, we also thank our publisher, Springer, and its publishing editor for their continuous support in the publication of this book.

Disclaimer

The authors of individual chapters are solely responsible for the ideas, views, data, figures, and geographical boundaries presented in the respective chapters of this book, and these have not been endorsed, in any form, by the publisher, the editor, and the authors of forewords, preambles, or other chapters.

Contents

Contributors

Zulfequar Ahmad Department of Civil Engineering, IIT Roorkee, Roorkee, India

T. Akhil National Centre for Earth Science Studies, Thiruvananthapuram, India

Wajahat Annayat Department of Civil Engineering, National Institute of Technology, Silchar, Assam, India

Kumar Ashwini Department of Civil Engineering, National Institute of Technology, Silchar, Assam, India

Arnab Banerjee U.T.D, Department of Environmental Science, Sarguja Viswavidyalaya, Ambikapur, Chattisgarh, India

Shiboram Banerjee Department of Conservation Biology, Durgapur Government College, Durgapur, West Bengal, India

Anindya Basu Department of Geography, Diamond Harbour Women's University, Sarisha, West Bengal, India

Sukla Basu Department of Rural Studies, West Bengal State University, Barasat, West Bengal, India

Suman Bera Department of Geography and Environment Management, Vidyasagar University, Midnapore, West Bengal, India

Adrija Bhattacharjee Department of Geography, Diamond Harbour Women's University, Sarisha, West Bengal, India

Soumik Bhattacharya Aerospatial Solution Pvt. Ltd., Thane, Maharashtra, India

Gouri Sankar Bhunia Randstad India Private Limited, New Delhi, India
Bihar Remote Sensing Application Centre, Bihar, India

Uday Chatterjee Department of Geography, Bhatter College, Dantan (Vidyasagar University), Paschim Medinipur, West Bengal, India

Aksha Chowdhary Department of Geography (Geoinformatics), Savitribai Phule Pune University (SPPU), Pune, Maharashtra, India

Pinakesh Das Department of Soil and Water Conservation and AICRP on Agroforestry, Bidhan Chandra Krishi Viswavidyalaya, Haringhata, West Bengal, India

Swarupa Das Department of Geography, Barjora College, Barjora, West Bengal, India

Anusree Dutta Department of Geography, Purwanchal Vidyamandir, Kolkata, India

Karishma Dutta Department of Geography, School of Human and Environmental Sciences, North Eastern Hill University, Shillong, Meghalaya, India

RupamKumar Dutta Department of Geography, Kultali Dr. B.R. Ambedkar College (University of Calcutta), Parganas, West Bengal, India

P.K. Garg Department of Civil Engineering, IIT Roorkee, Roorkee, India

MintuE. George Cochin University of Science and Technology, Ernakulam, India

Pranamee Gogoi Department of Geography, Rajiv Gandhi University, Itanagar, Arunachal Pradesh, India

Jayanta Gour Department of Geography, Sambhu Nath College, Birbhum, India

Anwesha Haldar Department of Geography, East Calcutta Girls' College, Kolkata, India

K. Justin George Agriculture & Soils Department, Indian Institute of Remote Sensing, Dehradun, India

KishorChandra Kandpal Uttarakhand Space Application Centre, Dehradun, India

Jayashree Karmakar Bankura Zilla Saradamani Mahila Mahavidyapith, Bankura, West Bengal, India

Nuruzzaman Kasemi Department of Geography, Raiganj University, Raiganj, India

N. Kasongi Department of Geography and Environmental Studies, University of Dodoma, Dodoma, Tanzania

Kaushik Kharghoria Department of Geography, School of Human and Environmental Sciences, North Eastern Hill University, Shillong, Meghalaya, India

Nasira Khatun Department of Geography, Women's College, Kolkata, India

Bappaditya Koley Department of Geography, Bakim Sardar College, Paraganas, West Bengal, India

Suresh Kumar Agriculture & Soils Department, Indian Institute of Remote Sensing, Dehradun, India

K.C. Lalmalsawmzauva Department of Geography & Resource Management, Mizoram University, Aizawl, India

Ramkrishna Maiti Department of Geography and Environment Management, Vidyasagar University, Midnapore, West Bengal, India

Sima Majumdar Department of Education, Government of Tripura, Agartala, India

Sushobhan Majumdar Department of Geography, Jadavpur University, Kolkata, India

Saikat Mandal Department of Zoology, Raghunathpur College, Raghunathpur, West Bengal, India

Durdanah Mattoo Department of Geography and Regional Development, University of Kashmir, Srinagar, Jammu and Kashmir, India

Swati Mollah Dumkal College, Murshidabad, West Bengal, India

Tahir Muntazari Department of Civil Engineering, National Institute of Technology, Srinagar, Jammu and Kashmir, India

Gardinia Nongbri Department of Environmental Studies, North-Eastern Hill University, Shillong, India

Adikanda Ojha GIS Analyst, ICZMP-SPMU, Bhubaneswar, Odisha, India

Debnath Palit Department of Botany, Durgapur Government College, Durgapur, West Bengal, India

Papia Nandy Palit Department of Geography, Barjora College, Barjora, West Bengal, India

Subhabrata Panda Department of Soil and Water Conservation and AICRP on Agroforestry, Bidhan Chandra Krishi Viswavidyalaya, Haringhata, West Bengal, India

S. Panneerselvam Water Technology Centre, Tamil Nadu Agricultural University, Coimbatore, Tamil Nadu, India

SantanuK. Patnaik Department of Geography, Rajiv Gandhi University, Itanagar, Arunachal Pradesh, India

Sellaperumal Pazhanivelan Department of Remote Sensing and Geographical Information System, Tamil Nadu Agricultural University, Coimbatore, Tamil Nadu, India

T.K. Prasad Kannur University, Kannur, India

MdFazla Rabby Department of Geography and Environment, Jahangirnagar University, Savar Dhaka, Bangladesh

M.K. Rafeeque University of Kerala, Thiruvananthapuram, India

Nusrat Rafique Department of Geography and Regional Development, University of Kashmir, Srinagar, Jammu and Kashmir, India

MdShahedur Rashid Department of Geography and Environment, Jahangirnagar University, Savar Dhaka, Bangladesh

Neelam Rawat Uttarakhand Space Application Centre, Dehradun, India

Rezaul Roni Department of Geography and Environment, Jahangirnagar University, Savar Dhaka, Bangladesh

Jajnaseni Rout Remote Sensing and GIS Consultant, Bhubaneswar, Odisha, India

AbiraDutta Roy Bankura Zilla Saradamani Mahila Mahavidyapith, Bankura, West Bengal, India

Bhaswati Roy Department of Geography, Raiganj University, Raiganj, India

KrishnaRao Sadasivuni Department of History and Archaeology, University of Dodoma, Dodoma, Tanzania

Venkadesh Samykannu Agro Climate Research Centre, Tamil Nadu Agricultural University, Coimbatore, Tamil Nadu, India

Ranajit Sardar Acharya Prafulla Chandra College, New Barrackpore, West Bengal, India

Lakshminarayan Satpati Department of Geography, UGC-HRDC, University of Calcutta, Kolkata, India

Pravat Kumar Shit Department of Geography, Raja N. L. Khan Women's College (Autonomous), Midnapore, West Bengal, India

BaiaroihunWar Shymbin Department of Geography, Mawsynram Border Area College, Mawsynram, India

BritiSundar Sil Department of Civil Engineering, National Institute of Technology, Silchar, Assam, India

Mst.Rebeka Sultana Department of Geography and Environmental Studies, University of Rajshahi, Rajshahi, Bangladesh

D.S. Suresh Babu National Centre for Earth Science Studies, Thiruvananthapuram, India

E.L. Temu Ministry of Culture, Dodoma, Tanzania

RowshonAra Toma Department of Geography and Environment, Jahangirnagar University, Savar Dhaka, Bangladesh

Ch. Udaya Bhaskara Rao Department of Geography and Resource Management, Mizoram University, Aizawl, India

Swapnil Vyas Department of Geography (Geoinformatics), Savitribai Phule Pune University (SPPU), Pune, Maharashtra, India

ShoibB. Wani Civil Engineering Department, B.S. Abdur Rahman University, Chennai, Tamil Nadu, India

Mohammad Zakwan Department of Civil Engineering, MANUU, Hyderabad, India

About the Editors

Gouri Sankar Bhunia is working at Randstad India Private Limited, New Delhi, India. He received a PhD from the Department of Geography, University of Calcutta, India, in 2015. He also worked as a visiting faculty in a private university (Seacom Skills University) in West Bengal, India. His PhD dissertation work focused on environmental control measures of infectious disease using geospatial technology. Currently, he is involved in smart city and urban planning project. He has received the Senior Research Fellowship Award from the Indian Council of Medical Research. His research interests include environmental modelling, health GIS, risk assessment, natural resources mapping and modelling, data mining, and information retrieval using geospatial technology. Dr Bhunia is associate editor and on the editorial boards of three international journals in health GIS and geosciences. Dr Bhunia has published more than 80 articles in various Scopus-indexed journals. He published four books with International Publishing Group. He is currently the editor of the GIScience and Geo-environmental Modelling (GGM) book series, Springer Nature.

Uday Chatterjee is an assistant professor in the Department of Geography, Bhatter College, Dantan, Paschim Medinipur, West Bengal, India, and an applied geographer with a postgraduate degree in applied geography from Utkal University and doctoral degree in applied geography from Ravenshaw University, Cuttack, Odisha, India. He has contributed various research papers published in various reputed national and international journals and edited book volumes. He has authored and jointly edited the book *Harmony with nature: Illusions and elusions from geographer's perspective in the 21st century*. He has also conducted (convener) one faculty development program on "Modern methods of teaching and advanced research methods" sponsored by the Indian Council of Social Science Research (ICSSR), Govt. of India. His areas of research interest cover urban planning, social and human geography, applied geomorphology, hazards and disasters, environmental issues, land use, and rural development. His research work has been funded by the West Bengal Pollution Control Board (WBPCB) Govt. of West Bengal, India. He has served as a reviewer for many International journals. Currently, Dr Chatterjee is the lead editor of the special issue (S.I) *Urbanism, Smart Cities and Modelling, Geojournal*, Springer.

K.C. Lalmalsawmzauva graduated from St. Edmund's College, Shillong, Meghalaya, and pursued MSc geography from North Eastern Hill University, and he was awarded doctorate degree (PhD) from the same University in the year 2012. He was the founder President of NEHU-Geographical Research Forum (NEHU-GRF). He joined Mizoram University in the year 2011 as Assistant Professor. To his credit, he published three books as a single author and co-authored four books. He completed projects by the Indian Council of Social Science Research (ICSSR) and involved himself in the project of the United Nations Food and Agriculture Organization (FOA) for agricultural development in the hill regions. He also Co-PI in the consultancy project of National Rural Livelihood Mission project, Government of Mizoram. He published more than 20 research papers in various journals and books. He has actively participated in national and international academic activisties. In his academic endeavor, he has visited Europe, America, and Africa.

Pravat Kumar Shit is an assistant professor in the PG Department of Geography, Raja N. L. Khan Women's College (Autonomous), West Bengal, India. He received his MSc and PhD degrees in geography from Vidyasagar University and PG diploma in remote sensing and GIS from Sambalpur University. His research interests include applied geomorphology, soil erosion, groundwater, forest resources, wetland ecosystem, environmental contaminants and pollution, and natural resources mapping and modelling. He has published 10 books (8 books with Springer) and more than 60 papers in peer-reviewed journals. He is currently the editor of the GIScience and Geo-environmental Modelling (GGM) book series, Springer Nature.

Chapter 1
An Introduction to Anthropogeomorphology and Geospatial Technology

Gouri Sankar Bhunia, Uday Chatterjee, Pravat Kumar Shit, and K. C. Lalmalsawmzauva

1.1 Introduction

Geomorphology is a scientific study of landforms, terrain and related processes that includes explanation of earth/planetary surfaces, substances, origins, nature and history. Geomorphological processes are the factors that alter the Earth, and others are the endogenetic mechanisms that reside inside the Earth; some forces placed beyond the Earth are the exogenetic processes. Moving water, wind, glaciers, karst and sea waves are powerful agents of degradation and aggradation that operate over a significant period of time and create gradual changes that lead to systematic development of landform. Landforms formed by dynamic anthropogenic processes are much less easy to identify, not smaller, but they do not require the introduction of a new path or strategies as much as the amplification of natural phenomena.

The concept of *anthropogeomorphology*, coined by Golomb and Eder (1964), is the study of the human role in forming landforms and changing the function of geomorphological processes. It therefore reflects throughout the Anthropocene on several key components of geomorphological processes. A significant aspect of the

G. S. Bhunia (✉)
Remote Sensing and GIS Expert, Randstad India Private Limited, New Delhi, Delhi, India

U. Chatterjee
Department of Geography, Bhatter College, Dantan (Vidyasagar University), Paschim Medinipur, West Bengal, India

P. K. Shit
PG Department of Geography, Raja N. L. Khan Women's College (Autonomous), Midnapore, West Bengal, India

K. C. Lalmalsawmzauva
Department of Geography & Resource Management, Mizoram University, Aizawl, Mizoram, India

G. S. Bhunia et al. (eds.), *Anthropogeomorphology*, Geography of the Physical Environment, https://doi.org/10.1007/978-3-030-77572-8_1

Anthropocene is physiographic modification, but its consequences can differ widely in space and time, and it is often underestimated in view on human interventions on the Earth's surface (Brown et al., 2013). Human behaviours impact the landscape and its natural features dramatically and cumulatively (Marsh, 1864). Geomorphological modifications arise from a variety of anthropogenic processes, namely, forest clearance, cultivation, ground drainage and filling, mining and quarrying, channelization, irrigation, and dam formation or other engineering development (Nir, 1983). Both the systematic erosion and accumulation of material and the unintended consequences of hydrological changes and the subsequent deforestation and sedimentation are implicated in these operations (Hooke, 2000).

1.2 Interaction Between Geomorphology and Human

The role of human being as a geomorphological agent is the topic of *anthropogeomorphology*. It has been presumed that the Earth's surface is a naturally occurring phenomenon capable of controlling human interaction, but that it is perhaps occasionally, if ever, strongly affected by human beings. Humans are now the most effective geomorphic entity in shaping and reforming the Earth's face, by transforming the physical climate. Exponentially the population growth has become faster, and the resources available for the demand has culminated in the widespread reworking of surface materials, which is projected to develop in the subsequent manner, at an even faster rate of population growth. In areas such as agriculture and mining, technological innovations will be designed and implemented, and growing population rates will result in more changes in land cover and in the utilization of natural resources. The early deforestation of the Neolithic slope in Central Europe, for example, may have been the most essential geomorphological process since the end of the Pleistocene, whereas in Dubai the coastline has altered during the last few centuries. Conversely, whenever significant anthropogenic modifications occurred, they have a direct global impact on the terrestrial ecosystem. Hooke et al. (2012) have estimated the terrestrial area amended by human activity in 2007 and recommend that more than 50% of the total area without ice has been altered by social existence. The Intergovernmental Panel on Climate Change (IPCC) has shown that global warming can dramatically alter biomes, contribute to major cryosphere modifications and enable sea levels to increase (IPCC, 2013). Wohl (2013) opines that geomorphologists can make a positive contribution to the management of what is now called the 'critical zone' in several contexts. This is the near-surface layer of the Earth from the tops of the trees down to the profoundest groundwater, with more and more human interactions with the surface of the Earth and the confluence of most geomorphological activity.

The dilemma of anthropogenic geomorphology is the identification of the broad and ever-expanding array of landforms on the surface, extremely varied in origin and function, generated by human activity. In a broader context, artificially generated landforms have diverse environmental effects (e.g. meso- and microclimate changes, morphology and so on) and alter natural processes. Natural environments that have been substantially changed as a result of direct management effects became anthropogenically varied systems. In geomorphology, a wide practice deals with the study of human impacts on river systems and other landscape structures (Thomas Jr., 1956). Direct anthropogenic practices, such as construction (e.g. spoil tips, embankments, sea walls), exploration (e.g. mining and quarrying), hydrological intervention (e.g. ponds and canals) and cultivation (e.g. terraces), generate numerous geomorphological features. The most common models of anthropogenic ecosystems involve agricultural fields, particularly cultivated land and pasture/grazing land and technological landscapes like urban landscapes and mining centre, etc. Slaymaker et al. (2009) pointed out the consequences of land cover modifications can be at least as significant as the changes that will be triggered by future climate change.

Nowadays, only in the sense of the impact of past human activities will the geomorphic consequences of continuing urban and suburban growth in the same communities be interpreted (Voli et al., 2013). Prolonged urbanisation is occurring globally and more quickly in the humid tropics' developing nations. The urban landscape features of LULC play a significant role in national, regional and local climate change. This accumulation will arise not only owing to variations in the patterns of susceptible and latent surface heat but also due to variation in surface albedo. The albedo is lower in an urban area in comparison to a rural region (Sailor, 2002). The lower albedo is recorded due to rooftops and asphalt roadways in the urban area. Much of the metropolitan landscape is marked by a large portion, or, if any, the proportion of urban surface coverage is disconcertingly low as a result of decreased supply of moisture. These elements of the urban landscape provide cities a much higher heat potential than natural surfaces.

Major driving forces of water, soil and air pollution are changes in land use/land cover characteristics. Mining activities can create contaminants from radioactive metals exposed in the operation. Agricultural pesticides are introduced into the soil and surface waters, including fertilizers, insecticides and pesticides, and, in some cases, persist as pollutants inside the region. Overtime, the deforestation degrades soil fertility and reduces soil suitability for potential agricultural use but also releases tremendous quantities of phosphate, nitrogen and sediments into waterways and other marine environments, with a kind of adverse effect (excessive sedimentation, turbidity, eutrophication and coastal hypoxia).

1.3 Role of Remote Sensing and GIS in Geomorphological Application

Geomorphological mapping and the study of different structures using advanced tools including remote sensing and geographic information system (GIS) serve as conceptual tools for inventory, exploration and governance of land resources and geomorphological and geological risk mitigation, along with generating baseline knowledge for other environmental research areas, such as landscape ecology, soil science, hydrology and forestry, etc. In several geomorphological research, GIS and remote sensing have been implemented to measure surface processes and land-masses. Using post classification, comparative analysis, standard image differentiation, employing image ratio, image regression and manual on-screen digitization of variations, main components evaluation and multi-temporal image classification, there are several systems designed in the earlier research work. Geospatial technology has been extensively used for classifying land type and landscape units with the continuous advancement of GIS and RS technologies, extracting those landform characteristics, quantifying process-landform interactions and defining geomorphic variations.

Punkari's (1987) study shows how imagery beyond the visible spectrum can be helpful in landform recognition. In this research, inter-drumlin regions displayed higher proportion of moisture that influenced land cover, leading to a lower reflectance in visible-near-infrared (VNIR) bands and thus better separating drumlins. In conjunction with shaded relief derived through DEMs, Jansson and Glasser (2005) found false-colour composites integrating with near infrared and thermal infrared bands to be the most effective for identifying landscape. Marchese et al. (2019) used Multispectral Instrument (MSI), Operational Land Imager (OLI), Sentinel-2A/2B and Landsat 8 satellites sensors data to test an original multichannel algorithm (normalized hotspot indices and normalized thermal index), which aims at mapping volcanic thermal anomalies at a global scale. Sensors including the Advanced Very-High-Resolution Radiometer (AVHRR) and Moderate Resolution Imaging Spectroradiometer (MODIS), which provide high temporal resolution data in the medium infrared (MIR) and thermal infrared (TIR) bands (up to 6 h in the case of AVHRR), have also been commonly used for operational monitoring of active volcanoes (Lombardo, 2016; Coppola & Cigolini, 2013). Thematic Mapper (TM) and Advanced Spaceborne Thermal Emission and Reflection Radiometer (ASTER) and image spectrometers such as Hyperion, providing medium-high spatial resolution near-infrared (NIR) and short-wave infrared (SWIR) data, are probably more useful for mapping lava flows and retrieving reliable volcanic feature information (Reath et al., 2019a, b; Davies et al., 2006). Satellite measurements of sparging, thermal variations and land breakage covering 17 years are used by Reath et al. (2019a) in 47 of the most active volcanoes in Latin America and associate these historical data with ground-based measurements documented by the Global Volcanism Program. This information offers to determine volcanic behaviour on a regional scale during, noneruptive, pre-eruptive, syneruptive and posteruptive periods. For the purpose of delineating overlapping lava flows near the summit through textural differences between flows, Smets et al. (2010) used individual SAR images and SAR pairs from ERS 1/ERS 2 (C-band), ENVISAT (C-band) and JERS (L-band: 23.5 cm) satellites. The snow has maximum reflectance accompanied by firn and ice as visible from the spectral response curve. Related reflectance of nearby rocks is observed in the debris layer on the glacier. An important remote sensing technique for distinguishing surface entities with various temperatures or emissivity is the thermal infrared (3–15 μm) (Lepparanta & Granberg, 2010).

The glacier's surface temperature is lower than that of the environment, and therefore the thermal data are being distinguished. The glacier is thermally active at a depth of just 10 m and can be detected throughout the season (Gareth & Pellikka, 2010). AVHRR, MODIS, Landsat and ASTER series are the most widely used thermal band sensors for the glaciological research. The key benefit of the microwave sensor in glacier tracking is the capability to encapsulate snow and ice at different depths through microwave signals and to provide information about the underlying configuration of the glacier. The L-band radar can be used to gather data about the inner stratigraphy of the glaciers. SAR data can be used for the evaluation of glacier faces, glacier stratigraphy and other metrics, including glacier thickness and acceleration, at high quality and high resolution. In particular, Landsat MSS, TM, AVHRR, MODIS, SPOT, ASTER and IRS VNIR bands are commonly used for mapping the snow cover zone worldwide. The Normalized Snow Difference Index (NDSI) is determined according to Dozier (1989) from a reflectance in band at wavelengths when the snow is bright (e.g. TM band 2 or MODIS band 1), where it is dark (e.g. TM band 5 or MODIS band 6) along with the band for threshold lights (e.g. TM band 4 or MODIS band 2). The Advanced Microwave Scanning System for Earth Observation (AMSR-E) is used since 2002 for providing a global Snow Water Equivalent (SWE) product (Nolin, 2010). SWE and snow depth recovery techniques use effective data from the microwave. SRTM and ASTER GDEM are the two global DEMs, publicly accessible and commonly used to build on the glaciers' topographical criteria. The glacier indices offer knowledge about the input of the avalanches from the adjacent to the glacier and impact the mass balance of the glacier. The glacier ratio is the proportional upsloping area and the downslope direction (Way et al., 2014).

GIS-based geomorphological applications span the entire spectrum of process fields and related landforms. On a range of spatial scales, anthropogenic landforms are generated. Under the general heading of 'engineering mapping', Fookes et al. (2007) aggregated several applied geomorphological mapping issues. Comparative maps of morphography, morphochronology, morphogenesis, tools and hazards made it possible to understand the important mapping criteria. More applications analyse the space-time patterns of geomorphic environments, multiscale characteristics and processes, scenarios of landscape transition, changes in disorder regimes and land destruction associated with natural and human forces, often in images or scientific visualization techniques. The convergence of GIS and RS with digital elevation models (DEMs) has, especially with the development of the early twenty-first century in LiDAR (light detection and ranging) and UAV (unmanned aerial vehicle) to get high-resolution DEMs, becomes one of the most prevalent strategies for geomorphologic exploration. Both the systematic extraction and accumulation of material and the unintended consequences of hydrological changes and subsequent degradation and sedimentation are implicated in these processes (Hooke, 2000). Geomorphological modifications arise from a variety of anthropogenic processes, involving forest clearance, agriculture, land mining and excavation, mining and quarrying, infusing, irrigation and dam building or other infrastructure properties (Nir, 1983). Global satellite positioning (GPS) technology is common in defining and incorporation of different data on geographical positions of landscape features and specific patterns. These emerging capabilities represent a significant improvement relative to conventional cartography in geomorphology. Conceptual and functional challenges that have a potential for geospatial solutions also need to be understood (Table 1.1).

1.4 Anthropogenic Landform and Intervention of Geospatial Technology

On a number of spatial scales, anthropogenic landforms are formed. Human activities like excavation, mining and quarrying have brought about dramatic changes in the landscape. Landforms of local scale emerge from excavation, cutting and levelling to change slopes and channel morphology and establish flat land for infrastructure for construction and transport infrastructure. Broader landforms of size are produced by mining and quarrying, as well as subsidence due to the exploitation of water or mineral resources. Many of the researchers warned about the destructions have been caused by human activity, but perhaps more harmful to life and property are disturbances induced by these practices (wastelands, scars arising from strip mining and open-pit quarrying). In regions with concentrated human occupation (e.g. in urban areas), the combined geomorphologic impacts of human activities are more prominent and arise mainly in the early-urban to mid-urban stages of growth (Chirico et al., 2020). Some of the human actions have induced slope displacement by steeping slopes, removing the support, removing protective cover, surplus stacking, drainage blockage, increased soil moisture and vibration. Geoscientific knowledge plays a critical role in evaluating resource capacity and suitability for urban planning, defining risks and advising management policies with increasing demand on urban areas (Fookes et al., 2005).

1.4.1 Mining, Quarrying and Geomorphological Change and Application of Geospatial Technology

Remote sensing methods have been successfully used globally in mineral extraction research (Fig. 1.1). While fine-resolution data has been used to analyse improvements in the scale of surface mining, owing to its global coverage, a vast number of studies are focused on Landsat imagery (Maxwell et al., 2014). Studies integrating mine recognition with multitemporal analysis can also be strengthened by using fine-resolution data to classify and establish mine borders while doing change analysis based on medium-resolution imagery (Koruyan et al., 2012). In semiarid areas, Schimmer (2008) used normalized difference vegetation index (NDVI), wetness and grain size homogeneity to establish a new metric unique to the detection of

Table 1.1 Spatiotemporal characteristics of geomorphological features

Order	Relief type	Spatial extension (km^2)	Temporal characteristics (in year)	Landforms	Satellite/sensor
0	Mega relief	$5.098*10^8$	$5*10^9$	Lithosphere, cryosphere, hydrosphere, atmosphere, ecosphere	Gravity Recovery and Climate Experiment, MODIS, NOAA-AVHRR, Landsat, ASTER, ESA's satellite SMOS, MIRAS
1		10^7–10^8	10^9–10^8	Continents, ocean basins and tectonic plates	InSAR, SPOT, ERS-2 radar, TERRASAR-X, JERS-1, ALOS, MOSART, IRS, AIRS
2	Macrorelief	10^7–10^5	10^8	Physiographic provinces, shields, large volcanoes, mountain ranges	AVHRR, MODIS, Landsat, ASTER, SPOT, CORONA, Sentinel-2, SEVIRI
3		10^5–10^3	10^8–10^7	Medium-scale tectonic units, mountain massifs, fault blocks, grabens	Landsat TM, ASTER, WorldView-2
4	Mesorelief	10^3–10^2	10^7	Small-scale tectonic units, fault blocks, sacking	
5	Microrelief	10^2–10^1	10^7–10^6	Large-scale erosional/depositional landforms, major valleys, piedmonts, deltas, landslides	SRTM, aerial photography, Landsat (MSS, TM, ETM+), SPOT, IRS, Radarsat, JERS-1, LDAR, CORONA satellite, RapidEye, Pleiades
6		10^1–10^{-1}	10^6–10^5	Medium-scale erosional/depositional landforms, cirques, moraines, floodplains, alluvial fans	ETM^+, LiDAR, SRTM, IRS, IKONOS, Digital Globe
7	Nanorelief	10^{-1}–10^{-3}	10^5–10^4	Small-scale erosion/depositional landforms, ridges, terraces, dunes, slump blocks, talus	MODIS – Terra/Aqua, ASTER, Landsat TM, ETM^+, aerial LiDAR, UAV, IKONOS, InSAR, IRS-1C PAN, SPOT-4, AVIRIS, RADARSAT
8	Picrorelief		10^4–10^3		

(continued)

Table 1.1 (continued)

Order	Relief type	Spatial extension (km^2)	Temporal characteristics (in year)	Landforms	Satellite/sensor
		10^{-3}–10^{-5}		Larger geomorphic process units, hillslopes, stream channel reaches, talus, small debris flows	Low-altitude unmanned aerial vehicle (UAV), terrestrial laser scanning (TLS), 3D Point Cloud (3DPC), aerial photograph, IKONOS, Quickbird, WorldView and GeoEye
9		10^{-5}–10^{-7}	10^3–10^2	Medium geomorphic process units, riffles/pools, river bars, slope facets, solution pits, gullies	SRTM, ASTER, RADAR, LiDAR, Google Map
10		10^{-7}–10^{-9}	$<10^2$	Smaller geomorphic process units, ripple marks, glacial striae, rills, raindrop impact pits	ICESat, CryoSat
11		10^{-9}–10^{-12}		Clast grain morphologies, clay mineral structure	

AIRS Atmospheric Infrared Sounder, *ASTER* Advanced Spaceborne Thermal Emission and Reflection Radiometer, *AVHRR* Advanced Very High Resolution Radiometer, *MIRAS* Microwave Imaging Radiometer with Aperture Synthesis, *InSAR* Interferometric Synthetic Aperture Radar, *MODIS* Moderate Resolution Imaging Spectroradiometer, *MOSART* Monitoring of Surface Deformation in Active Tectonic Zones, *GRACE* Gravity Recovery and Climate Experiment, Tropical Rainfall Measuring Mission, *SEVIRI* Spinning Enhanced Visible InfraRed Imager, *UAV* unmanned aerial vehicle

Source: modified after Napieralski et al. (2013)

copper mill tailings. In order to quantify changes in vegetation stemming from marble quarry expansion in a densely vegetated area, Koruyan et al. (2012) also used NDVI. To identify mining areas and assess recent mining expansion in Myanmar, LaJeunesse Connette et al. (2016) used freely available data (such as Google Earth) and open-source software. Open-cast mining causes piles of manure, tailings, solid waste and other issues impacting land use (Ozdemir & Kumral, 2019). Slope stability has introduced secret risks to mining areas (MA) of open-cast mines with a rise in mining depth and mining angle. The dynamic acquisition of multiple variables through UAVs equipped with different sensors at different phases in mining areas is an efficient method to provide constant monitoring of risk sources after quarrying and land reclamation planning (Jackisch et al., 2018; Xiao et al., 2019).

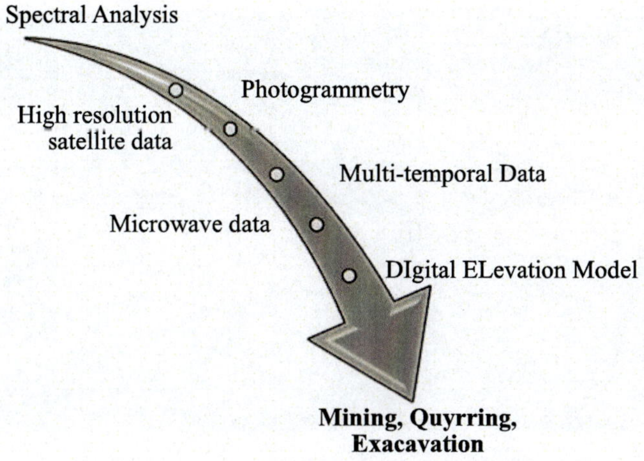

Fig. 1.1 Role of geospatial technology in mining, quarrying and excavation

1.4.2 Riverine Geomorphology and Human Intervention and Application of Geospatial Technology

Unintended side effects and future long-term legacies that can cause new complications in upstream or downstream sections are always taken into account in human interventions in riverine ecosystems (Hohensinner et al., 2018). Natural channel changes are superimposed by disruptions induced by humans that further enhance or curtail fluvial dynamics. A river channel's geometry represents the equilibrium or imbalance of erosion and depositional factors that configure the riverbed and the banks, respectively. Bandyopadhyay et al. (2013) reported that the elevation differences between narrow dried-up channels for brick factories to supply raw materials also cause the river course to change, alter the natural dynamics and eventually make the Haora River sick (Tripura, India). From the human viewpoint, the valley slope probably remains the same, and the supply of sediments are more susceptible and over shorter time frames respond to natural or human effects. Apart from specific impacts on human beings, such as changes in land cover or sand mining (Fig. 1.2), the fluvial morphology of the river rises to the most direct changes in river channelization. As the texture of the sand is very fine and it is of high demand for its manufacturing purposes, brick fields and some other building industries obtain sand from the riverbed. This sand extraction from both the riverbed and the riverbank induces numerous forms of hydrological changes within the river system (Saviour, 2012). Sediment quarrying from the riverbed is generally treated as a good activity that eliminates a river's dilemma of sedimentation. However, the sediments are randomly quarried, it poses another challenge for the flow, leaving the riverbed uneven and porous and generating more sedimentation to the lower river reaches (Bandyopadhyay & De, 2018). Most of the brick fields in this river basin accumulate mud by clumsily destroying the nearby low-elevated denudational uplands and

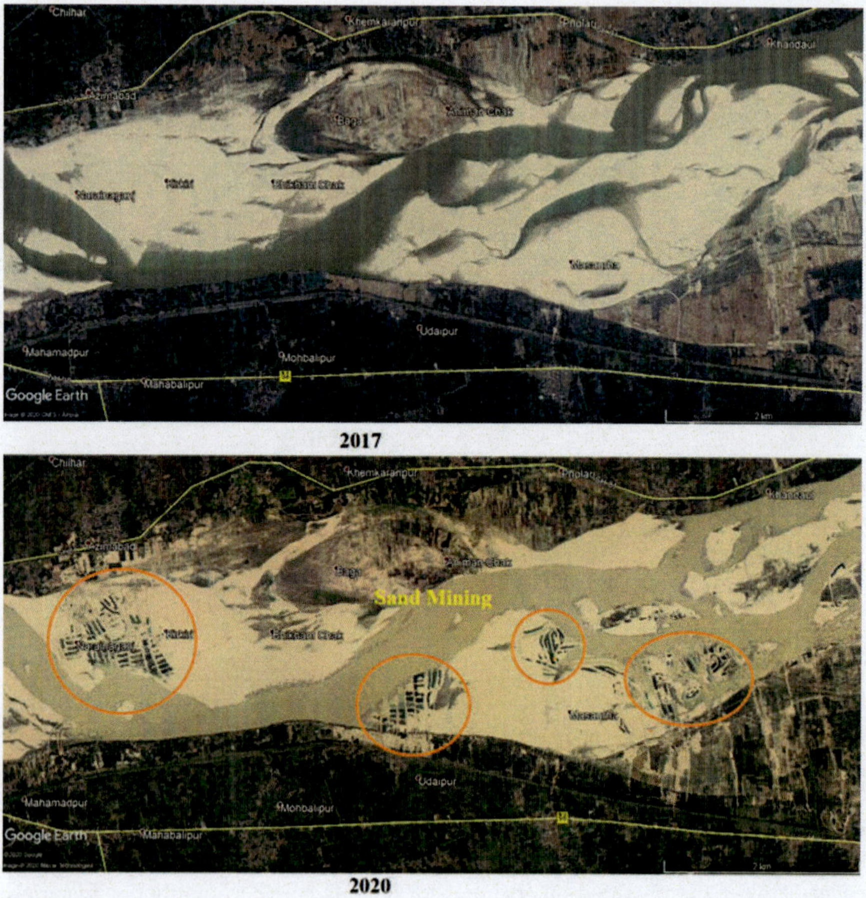

Fig. 1.2 Temporal view of anthropogenic intervention (sand mining) along the Sone river course in Bihar, India

transferring the left out loose debris to the river and creating unnecessary sedimentation (Fig. 1.3).

Moreover, brick fields situated along the river do not have sufficient road access in most instances. So, they have to move their goods over the river across those brick fields. Any temporary bridges across the river have been built by most of the brick fields, and they also often use the riverbed itself as their means of transport by dumping particles of bricks. This brick dumping inside the riverbed made the bed rough and affected the rugosity of the river.

Mining operations exacerbate coastal erosion not only by extracting sand directly but also by disrupting the sediment budget, frequently depriving regions of their sand input in the down-current direction. Despite the limited spatial scope of operations, the combination of optical images and Interferometric Synthetic Aperture Radar (InSAR) can be used to determine the environmental effects of black sand

Fig. 1.3 Temporal view of development of brick industry along Rupsa River course

mining (Chaussard & Kerosky, 2016). The L-band sensor on board the ALOS-2 and future NiSAR satellites will be especially suitable for ongoing surveillance and impact monitoring of mining activities.

The river's normal behaviour has been disrupted because of the unscientific development around the river. The river adjusts its behaviour to readjust the situation and has a great strain on those building systems that end with the fall of most of them. Bridge pier structures have certain morphological effects on rivers (Lane, 1955). Pier scouring happens as water levels discharge is raised and significant quantities of surface materials next to the bridge piers are swept away (Heidarnejad et al., 2010). Any of the removed soil particles are surrounded by turbidity currents along the bridge piers (Pasiok & Stilger-Szydło, 2010) and are accumulated as bars immediately downstream of the bridge.

Aher et al. (2012) describes the 35-year changes in the Pravara riverbank due to numerous natural and manmade activities such as floods, water intensity, sand extraction, plant cover destruction and rich soil excavation for the different proposals of the inhabitants of the immediate surrounding area using remote sensing data (SRTM) and GIS technology. Five different years (2010, 2003, 1989, 1980 and 1973) of Landsat Enhanced Thematic Mapper Plus (ETM+), Thematic Mapper (TM) and Multispectral Scanner (MSS) image and topographic maps of 1947 were used to identify and analysed vulnerability and morphological changes of Jamuna river course by Uddin et al. (2011).

1.4.3 Coastal Geomorphology and Human Intervention and Application of Geospatial Technology

At a global scale, delta morphologies are easily modified by the direct and indirect influences of human behaviour (Hoitink et al., 2020). Delta dynamics are largely governed by the availability of riverside sediments and subsequent reworking by tidal and wave driven, and variations in average sea level and in the distribution of sediments are influenced over longer periods of time. These environmental dynamics are detrimental to human activity, for example, the development of reservoirs in catchments on rivers has left many deltas starving (Kondolf et al., 2014; Schmitt et al., 2017). Delta planforms have been completely resolved by embankments (engineered levees) for purposes of land reclamation and flood control, thus competing between alternate land use categories (Giosan et al., 2013). Delta distribution channels usually terminate in mouth bar structures, where depths are limited, whether restricted by embankment or not (Fagherazzi et al., 2015). Erosion activities contribute to the emergence of deep pits in the channel beds in deltas of interconnected subsoil lithology, placing covered embankments at risk (Sloff et al., 2013). As a consequence of flow reconfiguration caused by human changes, vulnerable earthen dikes will explicitly collapse, including in the Ganges-Brahmaputra delta (Bain et al., 2019). In view of the projected rise in sea level, sand mining, groundwater extraction and human impact on delta soil, traditional delta landscape control approaches may become unsustainable (Schmitt et al., 2017).

High-resolution satellite data helps in continuos tracking of hydrodynamic and morphodynamic reactions to sudden and incremental shifts in boundary conditions (Fig. 1.4). A modern strategy to persistent flow and discharge control, for example, takes advantage of measurements from a horizontal acoustic Doppler profiler that can gather horizontal flow profiles through complex delta channels (Kästner et al., 2018). Particularly, under the surface of the water, airborne remote sensing techniques in turbid conditions associated with river deltas are confined. In shallow coastal and deltaic systems, this is a regular phenomenon. High-resolution digital elevation models (DEMs) interpolated from over 1000 measurements per square metre can be created by extensive use of light detection and ranging (LiDAR), motion structure and other techniques in terrestrial environments (Passalacqua et al., 2015). In clearwater systems, LiDAR and multispectral techniques will also work well to quantify bathymetric surfaces where measurements of up to 70-m depth are possible (Brock & Purkis, 2009; Wedding et al., 2008). The need to build automated software capable of collecting sensitive information without or with minimal user interaction is intensified by the vast data sources from remote sensing.

1.4.4 Human Intervention in Mountainous Region and Application of Geospatial Technology

Humans are able to preserve certain ecological resources. Therefore, effective interdisciplinary cooperation is required, requiring shared data sources that are open and clear to all interested populations. In the coupling of the ecological and human facets of delta processes, social sciences play an increasingly important role. Study activities have to become not just interdisciplinary but also transdisciplinary, that is, including stakeholders, for an efficient interface between delta science and delta management. A normal morphodynamic equilibrium usually no longer occurs,

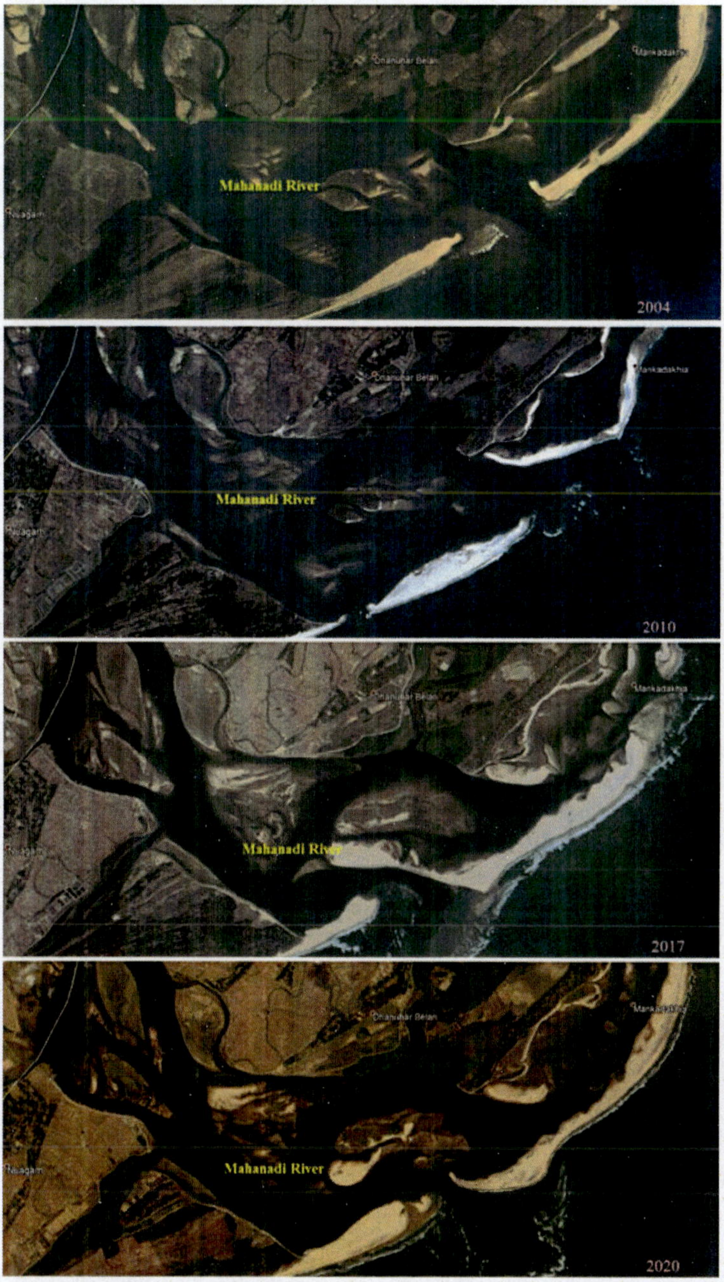

Fig. 1.4 Temporal view (2004–2020) of Mahanadi river delta

and deltas cannot achieve morphodynamic equilibrium as long as human actions take place due to continuous interventions. In addition to these, the stratigraphic record offers an extensive database of pristine delta behaviour that should be further studied in order to uncover unexplained natural morphological resilience mechanisms.

Anthropogenic triggering factors may decrease forest resistance to landslides through deforestation, forestry and mining or may increase vulnerability of forest areas to landslides through fragmentation caused by infrastructure construction, including extension of the transport network with the results of mass movements and slope failures (Shirvani, 2020). In the meantime, the impact of anthropogenic behaviour on the vulnerability of degraded forests to landslides can be revealed by a distinction between the influence of conditioning and triggering factors in protected and non-protected forests. Change in land use is a significant factor in the creation and movement of landslides caused by rainfall (Karsli et al., 2009). Landslides are specifically correlated to land use variations (Glade, 2003). The larger the volume of land cover, the less soil is destroyed by the rains. Since to recover a natural slope takes 30–35 years or more (Hung, 2009), slope land growth can enable a slope to drop its equilibrium, causing slope collapse during heavy downpours. Anthropogenic practises in mountain regions typically create volatile areas on hillsides in the Earth's material (Nefeslioglu et al., 2011). Human activities including deforestation have been described as a primary preparatory factor in a study area in Ethiopia for shallow landslides (> 3-m deep) by eliminating the stabilising forest greenery (Broothaerts et al., 2012). Recent scholars have investigated the existing generated mapping of landslide vulnerability in terms of applying common geo-environmental variables across various regions and periods (Reichenbach et al., 2018), taking into account fixed effects of a variable (Xiao et al., 2018) such as road distance and land use/land cover generated from the latest available images without analysing their dynamic changes. Shirvani (2020) investigated the significant conditioning and triggering factors that control the susceptibility of landslide using Landsat 8 multispectral images and digital elevation model (DEM) data.

1.4.5 Soil, Gully Erosion and Application of Geospatial Solution

Naturally, soil erosion happens by wind or extreme climatic conditions, but overgrazing, overcropping and deforestation are man-made practices, influencing soil erosion. *Overgrazing* happens when farmers store too many animals on their property, such as horses, cattle or goats. By chewing the plants and then scratching into damp soil or compacting dry soil with their hooves, the animals destroy the soil surface. *Overcropping* happens as the soil is continually harvested, and between crops it is not permitted to lay fallow. When it is continuously ploughed or stripped for crop growth, this continuous cultivation of the land decreases the capacity of the soils to provide useful humus for soil fertility. *Deforestation* is the chopping down of an open, uncovered ecosystem of vast tracts of trees. By leaving large areas vulnerable to excessive rainfall (which can cause leaching or flash floods) or wind erosion, deforestation accelerates soil erosion. When humans intervene with building, gardening, forestry and mining operations on the earth, the result is a deterioration of the Earth's surface layer, contributing to excessive wear and erosion. The eroded material is taken away or transferred and ultimately deposited by water, wind, etc. When water flows over land, it takes sand and other types of natural debris with it.

By raindrop splash, eroded soil is transported downslope. The erosion is termed rill erosion as drainage water focuses and passes into finger-like channels (rills) from upland areas containing soil particles, and the soil eroded from between rills is referred to as inter-rill erosion. When rill erosion starts, a sequence of deeply erodible head cuts usually advances up the slope (Shit et al., 2020). Gully is referred to as areas where accumulated drainage from a slope is adequate in volume and velocity to dig deep trenches or where concentrated water tends to cut in the same groove (such as rill) creating a deep incision of the surface. In terms of high-spatial-resolution satellite imagery and geographic object-based image analysis (GEOBIA) (Fourie, 2011), the development of remote sensing technologies provides the ability to map gullies with less effort, time and precision at an appropriate level. Previous experiments have shown that because of the spectral similarity with other non-erosion characteristics, supervised classification methods such as the maximum likelihood classification (MLC) algorithm do not express water erosion characteristics at an appropriate degree of precision (Pirie, 2009). Some human actions, such as agricultural practices, forest conversion to agriculture, etc., will increase the rate of erosion. Latest studies suggest that gully erosion is one of the most severe risks to agricultural lands and is an essential aspect of sediment in a number of environments, and gullies are effective connections for the transport of runoff and sediment from uplands to valley bottoms and permanent channels where the consequences of river erosion are exacerbated off-site. The prevalence and behaviour of gully erosion, however, has been constrained, and the geographical distribution of soil loss has been identified due to the constraints of small datasets in dynamic situations. Integrated erosion prediction models using remote sensing (RS) and geographical information system (GIS) not only provide estimated soil depletion but also provide spatial erosion distributions (Okalp, 2005). Through their drainage pattern, form, scale, colour and tone, gullies have been visually identified from multispectral remote sensing data. In the red and infrared (IR) bands, gullies with no vegetation appears bright. Pirie (2009) suggested that since they get embedded with the pixels, gullies smaller than 10 m cannot be observed using SPOT 5 imagery. It is also recommended that IO techniques, such as Quickbird, TerraSAR-X and IKONOS, be checked with better higher-spatial-resolution images. Bouaziz et al. (2011) investigated the possible contribution of the Advanced Space-borne Thermal Emission and Reflection Radiometer (ASTER) data and geomorphological parameters to the discernment of MER gully erosion trends and characteristics. In order to remove various gull shapes and patterns, maximum likelihood classification (MLC), support vector machine (SVM) and minimum distance (MD) classifiers have been used. The most productive data is provided by very-high-resolution satellite images (Desprats et al., 2013), traditional aerial photography (Maugnard et al., 2014) and nowadays also by images collected using unmanned aerial vehicles (UAVs) for precise tracking of damage in individual catchments or in agricultural fields (Eltner et al., 2014). Photogrammetry, primarily close-range photogrammetry (Wells et al., 2016) or UAV-based photogrammetry (Carollo et al., 2015) and terrestrial laser scanning (TLS) are the most frequently used noncontact techniques (Vinci et al., 2015). High-resolution data, such as point clouds, which enable digital surface models (DSMs) to be created and orthophotos of observed objects, such as erosion rills and gullies, to be acquired for more volumetric study, are an advantage of modern indirect methods. Gessesse et al. (2010) have used photogrammetric DSM time series to measure the change in tillage-induced microtopography on field plots due to inter-rill and rill erosion. Razavi-Termeh et al. (2020) have investigated the gully erosion susceptibility map (GESM) for the Abdanan region, Ilam province, Iran, using frequency ratio (FR), logistic regression (LR), an ensemble of radial basis function (RBF) and imperialist competitive algorithm (ICA) models. In a GIS platform, 12 factors influencing gully erosion were defined and planned to model the GESM, including height, slope angle, slope part, plan curvature, topographical wetness index (TWI), standardised differential vegetation index (NDVI), soil type, precipitation, river distance, road distance, lithology and land use in the study region. Thus, such strategies are effective where there has been great geomorphic change and high-quality historical evidence.

1.4.6 Urban Geomorphology and Application of Geospatial Solution

Urban growth is one of the world's biggest challenges. Urbanisation and the subsequent development of infrastructure have a high effect and often disrupt landforms completely (Reynard et al., 2017). Landform often borders between two geomorphological contexts that prohibit urbanisation in some directions, such as the borders of river or plateau. Some unique geomorphological backgrounds often contribute to significant urban development limitations. The slope is also a barrier to urbanisation and causes natural threats, such as landslides, rockslides or debris flows, particularly dangerous for settlements built without preparation on precarious slopes in humid climate conditions. Processes of geomorphology can lead to natural threats. Owing to the distribution of communities that maximise risk, damages and casualties in urban areas are significantly higher than in rural areas. Urbanization is a transformative vector of relief. This produces artificial types of land. These are often infills intended to remove pathways and make movement simpler. There are also hills made up of different waste materials. Finally, urbanisation is the aspect that affects landforms, like degradation. Many landforms, especially separate fillings,

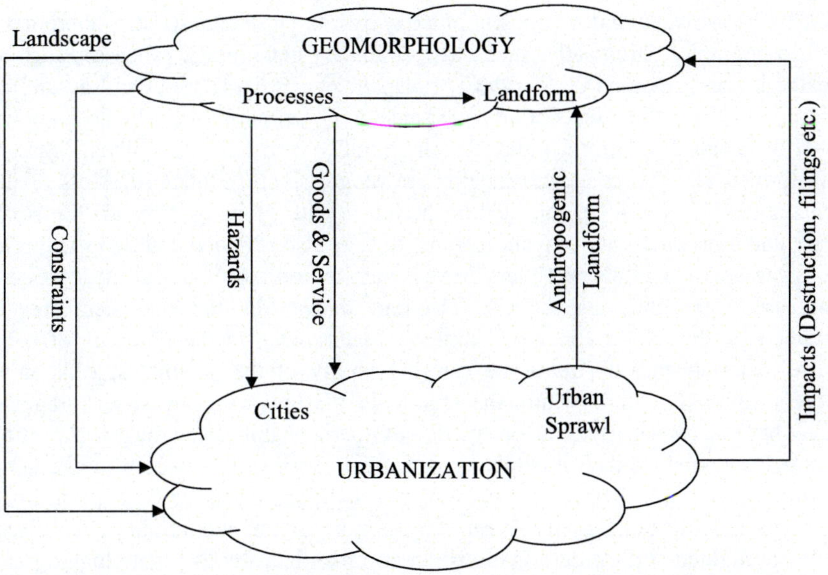

Fig. 1.5 Association between geomorphology and urbanization (Source: Reynard et al., 2017)

perforations (tunnels) or abrasions, are influenced by city planning. It contributes to the partial or complete invisibility of lands and sediments protected or degraded by urban facilities. As the urbanisation takes place, the formerly forested land is supplanted with impervious surface, distorting the size of waste and the sediment delivery to the stream network (Booth & Jackson, 1997).

Urban geomorphosites are places that illustrate the connections between geomorphology and urban growth in a classic sense (Fig. 1.5). Urban geomorphosites as well as other types of geomorphosites found in rural areas and natural areas should be considered, evaluated and controlled. They can be alone, or group of property, of complex type or geomorphological structures like all kinds of geomorphosites. Numerous settlements, with different systems, are in dynamic geomorphological settings, and the town site itself can be used as a geomorphosite (geomorphological system). Urban geomorphosites may be either dynamic (active geomorphosites) or inactive (i.e. indicative of no longer current historical morphodynamic conditions); others may be static geomorphosites emerging (Pelfini & Bollati, 2014). Urban geotourism has recently been developed. Urban areas also have fascinating contextual conditions depending upon geo-interpretation (Hose, 2012) for the production of geotouristic services (Pica et al., 2017).

The complex relationship between various facets of urban development, such as build-up area expansion, construction activities on natural features causing the diversion and depletion of aquifers and the unique geomorphic characteristics of the urban area, needs to be understood (Wolman, 1967). Sediments overload the waterways and thus modify the courses. The availability of groundwater from the urbanised region also declines and the frequency of runoff increases (Sala & Inbar,

1992). The growth and development of a city over the years are steadily altering the local topography. Ultimately, such changes impact geomorphic processes such as weathering and erosion (Viles, 1993). Mohapatra et al. (2014) used Landsat satellite image to evaluate the effects of urban expansion on geomorphology in the historical town of Gwalior in central India. Results initially revealed that residual hills are transformed and the denudational hills are progressively influenced. Most of the denudational hills will be impacted in the future with growing demands for urban space and nonavailability of plain regions. Results also revealed that two streams of the Swarnarekha and Morar Rivers flowed south to north through the city's western and eastern portions, respectively. The rain water from the catchment area is obtained by the Morar River by numerous nallas. Due to the dumping of solid waste materials that obstructs the carrying ability of the sediments, siltation is evident along its course within the region. In the low-lying areas, waterlogging issues have also been seen. The destruction of riparian habitats and the disconnection of stream channels to their floodplains are another detrimental effect of urbanisation (Segura & Booth, 2010). Haldar and Satpati (2018) enhanced the concept of urban geoforms in Kolkata Megacity (West Bengal, India) situated in humid tropics. They have highlighted surface deformations due to urban hazards like waterlogging and subsidence and also accidents due to uneven surfaces. However, numerous anthropogenic processes act to alter the topography of metropolitan environments. This may involve digging underground lines or pipes, cutting and filling surfaces, shifting of building materials and laying of roads and floors. In addition, indirect improvements include waste disposal, construction material disintegration, failure, shrinking or surface layer and the sedimentation degenerative streams/canals.

The developments in remote sensing and GIS technology have created a forum for space and time analysing landscape transformations (Estoque & Murayama, 2016). Landscape changes are regulated by anthropogenic behaviours in urban areas and are profoundly affected by spatial extension of built-up lands (Estoque & Murayama, 2015, 2016). LISS-III, Landsat TM, MSS and ETM+ satellite data have been used for 25 years (1977–2002) to study the urban structure of Ajmer city of Rajasthan, India (Jat et al., 2008). The spatial analysis of urban geography of Ranchi town, India, was studied by Ahmad and Goparaju (2016) using Landsat data from 1976, 2002 to 2015 and reported the prerequisite for unplanned urban growth tracking and sustainable urban development. Taubenböck et al. (2009) detected the temporal and spatial sprawl in cities, while Abd El-Kawy et al. (2011) proved the responsibility for land loss practices due to human activities. Geomorphology remains a major and productive method in allowing developers and policymakers to prepare for low-cost environmental planning and sustainable growth. In recent decades, for example, 68% of the coast of New England and the Central Atlantic region of the USA have suffered erosion for many reasons, mostly linked to anthropogenic influences (Goudie & Viles, 2016). In addition, it is a possible disaster, involving loss of life, to build and urbanise in/near sand dunes or in places where any other geomorphic risk, such as a base, being undermined by fragile bedrock. Therefore, the use of both high-spatial-resolution remote sensing (Unmanned aerial vehicles UAVs for 3D imaging) data and field research (ground truthing) to detect land use and land cover changes, such as the use of sample analysis of historical sites and soils to trace evidence and create/update a geomorphological map and the construction of a spatial database, can lead to the generation of an effective urban planning system.

1.5 Conclusion

Geomorphological maps are required on a wide range of scales, as various chemical, biological, meteorological and lithospheric processes are restricted and regulated by surface materials and topography. The most advanced approach to generate detailed geomorphological maps on multiple scales is still human analysis, even though subjectivity, reproductivity and validity continue to be a concern. Theoretical and philosophical advances propel advancement in research. Geomorphologists can now navigate a host of new data and technological features and can modify and interpret data using multifaceted processing sequences. The most powerful geomorphic entity of the universe is now human beings to shape and reform the surface by modifying the physical landscape. However, the advancements to digital geomorphologic mapping protocols in the current state of geospatial technology would rely upon combining geomorphological and GIS science expertise and on the person, who wants to cross the gap between two disciplines and contribute to computational geomorphology. This strategy often collectively acts as a way to create uniform forms of geographical territories. With global DEMs and satellite images accessible, geomorphologists must concentrate on map information material, vocabulary, setting the appropriate mapping requirements, evaluating objective methods and repeatable outcomes.

References

Abd El-Kawy, O. R., et al. (2011). Land use and land cover change detection in the western Nile delta of Egypt using remote sensing data. *Applied Geography, 31*(2), 483–494.

Aher, S. P., Bairagi, S. I., Deshmukh, P. P., & Gaikwad, R. D. (2012). River change detection and bank erosion identification using topographical and remote sensing data. *International Journal of Applied Information Systems, 2*(3), 1–7.

Ahmad, F., & Goparaju, L. (2016). Analysis of urban sprawl dynamics using geospatial technology in Ranchi City, Jharkhand. *Indian Journal of Environment and Geography, 9*(1–2), 7–13.

Bain, R. L., Hale, R. P., & Goodbred, S. L. (2019). Flow reorganization in an anthropogenically modified tidal channel network: An example from the southwestern Ganges-Brahmaputra-Meghna delta. *Journal of Geophysical Research: Earth Surface, 124*, 2141–2159.

Bandyopadhyay, S., & De, S. K. (2018). Anthropogenic impacts on the morphology of the Haora River, Tripura, India. *Géomorphologie: Relief, Processus, Environnement, 24*(2), 151–166.

Bandyopadhyay, S., Saha, S., Ghosh, K., & De, S. K. (2013). Channel planform change and detachment of tributary: A study on the Rivers Haora and Katakhal, Tripura, India. *Geomorphology, 193*, 25–35.

Booth, D. B., & Jackson, C. R. (1997). Urbanization of aquatic systems: degradation thresholds, stormwater detention, and the limits of mitigation. *Journal of the American Water Resources Association, 22*, 1–20. https://doi.org/10.1111/j.1752-1688.1997.tb04126.x

Bouaziz, M., Wijaya, A., & Gloaguen, R. (2011). Remote gully erosion mapping using aster data and geomorphologic analysis in the Main Ethiopian rift. *Geo-Spatial Information Science, 14*, 246–254. https://doi.org/10.1007/s11806-011-0565-1

Brock, J. C., & Purkis, S. J. (2009). The emerging role of lidar remote sensing in coastal research and resource management. *Journal of Coastal Research, 53*, 1–5.

Broothaerts, N., Kissi, E., Poesen, J., Van Rompaey, A., Getahun, K., Van Ranst, E., & Diels, J. (2012). Spatial patterns, causes and consequences of landslides in the Gilgel gibe catchment, SW Ethiopia. *Catena, 97*, 127–136.

Brown, C. J., Saunders, M. I., Possingham, H. P. & Richardson, A. J. (2013). Managing for interactions between local and global stressors of ecosystems. *PLoS One, 8*, e65765.

Carollo, F. G., Di Stefano, C., Ferro, V., & Pampalone, V. (2015). Measuring rill erosion at plot scale by a drone-based technology. *Hydrological Processes, 29*(17), 3802–3811.

Chaussard, E., & Kerosky, S. (2016). Characterization of black sand mining activities and their environmental impacts in the Philippines using remote sensing. *Remote Sensing, 8*, 100.

Chirico, P. G., Bergstresser, S. E., DeWitt, J. D., & Alessi, M. A. (2020). Geomorphological mapping and anthropogenic landform change in an urbanizing watershed using structure-from-motion photogrammetry and geospatial modeling techniques. *Journal of Maps*. https://doi.org/10.1080/17445647.2020.1746419

Coppola, D., & Cigolini, C. (2013). Thermal regimes and effusive trends at Nyamuragira volcano (DRC) from MODIS infrared data. *Bulletin of Volcanology, 75*, 744.

Davies, A. G., Chien, S., Baker, V., Doggett, T., Dohm, J., Greeley, R., Ip, C. R., Cichy, B., Rabideau, G., Tran, D., et al. (2006). Monitoring active volcanism with the autonomous sciencecraft experiment on EO-1. *Remote Sensing of Environment, 101*, 427–446.

Desprats, J. F., Raclot, D., Rousseau, M., Cerdan, O., Garcin, M., Le Bissonnais, Y., Ben Slimane, A., Fouche, J., & Monfort-Climent, D. (2013). Mapping linear erosion features using high and very high-resolution satellite imagery. *Land Degradation & Development, 24*(1), 22–32.

Dozier, J. (1989). Spectral signature of alpine snow cover from the Landsat thematic mapper. *Remote Sensing of Environment, 28*, 9–22.

Eltner, A., Baumgart, P., Maas, H., & Faust, D. (2014). Multi-temporal UAV data for automatic measurement of rill and interrill erosion on loess soil. *Earth Surface Processes and Landforms, 40*(6), 741–755.

Estoque, R. C., & Murayama, Y. (2015). Intensity and spatial pattern of urban land changes in the megacities of Southeast Asia. *Land Use Policy, 48*, 213–222.

Estoque, R. C., & Murayama, Y. (2016). Quantifying landscape pattern and ecosystem service value changes in four rapidly urbanizing hill stations of Southeast Asia. *Landscape Ecology, 31*, 1–27.

Fagherazzi, S., Edmonds, D. A., Nardin, W., Leonardi, N., Canestrelli, A., Falcini, F., Jerolmack, D. J., Mariotti, G., Rowland, J. C., & Slingerland, R. L. (2015). Dynamics of river mouth deposits. *Reviews of Geophysics, 53*, 642–672.

Fookes, P. G., Lee, E. M., & Milligan, G. (Eds.). (2005). *Geomorphology for engineers*. Whittles Publishing.

Fookes, P. G., Lee, E. M., & Griffiths, J. S. (Eds.). (2007). *Engineering geomorphology: Theory and practice*. CRC Press.

Fourie, C. (2011). *A one class object based system for sparse geographic feature identification, Stellenbosch, South Africa*. University of Stellenbosch, MSc thesis.

Gareth, R. W., & Pellikka, P. (2010). In P. Pellikka & G. Rees (Eds.), *Principles of remote sensing. Remote sensing of glaciers* (pp. 1–20). CRC Press.

Gessesse, G. D., Fuchs, H., Mansberger, R., Klik, A., & Rieke-Zapp, D. H. (2010). Assessment of erosion, deposition and rill development on irregular soil surfaces using close range digital photogrammetry. *The Photogrammetric Record, 25*(131), 299–318.

Giosan, L., Constantinescu, S., Filip, F., & Deng, B. (2013). Maintenance of large deltas through channelization: Nature vs. humans in the Danube delta. *Anthropocene, 1*, 35–45.

Glade, T. (2003). Landslide occurrence as a response to land use change: A review of evidence from New Zealand. *Catena, 51*, 297–314.

Golomb, B., & Eder, H. M. (1964). Landforms madeby man. *Landscape, 14*, 4–7.

Goudie, A., & Viles, H. (2016). *Geomorphology in the anthropocene*. Cambridge University Press.

Haldar, A., & Satpati, L. N. (2018). Urban geoforms: Concept and significance in Anthropogeomorphology. *Journal of Indian Geomorphology, 6*, 108–115.

Heidarnejad, M., ShafaiBajestan, M., & Masjedi, A. (2010). The effect of slots on scouring around piers in different positions of 180-degrees bends. *World Applied Sciences Journal, 8*(7), 892–899.

Hohensinner, S., Hauer, C., & Muhar, S. (2018). River morphology, channelization, and habitat restoration. In S. Schmutz & J. Sendzimir (Eds.), *Riverine ecosystem management* (Aquatic ecology series) (Vol. 8). Springer. https://doi.org/10.1007/978-3-319-73250-3_3

Hoitink, A. J. F., Nittrouer, J. A., Passalacqua, P., Shaw, J. B., Langendoen, E. J., Huismans, Y., & van Maren, D. S. (2020). Resilience of river deltas in the anthropocene. Grand Challenges in the Earth and Space Sciences. *Journal of Geophysical Research: Earth Surface, 125*(3), 1–24, e2019JF005201.

Hooke, R. L. (2000). On the history of humans as geomorphological agents. *Geology, 28*(9), 843–846.

Hooke, R. L., Martin-Duque, J. F., & Pedraza, J. (2012). Land transformation by humans: A review. *GSA Today, 22*, 4–10.

Hose, T. A. (2012). 3G's for modern geotourism. *Geoheritage, 4*, 7–24.

Hung, J. J. (2009). Slopeland disaster mitigation. *Sinotech, 46*, 7–16. (in Chinese).

IPCC. (2013). In T. F. Stocker, D. Qin, G.-K. Plattner, M. Tignor, S. K. Allen, J. Boschung, A. Nauels, Y. Xia, V. Bex, & P. M. Midgley (Eds.), *Climate change 2013: The physical science basis. Contribution of Working Group I to the Fifth Assessment Report of the Intergovernmental Panel on Climate Change*. Cambridge University Press. 1535 pp.

Jackisch, R., Lorenz, S., Zimmermann, R., et al. (2018). Drone-borne hyperspectral monitoring of acid mine drainage: An example from the Sokolov Lignite District. *Remote Sensing, 10*(3), 385–407.

Jansson, K. N., & Glasser, N. F. (2005). Palaeoglaciology of the Welsh sector of the British–Irish Ice Sheet. *Journal of the Geological Society, 162*(1), 25–37.

Jat, M. K., Garg, P. K., & Khare, D. (2008). Monitoring and modelling of urban sprawl using remote sensing and GIS techniques. *International Journal of Applied Earth Observation and Geoinformation, 10*, 26–43.

Karsli, F., Atasoy, M., Yalcin, A., Reis, S., Demir, O., & Gokceoglu, C. (2009). Effects of land-use changes on landslides in a landslide-prone area (Ardesen, Rize, NE Turkey). *Environmental Modeling and Assessment, 156*, 241–255.

Kästner, K., Hoitink, A. J. F., Torfs, P. J. J. F., Vermeulen, B., Ningsih, N. S., & Pramulya, M. (2018). Prerequisites for accurate monitoring of river discharge based on fixed-location velocity measurements. *Water Resources Research, 54*, 1058–1076.

Kondolf, G. M., Rubin, Z. K., & Minear, J. T. (2014). Dams on the Mekong: Cumulative sediment starvation. *Water Resources Research, 50*, 5158–5169.

Koruyan, K., Deliormanli, A. H., Karaca, Z., Momayez, M., Lu, H., & Yalçin, E. (2012). Remote sensing in management of mining land and proximate habitat. *Journal of the Southern African Institute of Mining and Metallurgy, 112*, 667–672.

LaJeunesse Connette, K. J., Connette, G., Bernd, A., Phyo, P., Aung, K. H., Tun, Y. L., Thein, Z. M., Horning, N., Leimgruber, P., & Songer, M. (2016). Assessment of mining extent and expansion in Myanmar based on freely-available satellite imagery. *Remote Sensing, 8*(11), 912. https://doi.org/10.3390/rs8110912

Lane, E. W. (1955). The importance of fluvial morphology in hydraulic engineering. *Journal of Hydraulics Division, ASCE, 81*, 1–17.

Lepparanta, M., & Granberg, H. B. (2010). Physics of glacier remote sensing. In P. Pellikka & G. Rees (Eds.), *Remote sensing of glaciers* (pp. 81–98). CRC Press.

Lombardo, V. (2016). AVHotRR: Near-real time routine for volcano monitoring using IR satellite data. *Geological Society of London, Special Publication, 426*, 73.

Marchese, F., Genzano, N., Neri, M., Falconieri, A., Mazzeo, G., & Pergola, N. (2019). A multi-channel algorithm for mapping volcanic thermal anomalies by means of Sentinel-2 MSI and Landsat-8 OLI data. *Remote Sensing, 11*(23), 2876. https://doi.org/10.3390/rs11232876

Marsh, G. P. (1864) (1965 edn). *Man and nature, or, physical geography as modified by human action (ed. Lowenthal, D.)*. Cambridge, mass.: Belknap Press of Harvard University Press, 472 pp.

Maugnard, A., Cordonnier, H., Degre, A., Demarcin, P., Pineux, N., & Bielders, C. L. (2014). Uncertainty assessment of ephemeral gully identification, characteristics and topographic threshold when using aerial photographs in agricultural settings. *Earth Surface Processes and Landforms, 39*(10), 1319–1330.

Maxwell, A. E., Warner, T. A., Strager, M. P., & Pal, M. (2014). Combining RapidEye satellite imagery and LiDAR for mapping of mining and mine reclamation. *Photogrammetric Engineering and Remote Sensing, 80*, 179–189.

Mohapatra, S. N., Pani, P, & Sharma, M. (2014). Rapid urban expansion and its implications on geomorphology: a remote sensing and GIS based study. *Geography Journal*, 10 p. https://doi.org/10.1155/2014/361459

Napieralski, J., Iestyn, B., Ulrich, K., & Matthieu K. (2013). Remote sensing and GIScience in geomorphology. *Treatise on Geomorphology, 3*, 187–227. https://doi.org/10.1016/B978-0-12-374739-6.00050-6

Nefeslioglu, H. A., Gokceoglu, C., Sonmez, H., & Gorum, T. (2011). Medium-scale hazard mapping for shallow landslide initiation: The Buyukkoy catchment area (Cayeli, Rize, Turkey). *Landslides, 8*, 459. https://doi.org/10.1007/s10346-011-0267

Nir, D. (1983). In D. Ashboren (Ed.), *Man, a geomorphological agent: An introduction to anthropogenic geomorphology*. Springer.

Nolin, A. W. (2010). Recent advances in remote sensing of seasonal snow. *Journal of Glaciology, 56*(200), 1141–1150.

Okalp, K. (2005). *Soil erosion risk mapping using geographic information systems: a case study on Kocadere creek watershed, Izmir*. M.Sc. Thesis. Department of Geodetic and Geographic Information Technologies, Natural and Applied Sciences of Middle East Technical University. Ankara, Turkey, pp. 20–21.

Ozdemir, B., & Kumral, M. (2019). A system-wide approach to minimize the operational cost of bench production in open-cast mining operations. *International Journal of Coal Science and Technology, 1*(6), 84–94.

Pasiok, R., & Stilger-Szydło, E. (2010). Sediment particles and turbulent flow simulation around bridge piers. *Archives of Civil and Mechanical Engineering, 10*(2), 67–79.

Passalacqua, P., Belmont, P., Staley, D. M., Simley, J. D., Arrowsmith, J. R., Bode, C. A., Crosby, C., DeLong, S. B., Glenn, N. F., Kelly, S. A., Lague, D., Sangireddy, H., Schaffrath, K., Tarboton, D. G., Wasklewicz, T., & Wheaton, J. M. (2015). Analyzing high resolution topography for advancing the understanding of mass and energy transfer through landscapes: A review. *Earth-Science Reviews, 148*, 174–193.

Pelfini, M., & Bollati, I. (2014). Landforms and geomorphosites ongoing changes: Concepts and implications for geoheritage promotion. *Quaestiones Geographicae, 33*(1), 131–143.

Pica, A., Reynard, E., Grangier, L., Kaiser, C., Ghiraldi, L., Perotti, L., & Del Monte, M. (2017). GeoGuides, urban geotourism offer powered by mobile application technology. *Geoheritage*. https://doi.org/10.1007/s12371-017-0237-0

Pirie, A. N. (2009). *Gully erosion mapping using SPOT 5 satellite imagery* (Honours research report). University of Pretoria.

Punkari, M. (1987). Glacial geomorphology and dynamics in the eastern parts of the Baltic Shield interpreted using Landsat imagery. *Photogrammetric Journal of Finland, 9*, 77–93.

Razavi-Termeh, S. V., Sadeghi-Niaraki, A., & Choi, S.-M. (2020). Gully erosion susceptibility mapping using artificial intelligence and statistical models. *Geomatics, Natural Hazards and Risk, 11*(1), 821–844. https://doi.org/10.1080/19475705.2020.1753824

Reath, K., Pritchard, M., Poland, M., Delgado, F., Carn, S., Coppola, D., Baker, S., Andrews, B., Ebmeier, S. K., Rumpf, E., et al. (2019a). Thermal, deformation, and degassing remote sensing time series (CE 2000–2017) at the 47 most active volcanoes in Latin America: Implications for volcanic systems. *Journal of Geophysical Research - Solid Earth, 124*, 195–2018.

Reath, K., Pritchard, M., Poland, M., Delgado, F., Carn, S., Coppola, D., Andrews, B., Ebmeier, S. K., Rumpf, E., Henderson, S., Baker, S., Lundgren, P., Wright, R., Biggs, J., Lopez, T., Wauthier, C., Moruzzi, S., Alcott, A., Wessels, R., Griswold, J., Ogburn, S., Loughlin, S., Meyer, F., Vaughan, G., & Bagnardi, M. (2019b). Thermal, deformation, and degassing remote sensing time series (CE 2000–2017) at the 47 most active volcanoes in Latin America: Implications for volcanic systems. *Journal of Geophysical Research: Solid Earth, 124*(1), 195–218.

Reichenbach, P., Rossi, M., Malamud, B. D., Mihir, M., & Guzzetti, F. (2018). A review of statistically-based landslide susceptibility models. *Earth Science Reviews, 180*, 60–91.

Reynard, E., Pica, A., & Coratza, P. (2017). Urban geomorphological heritage. An overview. *Quaestiones Geographicae, 36*(3), 7–20.

Sailor, D. J. (2002). *Urban Heat Islands, Opportunities and Challenges for Mitigation and Adaptation*. Toronto, Canada: North American Urban Heat Island Summit.

Sala, M., & Inbar, M. (1992). Some effects of urbanization in Catalan rivers. *Catena, 19*, 345–361.

Saviour, N. (2012). Environmental impact of soil and sand mining: a review. *International Journal of Science, Environment and Technology, 1*(3), 125–134.

Schimmer, R. (2008). A remote sensing and GIS method for detecting land surface areas covered by copper mill tailings. In *Proceedings of the Pecora 17—The future of land imaging...going operational, Denver, CO, USA, 18–20 November 2008*.

Schmitt, R., Rubin, Z., & Kondolf, G. (2017). Losing ground-scenarios of land loss as consequence of shifting sediment budgets in the Mekong Delta. *Geomorphology, 294*, 58–69.

Segura, C., & Booth, D. B. (2010). Effects of geomorphic setting and urbanization on wood, pools, sediment storage, and bank erosion in puget sound streams. *Journal of the American Water Resources Association, 46*(5), 972–986.

Shirvani, Z. (2020). A holistic analysis for landslide susceptibility mapping applying geographic object-based random forest: A comparison between protected and non-protected forests. *Remote Sensing, 12*(3), 434. https://doi.org/10.3390/rs12030434

Shit, P. K., Pourghasemi, H. R., & Bhunia, G. S. (2020). *Gully erosion studies from India and surrounding regions*. ISBN 978-3-030-23243-6.

Slaymaker, O., Spencer, T., & Embleton Hamann, C. (Eds.). (2009). *Geomorphology and global environmental change*. Cambridge University Press. An edited work that places anthropogeomorphology in the context of global change.

Sloff, C. J., Van Spijk, A., Stouthamer, E., & Sieben, A. S. (2013). Understanding and managing the morphology of branches incising into sand-clay deposits in the Dutch Rhine Delta. *International Journal of Sediment Research, 28*(2), 127–138.

Smets, B., Wauthier, C., & d'Oreye, N. (2010). A new map of the lava flow field of Nyamuragira (D.R. Congo) from satellite imagery. *Journal of African Earth Sciences, 58*, 778–786. https://doi.org/10.1016/j.jafrearsci.2010.07.005

Taubenböck, H., et al. (2009). Urbanization in India—Spatiotemporal analysis using remote sensing data. *Computers, Environment and Urban Systems, 33*(3), 179–188.

Thomas, W. L., Jr. (1956). *Man's role in changing the face of the earth*. University of Chicago Press.

Uddin, K., Shrestha, B., & Alam, M. S. (2011). *Assessment of morphological changes and vulnerability of river bank erosion alongside the river Jamuna using remote sensing*. Availabe at: https://www.isprs.org/proceedings/2011/gi4dm/pdf/op63.pdf

Viles, H. A. (1993). The environmental sensitivity of blistering of limestone walls in Oxford, England: A preliminary study. In D. S. G. Thomas & R. I. Allison (Eds.), *Landscape sensitivity* (pp. 309–326). Wiley.

Vinci, A., Brigante, R., Todisco, F., Mannocchi, F., & Radicioni, F. (2015). Measuring rill erosion by laser scanning. *Catena, 124*, 97–108.

Voli, M., Wegmann, K., Bohnenstiehl, D., Leithold, E., Osburn, C., & Polyakov, V. (2013). Fingerprinting the sources of suspended sediment delivery to a large municipal drinking water reservoir: Falls Lake, Neuse River, North Carolina, USA. *Journal of Soils and Sediments, 13*, 1692. https://doi.org/10.1007/s11368-013-0758-3

Way, R. G., Bell, T., & Barrand, N. E. (2014). An inventory and topographic analysis of glaciers in the Torngat Mountains, northern Labrador, Canada. *Journal of Glaciology, 60*(223), 945. https://doi.org/10.3189/2014JoG13J195

Wedding, L. M., Friedlander, A. M., McGranaghan, M., Yost, R. S., & Monaco, M. E. (2008). Using bathymetric lidar to define nearshore benthic habitat complexity: Implications for management of reef fish assemblages in Hawaii. *Remote Sensing of Environment, 112*(11), 4159–4165.

Wells, R. R., Momm, H. G., Bennett, S. J., Gesch, K. R., Dabney, S. M., Cruse, R., & Wilson, G. V. (2016). A measurement method for rill and ephemeral gully erosion assessments. *Soil Science Society of America Journal, 80*(1), 203–214.

Wohl, E. (2013). *Wide rivers crossed: The south platte and the illinois of the American prairie*. Boulder, CO: University Press of Colorado.

Wolman, M. G. (1967). A cycle of sedimentation and erosion in urban river channels. *Geografiska Annaler: Series A, 49*(2–4), 385–395.

Xiao, L., Zhang, Y., & Peng, G. (2018). Landslide susceptibility assessment using integrated deep learning algorithm along the China-Nepal highway. *Sensors, 18*, 4436.

Xiao, W., Chen, J., Zhao, Y., et al. (2019). Identify maize chlorophyll impacted by coal mining subsidence in high groundwater table area based on UAV remote sensing. *Journal of China Coal Society, 44*(1), 302–313.

Chapter 2
Braiding and Planform Pattern of Ganga

Zulfequar Ahmad, Mohammad Zakwan, and P. K. Garg

2.1 Introduction

The river morphology is concerned with the shapes of river channels and how they change over time (Leopold et al., 1964; Ahmad et al., 2017; Mukherjee & Pal, 2018; Zakwan et al., 2018). The morphology of a river channel is a function of several processes that may be natural or artificial and environmental conditions on multiple spatial and temporal scales (Wolman & Miller, 1960; Muzzammil et al., 2018; Ara & Zakwan, 2018; Aiyelokun et al. 2021; Nawaz et al., 2020). Planform index (*PFI*), sinuosity ratio (*SR*) and braiding ratio (*BR*) are commonly used parameters used to define planform pattern of the river (Ahmad et al., 2017). Remote sensing (*RS*) and geographic information system (*GIS*) have found multiple applications in tracing channel migration and braiding (Mukherjee & Pal, 2018; Akhter et al., 2019; Msaddek et al., 2019; Agnihotri et al., 2020). Brice (1981) analysed the meandering pattern of three reaches in the White River and concluded that the erosion along straight segments of a highly sinuous channel is negligible. Hooke (1987) stated that the channel planform is altered as a result of bank erosion (growth of meanders), deposition within the channel (braiding) or by cutoffs and avulsions that involve switching of channel position. Takagi et al. (2007) evaluated channel braiding and stability of the Brahmaputra Jamuna River using remote sensing images of 1960–1990. Akhtar et al. (2011) correlated braiding pattern in the Brahmaputra to the stream power of the river. Thakur et al. (2012) studied the impact of construction of Farakka barrage on its upstream and reported that with time the sinuosity and braiding index show significant positive trend in post–Farakka barrage period.

Z. Ahmad · P. K. Garg
Department of Civil Engineering, IIT Roorkee, Roorkee, Uttarakhand, India
e-mail: p.garg@ce.iitr.ac.in

M. Zakwan (✉)
Department of Civil Engineering, MANUU, Hyderabad, Telangana, India

G. S. Bhunia et al. (eds.), *Anthropogeomorphology*, Geography of the Physical Environment, https://doi.org/10.1007/978-3-030-77572-8_2

Analysing the satellite images of lower Brahmaputra Jamuna River, Bhuiyan et al. (2015) reported significant increase in river width, bar growth, meander bend and obvious planform changes resulting from natural processes and human interventions. Sinha et al. (2014) analysed the planform pattern of Kosi River, a tributary of Ganga, and found significant impact of planform pattern on the river avulsion. Yang et al. (2015) reported a decline in sinuosity in Lower Jingjiang River, China, with greater migration rates toward the left bank as compared to the right bank. Analysing the morphology of Ganga at Allahabad, Kumar et al. (2016) reported both sinuous and meandering pattern in the reach. Mukherjee and Pal (2018) asserted that Ganga River is undergoing significant channel migration in and around Malda region, India. Singh et al. (2018) analysed the braiding pattern in the middle Ganga plain and reported that the construction of hydraulic structure had significant impact on the morphology of Ganga. In the present chapter, imageries of Ganga River have been analysed using RS and GIS technique to study the changes in the planform pattern and braiding index from Devprayag to Farakka.

2.1.1 *Planform Pattern of the River*

Sinuosity Ratio (*SR*)

Based on planform pattern, rivers may be classified as straight, meandering or braided (Sharma, 2004). In natural rivers there exist a wide range of channel patterns, ranging from straight through meandering to braided. Sinuosity ratio (*SR*) is defined as the ratio of channel length to valley length or the ratio of valley slope or channel gradient as measured over the same length of the valley (Schumm, 1977). Based on the value of *SR*, channels are demarcated as straight or meandering channels; $SR = 1.5$ separates the sinuous river from the meandering river. The definition of the sinuosity ratio, as suggested by various investigators, are presented in Table 2.1; however, the definition proposed by Leopold and Wolman (1957) is commonly used.

Modified sinuosity parameter, *P*, defined by Friend and Sinha (1993) was utilized to define sinuosity in the present chapter:

$$P = \frac{L_{cmax}}{L_R} \tag{2.1}$$

Table 2.1 Definition of sinuosity ratio

SI	Sinuosity ratio	Source
1	$\frac{\text{Thalweg length}}{\text{Valley length}}$	Leopold and Wolman (1957)
2	$\frac{\text{Channel length}}{\text{Length of Meander belt axis}}$	Brice (1964)
3	$\frac{\text{Stream length}}{\text{Valley length}}$	Schumm (1963)

where L_R is the overall length of the channel belt reach measured along a straight line, or the valley length, and L_{cmax} is the mid channel length for the same reach or the stream length.

Planform Index

Three parameters which are commonly employed to characterize the braiding pattern are those proposed by Brice (1964), Rust (1978) and Friend and Sinha (1993):

$$\text{Brice Index } (BI) = 2\frac{\sum L_i}{L_r} \tag{2.2}$$

where $\sum L_i$ is the length of the islands or bars in a reach and L_r is the reach measured midway between the banks of channel. Factor 2 in the Eq. (2.2) accounts for the total length of the bars.

Braiding parameter proposed by Rust (1978) may be calculated as

$$RI = 2\frac{\sum L_b}{L_m} \tag{2.3}$$

where$\sum L_b$ is the sum of the braid lengths between the channel thalweg divergences and confluences in the reach and L_m is the average of meander wave lengths.

Braid-channel ratio (*BR*) is defined as

$$BR = \frac{L_{ctot}}{L_{cmax}} \tag{2.4}$$

where L_{ctot} is the sum of mid-channel lengths of all the segments of primary channels in a reach and L_{cmax} is the mid-channel length of the widest channel through the reach. BR measures the likelihood of the channel belt to be divided into the number of channels (Ahmad et al., 2017). Recently, Sharma (2004) proposed a Plan Form Index (*PFI*), cross-slope ratio and flow geometry index (*FGI*) to identify the braiding level in a highly braided river which may be expressed as

$$\text{Plan Form Index } (PFI) = \frac{\frac{T}{B} \times 100}{N} \tag{2.5}$$

$$\text{Flow geometry index} = \left[\frac{\sum d_i x_i}{RT}\right] \times N \tag{2.6}$$

$$\text{Cross} - \text{slope ratio} = \frac{\frac{B_L}{2}}{(\text{Bank level} - Av.\text{bed level})} \tag{2.7}$$

where T is the top flow width, B is the overall width, N is the number of braided channels, R is the hydraulic mean depth, d_i and x_i are the depth and width of the submerged sub-channel, respectively, and $B_L =$ transect length across river width. *PFI* reflects the river landform disposition with respect to water level. Low values of *PFI* indicate higher degree of braiding. According to Majumdar and Mandal (2020) and Raj and Singh (2021), $PFI < 4$ represents highly braided river reach, while $19 > PFI > 4$ represents moderately braided reach and $PFI > 19$ presents low-braided reach.

2.2 Study Area and Data

Ganga is the largest river in India, and the Ganga plain is densely populated with significant proportion of population directly or indirectly dependent on the Ganga River for their livelihood. With a limited rainfall input (June–September) and large area under agriculture, Ganga forms the backbone of agricultural economy of the country. The total length of Ganga River from Devprayag (30°10′10″ N, 78°05′00″ E, elevation: 430 m) to Farraka (24°19′29″ N, 88°36′30″ E, Elevation: 430 m) is 1824 km which was divided into 37 reaches for calculating morphological parameters. For the present study, toposheets with a scale of 1:50,000 were obtained from the Survey of India, and remote sensing digital data (Landsat MSS, IRS 1A LISS II, IRS 1C LISS III, IRS P6 LISS III) were obtained from GLCF website, USGS website and NRSC Hyderabad as shown in Table 2.2.

2.3 Methodology

Firstly, the Survey of India toposheets (1:50,000 scale) which cover the area of interest are procured. Then, these toposheets are geo-referenced and ortho-rectified using first-order polynomial and Lambert conformal conic (LCC) projection type with the help of ERDAS Imagine 2014 software, before initiating the analysis. After geo-referencing, clipping and mosaicking have been done. Relevant satellite images were procured from the National Remote Sensing Centre (NRSC), Hyderabad, and downloaded from the US Geological Survey (USGS) website. Preprocessing of the satellite images has been carried out using filters such as histogram equalization, matching, etc. to improve and equalize the brightness levels of similar features and to provide uniform information. After mosaicking the toposheets, all the satellite images were geo-referenced using image-to-image method by selecting several control points. The RMS error has been kept below the half pixel size. After geo-referencing, images have been mosaicked to cover the entire study area. For mosaicking the feather option of the software has been used to get the seamless boundaries between different images of the same year.

Table 2.2 Data used and their sources

Toposheets				
S. no.	Data		Data source	Scale
	Toposheets and year		SOI, Dehradun	1:50,000
1	**53**	G/-14 **(1974)**, 15 **(1977)** J/-4 **(1969)**, 8 **(1967)**, 12 **(1967)** K/-1 **(1972)**, 2 **(1970)**, 3 **(1970)**, 4 **(1974)** L/-1 **(1976)**, 2 **(1970)**, 3 **(1970)**, 6, 7 (1970), 8 **(1970)**, 12 **(1970)**, 16 **(1970)**		
	54	I/-9,13 M/-1, 2, 6,10,11, 12, 15 **(1976)**, 16 **(1976)**		
	63	A/-4 B/-1, 2, 5, 6, 7, 8, 11, 12, 16 F/-4 G/-1, 5, 6, 10, 11, 14, 15 K/-3 **(1973)**, 4 **(1972)**, 7 **(1973)**, 8 **(1973)**, 11 **(1976)**, 12 **(1978)**, 15 **(1974)**,16 **(1972)** O/-2, 3, 4, 7, 8, 10, 11, 14, 15		
	72	C/-1, 2, 5, 6, 9, 10, 13, 14 G/-2, 6, 7,10,11, 14, 15 K/-3 **(1984)**, 4 **(1981)**, 7 **(1980)**, 8 **(1981)**, 11 **(1980)**, 12 **(1982)**, 15 **(1980)**, 16 **(1983)** O/-1 **(1972)**, 2 **(1972)**, 3 **(1974)**, 4 **(1973)**, 7 **(1974)**, 8 **(1974)**, 11 **(1974)**, 12 **(1974)**, 15 **(1974)**, 16 **(1974)** P/-13 **(1975)**		

(continued)

Table 2.2 (continued)

Toposheets						
S. no.	Data			Data source		Scale
Remote sensing digital data						
S. no.	**Sensor**	**Data source**	**Year**	**Path**	**Row**	**Date**
1	Landsat MSS	GLCF and USGS website	1980	151	42	27/09/1980
				152	42	21/01/1980
				153	42	17/10/1980
				153	43	17/10/1980
				154	41	30/09/1980
				154	42	23/11/1980
				155	41	01/10/1980
				155	42	12/12/1980
				156	41	14/09/1980
				157	39	14/12/1980
				157	40	14/12/1980
				157	41	14/12/1980
2	IRS 1A LISS II (Each scene of LISS II contains four satellite images A1, A2, B1 and B2)	NRSC Hyderabad	1990	19	50	21/11/1990
				20	50	22/10/1989
				21	50	23/11/1990
				23	50	30/01/1991
				24	50	18/12/1990
				25	49	05/10/1989
				25	50	21/04/1990
				26	48	06/11/1990
				26	49	06/10/1989
				27	48	29/11/1990

(continued)

Table 2.2 (continued)

Toposheets						
S. no.	Data			Data source		Scale
				28	46	08/10/ 1989
				28	47	17/10/ 1990
				28	48	17/10/ 1990
3	IRS 1C LISS III	NRSC Hyderabad	2000	96	50	17/11/ 2000
				97	50	15/11/ 2000
				97	51	15/11/ 2000
				98	52	27/10/ 2000
				99	52	08/10/ 2000
				99	53	08/10/ 2000
				100	53	13/10/ 2000
				101	54	18/10/ 2000
				102	54	16/11/ 2000
				103	54	15/12/ 2000
				104	53	02/11/ 2000
				104	54	02/11/ 2000
				105	54	25/12/ 2000
				106	54	19/10/ 2000
				107	54	17/11/ 2000
4	IRS P6 LISS III	NRSC Hyderabad	2010	96	50	06/11/ 2010
				97	50	05/12/ 2010
				97	51	05/12/ 2010
				98	52	23/10/ 2010

(continued)

Table 2.2 (continued)

Toposheets						
S. no.	Data			Data source		Scale
				99	52	28/10/2010
				99	53	28/10/2010
				100	53	02/11/2010
				101	54	14/10/2010
				102	54	12/11/2010
				103	54	24/10/2010
				104	53	29/10/2010
				104	54	29/10/2010
				105	54	10/10/2010
				106	54	02/12/2010
				107	54	13/11/2010

The riverbank line has then been identified and delineated from mosaics developed using ERDAS Imagine 2014, for all the satellite images of years 1970, 1980, 1990, 2000 and 2010. The identified riverbank lines for the left, right and centre (main channel) have been digitized using ArcMap software. The left bank, right bank and centre line have been prepared for the years 1970, 1980, 1990, 2000 and 2010 (Fig. 2.1). The length of arcs of the left bank, right bank and centre line has been calculated using GIS software. Shifting of centre line, left bank and right bank of the river has been evaluated with respect to the centre line of year 2010. Sinuosity ratio and *PFI* index have been estimated using the satellite images of several years. The spatiotemporal variation of sinuosity ratio and *PFI* index for different reaches were analysed.

2.4 Results and Discussion

Sinuosity ratio, Planform Index and braiding ratio are commonly used parameters to define the planform pattern of the river. For calculating sinuosity ratio, Eq. (2.1) given by Friend and Sinha (1993) was utilized. The calculated values of sinuosity ratio for different reaches are given in Table 2.3 and graphically shown in Fig. 2.2.

Fig. 2.1 Mosaic of FCC of Ganga River of year 2010 for illustration (IRS P6 LISS III Images)

Based on the calculated values of sinuosity ratio of Ganga River (Table 2.3), it may be concluded that sinuosity ratios in the reaches 200–250 km (SR = 1.40), 600–700 km (SR = 1.82 and 1.84), 800–900 km (SR = 1.98 and 1.78) and 1750–1824 km (SR = 1.84 and 2.17) are relatively higher than the other reaches. As the average value of the sinuosity ratio for the whole reach of the river was 1.38, therefore, the Ganga shall be considered as sinuous river as per the classification laid by Leopold and Wolman (1957) except for the reaches identified above (M-1 to M-20) which can be classified as meander. In the whole reach of the river and for the years 1970, 1980, 1990, 2000 and 2010, the maximum, minimum and average sinuosity ratios are of the order of 2.2, 1.1 and 1.38, respectively. Negligible progressive change in the sinuosity ratio has been found from the year 1970 to 2010. Geometrical parameters of the prominent meanders such as axial wavelength, amplitude and radius of curvature as shown in Fig. 2.3 have been calculated and listed in Table 2.4. Prominent meanders are numbered as M-1 to M-20 and shown in Figs. 2.4, 2.5, 2.6, 2.7 and 2.8. From Table 2.4 and Figs. 2.4, 2.5, 2.6, 2.7 and 2.8, it may be concluded that except the meanders M-1(d/s Sultanganj), M-2 (Upstream of Sultanganj) and M-3 (Munger), all other meanders are stable, and no noticeable change in their geometry has been noticed from year 1970 to 2010. At the meanders M-1, M-2 and M-3, the river has left the meandering route and followed a relatively straight path over the years.

The *PFI* of Ganga River was calculated using the formula (Eq. 2.5) given by Sharma (2004). For calculating *PFI*, the 50-km reach of river was further subdivided into 2-km subreaches; the bed width, channel width and the number of channels of the river were computed. The *PFI* for each chainage was calculated, and the calculated values were graphically represented in Fig. 2.9a, b, c, d and e. From these plots, it may be concluded that the *PFI* values decreased in progressive decades

Table 2.3 Sinuosity ratios for Ganga River of years 1970, 1980, 1990, 2000 and 2010

Chainage (km)	Section	2010			2000			1990			1980			1970		
		L_{cmax}	L_R	SR-2010	L_{cmax}	L_R	SR-2000	L_{cmax}	L_R	SR-1990	L_{cmax}	L_R	SR-1980	L_{cmax}	L_R	SR-1970
0-50	1	50.00	38.68	**1.29**	49.60	38.64	**1.28**	46.34	38.13	**1.22**	410.18	38.23	**1.23**	48.80	310.55	**1.30**
50-100	2	50.00	310.21	**1.34**	46.66	310.82	**1.23**	45.71	310.48	**1.22**	49.57	310.57	**1.32**	48.91	39.18	**1.25**
100-150	3	50.00	33.68	**1.48**	48.50	34.45	**1.41**	45.61	33.97	**1.34**	40.09	34.98	**1.15**	46.49	34.97	**1.33**
150-200	4	50.00	40.26	**1.24**	43.48	39.13	**1.11**	45.95	39.28	**1.17**	46.16	40.88	**1.13**	60.17	40.16	**1.50**
200-250	5	50.00	35.64	**1.40**	63.95	38.10	**1.68**	50.40	310.07	**1.36**	75.44	34.89	**2.16**	75.28	36.04	**2.09**
250-300	6	50.00	41.61	**1.20**	45.37	39.79	**1.14**	44.44	40.33	**1.25**	50.77	42.68	**1.19**	51.14	42.27	**1.21**
300-350	7	50.00	40.90	**1.22**	48.60	41.57	**1.17**	410.41	41.26	**1.08**	50.96	43.73	**1.17**	49.37	42.08	**1.17**
350-400	8	50.00	43.88	**1.14**	51.42	43.14	**1.19**	52.90	43.12	**1.10**	49.42	42.29	**1.17**	49.35	42.91	**1.15**
400-450	9	50.00	44.74	**1.12**	49.09	44.51	**1.10**	61.70	44.62	**1.19**	60.14	44.67	**1.35**	61.27	44.97	**1.36**
450-500	10	50.00	40.80	**1.23**	50.21	41.30	**1.22**	54.47	41.22	**1.50**	51.21	41.52	**1.23**	52.02	41.93	**1.24**
500-550	11	50.00	39.47	**1.27**	50.45	38.99	**1.17**	48.96	39.05	**1.39**	48.48	38.61	**1.26**	49.91	38.42	**1.30**
550-600	12	50.00	38.82	**1.29**	48.98	39.25	**1.16**	49.73	38.48	**1.42**	49.25	38.92	**1.27**	50.29	38.16	**1.32**
600-650	13	50.00	210.43	**1.82**	50.30	210.22	**1.85**	53.54	210.56	**1.78**	50.21	210.32	**1.84**	50.01	210.28	**1.83**
650-700	14	50.00	210.17	**1.84**	50.44	210.08	**1.86**	50.28	210.04	**1.84**	49.24	210.08	**1.82**	50.47	210.06	**1.87**
700-750	15	50.00	39.40	**1.27**	49.86	39.51	**1.26**	50.50	39.57	**1.35**	50.54	39.46	**1.28**	51.22	39.46	**1.30**
750-800	16	50.00	33.83	**1.48**	49.42	33.96	**1.46**	49.19	33.91	**1.48**	410.94	33.84	**1.42**	49.17	33.96	**1.45**
800-850	17	50.00	25.28	**1.98**	50.65	25.23	**2.01**	50.22	25.30	**2.00**	410.88	25.24	**1.90**	50.14	25.24	**1.99**
850-900	18	50.00	28.09	**1.78**	49.79	210.78	**1.79**	49.62	210.54	**1.79**	48.26	210.49	**1.76**	48.89	26.80	**1.82**
900-950	19	50.00	34.77	**1.44**	48.57	33.02	**1.47**	53.42	33.91	**1.48**	51.66	33.45	**1.54**	53.66	34.99	**1.53**
950-1000	20	50.00	42.88	**1.17**	54.79	43.02	**1.27**	56.90	42.45	**1.34**	55.62	42.82	**1.30**	52.60	42.02	**1.25**
1000-1050	21	50.00	40.12	**1.25**	49.77	40.01	**1.24**	52.73	40.02	**1.32**	52.34	40.08	**1.31**	52.10	40.07	**1.30**
1050-1100	22	50.00	40.06	**1.25**	51.54	39.66	**1.30**	51.00	39.84	**1.28**	50.55	39.79	**1.27**	51.97	39.81	**1.31**

(continued)

Table 2.3 (continued)

1100-1150	23	50.00	410.36	**1.06**	52.82	48.10	**1.10**	51.84	410.46	**1.09**	54.04	410.69	**1.13**	53.41	410.96	**1.11**
1150-1200	24	50.00	42.42	**1.18**	48.26	42.42	**1.14**	48.20	42.41	**1.14**	51.32	42.45	**1.21**	510.03	42.47	**1.34**
1200-1250	25	50.00	41.82	**1.20**	48.22	41.74	**1.16**	46.38	41.79	**1.11**	48.54	41.74	**1.16**	51.01	41.75	**1.22**
1250-1300	26	50.00	42.70	**1.17**	48.04	41.98	**1.14**	51.73	42.02	**1.23**	50.38	43.28	**1.16**	50.22	39.69	**1.27**
1300-1350	27	50.00	40.44	**1.24**	49.19	40.87	**1.20**	44.71	310.93	**1.18**	50.26	38.82	**1.29**	52.39	40.33	**1.30**
1350-1400	28	50.00	41.90	**1.19**	50.35	42.31	**1.19**	53.76	43.24	**1.24**	51.84	42.61	**1.22**	56.33	43.53	**1.29**
1400-1450	29	50.00	42.70	**1.17**	50.90	42.95	**1.19**	52.95	42.52	**1.25**	55.76	42.01	**1.33**	50.55	41.65	**1.21**
1450-1500	30	50.00	42.24	**1.18**	49.87	40.85	**1.22**	48.62	41.97	**1.16**	49.83	41.55	**1.20**	50.48	41.61	**1.21**
1500-1550	31	50.0	41.72	**1.20**	48.47	41.96	**1.16**	50.21	41.11	**1.22**	49.27	41.75	**1.18**	53.25	41.97	**1.27**
1550-1600	32	50.0	40.31	**1.24**	50.67	40.31	**1.26**	53.14	41.19	**1.29**	50.80	40.71	**1.25**	51.94	40.21	**1.29**
1600-1650	33	50.0	310.30	**1.34**	50.92	35.75	**1.42**	45.32	310.04	**1.22**	410.40	310.46	**1.27**	410.15	310.24	**1.27**
1650-1700	34	50.0	35.85	**1.39**	48.87	35.22	**1.39**	46.29	36.86	**1.26**	410.07	36.55	**1.29**	55.73	35.33	**1.58**
1700-1750	35	50.0	41.93	**1.19**	52.87	41.74	**1.27**	51.53	41.01	**1.26**	53.68	40.83	**1.31**	55.96	41.95	**1.33**
1750-1800	36	50.0	210.19	**1.84**	49.30	26.47	**1.86**	49.65	26.58	**1.87**	48.84	26.51	**1.84**	50.01	210.31	**1.83**
1800-1824	37	24.0	11.06	**2.17**	24.22	11.03	**2.20**	24.12	11.06	**2.18**	23.98	11.07	**2.17**	24.33	11.02	**2.21**
Maximum				**2.17**			**2.20**			**2.18**			**2.17**			**2.21**
Minimum				**1.06**			**1.10**			**1.08**			**1.13**			**1.11**
Average				**1.36**			**1.36**			**1.37**			**1.38**			**1.42**

Shaded portion indicates high sinuosity ratio (SR>1.40)

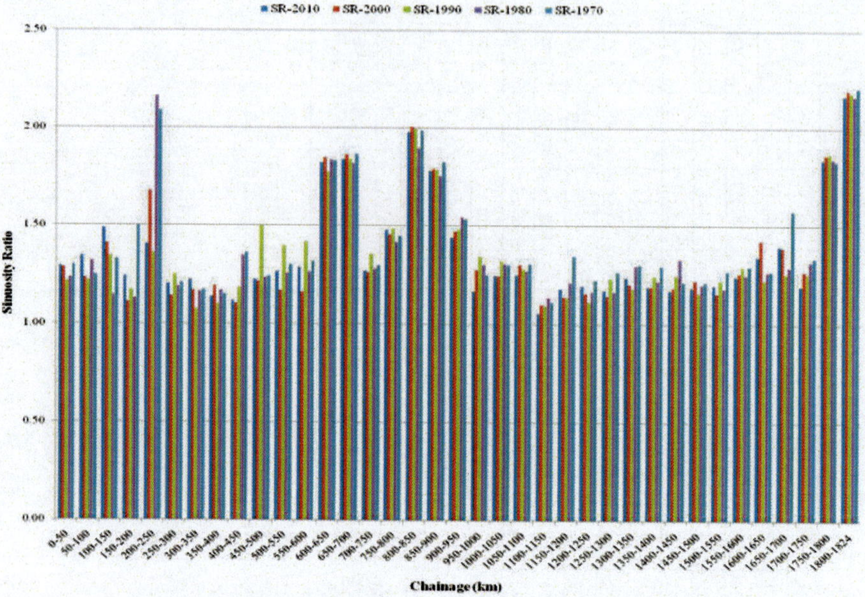

Fig. 2.2 Sinuosity ratio of Ganga River of years 1970, 1980, 1990, 2000 and 2010

Fig. 2.3 Geometrical parameters of a river meander (Sinha & Ghosh, 2012)

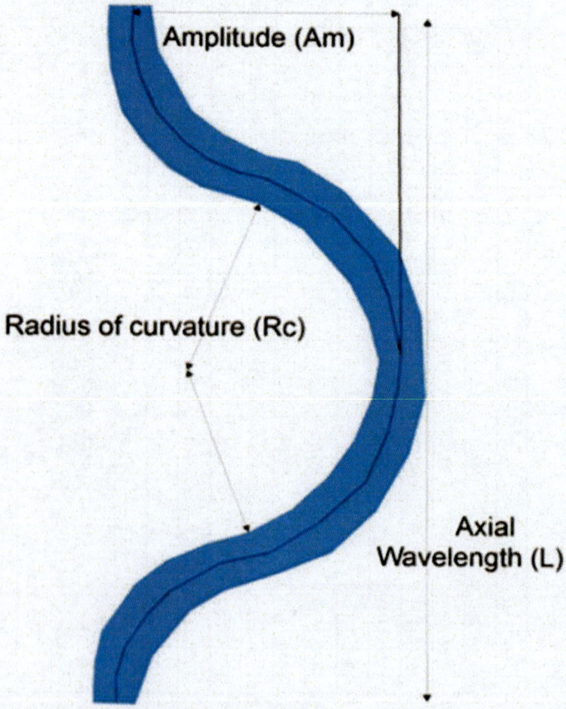

Table 2.4 Geometrical parameters of the prominent meanders in the Ganga River

Meander no.	Chainage (km)	Location	Year	Axis wave length (km)	Amplitude (km)	Radius of curvature (km)
M-1	195	D/S of Sultanganj	1970	10	8.5	4.5
			1980	SR~1.0		
			1990	SR~1.0		
			2000	SR~1.0		
			2010	SR~1.0		
M-2	220	U/S of Sultanganj at Fatehpur	1970	14	6.5	3.5
			1980	13.5	6.0	3.5
			1990	SR~1.0		
			2000	SR~1.0		
			2010	SR~1.0		
M-3	240	Munger	1970	14	13	7.5
			1980	14	13	7.5
			1990	10.7	4.5	4
			2000	12	11	3.5
			2010	SR~1.0		
M-4	630	U/S of Ghazipur	1970	17	15	2.5
			1980	17	15	2.5
			1990	17	15	2.5
			2000	17	15	2.5
			2010	17	15	2.5
M-5	680	At confluence point of Gomti river	1970	15.5	11.5	6.0
			1980	15.5	11.5	6.0
			1990	15.5	11.5	6.0
			2000	15.5	11.5	6.0
			2010	15.5	11.5	6.0
M-6	700	D/S of Varanasi	1970	18	10	5.0
			1980	18	10	5.0
			1990	18	10	5.0
			2000	18	10	5.0
			2010	18	10	5.0
M-7	715	At Varanasi	1970	17	6.5	4.5
			1980	17	6.5	4.5
			1990	17	6.5	4.5
			2000	17	6.5	4.5
			2010	17	6.5	4.5
M-8.	725	U/S of Varanasi	1970	15.5	7.0	4.0
			1980	15.5	7.0	4.0
			1990	15.5	7.0	4.0
			2000	15.5	7.0	4.0
			2010	15.5	7.0	4.0
M-9	740	D/S of Chunar	1970	18	5.0	3.0

(continued)

Table 2.4 (continued)

Meander no.	Chainage (km)	Location	Year	Axis wave length (km)	Amplitude (km)	Radius of curvature (km)
			1980	18	5.0	3.0
			1990	18	5.0	3.0
			2000	18	5.0	3.0
			2010	18	5.0	3.0
M-10	750	At Chunar	1970	31	10	4.5
			1980	31	10	4.5
			1990	31	10	4.5
			2000	31	10	4.5
			2010	31	10	4.5
M-11	785	D/S of Mirzapur	1970	16	11	4.5
			1980	16	11	4.5
			1990	16	11	4.5
			2000	16	11	4.5
			2010	16	11	4.5
M-12	795	At Mirzapur	1970	12	10	3.5
			1980	12	10	3.5
			1990	12	10	3.5
			2000	12	10	3.5
			2010	12	10	3.5
M-13	815	Near Gopiganj	1970	15	6.0	3.5
			1980	15	6.0	3.5
			1990	15	6.0	3.5
			2000	15	6.0	3.5
			2010	15	6.0	3.5
M-14	830	U/S of Gopiganj	1970	9.5	6.0	2.5
			1980	9.5	5.5	2.0
			1990	9.5	6.0	2.5
			2000	9.5	6.0	2.5
			2010	9.5	6.0	2.5
M-15	840	Kolainpur	1970	9.5	9.0	4.0
			1980	9.5	9.0	4.0
			1990	9.5	9.0	4.0
			2000	9.5	9.0	4.0
			2010	9.5	9.0	4.0
M-16	855	Katra	1970	8.5	15.5	2.0
			1980	8.5	15.5	2.0
			1990	8.5	15.5	2.0
			2000	8.5	15.5	2.0
			2010	8.5	15.5	2.0
M-17	875	Lachhagila	1970	18	9.5	4.5
			1980	18	9.5	4.5

(continued)

Table 2.4 (continued)

Meander no.	Chainage (km)	Location	Year	Axis wave length (km)	Amplitude (km)	Radius of curvature (km)
			1990	18	9.5	4.5
			2000	18	9.5	4.5
			2010	18	9.5	4.5
M-18	890	Sirsa	1970	10	5.5	3.0
			1980	10	5.5	3.0
			1990	10	5.5	3.0
			2000	10	5.5	3.0
			2010	10	5.5	3.0
M-19	970	Lalgopalganj	1970	10	5.5	3.5
			1980	10	5.5	3.5
			1990	10	5.5	3.5
			2000	10	5.5	3.5
			2010	10	5.5	3.5
M-20	1010	Manikpur	1970	8.0	4.0	2.5
			1980	8.0	4.0	2.5
			1990	8.0	4.0	2.5
			2000	8.0	4.0	2.5
			2010	8.0	4.0	2.5

as compared to 1970, indicating the increase in braiding intensities almost throughout the river except in the upper Ganga reaches (Rishikesh to Devprayag) where *PFI* value was consistently calculated to be close to 100. Thakur et al. (2012) too reported consistent increase in braiding index upstream of Farakka barrage. The eastern Ganga plain especially in the reach (0–500 km chainage, i.e. Farakka to Ballia) is highly braided, partly due to high looseness factor of Farakka barrage, and a number of tributaries (Kosi, Gandak, Ghagra) carrying significant sediment load join the Ganga River in this reach (Zakwan et al., 2018). Between 500 and 800 km chainage (Buxar and Mirzapur), the channel is almost unbraided. Again from around Allahabad till Haridwar, the Ganga River is highly braided, partly due to the presence of a large number of hydraulic structure in the reach and partly due to confluence of tributaries (Yamuna at Allahabad). Heavy braiding around Allahabad was also reported by Kumar et al. (2016). The confluence of tributaries carrying high sediment load and the presence of barrages with high looseness factors reduce the sediment transport capacity of the river. Examination of Fig. 2.9a, b, c, d, e and f and Fig. 2.10a, b, c, d, e, f, g, h and i reveals that Ganga is prominently braided downstream of Haridwar, Garhmukteshwar, Ramghat (d/s of Narora), Kachlabridge; upstream of Farrukhabad and upstream of Allahabad, Balia, Raghopur and Krusela. The braiding pattern of the Ganga at these locations is shown in Fig. 2.10a, b, c, d, e, f, g, h and i.

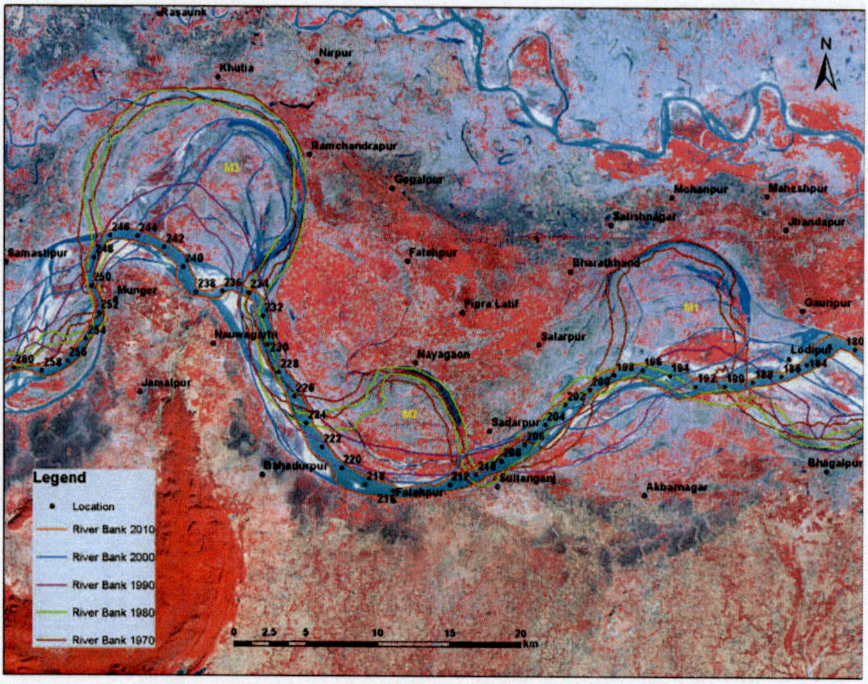

Fig. 2.4 Meandering pattern of Ganga River in the reach from chainage 180 km (Bhagalpur) to 260 km (Munger)

It has been observed that the important variables that affect the braiding of rivers are discharge, fluctuations in flow rate; the size distribution of the bed material and the rate and size distribution of sediment load, channel width, depth, slope, climatic and geologic factors (Garde, 2006; Zakwan, 2018; Pandey et al., 2018, 2020). It has been often observed that a given channel may change in a short distance from a braided to meandering and vice versa; such alterations are therefore attributed to the variation of local variables. It was also observed that the rivers dominated by braided as against the meandering channels possess higher flood peaks, higher total discharge and greater monthly discharge variation. Ganga is also characterized by very high discharge variability (Zakwan et al., 2018). Braiding was developed by sorting as the stream leaves behind those fractions of the load that it is incompetent to transport. If the river possesses the capability to transport all sediment sizes that it was transporting but was overloaded, aggradation may take place without braiding (Garde, 2006).

Lane (1957) analysed the planform of several rivers along with their history and found that basic reasons behind braiding were (i) overloading, i.e. the river has been overloaded with sediment load greater than it sediment transport capacity, thereby leading to deposition, and (ii) the presence of steep slopes leading to wide shallow channels, promoting formation of bars and islands. All the steep slope-type braided

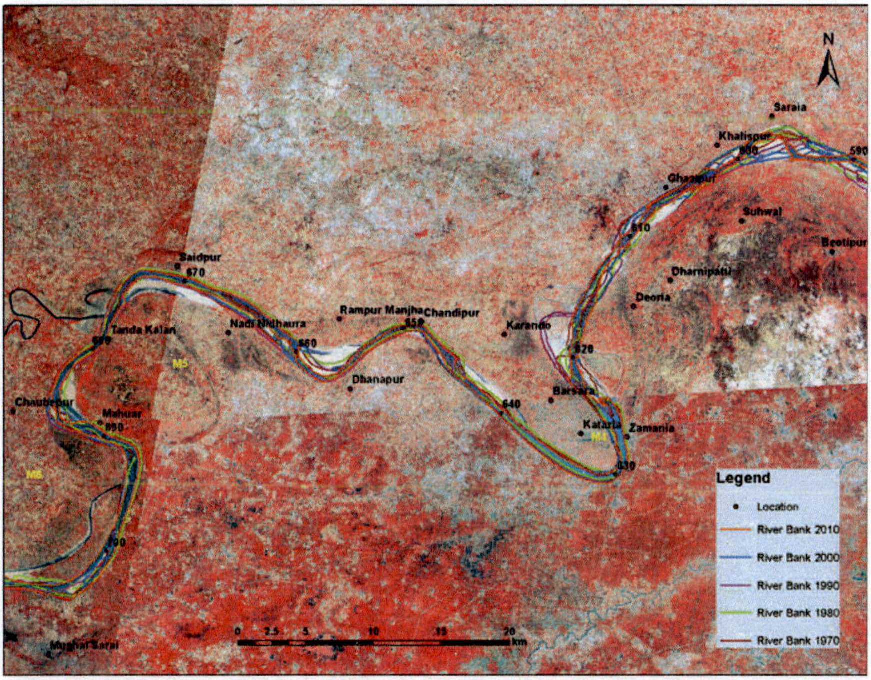

Fig. 2.5 Meandering pattern of Ganga River in the reach from chainage 600 km (Ghazipur) to 700 km (Mughalsarai)

channels have many characteristics in common apart from multiple channels; these are:

(i) Relatively straight course of main channel
(ii) Steep longitudinal slopes
(iii) Wide channels
(iv) Shallow depths
(v) Sand or coarse bed material
(vi) Usually high bed load

Since braided form can be due to steep slope or due to aggradation or combination of both, braided channels may be divided as:

I Braiding from steep slope	(a) Braiding in steep channel having degradation
	(b) Braiding in steep channel possessing equilibrium
II Braiding due to aggradation	(c) Braiding in steep channel having aggradation
	(d) Braiding in moderate slope channel having aggradation
	(e) Braiding in mild slope channel having aggradation

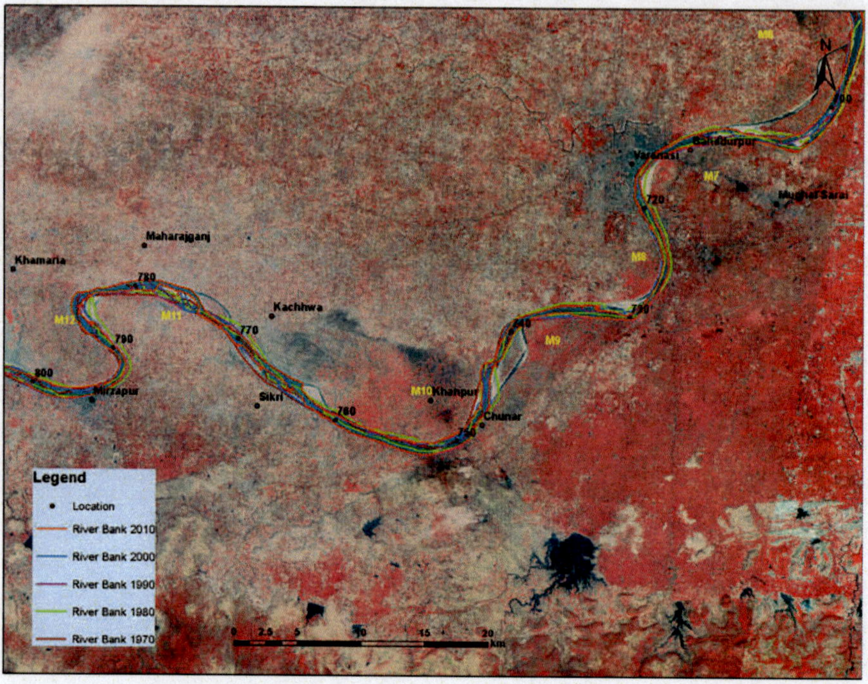

Fig. 2.6 Meandering pattern of Ganga River in the reach from chainage 700 km (Mughalsarai) to 800 km (Mirzapur)

High braiding in the Ganga River downstream of the Haridwar, upstream of Allahabad, Krusela, etc., was primarily due to aggradation which occurs as the river was incompetent to transport the sediment load that it or tributary has been bringing from the relatively steep reach. Further, construction/operation of a large number of barrages like Rishikesh barrage, Bhimgoda barrage, Chaudhary Charan Singh barrage, Ganga barrage, Farakka barrage and several other minor river works have contributed to the changing planform pattern in Ganga River. Changes in land use pattern also influences the landform pattern of the river.

2.5 Conclusion

In the present study on morphology of Ganga River, sinuosity ratio, braiding Index and *PFI* of various reaches, the rivers have been estimated using the satellite images for the years 1970, 1980, 1990, 2000 and 2010. The spatial variation in the sinuosity ratio from Devprayag to Farakka has been explored thereby identifying highly sinuous reaches of Ganga River as high sinuosity ratio represents chances of channel migration and landscape changes. Sinuosity ratios in the reaches 200–250 km, 600–700 km, 800–900 km and 1750–1824 km are relatively higher than the other

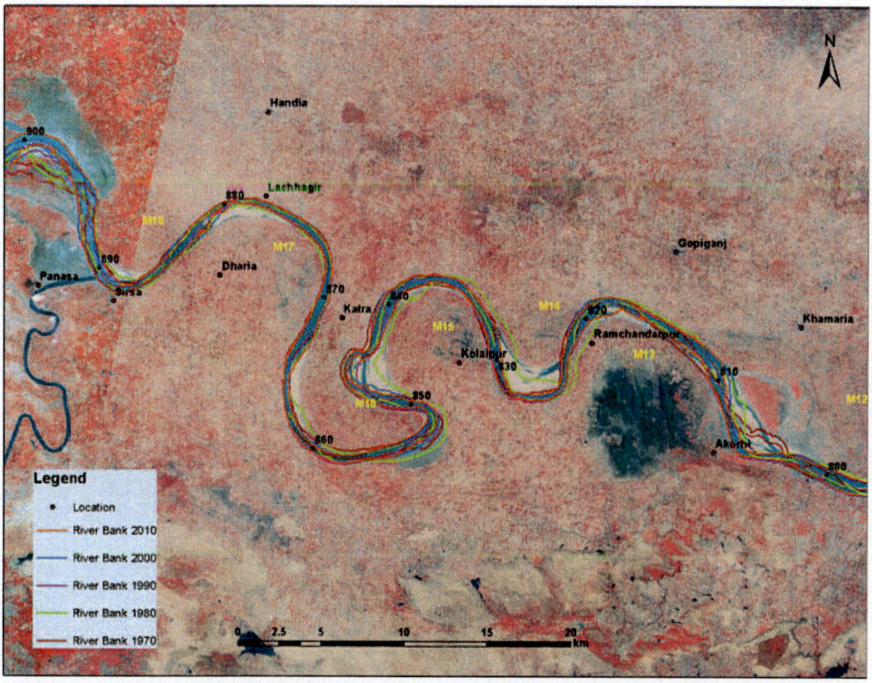

Fig. 2.7 Meandering pattern of Ganga River in the reach from chainage 800 km (Mirzapur) to 900 km (Saidabad)

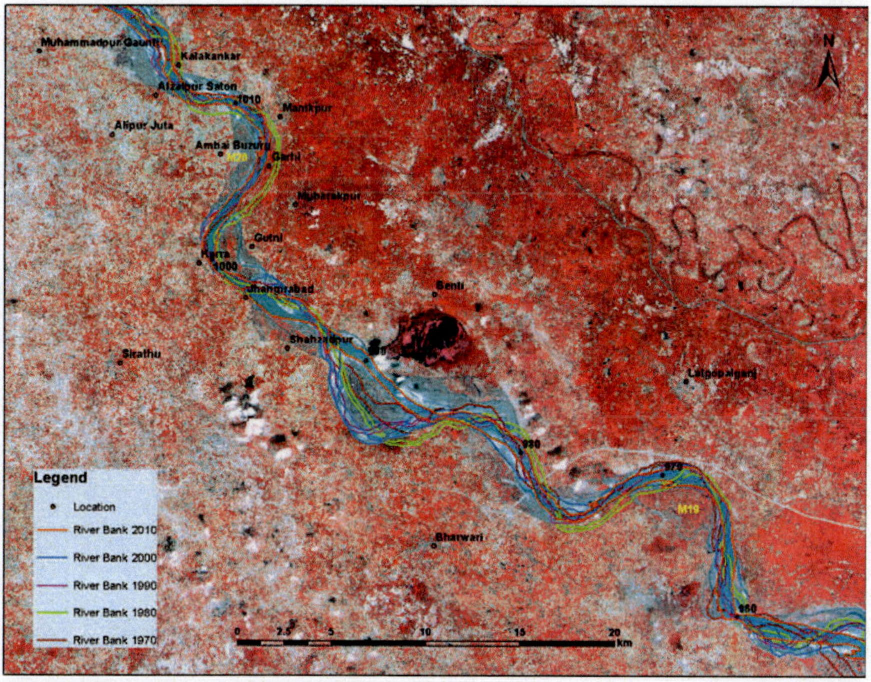

Fig. 2.8 Meandering pattern of Ganga River in the reach from chainage 960 km (Manauri) to 1015 km (Manikpur)

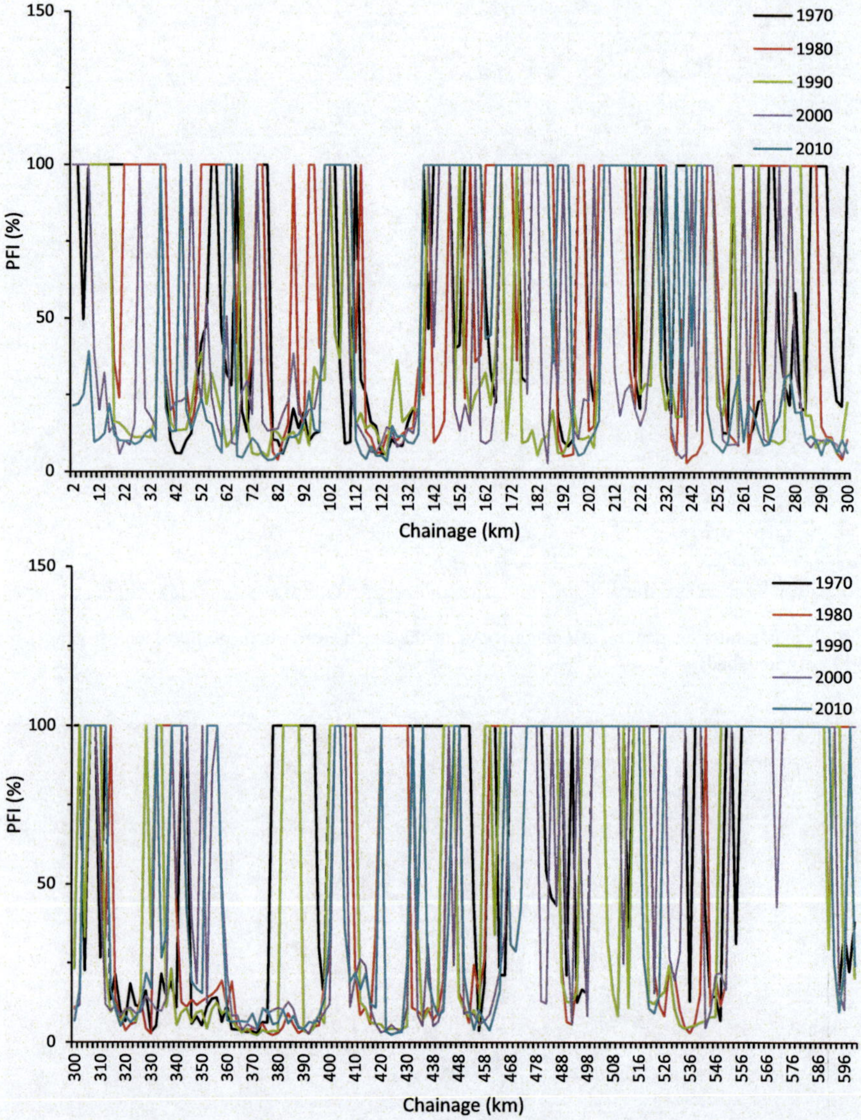

Fig. 2.9 (**a**) Planform Index of Ganga River from chainage 0–300 km. (**b**) Planform Index of Ganga River from chainage 300–600 km. (**c**) Planform Index of Ganga River from chainage 600–900 km. (**d**) Planform Index of Ganga River from chainage 900–1200 km. (**e**) Planform Index of Ganga River from chainage 1200–1500 km. (**f**) Planform Index of Ganga River from chainage 1500–1800 km

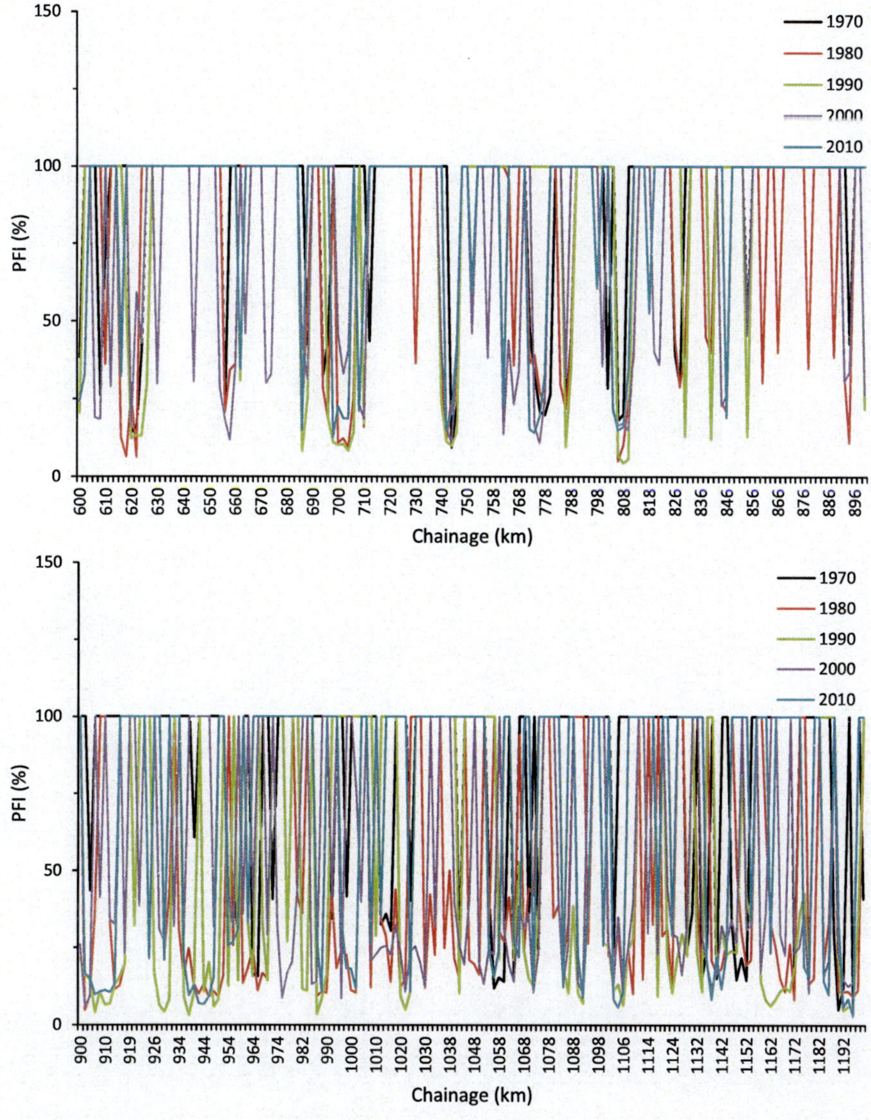

Fig. 2.9 (continued)

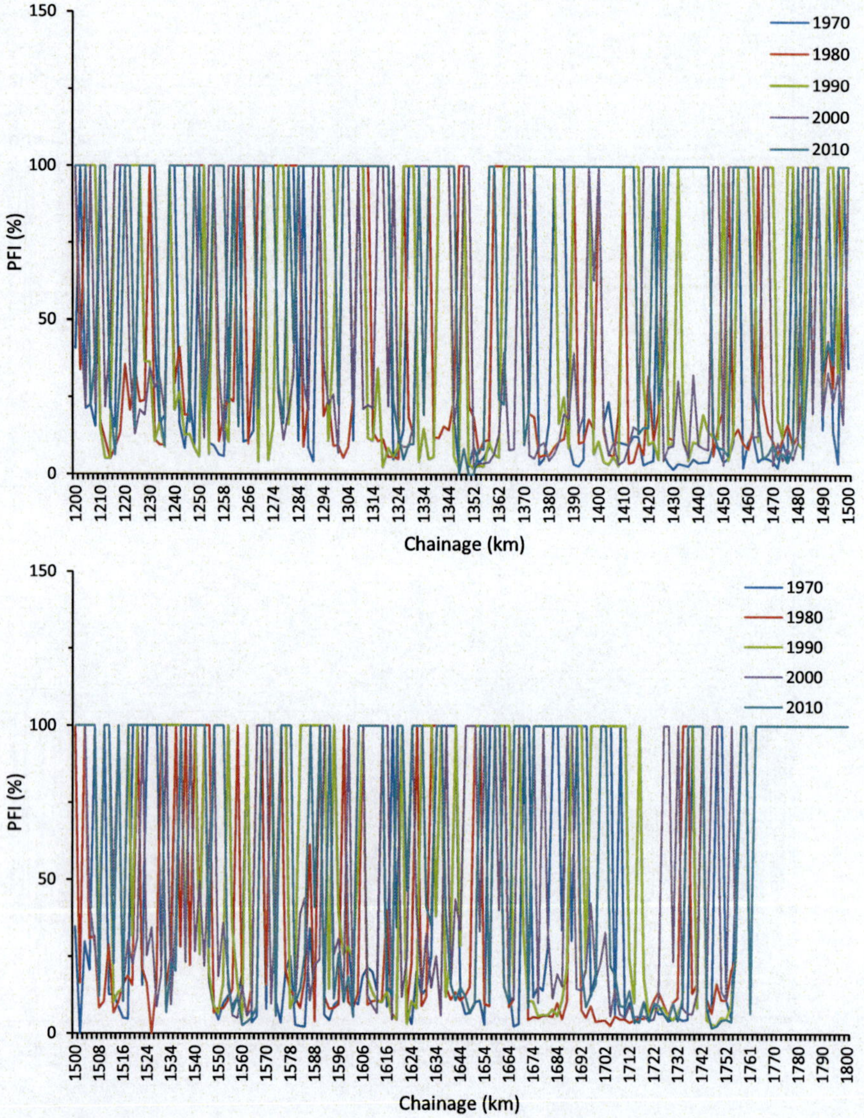

Fig. 2.9 (continued)

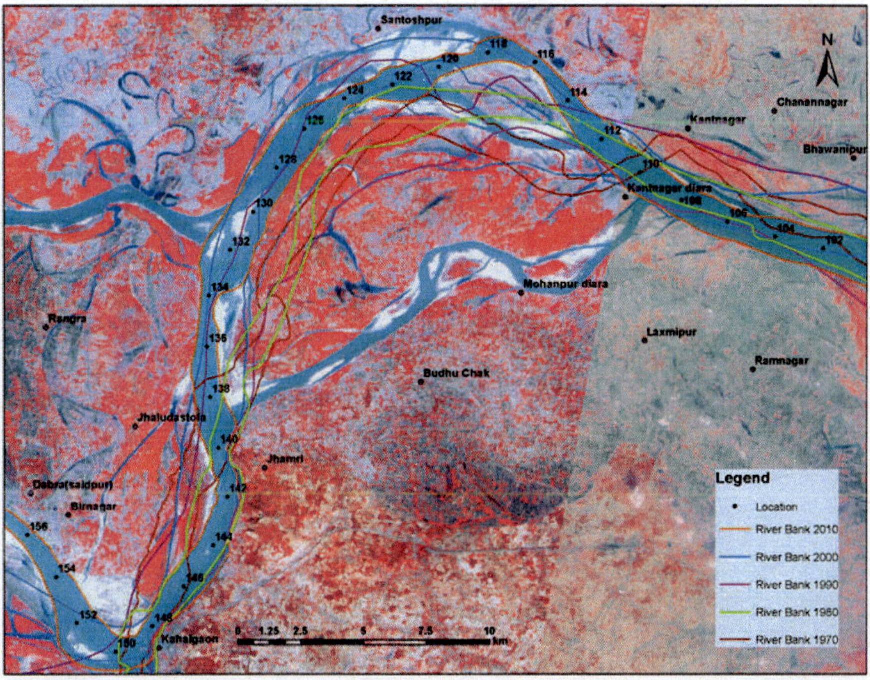

Fig. 2.10 (**a**) Braiding pattern of Ganga River in year 2010 from chainage 110–140 km near Krusela. (**b**) Braiding pattern of Ganga River in year 2010 from chainage 360–400 km near Raghopur. (**c**) Braiding pattern of Ganga River in year 1990 from chainage 520–550 km near Ballia. (**d**) Braiding pattern of Ganga River in year 2010 from chainage 930–950 km u/s of Allahabad. (**e**) Braiding pattern of Ganga River in year 1990 from chainage 1340–1360 km u/s of Farrukhabad. (**f**) Braiding pattern of Ganga River in year 1970 from chainage 1400–1440 km near Kachhla bridge. (**g**) Braiding pattern of Ganga River in year 1970 from chainage 1440–1480 km near Ramghat (u/s of Narora). (**h**) Braiding pattern of Ganga River in year 1970 from chainage 1550–1590 km near Garhmukteshwar. (**i**) Braiding pattern of Ganga River in year 2010 from chainage 1710–1740 km downstream of Haridwar

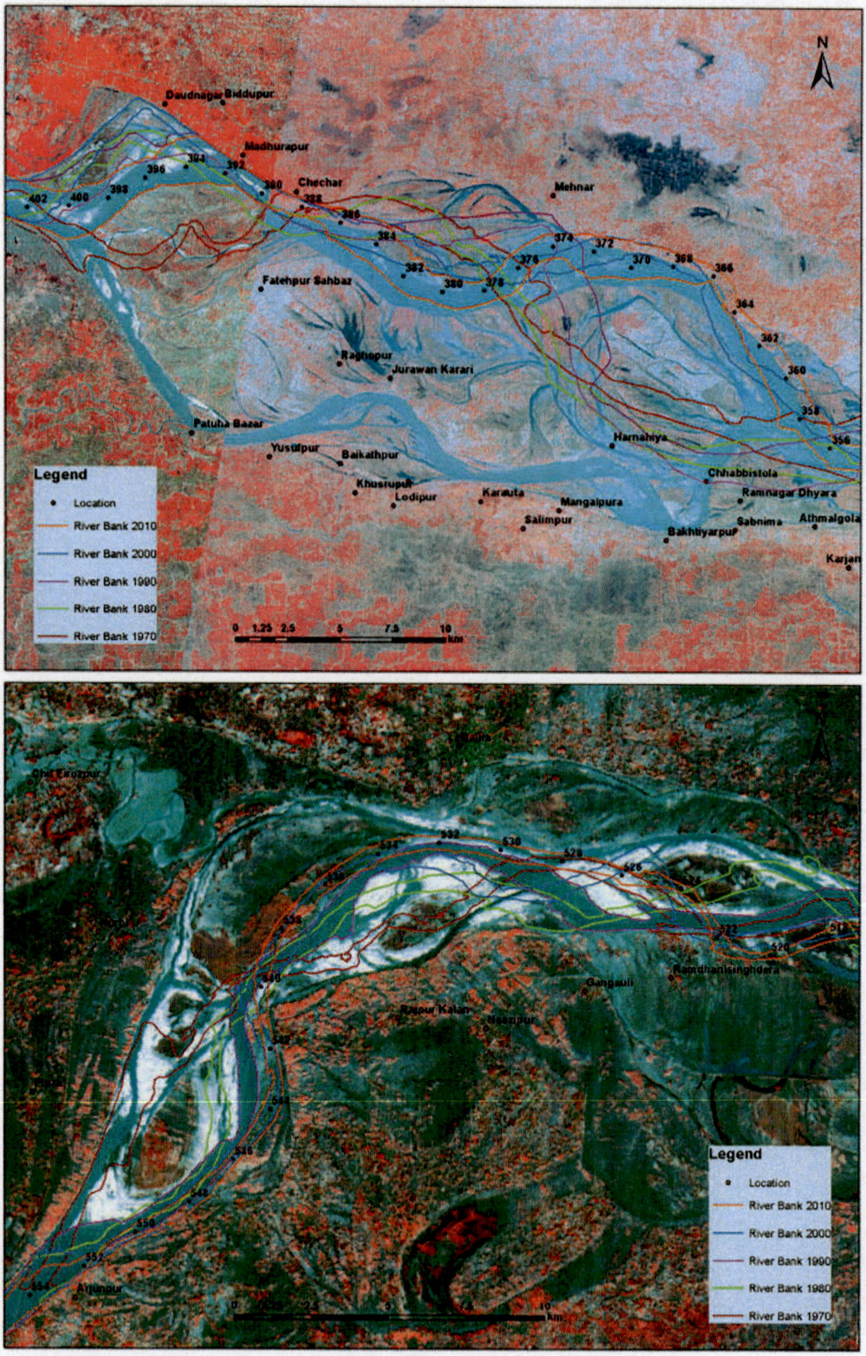

Fig. 2.10 (continued)

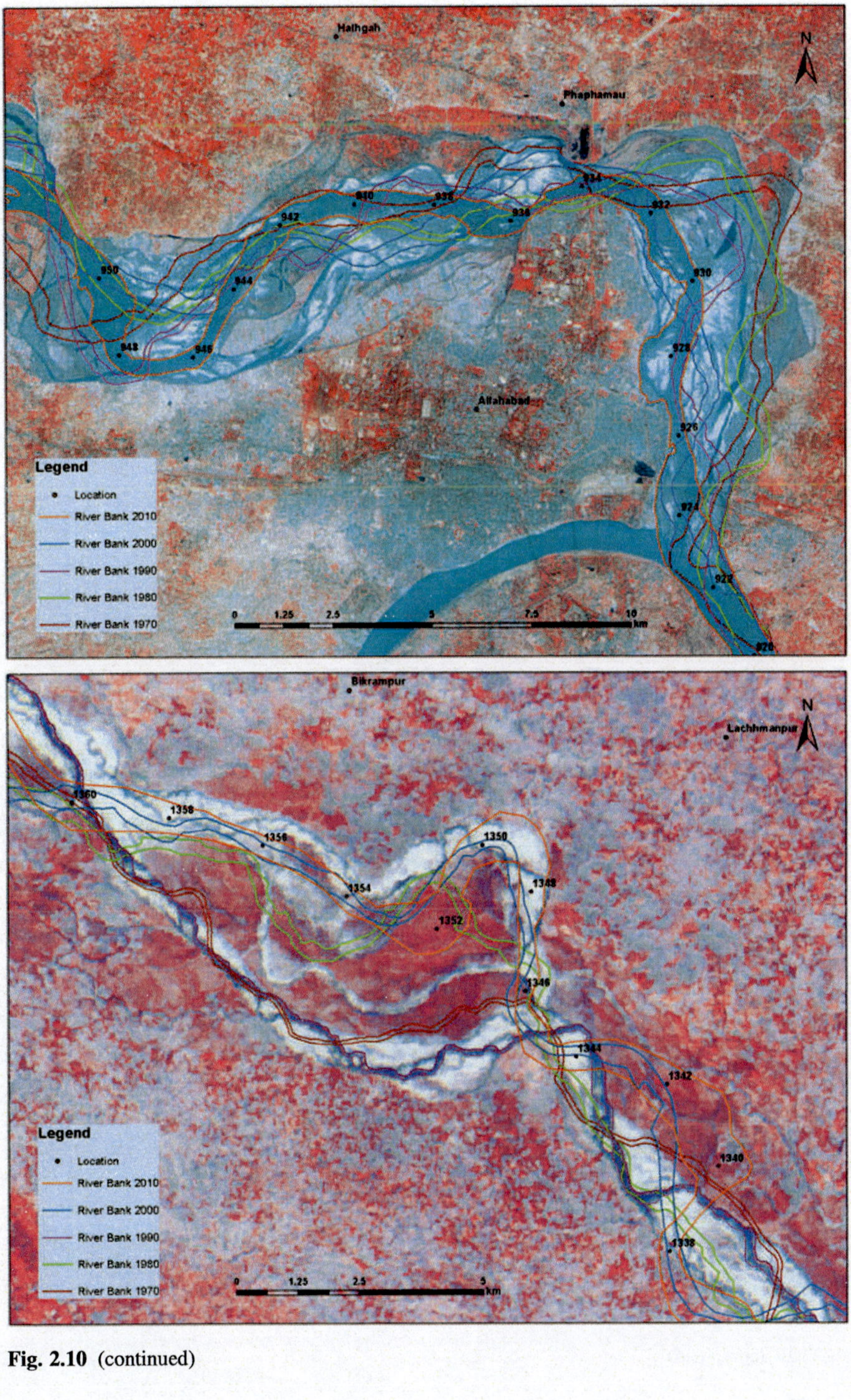

Fig. 2.10 (continued)

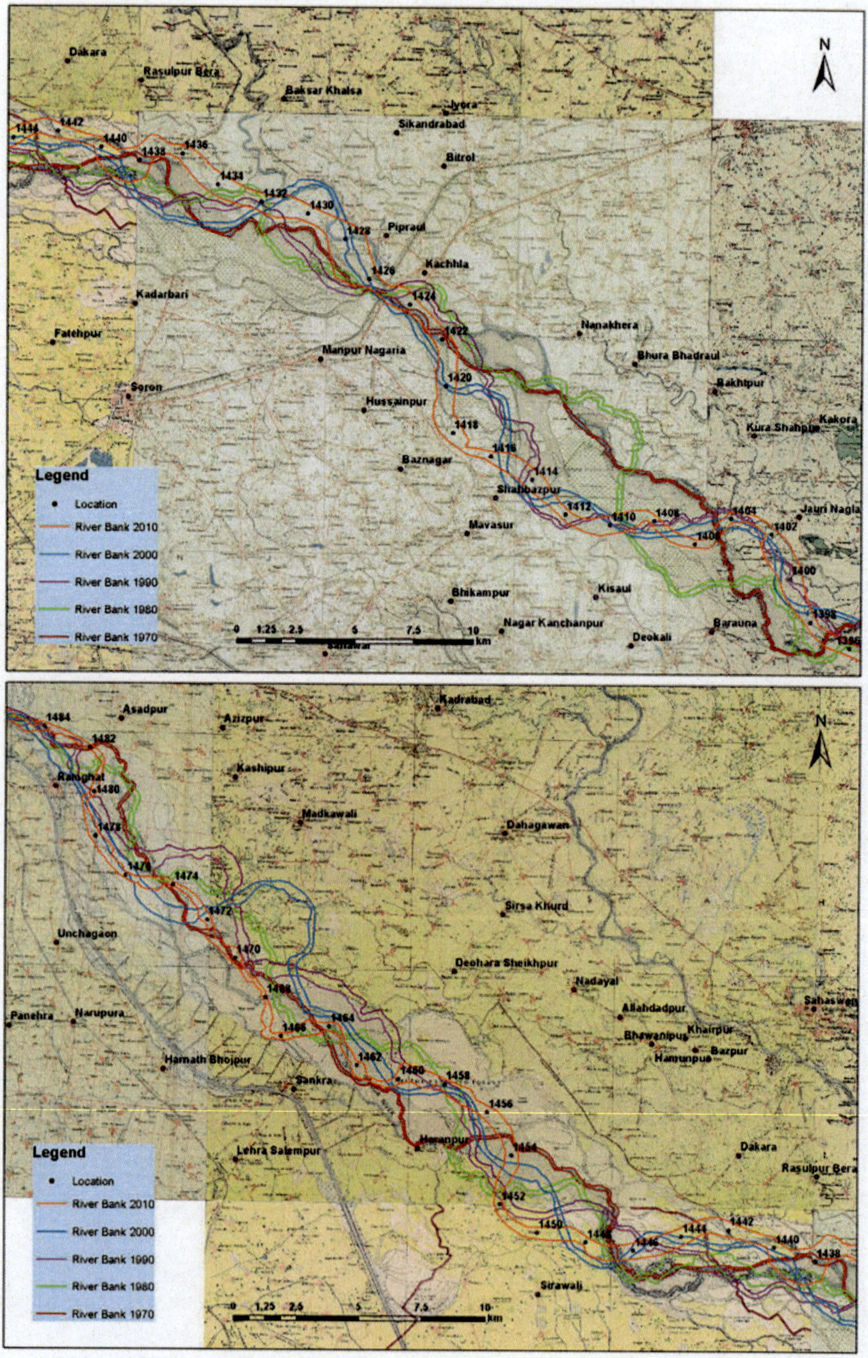

Fig. 2.10 (continued)

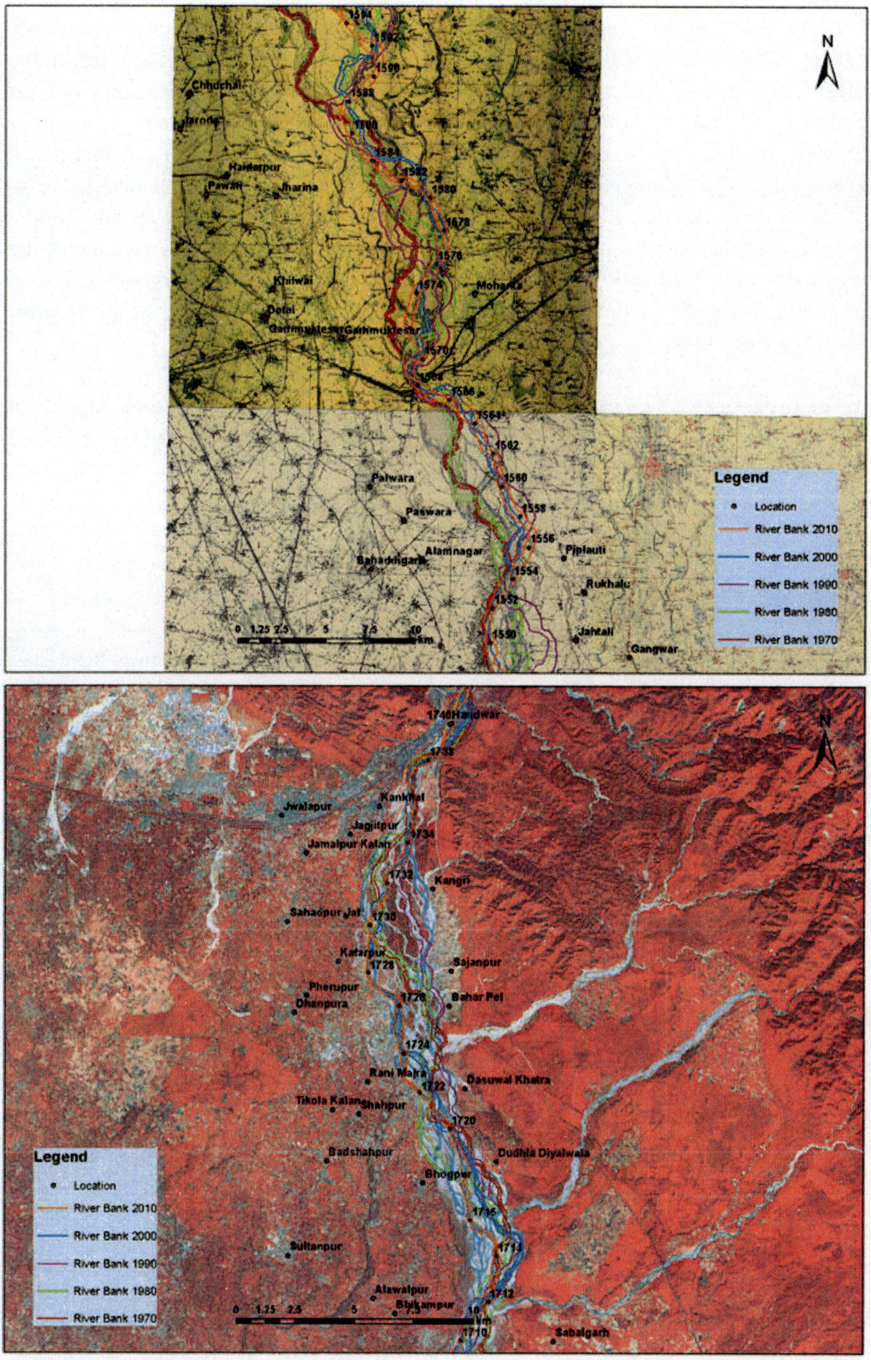

Fig. 2.10 (continued)

reaches. Based on the present analysis, Ganga shall be considered as a sinuous river except few reaches in which the river may be treated to be a meandering river. In the meandering reach, most of the meanders are stable, and no noticeable change in their geometry has been noticed from years 1970 to 2010. Ganga River is a prominently braided downstream of Haridwar, Garhmukteshwar, Ramghat and Kachlabridge; upstream of Farrukhabad, upstream of Allahabad, Ballia, Raghopur and Krusela. High looseness factors of barrages constructed across the Ganga River and contribution of significant sediment load from tributaries could be identified as one of the major causes behind braiding. In most of the reaches, Ganga River has been found to be braided. The *PFI* index has been found to be decreasing in most of the reaches resulting from both natural and anthropogenic factors.

Acknowledgement The authors would like to thank the Central Water Commission, Ministry of Water Resource, India, for providing funds for carrying this study. Contributions of Prof. R D Garg and Prof. P K Sharma, Dept. of Civil Eng., IIT Roorkee, are also acknowledged.

References

Agnihotri, A. K., Ohri, A., & Mishra, S. (2020). Channel planform dynamics of lower RamGanga River, Ganga basin, GIS and remote sensing analyses. *Geocarto International, 35*(9), 934–953.

Ahmad, Z., Garg, P. K., Garg, R. D., & Sharma, P. K. (2017). *Morphological study of Ganga River using remote sensing techniques*. CWC, Ministry of Water Resources.

Aiyelokun, O., Pham, Q. B., Aiyelokun, O., Malik, A. & Zakwan, M. (2021). Credibility of design rainfall estimates for drainage infrastructures: extent of disregard in Nigeria and proposed framework for practice. *Natural Hazards*. Article https://doi.org/10.1007/s11069-021-04889-1

Akhtar, M. P., Sharma, N., & Ojha, C. S. P. (2011). Braiding process and bank erosion in the Brahmaputra River. *International Journal of Sediment Research, 26*, 431–444.

Akhter, S., Eibek, K. U., Islam, S., Islam, A. R. M. T., Chu, R., & Shuanghe, S. (2019). Predicting spatiotemporal changes of channel morphology in the reach of Teesta River, Bangladesh using GIS and ARIMA modeling. *Quaternary International, 513*, 80–94.

Ara, Z., & Zakwan, M. (2018). Estimating runoff using SCS curve number method. *International Journal of Emerging Technology and Advanced Engineering, 8*, 195–200.

Bhuiyan, M. A., Kumamoto, T., & Suzuki, S. (2015). Application of remote sensing and GIS for evaluation of the recent morphological characteristics of the lower Brahmaputra-Jamuna River, Bangladesh. *Earth Science Informatics, 8*(3), 551–568.

Brice, J. C. (1964). *Channel patterns and terraces of the Loup Rivers in Nebraska*. U.S. Geological Survery Professional Paper 422-D, 41 pp.

Brice, J. C. (1981). Meander pattern of the White River in Indiana: A analysis. In M. Morisawa (Ed.), *River geomorphology* (pp. 178–200).

Friend, P. F., & Sinha, R. (1993). Braiding and meandering parameters. *Geological Society, London, Special Publications, 75*(1), 105–111.

Garde, R. J. (2006). *River morphology*. New age International Publishers.

Hooke, J. M. (1987). Changes in meander morphology. In V. Gardiner (Ed.), *International geomorphology 1986 part I* (pp. 591–609). Wiley.

Kumar, M., Denis, D. M., & Gourav, P. (2016). Study of meandering of river Ganga near Allahabad (India) using remote sensing and GIS techniques. *Asian Journal of Environmental Science, 11* (1), 59–63.

Lane, E. (1957). *A study of the shape channels formed by natural streams flowing in erodible material.* U.S. Army Eng. Division.

Leopold, L. B., & Wolman, M. G. (1957). *River channel patters: Braided, meandering, straight, physiographic and hydraulic studies of rivers*. U.S. Government Printing Office.

Leopold, L. B., Wolman, M. G., & Miller, J. P. (1964). *Fluvial processes in geomorphology* (p. 522). W.H. Freeman.

Majumdar, S., & Mandal, S. (2020). Assessment of relationship of braiding intensities with stream power and bank erosion rate through plan form index (PFI) method: A study on selected reaches of the upstream of Ganga river near Malda district, West Bengal, India. *Sustainable Water Resources Management, 6*(6), 1–18.

Msaddek, M. H., Moumni, Y., Chenini, I., & Dlala, M. (2019). Applicability of developed algorithm for semi-automated extraction and morphotectonic interpretation of lineaments using remotely sensed data, southwestern Tunisia. *Remote Sensing in Earth Systems Sciences, 2*(4), 292–307.

Mukherjee, K., & Pal, S. (2018). Channel migration zone mapping of the River Ganga in the Diara surrounding region of eastern India. *Environment, Development and Sustainability, 20*(5), 2181–2203.

Muzzammil, M., Alam, J., & Zakwan, M. (2018). A spreadsheet approach for prediction of rating curve parameters. In *Hydrologic modeling* (pp. 525–533). Springer. https://doi.org/10.1007/978-981-10-5801-1_36

Nawaz, A. R., Zakwan, M., Khan, I., & Rahim, Z. A. (2020). Comparative analysis of variants of Muskingum model. *Water and Energy International, 63*(7), 64–73.

Pandey, M., Zakwan, M., Sharma, P. K., & Ahmad, Z. (2018). Multiple linear regression and genetic algorithm approaches to predict temporal scour depth near circular pier in non-cohesive sediment. *ISH Journal of Hydraulic Engineering, 26*(1), 96–103. https://doi.org/10.1080/09715010.2018.1457455

Pandey, M., Zakwan, M., Khan, M. A., & Bhawe, S. (2020). Development of scour around a circular pier and its modelling using genetic algorithm. *Water Supply, 20*, 3358.

Raj, C., & Singh, V. (2021). Spatial and temporal variation of fluvial islands and sandbars in River Ganga from Bhagalpur to Farakka during 1955–2019. *Sustainable Water Resources Management, 7(3)*, 1-16.

Rust, B. R. (1978). A classification of alluvial channel systems. *Canadian Society of Petroleum Geologists, Memoir, 5*, 187–198.

Schumm, S. A. (1963). Sinuosity of alluvial rivers on the great plains. *Geological Society of America Bulletin, 74*, 1089–1100.

Schumm, S. A. (1977). *The fluvial system.* Wiley.

Sharma, N. (2004). *Mathematical modelling and braid indicators – The Brahmaputra Basin water resources* (Vol. 47, pp. 229–260). Kluwer Academic Publishers.

Singh, P., Patil, R. G., & Singh, A. (2018). Assessment of recent changes in planform of river Ganga from Mirapur Khadar to Narora barrage, Uttar Pradesh, India. *Sustainable Water Resources Management, 5*, 575. https://doi.org/10.1007/s40899-018-0222-z

Sinha, R., & Ghosh, S. (2012). Understanding dynamics of large rivers aided by satellite remote sensing: A case study from Lower Ganga plains, India. *Geocarto International, 27*(3), 207–219.

Sinha, R., Sripriyanka, K., Jain, V., & Mukul, M. (2014). Avulsion threshold and planform dynamics of the Kosi river in North Bihar (India) and Nepal: A GIS framework. *Geomorphology, 216*, 157–170.

Takagi, T., Oguchi, T., Matsumoto, J., Grossma, M. J., Sarker, M. H., & Matin, M. A. (2007). Channel braiding and stability of the Brahmaputra River, Bangladesh, since 1967: GIS and remote sensing analyses. *Geomorphology, 85*, 294–305.

Thakur, P. K., Salui, C., & Aggarwal, S. P. (2012). River bank erosion hazard study of river Ganga, upstream of Farakka barrage using remote sensing and GIS. *Natural Hazards, 61*, 967–987.

Wolman, M. G., & Miller, J. P. (1960). Magnitude and frequency of forces in geomorphic processes. *Journal of Geology, 68*, 54–74.

Yang, C., Cai, X., Wang, X., Yan, R., Zhang, T., Zhang, Q., & Lu, X. (2015). Remotely sensed trajectory analysis of channel migration in lower Jingjiang reach during the period of 1983–2013. *Remote Sensing, 7*(12), 16241–16256.

Zakwan, M. (2018). Spreadsheet-based modelling of hysteresis-affected curves. *Applied Water Science, 8*(4), 101–105. https://doi.org/10.1007/s13201-018-0745-3

Zakwan, M., Ahmad, Z., & Sharief, S. M. V. (2018). Magnitude-frequency analysis for suspended sediment transport in the Ganga River. *Journal of Hydrologic Engineering, 23*(7), 05018013.

Chapter 3
SWAT-Based Analysis of Ganga Water Availability at Farakka

Suman Bera and Ramkrishna Maiti

3.1 Introduction

The spatial and temporal variations of water flow at the basin level are more essential to understand the water resource availability. This availability becomes more crucial when river water is shared between two or more countries. The land use and climatic change (i.e., rainfall trend) are the significant components to determine the hydrological response (Vorosmarty et al., 2000; Manjan & Aggarwal, 2014) and water availability of any river basin. In this circumstance, FAO categorized the per capita water availability. The value with less than 2000 cubic meter per year and less than 1000 cubic meter per year is called "water-stressed" and "water-scarce condition," respectively. In India, the water availability has significantly dropped from 5177 cubic m/year (1951) to 1588 cubic m/year (2010). Furthermore, 14 river basins in India also face water-stressed conditions. The Ganga river basin (water availability is 1039 cubic m/year in 2010) is one of them (CWC, 2010; Gaur & Amarsinghe, 2011) and more significant for water sharing issues. Moreover, it is apparent from the report that fluctuation and reduction of water availability are attributed to irregular rainfall. Similarly, the availability of water at Farakka also fluctuates which has intensified the issue of water conflict. Several studies have been done on this issue and concluded that the water flow at Farakka is reduced mainly at lean season (Rao, 1979; Montu, 1980; Iyer, 1999; Nishat & Pasha, 2001; Haque, 2002; Chakraborty & Serageldin, 2004; Kamruzzaman & Zuppi, 2009; Parua, 2010). Numerous studies exposed that dry climate and land use change in the upper catchments area are responsible for declining of water availability at Farakka (Parua, 2010). Contrary, a number of research work made upstream obstruction responsible for water reduction at Farakka as well as in Bangladesh (Engineers Association of Bangladesh, 1997;

S. Bera (✉) · R. Maiti
Department of Geography and Environment Management, Vidyasagar University, Midnapore, West Bengal, India

G. S. Bhunia et al. (eds.), *Anthropogeomorphology*, Geography of the Physical Environment, https://doi.org/10.1007/978-3-030-77572-8_3

Mirza, 2004; Faisal, 2002; Khalid, 2010; etc.). Additionally, the unavailability of water flow data at Farakka point leads to the scope of confusion about the availability of water. Therefore, the analysis of spatial distribution and temporal variation of the surface runoff at the regional scale is required to validate the claim of water reduction at Farakka and also to understand the magnitude of water reduction.

World over, among 276 river basins, the Ganga river basin is one of the major transboundary rivers. Though the Ganga is shared by four countries, the water conflict exists between India and Bangladesh which deteriorates their neighboring relationships. A number of treaties in 1975, 1977, 1982, 1985, and 1996 have been concluded between them to devise a formula for sharing waters at Farakka for lean periods (January to May). So far, those formulas are not proven to be effective due to inadequate water at Farakka (shown in Fig. 3.1a), and the treaties are unsuccessful.

However, to measure the runoff, field techniques are very useful, but sometimes it is very expensive, time-consuming, and difficult; so rainfall-runoff models are applicable to compute it. In this context, the Soil Conservation Service (SCS, 1985) curve quantity technique may be very famous and extensively utilized by a number of researchers to decide the rainfall-runoff dating with minimal fact inputs (USDA, 1986; Schulze et al., 1992; SCS, 1972; Chatterjee et al., 2001; Bhuyan et al., 2003; Stuebe & Douglas, 1990; Sharma et al., 2001; Sharma & Kumar, 2002; Mishra et al., 2004; Pandey & Dabral, 2004; Pandey et al., 2005; Mishra et al., 2005; Jain et al., 2006; Dwivedi et al., 2017; Gupta & Panigrahy, 2008, Kumar & Rajpoot, 2013). Recently this SCS technique has been applied in numerous hydrological, erosion, and water quality models as CREAMS (Kinsel, 1980), EPIC (Sharpley & Williams, 1990), AGNPS (Young et al., 1989), and SWAT (Arnold et al., 1998). Lately, the Soil and Water Assessment Tool (SWAT) model (Arnold et al., 1998; Arnold & Fohrer, 2005) has been diagnosed to be a powerful and dependable device for assessing water aid availability and environmental situations across the world (Arnold et al., 1998; Arnold & Fohrer, 2005; Neitsch et al., 2005; Gassman et al., 2010).

Several researchers have applied the SWAT model on the Indian River to examine basin hydrology (Kaur et al., 2003; Gosain et al., 2006; Immerzeel et al., 2008; Kumar & Joshi, 2015). Santra and Das (2013) have exerted this model to estimate the runoff from an agricultural watershed in Chilika Lake, Odisha. Madhusudhan and Shivpuri (2016) have utilized the SWAT model over the Bennihalla river basin, a tributary of the Krishna River. Similarly, Sowjanya et al. (2015) used the model on the upper Krishna river basin. Kumar and Joshi (2015) have studied the upper catchment of the Subarnarekha River by using the SWAT model. Even to predict the sediment yield of the Ngawa watershed, SWAT is applied (Singh et al., 2012). In southern India, water availability in a semiarid catchment is assessed by SWAT (Perrin et al., 2012). To investigate the impacts of climate change on the runoff of the lower Brahmaputra river basin, the SWAT model has been used by Gain et al. (2011). Over the Bagmati River, Sharma and Shakya (2006) applied the model to study hydrological changes, and Manjan and Aggarwal (2014) exercised for water resource management. Khare et al. (2014) have done the hydrological modelling of Barinallah catchment in Chambal district, India. The

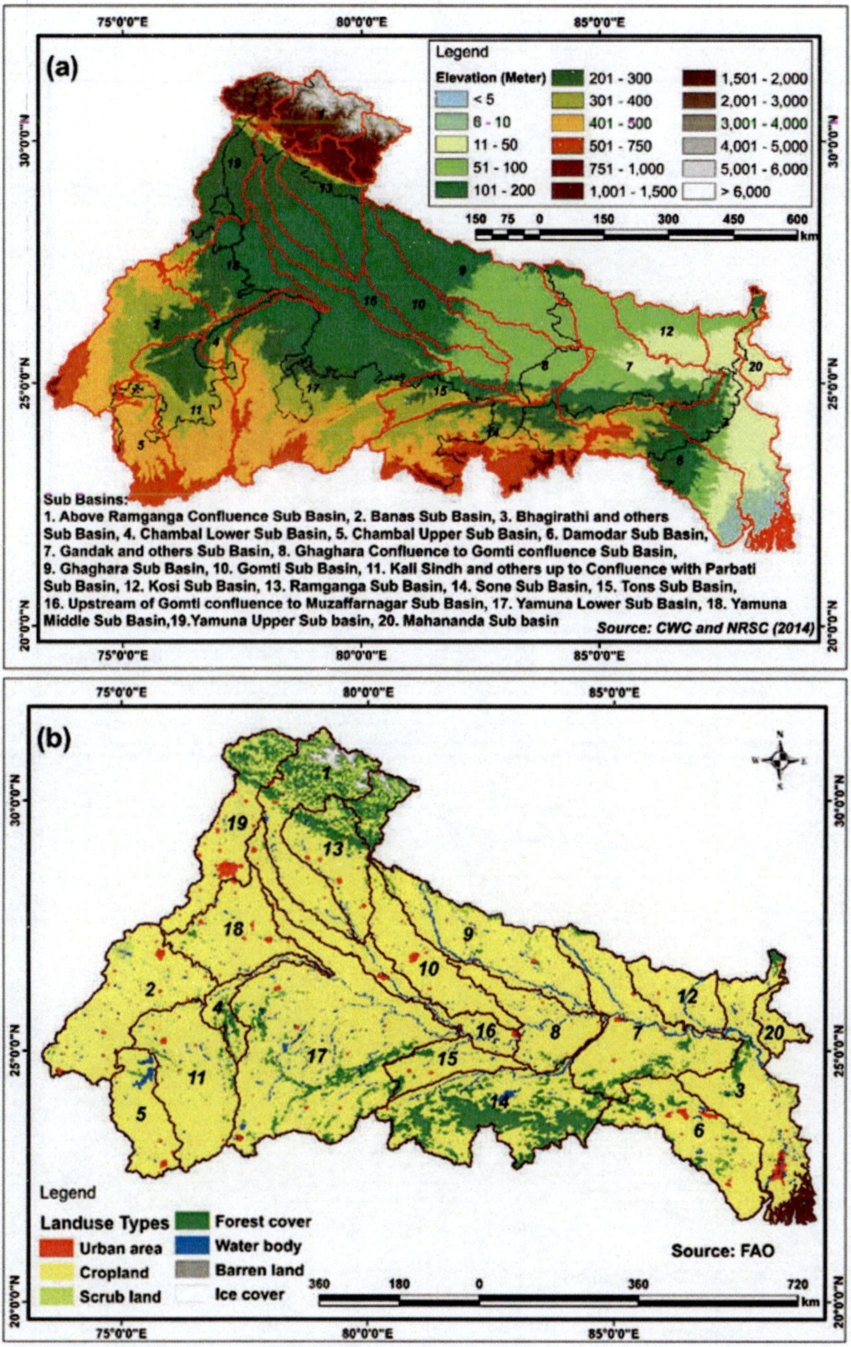

Fig. 3.1 Model input data for SWAT: (**a**) DEM, (**b**) land use and land cover, (**c**) soil groups, (**d**) location of weather stations

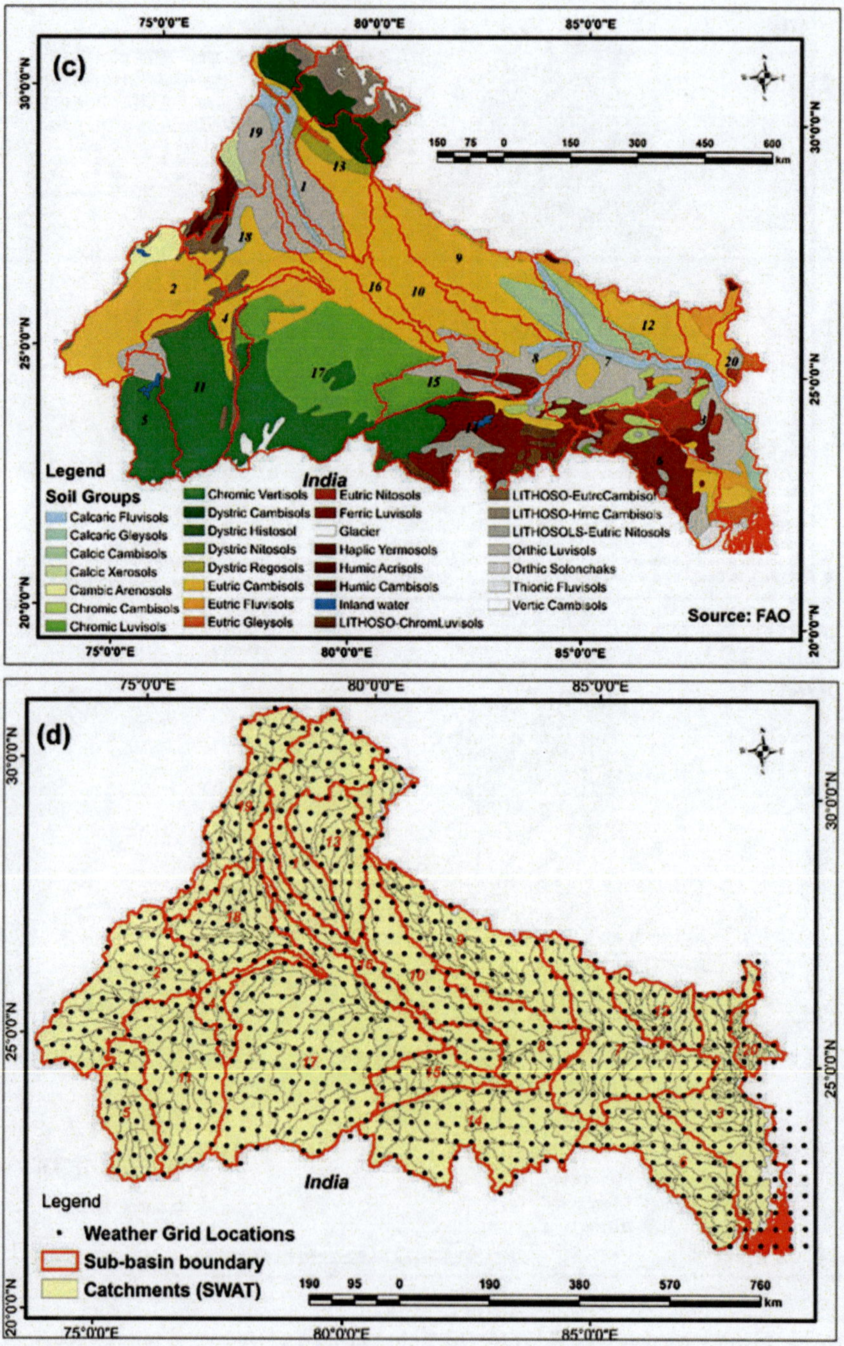

Fig. 3.1 (continued)

same model has been simulated over the Subansiri river basin by Chanda and Borah (2016) and over Vamsadhara River of south India by Jain and Sharma (2014). Recently, Bhunia (2016) applied this SWAT model on Jaipanda river basin for the estimation of surface runoff and water resources for effective management.

In the context of water sharing issues, limited research works are present on the Ganga river basin. Rao (1975) reported that the total annual runoff of the Ganga basin in India is 55 Mha-m, where Chaturvedi (1976) estimated as 49.3 Mha-m and Zade et al. (2005) calculated 58.93 Mha-m. Gupta and Panigrahy (2008) used SCS model to estimate the annual runoff of all river basins in India and found that Ganga river basin has produced 37.6 Mha-m in 2004 which is comparatively lower than the previous years. Further, Aggarwal et al. (2013) have assessed the annual runoff of the Ganga basin as 83.40 Mha-m. Ganguly et al. (2011) have run this SWAT model over the Ganga basin in India and obtained the declining trend of simulated water flow in annual, monsoon, and dry season at Farakka.

This paper represents an overview of the water resource availability of the Ganga basin with its sub-basins in India. According to CWC and NRSC (2014), Ganga basin in India consists of 19 sub-basins. Here Mahananda as the 20th sub-basin has been separated from Bhagirathi sub-basin. The analysis has been restricted from 1990 to 2014. Firstly, the water yield capacity of each, and every catchment under the basin has been estimated through the SWAT model in ArcGIS software. After that, discharge at every mouth of the catchment has been calculated for understanding the season-wise water availability of Ganga basin.

3.2 Materials and Methods

SWAT 2009 model has been successfully worked out through ArcGIS version 9.3.1. This model subdivided the DEM into sub-basins, and the hydrological characteristics are estimated based on topography, land use, soil, and other factors of the watershed (Olivera et al., 2006). Land use, soil types, and slope conditions are taken into consideration for further division of the basin into hydrological response units (HRUs) in this model (Wu & Xu, 2005). Finally, this model develops the daily, monthly, or yearly estimated runoff for a long time period. The chief benefit of this SWAT model is that it is used over watersheds where the gauge station is not present or the basin where hydrological data is restricted or where hydrological parameters are not available. So, this model is more appropriate for the Ganga basin due to hydrological data restriction.

The SWAT 2009 model required some essential databases like DEM, land use data, soil data, and meteorological data. Details of the data are as follows.

3.2.1 Digital Elevation Model (DEM)

Digital elevation model (DEM) is the main input, required for SWAT model to generate topographical characteristics of the Ganga basin. SRTM DEM of 90-m resolution is used to delineate the basin boundary of the river. The hydrological features like stream network, longest path, etc. of the basin are generated from the prepared DEM (Fig. 3.1a).

3.2.2 Land Use and Land Cover

Land use and land cover (LULC) is essential in SWAT model simulation because it affects the surface, subsurface runoff, soil erosion, and evapotranspiration (Mango et al., 2011; Neitsch et al., 2005). The LULC (FAO) of 1992–1993 has been used in this model (Fig. 3.1b). The dominant land use of the Ganga basin is cropland which accounts for 65.7% of this basin in the Indian part.

3.2.3 Soil Data

The soil layer with its textural and physicochemical properties, bulk density, water capacity, saturated hydraulic conductivity, organic carbon content, and electric conductivity for every soil category is required for the modelling (Setegn et al., 2009). Here the soil data has been collected from FAO (1992–1993) and has been converted into hydrological soil group (HSG) through maintaining with the HSG soil characters (Fig. 3.1c).

3.2.4 Meteorological Data

Besides the DEM, LULC, and soil data, meteorological information are required for SWAT simulation. This meteorological data has been collected from the Global Weather Data for SWAT which contains daily data of precipitation, temperature, humidity, solar radiation, and wind velocity for the period of 01/01/1979 to 31/07/2014. This meteorological data distributes at a resolution of 0.5×0.5 latitude and longitude grid points (Fig. 3.1d).

3.2.5 Model Setup

Firstly, the DEM, which is projected to WGS 1984 UTM Zone 45N, has been used for watershed delineation in the Arc SWAT model. Flow direction, flow accumulation, and stream network has been prepared, and outlet has been selected. In the next step, projected LULC, soil, and slope data have been exerted to obtain hydrological response units (HRUs). Daily precipitation, minimum and maximum temperature, relative humidity, solar radiation, and wind speed have been imported into the model and are available from Global Weather Data for SWAT. The hydrological processes of the selected basin are estimated by the following water balance equation (Eq. 3.1).

$$SW_t = SW + \sum_{i=1}^{t} (R_{it} - Q_i - ET_i - P_i - QR_i) \tag{3.1}$$

where *SW* is defined as the soil water content minus the wilting point water content and *R*, *Q*, *ET*, *P*, and *QR* are defined as daily precipitation (mm), runoff, evapotranspiration, percolation, and ground water flow, correspondingly.

3.2.6 Surface Runoff Estimation

Surface runoff means the volume of water which simply flows over the Earth's surface, following the slope condition. Actually, SWAT 2009 version takes land use-land cover, soil, antecedent moisture condition, and climatic parameters as inputs and simulates the floor runoff (Arnold et al., 1998) with the aid of using the usage of Soil Conservation Service (SCS) curve variety method (USDA & Soil Conservation Service, 1972). The equation of runoff estimation is as follows:

$$Q_{\text{surf}} = \frac{(R_{\text{day}} - I_a)^2}{(R_{\text{day}} - I_a + S)} \tag{3.2}$$

where Q_{surf} is the accrued runoff or rainfall excess (mm of H_2O) in m^3/s, R_{day} means rainfall intensity for the day (mm of H_2O), I_a is initial abstractions which includes surface storage interception and infiltration previous to runoff, and *S* denotes retention factor (mm of H_2O).

Soil retention parameter is determined from the curve number which differs in spatial and temporal scales. In the basin area, the variation of soil, water contents, soil types, land use, land management, and slope condition has significant effects on moisture retention capacity. It is determined by

$$S = 25.4\left(\frac{1000}{CN} - 10\right) \tag{3.3}$$

where CN means the curve number, a basic parameter that includes the areas, hydrologic soil group, land use, and hydrologic conditions (Williams, 1995).

Generally initial abstractions, I_a, are approximated as 0.2S, and after that Eq. (3.2) became (USDA & Soil Conservation Service, 1972) Eq. (3.4).

$$Q_{\text{surf}} = \frac{\left(R_{\text{day}} - 0.2S\right)^2}{\left(R_{\text{day}} + 0.8S\right)} \tag{3.4}$$

3.2.7 Calibration and Validation

The simulated output data are verified with NATCOM data from IIT, Delhi, for better accuracy. The calibration and validation of the datasets are made through two standard evaluation techniques.

3.2.8 Model Performance Evaluation

Nearly 20 statistical techniques are available to check the accuracy between the observed data and modelled simulated data (Coffey et al., 2004; Kumar & Joshi, 2015). Nevertheless, the popular techniques for assessing the performance of the model are the coefficient of determination (R^2) and Nash-Sutcliffe efficiency (*NSE*).

R^2 denotes the degree of correlation between the observed and simulated data (Eq. 3.5) ranging from 0 to 1. For this model, if the value of the R^2 becomes greater than 0.5, it is considered as the accepted value (Santhi et al., 2001; Van Liew et al., 2003; World Bank, 2011).

$$R^2 = \left[\frac{\sum_{i=1}^{n}\left(O_i - \overline{O}\right)\left(S_i - \overline{S}\right)}{\sqrt{\sum_{i=1}^{n}\left(O_i - \overline{O}\right)^2}\ \sqrt{\sum_{i=1}^{n}\left(S_i - \overline{S}\right)^2}}\right]^2 \tag{3.5}$$

The Nash-Sutcliffe efficiency (*NSE*) is some other statistical method (Eq. 3.6) which outlines the relative magnitude of residual variance on the subject of measured data (Nash & Sutcliffe, 1970; World Bank, 2011):

$$NSE = \left[\frac{\sum_{i}^{n} \left(Y_I^{obs} - Y_i^{sim}\right)^2}{\sum_{i}^{n} \left(Y_I^{obs} - Y^{mean}\right)^2} \right] \tag{3.6}$$

where Y_I^{obs} is the ith observed value, $Y_i^{sim}$is the ith simulated value, and Y^{mean} is the mean of the observed data set for evaluation. The *NSE* value varies from $-\infty$ to 1.0 where the resulted values within 0–1 are treated as acceptable only (Moriasi et al., 2007; World Bank, 2011).

3.3 Result and Discussion

The modelled annual surface flow has been estimated for temporal change of the annual and lean season over the basin. The simulation revealed, the maximum portion of the basin, i.e., 47–63.8%, has generated above 500 m^3/s surface flows, ensuring the water-rich nature of the basin annually.

More precisely, the highest flow ranges from 18530.82 to 64802.9 m^3/s, which has been observed in the lower catchments of Yamuna Lower, Sone, Gandak, Ghaghara, and Kosi sub-basins. Thus, it can be said that these sub-basins are a good contributor to the annual flow to the mighty Ganga. Therefore, the annual water flow at Farakka is enough because of more water flow contribution from the upper catchments. But, during the lean season, only 1.1–1.5% areas of the basin, situated in Gomti, Ghaghara, Ramganga, and Gandak sub-basins, have produced above 500 cumec surface flows. In a nutshell, Farakka has been facing the reduction of water flow resulting from meagre amount of water generation in the upper riparian sub-basins. Hence, water availability is a critical issue of this river basin throughout the treaty periods or lean season (January to May). Here the spatiotemporal changes of annual and lean season surface water availability of two different years (1991 and 2011) have been exhibited in Fig. 3.2a, b, c, and d.

In this circumstance, the lean season analysis is a very necessary or precise study. So, the line graph of Fig. 3.3 shows a very high variation in their surface water behavior. The simulated monthly discharge data (1990–2014) at Farakka outlet expressed that annually it holds a flow of 1073314 m^3/s on average. Nevertheless, the year-wise variation is quite high ranging from 36003 to 2171189 m^3/s. As expected, the high discharge is seen in the monsoon months in August and September.

In the lean season, the minimum and maximum simulated total discharges are 8092 m^3/s and 55059.2 m^3/s, respectively (Fig. 3.3). The maximum discharge of 22960 m^3/s is recorded in 1998. Contrary, the average lower values are recorded in the year 1996 (1237.6 m^3/s in May), 2000 (1120.7 m^3/s in April), 2001 (506.8 m^3/s in March), 2002 (1450.3 m^3/s in April), 2003 (1681.5 m^3/s in May), 2004 (621.4 m^3/s in March), 2006 (268 m^3/s in February), 2008 (766.9 m^3/s March), 2009 (886.1 m^3/

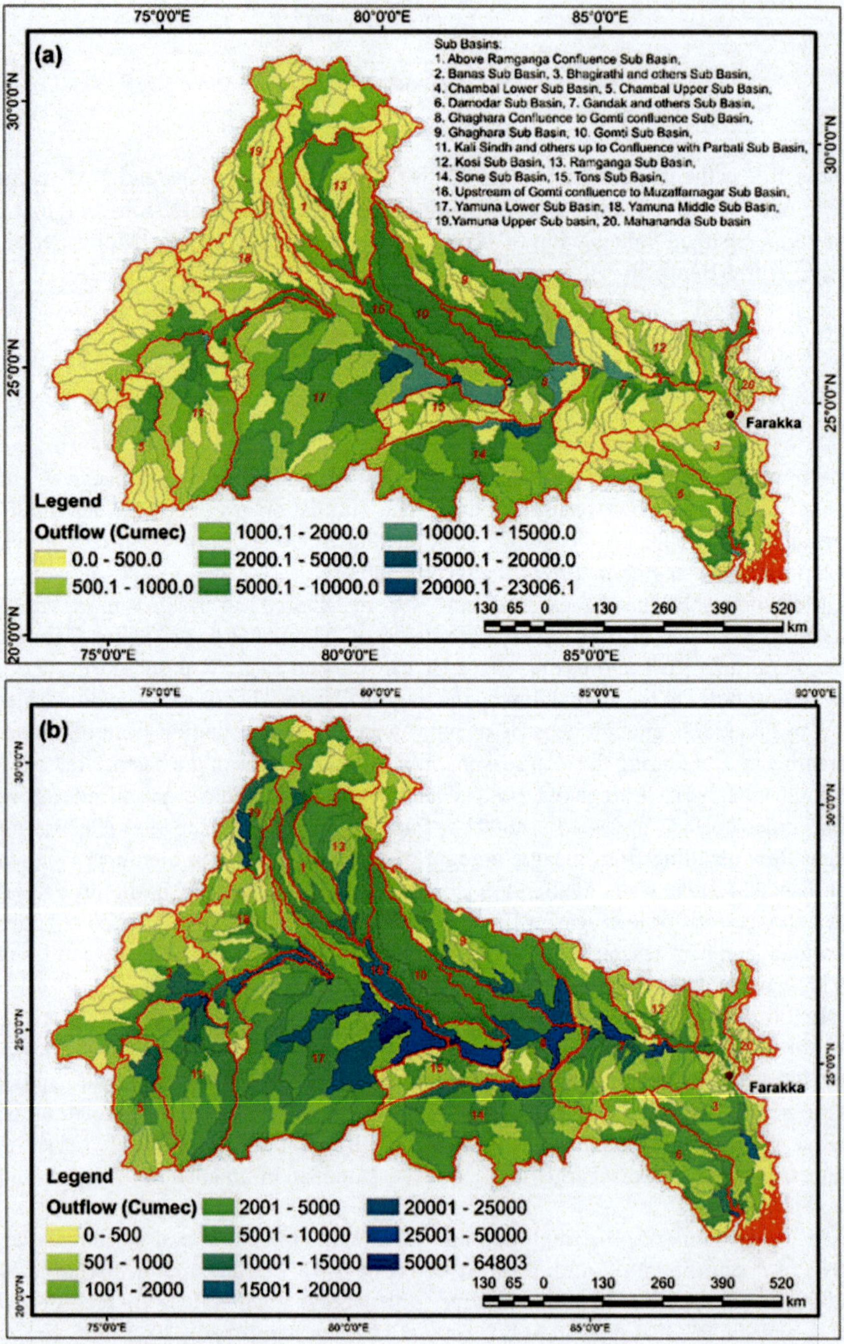

Fig. 3.2 Spatiotemporal pattern of surface water for (**a**) annual, 1991; (**b**) annual, 2011; (**c**) lean season, 1991; and (**d**) lean season, 2011

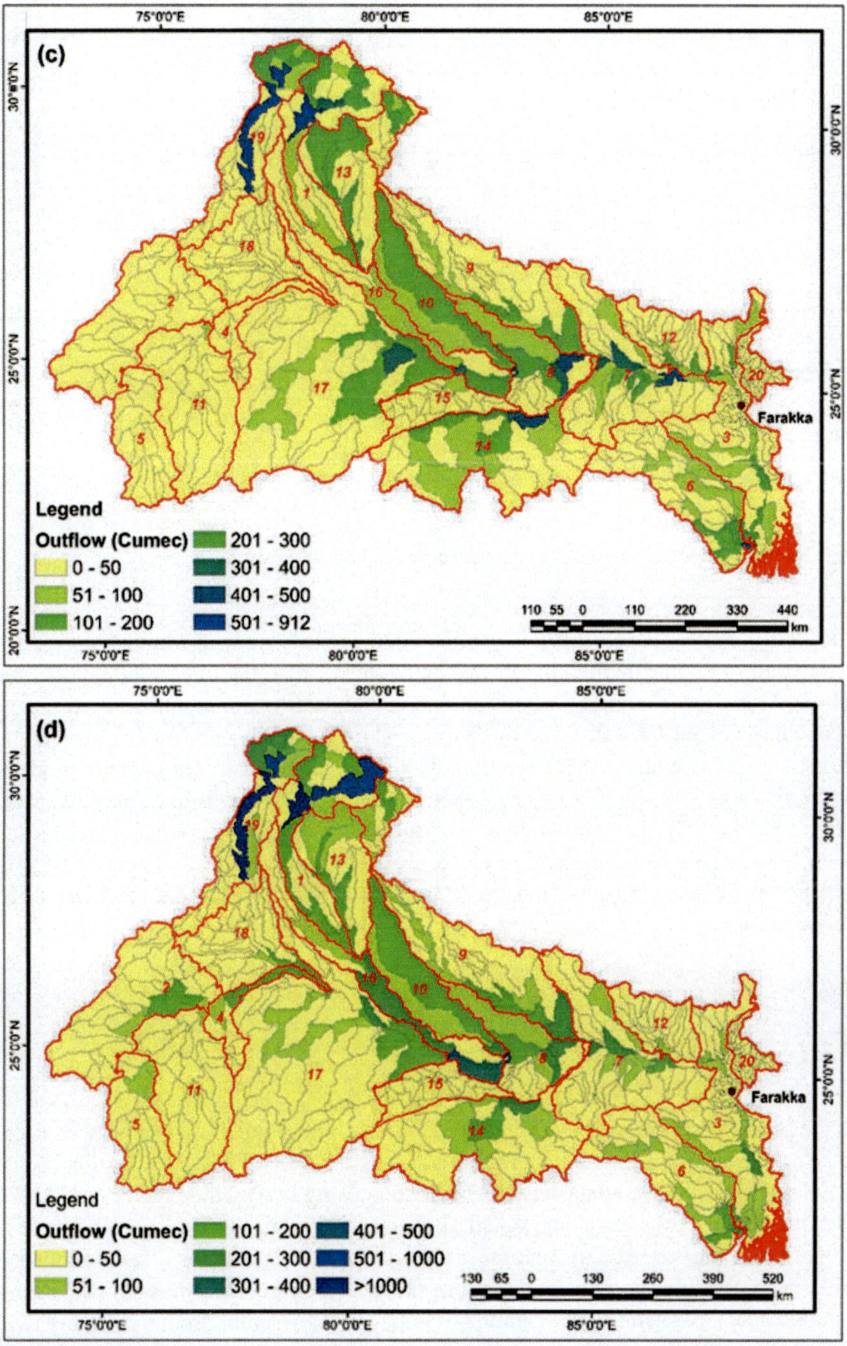

Fig. 3.2 (continued)

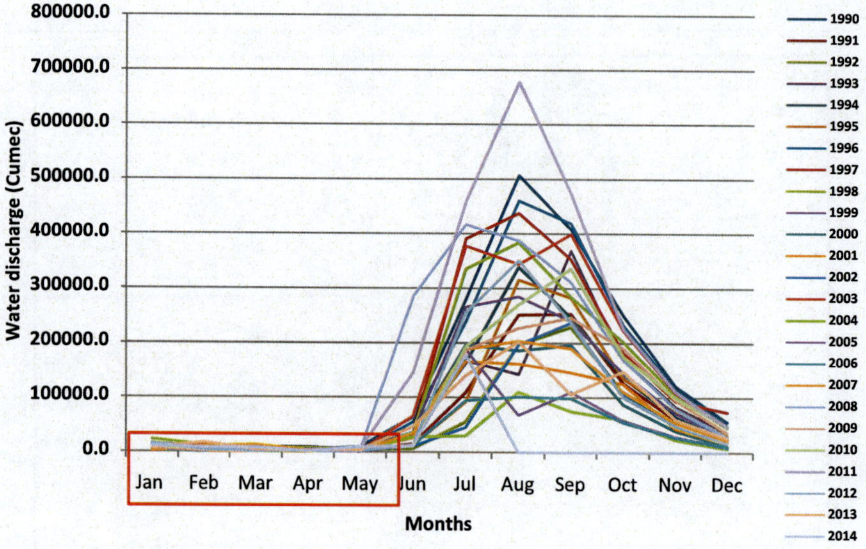

Fig. 3.3 Mean monthly variations of simulated stream flow at Farakka

s in April), 2010 (1084.4 m^3/s in April), 2012 (1615.8 m^3/s in March), and 2014 (1601.6 m^3/s in April) when the flow is much lower than the minimum water level required for water sharing (1982 m^3/s). Among the lean season months, April seems to be more crucial for water sharing as, in most cases, low water flow is observed in this month. It is evident that the water flow at Farakka fluctuates in the period of water sharing. Actually, the physical factor, i.e., negative trends in rainfall (Bera & Maiti, 2015), and the human intervention, i.e., changing patterns of land use, contribute largely to the reduction of the flow at Farakka barrage (Bera, 2019). It appears to be serious constraints in water sharing that further influence regional cooperation.

3.3.1 SWAT Model Performance

In the present study area, the discharge data of ten hydrological sites have been used to be calibrated and validated with the SWAT result. The statistical evaluation of the simulated and observed data sets has been done using two methods as R^2 and *NSE*. The simulated results can be accepted as satisfactory when $R^2>0{\cdot}5$ and $NSE>0{\cdot}5$.

Table 3.1 registers the evaluation results during the time series. The catchments have been carefully chosen mainly from the upper riparian sub-basins, i.e., Ramganga, Sone, Ghaghara, Kosi, Yamuna, Banas, and Kali Sindh. A comparison of two sets of discharge data designates a good performance.

Table 3.1 Performance criteria of the simulated data

Sl. no.	Calibration 2000–2002	*NSE*	R^2	Sl. no.	Validation 2003–2005	NSE	R^2
1	Dhasan_20316	0.80	0.82	1	Dhasan_20316	0.26	0.45
2	Gorara_2032101	0.36	0.75	2	Gorara_2032101	0.26	0.55
3	Baghain_2032015	0.62	0.75	3	Baghain_2032015	0.45	0.51
4	Tons_2060505	0.50	0.90	4	Tons_2060505	0.42	0.66
5	Son_2051201	0.83	0.83	5	Son_2051201	0.73	0.77
6	Katna_2051211	0.56	0.61	6	Katna_2051211	0.46	0.63
7	Kasai_2010801	0.40	0.63	7	Kasai_2010801	0.46	0.57
8	Ghagra_2020602	0.66	0.71	8	Ghagra_2020602	0.38	0.54
9	Sapt kosi_2070902	0.46	0.50	9	Sapt kosi_2070902	0.38	0.49
10	Kamla_2080105	0.61	0.69	10	Kamla_2080105	0.40	0.78

3.4 Conclusion

Hydrological data availability is a serious problem for the Ganga basin because of an international issue. For this reason, the well-appreciated SWAT model has been applied for simulating the water discharge to understand water resource availability. The simulated model data are matched with the data of ten hydrological sites within the Ganga basin. In the selected catchments, the calibration and validation with two methods show acceptable performance. The annual outflow is higher over the entire Ganga basin of India. According to the result, the average annual water flow at Farakka is 1073314 m^3/s. But during the lean season (January to May), the entire Ganga basin suffers from very low discharge. Concisely, the lean season surface flow of upper catchments of Farakka point is very meager with which water availability responses to shrink. Consequently, the actual water sharing has been failed. Hence it is essential to go to catchment-wise integrated water management that will develop new strategies to augment water resources.

References

Aggarwal, S. P., Garg, V., Gupta, P. K., Nikam, B. R., Thakur, P. K., & Roy, P. S. (2013). Run-off potential assessment over Indian landmass: A macro-scale hydrological modelling approach. *Current Science, 104*(7), 950–959.

Arnold, J. G., & Fohrer, N. (2005). SWAT2000: Current capabilities and research opportunities in applied watershed modelling. *Hydrological Processes, 19*(3), 563–572.

Arnold, J. G., Srinivasan, R., Muttiah, R. S., & Wiliams, J. R. (1998). Large area hydrological modeling and assessment part I: Model development. *Journal of the American Water Resources Association, 34*(1), 73–89.

Bera, S. (2019). *Cross boundary water conflict- A study on River Ganga.* Unpublished Thesis, Vidyasagar University.

Bera, S., & Maiti, R. (2015). Spatial and temporal variability of pre-monsoon rainfall during 1950–2000 in Ganga Basin, India and Bangladesh. *Indian Journal of Geography & Environment, 14*, 53–62.

Bhunia, P. (2016). *Estimation of water resources for effective management in Jaipanda River basin: A study through system approach*. Department of Geography and Environment Management, Vidyasagar University, Midnapore, 721102. [Thesis].

Bhuyan, S. J., Mankin, K. R., & Koelliker, J. K. (2003). Watershed-scale AMC selection for hydrologic modelling. *Transactions of ASAE, 46*, 237–244.

Chakraborty, R., & Serageldin, I. (2004). Sharing of river waters among India and its neighbors in the 21st century: War or peace? *Water International, 29*(2), 201–208.

Chanda, S., & Borah, T. (2016). Development of rainfall-runoff modelling using ArcSWAT in the Subansiri Basin. *Journal of Transactions on Engineering and Sciences, 4*(2), 70–88.

Chatterjee, C., Jha, R., Lohani, A. K., Kumar, R., & Singh, R. (2001). Runoff curve number estimation for a basin using remote sensing and GIS. *Asian Pacific Remote Sensing and GIS Journal, 14*, 1–7.

Chaturvedi, M. C. (1976). *Second India studies: Water* (pp. 12–22). MacMillion Company of India Ltd.

Coffey, M. E., Workman, S. R., Taraba, J. L., & Fogle, A. W. (2004). Statistical procedures for evaluating daily and monthly hydrologic model predictions. *Transactions of ASAE, 47*(1), 59–68.

CWC. (2010). *Central Water Commission: Preliminary consolidated report on effect of climate change on water resources, Government of India*. Available at: http://cwc.gov.in/main/downloads/Preliminary_Report_final.pdf. Accessed on 2 Dec 2017.

CWC, & NRSC. (2014). *Government of India: Ganga Basin. Version 2*. Ministry of Water Resources.

Dwivedi, P., Mishra, A., Karwariya, S., Goyal, S., & Thomas, T. (2017). SCS-CN method for surface runoff calculation of agricultural watershed area of Bhojtal. *SGVU Journal of Climate Change and Water, 1*(2), 9–12.

Engineers Association of Bangladesh. (1997). *Ganges water sharing treaty*. Press Club.

Faisal, I. M. (2002). Managing common waters in the Ganges-Brahmaputra-Meghna region: Looking ahead. *SAIS Review, Summer-Fall 2002, The Johns Hopkins University Press, 22*(2), 309–327.

Gain, W. W., Immerzeel, F. C., Sperna, W., & Bierkens, M. F. P. (2011). Impact of climate change on the stream flow of the lower Brahmaputra: Trends in high and low flows based on discharge-weighted ensemble modelling. *Hydrology and Earth System Sciences, 15*, 1537–1545.

Ganguly, T., Dubey, A. K., Dutta, S., Kartha, S. A., & Kumar, B. (2011). *Hydrological trend analysis of the Ganga Flow at Farakka Barrage, Conference Paper, February*. https://www.researchgate.net/publication/270957691

Gassman, P. W., Arnold, J. G., Srinivasan, R., & Reyes, M. (2010). The worldwide use of the SWAT model: Technological drives, net-working impacts and simulation trends. In *Proceeding 21st century watershed technology: Improving water quality and environment* (ASABE publication no 701P0210cd.St. Joseph, Mich.: ASABE. Costa Rica, February 2010). American Society of Agricultural and Biological Engineers.

Gaur, A., & Amarsinghe, P. (2011). A river basin perspective of water resources and challenges. In *India infrastructure report 2011, water: Policy and performance for sustainable development*. Oxford University Press.

Gosain, A. K., Rao, S., & Basuray, D. (2006). Climate change impact assessment on hydrology of Indian river basins. *Current Science, 90*(3), 346–353.

Gupta, P. K., & Panigrahy, S. (2008). *Geo-spatial modeling of runoff of large land mass: Analysis, approach and results for major river basins of India* (pp. 63–68). The International Archives of the Photogrammetry, Remote Sensing and Spatial Information Sciences, 37(B2), Beijing.

Haque, M. A. (2002). Treaty on sharing the Ganges waters: A milestone towards regional cooperation for development and management of water resources, regional hydrology: Bridging the gap between research and practice. In *Proceeding of the 4th international FRIEND conference held at cape town, South Africa, march 2002* (IAHS publ. no. 274) (pp. 307–313).

Immerzeel, W. W., Gaur, A., & Zwart, S. J. (2008). Integrating remote sensing and a process-based hydrological model to evaluate water use and productivity in a south Indian catchment. *Agricultural Water Management, 95*(1), 11–24.

Iyer, R. R. (1999). The fallacy of 'augmentation' demands on ganga waters. *Economic and Political Weekly, 34*(33), 2296–2297.

Jain, M., & Sharma, S. D. (2014). Hydrological modeling of Vamsadhara River basin, India using SWAT. In *International Conference on emerging trends in computer and image processing (ICETCIP'2014) December 15–16, 2014 Pattaya (Thailand)* (pp. 82–86).

Jain, M. K., Mishra, S. K., & Singh, V. P. (2006). Evaluation of AMC-dependent SCS-CN based models using watershed characteristics. *Journal of Water Resources Research, 35*, 3103–3114.

Kamruzzaman, M., & Zuppi, G. M. (2009). The root of the water paradox in Bangladesh: A case study in Ganges River basin. In *Thirteen International Water Technology Conference, IWTC 13 2009, Hurghada, Egypt* (pp. 797–812).

Kaur, R., Srinivasan, R., Mishra, K., Dutta, D., Prasad, D., & Bansal, G. (2003). Assessment of a SWAT model for soil and water management in India. *Land Use Water Resources Research, 3* (4), 1–7.

Khalid, I. (2010). Bangladesh water concern. *South Asian Studies, A Research Journal of South Asian Studies, 25*(1), 73–87.

Khare, D., Singh, R., & Shukla, R. (2014). Hydrological modelling of Barinallah watershed using Arc-Swat model. *International Journal of Geology, Earth and Environmental Sciences, 4*(1), 224–235.

Kinsel, W. G. (1980). *CREAMS: A field-scale model for chemical, runoff and erosion from agricultural management systems* (Conservation research report no. 26). South East Area, US Department of Agriculture.

Kumar, P., & Joshi, V. (2015). Applications of hydrological model SWAT on the upper watershed of river Subarnarekha with special reference to model performance and its evaluation. *Journal of Basic and Applied Engineering Research, Krishi Sanskriti Publications, 2*(13), 1128–1134.

Kumar, A., & Rajpoot, P. S. (2013). Assessment of hydro-environmental loss as surface runoff using CN method of Pahuj River basin Datia, India. *Proceeding of the International Academy of Ecology and Environmental Sciences, 3*(4), 324–329.

Madhusudhan, M. S., & Shivpuri, A. V. (2016). Application of SWAT model in generating surface runoff for Bennihalla River basin. *International Research Journal of Engineering and Technology, 3*(6), 1093–1097.

Mango, L. M., Melesse, A. M., McClain, M. E., Gann, D., & Setegn, S. G. (2011). Land use and climate change impacts on the hydrology of the upper Mara River basin, Kenya: Results of a modeling study to support better resource management. *Hydrology and Earth System Sciences, 15*, 2245–2258.

Manjan, S. K., & Aggarwal, S. P. (2014). Hydrological modeling of Bagmati River basin, Nepal. *International Journal of Current Engineering and Technology, 4*(6), 3978–3984.

Mirza, M. M. Q. (2004). The Ganges water-sharing treaty: Risk analysis of the negotiated discharge. In M. M. Q. Mirza (Ed.), *The Ganges water dispersion: Environmental effects and implications* (pp. 287–303). Kluwer Academic Publishers.

Mishra, S. K., Jain, M. K., & Singh, V. P. (2004). Evaluation of the SCS-CN-based model incorporating antecedent moisture. *Water Resources Management, 18*, 567–589.

Mishra, S. K., Jain, M. K., Bhunya, P. K., & Singh, V. P. (2005). Field applicability of the SCS-CN-based Mishra-Singh general model and its variants. *Water Resources Management, 19*, 37–62.

Montu, K. (1980, July 5). Farakka dispute. *Economic and Political Weekly, XV*(27), 1132.

Moriasi, D. N., Arnold, J. G., Van Liew, M. W., Bingner, R., Harmel, R. D., & Veith, T. L. (2007). Model evaluation guidelines for systematic quantification of accuracy in watershed simulations. *Transactions of the ASABE, 50*(3), 885–900.

Nash, J. E., & Sutcliffe, J. V. (1970). River flow forecasting through conceptual models part I- A discussion of principles. *Journal of Hydrology, 10*(3), 282–290.

Neitsch, S. L., Arnold, J. G., Kiniry, J. R., Srinivasan, R., & Wiiliams, J. R. (2005). *Soil and water assessment tool input/output file documentation- version 2005*. Blackland Research Centre, Texas Agriculture Experiment Station.

Nishat A. and Pasha M. F. K.: A review of the Ganges treaty of 1996. Globalization and water resources management: The changing value of water, University of Dundee International Specialty Conference, August 6–8, (2001).

Olivera, F., Valenzuela, M., Srinivasan, R., Choi, J., Cho, H., Koka, K., & Agrawal, A. (2006). ArcGIS-SWAT: A geodata model and GIS interface for SWAT. *Journal of the American Water Resources Association, 04087*, 295–309.

Pandey, A., & Dabral, P. P. (2004). Estimation of runoff for hilly catchment using satellite data. *Journal of the Indian Society of Remote Sensing, 32*(2), 235–240.

Perrin, J., Ferrant, S., Massuel, S., Dewamdel, B., Marechal, J. C., Aulong, S & Ahmed, J. (2012). Assessing water availability in a semi-arid watershed of southern India using a semi-distributed model. *Journal of Hydrology, 460–461*, 143–155.

Pandey, A., Chowdary, V. M., Mal, B. C., & Dabral, P. P. (2005). Estimation of surface water potential of agricultural watershed using geographic information system. *Asian Journal of Geoinformatics, 5*(4), 29–36.

Parua, P. K. (2010). *The Ganga: Water use in the Indian subcontinent* (Water science and technology library-64). Springer.

Rao, A. L. (1975). *India's water wealth: Its assessment, uses and projections*. Orient Longman Limited.

Rao, K. L. (1979). *India's water wealth*. Orient Longman.

Santhi, C., Arnold, J. G., Williams, J. R., Dugas, W. A., Srinivasan, R., & Hauck, L. M. (2001). Validation of the SWAT model on a large river basin with point and nonpoint sources. *Journal of the American Water Resources Association, 37*(5), 1169–1188.

Santra, P., & Das, B. S. (2013). Modeling runoff from an agricultural watershed of western catchment of Chilka lake through ArcSWAT. *Journal of Hydro-Environment Research, 7*(4), 261–269.

Schulze, R. E., Schmidt, E. J., & Smithers, J. C. (1992). *SCS-SA user manual, PC based SCS design flood estimates for small catchments in southern Africa* (Report no. 40). Department of Agricultural Engineering University of Natal.

SCS. (1972/1985). Soil conservation service: Hydrology. In *National engineering handbook, supplement A, section 4, chapter 10*. USDA.

Setegn, S. G., Srinivasan, R., Dargahi, B., & Melesse, A. M. (2009). Spatial delineation of soil erosion vulnerability in the Lake Tana Basin, Ethiopia. *Hydrological Processes, 23*, 3738–3750.

Sharma, D., & Kumar, V. (2002). Application of SCS model with GIS data base for estimation of runoff in an arid watershed. *Journal of Soil and Water Conservation, 30*(2), 141–145.

Sharma, R. H., & Shakya, N. M. (2006). Hydrological changes and its impact on water resources of Bagmati watershed, Nepal. *Journal of Hydrology, 327*(3–4), 315–322.

Sharma, T., Satya Kiran, P. V., Singh, T. P., Trivedi, A. V., & Navalgund, R. R. (2001). Hydrological response of watershed to Landuse changes: Remote sensing and GIS approach. *International Journal of Remote Sensing, 22*(11), 2095–2108.

Sharpley, A. N., & Williams, J. R. (1990). *EPIC—Erosion/productivity impact calculator: 1. Model documentation* (US Department of Agriculture technical bulletin no. 1768). US Government Printing Office.

Singh, M. I., Imtiyaz, M., Isaac, R. K., & Denis, D. M. (2012). Comparison of soil and water assessment tool (SWAT) and multilayer perception (MLP) artificial neutral network for predicting sediment yield in the Nagwa agricultural watershed in Jharkhand, India. *Agricultural Water Management, 104*, 113–120.

Sowjanya, P., Venkat, R. K., & Shashi, M. (2015). Climate change effect on water resources using climate model data of Indian river basin. *International Journal of Research in Engineering and*

Technology, *4*(11), 150–154. Retrieved from https://pdfs.semanticscholar.org/b462/6cea369fcd7826f0c423f129898cb28a3301.pdf

Stuebe, M. M., & Douglas, M. J. (1990). Runoff volume estimation using GIS techniques. *Water Resources Bulletin, 26*(4), 611–620.

USDA, & Soil Conservation Service. (1972). *National engineering handbook, section 4: Hydrology*. USDA.

USDA. (1986). *Urban hydrology for small watersheds. Technical release*. National Technical Information Service.

Van Liew, M. W., Arnold, J. G., & Garbrecht, J. D. (2003). Hydrologic simulation on agricultural watersheds: Choosing between two models. *Transactions of the ASABE, 46*(6), 1539–1551.

Vorosmarty, C. J., Green, P., Salisbury, J., & Lammers, R. B. (2000). Global water resources: Vulnerability from climate change and population growth. *Science, 289*, 284–288.

Williams, J. R. (1995). The EPIC model. In V. P. Singh (Ed.), *Computer models of watershed hydrology*. Water Resources Publications.

World Bank (2011). *Water systems modeling for Ganga basin: Final report (English)*. World Bank Group. Retrieved from http://documents.worldbank.org/curated/en/661041468197387099/Water-systems-modeling-for-Ganga-basin-final-report

Wu, K., & Xu, J. (2005). Applicability of SWAT for three coastal watersheds in Louisiana. Watershed management to meet water quality standards and emerging TMDL (Total maximum daily load). In *Proceedings of the Third Conference 5–9 March 2005. Atlanta, Georgia, USA*.

Young, R. A., Onstad, C. A., Bosch, D. D., & Anderson, W. P. (1989). AGNPS: A nonpoint-source pollution model for evaluating agricultural watersheds. *Journal of Soil and Water Conservation, 44*, 168–173.

Zade, M., Ray, S. S., Dutta, S., & Panigrahy, S. (2005). Analysis of runoff pattern for all major basins of India derived using remote sensing data. *Current Science, 88*(8), 1301–1305.

Chapter 4
Bank Erosion and Sediment Deposition in Teesta River: A Spatiotemporal Analysis

Mst. Rebeka Sultana

4.1 Introduction

The downstream location of Teesta River is susceptible to bank erosion and sediment deposition. Originating from Himalaya, the Teesta flows through India and Bangladesh and meets the Brahmaputra at Gaibandha district of Bangladesh. The banks of Teesta are characterized by its massive erosion proneness. Similarly, gentle slope along with loose soil compactness increases the vulnerability of bank erosion of the river. The river brings huge sediment load during floods. Large-scale suspended sediments such as silt and clay are carried by the rivers of Bangladesh in flooding period (Sikder, 2008). The Teesta river transports 31.4 million ton of sediment per year, and the mean annual runoff of the river is 25.2 billion m^3 (Shi et al., 2019). Moreover, riverbank erosion also increases the depositional activities of rivers. Likewise, rill and gully erosion in the upstream location and in the floodplains also acts as source materials for deposition of sediment in riverbanks and bed as well. The sediments of Teesta are composed of silt, sand, and mud. Predominantly the sediments of Teesta River are silt and clay, but sand reduces in the downstream location and the concentration of mud accelerates (Saha et al., 2017). The driving forces of bank erosion in the country are turbulence and vortices of water in the monsoon, non-cohesive soil texture, and anthropogenic impacts in the river. Deforestation, urbanization, and various anthropogenic activity regulate soil erosion and deposition of river basins (Suif et al., 2016). Fluctuation of water discharge in dry and wet seasons after dam and barrage construction plays a vital role in bank erosion and sediment deposition of Teesta as well. Water diversion, irrigation projects, barrage construction, etc. effects on river stability and drought occur for reduced

M. R. Sultana (✉)
Department of Geography and Environmental Studies, University of Rajshahi, Rajshahi, Bangladesh
e-mail: rebeka.sultana@ru.ac.bd

G. S. Bhunia et al. (eds.), *Anthropogeomorphology*, Geography of the Physical Environment, https://doi.org/10.1007/978-3-030-77572-8_4

water flow in dry season in Teesta (Khan & Islam, 2015). Water discharge capacity of Teesta River decreases abruptly during the dry and monsoon periods (Mondal & Islam, 2017). Intensive rainfall in the upper catchments follows huge water discharge in the Teesta River, and inundation occurs in the floodplains of Bangladesh which initiates riverbank erosion in monsoon and post monsoon periods, and also high yield of sediment transportation occurs. Onset of monsoon carries more that 200–600 mm rainfall in Teesta catchment in each year, and the river transports the highest sediment load than the other rivers of the Himalayan belt (Pal et al., 2016). The extent of flood determines the prevalence of erosion and deposition of sediment. Low magnitude flood stores fine deposits in the depressions, but in the case of high extent of flood, sand accumulation occurs in the main stream channel (Middelkoop, 2006). Likewise, sediment deposition and erosion are exposed to change in stream channel. The Teesta River faces extreme channel change problem due to erosion and sedimentation. Stream channel migration is visible in the banks of a river where erosion takes place and the river deposits sediments in the previous channels (Sarma et al., 2007).

The scientific study of erosion and sediment deposition of different river reaches has been investigated by geographers, river engineers, environmentalists, and researchers of different fields. Recently, the trend of river research accelerated with GIS and remote sensing techniques. Alekseevskiy et al. (2008) illustrated the fluvial erosion and sediment conveyance along with accumulation in a river system. The study also explored the channel process in spatiotemporal scale. Kotoky et al. (2015) conducted a research on the erosional and depositional activity of Brahmaputra River in Assam, India. Ety and Rashid (2020) detected the spatiotemporal changes of erosion and deposition of Ganges River in Bangladesh. Moreover, the paper explored the comparison of James Rennell's map of Ganges river basin from a long historical perspective from 1760–2015. Correspondingly, Billah (2018) examined the changes in riverbanks of Padma and its erosion-accretion dynamics through GIS and RS techniques. Chalov et al. (2015) carried out a study on the spatiotemporal analysis of sediment transport of Selenga River. Laskar and Phukon (2012) investigated the erosion variability of Barak River with satellite data and topomaps. The study analyzed 85 years of erosional activities of the river spanning from six spatiotemporal data sets. Petropoulos et al. (2015) analyzed the spatiotemporal changes of Axiosand and Aliakmonas river deltas. The study emphasized on the erosion and deposition of Mediterranean delta environment. Chakraborty and Datta (2013) provided a GIS- and remote sensing-based approach on the channel changes of Jaldhaka-Diana River. Thakur et al. (2012) carried out a study on bank erosion of Ganges River using morphological parameters. The study is based on satellite imagery and highlighted the left bank erosion of the river in Malda district from 1955 to 2005 with increased sinuosity. Pal et al. (2017) used the multi-temporal satellite images to determine the condition of erosion and accretion of Dudhkumar River of northern Bangladesh. Mukherjee et al. (2017) assessed the bank dynamics of lower Ramganga River and explained the river planform pattern change with erosion and accretion mechanism. Aher et al. (2012) identified bank erosion of

Pravara River and also detected channel change with remote sensing technique and topographical sheet study.

Channel morphology change, sedimentation, planform pattern, and erosion of Teesta River in both Bangladesh and Indian portions have been explored by many researchers of geography, geomorphology, and other disciplines. Chakraborty and Ghosh (2010) reviewed a study on the geomorphological and sedimentological aspects of Teesta mega fan. The study elaborated the process of Teesta mega fan building. Tarannum et al. (2018) examined the change in Teesta River morphology. The study used the data of morphological survey of Bangladesh Water Development Board from 2001 to 2014 and identified net erosion and deposition of Teesta River in some selected points. Froehlich and Walling (2006) investigated the over bank sedimentation of Teesta River along with sedimentation pattern in the floodplains. Akhter et al. (2019) predicted the change in Teesta river channel with GIS-based technology and also applied the ARIMA model in their study. Ghosh (2014) explored the planform pattern of Teesta and narrated the PFI, bar development, and channel sinuosity of the river.

Hence, the present study recognizes bank erosion and sediment deposition are such an event that impacts on the community and the downstream reaches of the Teesta river basin every year. Considering the previous studies, the present study explored the fact that the whole lower course of Teesta River in Bangladesh is unexplored in the context of the district-level erosion and deposition. The present study documents a detailed analysis of erosion and sediment deposition of Teesta Fiver from 1975 to 2017. To illustrate the erosion and deposition of Teesta River, the study has been carried out in the five districts of Teesta river course in Bangladesh. Therefore, the paper aims to examine the bank erosion and sediment deposition in the five districts of Teesta river flow in Bangladesh. Likewise, it also seeks to explore the process of bank erosion and deposition of the river. The focal theme of the existing research is to identify the most eroded and deposited district of Teesta riverbank in Bangladesh. Different research exemplifies the erosion and deposition status of the three main rivers, i.e., the Padma, the Meghna, and the Brahmaputra-Jamuna of Bangladesh. But the Teesta has not received much devotion from the researchers at home and abroad in the context of district-level erosion and deposition in temporal scale, though the Teesta is geostrategically important for Bangladesh. With this backdrop, the present research attempts to explore the spatiotemporal analysis of bank erosion and sediment deposition of Teesta River in Bangladesh.

4.2 Methodology and Data Sources

The objective of the present research is to explore the riverbank erosion and sediment deposition and identify the most eroded and deposited district of Teesta riverbank in Bangladesh. The study area lies beneath the Teesta mega fan which is an important physiographic unit. This physiographic unit is mostly floodplain, and the area is highly dynamic due to intermittent erosion and sediment depositional activities. The Teesta floodplain is also occupied by several rivers such as the mighty Brahmaputra, Karatoa, Attrai, and Punarvaba. The study area is a transitional region between the Ganges floodplain and the Brahmaputra-Jamuna floodplain. Physiographically, the Teesta floodplain is banded with plain lands, ridges, and natural levees. The study investigates the spatial analysis of bank erosion and sediment deposition of Teesta River from 1975 to 2017 in the five districts (Nilphamari, Rangpur, Kurigram, Lalmonirhat, and Gaibandha) of Bangladesh. These five districts are directly influenced by the severe bank erosion and sediment deposition of Teesta River followed by floods in each year. Therefore, the districts have been chosen as study area for field data collection and satellite image analysis.

The methodology of the research includes both quantitative and qualitative representations of statistics and information obtained from satellite images, field observation, field survey, and existing literature review. Field survey has been carried out to understand the mechanism of bank erosion and sediment deposition of Teesta River. Field survey was conducted during and after floods in 2017 in Teesta River to monitor the condition of erosion and deposition. Moreover, relevant literatures were reviewed regarding Teesta and other river erosion and deposition.

The study used multi-temporal Landsat images of MSS, TM, and OLI_TIRS to determine the bank erosion and sediment deposition of Teesta River. The study analyzed the decadal variation of erosion-deposition from 1975–1987, 1987–1997, 1997–2006, and 2006–2017. All the Landsat images have been collected from USGS Earth Explorer. The collected images have been classified through ERDAS Imagine 2014. The study used Arc GIS 10.2.2 software to detect the erosion and sedimentation in the study districts. Excel software was used in the study for data interpretation. Red, blue, green, infrared, and near infrared bands of Landsat imageries have been used for false color composite (FCC). The projection method was Universal Transverse Mercator (UTM) 46N. Classified images were transferred into GIS layer to mapping interpretation and area measurement. Then bank line of the river and classified images were overlapped, and the area of erosion-deposition was measured. Table 4.1 shows the list of imageries used in the study.

The satellite image processing and analysis include the process of image collection, geometric and radio metric correction, digitization, classification, and change detection. Figure 4.1 represents the flow diagram of satellite image processing, analysis, and interpretation.

Table 4.1 List of imageries used in the study

No.	Images	Path/row	Sensor ID	Resolution	Date of acquisition
1.	Landsat 1975	148/42	MSS	60 m	17 and 18 November 1975
		149/42			
2.	Landsat 1987	138/42	TM	30 m	24 November 1987
3.	Landsat 1997	138/42	TM	30 m	1 November 1997
4.	Landsat 2006	138/42	TM	30 m	1 January 2006
5.	Landsat 2017	138/42	OLI_TIRS	30 m	8 November 2017

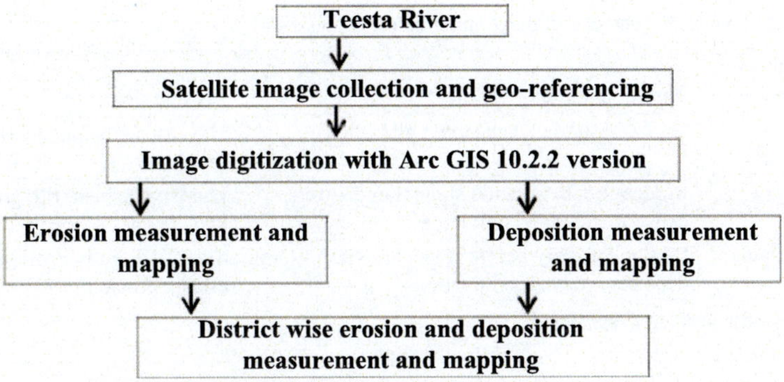

Fig. 4.1 Flow chart of satellite image processing, analysis, and interpretation

4.3 Results and Discussion

4.3.1 The Process of Erosion and Deposition

Sediment deposition and erosion are the key determinants of change in river morphology. River morphology and channel pattern alters through bank erosion, and riparian environment develops with the sediment originated from bank collapse (Florsheim et al., 2008). The present paper tries to investigate the erosion and deposition process of Teesta River. Over the study decades, the Teesta River widened its channel through bank erosion. During field investigation, the inhabitants of Teesta riverine environment informed that formerly the river was only 2–3-km wide, but at present the river valley widened ranging from 7 to 10 km owing to its erosional activities. Subsequently the eroded bank material decreases the valley depth. Therefore, the process of erosion involved in sediment accumulation. Erosion is the process by which water is the main transporting agent of weathered rocks in a river basin (Ward, 1975). Removable sediment slips from upper banks of the river and is set down on the bottom of stream banks (Prosser et al., 2000). Erosion occurs in the outer meander bend (concave bank), and sediment deposition takes place in the inner meander bend (convex bank) of an alluvial river. Accretion or sediment deposition in the riverbank and bed is determined by different natural and manmade causes. Flood, erosion, and landslide in the upper catchment of Teesta cause sediment deposition in the downstream valleys. The share stress of influx of water and stream bed is the factor responsible for the movement of river water and transportation of suspended sediment load of 13 million tons/day (Sikder, 2008). The sediment source regions and erosion processes of Teesta are shown in Table 4.2.

The movement of water in the riverbank, type of bank material, and slope aspect intensifies the bank failure of alluvial rivers like the Teesta. The riverbank erodes after the subsequent retreat of water from a bank full stage of flood. Riverbank erosion and accretion are a complex process in which hydroclimatic and geomorphological processes are involved to regulate the river morphodynamics. Figure 4.2

Table 4.2 Source regions of sediment and erosion processes

Sediment source region	Terrain and major features	Erosion processes
Sikkim Himalaya	High elevation, steep slope, glaciation	Hill slope erosion, debris flow, rock fall
Himalayan foothills	Moderate relief, gentle slope, fan deposits, dense forest, and cultivation	Landslide, sheet, rill, gully, and channel erosion
Teesta plains	Low relief, very gentle slope, alluvial deposits, grazing, cultivation	Sheet, rill, gully erosion, riverbank erosion

Source: Modified from Wasson (2003)

Fig. 4.2 Riverbank erosion and deposition process. (Source: Modified from Hung, 2006)

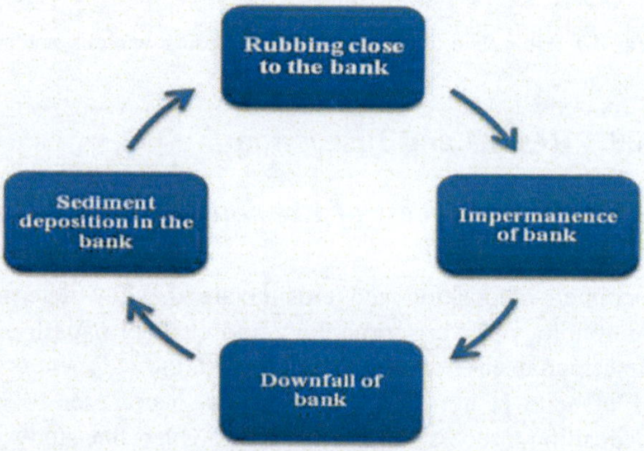

Fig. 4.3 Bank erosion and sediment deposition of Teesta. (Source: Field survey, 2017)

illustrates the process of erosion and deposition of a river (modified from Hung, 2006). The Teesta riverbank dwellers experience substantial damage of property and loss of crop production and livelihoods caused by erosion and sediment deposition after flood. The condition of bank erosion and sediment deposition has been illustrated in Fig. 4.3. The photos were captured after Teesta floods in August 2017.

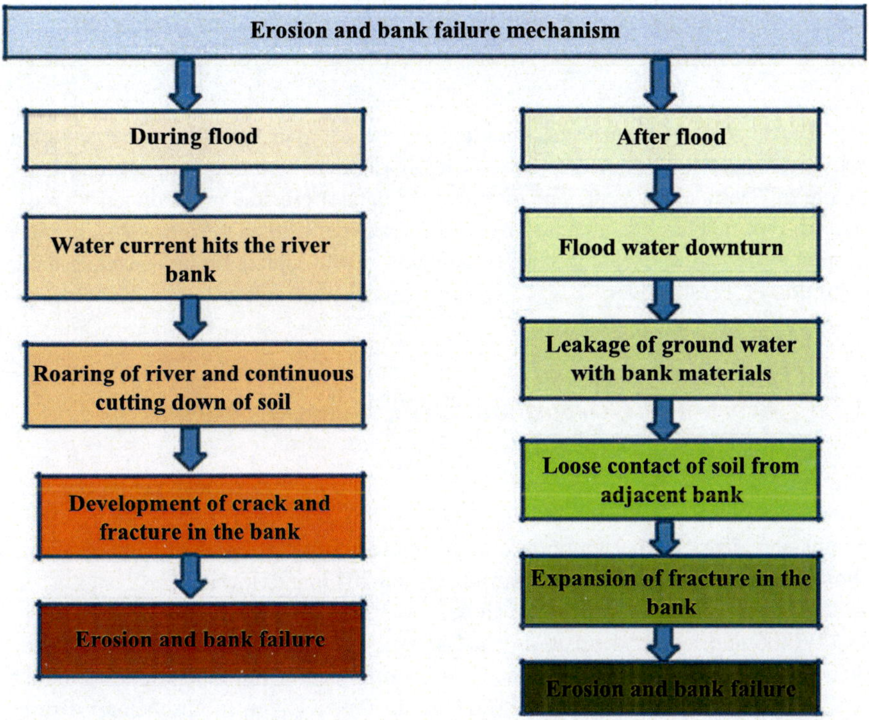

Fig. 4.4 Erosion and bank failure mechanism of an alluvial river (Developed by the researcher)

The erosion of the Teesta River follows two distinct phases: erosion during flooding period and erosion after flood water recession. The process of riverbank erosion and bank failure has been illustrated in Fig. 4.4. During the flood the stream current hits the bank, and the bank materials collapse. Bank erosion occurs suddenly in the monsoon, but bank collapse and soil erosion occur slowly in post flood situation. The water contamination in the riverbank loses the compactness of soil bonding, and bank failure takes place. Seepage of ground water into the river is responsible for post flood erosion (Mohiuddin et al., 2007). Rainfall, flood, mass movement, soil composition, deforestation, topography, tectonic movement, and human interference on river are the controlling factors of riverbank erosion, and thus the river changes its morphology. Riverbank erosion is a dynamic fluvial process. Bank erosion mechanism is influenced by scouring in the lower slope and sliding in the upper slope after flood water recession (Klaassen et al., 2005). Riverbank erosion occurs during June–November (monsoon-post monsoon season) in Teesta River. The intensity of erosion varies with the change in the variables of channel morphology and hydrology. Riverbank erosion involves with corrosion, and that is controlled by water flow as well as slumping which is governed by soil moisture content (Hooke, 1979).

4.3.2 Spatiotemporal Analysis of Bank Erosion and Sediment Deposition

The Teesta River demonstrates braiding stream channel through its depositional activities in the study area. All the five study districts face huge erosion and bank failure of Teesta every year. The district level data of erosion and sediment deposition of Teesta river accompanied by erosion and deposition rate in per sq.km has been analyzed in the present study. The study result depicts highly unstable bank condition of Teesta River.

4.3.3 Erosion and Deposition in Nilphamari District

The transboundary Teesta River enters into the plains of Bangladesh via Nilphamari district. Table 4.3 shows the erosion and deposition (in sq.km) of Teesta River in Nilphamari district. The temporal change in erosion pattern of Nilphamari district shows highest erosion (7.41 sq.km) in 2006–2017 and lowest (2.47 sq.km) in 1975–1987. The analyzed data shows highest deposition (24.35 sq.km) in 1997–2006 and lowest deposition (6.84 sq.km) in 1987–1997. The table indicates that during 1975–2017 in Nilphamari district, sediment deposition is higher (46.21 sq.km) than erosion (6.47 sq.km) in the Teesta River. In Nilphamari district highest erosion rate/year is 0.67 sq.km in 2006–2017, and highest deposition rate is 2.70 in 1997–2006. Figure 4.5 illustrates the condition of erosion and deposition of Teesta in Nilphamari district.

4.3.4 Erosion and Deposition in Rangpur District

The spatiotemporal variation of erosion and deposition of Teesta River in Rangpur district designates that during 1975–2017 deposition (66.48 sq.km) is four times higher than erosion (15.32 sq.km). Table 4.4 shows highest erosion (18.08 sq.km) in 1975–1987, and the erosion rate is 1.50 sq.km per year. The decade 1997–2006

Table 4.3 Erosion and deposition (sq.km) in Nilphamari district

Year	Erosion (sq.km)	Erosion rate/year (sq.km)	Deposition (sq.km)	Deposition rate/year (sq.km)
1975–1987	2.47	0.20	12.84	1.07
1987–1997	4.71	0.47	6.84	0.68
1997–2006	3.23	0.35	24.35	2.70
2006–2017	7.41	0.67	13.56	1.23
1975–2017	6.47	0.15	46.21	1.10

Source: Calculated from satellite image

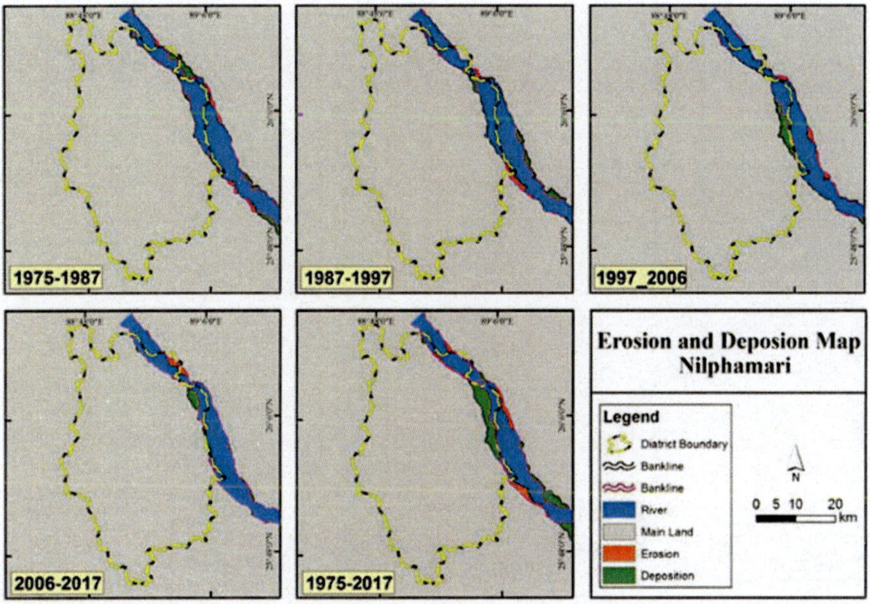

Fig. 4.5 Erosion and deposition of Teesta in Nilphamari

Table 4.4 Erosion and deposition (sq.km) in Rangpur district

Year	Erosion (sq.km)	Erosion rate/year (sq.km)	Deposition (sq.km)	Deposition rate/year (sq.km)
1975–1987	18.08	1.50	30.60	2.55
1987–1997	17.17	1.71	17.15	1.71
1997–2006	4.89	0.54	19.03	2.11
2006–2017	6.06	0.55	30.58	2.78
1975–2017	15.32	0.36	66.48	1.58

Source: Calculated from satellite image

indicates lowest erosion (4.89 sq.km) in the district than the other decades. Highest sediment deposition in Rangpur district has been observed to be 30.58 sq.km during 2006–2017, and lowest deposition has been detected to be 17.15 sq.km in the time period of 1987–1997. Figure 4.6 shows the condition of erosion and deposition of Teesta in Rangpur district.

4.3.5 *Erosion and Deposition in Lalmonirhat District*

Figure 4.7 displays the situation of erosion and deposition of Teesta River in Lalmonirhat district. The erosion and deposition of Teesta River during 1975–1987 and 1987–1997 shows nearly parallel pattern. But erosion is almost

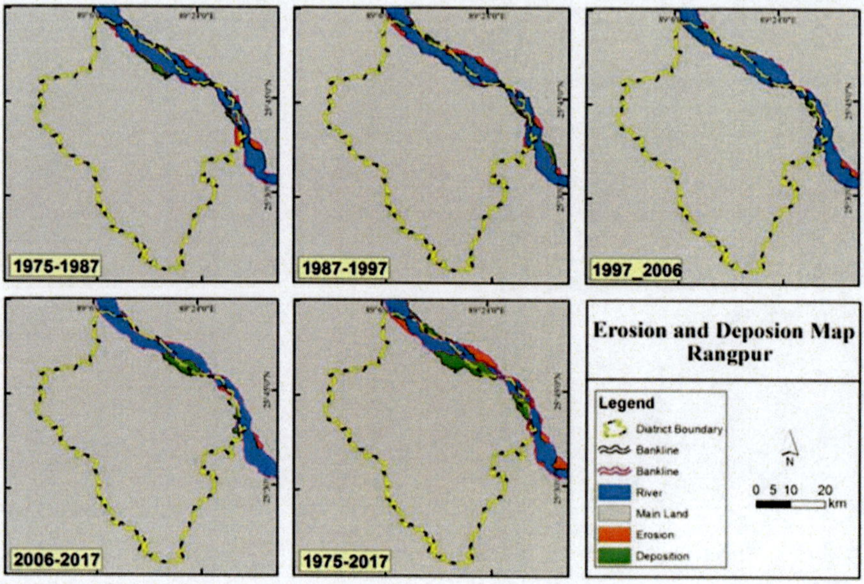

Fig. 4.6 Erosion and deposition of Teesta in Rangpur

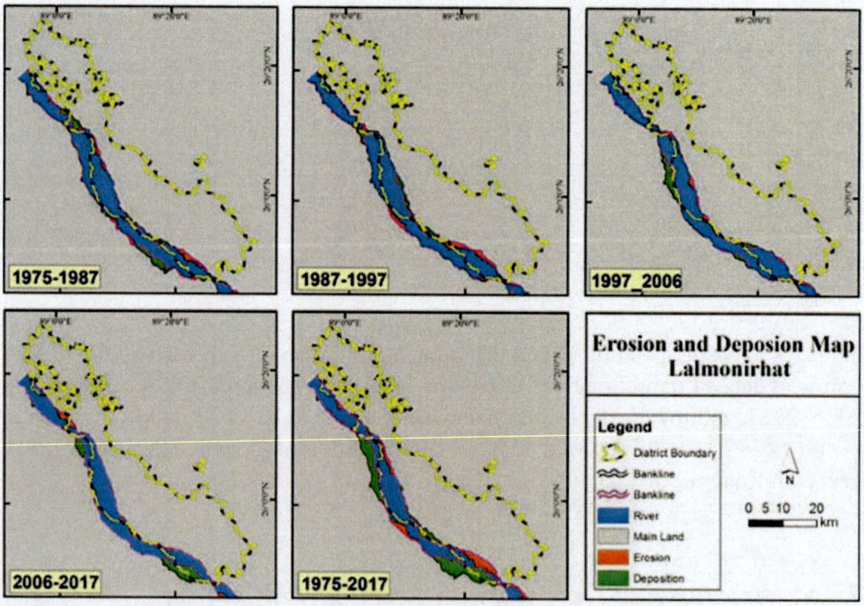

Fig 4.7 Erosion and deposition of Teesta in Lalmonirhat

two times higher (15.87 sq.km) than deposition (8.56 sq.km) during 1997–2006, and deposition is also more than two times higher (21.67 sq.km) than erosion (9.32 sq. km) during 2006–2017. Highest erosion (18.89 sq.km) has been observed in 1975–1987, and lowest erosion (9.32 sq.km) has been detected in 2006–2017. Simultaneously highest deposition (21.67 sq.km) has been identified in 2006–2017, and lowest deposition (8.56 sq.km) has been noticed in 1997–2006. The study result of 42 years (1975–017) of erosion and deposition of Teesta River in Lalmonirhat district displays that erosion occurred at 31.72 sq.km and deposition has been happened in 40.17 sq.km (Table 4.5).

4.3.6 *Erosion and Deposition in Kurigram District*

The erosional and depositional status of Kurigram district highlights that erosion is higher than the deposition during the study decades except the decade of 1987–1997 when deposition (14.07 sq.km) is higher than erosion (12.01 sq.km). During the whole study period (1975–2017), erosion is four times higher (29.11 sq.km) than deposition (7.39 sq.km) in Kurigram district. Highest erosion (14.26 sq.km) has been noticed in 1975–1987, and lowest erosion (9.23 sq.km) has been identified in 2006–2017. On the other hand, highest deposition (14.07) has been detected in 1987–1997, and lowest deposition (1.65 sq.km) has been observed in 2006–2017 (Table 4.6). Figure 4.8 portrays the condition of erosion and deposition of Teesta River in Kurigram district.

Table 4.5 Erosion and deposition (sq.km) in Lalmonirhat district

Year	Erosion (sq.km)	Erosion rate/year (sq.km)	Deposition (sq.km)	Deposition rate/year (sq.km)
1975–1987	18.89	1.57	20.17	1.68
1987–1997	17.12	1.71	19.42	1.94
1997–2006	15.87	1.76	8.56	0.95
2006–2017	9.32	0.84	21.67	1.97
1975–2017	31.72	0.75	40.17	0.95

Source: Calculated from satellite image

Table 4.6 Erosion and deposition (sq.km) in Kurigram district

Year	Erosion (sq.km)	Erosion rate/year (sq.km)	Deposition (sq.km)	Deposition rate/year (sq.km)
1975–1987	14.26	1.18	4.38	0.36
1987–1997	12.01	1.20	14.07	1.40
1997–2006	11.17	1.24	4.86	0.54
2006–2017	9.23	0.83	1.65	0.15
1975–2017	29.11	0.69	7.39	0.17

Source: Calculated from satellite image

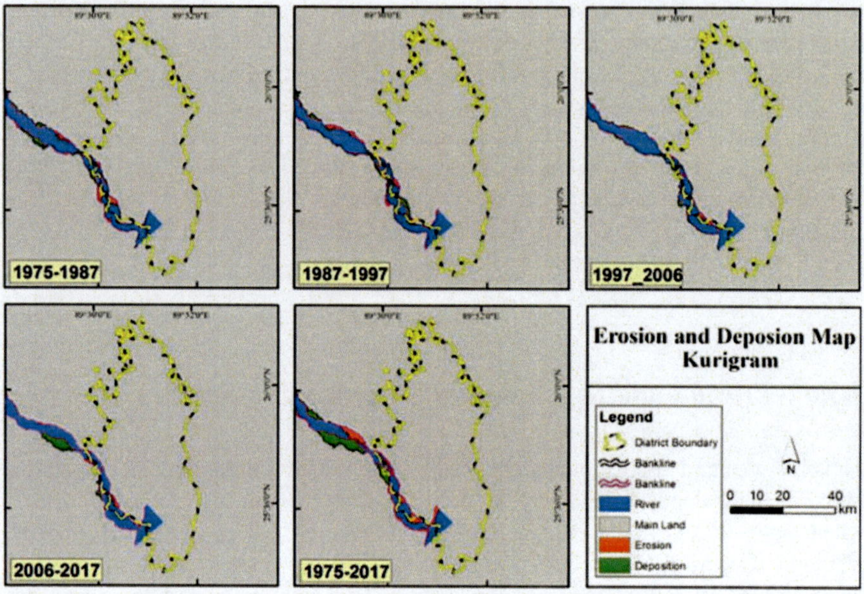

Fig. 4.8 Erosion and deposition of Teesta in Kurigram

Table 4.7 Erosion and deposition (sq.km) in Gaibandha district

Year	Erosion (sq.km)	Erosion rate/year (sq.km)	Deposition (sq.km)	Deposition rate/year (sq.km)
1975–1987	12.12	1.01	2.18	0.18
1987–1997	7.8	0.78	2.85	0.28
1997–2006	5.78	0.64	6.07	0.67
2006–2017	4.23	0.38	2.77	0.25
1975–2017	21.30	0.50	5.26	0.12

Source: Calculated from satellite image

4.3.7 Erosion and Deposition in Gaibandha District

Table 4.7 represents the condition of erosion and deposition (in sq.km) in Gaibandha district. Among the study decades, erosion is highest (12.12 sq.km) during 1975–1987 and lowest (4.23 sq.km) during 2006–2017 in Gaibandha district. The district follows lowest deposition (2.18 sq.km) in the study period 1975–1987 and highest deposition (6.07 sq.km) during 1997–2006. During the 42 years, (1975–2017) riverbank change in Gaibandha district erosion is four times higher (21.30 sq.km) than deposition (5.26 sq.km). The condition of erosion and deposition of Teesta in Gaibandha district has been depicted in Fig. 4.9.

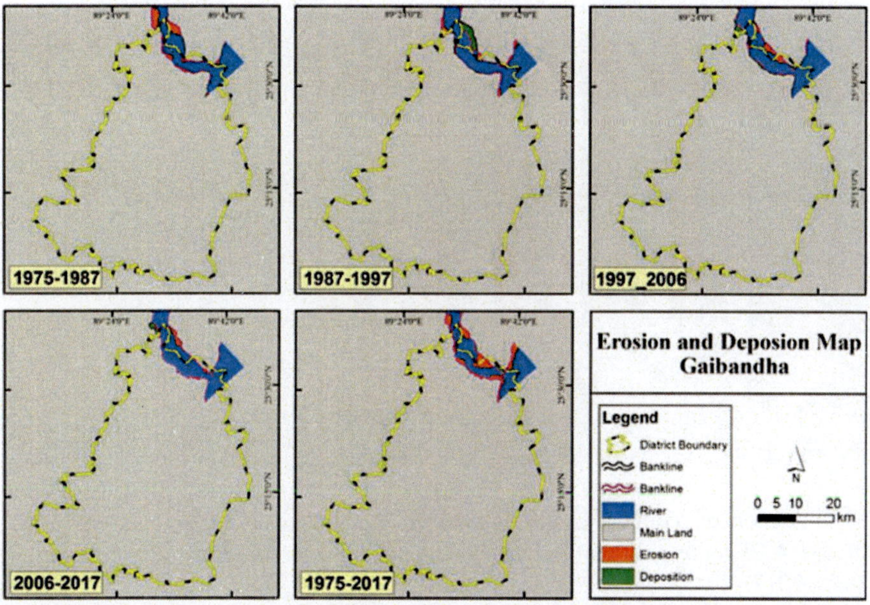

Fig. 4.9 Erosion and deposition of Teesta in Gaibandha

4.3.8 Comparison of Erosion and Deposition Among the Study Districts

Satellite image analysis of the five districts of Teesta River illustrates that the erosional and depositional status of the studied river has an equilibrium. In Bangladesh, the upstream districts like Rangpur (66.48 sq.km) shows the highest, and Nilphamari (46.21 sq.km) displays second highest deposition than erosion during 1975–2017. The third highest deposition has been detected in Lalmonirhat district (40.17 sq.km). Simultaneously, the downstream districts within Bangladesh followed the opposite condition and depicted higher erosion than deposition. But the district of Lalmonirhat portrays almost balanced erosion (31.72 sq.km) and deposition (40.17 sq.km) than the other study districts during the study period. The highest erosion has been observed in Lalmonirhat district among the five districts. The second and third highest eroded districts are Kurigram (29.11 sq.km) and Gaibandha (21.30 sq.km), respectively. Field survey result identified the causes behind erosion in the downstream districts and deposition in the upstream districts of Teesta River. The study finds out that the causes of high erosion in the downstream districts of Bangladesh are weak soil condition, dam and barrage construction, stone collection from the riverbed, flood, and excessive water discharge from the upstream in the monsoon. On the other hand, the study also explores flood, erosion, human activities, and sediment transportation by Teesta River from India cause high

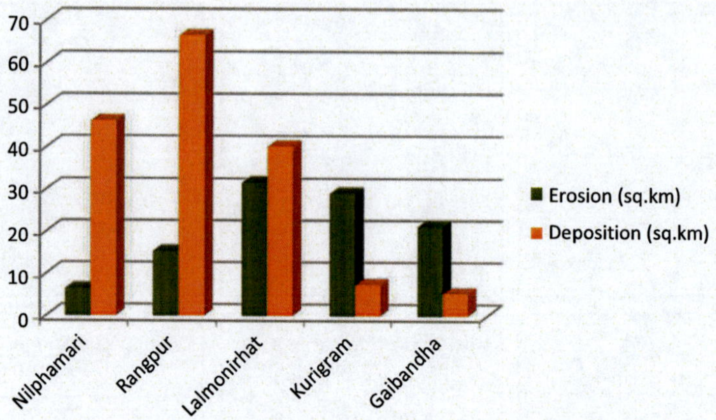

Fig. 4.10 Comparison of erosion and deposition in the study districts (1975–2017)

accumulation of sediment in the districts of Nilphamari, Rangpur, and Kurigram. Figure 4.10 represents the erosion and deposition of the study districts during 1975–2017.

4.4 Recommendations

The study detected erosion and sediment deposition in all the study districts in varied magnitude which causes bank failure and multithreaded channel development in Teesta River. Enormous bank erosion and excessive sediment deposition of Teesta River are the major issues of northern Bangladesh in socioeconomic and geo-political perspectives. Correspondingly, the change in the socio-economy of bank dwellers brings huge sufferings after erosion occurrences and sand deposition in agricultural lands. Therefore, adoption of technology-based riverbank erosion forecasting and disseminating the message to mass people will potentially reduce the economic loss of the community of Teesta river basin. Moreover, the research recommends constructing and repairing embankments in the most erosive districts to protect the riverbanks from massive erosion. Similarly, immediate government intervention is needed for dredging in Teesta to manage tremendous sediment load of the river. In addition, the government should give emphasis on the maintenance of a channel in Teesta River rather than braided channel developed by sediment deposition.

4.5 Conclusion

Bank erosion and sediment deposition modifies riverine environments, and these dynamic processes change river morphology. Riverbank erosion and sediment deposition is geomorphic process that involves with flood plain development (Hooke, 1979). The multi-temporal data analysis of the present study depicts spatial variation in riverbank erosion and sediment accumulation in the study districts within Bangladesh. The satellite image analysis result also displays that spatial variation of bank erosion and sediment deposition is unequal and becomes a growing concern in the lower course of Teesta River in Bangladesh. The study result shows high erosive activities of Teesta in Lalmonirhat, Kurigram, and Gaibandha. Simultaneously, the fluvial activity of Teesta displays high sediment deposition activities in Rangpur, Nilphamari, and Lalmonirhat districts. Therefore, the study depicts Rangpur is highest deposited district and Lalmonirhat is highest erosive district. Likewise, Lalmonirhat district is most vulnerable in terms of erosion and deposition in the study period. Hence, the Teesta riverbank erosion issue needs special attention toward policy formulation and implementation. Similarly, accelerated sediment deposition by human development works should be taken into consideration to prevent increase

in riverbed height. Therefore, the government should give emphasis on the hydrological issues through international collaboration to mitigate the Teesta River water dispute. Consequently, the government should give prominence toward Lalmonirhat district for bank erosion protection, and sediment control measures should be implemented in Rangpur district immediately.

References

Aher, S. P., Bairagi, S. I., Deshmukh, P. P., & Gaikwad, R. D. (2012). River change detection and bank erosion identification using topographical and remote sensing data. *International Journal of Applied Information Systems, 2*(3), 1–7.

Akhter, S., Eibek, K. U., Islam, S., Islam, A. R. M. T., Chu, R., & Shuanghe, S. (2019). Predicting spatiotemporal changes of channel morphology in the reach of Teesta River, Bangladesh using GIS and ARIMA modeling. *Quaternary International, 513*, 80–94.

Alekseevskiy, N. I., Berkovich, K. M., & Chalov, R. S. (2008). Erosion, sediment transportation and accumulation in rivers. *International Journal of Sediment Research, 23*(2), 93–105.

Billah, M. M. (2018). Mapping and monitoring erosion-accretion in an alluvial river using satellite imagery–the riverbank changes of the Padma river in Bangladesh. *Quaestiones Geographicae, 37*(3), 87–95.

Chakraborty, S., & Datta, K. (2013). Causes and consequences of channel changes–a spatio-temporal analysis using remote sensing and GIS—Jaldhaka-Diana River system (lower course), Jalpaiguri (Duars), West Bengal, India. *Journal of Geography and Natural Disasters, 3*(1), 1–13.

Chakraborty, T., & Ghosh, P. (2010). The geomorphology and sedimentology of the Tista megafan, Darjeeling Himalaya: Implications for megafan building processes. *Geomorphology, 115*(3–4), 252–266.

Chalov, S. R., Jarsjö, J., Kasimov, N. S., Romanchenko, A. O., Pietroń, J., Thorslund, J., & Promakhova, E. V. (2015). Spatio-temporal variation of sediment transport in the Selenga River basin, Mongolia and Russia. *Environmental Earth Sciences, 73*(2), 663–680.

Ety, N. J., & Rashid, M. S. (2020). Spatiotemporal variability of erosion and accretion in Ganges River using GIS and RS: A comparative study overlapping Rennell's map of 1760s. *Environment, Development and Sustainability, 22*(4), 3757–3775.

Florsheim, J. L., Mount, J. F., & Chin, A. (2008). Bank erosion as a desirable attribute of rivers. *Bioscience, 58*(6), 519–529.

Froehlich, W., & Walling, D. E. (2006). The use of^ 1^ 3^ 7Cs and^ 2^ 1^ 0Pbex to investigate sediment sources and overbank sedimentation rates in the Teesta River basin, Sikkim Himalaya, India. *IAHS Publication, 306*, 380–388.

Ghosh, K. (2014). Planform pattern of the lower Teesta River after the Gazaldoba barrage. *Indian Journal of Geography and Environment, 13*, 127–137.

Hooke, J. M. (1979). An analysis of the processes of river bank erosion. *Journal of Hydrology, 42* (1–2), 39–62.

Hung, N. N. (2006). *Bank erosion prediction and mitigation measures along the lower Mekong river* (M. Sc. Thesis, Nguyen Nghia Hung at Unesco-IHE Delft).

Khan, S. S., & Islam, A. R. M. T. (2015). Anthropogenic impact on morphology of Teesta River in northern Bangladesh: An exploratory study. *Journal of Geosciences and Geomatics, 3*(3), 50–55.

Klaassen, G. J., Pilarczyk, K. W., & San, D. C. (2005). River bank erosion and mitigation strategies in Vietnam. In V. Alphen, V. Beek, & Tal (Eds.), *Floods, from defense to management* (pp. 269–279). Taylor & Francis.

Kotoky, P., Bezbaruah, D., & Sarma, J. N. (2015). Spatio-temporal variations of erosion-deposition in the Brahmaputra River, Majuli—Kaziranga sector, Assam: Implications on flood management and flow mitigation. In *Environmental management of River Basin ecosystems* (pp. 227–251). Springer.

Laskar, A. A., & Phukon, P. (2012). Erosional vulnerability and spatio-temporal variability of the Barak River, NE India. *Current Science, 103*(1), 80–86.

Middelkoop, H. (2006). Floodplain sedimentation–Methods, patterns, and processes: A review with examples from the Lower Rhine, the Netherlands. In *Encyclopedia of hydrological sciences.* Retrieved from https://scholar.google.com/scholar?hl=en&as_sdt=0%2C5&q=Floodplain+Sedimentation%E2%80%93Methods%2C+Patterns%2C+and+Processes%3A+A+Review+with+Examples+from+the+Lower+Rhine%2C+the+Netherlands&btnG=

Mohiuddin, F. A., Kabir, M. M., & Hye, J. M. A. (2007). Forecasting of the Padma river bank erosion and its mitigation measures. In *International conference on water and flood management (ICWMF)* (Vol. 2, pp. 709–717). Institute of Water and Flood Management, BUET.

Mondal, S. H., & Islam, S. (2017). Chronological trends in maximum and minimum water flows of the Teesta River, Bangladesh, and its implications. *Jàmbá: Journal of Disaster Risk Studies, 9* (1), 1–11.

Mukherjee, R., Bilas, R., Biswas, S. S., & Pal, R. (2017). Bank erosion and accretion dynamics explored by GIS techniques in lower Ramganga river, Western Uttar Pradesh, India. *Spatial Information Research, 25*(1), 23–38.

Pal, R., Biswas, S. S., Mondal, B., & Pramanik, M. K. (2016). Landslides and floods in the Tista Basin (Darjeeling and Jalpaiguri districts): Historical evidence, causes and consequences. *Journal of Geophysics Union, 20*(2), 209–215.

Pal, P. K., Rahman, A., & Yunus, A. (2017). Analysis on river bank erosion-accretion and bar dynamics using multi-temporal satellite images. *American Journal of Water Resources, 5*(4), 132–141.

Petropoulos, G. P., Kalivas, D. P., Griffiths, H. M., & Dimou, P. P. (2015). Remote sensing and GIS analysis for mapping spatio-temporal changes of erosion and deposition of two Mediterranean river deltas: The case of the Axios and Aliakmonas rivers, Greece. *International Journal of Applied Earth Observation and Geoinformation, 35*, 217–228.

Prosser, I. P., Hughes, A. O., & Rutherfurd, I. D. (2000). Bank erosion of an incised upland channel by subaerial processes: Tasmania, Australia. *Earth Surface Processes and Landforms: The Journal of the British Geomorphological Research Group, 25*(10), 1085–1101.

Saha, S., Roy, M., & Reza, A. S. (2017). Textural characteristics of the sediments of the Tista River, Rangpur, Bangladesh. *Journal of Life and Earth Science, 12*, 73–79.

Sarma, J. N., Borah, D., & Goswami, U. (2007). Change of river channel and bank erosion of the Burhi Dihing river (Assam), assessed using remote sensing data and GIS. *Journal of the Indian Society of Remote Sensing, 35*(1), 93–100.

Shi, H., Ji, Z., Lu, Q., & Wang, D. (2019, May). Study on fluvial processes and sediment transport capacity of Teesta River in Bangladesh. In *World environmental and water resources congress 2019: Hydraulics, waterways, and water distribution systems analysis* (pp. 399–410). American Society of Civil Engineers.

Sikder, M. D. H. (2008). Hydrodynamic characteristics of the coastal areas of Bangladesh. In S. M. N. Amin (Ed.), *Studies on coastal environments in Bangladesh* (pp. 15–28). A H Development Publishing House.

Suif, Z., Fleifle, A., Yoshimura, C., & Saavedra, O. (2016). Spatio-temporal patterns of soil erosion and suspended sediment dynamics in the Mekong River basin. *Science of the Total Environment, 568*, 933–945.

Tarannum, T., Bhuyan, A., & Badhon, F. F. (2018, March). Morphological changes of river Teestaduring the last decade. In *Paper presented at 1st National Conference on water resources engineering*. Chittagong University of Engineering and Technology.

Thakur, P. K., Laha, C., & Aggarwal, S. P. (2012). River bank erosion hazard study of river ganga, upstream of Farakka barrage using remote sensing and GIS. *Natural Hazards, 61*(3), 967–987.

Wasson, R. J. (2003). A sediment budget for the ganga–Brahmaputra catchment. *Current Science, 84*, 1041–1047.

Ward, R. C. (1975). *Principles of Hydrology*. London: McGraw-Hill book company limited.

Chapter 5
Remote Sensing and GIS Application in Flood Management: A Case Study of the Jiadhal River Basin of Dhemaji District, Assam, India

Pranamee Gogoi and Santanu K. Patnaik

5.1 Introduction

Flood is one of the common destructive acts of nature which is spread globally and causes extensive loss to lives and properties. Riverine floods are most widespread due to heavy, prolonged rainfall, rapid snowmelt in upstream watersheds or the regular spring thaw (1991). The area affected by flood every year covers a large portion where the density of population is very high, and that is why flood hazards throughout the world are responsible for causing heavy damage to the human lives (Das, 2019). Floods have a direct consequence which leads to loss of lives and property and damages to infrastructures, ecosystems, and also to historical and cultural values (Jonkman, 2005), and the other effects includes outbreaks of diseases as well as tertiary effects like loss of soil fertility, famine, and poverty (Opolot, 2013).

Like all other natural hazard, flood is also very difficult to control, and the drainage problem arises concurrently if the flood is for a longer duration of time. In India, the rivers flowing from the Himalayan causes flood problems in the Ganga-Brahmaputra region owing to the high discharge of sediment which blocks the drainage arteries and also causes changes in the river courses and braiding of channels (Murthy & Prasad, 2004). In the Brahmaputra floodplain region of Assam, the event of flood is of frequent occurrence, and because of this the region has to face the worst situation of flood every year. The state has a chronic history of devastating flood, and the situation has not changed much over the years (Saikia & Das, 2002). The Jiadhal River in the Dhemaji district of Assam, a subtributary of the Brahmaputra River creates a devastating hazard to the people living in that area. It is

P. Gogoi (✉) · S. K. Patnaik
Department of Geography, Rajiv Gandhi University, Itanagar, Arunachal Pradesh, India
e-mail: santanu.patnaik@rgu.ac.in

G. S. Bhunia et al. (eds.), *Anthropogeomorphology*, Geography of the Physical Environment, https://doi.org/10.1007/978-3-030-77572-8_5

one of the most frequently flooded rivers of the district. Flood occurs in the river due to the overflow of water resulting in embankment failure and later on shifting of the river. So, a flood hazard map for this area is very much essential to identify the flood-affected areas of the region for proper planning and development of the region. Remote sensing and geographical information system (GIS) application in regard to flood management have been considered as an important tool. Remote sensing and GIS help to gather information for floodplain areas and also to compare the data collected before and after the flood. With the help of remote sensing and GIS, a flood hazard zone map can be generated using satellite imagery and digital elevation model (DEM). The main objective of the flood hazard zone map is to prevent the loss of lives, properties, and infrastructures and to enhance awareness among the people about the importance of flood disaster preparedness (Thakuriah & Saikia, 2015).

This study takes into account the Jiadhal River due to its devastating effect of flood during the monsoon season. The roads, railway lines, and embankment in the region have been subjected to breaching during the flood season, and the risks have been increasing every year, and the left embankment has undergone a numerous breaching event within the last few decades compared to the left embankment (Borgohain et al., 2016). The study aims to generate a flood hazard zone map of the river and the nearby areas with the help of weighted overlay method in ArcGIS 10.3 to identify the areas which have higher potential for hazard. The study is limited to the factors like attitude, slope, drainage network, drainage proximity, land use/ land cover, and geomorphology. Therefore, the paper describes about the utilities of remote sensing and GIS in flood hazard mapping.

5.2 Study Area

The Jiadhal River situated in the Dhemaji district of Assam is a north bank subtributary of the Brahmaputra River. The basin extended from 27°8′24″ N to 27 °45′0″ N latitudes and from 94°15′25″ E to 94°37′22″ E longitudes (Fig. 5.1). The total area of the catchment is 1124.65 km^2. It originates in the West Siang district of Arunachal Pradesh where the three rivers Siri, Sido, and Sika join together and flow down as the river Jiadhal. These three rivers contribute most of the runoff and sediment discharge in the main stream. The river passes through a narrow gorge in Arunachal Pradesh and enters into the plains of Assam in a place called Jiadhalmukh. The name Jiadhal derived from the two words "Jia" which means "alive" and "dhal" means "flash flood" as the river has a great tendency of flash flood during the monsoon season (Borgohain et al., 2016; Sharma & Sarma, 2018). In Assam the upper reach of the river is known as Jiadhal, and the lower reach is known as Kumotiya. According to the Water Resource Department, government of Assam, the morphology of the River Jiadhal has been changed drastically, and the frequent changed in the river course created problem in understanding the channel

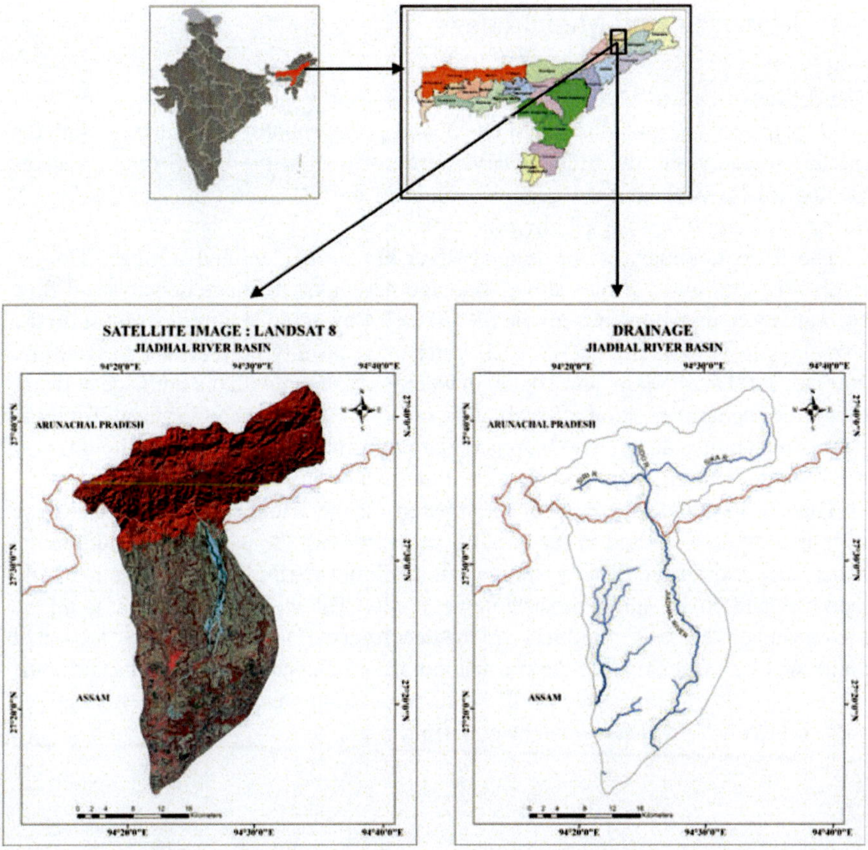

Fig. 5.1 Location map of the study area

morphology. The bed of the river has been rising continuously due to the sediment load carried from the upstream, and as a result, the river water overflows and flood occurs.

The Jiadhal River basin has two distinct geological division – (i) the upper part of the basin comes under the Siwalik formation, and (ii) the lower part of the basin consists of the alluvial sediments. The climate of the basin is typically characterized by hot and humid climatic condition. The rainfall occurs during the monsoon period, i.e., from June to September

5.3 Materials and Methodology

The datasets that was used in the study are Landsat 8 satellite imagery of the year 2017 (path 135 and row 41), SRTM DEM, and geomorphological unit map. Both the satellite imagery and the SRTM DEM were taken from the USGS (United States Geological Survey), and the geomorphological unit is taken from the Geological Survey of India (bhukosh.gsi.gov.in).

The basin boundary of the Jiadhal River has been identified from the Google Earth. The altitudinal zones, slope, drainage networks, drainage density, and flow accumulation are prepared from the SRTM DEM by using Hydrology tool from the spatial analysis tools in ArcGIS 10.3. Later on proximity buffers of 1 m, 1000 m, 2000 m, 3000 m, 4000 m, and 10,000 m were made along with the drainage network to get the scenario of flood affected areas closer to the river and far away from the river. The land use/land cover map is prepared from the Landsat 8 satellite imagery using the maximum likelihood classification method in ArcGIS.

Thus, a flood hazard zone map is prepared for the Jiadhal River basin using the weighted overlay method in the ArcGIS software with the parameters – altitudinal zones, slope, drainage density, flow accumulation, proximity to drainage network, land use/land cover, and geomorphological units. The weightage is given to all the parameters on the basis of influence in the occurrence of flood. The highest weighted value will be considered as the most flood-affected zone or the high hazard zone where the risk of vulnerability will be more as compared to the lowest weighted value which is a flood-free zone or no hazard zone.

5.3.1 Land Use/Land Cover

The land use/land cover is an important factor for flood hazard mapping. To generate the thematic layer, Landsat 8 OLI-TIRS multispectral satellite imagery of 2017 was downloaded from USGS with a minimum percentage of cloud coverage for the study. The supervised classification has been used in the study to classify the different land use/land cover categories in the ArcGIS 10.3 software. Later on, the land use/land cover classes of the Jiadhal River basin have been categorized into six classes: (i) dense forest (333.23 km^2), (ii) sparse vegetation (354.45 km^2), (iii) water bodies (7.60 km^2), (iv) agricultural land (213.35 km^2), (v) barren land (33.27 km^2), and (vi) sandbars (10.09 km^2). It is seen from Fig. 5.2 that the maximum part of the dense forest falls in the upper catchment area of the river basin, i.e., in the state of Arunachal Pradesh, and a little part of the dense forest can be seen in the foothills. The lower part of the catchment mostly comprises of sparse vegetation which covers the maximum area of the basin followed by agricultural land, barren land, sandbars, and water bodies. The settlement class was not included in the classification because of less number of pixels for the same and was creating difficulties in the image classification process. Most of the areas in the lower catchment are suitable for

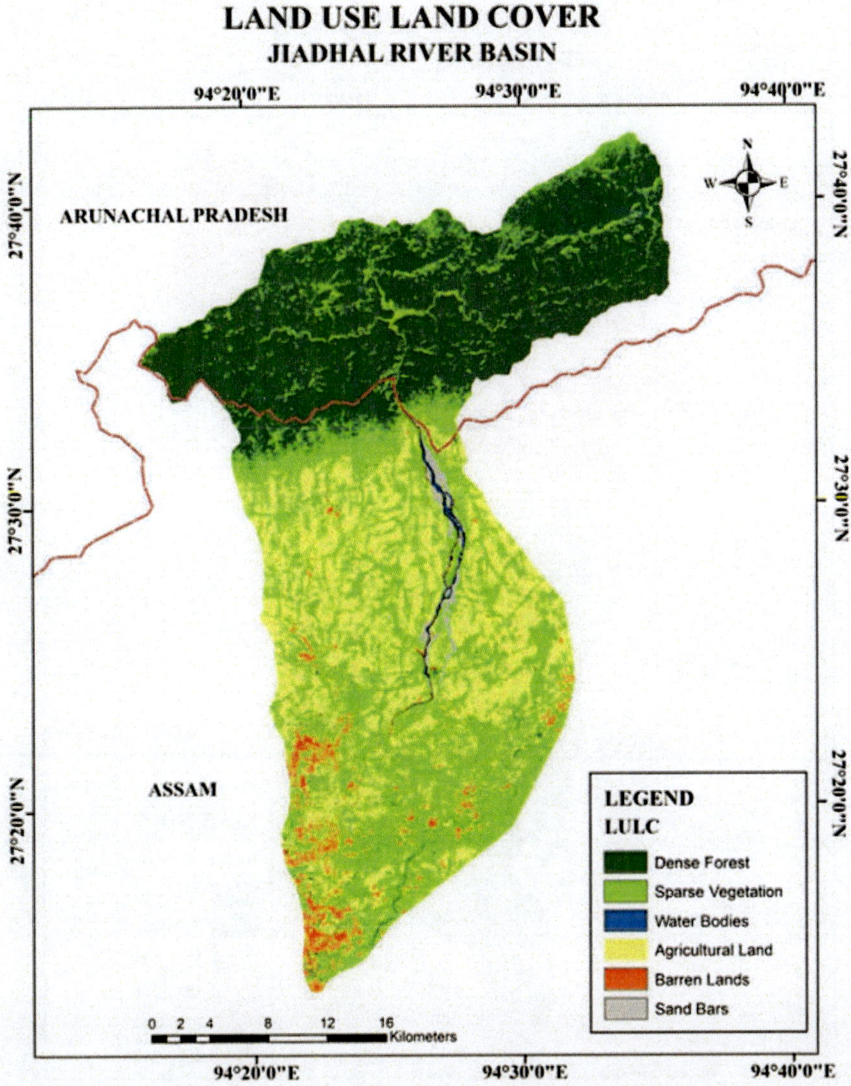

Fig. 5.2 Land use/land cover

agricultural purposes because of the alluvial soil, and also the double-cropping method is practiced in the area. The sparse vegetation is used as grazing land for domestic animals. The water bodies include rivers, ponds, lakes, and wetlands locally called as beels.

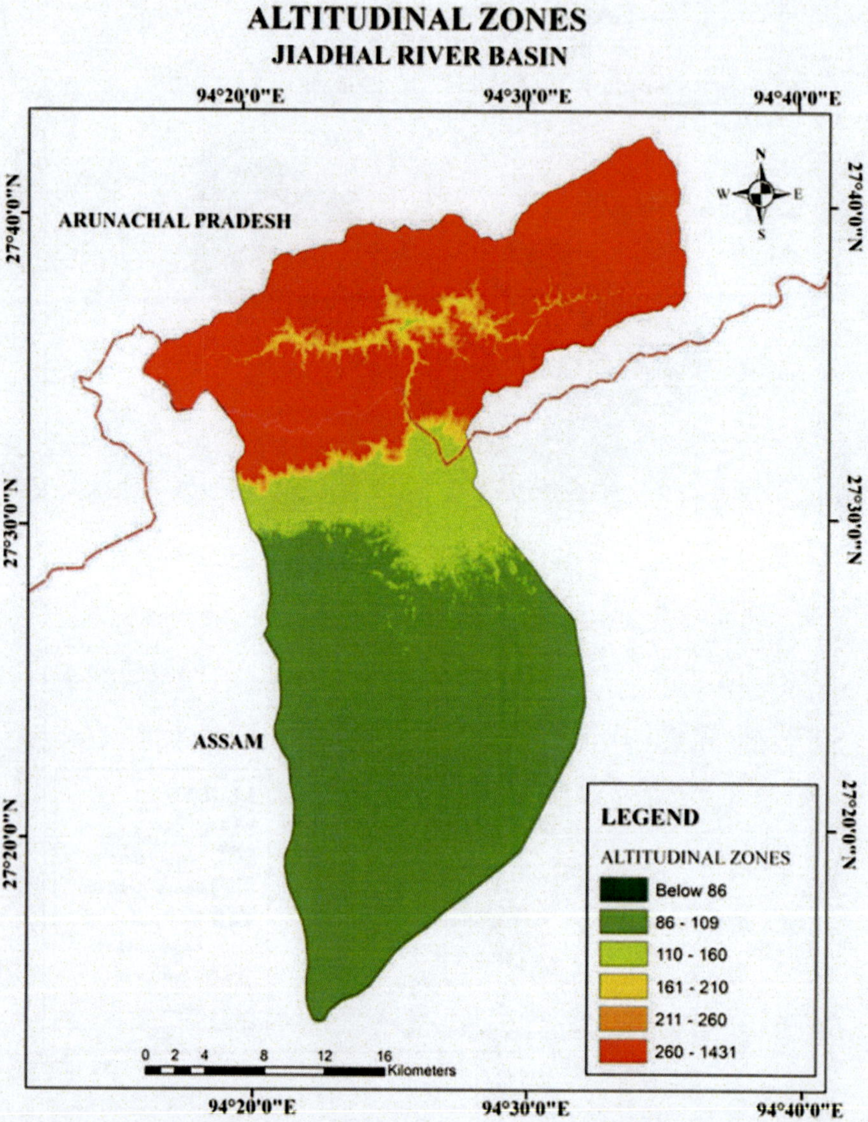

Fig. 5.3 Altitudinal zones

5.3.2 Altitudinal Zones

The altitude or the elevation plays an important role in flood hazard. Flood mostly occurs where the altitude is very low, and the high altitudinal zones are free from flood havoc. The altitudinal zone map of the Jiadhal River basin was prepared using the Shuttle Radar Topography Mission (SRTM) DEM of 30-meter resolution in ArcGIS 10.3 shown in Fig. 5.3. The highest point of altitude is 1431 m above mean sea level, and the lowest point is 86 m above mean sea level of the river basin. This difference in the altitude creates a great variation in the gradient of the river and is considered as one of the factors for which the river follows down changing its course continuously. This continuous changing pattern of the river causes flood in the plain areas which leads to embankment breaching and sand deposition to the new areas of the river basin (Bormudoi & Nagai, 2016).

5.3.3 Slope

Slope is another effective element of flood, as the slope increases, the danger of flood also increases (Thakuriah & Saikia, 2015). The flow velocity of the river also increases with the increase of river slope (Thakuriah & Saikia, 2015). The slope map for the study area was prepared using the SRTM DEM. The slope was calculated in terms of degree and the divided into six categories or classes. The slope ranges from 0 to 73 degree. The slope of the study area ranges between 0 and 73. When the slope is plain or low, there are more possibilities of flood occurrence.

5.3.4 Flow Accumulation

The flow accumulation is always created by using the flow direction, and if the valuc of flow accumulation is greater, then there is a chance for runoff in the river. The outlets of large rivers will have large values, and small rivers will have small values.

5.3.5 Drainage Network

The drainage delineation was carried using Hydrology tools in ArcGIS 10.3 using SRTM DEM. There are three major tributaries of the Jiadhal River, i.e., Siri, Sido, and Sika, in the upper catchment. After the convergence the river is known as Jiadhal and flows southward through a narrow gorge and enters into the plains of Assam.

5.3.6 Drainage Proximity

Drainage proximity or drainage buffer was prepared using the drainage network. Proximity buffers of 1 m, 1000 m, 2000 m, 3000 m, 4000 m, and 10,000 m were made along with the drainage network as shown in Fig. 5.4 to get the scenario of flood-affected areas closer to the river and far away from the river. It is seen from the

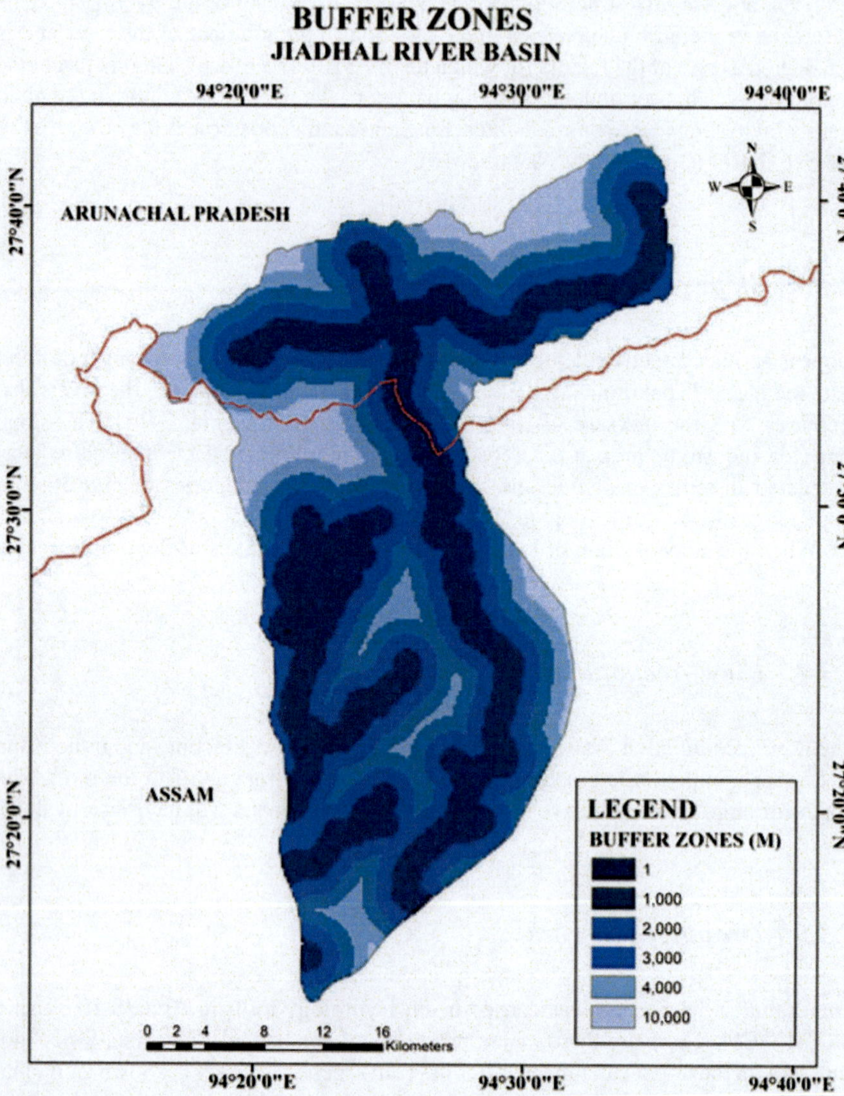

Fig. 5.4 Buffer zones

map that the occurrence of flood along the river in 1 m distance and 1000 m distance is more than that of 10,000 m distance.

5.3.7 Geomorphological Units

Since the Jiadhal River originates in Arunachal Pradesh and flows down to Assam, so it has mainly two physiographic divisions, i.e., the mountainous region and the plain region. The plain region covers a major part of the basin than the mountainous region. Further on the basin area can be divided into eight geomorphic units – (a) alluvial plain, (b) flood plain, (c) highly dissected hills and valleys, (d) low dissected hills and valleys, (e) mass wasting, (f) moderately dissected hills and valleys, (g) piedmont slope, and (h) water body (river). Highly dissected, moderately dissected, and low dissected hills and valleys are part of the Eastern Himalaya. It covers the entire northern part of the basin. There are few patches of mass wasting that cover a very little portion of the basin. Piedmont slopes connect the foothills and the alluvial plain. A small section of the flood plain is seen in the mountain region in the confluence of the three rivers Siri, Sido, and Sika where it forms the Jiadhal River. The maximum part of the lower basin is covered by flood plain areas. A major portion of it is seen along the river with extensive stretches of sand deposits. There are also some small ponds or lakes locally known as *beels* seen within alluvial plains and the flood plains.

5.4 Results

5.4.1 Flood Hazard Zones

Flood hazard mapping is an important component of land use planning in flood-affected areas as it helps in identifying the risk zone in the basin and prioritizes their mitigation effects (Forkuo, 2014). Flood hazard is defined as some threat, natural, technological, or civil to people, property, and environment, and is viewed as the probability of the occurrence of a potentially damaging flood event of a certain magnitude in a given area within a specific period of time (Sanyal & Lu, 2004). The concept of flood hazard is very important in flood management especially in the determination of flood risk (Opolot, 2013). The flood hazard zone map for the Jiadhal River basin has been created using the weighted overlay method in ArcGIS 10.3 (Table 5.1). The parameters taken are altitudinal zones, slope, flow accumulation drainage proximity, geomorphological units, and land-use/land-cover. The weightage was given to all the parameters on the basis of influence in the occurrence of flood and were assigned on a scale of 1 to 5. The flood hazard zone map that has been prepared for the Jiadhal River has five hazard zones, i.e., high, moderated, low, least, and no hazard zones. There are a total of 277 villages in the study area, out of which 260 villages are under moderate hazard zones, and the total area covered by this zone is 536.37 sq. km. The high hazard zone occupies 17.14 sq. km., whereas the moderate hazard zone covers 536.37 sq. km.of the total area shown in Table 5.2.

Table 5.1 Weightage of the parameters

Parameters	Weightage	Subsets	Scale value (weight)	Occurrence of flood
Altitudinal zones	25	Below 76	5	High hazard
		76–110	3	Low hazard
		110–160	2	Least hazard
		160–210	1	No hazard
		210–270	1	No hazard
		Above 270	1	No hazard
Slope	20	0–7	5	High hazard
		7–16	4	Moderate hazard
		16–25	2	Least hazard
		25–33	2	Least hazard
		33–42	1	No hazard
		42–73	1	No hazard
Flow accumulation	10	1	1	No hazard
		2	1	No hazard
		3	3	Low hazard
		4	5	High hazard
		5	5	High hazard
Buffer zones	15	1	5	High hazard
		1000	5	High hazard
		2000	4	Moderate hazard
		3000	3	Low hazard
		4000	2	Least hazard
		10,000	1	No hazard
Land use/land cover	15	Dense forest	1	No hazard
		Sparse vegetation	3	Low hazard
		Water bodies	5	High hazard
		Agricultural land	4	Moderate hazard
		Barren land	5	High hazard
		Sandbars	5	High hazard
Geomorphology	15	Moderate hills	1	No hazard
		Low hills	1	No hazard
		Piedmont slope	2	Low hazard
		Mass wasting	2	Low hazard
		Alluvial plain	4	Moderate hazard
		Flood plain	5	High hazard
		Water bodies	5	High hazard
		High hills	1	No hazard

Table 5.2 The total area, percentage, and the number of villages that comes under different hazard zones

	Hazard zones	Area in km^2	Area in percentage	No. of village affected
1	No hazard zone	117.97	12.47	4
2	Least hazard zone	225.73	23.87	6
3	Low hazard zone	48.41	5.12	3
4	Moderate hazard zone	536.37	56.72	260
5	High hazard zone	17.14	1.81	4

5.4.2 *Flood Management*

Flood management does not strive to eliminate flood hazard but to mitigate them (Tingsanchali, 2012). Flood management can be of both structural measures and nonstructural measures. Flood hazard zone mapping is considered as a means of nonstructural measures in flood management process to manage the flood plain development measures. Flood and drainage congestion is a major problem in the Jiadhal River basin. Due to heavy rainfall during the monsoon season, the lack of adequate gradient to drain out the high water and sediment discharge of the river and breaching of embankment flood occur in the basin area.

The communities living in the region has adopted with the flood problem and has learnt to respond positively toward flood hazard reduction. People of the Jiadhal basin have their own traditional and cultural believe to cope with the changing scenario of flood. The unique adaptation that has been taken in response to flood disasters is the stilt houses (chang-ghar) made from bamboo poles and using wooden boat as a means for transportation. For cultivation also the people adopted crop diversification and mixed cropping method as one of the critical mechanisms to cope up against heavy losses. The store houses locally called as Bhoral are higher than the houses to protect the grains from flood. The Jiadhal River is a frequently avulsing river and indigenous communities have been using bamboo porcupine to divert the flow of the river from residential areas (Das, 2015).

5.5 Conclusion

Remote sensing technology along with geographical information system (GIS) become the key tool for flood monitoring in recent years (Sanyal & Lu, 2004). But in spite of the great potential that remote sensing and GIS offer in flood management, their uses has been limited to some extent, and the presence of cloud cover during the flooding periods has been a major challenge in the use of optical remote sensing in the flood management (Sanyal & Lu, 2005). Floods of different intensity have been occurring in the flood plains since time immemorial, but with the increasing settlement in the flood plains, the loss of life and properties are also increasing. Remote sensing and GIS techniques have been regarded as a powerful tool in dealing with the various aspects of flood management like prevention, preparedness, and relief management of flood disaster, and the extensive use of these technologies has great prospect in creating long-term database on flood proneness, risk assessment, and relief management. In studies dealing with the application of remote sensing in inundated area delineation and flood risk assessment, DEM is used to visualize the interface of flood water with the terrain (Sanyal & Lu, 2004).

The main reason for heavy flood in the study area is because of the heavy rainfall during the monsoon season in the hilly part of the catchment as well as in the adjoining foothills. Flood hazard zoning and proper management of flood are very important in order to limit the damages caused by flood every year. In this study flood hazard zone map was generated based on appropriate parameters and are useful in identifying different zones that are prone to flood. The outcome of the study shows that 56.72% of the basin area comes under moderated hazard zone and 12.47% of the area comes under high hazard zone. The no hazard zone covers the major portion of the upper catchment area of the river basin. Thus, GIS has provided significant value in understanding sub-basin drainage characteristics with respect to flooding and also assists flood mitigation and land use planning.

References

Borgohain, S., Das, J., Saraf, A. K., Singh, G., Baral, S. S., & Sharma, K. (2016). Flood hazard assessment of the Jiadhal Fan, Assam, India. In *International conference on Climate Change Mitigation and Technologies for Adaptation (IC3MTA 2016), Shillong, Meghalaya, 23–26 June*.

Bormudoi, A., & Nagai, M. (2016). A remote-sensing-based vegetative technique for flood hazard mitigation of Jiadhal basin, India. *Natural Hazards, 83*(1), 411–423. https://doi.org/10.1007/s11069-016-2321-1.

Das, L. M. (2015). Channel Avulsion in Jiadhol River of Brahmaputra Basin Luna Moni Das Geography. *International Journal of Scientific Research, 4*(12), 274–276.

Das, L. M. (2019). Flood hazard zonation in Jiadhol River Basin of Assam. *International Journal of Science and Research (IJSR), 8*(1), 297–300.

Forkuo, E. K. (2014). Flood hazard mapping using Aster image data with GIS flood Hazard mapping using Aster image data with GIS. *International Journal of Geomatics and Geosciences, 1*(4), 932–948.

Jonkman, S. N. (2005). Global perspectives on loss of human life caused by floods. *Natural Hazards, 34*, 151–175.

Murthy, V. B., & Prasad, V. H. (2004). Remote sensing and management GIS in flood. *Water Science and Technology Library, 47*, 559–560.

Opolot, E. (2013). Application of remote sensing and geographical information systems in flood management: A review. *Research Journal of Applied Sciences, Engineering and Technology, 6*(10), 1884–1894.

Saikia, B. K., & Das, P. K. (2002). Application of fuzzy logic in modeling river catchment. *Proceeding of the National Seminar on Mathematics*, 139–151, *Dibrugarh University*.

Sanyal, J., & Lu, X. (2004). Application of remote sensing in flood management with special reference to monsoon Asia: A review. *Natural Hazards, 2014*, 283–301. https://doi.org/10.1023/B:NHAZ.0000037035.65105.95.

Sanyal, J., & Lu, X. X. (2005). Remote sensing and GIS-based flood vulnerability assessment of human settlements: A case study of Gangetic West Bengal, India. *Hydrological Processes, 19*, 3699–3716. https://doi.org/10.1002/hyp.5852.

Sharma, D., & Sarma, B. (2018). Prioritization of Jiya Dhol Basin using GIS. *International Journal of Science and Research (IJSR), 7*(7), 743–749. https://doi.org/10.21275/ART201962.

Thakuriah, G., & Saikia, R. (2015). GIS-based food hazard mapping: A case study in Krishnai river basin, India. *Research Journal of Recent Sciences, 5*(January), 50–59.

Tingsanchali, T. (2012). Urban flood disaster management. *Procedia Engineering, 32*, 25–37. https://doi.org/10.1016/j.proeng.2012.01.1233.

Chapter 6
Geomorphic History and Analysis of Deterioration of Quaternary Red Sands of Visakhapatnam, East Coast of India

Ch. Udaya Bhaskara Rao

6.1 Introduction

The Visakhapatnam coastal region is a part of the Eastern Ghats mobile belt, located on the east coast of India. This coastal tract on the east coast of India is characterized by a series of crescent-shaped bays and headlands as the hills are projecting into the Bay of Bengal in east-west direction deviating the normal trend (NE-SW) of Eastern Ghats. It presents remarkable red-looking sediments and awe-inspiring badland topography with a wide range of geomorphic features.

The red sediments of Visakhapatnam coastal area are the remnants of eroded sands believed to be formed during the Post Pliocene time or isolated sand banks formed around the sunken hills (King, 1886). Mahadevan and Sathapathi (1949) inferred that the red-looking sediments were formed by the cumulative work of wind and running water. Vishnuvardhana Rao and Durga Prasad Rao (1968) considered that the red sediments in this coastal region are dune sands based on grain size analysis. The red sediments of Visakhapatnam were believed to be formed by the mechanical weathering of the source rocks in an arid phase and transported to the present places in humid periods by fluvial agencies (Rao, 1978). Rao (1978) classified the red sands of Visakhapatnam into coastal and inland red sediments. The gravel-silt zones present in these sediments were believed to be deposited during several episodes of fluvial phases. The archeological and climate change studies by Rao (1978) suggest that the red sediments were formed around 11,000 years B.P. Raman and Rao (1980) have suggested fluvial origin for the origin of red sediments based on textural analysis. Srihari (1980) inferred dune origin for the red sediments based on textural analysis, and the pebbly horizons at various levels are the deposits of storm periods. Prudhvi Raju et al. (1985) have classified red

C. Udaya Bhaskara Rao (✉)
Department of Geography and Resource Management, Mizoram University,
Aizawl, Mizoram, India

G. S. Bhunia et al. (eds.), *Anthropogeomorphology*, Geography of the Physical Environment, https://doi.org/10.1007/978-3-030-77572-8_6

sediments into the bottom white to yellow sands with bouldery pebbly gravel as beach deposit, white to yellowish sands as dune deposits, and the top red-looking sands as beach deposits. The coastal red sediments of Visakhapatnam-Bhimunipatnam region are broadly classified by Rao et al. (1993a, b) based on granulometric studies into top fluvial red sediments with intermittent pebbly zones, brownish-red dune sands with calcium carbonate concretions, reddish-brown dune sands, and the bottom yellow sand with pebbly zones under the cumulative effect of the fluvial and aeolian processes. Nageswara Rao et al. (1993) based on C14 dating of calcium carbonate concretions in brownish-red sands and archaeological inferences assigned an age of 6000–3000 years BP to the red sand deposits. Srinivas (1995) classified the red sediments into three broad categories as khondalite basement (Archaean meta-sedimentaries), 75 cm thick pebble bed of fluvial origin, the upward sequence of 1–5 m thick coarse to fine-grained sandstones, 1.5 m thick red and yellow siltstones (Pleistocene), and 75 cm thick well-rounded pebbly gravel and the top hematite-coated sands (Holocene). The presence of lineaments and double cone structure and the calcium carbonate columnar concretions as inferred by Srinivas (1995) indicate the neotectonic activity around 6000 years BP. Nageswara Rao et al. (2006) have inferred six layers such as khondalite basement, 2.5 m thick weathered sandstone, 10 m thick yellow sand with graded and cross bedding, 1550 cm-thick duricrust, and 25 to 60 cm thick pebble beds, 15 m thick reddish-brown concretion bearing sands, 15 m thick brick-red sands, and 1 m top light-yellow sand. The bottom yellow sand was inferred as fluvial origin, reddish-brown concretion bearing sand as dune environment, brick-red sand as dune environment, and the top light-yellow sand as dune sands. These sediments are believed to be late Pleistocene origin (Nageswara Rao et al., 2006). The pebbly gravel beds present in the bottom yellow sand and the brick-red sand indicate several phases of fluvial transportation from the source regions in the vicinity of the study area. The duricrust horizons (20,000–30,000 years BP) which are embedded in the thick red and yellow sand deposits indicate the prolonged dry phases and time lags around 20,000–30,000 years BP (Rath, 1996) between the deposition of sediments under diverse environmental conditions that prevailed in the area. The red sands occurring in this area are similar to the red sands (Teris) which occur along the Tamil Nadu coast in the southeast coastal India and Sri Lanka (Gardner, 1995).

All the earlier studies were based on granulometric and archaeological evidences to infer their mode of origin. However, the present study is an attempt to understand the Quaternary geomorphic history and to analyze the causes for the deterioration of 30–40 m thick Quaternary red sediments by field mapping, remote sensing, and the techniques of geographical information systems.

6.2 Materials and Methods

Survey of India topographical maps on 1:50,000 scale have been used to delineate study area boundary through on-screen digitization with the help of ArcGIS software tools. SPOT FCC geocoded satellite image of the year 1990 and Google Earth image of the year 2020 have been used to understand the gully morphology and to demarcate the extent of gullies. Aerial photographs of the year 1973 have been used to study the dune morphology. Geomorphic features have been mapped using Survey of India topographical maps and satellite imagery by visual interpretation techniques. Field mapping has been carried out to identify all microrelief features. ASTER Global Digital Elevation Model (GDEM) with 30 m spatial resolution has been downloaded from NASA Earthdata website. Slope and flow direction layers have been generated from ASTER GDEM using the tools of ArcGIS spatial analyst. Slope, flow direction, and gully extent layers have been assigned weight of 50%, 25%, and 25%, respectively, and evaluated by multi-criteria evaluation technique using raster calculator of spatial analyst to delineate the extent of different forms of erosion. Four types of erosion such as severe gully erosion, severe sheet erosion, moderate sheet erosion, and low sheet erosion have been delineated based on the degree of slope, flow direction, and the gully extent.

6.3 Background of the Study Area

The area lies between 17° 51' and 17° 53’ latitudes and 83° 23' 36" and 83° 24' longitudes and located at about 24 km north of Visakhapatnam City (Fig. 6.1). The area is bounded on the north by a narrow to broad-crested hill range, by Pedda Gedda River on the south, on the east by the Bay of Bengal, and on the west by Chitti Gedda, a tributary to the Pedda Gedda River. The study area falls in semi-arid

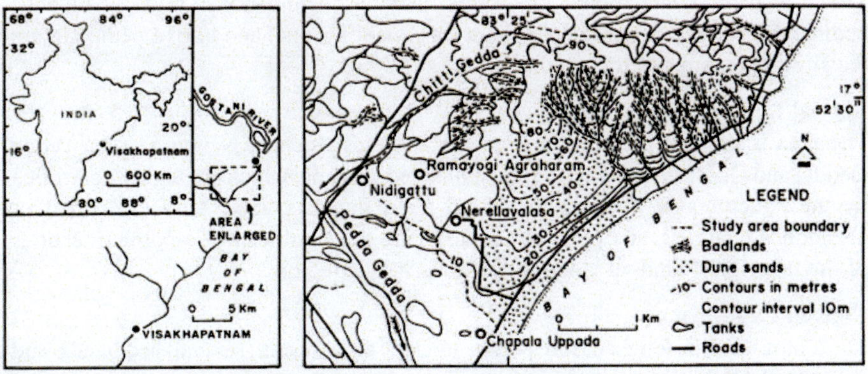

Fig. 6.1 Location of the study area

climate with an annual rainfall about 1000 mm. Temperature ranges between 18 and 34 °C.

The Visakhapatnam region is composed mostly by granulite grade rocks such as garnet-sillimanite gneisses (khondalites) and garnet-biotite gneisses (leptynites) with three generations of folds (Rao et al., 1993a, b). The hills in this area are composed of khondalites, leptynites, charnockites, pyroxene granulites, and pegmatites and granites. Quartzites also occur as thin veins in khondalites and leptynites at many places. Charnockites also occur as discontinuous rock groups and show an intrusive relationship with khondalites with a characteristic feature of spheroidal weathering (Rao et al., 1995).

Laterites like in-situ and colluvial type are seen at higher altitudes ranging from +10 to +42 m above MSL and 2 to 3 km inland.

6.4 Results and Discussion

Geomorphic Environment
A systematic interpretation of the satellite image pertaining to the study area reveals the prominent geomorphic features belonging to different environments.

Structural Landforms
The area presents a variety of structural landforms such as narrow to broad-crested hills; well-rounded and flat-topped hills are present at a number of places along this coastal tract (Fig. 6.2). A prominent structural dome with conspicuous characteristic features like plunging anticline and nose of the dome are present in this region. Linear to curvilinear structural valleys which are controlled by faults and fractures are quite obvious. Many prominent streams and rivers are controlled by the structures like faults and fractures developed by the ongoing tectonic activity in this region.

Denudational Landforms
The prominent denudational landforms, such as denudational hills, pediments, pediment inselberg complex, piedmont fans, debris slope, and buried pediplain are the striking features of this region.

Fluvial Landforms
The area is drained by many major river systems, such as Narava Gedda, Pedda Gedda, and Gostani, along with a few unnamed streams, which are highly controlled by the structures of this region. Flood plain deposits, river terraces, buried or abandoned channels, and point bar deposits are the significant geomorphic features of fluvial origin found all along the streams in this region.

Coastal Landforms
The entire 24-km-long coastal part is rocky; as a result, several headlands and crescent-shaped bays are conspicuous. The remarkable erosional coastal landforms like wave-cut terraces, sea stacks, sea cliffs, and depositional forms such as beach,

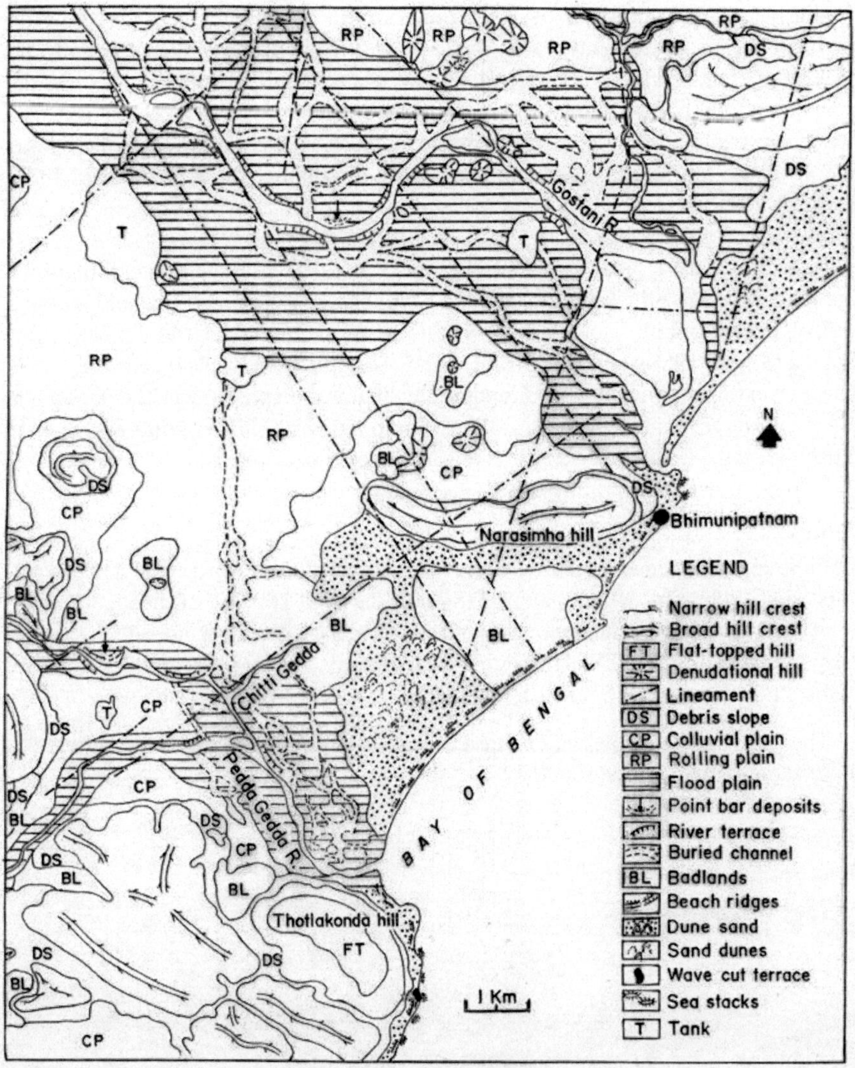

Fig. 6.2 Geomorphic features in the Visakhapatnam coastal region

sand dunes, and beach ridges are seen along the coast. There are about five such rock promontories along this coast. In fact, the Eastern Ghats run parallel to the coast in NE-SW direction. But along the Visakhapatnam coast, some hills are directly jetting into the sea in E-W direction at five locations which are prone to erosion as these are the points of wave attack. As a result, a number of sea cliffs, wave-cut terraces, and sea stacks have developed along the coast.

Similarly, several landforms of depositional origin can also be seen all along the 30 to 60 m wide crescent beach. Dune sand fields and sand dunes occupy the backshore zone up to 2–4 km inland. A few sets of stabilized dune ridges are also seen behind the beach. Three sets of such beach ridges at about 5–6 m above MSL which run parallel to the coast are conspicuous. These were inferred as the continuation of the Holocene strandlines that occur throughout the east coast of India. These ridges indicate the Holocene sea level up to 5–6 m above the present sea level (Nageswara Rao & Sadakata, 1993).

In addition, the typical badland topography that developed with dendritic drainage pattern in these thick unconsolidated sediments resembles the badland topography of Perth Amboy, New Jersey (Schumm, 1956); National Monument, South Dakota (Smith, 1958); and Lockwood Valley, California badlands (Caraman, 1958). The area exhibits typical badland topographic features displayed worldwide, such as sharp interfluves, narrow ridges, rolling, slumping, earth pillars, knife edges, earth pillars, flutings, hoodoods, septum like sub-ridges, steep vertical head sections with distal determinations, and haystacks (Fig. 6.3).

Gullies

Gullies are the ephemeral features that exist in the badlands which play a major role in erosion (Schumm, 1956). About 218 ha of the total 1000 ha of the study area is composed of gullies. Gully erosion is considered as the most prevalent form of erosion in this area. Gullies exhibit dendritic pattern. There are 16 prominent gullies in the study area which have different catchments (Fig. 6.4 and Table 6.1).

The gully heads in the area exhibit 10–20 m high cliffs with shapes like cave, inclined, armchair, and vertical.

Fig. 6.3 Badland topography in the thick unconsolidated red-looking sands

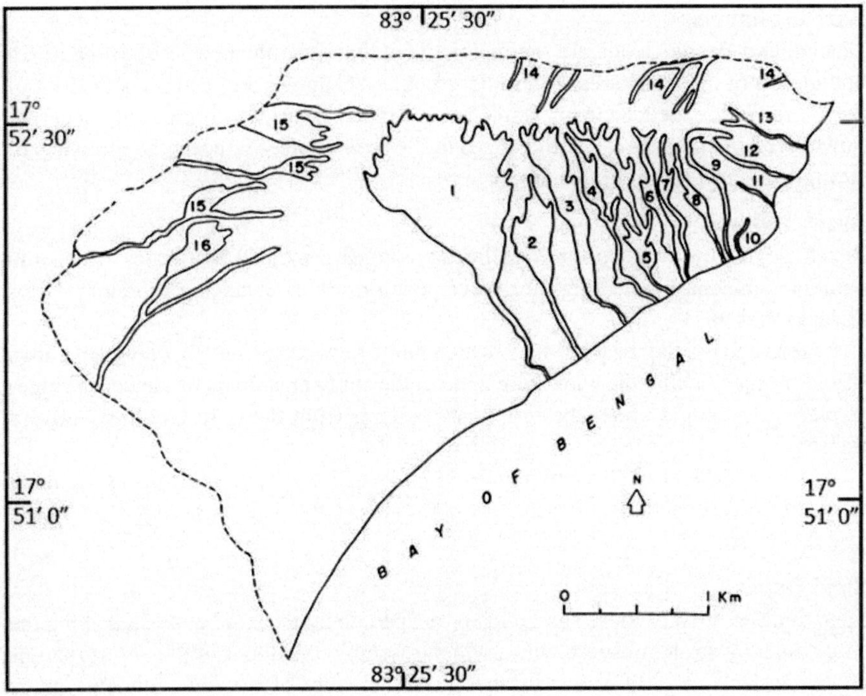

Fig. 6.4 Extent of gullies

Table 6.1 Morphometric characteristics of the gullies

Gully	Length (km)	Area in hectares	No. of tributaries First order	Second order	Third order	Fourth order
1	2.40	75.0	23	7	2	1
2	1.25	14.3	4	1	–	–
3	1.50	17.5	5	1	–	–
4	1.38	6.8	3	1	–	–
5	1.38	12.5	7	1	–	–
6	1.00	6.3	3	1	–	–
7	1.00	4.4	2	1	–	–
8	0.56	3.8	2	1	–	–
9	1.00	3.1	1	–	–	–
10	0.13	0.5	1	–	–	–
11	0.13	0.63	1	–	–	–
12	0.63	3.1	1	–	–	–
13	0.63	1.6	2	1	–	–
14	2.50	5.5	8	2	1	–
15	0.70	46.0	11	3	–	–
16	1.40	16.2	4	1	–	–

Buried Channels

Gostani and Pedda Gedda are the major rivers that form the northern and southern boundaries of the study area. A number of buried channels are seen on both sides of the Chitti Gedda, Pedda Gedda, and Gostani river courses which indicate the earlier flow directions in the area. The changes in the river courses indicate the influence of strong tectonic activity in this area.

Beach Ridges

Beach ridges are low, narrow, and linear mounds separated by swales which run parallel to the coast indicating successive positions of an advancing shoreline (Bates & Jackson, 1987).

Three such parallel beach ridges, which range between 3 and 6 m elevation above MSL, are seen behind the backshore zone in the study area. Parts of the beach ridges at places are buried under the sediments coming from the upstream areas toward the sea.

The beach ridges of this area represent Holocene sea level positions as the inner most ridge dates back to approximately 6000 years BP (Nageswara Rao & Sadakata, 1993).

Sand Dunes

Sand dunes consist of loose sand piled or heaped up by the wind deposited along the shore zone by strong onshore winds which resemble the dunes of the desert regions (Bates & Jackson, 1987). Sand dunes in the vicinity of the red sediments range between 2 and 10 m progressing toward the sea as the area is influenced by the strong onshore winds and associated aeolian activity. Some stabilized dunes which reach up to 60 m and up to 2 km inland from the present shoreline are also peculiar in this region. The dunes are mostly parabolic type, as seen on aerial photographs of the year 1973, which were later modified as flat lands by the agricultural activities in the area.

Lineaments

Lineaments are straight or gently curved linear features of the surface representing fractures, joints, or faults in the bedrock (Gary et al., 1972).

A number of prominent lineaments pass across as the area is tectonically active. There are four such prominent active lineaments identified in the area. The lineament one which passes through northern hill range and the southern distributaries of the Gostani River is significant in terms of its intensity which coincides with the major gully 1 (Fig. 6.5). Another prominent lineament passes through Chitti Gedda which exhibits a 30 m wide and about 1000 m long straight course in SW-NE direction (Fig. 6.6).

Two more major lineaments pass through the western flank of the northern range which shows the connectivity between a paleochannel of the Gostani River and the main active gully 1 and gully 2, respectively. The movement of ground water along the tributary of the gully 1 indicates the existence of a prominent fracture formed by neotectonic activity, as observed by Srinivas (1995). This lineament appears to be active even at present.

Fig. 6.5 A fracture-controlled major gully (gully 1) running in NE-SW direction

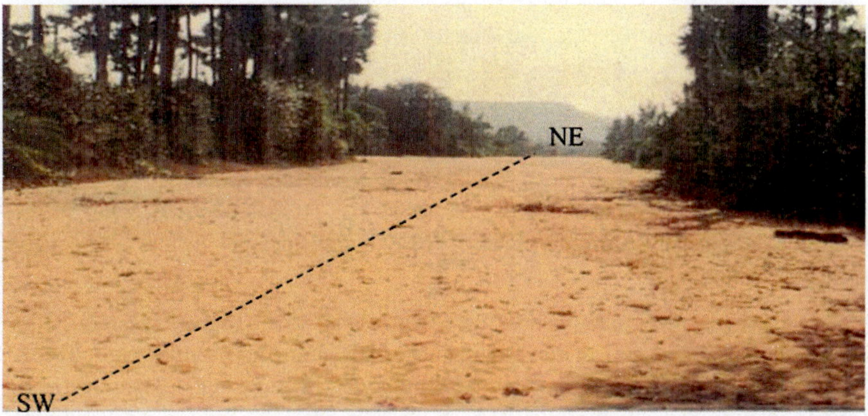

Fig. 6.6 A fracture-controlled channel of the Chitti Gedda

Yet, one more prominent lineament passes through the western part of the study area which coincides with a paleochannel of the Gostani River and Chitti Gedda stream in a N-S direction. More interestingly, all the gullies in the eastern part of the area are oriented in N 30° W direction which is a strong evidence of the existence of several fractures most probably in the subsurface bedrock.

Moreover, gullies in the eastern side are mainly oriented in one preferred direction of N 30° W. It is a clear field evidence of the existence of fractures. Particularly, the headward ends of the gullies end abruptly in this region following the general strike in E-W direction.

Knickpoints

The longitudinal profiles of the major gullies show the presence of waterfalls developed by the knickpoints in the bedrock and also due to the presence of hard

Fig. 6.7 A 5 m steep waterfall developed in the bottom yellow sand due to the presence of a 25 cm thick duricrust layer

indurated duricrust (laterites) and thick pebble beds at various levels in these thick sedimentary horizons. A sequence of knickpoints are seen all along the gullies at various levels ranging from 6 to 45 m above the present MSL. Among all the waterfalls developed in the area, the one formed by the thick pebble bed is a peculiar feature of the area.

A 2 m deep gorge is also seen in the bedrock from where it gives rise to a 1.5 m deep waterfall at about 11 m elevation at about 850 m inland from the shoreline. A 5 m steep waterfall is formed at about 25 m above MSL at about 1.3 km upstream from the shoreline due to presence of a 25 cm thick ferricrete layer (duricrust) in the bottom yellow sand (Fig. 6.7). A 4 m-high waterfall created by the khondalite bedrock at about 30 m elevation above MSL at about 1.35 km inland is seen in the major tributary of the gully 1. Another 2 m deep waterfall is present at 33 m elevation above MSL at about 1.5 km inland from the shoreline due to presence of a 2 m thick duricrust in gully1.

Three 2 m deep waterfalls at elevations 38 m, 39 m, and 42 m, respectively, are seen in the gully 15 in the western part of the study area.

The most striking 2 m and 60 cm-high waterfalls were created by the presence of 25 and 60 cm thick quartzite pebble beds at about +38 m and +40 m above MSL at about 1.74 km and 1.8 km inland, respectively, from the shoreline in the gully 1 (Fig. 6.8).

All the waterfalls seen in the area are characterized by spectacular plunge pools which vary from 30 to 75 cm in depth with up to 3 m in diameter. The presence of pebble beds here indicates the existence of a major river during Pleistocene. The well-rounded nature of the pebbles in the pebble beds indicates the existence of

Fig. 6.8 About 60 cm thick cemented well-rounded pebbly-gravel bed

major river systems which have flown across this area during Pleistocene and brought huge volumes of sediments from the dunes stabilized inland (Table 6.2). The two possible rivers in this are Gostani and Pedda Gedda which might have flown in this area as indicated by the orientation of their paleochannels toward this area.

The roundness index of the pebbles was measured (Cailleux, 1945; Kunen, 1956). The rounded nature of the pebbles in this area indicates longer abrasion history (Wentworth, 1922; Butzer, 1964). The orientation of the pebbles ranges between 16° and 36°, which is one of the characteristics of fluvial gravels (Cailleux, 1945). Moreover, the inclination of the measured pebbles is parallel to the flow direction (White, 1952) and shows an up-current dip (Pettijohn, 1957). All these observations suggest the sediments in the area were reworked by an active fluvial system in the past. It is noteworthy that the initial deposition might have taken place around +10 m above MSL as the bed levels of the two active river systems in the area, Gostani and Pedda Gedda, were around +10 m.

Table 6.2 Morphometric characteristics of pebbles in the pebble beds

Class	Shape of the pebble	Texture	Material	Percentage
A	Well-rounded	Fine	Quartzite	15
B	Sub-rounded	Fine	Quartzite	47
C	Elongated	Fine	Quartzite	10
D	Fine	Fine	Quartzite	8

The occurrence of these fluvial pebble beds between +38 and +42 m above MSL indicates the strong post depositional tectonic uplift, though the initial deposition had taken place around +10 m above MSL.

Wave-Cut Terrace

Wave-cut terrace is a gently sloping surface produced by wave erosion which indicates relative drop in the sea level (Bates & Jackson, 1987). A khondalite platform exists at +12.10 m above MSL at about 970 m inland from the shoreline. This platform spreading in an area of about 285 m^2 is dipping 10° toward the sea which is buried under the thick pile of sediments of about 30 m. Based on the existence of the discontinuous flat pebbly conglomerate with about 95% flat pebbles and the nature of the surface as well its orientation towards the coast confirms its marine origin (Cailleux, 1945).

Earlier studies made by Prudhvi Raju and Vaidyanadhan (1978) reveal that this platform is at about +7 m above MSL. ^{18}O and ^{16}O ratio measurements of benthonic and planktonic foraminifers on the ocean floor by Shackleton and Opdyke (1973) resolved that the sea level at 120,000 years BP was about +15 m. The submerged shorelines and shelves in Hawaiian islands during the last interglacial period was about +6 m above MSL (Streams, 1974). During the late Quaternary period between 120,000 and 130,000 years BP in the coral terraces in Bermuda, Florida, Oahu, and Hawaiian islands, the sea level was estimated at about +9 m above MSL (Gascoyne et al., 1979). The U-series isotope studies carried out by Moore (1982) reveal that the sea levels were oscillated between +2 and +7 m during the late Quaternary period between 120,000 and 140,000 years BP. The last interglacial shorelines in New South Wales was estimated by Marshall and Thom (1976) between +4 (younger) and + 6 (older) above MSL by $^{230}Th/^{230}U$ dating of corals. In the last interglacial phase at about 122,000 years BP, the highest sea level positions along the Barbados III coast based on Th^{230}, U^{230}, Pa^{231}, and U^{235} dating were estimated at +6 m by Broecker et al. (1968). Similar studies of $^{230}Th/^{234}U$ of Barbados III by Chapell (1974) and Bloom et al. (1974) revealed the sea level position during the late interglacial period as +6 m above MSL.

Based on the multiple paleoenvironmental criteria, Cronin et al. (1981) have predicted the higher sea levels during the last interglacial period along the South Carolina coast at +7.5 m ± 1.5 m around 120,000 years BP and +6.5 m ± 3.5 m around 94,000 years BP, respectively. Kupang, West Timor, had experienced the sea levels between +5 and +6 m above MSL during the last interglacial period, and +39 m neotectonic uplift was estimated with a mean rate of 0.3 mm/year by Jouannic et al. (1988). The Bermuda had experienced sea levels between +2 and +6 m around 125 kyrs. BP as revealed by Harmon et al. (1981) based on $^{230}Th/^{234}U$ dating of corals, amino acid racemization, speleothems, and mollusks. The Th^{230}/U^{238} and U^{234}/U^{238} dating of coral reefs along the Pacific and Indian Oceans revealed the late Quaternary sea level between 90,000 and 140,000 years BP were at +2 and +9 m, respectively (Veeh, 1966). The fossil dating of corals along the shoreline of Waimanalo of Oahu by Ku et al. (1974) reveals the last interglacial sea level around 120 ka. BP at +7.6 m above MSL. The last interglacial sea level of Barbados around 120 ka was between +2 and +8 m as revealed by Gallup et al. (1994) based on ^{234}U and $^{230}Th/^{238}U$ dating of the coral terraces. Selivanov (1991) analyzed about 86 radiometrically dated terraces and inferred the global mean sea level around 125 ka during the last interglacial period was between +5 and +8 m above the present mean sea level.

Studies made in several parts of India also prove that the last interglacial period (125 ka) has higher sea levels according to Gupta (1972). Gupta and Amin (1974) have identifed few terraces which range elevation between +2 and +6 m above MSL along the Kathiawar Peninsula coast based on the molluskan dating. Lele (1989) and Merh (1992) inferred the sea levels of Saurashtra Peninsula coast, western India, during the Quaternary period were between +2 and +20 m above MSL. The last interglacial sea level along the Saurashtra coast was inferred by a detailed study of raised terraces by Pant and Juyal as +7 m above MSL.

For the $^{230}Th/^{234}U$ radiocarbon dating of oysters and clam shells, Juyal et al. (1995) inferred the last interglacial sea level around 126.2 ± 5.7 kyrs as +7 m. The last interglacial sea level stand along the Tamil Nadu coast, southeast India, was about +8 m.

Particularly, along the Visakhapatnam coast, sea caves were reported at +7 m by Mahadevan and Sathapathi (1949) in the rock promontories of the hill ranges on the north and south of Visakhapatnam City. Wave-cut terraces at about +7 m above the present level were identified near Gangavaram and Pudimadaka at about 11 km and 29 km, respectively, in the southern part of Visakhapatnam by Bhaskara Rao and Vaidyanadhan (1975) and inferred them as of Holocene origin. The higher sea levels along the east coast of India of the last interglacial period (Sangamon) were between +2 and +8 m above MSL as inferred by Bruckner (1988).

A few similar terraces mapped in the further south of the red sands range between +6.5 and +7 m above MSL.

The +7 m wave-cut terrace at about 10 km south of these red sands must be older than the Lower Paleolithic culture (300,000 to 150,000 years BP) as inferred by Thimma Reddy et al. (1995). Based on the occurrence of the Paleolithic cultural material on this terrace at +7 m, it is inferred by Mohana Rao and Rao (1994) as pre-Holocene origin.

The relation between +7 sea level position and Acheulian industry (Subrahmanyam & Sireesha, 1989) supports the last interglacial higher sea levels along the east coast of India. As conjectured by Thimma Reddy (1994) based on the existence of split pebble industry on the wave-cut terrace found at Rushikonda 10 km south of red sands is older than 125,000 years (Nageswara Rao & Udaya

Bhaskara Rao, 1999). The terrestrial and aquatic pollen dating indicates an age of 8.0 to 4.9 cal ka BP for the middle Holocene sea level rising along the east coast of India (Nageswara Rao et al., 2020).

The appraisal of these studies prove that the terrace in the present study area is at the age of 125 ka BP, but the present elevation of the wave-cut terrace at +12 above the present mean sea level may be considered as tectonic uplift of the area as the area is tectonically active.

The geomorphic evidences reveal the Quaternary geomorphic history of the area. Evidences suggest that the major river systems like Gostani or Pedda Gedda might have flown across the area probably during the last Sangamon interglacial period when the terrain was at much lower elevation than at present and the sea was also far away from the present position. The pebble beds which occur at various levels in these thick sediments have brought by the paleo-fluvial systems initially deposited along the then meandering courses. Probably, the wave-cut terrace in the area was formed initially at +7 m above MSL during the last interglacial period around 125 ka BP as archeological evidences suggest. The area experienced a drier phase due to climatic change possibly at the beginning of the last major glacial stage of the Pleistocene, and the dune formation took place, burying the preexisting pebbly riverbed. The dune reddening was taken place in the subsequent worldwide lowering of the sea level. In the subsequent periods, the area uplifted by neotectonic activity led to domal upwarp of the area as the wave cut-terrace occurs around +12 m above the mean sea level (Nageswara Rao et al., 2006). The dune formation was continued during the Holocene, and dune reddening has also taken place as the sea levels dropped further. The ongoing tectonic activity led to the formation of a series of faults which results in the displacement of various sedimentary horizons both vertically and horizontally in the area as we see at present.

6.5 Human Intervention and Spatio-temporal Changes

The human interference by inappropriate agricultural practices by the local farmers in this area might have enhanced erosion in multiple dimensions in addition to the natural erosion processes. As the area is composed of loose and unconsolidated sediments, erosion in different forms is prevalent. Nageswara Rao et al. (2008a, b) have estimated the total volume of the sediment in this area based on geophysical and GIS analyses as 145 million cubic meters. It was estimated based on the dimensions of the gullies in the upper and middle and lower sections using remote sensing and field mapping techniques along with the rate of erosion along the active gully fronts. The rate of erosion as estimated in this study was 0.043 million cubic m. It is inferred that the initiation of gully erosion started around 1860 AD (Nageswara Rao et al., 2008a, b). Though, erosion may not be uniform in the area as it is influenced by several physical factors like the degree of slope, aspect, and storm intensity. It is estimated in this study based on the integration of the slope, flow direction, and the existing gullies about 2.18 km^2 area is prone to severe gully

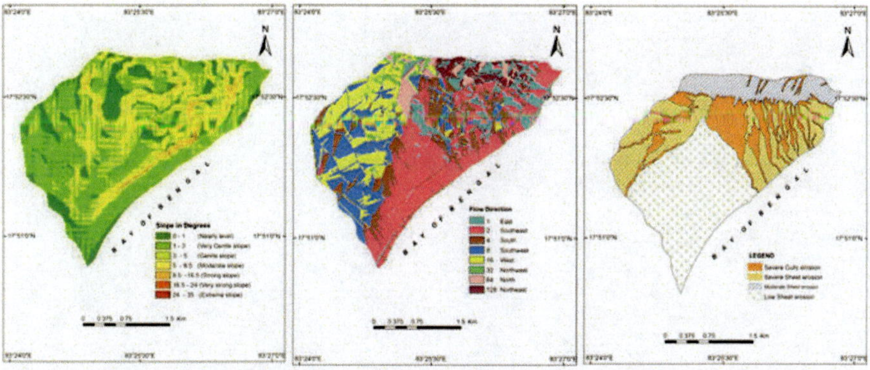

Fig. 6.9 (**a** and **b**) Slope and flow direction and (**c**) the nature of erosion

Fig. 6.10 SPOT FCC, image – 1990

erosion, 1.48 km^2 to severe sheet erosion, 1.57 km^2 to moderate sheet erosion, and about 4.77 km^2 area to very low amount of sheet erosion (Fig. 6.9).

The spatio-temporal analysis of the multi-date satellite images of the area reveals significant changes in the vegetation density. It is inferred based on gully dimensions that changes have been observed from SPOT FCC satellite image (Fig. 6.10) of the year 1990 and Landsat Google image (Fig. 6.11) of 2020 in the morphology of the gullies in the area during the period of 30 years.

Some efforts have already been made toward the implementation of massive afforestation programs in this area to stabilize the sands. The massive afforestation implemented in the gullies seems to be working well to some extent. The area is still experiencing severe gully and sheet erosion due to the faults and fractures developed by the ongoing tectonic activity more importantly due to unconsolidated sandy nature of the sediment. The area needs gully erosion control measures as erosion is going on at places along the steep gully walls.

Fig. 6.11 Google Earth (Landsat), image – 2020

6.6 Remedial Measures

The sand erosion should be checked by adopting appropriate techniques to the maximum possible extent as it preserves the rich past geomorphic history of millions of years.

Wire-gabion structures seem to be the suitable structures that should be constructed at multiple locations all along gullies to check gully erosion and particularly to control widening of gullies. Moreover, gully banks should be stabilized with retaining walls at selected locations which can prevent widening of the walls. Eco-development programs like massive afforestation should be adopted to prevent severe sheet and gully erosion. Locally available khondalite stones can be put across gullies in the upstream areas to arrest sediment load towards the lower reaches which can prevent gully widening also. Gully banks in the upper, middle, and lower reaches should be planted by quickly adaptable plant species like *Shorea robusta*, *Jatropha curcas*, *Salix* species, *Euphorbia globules*, *Casuarina equisetifolia*, *Acacia nilotica*, *Azadirachta indica*, etc. In addition, on the surface of the sediments, economically viable plants such as *Anacardium occidentale*, *Mangifera indica*, etc. should be planted extensively, though there will be erosion to some extent in the initial stage.

6.7 Conclusions

The geomorphic features present in the red sand deposits revealed the Quaternary geomorphic history and sea level changes that occurred in this part on the east coast of India. Further, the anthropogenic causes like inappropriate agricultural practices have accelerated soil erosion, thereby leading to the deterioration of the red sand deposits of the Quaternary period on the Visakhapatnam coast.

References

Bates, R. L., & Jackson, I. A. (1987). *Glossary of geology* (3rd ed.). American Geological Institute. 788 p.

Bhaskara Rao, V. U., & Vaidyanadhan, R. (1975). Photogeomorphic studies of coastal features between Visakhapatnam and Pudimadaka. *Photonirvachak, 3*, 43–46.

Bloom, A. L., Broecker, W. S., Chappell, J. M. A., Matthews, R. K., & Mesolella, K. J. (1974). Quaternary sea level fluctuations on a tectonic coast New ^{230}Th / ^{234}U dates from the Huon Peninsula, New Guinea. *Quaternary Research, 4*, 185–205.

Broecker, W. S., Thurber, D. L., Goddard, J., Ku, T. L., Matthews, R. K., & Mesolella, K. J. (1968). Milankovitch hypothesis supported by precise dating of coral reefs and deep-sea sediments. *Science, 159*, 279–300.

Bruckner, H. D. (1988). Indicators for formerly higher sea levels along the east coast of India and on the Andaman Islands. *Hamburger Geographische studien on Hett, 44*, 47–72.

Butzer, K. W. (1964). *Environment and archaeology (An introduction to pleistocene geography)* (pp. 151–172). Methunen & Co. Limited.

Cailleux, A. (1945). Distinction des galets marins et fluviatiles. *Bulletin Geological Society of France, 15*(5), 375–404.

Caraman, M. F., Jr. (1958). Formation of badland topography. *Bulletin Geological Society of America, 69*, 789–790.

Chappell, J. (1974). Geology of coral terraces. Huon Peninsula, New Guinea. A study of Quaternary tectonic movements and sea level changes. *Bulletin Geological Society of America, 85*, 553–570.

Cronin, T. M., Szabo, B. J., Ager, T. A., Hazel, J. E., & Owens, J. P. (1981). Quaternary climates and sea levels of the U.S Atlantic coastal plain. *Science, 211*, 233–240.

Gallup, C. D., Edwards, R. L., & Johnson, K. G. (1994). The timing of high sea levels over the past 200,000 years. *Science, 263*, 796–800.

Gardner, R. A. M. (1995). Red dunes and Quaternary Palaeoenvironment in India and Sri Lanka. *Memoirs Geological Society of India, 32*, 391–404.

Gary, M., McA Fee, R., Jr., & Woli, C. L. (Eds.). (1972). *Glossary of geology*. American Geological Institute. 805 p.

Gascoyne, M., Benjamin, G. J., Schwarcz, H. P., & Ford, D. C. (1979). Sea level lowering during the Illinoin Glaciation: Evidence from a Bahama "Blue Hole". *Science, 205*, 806–808.

Gupta, S. K. (1972). Chronology of the raised beaches and inland coral reefs of the Saurashtra coast. *Journal of the Geological Society of India*, 351–361.

Gupta, S. K., & Amin, B. S. (1974). I o/ U ages of corals from Saurashtra coast. *Marine Geology, 16*, 79–83.

Harmon, R. S., Land, L. S., Mitterer, R. M., Garrett, P., Schwarcz, H. P., & Larson, G. J. (1981). Bermuda sea level during the last interglacial. *Nature, 289*, 481–483.

Jouannic, C., Hoang, C. T., Hantoro, W. S., & Delinom, R. M. (1988). Uplift rate of coral reef terraces in the area of Kupang, West Timor: Preliminary results. *Palaeogeography Palaeoclimatology Palaeoecology, 68*, 259–272.

Juyal, N., Pant, R. K., Bhushan, R., & Somayajulu, B. L. K. (1995). Radiometric dating of late Quaternary sea levels of the Saurashtra coast, western India: AN experiment with Oyster and Clam shells. *Memoirs Geological Society of India, 32*, 372–379.

King, W. (1886). The geological sketch of Visakhapatnam district. *Rec Geological Survey of India, 19*, 143–156.

Ku, T. L., Kimmel, M. A., Easton, W. H., & O'Neil, T. (1974). Eustatic sea level 120,000 years ago on Oahu, Hawaii. *Science, 183*, 959–962.

Kuenen, P. H. (1956). Experimental abrasion of pebbles 2. Rolling by current. *Journal of Geology, 64*, 336–368.

Lele, V. S. (1989). Quaternary formations in the Bhadar Valley-Western India 2. *Bulletin of the Deccan College Post-Graduate and Research Institute, 49*, 165–205.

Mahadevan, C., & Sathapathi, N. (1949). The origin of the Waltair highlands. *Geography Journal, India, 24*, 1–26.

Marshall, J. F., & Thom, B. G. (1976). The sea level in the last interglacial. *Nature, 263*, 120–121.

Merh, S. S. (1992). Quaternary sea level changes along Indian coast. *Proceedings of the Indian National Academy of Science, 58*, 461–472.

Mohana Rao, K., & Rao, T. C. S. (1994). Holocene sea levels of Visakhapatnam shelf, east coast of India. *Journal of the Geological Society of India, 44*, 685–689.

Moore, W. S. (1982). Late Pleistocene sea-level history. In *Uranium series disequilibrium applications to environmental problems* (pp. 481–495). Clarendon.

Nageswara Rao, K., & Sadakata, N. (1993). *Holocene evolution of deltas on the east coast of India.* In Proceedings of the 8th symposium on coastal and ocean management, 19–23 July, New Orleans, Louisiana, pp. 1–15.

Nageswara Rao, K., & Udaya Bhaskara Rao, C. (1999). Geomorphic evidence of Pleistocene higher sea levels near Visakhapatnam, east coast of India. *Zeitschrift für Geomorphologie, 43*(1), 19–25.

Nageswara Rao, P., Rao, A. T., Deva Verma, D., & Purnachandra Rao, K. (1993). Calcretes in the red sediments of Visakhapatnam. *Indian Journal of Landscape Systems and Ecological Studies, 16*(1), 92–100.

Nageswara Rao, K., Udaya Bhaskara Rao, C., Vijaya Prakash, P., & Thimma Reddy, K. (2006). Morphostartigraphy and evolution of quaternary 'Red Sands' near Bhimunipatnam, East Coast of India. *Journal of the Geological Society of India, 68*, 857–873.

Nageswara Rao, K., Udaya Bhaskara Rao, C., & Venkateswara Rao, T. (2008a). Estimation of sediment volume through Geophysical and GIS analysis – A case study of the red sand deposit along Visakhapatnam coast. *Journal of the Indian Geophysics Union, 12*(1), 23–30.

Nageswara Rao, K., Udaya Bhaskara Rao, C., & Sadakata, N. (2008b). Time frame for the initiation of gully erosion in coastal red sand formations near Visakhapatnam, east coast of India. *Annals of the Yamaguchi Geographical Union (Area Yamaguchi), 37*, 35–44.

Nageswara Rao, K., Saito, Y., Nagakumar, K. C. V., Kubo, S., Pandey, S., Li, Z., Demudu, G., & Rajawat, A. S. (2020). Holocene evolution and Anthropocene destruction of the Krishna Delta on the east coast of India: Delta lobe shifts, human impacts, and sea-level history. *Marine Geology, 427*, 106229.

Pettijohn, F. J. (1957). *Sedimentary rocks*. Oxford & IBH Publishing Co. 701 p.

Prudhvi Raju, K. N., & Vaidyanadhan, R. (1978). Geomorphology of Visakhapatnam, Andhra Pradesh. *Journal of the Geological Society of India, 19*(l), 26–34.

Prudhvi Raju, K. N., Mahalakshmi, K. B., Prasada Raju, P. V. S., Krishna, K., Bhagavan, S. V. B., & David Emmanuel, B. (1985). Geomorphic processes in the formation of (Red) sands of Bhimilipatnam, Visakhapatnam District, Andhra Pradesh. *Journal of the Geological Society of India, 26*, 336–344.

Raman, C. V., & Rao, A. T. (1980). Textural analysis of red sediments from Visakhapatnam district, Andhra Pradesh. *Journal Geological Society of India, 21*, 48–53.

Rao, A. T. (1978). Red sediments of Visakhapatnam, Andhra Pradesh. *Journal Geological Society of India, 19*(2), 79–82.

Rao, A. T., Rao, J. U., & Yoshida, M. (1993a). Geochemistry and tectonic evolution of the pyroxene granulites from Visakhapatnam area in the Eastern Ghats granulite belt, India. *Journal of Geosciences. Osaka City University, 36*, 135–150.

Rao, A. T., Rao, P. N., Deva Varma, D., & Purnachandra Rao, K. (1993b). Coastal sediments along Visakhapatnam – Bhimunipatnam region, Andhra Pradesh. *Journal of Indian of Association of Sedimentologists, 12*, 1–9.

Rao, A. T., Divakara Rao, V., Yoshida, M., & Arima, M. (1995). Geochemistry of Charnockites from the Eastern Ghats granulite belt: Evidence for possible linkage between India and Antarctica. *Memoirs Geological Society of India, 34*, 273–291.

Rath, A. (1996). *Quaternary environment and Prehistoric cultural development: A micro level study on the east coast of India*. Unpublished Ph.D thesis, Andhra University, Visakhapatnam, 356 p.

Schumm, S. A. (1956). Evolution of drainage systems and slopes of badlands of Perth Amboy, New Jersey. *Bulletin Geological Society of America, 67*, 597–646.

Selivanov, A. O. (1991). *Spatial analysis of sea level during the certain Late Pleistocene and geoidally-induced sea level changes*. Annual report of IGCP Project #274 and News Letter, pp. 144–145.

Shackleton, N. J., & Opdyke, N. D. (1973). Oxygen isotope and palaeomagnetic stratigraphy of equatorial Pacific core, isotope temperatures and ice volumes on a 105 year and 106 year scale. *Quaternary Research, 3*, 39–55.

Smith, K. G. (1958). Erosional processes and landforms in badlands, National Monument, South Dakota. *Bulletin Geological Society of America, 69*, 975–1008.

Srihari, Y. (1980). *Origin of Visakhapatnam red sediments*. Unpublished Ph.D thesis, Andhra University, Visakhapatnam, 152 p.

Srinivas, M. (1995). Possible indicators of neotectonic activity near Bhimunipatnam coast, Andhra Pradesh. *Current Science (India), 68*(5), 552–555.

Stearns, H. T. (1974). Submerged shorelines and shelves in the Hawaiian islands and a revision of some of the eustatic emerged shorelines. *Bulletin Geological Society of America, 85*, 705–804.

Subrahrnanyam, B. K., & Sireesha, E. J. (1989). Fossil beaches and palaeolithic investigations in coastal Andhra Pradesh. *Journal of the M.S. University of Baroda, 37–38*(1), 77–86.

Thimma Reddy, K. (1994). Coastal ecology and archaeology: Evidence from the east coast of India. *Man and Environment, XIX*(1–2), 43–55.

Thimma Reddy, K., Vijaya Prakash, P., Rath, A., & Udaya Bhaskara Rao, C. (1995). A pebble tool assemblage on the Visakhapatnam coast. *Man and Environment, XX*(1), 114–118.

Veeh, H. H. (1966). Th^{230}/U^{238} and U^{234}/U^{238} ages of Pleistocene high sea level stand 3. *Journal of Geophysical Research, 7l*(14), 3379–3386.

Vishnuvardhana Rao, M., & Durga Prasada Rao, N. V. N. (1968). A note on the origin of Waltair high-lands. *Current Science, 37*, 438–439.

Wentworth, C. K. (1922). *The shapes of beach pebbles. U.S.* (Geological Survey Project paper 131-C, pp. 74–83).

White, W. S. (1952). Imbrication and initial dip in a Keweenawan conglomerate bed. *Journal of Sedimentary Petrology, 22*, 189–199.

Chapter 7
Recent Disturbances in the Geomorphic Processes Due to Human Interventions Along the West Coast of India

M. K. Rafeeque, T. Akhil, Mintu E. George, D. S. Suresh Babu, and T. K. Prasad

7.1 Introduction

The western coastal zone of India with diverse coastal morphology due to backwaters and barrier reefs of recent formation narrow cliffs and pocket beaches, promonotories, strand plains, and wide extensive coastal mudflats having tidal range from 0.5 m at *Kanyakumari* coast to greater than 8.5 m in the Gulf of *Khambhat* in *Gujarat* coast (Mukhopadhyay and Karisiddaiah, 2014; V. Sanil Kumar et al., 2006; M. M. Nair, 1987). This coastline is more stable than the east coast of India, even though being a high-energy coast. Prominent geomorphic features that are active on this coast include marine, estuarine, fluvial, and denudational processes. The southern part of this coast is reported as younger coastal plain of recent formation, and the age of the shoreline gradually increases toward the northward of this coast.

M. K. Rafeeque (✉)
National Centre for Earth Science Studies, Thiruvananthapuram, India

University of Kerala, Thiruvananthapuram, India
e-mail: rafeeque.mk@ncess.gov.in

T. Akhil · D. S. Suresh Babu
National Centre for Earth Science Studies, Thiruvananthapuram, India
e-mail: dss.babu@ncess.gov.in

M. E. George
National Centre for Earth Science Studies, Thiruvananthapuram, India

Cochin University of Science and Technology, Ernakulam, India

T. K. Prasad
University of Kerala, Thiruvananthapuram, India

Kannur University, Kannur, India

G. S. Bhunia et al. (eds.), *Anthropogeomorphology*, Geography of the Physical Environment, https://doi.org/10.1007/978-3-030-77572-8_7

The paleo-shoreline of the *Kerala* state of the southern region of this coast is demarcated from a few hundred meters to several kilometers landward of the present shoreline at many locations. Young coastal geomorphological features like lagoons and backwaters, coastal barriers, swale, reefs, and coastal plains are ubiquitous in this coastal belt. The young coastal landforms (Nair NJK et al., 1989; Samsuddin M et al., 2008) of this coast demonstrate its accretion trend during the recent geological history. As per the *Shoreline Change Atlas of the Indian Coast* published by the Central Water Commission of India, nearly 40% of the total coastline undergoes erosion, while 30% of the coastline is protected by natural accretion, and the remaining 30% is represented as a stable coast (Table 7.1). The characteristics of wave and beach dynamism of the *Kerala* coast indicate that the direction of littoral drift is toward the south in June–July and toward the north in other seasons (KERI, 1978; Suchindan et al., 1987). Reddy and Varadachari (1973), Nair et al. (1973), and Murthy et al. (1980) have examined the sediment movements related to coastal erosion and accretion along the west coast of India in detail.

Nearly 50% of the *Kerala* coast is regulated with different coastal structures, including ports/harbors, breakwaters, groins, and seawalls (Sheela et al., 2018). Due to a very high density of coastal population of 2931 person/sq km along Kerala coastal plain (Rafeeque et al., 2021) and continuous parallel settlements along the Kerala coastal belt, the end/side erosion of already constructed seawalls or other coastal structures will frequently necessitate the construction of new structures or extend the existing one toward further extremes (Sheela et al., 2018). The shoreline changes associated with coastal protection measures constructed along *Kerala* coast were studied by Murthy et al. (1980). Sreekala et al. (1997) studied the long-term shoreline changes of *Kerala* coast from multi-dated satellite images, aerial photographs, and topographic maps. But, due to the different anthropogenic activities and natural factors, *Kerala* coast has been reported for drastic changes in the locations of erosion/accretion zones over the periods (Sreekala et al., 1997; Narayana & Priju, 2006; Kurian et al., 2007; Sheela Nair, 2014; Prasad et al., 2016). This is mainly due to the construction of ill-conceived shore protection structures, beach sand mining, port, and harbor development, etc. in addition to the natural factors such as the 2004 tsunami event, coastal flooding, mudbank formations, etc. (Kurian et al., 2006; Prasad et al., 2016; Sheela et al., 2018). Forty-five percent of the shoreline along

Table 7.1 Shoreline trend of states along SW coast of India

Sl. no.	State	Stable (%)	Erosion (%)	Accretion (%)
1	Gujarat, Daman, Diu	47.075	32.820	20.105
2	Maharashtra	6.506	60.558	32.936
3	Goa	52.371	17.395	30.234
4	Karnataka	24.594	35.601	39.805
5	Kerala	12.568	37.227	50.205
	Total	30.104	39.768	30.962

Source: *Shoreline Change Atlas of the Indian Coast*, Central Water Commission, INDIA, 2014 (Assessment Year 2006)

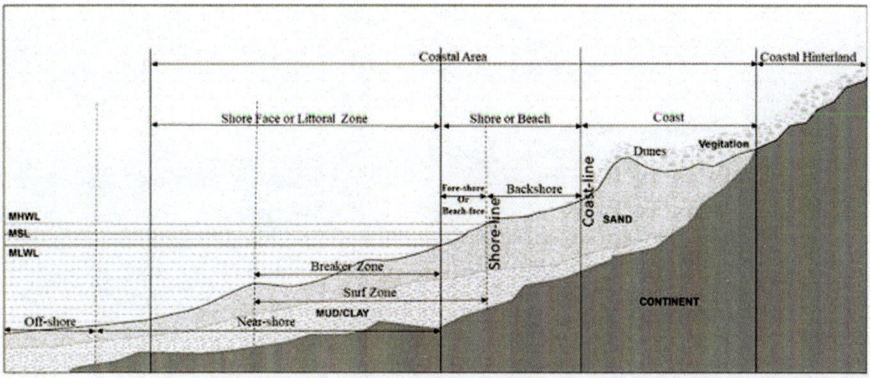

Fig. 7.1 A typical cross-section of a coast showing major morphological features and zones, which are also visible in the study area. (Source: conceptualized by author/s from different schema)

Kerala coast is suffering from coastal erosion, while 29% is recorded for coastal accretion and 26% of the shoreline is marked as stable (Sheela et al., 2018). The erosive trend of *Kerala* coast is boosted after 2004–2006 from 37% to 45% in 2014. The artificial coastal structures due to increased demand of seawalls and groins from inhabitants of the coastal zone after the 2004 tsunami rediverted the coastal dynamism and the shoreline movement (Fig. 7.1).

Nearly half of the coastline between *Kadalur* Cape in the north and *Beypur* estuary in the south of *Kozhikode* district of *Kerala* exists as stable coast as per the current study. The reported erosion in this segment is mainly associated with different artificial/natural coastal structures. The important coastal features of this coastal zone include three estuaries, three harbors, two main capes, and many micro-geomorphic and artificial features like promontories, sea pier, and groins. Almost all artificial coastal structures were constructed within the last 20 years, except sea piers (Fig. 7.2) and *Elathur* harbor, while piers were constructed way back in the time of colonial dynasty and *Elathur* fishing harbor was constructed in the 1990s. Accordingly, the rate of coastal erosion had increased during the last 20 years.

7.1.1 Study Area

The spatiotemporal analysis was focused on the 47-km-long coastal belt of *Kozhikode* city area bounded by *Kadalur* headland in the north and *Chaliyar* river basin in the south, which forms part of the coastal zone of SW Indian subcontinent lying between the Lakshadweep Sea and Western Ghats. The spatial extent can be demarcated with graticules of 11° 28′ N latitude and 75° 38′ E longitude in the north and 11° 09′ N latitude and 75° 48′ 37″ E longitude in the south. This coastal plain has a rich variety of morphological features such as sandy beaches, complex cliffs, headlands promontories, estuaries, creeks, barrier reefs, etc.

Fig. 7.2 Aerial view of *Kozhikode* beach. The remains of first artificial (sea pier) and recent tourism structures are also seen. (Photograph taken by author/s)

Geology of the Area Geologically *Kerala* is attached to the southern Indian shield, and the important geological formations are (i) crystalline rocks of the Precambrian, (ii) sedimentary rocks of the Tertiary age, (iii) laterites capping on crystalline and sedimentary formations, and (iv) recent and sub-recent sediments covered over low-lying areas, coastal zones, and river valleys (Geological Survey of India, 2005; Chattopadhyay, 2017). According to the "District Survey Report of Minor Minerals," prepared by the "Department of Mining and Geology" (GoK, 2016), *Kozhikode* is divided into three geological zones such as (i) a linear NW-SE trending gneissic belt, along the middle extending from north to south; (ii) a charnockite belt occupying areas in the northeast and south, extending to the adjacent districts, and also occurring as pockets within the gneissic terrain; and (iii) a narrow coastal belt. The structural configuration of the Kerala coastline is controlled by several faults and lineaments. Lineaments trending NW-SE, NNW-SSE, ENE-WSW, and WNW-ESE directions have been recognized (Prabhakara Rao et al., 1985), and the ones along NW-SE and NNW-SSE, trending parallel to the coastline, are found to be neotectonically active. Off-setting of the straight-line coast, acute bends in stream courses and lagoons, and minor differences in elevation, on either side of the beach ridges, are cited as shreds of evidence in favor of neotectonism (Prabhakara Rao et al., 1985).

Geomorphology of *Kozhikode* The important geomorphological features found in *Kozhikode* district are alluvial plains, flood plains, ridges, residual hills, denudational hills, lower plateaux, hillcrest, and rocky slope (scarp face). The coastal belt of *Kozhikode* is dominated by the young coastal plain of Holocene origin. The barrier reefs and ridges with intermediate swales are the dominant morphological structures of *the Kozhikode* coast. The northern side of *the Kozhikode* coastline has several rock outcrops of more than 10-m heights. The *Palakkulam* hill is the highest among them with a maximum height of 58 m above MSL. Other rocky structures are *Kadalur* hills (26 m), *Vellarakkad* (26 m), *Parappalli* (25 m), *Thoovappara* (15 m), and *Elathur* (11 m). These outcrops are located within 25 km of the northern portion of the project area, between *Kadalur* headland in the north and *Elathur* headland in the south. This coastal belt has a parallel drainage system of *Korapuzha* along its landward side which debouches into the Lakshadweep sea through the northern bank of *Elathur* headland, through a 4-km-long backwater system jointly with *Agalapuzha*, aligned in the landward side of *Elathur* headland. To the south of the *Elathur* headland, *the Kozhikode* coastline is a low-laying continuous coastal plain with less than 3-m elevation. Two rivers, *the Kallayi* River and the *Chaliyar* River, are discharging into the Lakshadweep sea through this coastal belt. The *Korapuzha* backwater system is connected with *the Kallayi* River and *Chaliyar* River with a coast-parallel artificial canal (*Canolly Canal*) constructed during the colonial era.

Sociocultural Setting Up The study area consists of coastal areas of two urban centers, viz., *Kozhikode* Municipal Corporation and *Koyilandi* Municipality and four Grama Panchayaths of *Moodadi*, *Chengottukavu*, *Chemancheri*, and *Kadalundi*. *Kozhikode* is categorized as an urban agglomeration by the Government of India census report 2011. The total population of the *Kozhikode* Municipal Corporation is 608,255 with a population density of 5129 persons per sq km. The local bodies coming under the study area and its population structure are given in Table 7.2. The densely populated settlement pattern in *Kozhikode* coast also faces severe coastal vulnerability due to many factors.

7.1.2 Oceanographic Condition in the Vicinity

Nearshore Waves The average wave height of *the Kozhikode* coast is less than 1 m. However, high waves with a significant wave height of 2.5 m or even higher

Table 7.2 Coastal local bodies and population scenario

Sl. no.	Name of local body	Type of local body	Population in person (2011)	Area in sq km	Density in per/sq km
1	Moodadi	Grama Panchayath	27,653	16.02	1726
2	Koyilandi	Municipality	71,873	29.1	2474
3	Chengottukavu	Grama Panchayath	23,702	13.6	1743
4	Chemancheri	Grama Panchayath	31,326	16.76	1869
5	Kozhikode	Municipal Corporation	608,255	118.59	5129
7	Kadalundi	Grama Panchayath	35,171	11.83	2973

Source: Census India Report (2011)

Table 7.3 Tidal levels referred to datum of sounding

Place	Lat. N	Long. E	Heights in meters above datum				
			MHHW	MLHW	MHLW	MLLW	MSL
Calicut	11° 15′	75° 46′	1.4	1.2	0.8	0.4	1.0
Beypur	11° 10′	75° 48′	1.3	1.1	0.8	0.4	0.9

Source: Hydrographic India Chart No: 2053

have been reported during the SW monsoon season. Such seasonal high waves are the causative factors overtopping and coastal erosion. The wave breaker zone of *Kozhikode* coast is also dynamic in nature as a result of variations in the nearshore wave heights. The plunging and/or spilling type of wave breakers is dominant in this coastal zone during the rough season, and the plunging and collapsing types of wave breakers are more active during the fair season. The surging type of wave breaker is also reported where every beaches are very steep. Due to inter- and intra-seasonal variations in the coastal morphology, particularly the beach characteristics, there is a corresponding change in the breaker characteristics which in turn affects the beach state all along the *Kerala* coast (Baba & Kurian, 1988).

Longshore Currents The longshore currents, i.e., the movement of water parallel to the coastline with the influence of variations in wave breaking heights and the oblique approach of the wave to the coast, along the *Kozhikode* coast, also change its direction seasonally. The longshore currents are predominantly northerly during the fair season, whereas it is mostly toward the south during the monsoon season. The post-monsoon season experiences relatively higher current compared to the pre-monsoon season (Hameed et al., 2007). Even the direction of longshore current is uniform all over the coast in a particular season, the magnitude of the current will not be necessarily uniform everywhere.

Tide Being part of the southwest coast of India, the *Kozhikode* coast is categorized as the coast of the micro-tidal range. This micro-tidal range is caused by the narrow shelf region, and the coastal geomorphological features of the area is determined by the tidal range. According to the Survey of India Tide Table (2020), the tide along the Kozhikode coast ranges from 0.04 m below MSL to 1.67 m above MSL. The details of tide level and maximum highest high water to maximum lowest low water reported along this coast are given in Table 7.3.

7.2 Materials and Methods

7.2.1 Mapping Coastal Geomorphology

High-resolution WorldView-3 satellite images of 2011 procured from the Digital Globe were used as the base map for demarcating the coastal structures and beach morphology. High-precision ground control points, collected with differential positioning survey method using Leica Global Navigation Satellite System receivers, were used to bring different raster datasets into a common geospatial reference system. The standard Universal Transverse Mercator geo-coordinate system with WGS 84 reference datum was used for geo-referencing. The recent timeline imageries provided by Google Earth were downloaded and overlaid on the base map with common GCPs and was used for the latest update of coastal morphological features. The different temporal multiresolution satellite images of 1984 to present such as Landsat, Cartosat, and WorldView satellites were analyzed for understanding the morphological processes and landform changes over the years.

Hydrographic Chart No. 2053, published by the National Hydrographic Office, Dehradun, in a 1:50,000 scale by incorporating the updated survey details of 2002, was used to generate nearshore morphology. By interpolating depth points and contours with ArcGIS spatial analysis tool, we generated the contour of a 1-m interval for nearshore bathymetry up to 10-m depths, 2-m interval up to 20-m depths, and 5-m interval beyond 20 m. A Digital Terrain Model of Kozhikode coast as derived from the processed hydrographic chart data and nearshore bottom features of this coastal zone was prepared in a 1:10,000 scale and validated through local information collected from fishermen and coastal community during field visits.

7.2.2 *Shoreline Analysis*

Shorelines of the different periods from 1968 to 2018 were extracted from different sources like topographic sheets, multi-resolution multi-dated satellite images, and open-source online satellite imageries. The details of datasets used for shoreline extraction are given in Table 7.4. The Google Earth images were downloaded with maximum resolution and scale and resampled into the base map for geo-coding. The scale disparity was considered for different data sources, and appropriate uncertainty value was assigned at the time of processing. The high tide line (HTL) identified with permanent vegetation or the landward side of active seawall/hard structures were taken into consideration as shoreline for change analysis. The limitation of this method is that it is difficult to represent the erosion along the coast protected with seawall. Many locations are experiencing overtopping or destruction of seawalls partly or fully during high-energy waves, and the seawall may be restored during the consequent fair season. But, the shoreline will be demarcated with the same line of seawall over the years, and the coastal erosion experienced there will not be reflected. Such a kind of overtopping is noticed at *Thottumukkam*, *Kannankadavu*, *Azheekkal*, *Konad*, *Kothi*, *Kuttichira*, *Marad*, and *Gotheeswaram*. The seawalls along these coastlines are restored within few years for protecting the inhabited population regularly. The Digital Shoreline Analysis System (DSAS), an “add-on” tool to ArcGIS software of the Environmental Systems Research Institute (ESRI) and developed by the United States Geological Survey (Thieler et al., 2009), was used to analyze the shoreline changes. The statistics of end point rate (EPR) and least regression rate (LRR) were chosen for shoreline analysis. The positive sign of EPR

Table 7.4 Datasets used for shoreline mapping from 1968 to 2018

Sl. no.	Shoreline year	Data source	Publisher	Year of survey	Scale/ resolution
1	1968	Topographic sheet	Survey of India	1963–65	1:50,000
2	1979	Topographic sheet	Survey of India	1977–1978	1:250,00
3	1984	Google Earth	Google LLC	1984	–
4	1990	Landsat TM mosaic	Global Land Cover Facility	1990	30 m
5	2000	Landsat 7	Global Land Cover Facility	1999	15 m
6	2007	CARTOSAT	National Remote Sensing Centre	2007	2.5 m
7	2010	CARTOSAT	NRSC	2010	2.5 m
8	2011	WorldView-2	Digital Globe	2011	0.5 m
9	2014	CARTOSAT	NRSC	2014	2.5 m
10	2018	Google Earth	Google LLC	2018	< 1 m

and LRR values indicates the shoreline accretion, and the negative values indicate the erosion intensity. The EPR and LRR values derived from DSAS analysis were then compiled to examine the coastal stability of the Kozhikode coast. For ease of understanding the erosion/accretion pattern, the derived shoreline change values were categorized into seven classes as mentioned in Table 7.3 (Sheela et al. 2018).

7.3 Results and Discussion

7.3.1 Coastal Geomorphology and Nearshore Landforms

Coastal Geomorphology The current coastline of Kozhikode was formed during the Holocene period. The width of the young coastal plain of recent origin along *Kozhikode* coast is identified 2.5–5 km in the *Beypur–Kallayi* sector, 1–2 km in the *Kallayi-Elathur* sector, and 1–2.5 km in the *Elathur-Koyilandi* sector. Paleo channels of this area changed its direction in many places during the Holocene-Pleistocene period under the fluvio-tidal influences (NJK Nair et al., 1989; V. Ambili & A.C. Narayana, 2014). The promontories/hills of weathered rock formation like *Kadalur* hills, *Thoovappara*, and *Elathur* hills regulate the coastal dynamism and control the fluvio-marine sediment movements (Figs. 7.3a and 7.3b). North of the study area witnessed as barrier reef formations, while the south portion retained as a young coastal plain due to the combined action of fluvial and marine forces. Three river systems, viz., *Kallayi*, *Korapuzha,* and *Chaliyar,* are debouching into the Lakshadweep Sea along the 40-km-long coast under consideration. The two main harbors (*Koyilandi* and *Puthiyappa*) and one minor harbor (*Vellayil*) play a key

Fig. 7.3a Coastal structure; natural vs anthropogenic: A view from Gotheeswaram. (Photograph taken by author/s)

Fig. 7.3b Coastal structure: Thoovappara promontory, North Kappad, Kozhikode. (Photograph taken by author/s)

role in the fishing activity of the region, along with the major estuaries, and they also influence the genesis of microgeomorphology of the area.

Based on the landform signatures, the geomorphic dynamism of the entire area may be categorized under three different geographical entities. The northern portion –

Koyilandi region extends from *Kadalur* headlands to the *Elathur* headland, the central portion *Kozhikode* region belongs to the area between *Elathur* headland, and the north of *Kallayi* River and the southern portion – *Beypur* region – is located between *the Kallayi* River basin and *Chaliyar* river basins. The *Koyilandi* region is marked as the zone of parallel barriers and ridges with the fluvio-marine formation, and the *Beypur* region is recorded as a zone of barriers and vast coastal plain with low-lying swale and flood plain, while the *Kozhikode* region (central portion) is made up of marine deposits along the hinterland region of the tertiary formation. The significant characteristics of paleo channels traced from coastal plains of the north *Koyilandi* region and south *Beypur* region support the concepts of the geomorphic evolution of this coast. A very dynamic moving spit of more than 600 m in length with an average width of 65 m has been present at the northern side of *Korapuzha* estuary of *the Koyilandi* region. The analysis of different timeline satellite imagery shows that this spit started disintegrating in 2015 and disappeared completely during the SW monsoon season of that year. Sediment deposition initiated the southern side of the estuary along with the *Elathur* headland, and it grows gradually and continuously. It indicates that there is a drastic change in the coastal dynamism.

Nearshore Landforms Nearshore bottom features of the area are diversified with parallel and transverse bars, reefs, exposed and buried rocks, stacks, etc. The major nearshore features are demarcated as *Kadalur* cape, *Thoovappara*, *Elathur* cape, *Thikkodi* reef, *Kadalur* reef, *Anchorage* reef, *Coote* reef, *Calicut* reef, *Rocky It*, *Gilham* rocks, *Rocky* points, *Black* rock, and *Puthiyangadi* bay (Fig. 7.4). The zone of shoreface has the maximum diversity in nearshore bottom features when compared to other zones. The beach morphology of the area shows that the backshore of the area is a little wider than the foreshore zone. The breaker zone or littoral zone has an abrupt slope, and the bathymetric contours become wider when it moves further from the offshore region. Three estuarine regions discharge a remarkable quantity of sediment to the sea, and it determines the sediment movements of the coast. Three harbors, two main headlands, and many small promontories and rock exposures of the area also influence the sediment transport. The *Elathur* headland divides the targeted coast into two major sediment cells, and the natural minor sediment cells are disturbed due to different artificial coastal structures such as harbors, breakwaters, and seawalls. Even if the sediment drift is reversing its direction twice in a year according to the monsoon, the net sediment transport is resulting from south to north. One of the major sediment cells of the area formed between *Kadalur* headland and *Elathur* headland and one between *Elathur* headland and *Beypur* estuary. The northern sector of the area was reported for frequent mudbank formation. The northern *Kadalur* headland blocks the movement of sediments from south to north, which could be the reason for finer sediments to come into suspension and develop mudbank phenomenon.

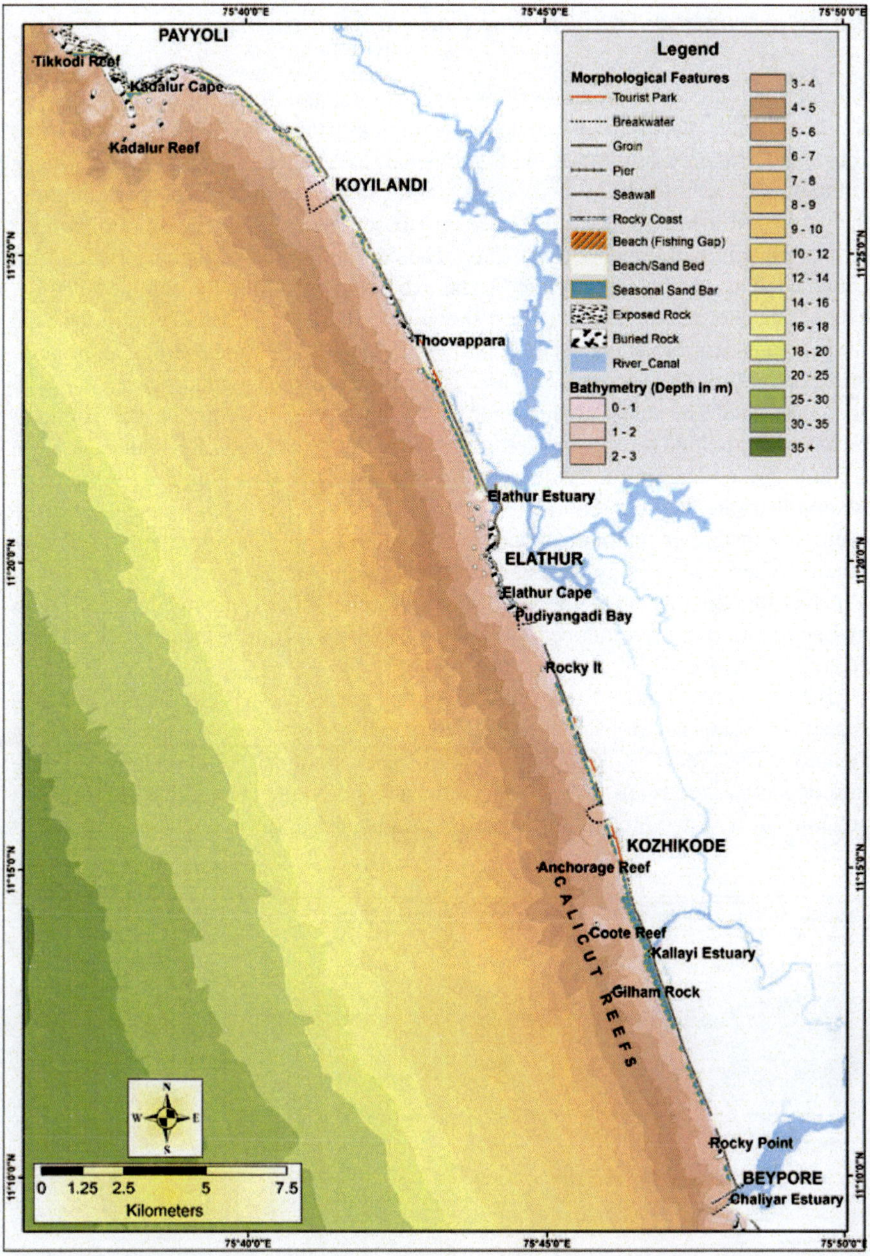

Fig. 7.4 Morphology of Kozhikode coast – with near shore bottom features associated to artificial and natural coastal structures. Bathymetry data are visible

7.3.2 *Shoreline Analysis and Coastal Modification*

Among the total coastal stretch of this coast, *the Koyilandi* region is demarcated as the most stable when compared to other regions (Fig. 7.5). The erosion of this coastline is mainly associated with rocky coasts and end erosion of man-made structures. Even though the coastal protection measures are persistent throughout the total coastline except for a few pockets, the southern *Beypur* region is known for coastal erosion with moderate intensity. The central portion of the city region also demarcated as a stable and accreted coast except for 1.7 km in the south of *Vellayil* beach. The current data interpretation shows that 48.21% of the total coastline of the study area is demarcated as a stable coast, while only 35.79% is recorded as erosion-prone area (Table 7.5) and 15.99% area is reported as beach gaining area. Very high accretion of LRR value greater than 5 is reported just south of the southern breakwater of *Chaliyar* River inlet, and the high accretion of LRR value between 2.5 and 5 is noticed in the southern side of breakwaters of *Chaliyar* (*Beypur*) and Elathur harbor. The accretion of moderate intensity has been found along the southern coastline of major breakwaters of *Koyilandi* and *Elathur* harbor and *Beypur* estuary. There are no major erosion hot spots with erosion intensity of more than 5 m/year that are evident from the present investigation. Few erosive locations with moderate intensity were noticed either along rocky coast or associated with major artificial structures.

The shoreline studies of this coast show that the increase in the artificial coastal protection measures impacted the beach gaining processes of the coastline in general. The local government bodies and political representatives were misconceived as seawalls, and groins are the only and immediate solution for coastal erosion. Most of the protection measures are planned as an immediate solution for

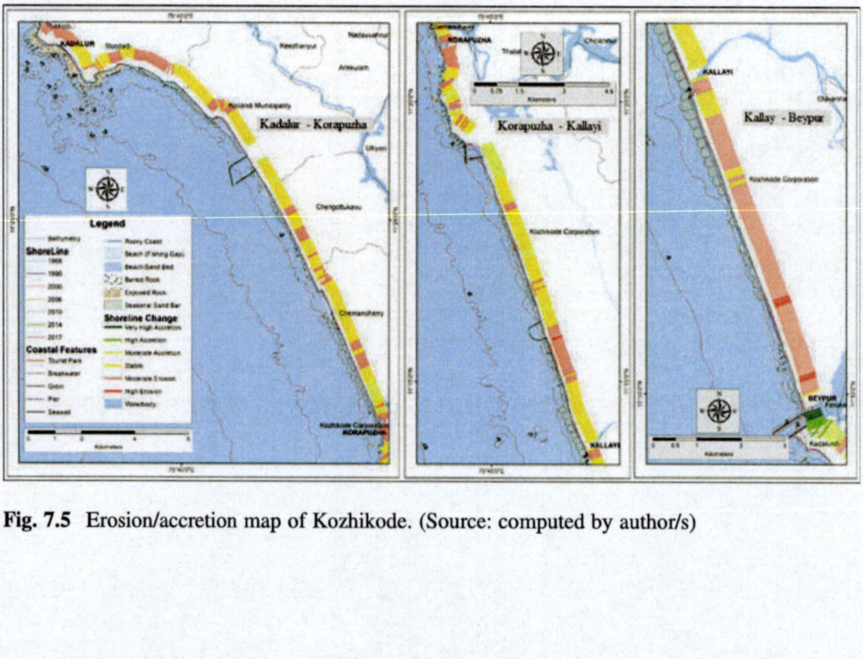

Fig. 7.5 Erosion/accretion map of Kozhikode. (Source: computed by author/s)

Table 7.5 Shoreline change classification based on EPR and LRR values

		Count of transects at each class	
Sl. no.	Class	EPR %	LRR %
1	Very high erosion	0	0
2	High erosion	0.80	0.92
3	Moderate erosion	34.98	39.12
4	Stable	48.21	48.68
5	Moderate accretion	14.84	10.12
6	High accretion	0.80	0.69
7	Very high accretion	0.34	0.46

Classification of coast based on EPR/LRR value: < -5, very high erosion; -5 to -2.5, high erosion; -2.5 to -0.5, moderate erosion; -0.5 to 0.5, stable; 0.5 to 2.5, moderate accretion; 2.5 to 5, high accretion; >5, very high accretion

the coastal erosion during the monsoon/erosive seasons, without a proper study of the coastal dynamism of the specific area. The pressure of high population density and political reasons have enforced the authorities to find out an immediate solution for the coastal erosion for each location. The construction of any such artificial protection measures, especially without a proper interpretation of coastal processes will disturb the coastal dynamism, and it may lead to the abnormal behavior of the coastline. Severe coastal erosion is reported to the south of the seawall constructed along *Konad* beach to protect the densely populated coastal belt (*Shanthinagar*) as a result of the end erosion. Such kind of end erosions is also reported from many locations along the *Kozhikode* coast on either side of the seawalls. Fishing gaps in between the protected coasts, which is designated for the fishing and landing facility of local fishermen, are the best examples of such artificial-induced coastal erosion. Besides the seawalls and groins, the construction of harbors and breakwaters to facilitate the coastal activities is also highly redefining the coastal dynamism of the impacted coastline.

The construction of long (more than 850 m) breakwaters at the mouth of *Chaliyar* River (*Beypur* estuary) on both sides choked the sediment supply of the coast, as the *Chaliyar* River was one of the most important sediment sources of this coast. Besides, the frequent dredging of this river mouth to facilitate the port constructed inside the estuary is also causing to decrease in the sediment availability of the coast. Due to the northward direction of net sediment transport of this coast, the breakwater constructed at the south side of the estuary blocks the sediment transport from further south. It causes to settle down the sediments and coastline of this area experience an erosive trend. Likewise, the north breakwater blocks the further northward transport of the sediments, and it causes the deficiency in sediment budgeting of sediment cells and leads to coastal erosion along *Gotheeswaram-Marad* area. A similar dynamism is also happening to the south side of *Puthiyappa* harbor. The beach to the south of this harbor is reported as beach gaining area with an annual rate up to 2.5 m/year. There is intensive sand mining from this beach due to its depositional nature unless the rate of accretion will be very high. This harbor is constructed in the *Puthiyappa* bay to the south of *Elathur* headland. Due to the rocky structures of the headland at the north side of the harbor, the coastline is almost stable, and there is no noticeable change in the coastal dynamism. The recently constructed, comparatively small, *Vellayil* harbor along the *Kozhikode* city region is also showing the same phenomenon of deposition at the south side and erosion at the north side of the harbor. When compared to other major structures, the *Koyilandi* harbor doesn't have modified the coastal dynamism of the area, except the further north of the harbor. There are reports of significant coastal erosion and overtopping to the north of *Valiyakath* mosquee, which may be the influence of the harbor and associated seawall.

The southward spit of the northern side of *Korapuzha* estuary, which was present up to 2015, had disappeared during that monsoon season (Fig. 7.6). Similarly, sediments are found to be deposited at the southern side of the estuary, which is the northward side of *Elathur* headland. The differential timeline imageries show that this spit started disappearing in March 2015 and completely washed out in August 2015. The initiation of sediment deposition was noticed at the conjunction of river and sea, just landward side of the rocky coast of *Elathur* headland. The sediment deposition at the southern bank is now well established and has more than 1 km of length and 100 m of average width. From the recent satellite images, it is clear that the beach formed on the southern side is still growing up, and continuous sedimentation is going on. The different temporal satellite images from 1984 show that this spit was actively present up to 2015 with more than 600-m length and an average width of 65 m. The reason for these conspicuous changes in the sediment movement system has to be examined in detail. A seawall was constructed in the northern tip of the spit in the year 2005, and there was a wide beach in the seaward side till 2015. This active beach has also been washed out along with the spit during the 2015 phenomenon, and the shoreline is realigned landward to the line of seawall, which was inactive for the last 10 years.

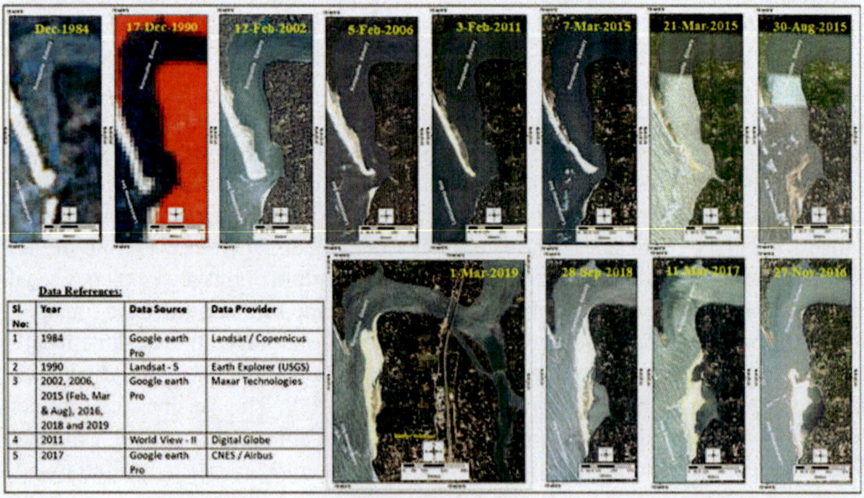

Sl. No:	Year	Data Source	Data Provider
1	1984	Google earth Pro	Landsat / Copernicus
2	1990	Landsat - 5	Earth Explorer (USGS)
3	2002, 2006, 2015 (Feb, Mar & Aug), 2016, 2018 and 2019	Google earth Pro	Maxar Technologies
4	2011	World View - II	Digital Globe
5	2017	Google earth Pro	CNES / Airbus

Fig. 7.6 The morphological change of Korapuzha spit from 1984 to 2019. (Source: compiled by author/s)

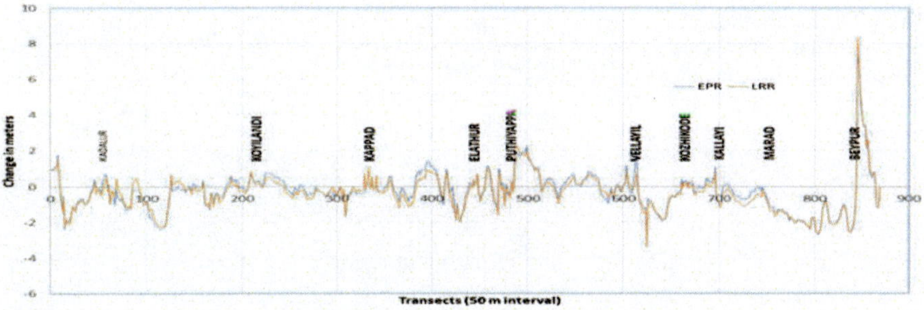

Fig. 7.7 Shoreline change along Kozhikode coast from 1968 to 2018. (Source: computed by author/s)

Fig. 7.8 Coastal protection measures and associated coastal erosions; **(a)** *Gotheeswaram* beach, **(b)** *Kothi*, **(c)** *Kuttichira* beach, and **(d)** *Konad*. (Photograph taken by author/s)

Figure 7.7 shows that the shoreline of *the Kozhikode* coast is nearly stable, and not much changes took place during the last half-century. The EPR/LRR value exceeds moderate intensity of accretion and erosion, i.e., EPR/LRR value of > 2.5 or < −2.5 has been reported marginally, and wherever it has been reported is associated with anthropological activities (Fig. 7.8).

7.3.3 Coastal Features and Coastal Stability

The very high population density of the *Kerala* coastal belt compelled the authorities to construct different hard engineering structures of coastal protection measures and

Table 7.6 Morphological structures along Kozhikode coast

Sl. no.	Structure	Quantity	Percentage
Natural structures			
1	Open beach	5.97 km	12.70
2	Rocky coast	6.36 km	13.53
3	Estuary	3 Nos	–
Artificial structures			
4	Seawall	31.45 km	66.91
5	Tourism structures	2.70 km	5.74
6	Groins	18 Nos	–
7	Breakwater	10 Nos	–
8	Harbor	3 Nos	–
9	Port	1 (estuary)	–
10	Sea pier	2 Nos	–

coastal facilities during the last few decades, mainly to satisfy the increased needs of inhabitants. Construction of such high-intense coastal structures disturbed the wave and long-shore current activities and reduced the net sediment transportation, and it gradually turns the accreting beach into the erosional one.

Being a young coastal plain of Holocene origin, the *Kozhikode* coast should not have been facing serious coastal hazards and calamities. The archival literature and historical documents suggest that this coast was generally accreting except at few seasonal erosion points. But the current study shows that the accretion trend of this coast was retarded, and beach gaining processes is restricted to few locations. The two sea piers constructed during the mid-twentieth century in *Kozhikode* beach form the first artificial structures along this coast. Now, more than 74% of the total coastline is regulated by different coastal structures, and out of the remaining 26%, 13.69% is suffering from the complex nature of the rocky coast (Table 7.6). The major sediment cells of this coast are defined by the *Kadalur* and *Elathur* headlands and *Kallayi* and *Beypur* estuaries. Besides, a number of microscale sediment cells were formed by the influence of small promontories and major breakwaters. According to the Indian monsoon data, the sediment cells of this region are more active during SW monsoon (Sanil Kumar et al., 2006). The reversing sediment drift along with sediment budgeting and micromorphological features intensifies the net sediment movements and causes to coastal erosion/accretion as per the wave action.

The natural stability of the *Kozhikode* coast is affected by the construction of artificial structures for facilitating the human needs and as the protection measures of the coastline from the seasonal and/or the short-duration erosional phenomenon. The former sediment cells of the coastline are disturbed due to the breakwater construction for the harbors and estuarine ports. The major sediment sources of the area are the rivers of *Korapuzha*, *Kallayi*, and *Chaliyar* and the further south estuary of *Kadalundi* River. Among these rivers, 169-km-long *Chaliyar* debouches directly to the Lakshadweep sea after draining an area of 2539.82 sq km (Kerala Irrigation Department). The 46-km-long *Korapuzha* debouches into the sea after making a backwater system in the coastal plain, and hence the sediment load is reduced when it reaches the Lakshadweep sea. *Kallayi* is a small river with only 22 km and 87.52 sq km area of drainage systems. It shows that the *Chaliyar* River is the major sediment source of the *Kozhikode* coastal area. Even though the current of the *Kozhikode* coast is changing their direction according to seasons, the net transport of this region is reported as northward (Sheela, 2014).

Construction of breakwaters at river mouths and harbors redefined the sediment cells of the area, and new microsediment cells are formed to either side of the breakwaters. The remarkable beach accretions are reported to the south of breakwaters constructed for harbors of *Puthiyappa* and *Vellayil* and at *Chaliyar* river mouth. The accreting rate of the beach at the south of *Chaliyar* breakwater is calculated as more than 8 m/year, while the accreting rate to the south of *Puthiyappa* harbor is controlled intensive sand mining, and the accreting rate of this beach is reported as 2.5 m/year. The *Vellayil* harbor is newly constructed, and deposition has also started at the southward of breakwater. Correspondingly, a remarkable coastal erosion is reported to the north of these breakwaters, and protection measures such as seawall are constructed to protect these coasts from erosion, except for *Puthiyappa* harbor (for Puthiyappa harbor the natural rocky coasts are preventing the coast from erosion). *Koyilandi* harbor have not modified the coastal stability due to the locational and structural specialty of the breakwaters when compared to other major breakwaters of the area. The minimized sediment availability is also influencing the sediment transport system of the region.

7.4 Conclusions

The modifications in the coastal geomorphology of the *Kozhikode* were investigated by analyzing the coastal and nearshore bottom morphological features and shoreline changes pertaining to over the last 50 years. Different datasets such as the Survey of India topographic sheets, multi-resolution satellite images, and the hydrographic survey charts of different temporal resolutions starting from 1968 to 2018 were analyzed and precisely geocoded them onto a common GCPs for the detection of temporal changes. Micromorphological features and shoreline structures were mapped on a 1:10,000 scale accuracy in the field. Our analysis shows that the coastal dynamism of the entire coastal stretch of *Kozhikode* is currently in a transformation stage from the beach gaining processes to a beach loosing stage. For understanding the long-term shoreline changes of the area, the end point rate (EPR) and least regression rate (LRR) of 1968–2018 period were derived from DSAS software developed by the USGS. The shoreline change analysis conveyed that the coastal belt of *the Beypur* region, which has been protected most by different engineering structures, affected significant coastal erosion. On the contrary, the least structured coastline of *the Koyilandi* region is reportedly the most stable coastline within the area under investigation. A total of 87% of the *Kozhikode* coastline has been modulated by either artificial structures or rocky coasts. These hard coastal structures, breakwaters, and cliffs disturb the natural sediment movements and redefine the sediment cells of shallow area.

Any modifications occurring in the coastal dynamism of wave-current-sediment processes affect the equilibrium status of the coast, leading to coastal instability. The areas of severe coastal erosion reported along *Kozhikode* coasts, such as *Kothi-Nainamvalappu*, *Payyanakkal*, *Gotheeswaram*, *Konad*, and *Kappad-Azheekal*, are concentrated along the coastal belts which are the zone of high population density and artificial coastal protection measures. After the construction of seawalls, the coastline experiences the intensified wave action and overtopping during extreme waves, but the shoreline has been continuously maintained by repairing the seawall damaged, and hence the effective shoreline change will be measured as the minimum (Fig. 7.9). The local inhabitants certify that the frequency and intensity of high waves and associated coastal hazardous, wherever reported, were increased after the construction of seawalls. It is concluded that the stability of *the Kozhikode* coast was affected due to anthropogenic activities such as the construction of coastal structures, sand mining, dredging, and choking the river sediments along with its course. As a result and being a high-energy coast, the occurrence of events associated with the artificial coastal protection measures such as end erosion, overtopping, and soil leaching have been increased during the last few decades.

Fig. 7.9 Maintaining the *coastal stability*; tetra-pods are deployed to protect the coast after eroding the previous gabions – view from Koyilandi. (Photograph taken by author/s)

References

Ambili, V., & Narayana, A. C. (2014). Tectonic effects on the longitudinal profiles of the Chaliyar River and its tributaries, southwest India. *Geomorphology, 217*, 37–47.

Baba, M., & Kurian, N. P. (1988). *Ocean waves and beach processes of southwest coast of India and their prediction* (p. 249). Centre for Earth Science Studies.

Chattopadhyay, S. (2017). *Geomorphological field guide book on laterites and backwaters of Kerala*. In: Amal Kar (Ed.). Indian Institute of Geomorphologists, Allahabad.

Geological Survey of India. (2005). *Compiled Report. Geology and Mineral Resources of Kerala*. Miscellaneous Publication No 30, part 9, 2nd revised edition, 83p

Hameed, T. S. S., Kurian, N. P., Thomas, K. V., Rajith, K., & Prakash, T. N. (2007). Wave and current regime off the southwest coast of India. *Journal of Coastal Research, 23*(5), 1167–1174.

KERI. (1978). *Simultaneous observation of coastal data, Kerala coast*. Kerala Engineering Research Institute Annual Report.

Kurian, N. P., Prakash, T. N., Baba, M., & Nirupama, N. (2006). Observations of tsunami impact on the coast of Kerala, India. *MarineGeodesy, 29*, 135–145. https://doi.org/10.1080/01490410600748301

Kurian, N. P., Ramachandran, K. K., Sheela Nair, L., Thomas, K. V., Hameed, T. S. S., & Ramana Murthy, M. V. (2007). *Long-term shoreline change along the south-central Kerala coast*. In Proceedings of the Fourth Indian National Conference on Harbour and Ocean Engineering, 12–14 December 2007. NIIT-K, pp. 761–768.

Mukhopadhyay, R., & Karisiddaiah, S. M. (2014). The Indian coastline: Processes and landforms. In V. Kale (Ed.), *Landscapes and landforms of India. World geomorphological landscapes*. Springer. https://doi.org/10.1007/978-94-017-8029-2_8

Murthy, C. S., Sasthry, J. S., & Varadachari, V. V. R. (1980). Shoreline deformation in relation to shore protection structures along the Kerala coast. *Indian Journal of Marine Sciences, 9*, 77–81.

Nair, M. M. (1987). Coastal geomorphology of Kerala. *Journal Geological Society of India, 29*, 450–458.

Nair, R. R., Varma, P. U., Pylee, A., & Varadachari, V. V. R. (1973). Studies on the sediment transport in the Mopla Bay. *Proceedings of the Indian Geophysics Union, 10*, 193–197.

Nair, N. J. K., Nalinkumar, S., Reghunathan Pillai, K., & Suresh Babu, D. S. (1989). *Environmental geomorphic Atlas: Kerala Coastal zone*. CESS.

Narayana, A. C., & Priju, C. P. (2006). Landform and shoreline changes inferred from satellite images along the central Kerala coast. *Journal Geological Society of India, 68*, 35–49.

Prabhakara Rao, P., Nair, M. M., & Raju, D. V. (1985). Assessment of the role of remote sensing techniques in monitoring shoreline changes: A case study of the Kerala coast. *International Journal of Remote Sensing, 6*(3/4), 549–558.

Prasad, R., Sheela Nair, L., Kurian, N. P., Prakash, T. N., & Varghese, T. I. (2016). Erosion and heavy mineral depletion of a placer mining beach along the southwest coast of India: Part III–Short and long term morphological changes. *Natural Hazards, 83*, 823–847. https://doi.org/10.1007/s11069-016-2346-5

Rafeeque, M. K., Ramehan, M., & Sreeraj, M. K. (2021). Measuring the vulnerability of coastal ecosystems in a densely populated west coast landscape, India – A remote sensing perspective. In R. Meenu, P. Kumar, S. Kaliraj, R. Sufia, & S. Haroon (Eds.), *Remote sensing of ocean and coastal environments* (pp. 203–224). Elsevier. https://doi.org/10.1016/B978-0-12-819604-5.00013-5

Reddy, M. P. M., & Varadachari, V. V. R. (1973). Sediment movement in relation to the wave refraction along the West Coast of India. *Proceedings of the Indian Geophysics Union, 10*, 169–191.

Samsuddin, M., Ramachandran, K. K., Mathai, J., Jayaprasad, B. K., & Neelakandan, V. N. (2008). *Natural resources and environmental atlas of Kerala; Kozhikode District*. CESS.

Sanil Kumar, V., Pathak, K. C., Pednekar, P., Raju, N. S. N., & Gowthaman, R. (2006). Coastal processes along the Indian coastline. *Current Science, 91*(4), 530–536.

Sheela Nair, L. (2014). *Sediment dynamics along the coast of Kerala, India*. Ph.D. thesis, Department of Ocean Engineering, Indian Institute of Technology, Madras.

Sheela Nair, L., Prasad, R., Rafeeque, M. K., & Prakash, T. N. (2018). Coastal morphology and long-term shoreline changes along the southwest coast of India. *Journal Geological Society of India, 92*, 588–595.

Shoreline Change Assessment for Kerala Coast (Report): National Centre for Sustainable Coastal Management, Society of Integrated Coastal Management, Ministry of Environment and Forest, Government of India.

Shoreline Change Atlas of the Indian Coast (Volume 1–3): Space Applications Centre (ISRO) and Coastal Erosion Directorate, Central Water Commission, Government of India, New Delhi (2014).

Sreekala, S. P., Baba, M., & Muralikrishna, M. (1997). Shoreline changes of Kerala coast using IRS data and aerial photographs. *Indian Journal of Marine Sciences, 27*, 144–148.

Suchindan, G. K., Samsuddin, M., & Thrivikramji, K. P. (1987). Coastal geomorphology and beach erosion and accretion in the Northern Kerala Coast. *Journal Geological Society of India, 29*, 379–389.

Thieler, E. R., Himmelstoss, E. A., & Zichichi, J. L., Ergul, A. (2009). Digital Shoreline Analysis System (DSAS) version 4.0 – An ArcGIS extension for calculating shoreline change: U.S. Geological Survey Open- File Report 2008-1278.

Vaidyanadhan, R. (1987). Coastal geomorphology in India. *Journal Geological Society of India, 29*, 373–378.

Census of India 2011, District Census Handbook Kozhikode, Directorate of Census Operations Kerala, Series -33 Part XII-B. https://censusindia.gov.in/2011census/dchb/3204_PART_B_KOZHIKODE.pdf

Kerala Irrigation Department. http://www.irrigation.kerala.gov.in/index.php/resources/water-bodies/rivers. Last accessed 19 Nov 2020.

LSGD page. http://lsgkerala.gov.in/index.php/ml/lsgd/localbody-list. Last accessed 7 Sept 2020.

LSGD Moodadi GP home page. http://lsgkerala.in/moodadipanchayat/. Last accessed 7 Sept 2020.

LSGD Chengottukave GP home page. http://lsgkerala.in/chengottukavupanchayat/. Last accessed 7 Sept 2020.

LSGD page Chemancheri GP home. http://lsgkerala.in/chemancherypanchayat/. Last accessed 7 Sept 2020.

LSGD page Kadalundi GP home. http://lsgkerala.in/kadalundipanchayat/. Last accessed 7 Sept 2020.

LSGD page Kozhikode Corporation home. https://kozhikodecorporation.lsgkerala.gov.in/. Last accessed 7 Sept 2020.

Chapter 8
River Dynamics of the Ganga and Its Tributaries in the Siwalik-Tarai Region of Haridwar District, Uttarakhand, India

Rupam Kumar Dutta

8.1 Introduction

Himalayan rivers occupy a highly dynamic environment with extreme variability in discharge and sediment load. Therefore, shifting of river courses and scouring of bed and banks are some of the most distinguishing characteristics of the Himalayan rivers (Kale, 2002). Except the Siwalik ridge, the entire study area of Haridwar district falls under the geomorphic region of Ganga alluvial plain. Several scientists have contributed their extraordinary works on the geological and geomorphological significance of the Ganga alluvial plan. Geomorphological characteristics of the region have been vividly discussed by many scientists (Shukla & Bora, 2003). The understanding of geological significance and tectonics of the area is very helpful to study the dynamic natures of the Ganga, and its tributaries have been analyzed by several renowned workers (Srivastava et al., 2003; Uniyal et al., 2010). The dynamic natures of rivers in the Ganga plains have been known to engineers and earth scientists for a long time. The most common types of changes identified in the Ganga plains are shifting in channel position due to avulsion, alternation of the flow direction because of cut-off, and channel widening in response to bank erosion and/or bar development. Apart from geology and tectonics, the geomorphic characteristics of an area play a crucial role in influencing the dynamic nature of the rivers in the study area. Jones and Schumm (1999) have nicely illustrated avulsion classification. Mukhopadhyay (2010) also described the mechanism of channel avulsion in floodplain surface. Most of the scientists (Srivastava et al., 2003) have established the concept of different existing geomorphic surfaces of the Ganga plain including fan surfaces and their development in different geological periods. Different parts of the Ganga valley exhibit various salient geomorphic signatures

R. K. Dutta (✉)
Department of Geography, Kultali Dr. B.R. Ambedkar College (University of Calcutta), Parganas, West Bengal, India

G. S. Bhunia et al. (eds.), *Anthropogeomorphology*, Geography of the Physical Environment, https://doi.org/10.1007/978-3-030-77572-8_8

through its channel behavior. Depending on the morphological characteristics, the Ganga valley has been classified into seven segments by some scientists (Singh & Singh, 1992). From the previous study it is found that the Himalayan foothill, over a long time periods channels must shift over its entire surface to maintain fan form (Field, 2001). The diversity in channel morphology is caused due to variability in water discharge, sediment load, ground slope, human activities, and tectonic influences (Valdiya, 2003). River channel dynamics are natural autogenic occurrences for rivers with influences from geological, geomorphological, and climatic factors including human modification. Kumar (1995) has showed long-term changes (1922–2000) of channel widening in all rivers of the Rajaji National Park, Haridwar, in his remarkable works on the sustainability of the park. Dutta (2017) has also analyzed controlling factors of salient changes in Ganga River channel near Haridwar city and its adjacent areas based on intensive field survey.

Several articles and reports by earlier scientists have been consulted to study the trend of research on Ganga River dynamics and find the gap in the research of the earlier works. The present work has uniquely focused on several issues regarding Ganga River channel characteristics of Haridwar, which were hardly discussed in the previous studies. In this chapter, the shifting pattern of the entire Ganga River of the district in more than three decades (1967–2005) has been illustrated. Comprehensive understanding about the shifting pattern is important for proper planning on land use as well as pre-hazard management along the vulnerable zones of riverbanks in significant city, such as Haridwar.

Previous scientists have established the concept of different existing geomorphic surfaces of the vast Ganga plain, but the present chapter specially highlights the four physiographic divisions of the study area to distinguish salient channel characteristics under different physiographic zones. To analyze the role of physiography on channel forms, proper ideas about physiographic divisions of an area are very crucial.

The earlier scientists have mainly focused on major physical factors (geology, tectonics, geomorphology, etc.) in their research, but the present work has not only discussed the physical controlling factors of river channels but also distinctively stated the influence of anthropogenic effects (Upper Ganga Canal and Bhimgoda barrage, channel mining, religious activities) on channel instability.

The objective of the chapter is to study the dynamic natures of the Ganga River and its other major tributaries with special emphasis on diverse fluvio-geometrical parameters (channel width, depth, pattern, area, the number of channel bars, etc.) under different physiographic zones like Siwalik, Bhabar, and Terai of Haridwar district, part of the upper Ganga plain, in more than three decades. The present study has focused on different prime-controlling factors (physical and anthropogenic) of aggrading, widening, shifting, and scouring nature of the present Ganga River channel including different types of channel avulsion and abandonments.

8.2 Materials and Methods

This chapter illustrates river channel dynamics of the Ganga and its tributaries of Haridwar district with modern methods including field techniques.

A broad drainage map of the study area has been prepared with the help of toposheets (SOI, 1967) and satellite images (LISS-III, 1997) to get the idea about drainage pattern of the Ganga and its tributaries in Haridwar.

The shifting trend of Ganga River (1967, 199, and 2005) has been illustrated with the help of toposheets (SOI, 1967) and satellite images (LISS-III, 1997 and 2005) (Table 8.1). IRS LISS-III,1997 & 2005 and LISS-IV, 2005 have been purchased by the author from the National Remote Sensing Centre (NRSC). The change detection map clearly reveals maximum tendency of channel shifting in Terai zone. Major physical factors (geology, tectonic setup, geomorphic surfaces, variability of water discharge, sedimentation) and anthropogenic factors (Upper Ganga Canal and Bhimgoda barrage, channel mining, religious activities) have been identified and discussed by the author with the help of intensive field documents (field survey from 2006 to 2017) and profiles (1445 m cross-section of the Ganga River, longitudinal profile from Sukhrao valley of extreme north to Ganga River of extreme south) as the causes of shifting and dynamic nature of the Ganga and its tributaries. In order to relate river channel pattern with geological structure and geomorphology of the area, broad geomorphological and geological maps have been prepared by the author with the help of the geological map of Saharanpur (GSI, 1997) and the Survey of India toposheets (no: 53 K/1, 53 K/2, 53 K/5,53 J/4, 53 G/13, 53 G/14) of the years 1967 and 1971. Physiographic division map of the study area has been prepared to differentiate the variation of channel characteristics in Siwalik, Bhabar, and Terai zones. Few sites have been identified like Nildhara area (near Chandi bridge), Kangri, Tatwala, Gajiwali-Shyampur villages, etc., which investigate channel shifting; bank erosion and avulsion events and their impacts on settlement, depending on field observation; GPS tracking over satellite image (Google Earth, 2016 and 2020) (Table 8.1); and the study of channel cross-section. Time series analysis has been done of Chillawali and Sukh rivers to find out the trend of channel widening from 1922 to 2000. The increasing trend of river channel width has been studied (Table 8.2) to establish the dynamic natures of the tributaries. Average annual increasing rate of channel width in different tributaries (in Siwalik zone) has been calculated by the author to show the channel instability. In order to explain the impacts of sedimentation on channel bars and the aggrading attribute of braided channel (in Bhabar-Terai), the area of a mid-channel bar of Ganga River, near Chandi bridge, has been measured to study its spatiotemporal changes

Table 8.1 Details of satellite data used in the chapter

Satellite sensor/product	Date of image
IRS LISS III	30 November 1997
IRS LISS III	24 November 2005
IRS LISS IV	27 November 2005
Google Earth	March 2016 and May 2020

Table 8.2 Average rate increase of the river width in Rajaji National Park (Kumar, 1995)

River name	Channel width (1922)	Channel width (2000)	Changed width within 78 years	Average annual increasing rate of width of the rivers (meter/year)
Malowali River	62.18	152.4	90.22	1.1567
Beenj River	112.47	215	102.53	1.3145
Sukh River	139.92	309	169.08	2.1677
Andheri River	42.06	314	271.4	3.4795
Gaj River	87.78	342	254.22	3.259
Gholana River	10.97	349.7	338.73	4.3427
Bam River	59.44	374.8	315.36	4.0431
Mohand River	182.88	492	309.12	3.9631
Chillawali River	291.69	590	298.31	3.8245
Betban River	307.24	968.8	661.56	8.4815

(1967–2010). Emphasis has been given on field observation for understanding the impact of flood discharge on river channel morphology (2010, 2013). Besides these, special attention has been provided on anthropogenic interferences with river channel like the construction of Bhimgoda barrage and Upper Ganga Canal based on Wolman's (1967) model. In order to elaborate the discussion, relevant primary and secondary data have been analyzed in different qualitative and quantitative approaches. Depending on the comprehensive study, the author has drawn conclusion emphasizing a few significant findings.

8.2.1 Area and Location

From Haridwar, the Ganga River flows through into the regional surface of the Ganga alluvial plain. Therefore, Haridwar, the study area, falls under part of upper Ganga plain. The Haridwar district consists of three Tahesils, namely, (1) Haridwar, (2) Roorkee, and (3), Laksar, including six blocks, namely, (1) Bhagwanpur, (2) Roorkee, and (3) Narsan, (4) Bahadrabad, (5) Laksar, and (6) Khanpur. The study area extends from 29°35'37"N to 30°13'29"N latitude and 77° 52' 52"E to78° 21' 57"E longitude, covering an area of 1883 sq km. The main noticeable characteristic of the study area (Fig. 8.1) is its numerous tributaries, (Raos) Malowali, Beenj, Mohand, Sukh, Chillawali, Chikna, Gaj, Andheri, Binj, Dholkhand, Sampuwali, Mallawali, Bam, etc., flowing from northeast to southwest direction according to the slope. The northern portion of the area includes young folded tertiary Himalayan mountain, and the southern part falls under the Ganga-Yamuna

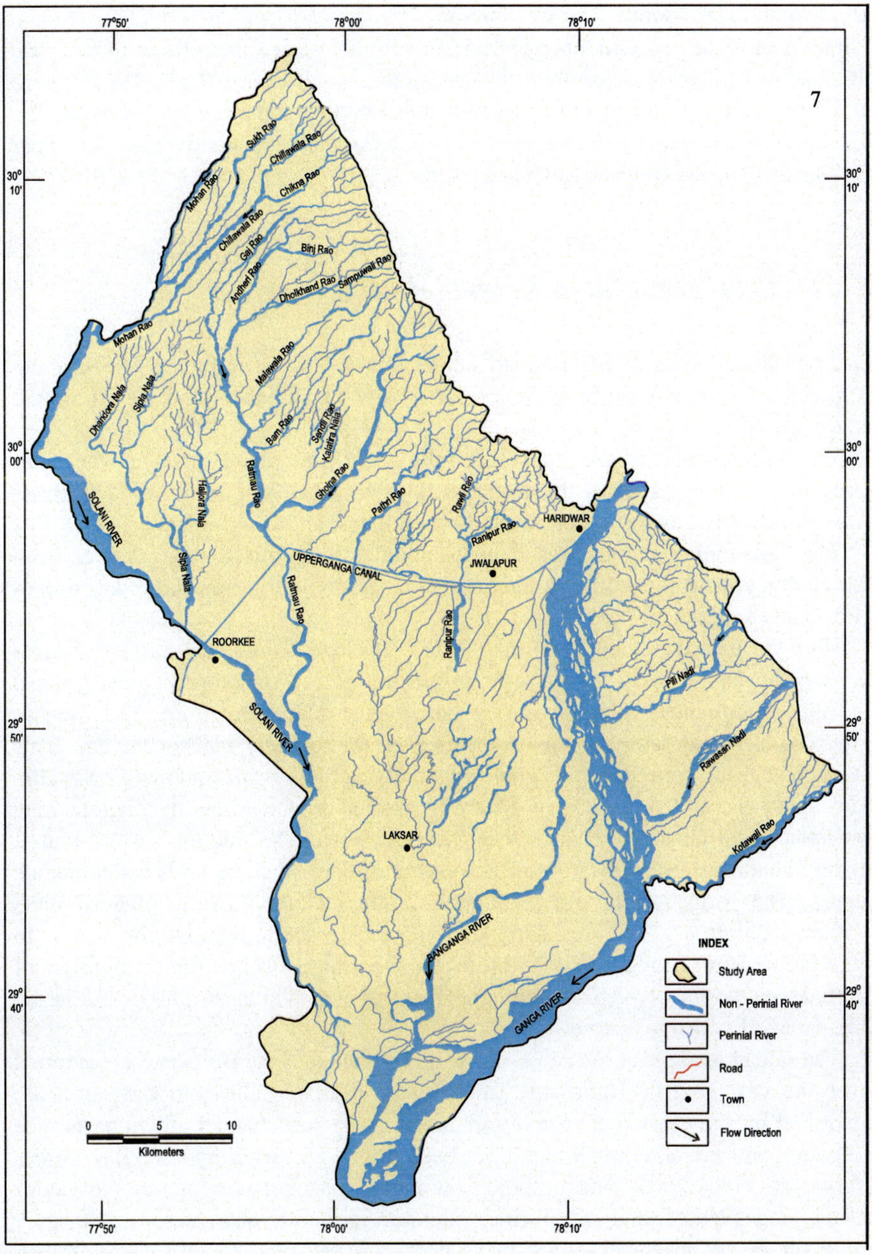

Fig. 8.1 The study area showing broad river network. (Source: Prepared based on toposheets, SOI, 1967)

Doab. In the north-south direction, the study area extends from Siwalik Himalaya to Terai. Along the western boundary of the study area, Solani River flows, and ultimately at the extreme south of the area, it joins with Ganga River. The northern and eastern parts of the area maintains administrative boundary of Haridwar district. The study area consists of enough variation of topographic distinctiveness like upper and lower Siwalik, Bhabar, and Terai plain.

8.2.2 Geological Characteristics of the Study Area

Geological structure is a dominant controlling factor to influence the channel behavior in different physiographic zones (Siwalik, Bhabar, and Terai) of the study area.

The northern part of the study area falls under the outer Himalaya. The outer Himalayan belt consists of the recently uplifted Siwalik of Miocene-Pleistocene detrital sediments, namely, coarse sandstones, clays, and conglomerates.

The Ganga alluvial plain represents the alluvial filling of sediments derived from the Himalaya and the northern Indian craton regions during the late Pleistocene-Holocene (Singh et al., 2007).

With the help of a geological map (Fig. 8.2) of the Geological Survey of India (1997), the entire study area is categorized with three major geological groups: Siwalik group, older alluvium group, and newer alluvium group. The Siwalik group of the study area belongs to the Pleistocene to Early Pleistocene age. This geological group consists of two geological units: upper Siwalik and middle Siwalik. The upper Siwalik unit is very distinctly noticed in the form of conglomerate, sandstone, and claystone sequence in the study area. The middle Siwalik unit is mainly characterized with gray micaceous sandstone, siltstone with conglomerate lenses. The formation of older alluvium group including two geological units Varanasi alluvium 1 and Varanasi alluvium 2 took place in between the Middle to Late Pleistocene. Brownish silt and sand are the prominent lithological imprints of Varanasi alluvium geological unit. The formation of Varanasi alluvium 2 is the comparatively recent deposits.

The salient features of the newer alluvium formation of the Holocene age is found over the vast area of Haridwar. Three types of newer alluvium formation are identified by the Geological Survey of India. These are channel alluvium, terrace alluvium, and fan alluvium which have been discussed earlier by several scientists (Srivastava et al., 2003). Main lithological units of channel alluvium are micaceous sands, gravel lenses, and silt. Terrace alluvium is mainly characterized with gray sand, silt, clay, and gravel lenses. major portion of the area is covered with gravel, sand, brown silt, and clay as the lithological expression of fan alluvium geological unit. In the extreme northern parts of the study area, due to structural control on river courses, channels are comparatively confined, but as the rivers flow toward the southern direction of the district, river channels become gradually unconfined, and at the extreme south mainly in the Terai zone, the Ganga River channel becomes extended with braiding nature.

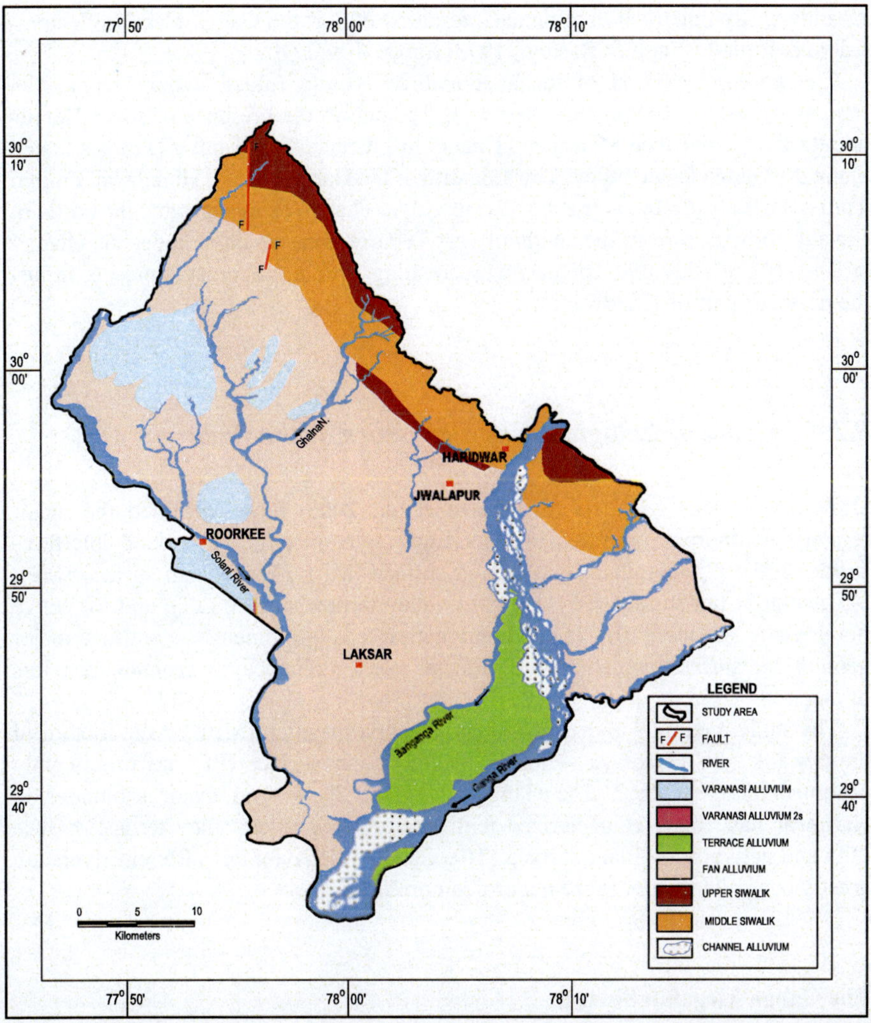

Fig. 8.2 Geological map. (Source: Prepared based on District Resource map of Saharanpur, GSI, 1997)

8.2.3 *Tectonic Setup of the Area*

The evolution of the Ganga plain foreland basin is related to the collision of the Indian and Asian plates during the early Miocene and the closure of Indus-Suture during the Late Cretaceous. The Ganga basin expanded and attained its present configuration during the Late Quaternary (Singh, 1996a, b).

So, the Ganga plain is an area of compressional stresses and active tectonics. Many of the geomorphic features like the alignment of river channels, distorted meanders, asymmetrical escarpments, terraces, etc. of the Ganga plain are tectonically controlled (Singh & Rastogi, 1973; Singh, 1996a, b).

The northern portion of the area includes young folded tertiary Himalayan mountain which is tectonically very sensitive due to the presence of several faults and thrusts. In between Mansa and Chandi hill, Ganga Tear Fault (GTF) is situated along the Ganga River valley. Toward north of the Haridwar city, Himalayan Frontal Thrust (HFT) is located along the Siwalik ridge. The HFT has separated the northern Siwalik formation from the southern vast Terai region, which is under the characteristics of fan alluvium. The east-west trending Bhimgoda thrust is present in just the northern part of Haridwar city.

8.2.4 Geomorphological Characteristics of the Area

Different eminent scientists (Srivastava et al., 2003) have identified the major regional geomorphic surfaces in the Ganga plain, like the (1) upland interfluve surface (T2); (2) marginal plain upland surface (MP); (3) megafan surface (MF); (4) piedmont fan surface (PF); (5) river valley terrace surface (T1); and (6) active flood plain surface (T0). The present author has prominently identified major geomorphic surfaces which have significant role to influence the channel behaviors of the area.

The study area falls under the Ganga-Yamuna megafan (MF). At the foothill of the Siwalik, the Himalaya series of piedmont fan surface (PF) are prominently observed by the author. The Ganga River and its several major tributaries in Haridwar have been characterized with well-marked river valley terrace surface (T1) and active flood plain surface (T0). Channel behaviors of different rivers are greatly controlled under those major geomorphic surfaces.

The Ganga Megafan Surface

Except Siwalik, the entire study area is a part of the Ganga megafan (Fig. 8.2). The megafan surface shows several N-S aligned drainage channels (Shukla et al., 2001). The fan surface is most deeply incised by the active Ganga River and other rivers like Ranipur Rao, Solani, Ratmao Rao, etc. near the Himalayan foothill. Prominent channel scarps of those rivers indicate active incision of channels over the megafan surface. At present, the megafans are relict features, and the surface is being modified by sheet erosion and minor channels as observed by the author near Sureswari temple area, in Ranipur forest range.

River Valley Terrace Surface (T1)

Terraces are the remnants of the earlier riverbed. The major rivers of the Ganga plain display abroad alluvial terrace surface (T1) (Fig. 8.3). In Haridwar the active channel of Ganga and other major rivers have developed paired and unpaired terraces, which

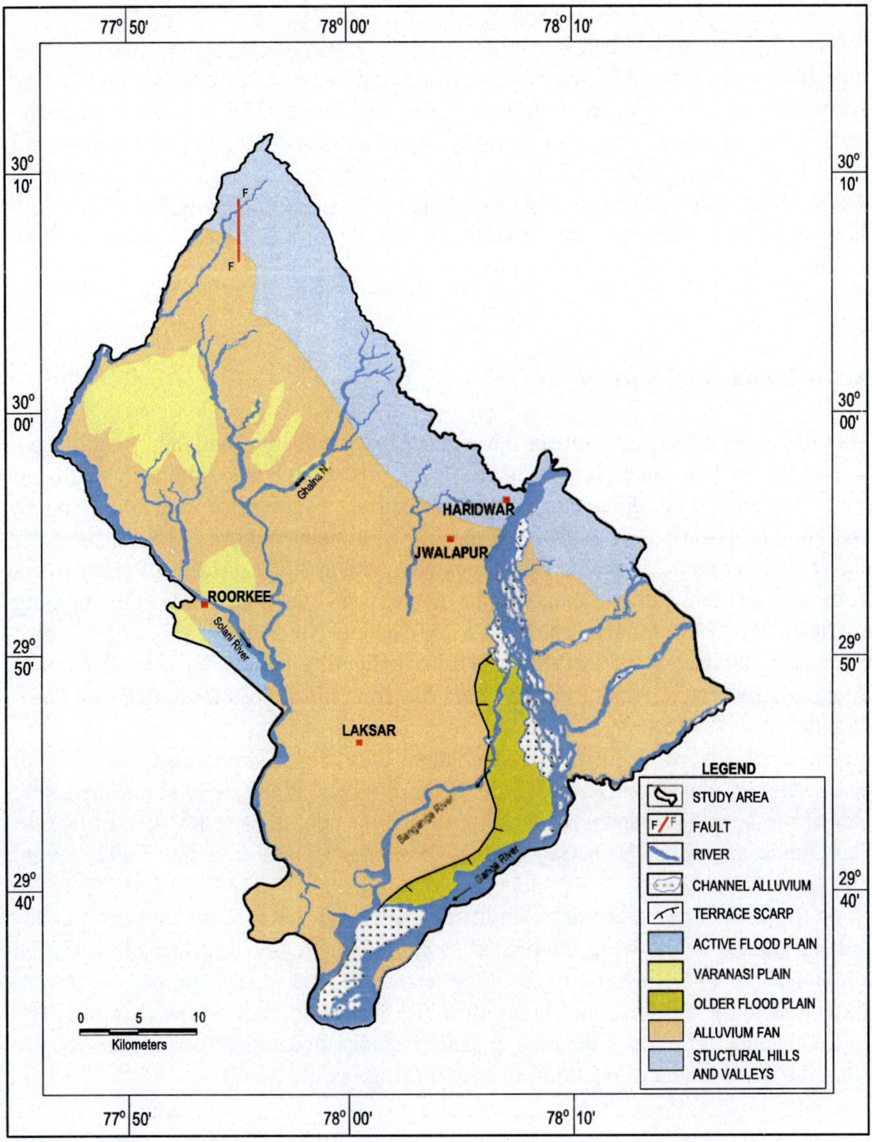

Fig. 8.3 Geomorphological map of the study area showing different geomorphic surfaces (Source: prepared by the author based on DRM, Saharanpur, 1997)

are the evidences of strong incision over the earlier terrace surfaces. Channels over the T1 surface are Yazoo type (Fig. 8.1). The surface exhibits features like abandoned channels and meander cut-off (Singh, 1996b).

Piedmont Fan Surface (PF)

PF is a 10–30-km broad belt formed by coalescing gravely piedmont fans. The PF is superimposed on the MF. Numerous small and braided, parallel to subparallel, ephemeral streams drain the surface. The major rivers, which flow through the surface, are deeply incised. The gravels are subangular to angular and are derived from the Himalaya and from the upper Siwaliks. The surface is mostly covered by silty sediments and is densely vegetated. The PF is cut by ephemeral streams, which carry well-rounded pebbles and possess terrace deposits of rounded pebbles. They are the deposits of the Late Holocene (Singh, 1996b).

Active Flood Plain Surface (T0)

This surface essentially comprises the presently active channel and their flood plains. Due to entrenched nature of most of the rivers, this surface is narrow with small and poorly developed floodplains (Fig. 8.3). The surface is subjected to annual flooding and exhibits a variety of landforms, namely, channels, sandbars, levees, meander cut-off, and swamps. In recent years, the active channel (T0 surface) of many rivers of the Ganga plain shows increase in the area of active sandbars, indicating ongoing aggradations. The author has analyzed a mid-channel bar (Dutta, 2017) which is also increasing its size in the last three decades indicating significant river dynamics. Such an accelerated siltation in the rivers has sometimes been attributed to anthropogenic activities (Singh, 1996a).

On the basis of distinctive assemblages of valley morphology and channel characteristics, the Ganga River valley (GRV) has been divided into seven segments. Out of the seven segments, the study area mainly falls under GRV-I: Haridwar to Garhmuktesar and GRV-0: Gaumukh to Haridwar (Singh & Singh, 1992; Singh, 1996a, b).

At the GRV-I downstream of Haridwar, the Ganga River flows through a 2–35-m-deep and 2–40-km-wide alluvial valley cut into the regional surface of the Ganga alluvial plain. In GRV-I the river valley terrace (T1) is prominent, located 2–8 m above the active channel, and shows swamps, lakes, and Yazoo-type channels. The active channel is 1–3 km wide in flood discharge, braided with huge sandbars mostly 100–500 m wide and 0.5–5.0 km in length (Singh et al., 2007).

8.3 Result and Discussion

8.3.1 Analysis of Channel Characteristics

In Haridwar, the main Ganga River and its several tributaries have developed Yazoo types of drainage pattern (Fig. 8.1) (Uniyal et al., 2010). The present researcher has divided four physiographic divisions depending on major breaks in the longitudinal profile drawn from the extreme north to south of the district (Fig. 8.10) like the (1) higher Siwalik ridge zone, (2) dissected lower Siwalik, (3) Bhabar, and (4)Terai to differentiate the channel characteristics in different physiographic zones.

Altitudinal Variation and River Dynamics

Higher Siwalik Ridge Zone (More Than 500 m)
From 500 meter to 900 meters, the physiographic zone has been considered by the author as higher Siwalik ridge (Fig. 8.7) which is extended along the northern limit of the study area. The main noticeable characteristics of the zone are the numerous tributaries (Raos) Malowali, Beenj, Mohand, Sukh, Chillawali, Chikna ,Gaj, Andheri, Binj, Dholkhand, Sampuwali, Mallawali, Betban, Bam, etc. flowing from north-east to south-west direction according to slope along with the major distinct breaks of slope forming narrow gorges (Fig. 8.4). The channel of these rivers in this zone is comparativcly steeper and mainly covered with bigger size of boulders. Active fluvial incision is prominently observed along the river channel in the higher Siwalik Himalayan zone of the area.

Dissected Lower Siwalik Zone (300–500m)
In between 300 and 500 meters (Fig. 8.7), the author has identified dissected lower Siwalik zone. In this zone a series of micro-alluvial fans have been developed under the process of fluvial deposition. Earlier scientists (Srivastava et al., 2003; Shukla

Fig. 8.4 Mohand Rao flowing through well-marked gorge in higher Siwalik range near Mohand, along NH-72A. (Source: Field survey, October, 2008)

Fig. 8.5 Deeply incised left bank with wide, flat boulder channel of Mohand Rao in dissected lower Siwalik near NH-72A. Gate no. 2, Mohand, Rajaji National Park, Chillawali Range Mohand. (Source: Field survey, October, 2008)

et al., 1999) had already stated those fans as piedmont fan in their works. The Haridwar city is located at the transitional part of the dissected lower Siwalik and Bhabar zones. As there is an abrupt decrease in the gradient of the slope in topography from higher Siwalik to dissected lower Siwalik, stream velocity suddenly declines and fluvial materials are deposited over the riverbed. So, wide, flat boulder channels (Fig. 8.5) are observed in these zones. Several temporary channel bars are the characteristic features of the rivers. In these physiographic zones, channels are covered with boulders, cobbles, and pebbles. During torrential rainfall of monsoon period, flash floods occur in these channels. So at the junctional point between higher Siwalik and dissected Siwalik zones, the channel widths of all tributaries are suddenly increased. Depending on the data provided by Kumar (1995) (Table 8.2), it has been found that the average annual increasing rate of channel width in different tributaries varies from about 2–9 meter/year. Time series graph shows positive trend of increasing width with time (Fig. 8.13) of Chillawali Rao and Sukh Rao (Fig. 8.13). Apart from this, in this zone, channel aggradation in all rivers is evidently noticed. During flooding condition the channels become unconfined, and active avulsion occurs in the channel courses of the rivers. Continuous depositional works by those rives are responsible for the development of recent piedmont fans. Strong topographic influences and excessive sedimentation processes have developed fast widening (Table 8.2) and scouring nature of several parallel river channels.

Bhabar Zone (200–300m)

At the southern part of dissected lower Siwalik zone, Bhabar area is located. It is comparatively a narrow tract situated in between lower Siwalik and Terai zone (Fig. 8.7). In this zone Ganga channel is characterized with boulders, cobbles, pebbles, and sands. The author has witnessed different types of channel behaviors in this zone like dry river channel, channel avulsion, channel shifting, flash flood (during monsoon), etc. In the rivers of the physiographic zone due to frequent diversion of channel, villages located along the riverbank face channel shifting-

Fig. 8.6 Left side photograph indicates active soil and gully erosion along the non-perennial channel in Bhabar zone near Sureswari temple and Ranipur forest range. Right side photograph shows very wide and unconfined braided channel of Ganga near Ballawali. (Source: field survey, May 2008)

associated problems. The author has also identified active soil and gully erosion (Fig. 8.6) along some parts of different riverbanks like Ranipur Rao, Ganga, etc.

Terai Zone (Less Than 200m)
Terai is the southern extensive physiographic division of the study are (Fig. 8.7). Laksar city is located in the zone within 200 meter altitude, where a vast wide-braided channel of the river Ganga is observed. Here Solani river joins with Ganga River, and a vast wetland has been developed at the Solani-Ganga confluence. Near Bishanpur a different branch of Ganga River is observed, and after its 45-km journey, it joins again with the Ganga. The confluence zone of Banganga and Ganga is basically a vast water-flooded area. Hence, a riverine wetland ecosystem prevails here. This wetland ecosystem is situated in Laksar range, and it is called Banganga wetland. Near Ballawali the Ganga River channel is characterized with an extensive braided pattern (Fig. 8.6) mainly filled with coarser and finer sands and silt. In this zone floods occur every year during the monsoon period. Besides flood, shifting of channel and channel avulsion are the major geomorphic hazards of the area. Villages situated in these areas face huge loss of property due to these hazards. Roorkee city is also situated at the transitional zone of the Bhabar and Terai.

8.3.2 Aggrading Characteristics of the Ganga and Its tributaries

As the study area is a part of upper Ganga plain, channel sedimentation is a common feature in Ganga and its other tributaries. In Bhabar and Terai zones, the formation of several channel bars is the strong geomorphic evidence of aggrading nature of channel. At the southern part of Haridwar district (Terai zone), a prominent braided

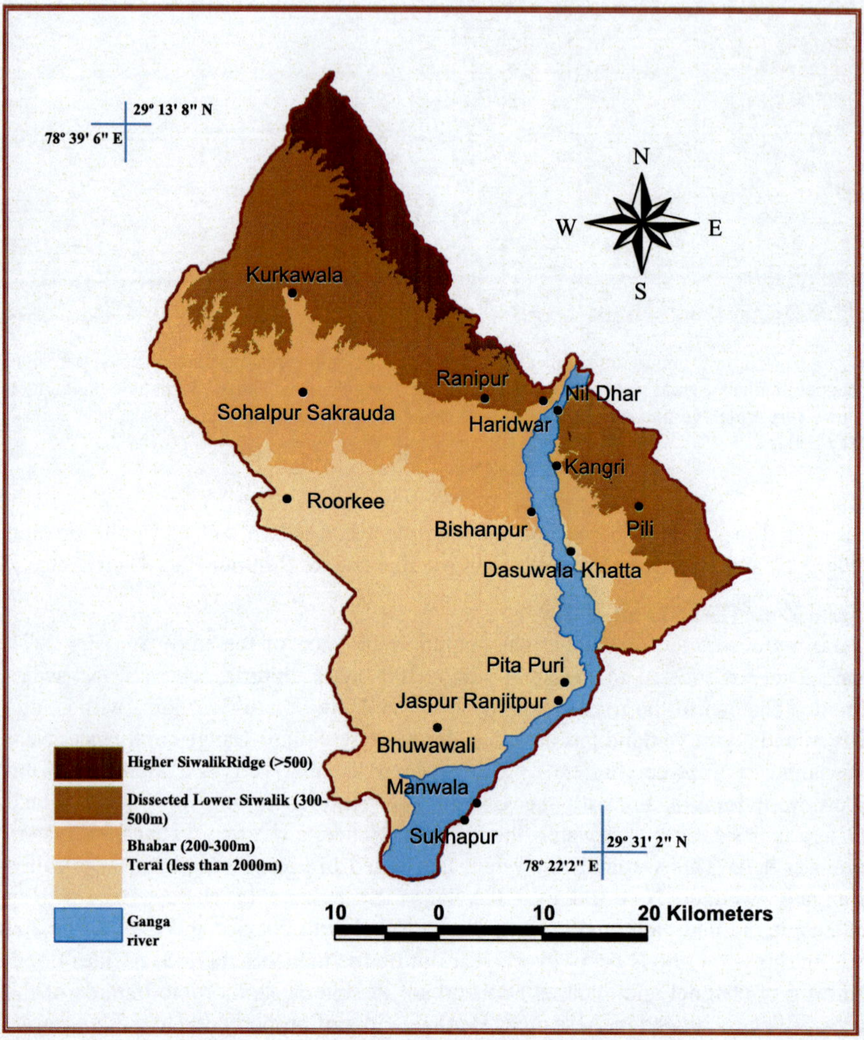

Fig. 8.7 Major physiographic divisions of the study area. (Source: prepared by the author)

channel is observed from the satellite image (Fig. 8.8). The braided channel pattern stands for aggrading Ganga River system. Aggrading Ganga river channel in Haridwar district is one of the causes to enhance channel dynamics.

Impacts of Sedimentation on Channel Bars

The rivers draining the study area (Siwalik to Terai) exhibit significant channel sedimentation process that are reflected in the form of numerous channel bars in the

Fig. 8.8 LISS IV image of Ganga River indicates extensive braided channel in downstream (near Laksar) and in upstream (near Haridwar) comparatively confined channel observed due to structural control. (Source: LISS-IV, 2005, NRSC)

entire Ganga River course of Haridwar district (Sinha, 2005). The researcher has vigorously studied the impact of sedimentation on Ganga River channel bars. The number of channel bars in Ganga River counted by author from topographical sheets in 1967 (SOI, 1967) were 180, and in 2005, the number of channel bars identified from the satellite image (LISS III, 2005) were 134. So it is very clear that the number of channel bar is declining from 1967 to 2005, and the researcher has also studied that the area of most of the bars are gradually increasing with their size. So, from this investigation, it is very clear that the declining number and gradually increasing size of the bars clearly indicate rapid channel deposition. Sandbars along the Ganga River are observed with occasional changes due to floods. The dynamic nature of the channel bars has a great influence on the flow properties of several primary and secondary channels of the Ganga River.

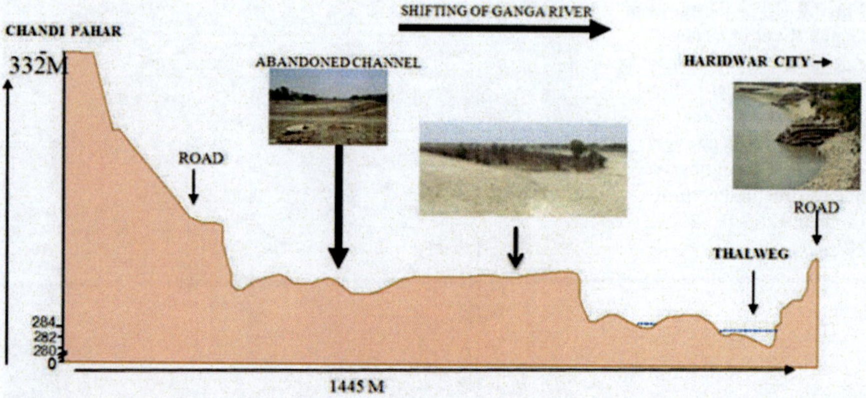

Fig. 8.9 Cross-section of Ganga River channel near Chandi bridge, showing abandoned channel, mid-channel bar, terrace, and sifting nature (toward the right bank) of the active Ganga channel. (Source: field survey by the author, May 2008)

Channel Forms in the Cross-Section of Ganga River

For the vivid analysis about channel dynamics of Ganga River, a cross-sectional (Fig. 8.9) study is very relevant. Therefore, a vast cross-section (about 1500 m) of Ganga River channel in between Mansa and Chandi hill valley has been studied. Geomorphologically, the entire cross-sectional area is under a vast river valley characterized with abandoned channel (Fig. 8.9), dry channel, lower alluvial terraces, stable mid-channel bar, unstable channel bars, thalweg, etc. In the present channel, stable and unstable bars are very evidently observed. The thalweg line of Ganga River is shifting toward the west direction. At present the thalweg line is in the extreme west side of the channel. The shifting of thalweg line toward Haridwar city is a serious issue. Man-made embankment with boulders was constructed to protect the city, but the effort is not enough. Serious damages in some parts of the embankment are caused by riverbank erosion.

Channel Abandonment and Development of Stable Mid-Channel Bar in Ganga River

An abandoned channel in a cross-section is a strong geomorphological evidence of the shifting nature of Ganga channel. It is situated below the Chandi bridge, and it is located on the left bank (Nildhara area) of the river Ganga (Fig. 8.9). Along the abandoned channel, avulsion occurs during every monsoon.

In the abandoned channel area, a very interesting relationship has been noticed by the researcher between channel morphology and riverbank erosion. Along the left bank of the Ganga, severe riverbank erosion is noticed, but on the right bank of the Ganga River, bank erosion problem is not so serious as some protective engineering measures have been taken. If the erosion is not properly managed, in the future, the problem of erosion on the right bank may be a more serious issue than the bank erosion problem of the left bank, because the main Haridwar city is located on the right side (Fig. 8.9) of the Ganga River and at present the shifting trend of the active channel is toward the city.

Bank erosion scenario along the left bank of the Ganga River is totally different than its right bank. In spite of the excessive erosion along the left bank during monsoon period, no effective preventive measures have been taken by the local authority from 2006 to 2017. Every monsoon, a serious erosion takes place due to shifting and avulsion of channel over the left bank.

The investigator surveyed the study area in the month of October 2010, after a flood, and found some remarkable impact of flood on riverbank erosion. Along the left bank, more than 500 meters south from the Chandi bridge, the top soil layer is totally washed away, and about 40 meters along the bank, it is totally eroded away.

The author has identified a crucial reason of riverbank erosion on the left bank. In this part due to rapid channel sedimentation, a vast mid-channel bar (Fig. 8.9) has been developed, which is increasing in size very rapidly and becoming a stable bar covered with trees and bushes.

The growing size of the mid-channel bar is an important reason to enhance the bank scouring and avulsion nature of Ganga channel. The researcher has analyzed the dynamic nature of the mid-channel bar with the help of primary and secondary data. In 1971 (toposheet, SOI), the area of the bar was 6.31 ha. In the year 1997 (LISS-III), the area of the bar has increased to 11.68 ha. Again the area of the mid-channel bar was measured from the satellite image (LISS-III, 2005), and it was 12.65 ha in the year 2005. The researcher has surveyed the mid-bar with a GPS receiver in the year 2010, October, and then the area has increased to 12.9 ha (Dutta, 2017).

However, from the change detection study, it can be depicted that both size and area of the bar are increasing very rapidly, indicating an active sedimentation on the riverbed.

Sinha (2005) already discussed the growing nature of Ganga channel bars under active sedimentation. Singh et al. (2007) have also mentioned in Haridwar Ganga downstream braided channel with huge sandbars mostly 100–500 m wide and 0.5–5.0 km in length.

As a result, the hydraulic radius of the channel of Ganga is being reduced, and during monsoon excessive water discharge creates unprecedented bank scouring, and the entire events lead to the dynamic nature of the river channel (Fig. 8.10).

8.3.3 *Shifting and Avulsion Nature of River Channels*

Shifting and avulsion of channel are significant fluvial geomorphic characteristics of Ganga and its tributaries. In sedimentary geology and fluvial geomorphology, avulsion is the rapid abandonment of a river channel and the formation of a new

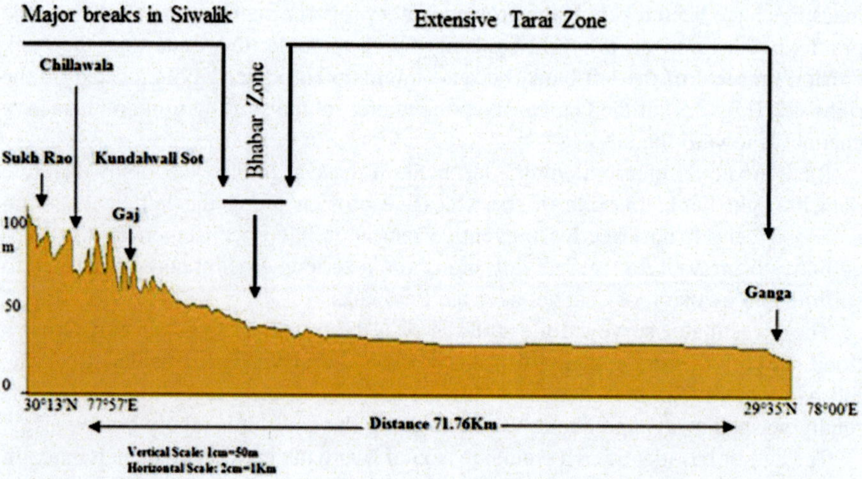

Fig. 8.10 Longitudinal profile from the extreme north (Sukh Rao valley) to the extreme south (Ganga River) of the study area reveals major breaks of slope in the northern Siwalik zone and extensive Terai zone in the south. (Source: prepared by author depending on the toposheets, SOI (1967 and 1971)

river channel. In braided stream mainly three types of avulsions are found: *choking avulsion, constriction avulsion, and apex avulsion* (Jones & Schumm, 1999).

Different Sites of Avulsion

The author has visited (2006–2017) different parts of the area and identified several types of channel avulsion in Ganga, Solani, Ratmao, Pathri, etc. During monsoon channel avulsion becomes very active mainly in the southern part of the study area. Along the Ganga River, nodal and random types of channel avulsion (Dutta, 2017) are observed. Near Nildhara area nodal avulsion takes place during every monsoon season. This type of avulsion is considered as constriction avulsion that is produced by deflection, confinement, and subsequent diversion of the flow by a bar formation.

Near the village of Tatwala, random avulsion is noticed. This type of avulsion is observed when there is erosion at the outside of sinuous thalweg and confined meander bends.

At the extreme southern part, near Banganga wetland area, some imprint of regional avulsion (Jones & Schumm, 1999) is observed by the author.

Near Kangri (Bhabar zone), Gajiwali-Shyampur villages (Fig. 8.11), minor random avulsion is observed in the wider Ganga River channel. This type of avulsion is a choking avulsion which is caused by blockage of one channel by a sediment lobe. All types of avulsions in the study area take place during the monsoon season.

Fig. 8.11 From Google Earth satellite image. An extensive dynamic braided pattern of Ganga River shows several random avulsion courses near flood-prone Kangri and Gajiwali-Shyampur villages. (Source: field survey with GPS receiver on 19 February 2017)

Change Detection Study on Shifting and Avulsion of Ganga and Other Rivers

The study area is characterized with channel shifting and channel avulsion in major rivers such as Ganga (Fig. 8.12b), Solali, Ratmao, Patri, etc. (Fig. 8.1). During monsoon, channel avulsion becomes very active, mainly in the southern part of the study area of Ganga River. For the better understanding of channel shifting of Ganga River, the present researcher has studied the river channels from 1967 (toposheet, SOI) to 2005 (LISS-III). From the change detection map (Fig. 8.12a), it is very evident that in the southern portion of the study area (Terai), the tendency of channel shifting and avulsion is very high in all the abovementioned rivers. The riverbank-eroded area is measured by the author to be about 49.30 sq km (Fig. 8.12b) from 1967 to 2005. That reveals a maximum channel instability in the Terai zone compared to the Siwalik area of Haridwar. This sharp channel instability causes great loss to agricultural land and road network of Terai zone. Based on recent satellite image (May 2020), it is evident that (Fig. 8.12c) several secondary channels and braided bars appeared along Ganga River in Haridwar. The area of the braided Ganga River in 2020 is measured to be about 102 sq km. In that image, the width of

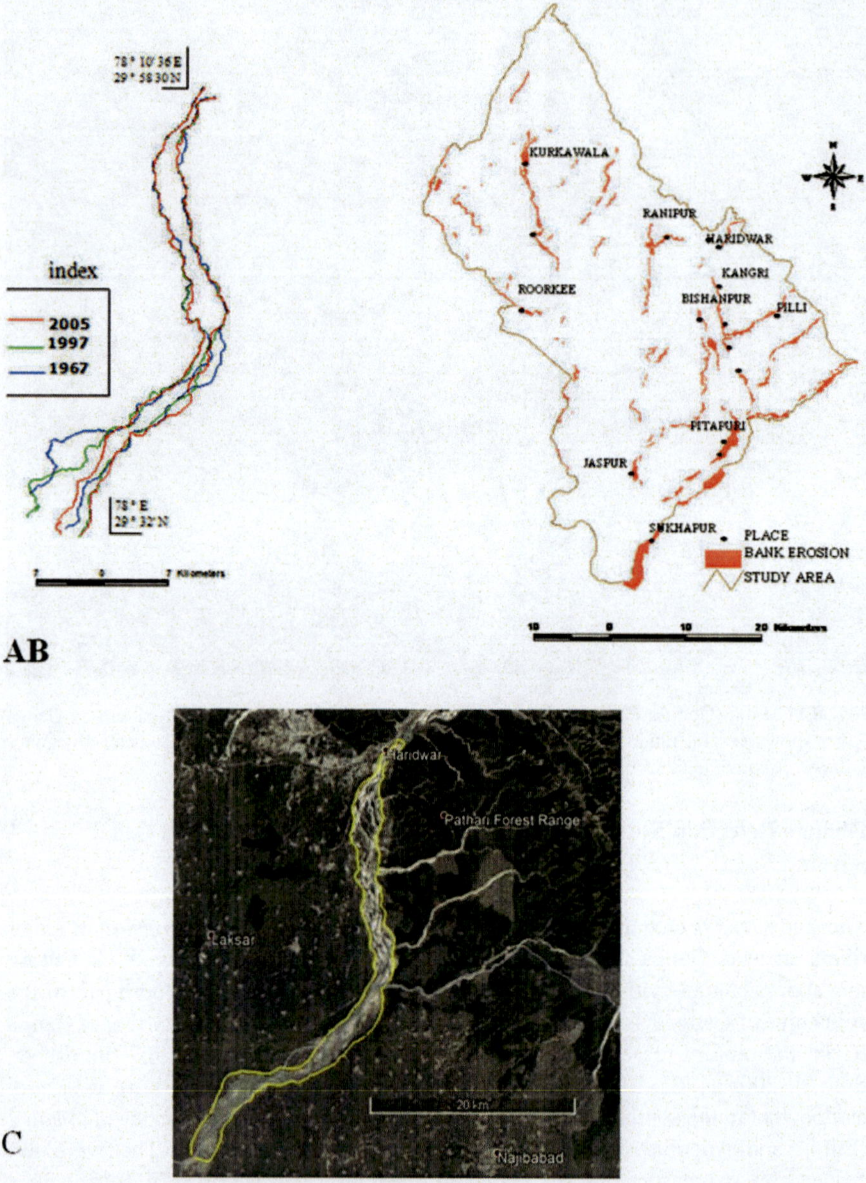

Fig. 8.12 The left one (**a**) expresses the shifting nature of the entire Ganga River channel (from 1967 to 2005) in Haridwar district, emphasizing unconfined channel in Terai and, on the other hand, Siwalik structure and human-controlled narrow channel in Siwalik zone. The right one (**b**) indicates a bank-eroded area (about 49.30 sq km) by Ganga and its tributaries (from 1967 to 2005). (Source: toposheets-1967, LISS-III, 1997 and 2005). The yellow line on the satellite image (**c**) shows the outline of Ganga expressing prominent braided river channel. (Source: Google Earth image, May 2020)

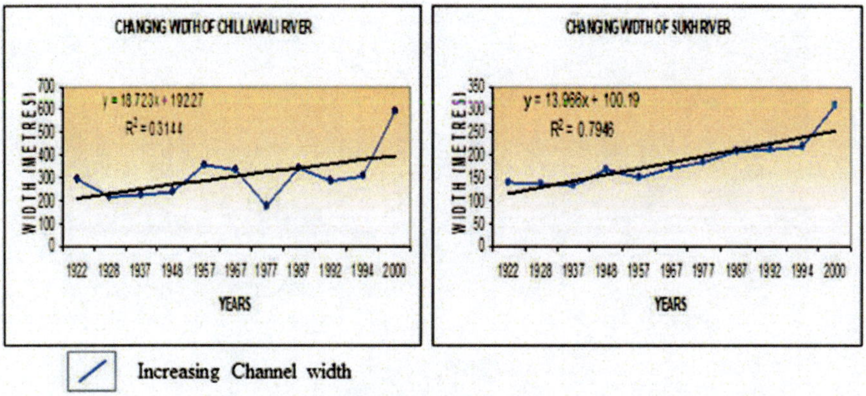

Fig. 8.13 Time series graph and trend lines indicate increasing channel width (1922–2000) in Chillawali and Sukh rivers in Rajaji National Park, Haridwar. (Source: based on data by Kumar, 1995)

the river at the extreme south is comparatively lesser (May) than the 2005 LISS-III image (November). So, the channel width of Ganga River varies with season (pre- and post-monsoon) (Fig. 8.13).

8.3.4 *Observation of Flood-Induced Channel Avulsion*

During field verification, the researcher has observed flood-induced channel avulsion in 2010 and 2013. It should be understood that the instability of a channel (i.e., avulsion threshold) caused by different factors (excessive sedimentation, meandering, braided channel course, variability, magnitude of discharge, etc.) does not necessarily induce avulsion unless a triggering event like flood or tectonic tremor occurs (Heller & Paola, 1996). It, therefore, follows that avulsion frequency would depend on how the processes inducing channel instability interact with the triggering events (Jones & Schumm, 1999). Mukhopadhyay (2010) rightly stated that the fundamental cause of avulsion is the relative loss of sediment-fluid carrying capacity of a channel to such an extent that the existing path becomes unfavorable for sustaining the flow, and the channel then finds a new course along a favorable gradient. There is no doubt that flood is the triggering factor for random avulsion in the braided channel of the Ganga River's course toward the southern part of Haridwar, because it is observed by the researcher during field verification that the occurrences of avulsion took place in flood situation of 2010 and 2013. During 2010, flood-induced channel avulsion seriously damaged a permanent road below the Chandi bridge where an abandoned channel have been identified by Dutta (2017). In 2010 and 2013, due to a flood, new micro-channels (Figs. 8.14 and 8.16) have been developed along the left bank of Ganga, below of Chandi bridge. In the 2009–2010 flood, about 69 villages were badly damaged.

Fig. 8.14 Development of a new avulsion channel, below Chandi bridge during the 2010 flood (field survey, October 2010)

Variability of Water Discharge and Flooding Nature of the Rivers

The variability of water discharge is one of the controlling factors of the dynamic nature of river channels of the area. In Haridwar riverbank erosion is caused by a heavy fluctuation of stream discharge. Except Ganga, all of the rivers are non-perennial in nature, and that is why during dry season in many rivers, water flow does not appeared, but in the period of monsoon, water discharge suddenly increases many times, and the excessive water flow enhances fluvial erosion power. As a consequence, sudden unprecedented flow induced bank scouring; shifting and avulsion are prominently noticed. The variation in water discharge (maximum and minimum water discharge at Rishikesh) in Ganga River is observed with seasonal change also. The maximum water discharge observed during July–August 2001 is about 3500 cumecs, and the minimum water discharge noticed during December–February is below 500 cumecs.

During monsoon period, flood occurs, and sometimes prolonged rainfall creates flash flood in mountainous streams. In the year 2010, during July–August, excessive monsoon rainfalls (1516 mm. up to 25 September) caused a major flood in the area. Heavy and continuous rainfall took place in the district for 3 days. High flood was observed in Ganga and in other local/regional rivers and rivulets. Breach in embankments of rivers, rivulets, and canals was noticed. More than 6 lakh cusecs of water discharge by Ganga River was reported in Laksar continuously for 35 hrs. The water discharges from different rivulets/rivers were as follows: Sonali river (25,000 cusecs), Ratmao (55,000 cusecs), Pelli (30,000 cusecs), and Pathari Roa (25,000 cusecs). Even in Ganga River, a massive flood took place in different years like 1910, 1924, 1978, 1995, 1997, 2010, and 2013. In those years, the maximum flood peaks were more than 4 lacks. During lean period the channels of almost all tributaries are observed as dry channel. So extreme variability of discharge in Ganga River and its tributaries is the main influencing factor of river dynamics in the area.

Apart from the variability of water discharge, the magnitude of water flow directly influences channel dynamics of Ganga and its other tributaries. It is said that when water level rises more than 1 meter than the danger level (DL) of a gauge site, then the flood is considered as a major flood (Dhar & Nandargi, 2004). Based on collected flood data from the DM office, Roshnabad, Haridwar, it is found that from 1894 to 2010, nine major flood peaks in Ganga River have been observed in the area, but since the year 1995, the occurrence of the major flood is observed every 1 or 2 years interval. It means the occurrence of major flood is rising gradually. The danger level for the gauge site of Haridwar is 294.00 meters. On 3 August 1997, a major flood occurred in Haridwar with the flood level of 295.35 meter. The difference between danger level and flood level was 1.35 meter (Dhar & Nandargi, 2004). At Haridwar, the river annually discharges 21,393 × 106 m^3, of which 75% flows during the monsoon period (June to September) and 25% is distributed during the remaining part of the year. The average discharge of the river is over 1523 m^3/s during the monsoon period and decreases to 282 m^3/s in winter. In 2009–2010, about 69 villages of Haridwar district were flooded. This year, the Ganga River flooded the villages of Kangri, Gajiwali-Shyampur, Sajanpur Pili, Rasulpur, Mithiberi, Tatwala, Bishanpur, etc.

Impact of the 2013 Flood on Channel Geomorphology

The present author has visited the area on October 2013 and observed some remarkable changes about channel forms. Deposition of fresh sediments on the riverbed (Fig. 8.15) is distinctly observed in Ganga River. Numerous uprooted trees have blocked some parts of the channel of Ganga and other tributaries. Near Nildhara area, a number of new channel bars have been developed under the influence of fresh channel sedimentation. During the high flood, massive water flow has created serious soil and bank erosion. The settlement and transport system were also badly affected by flood and flood-associated bank erosion (Fig. 8.16).

Fig. 8.15 Fresh sediment supply in Ganga channel after the 2013 flood. (Source: Field survey, October, 2013)

Fig. 8.16 A newly developed channel along the left bank, at Haridwar, after the flood hazard in Ganga River, Nildhara area. (Source: Field survey, October, 2013)

8.3.5 Anthropogenic Impacts on Ganga River Channel

The construction of the upper Ganga canal and Bhimgoda barrage has created control on channel width (Fig. 8.17) in Mansa and Chandi hill valley. This man-made-confined channel has increased channel instability in downstream Ganga.

Fig. 8.17 Bhimgoda barrage at Haridwar to divert Ganga water into the upper Ganga canal, influencing excessive channel sedimentation and interruption in channel equilibrium. (Source: field survey, February, 2010)

Wolman (1967) showed that the changes of land use can affect sediment yield in a river. Urbanization of the part of the catchment can increase sediment load by several hundredfold. The author reasons that reservoir can greatly influence the longitudinal profile of a river. The aggradational characteristics are seen in the longitudinal profile above the reservoir and erosion immediately below the sluice gates. Aggradation appears again further downstream of the barrage. Morisawa (1985) also stated that the indication of adjustment is seen by the hydraulic geometry which is different from the above and below the dam.

It has been found that Bhimgoda barrage and upper Ganga canal have greatly influenced the normal equilibrium of the Ganga River. The construction of upper Ganga canal was conceived and constructed by Proby T. Cautley during the period 1840–1854 (http://planningcommission.nic.in/reports/sereport/ser/ugc/ch1.pdf). In the beginning one of the branches of the river – a natural channel flowing near Haridwar – was used to divert practically the entire winter flow by constructing temporary obstructions across other branches. The upper Ganga canal takes off from the right flank of Bhimgoda barrage which replaced the old weir at Haridwar in 1991–1992.

The barrage (Fig. 8.17) (29° 57′ 23′′N, 78° 10′ 49′′E) is 455 m (1493 ft) long and has the head of a 23,000 km^2 (8880 sq mi) catchment area. It contains 15 spillway gates and 7 under sluice gates, 18 m (59 ft) wide all in all. The flood discharge of the barrage is 19,300 m^3/s (681,573 cu ft/s). Adjacent to it, on the right bank of the river, the barrage diverts water into the upper Ganga canal. The canal system is immense, consisting of 6450 km (4008 mi) of main canal and branches, providing irrigation for up to 2,023,000 ha (4,998,942 acres) land.

The photographic evidences of Bhimgoda barrage clearly indicate excessive siltation at upstream and downstream courses of the barrage (Fig. 8.17). Undoubtedly, the barrage has disturbed the natural longitudinal profile of the Ganga channel. The upstream and downstream of the barrage channel aggradation in the Ganga River has enhanced channel instability.

Besides this, channel materials (boulders, pebbles, sands), mining in Ganga, and other tributaries have significant impacts on channel behaviors. Sometimes unscientific channel excavation increases channel roughness that creates discontinuity of channel flow or eddies. These types of anthropogenic intervention directly and indirectly enhance the dynamic nature of Ganga River.

Ganga River channel courses and a number of channel bars are intensively utilized for the temporary tent for the pilgrims (Fig. 8.18) during Kumbh Mela and other religious festivals. Therefore, the natural vegetation of the channel bars are totally obliterated, and the stability of those channel bars is disturbed by anthropogenic effects. However, the human-induced wastages are removed into Ganga River unintentionally. These wastage materials seriously accumulate the human-induced sediments into the Ganga River. If it continues, there is no doubt the channel instability would increase day by day.

Fig. 8.18 Temporary tents and shelters for pilgrims over Ganga channel course, near Chandi bridge. (Source: field survey, February, 2010)

8.4 Conclusion

From the above discussion, the authors have come to the following conclusions.

River dynamics are evidently expressed in various channel behaviors like channel sedimentation, avulsion, channel abandonment, meandering and braiding, channel shifting, bank erosion, etc. of Ganga and its tributaries.

The varied channel behaviors are influenced by several physical factors like geology and tectonics, geomorphology, occurrence of flood, channel sedimentation, etc. and anthropogenic factors like Bhimgoda barrage, upper Ganga canal, channel material mining, Kumbh Mela, and other festivals.

Different physiographic divisions have distinct impacts on channel forms.

The area consists of significant landforms (alignment of river channels, channel escarpments, incised meander, river terraces, etc.) under megafan, piedmont fan, river valley terrace, and active flood plain surfaces that indicate strong tectonic influence on the channels.

The change detection map of the Ganga from 1967 to 2005 and the outline of the Ganga (2020) clearly reveal that toward the southern direction (mainly Terai zone), the tendency of shifting and avulsion is comparatively higher than the northern Siwalik zone and the fluvial processes varies on pre- and post-monsoon condition. In the study area, channel instability is increased in the downstream area. As a consequence, villages located in the downstream area face different river channel instability-associated hazards like avulsion, shifting, bank erosion, flood, etc.

The construction of Bhimgoda barrage and upper Ganga canal has greatly influenced the longitudinal profile of the Ganga River. Those constructions have confined the Ganga River channel width in Mansa and Chandi hill valley, resulting in channel instability in the downstream. The barrage accelerates channel aggradational rate, enhancing channel instability. Excessive excavation of materials from the riverbed has also disturbed the natural equilibrium of the Ganga River channel.

References

Dhar, O. N., & Nandargi, S. (2004). Flood in North Indian river systems. In *Copping with natural hazards: Indian Context* (pp. 104–111). Orient Longman private Limited.

Dutta, R. K. (2017). Controlling factors of channel shifting and avulsion in Haridwar district, Uttarakhand, Geographical Review of India. *Geographical Society of India, Kolkata, 79*(20), 153–167.

Field, J. (2001). Channel avulsion on alluvial fans in southern Arizona. *Geomorphology, Elsevier, 37*, 93–104.

Heller, P. L., & Paola, C. (1996). Downstream changes in alluvial architecture: an exploration of the control on channel stacking pattern. *The Journal of Sedimentary Research, Society of Sedimentary Geology, 66*(2), 297–306.

https://www.nrsc.gov.in/EOS_dissemination_Order

Jones, L. S., & Schumm, S. A. (1999). Causes of Avulsion :An overview. *The International Association of Sedimentologists, 28*, 171–178.

Kale, V. S. (2002). Fluvial geomorphology of Indian rivers: An overview. Progress in Physical Geography. *Arnold, 26*(3), 423–456.

Kumar, D. (1995). Rajaji National Park, Dehra Dun: Why it is not in sustainable position-remedies for its sustainability . Himalayan Geology. *Wadia Institute of Himalayan Geology, 6*(2), 95–110.

Morisawa, M. (1985). *Rivers – Form and Process*. Longman Inc., New York.

Mukhopadhyay, S. C. (2010). Fluvial processes with special reference to avulsion in the flood plains of the Unstable Mahananda-Sankosh interfluvial areas, West Bengal. *Indian Journal of Landscape System and Ecological Studies, ILEE, 32*(2), 521–538.

Shukla, U. K., & Bora, D. S. (2003). Geomorphology and sedimentology of Piedmont zone, Ganga Plain, India. *Current Science, 84*(8), 1034–1039.

Shukla, U. K., Singh, I. B., Srivastava, P., & Singh, D. S. (1999). Paleocurrent patterns in braid-bar and point-bar deposits: Examples from the Ganga river. *International Journal of Sediment Research, 69*, 992–1002.

Shukla, U. K., Singh, I. B., Sharma, M., & Sharma, S. (2001). A model of alluvial megafan sedimentation: Ganga Megafan. *Sedimentary Geology, 14*, 243–262.

Singh, I. B. (1996a). Late Quaternary sedimentation of Ganga plain foreland basin. Proc. Symp. NW Himalaya and foredeep. *Geological Survey of India Special Publication, 21*, 161–172.

Singh, I. B. (1996b). Geological evolution of Ganga plain-an overview. *Journal of the Palaeontological Society of India, 41*, 99–137.

Singh, I. B., & Rastogi, S. P. (1973). Tectonic framework of Gangetic alluvium, with special reference to Ganga River in Uttar Pradesh. *Current Science, 42*, 305–307.

Singh, M., & Singh, I. B. (1992). The Ganga River Valley: Alluvial valley in an active foreland basin. In *29th International Geological Conference, Kyoto, 2* (p. 30). Japan.

Singh, M., Singh, I. B., & Müller, G. (2007). Sediment characteristics and transportation dynamics of the Ganga River. *Science Direct, Elsevier, Geomorphology, 86*, 144–175.

Sinha, R. (2005). Why do Gangetic rivers aggrade or degrade? *Current Science, 89*(5), 836–839.

Srivastava, P., Singh, I. B., Sharma, M., & Singhvi, A. K. (2003). Luminescence chronometry and Late Quaternary geomorphic history of the Ganga Plain India. *Palaeogeography, Palaeoclimatology, Elsevier Palaeoecology, 197*, 15–41.

Uniyal, A., Ravindran, K. V., & Prasad, C. (2010). Tectonic control on alluvial fans, piedmont streams and Ganga River in western Ganga Plain (India) using satellite remote sensing data. *Current Science, 99*(1), 91–97.

Valdiya, K. S. (2003). Reactivation of Himalayan frontal fault. *Current Science, 85*(7), 1031–1040.

Wolman, M. G. (1967). Two problems involving river channel changes and background observations. *Northwest University Studies in Geography, 14*, 67–107. http://planningcommission.nic.in/reports/sereport/ser/ugc/ch1.pdf; https://en.wikipedia.org/wiki/Bhimgoda_Barrage

Chapter 9
Tidal Mechanism and the Changing Fluvio-Geomorphological Environment of the Interfluves in Indian Indian Sundarbans: A Geospatial Technology-Based Approach

Jayanta Gour

9.1 Introduction

Extending between 21°32′N and 22°40′N latitudes and between 88°05′E to 89°00′E longitudes, Sundarbans, the largest mangrove delta, is one of the most complex as well as hydro-geomorphologically the most dynamic tide-dominated site of the tropical world. Although, along with the terrestrial river falling into any bay, sea, or ocean, tide plays a key role in updating the configuration of the islands, interfluves of almost every delta. But the tidal mechanism working behind the building and changing drainage, pedological, ecological, and fluvio-geomorphological character of entire deltaic part, is least discussed all over the world. Hence, there is a need for a detailed study to collect, evaluate, and analyze the tidal data from each and every corner of any delta as the hidden dynamic tidal mechanism if revealed may help geomorphologists a wee bit in mitigating and managing the rising environmental and sea level changing issues.

The rich biospheric assets of Sundarbans in India has been experiencing an intangible change in its drainage, salinity, erosional, and accretional character since the last centuries. At present, Indian Sundarbans comprises of 19 blocks with 16 police stations of North 24-Parganas (Haroa, Hasnabad, Hingalganj, Minakhan, Sandeshkhali-I, and Sandeshkhali-II blocks) and South 24-Parganas (Basanti, Canning-I, Canning-II, Gosaba, Joynagar-I, Joynagar-II, Kakdwip, Kultali, Mathurapur-I, Mathurapur-II, Namkhana, Patharpratima, and Sagar blocks) Districts of the State of West Bengal. In South 24-Parganas District, there are 54 islands in this region. The land area measures about 9629 sq. km., of which 4493 sq. km is inhabited by people and the rest is forest reserve. The total number of mouzas under

J. Gour (✉)
Department of Geography, Sambhu Nath College, Birbhum, West Bengal, India

G. S. Bhunia et al. (eds.), *Anthropogeomorphology*, Geography of the Physical Environment, https://doi.org/10.1007/978-3-030-77572-8_9

the region is 1093. Perhaps, it was quite natural and was growing rhythmically before the British invasion and reclamation of its natural biota. Since the 1770s, the nature of tides not only changed but are being more complex because of which the entire social, economic, and environment of the habited and uninhibited islands are becoming endangered. Albeit, the global warming and rise in sea level may be considered as one of the prime factors.

It will be noteworthy to survey the previous literatures in global, national, regional, and local level as lots of international, national, state-level, and village-level government or private works have been done so far on the fluvio-geomorphological environment of different deltas and mangrove regions and their impacts on the land use in those regions alike the Matla-Bidyadhari interfluve in Indian Sundarbans. The following works have been taken into consideration, which correlates very much both from areal and thematic point of view with the research topic. The reason behind this is the weightage of the theme of the research increases as one comes down narrowing the area.

A brief account of the international works on deltas and mangroves has been reviewed to get an idea of the research methodologies.

Wolanski et al. (1980) worked on hydrodynamics of a tidal creek-mangrove swamp system in Australia. Ridd et al. (1990) carried out work on dynamics, flushing, and trapping in Hinchinbrook Channel, a giant mangrove swamp in Australia. Khalil (1999) worked on the economic valuation of the mangrove ecosystem along the Karachi coastal areas in the Indus Delta of Pakistan where it was noticed that a broad array of goods and services is provided by the mangroves and uses market data to estimate the economic value of a few of them. It then argues for the importance of more thorough mangrove valuation studies as a crucial input into policy decisions, which will affect the viability of mangrove ecosystems in the future. Kjerfve et al. (1999) worked on the hydrological and hydro-geochemical variations in mangrove ecosystems in the Itacurusa Experimental Forest, Baía de Sepetiba, Brazil. Work on coastal erosion due to long-term human impact on mangrove forests has also been done by Aucan and Ridd (2000). Bach et al. (2005) worked on restoring large-scale functional lake-fringe and floodplain wetlands at the Williamson River Delta, Oregon, USA. A hydraulic model of the delta was developed by them to identify and design levee breaches and potential riverine restoration options, as well as to evaluate the geomorphic and hydraulic effects of proposed restoration alternatives. Santos de Farias et al. (2005) worked on the land use and changes in the river Ipojuca, Pernambuco, to the northeast of Brazil. They showed the ambient characteristics through the land use and polluting sources and uses of the water in the basin of the river Ipojuca, located in the state of Pernambuco, northeast of Brazil. As well as, by the analysis of the quality of the water, the natural and socioeconomic factors have identified that influence the environment of the river Ipojuca. Ratna (2005) of Tata Institute of Social Sciences, Mumbai, India, worked on the restoration and management of mangrove forests through people's participation and experiences from India. The work highlighted the need of implementing proactive five-pronged approaches comprising education, advocacy, collaboration, conservation, and rehabilitation based on sustainable community-based development programs for effective restoration and long-term preservation of mangroves. Islam and Gnauck (2007) worked on the water sharing conflict in transboundary catchment of Ganges River and degradation of mangroves wetland ecosystem in the Sundarbans of Bangladesh. In the thesis, it was found that upstream freshwater supply is necessary and emergent for the protection of cultural landscapes and mangrove wetlands ecosystems in the Sundarbans region. Amjad et al. (2007) worked on the degradation of Indus Delta Mangroves in Pakistan where they have opined that the reduction in inflow of freshwater from Indus on account of diversion of water for other purposes, inflow of pollutants from industries, navigational activities, and intermix of industrial effluent. Human and livestock population pressure for fuel wood and fodder collection has exposed this complex ecosystem to severe environmental and social stresses in the form of loss of habitat and biodiversity, decline in fish productivity, and social problems for coastal communities. A study on the patterns of nutrient exchange in a riverine mangrove forest in the Shark River Estuary in Florida in the USA was carried out by Rivera-Monroy et al. (2007). A project being carried out by the Deutsche Gesellschaft für Technische Zusammenarbeit (GTZ) GmbH (German Technical Cooperation) in the Mekong Delta of Vietnam on behalf of the German Federal Ministry for Economic Cooperation and Development is for protecting the mangroves and for the sustainable use of coastal wetland in the southern province of Soc Trang (Eschborn, 8 May 2008). The project work has revealed that in Vietnam the deforestation of mangrove forests is leading to the loss of natural coastal protection in the Mekong Delta, increasing the danger of flooding. Sidik (2008) worked on the complex social-environmental pattern of linkages in resources utilization in Mahakama Delta, Indonesia. Yu et al. (2011) worked on the impacts of climate change on agriculture and assessing the impacts of climate change on irrigation systems in Vietnam, using a modeling approach. Works on improving rice tolerance of submergence and salinity stress to cope with climate change in coastal areas of Vietnamese deltas has been done by Le-Hung Linh et al. (2011).

Lots of research works have been done on the mangrove program restoration of mangrove ecosystems (Hong et al., 1996) – water flow and tidal influence in the mangrove delta system (Loon, 2005) of the deltaic region of Vietnam. Delta management programs in response to climatic changes of California and the Netherland has been discussed.

Paul et al. (1986) have worked on the morpho-ecological variations in Sundarbans mudflats where they opined that mudflat control the food chain in estuarine ecosystem, and these provide a platform for the species replenishment of the mangrove flora in the tropical wet coasts and estuaries. According to them excessive reclamation, however, poses the greatest single threat to shallow estuarine mudflats of the western and northern Sundarbans as it destroys forever the natural habitat and ecosystem. Paul (1987) worked on the coastal dune systems of the Sundarbans and effective management strategies for the coast of West Bengal, respectively, and established the fact that increased human uses of the coast are partly responsible for the high rate of coastline erosion and destabilization of coastal sand dunes. He proposed a special plan for Digha areas and the reclaimed

Sundarbans, because they are highly used by man for his socioeconomic development. Parua (1990) worked on the riverbank erosion, a potential threat to land resources. A work on the exploration of geophysics for hydrocarbons was carried out by Biswas and Neogi (1988). Banerjee (1998) dealt with work on environment, population, and human settlement of Sundarbans delta. Currey and Munasinghe (1989) worked on the intraplate deformation in the northeast Indian Ocean. Sil (1990) did another worthy work on the adaptation of the migrant Oraon in their present habitat in Sundarbans. Biswas and Agarwal (1992) did remarkable work on the tectonic evolution of the Bengal foreland basin since the Early Pliocene and its implication on the development of Bengal fan. A work on the new dates of Pleistocene-Holocene suburb samples from South Bengal, India, was dealt by Hait et al. (1996). Ghosh et al. (2001) worked on the assessment of land use/land cover dynamics and shoreline changes of Sagar Island through remote sensing. Those scientists again in 2002 carried out a work on the coastal erosion and flooding in Ghoramara and adjoining islands of Sundarbans deltaic system. Hazra et al. (2001) explained the sea level and associated changes in the Sundarbans. Chakrabarti and Bhandari (2001) worked on the improvement of bearing capacity using compacted PFA and geo-jute reinforcement over soft clay layer and coastal erosion and flooding in Ghoramara and adjoining islands of Sundarbans deltaic system, respectively. Ghosh et al. (2003) also worked on the application of a bio-engineering technique to protect Ghoramara Island (Bay of Bengal) from severe erosion. A preparatory assessment of vulnerability of the ecologically sensitive Sundarbans island system, West Bengal, in the perspective of climate change was carried out by Hazra et al. (2003). Hazra and Baksi (2003) dealt with the environmental refugees from vanishing islands. Chakrabarti (2005) worked on the Sundarbans, viz., its terrain, legends, gods, and myths. Works have been done on the reproductive behavior of the population confined only to Sagar District in the Sundarbans by Shukla (2007) focusing on the tremendous growth of population in that area. Aitch et al. (2005) had a research on a study on bank line migration of Bhagirathi-Hooghly River by GIS.

A limited work on the ground contamination in and its consequences, case studies from deltaic West Bengal, has been done by Bhattacharya (2008). Das (2008) carried out a detailed work on the evidence of some erosional features along the riverbank of Sundarbans. The research reveals the fact that this process of riverbank erosion and deposition leading to shifting of their courses all over the deltaic plains of Sundarbans might have relations to the neo-tectonism of the Bengal basin. Das Gupta (2008), Chief Engineer (Retd.), Irrigation and Waterways Department of the Government of West Bengal, did sound work on the flood protective embankment in West Bengal. Mukhopadhyay et al. (1997) did work on a simple model to study wave-surge interaction which helps in analyzing the cyclonic storm surges and wave-surge interaction in Sundarbans.

A little bit of lithological properties of the Sundarbans can be understood from the work of Mukherjee and Hazra (1997) on the changing paradigm of petroleum exploration in Bengal basin. Bhandari et al. (1997) worked on erosion control with geosynthetics. Hazra with Banerjee (2000) did an outline of an integrated exploration strategy for oil and gas and allied nonconventional energy resources of West Bengal where the stratigraphy and the geology of the deltaic Bengal has been discussed. Sen et al. (2002) carried out a work on evaluation of water quality parameters leading to pollution hazard from east of Sagar Island. Hazra et al. (2003) dealt with the vulnerability assessment of Sundarbans delta. Bhandari (2005) did significant work on probable causes of failure of existing embankments at Sundarbans and its remedial measures. Dasgupta and Hazra (2005) worked on the fishing scenario of Sundarbans. Mukherjee et al. (2006) had dealt with the physico-chemical properties of water and fish availability at the Muriganga estuary (adjoining the Bakkhali region of western Indian Sundarbans). Hazra and Mukherjee (2007) carried out another work on the coastal disaster management and fisher folk in Sundarbans wetlands. Mitra et al. (2006) did a case study on mangrove conservation efforts and the local economy.

Mitra et al. (2009) had observed changes in water mass properties in the Indian Sundarbans (northwestern Bay of Bengal) during 1980–2007. Lots of papers were presented in the international seminar held at Hanoi, Vietnam, 2011. In that particular seminar, Dhar and Garg (2011) discussed on the implications of climatic changes for the rural population residing in the vicinity of Piyali River, Sundarbans, India; Bardhan and Banerjee (2011) on mitigating disaster with salt tolerant paddy IR72046 in "Aila" affected coastal lands of West Bengal, India; and Sarkar et al. (2011) on the extent of tolerance of multiple abiotic stresses in rice landraces adapted to coastal zones of the Ganges and Mahanadi deltas, India, and prospects for use in breeding.

Other than the scientific journals, there are a huge number of valuable articles and research notes in different locally published magazines and books, and the contribution of them in the research progress cannot be ignored. K. Sarkar (2007), Coordinator of Tagore Society, Rangabelia Project, Gosaba, worked for one decade on the history of Sundarbans. Padma Shri Tushar Kanjilal, founder of Tagore Society for Rural Development at Rangabelia, Gosaba carried out an extensive work incorporating the local people of Gosaba for developing the livelihoods with the tides in Sundarbans since 1969 and introduced different cooperative programmes to develop the socioeconomic conditions in the remote islands of Indian Sundarbans. His views in different journals like- Compass, Shudhu Sundarban 'Nimnogangeyo Sundarban Sanskriti Patro are still very much relevant to young researchers.

On account of the previous literature, it can be stated that all the referred works mainly are restricted to the Sagar Island and coast-oriented geomorphological works. Though lots of works have been carried on the socioeconomic- and population-related issues of Sundarbans, least attention has been given to the tidal dynamisms acting upon entire Sundarbans. This longtime research gap has been found in this context. The fluvio-geomorphological attributes of the major rivers from west to east have merely been carried out in relation to changing tidal nature since the last few decades. To minimize this research gap, this chapter emphasized on the tidal dynamic and the changing fluvio-geomorphological environment of the interfluves in Indian Sundarbans through geospatial technology-based approach.

9.2 Objectives

The objectives of this work are as follows

1. To search for major tidal rivers and their hydro-morphological character in compliance with the tidal mechanism
2. To correlate tidal bore range of last decade to the present topographic profiles
3. To find out reasons behind intangible tidal dynamism acting upon the fluvio-geomorphological environment in Indian Sundarbans
4. To analyze and interpret the major findings of tidal dynamism in Indian Sundarbans

9.3 Methodology

For the convenience of discussion, previous publications and historical records along with recent articles or chapters in books, journals, magazines, and newspapers related with Sundarbans have been consulted before framing the entire work. Topographical maps of the Survey of India bearing the numbers like 79 B/10, 79 B/11, 79 B/12, 79 B/15, 79 B/16, 79 C/9, and 79 C/13 and other available topographical maps of RF 1: 50000 and topographical maps of Khula and Putney Island of 1924 with RF 1: 250000 have been consulted in the pre-field stage to have detailed knowledge on spatiotemporal changes in hydro-geomorphological environment of the study area. For the measurement and analysis of fluvio-geomorphological attributes, IRS P6 LISS-III satellite images and DEM have been applied. NATMO maps, census maps, and maps published in the Statistical Handbooks of the Districts of North and South 24-Parganas, West Bengal, India, have been taken into consideration. Mathematical and statistical techniques have been applied to frame rules to conduct both fieldwork and post-fieldwork analysis.

Primary data from major river sites of the study area, block and panchayat offices falling under the territory of Indian Sundarbans, have been collected during the field study since 2007. Detailed field surveys have been conducted during the pre-monsoon, monsoon, and post-monsoon seasons to collect the ground-level data. Careful fieldwork has been done incorporating the measurement of the height and depth of the riverbeds; embankments; locational attributes of the settlements; spatiotemporal changes in the courses of tidal rivers, channels, khals, creeks, and areal extension such as shrinkage of forests and wetlands and increase or decrease in the backswamps, marshy, and swampy areas; and identifying wastelands, charas, and bheris. Profiles have been prepared with the help of both filed survey instruments and computer software when required. Lots of face-to-face and telephonic interviews have been carried on with both structured and unstructured questionnaires, and sometimes opinions and perceptions of the native people has been taken into consideration for getting in-depth knowledge regarding the changing nature of the islands, rivers, and embankments. Soil samples were collected and tested with soil kits considering the pH, salinity, turbidity, etc. to get the seasonal and yearly changes in the soil characteristics to incorporate it during LULC analysis of the study. Photographs and video recordings were also taken from time to time as required for justification, correlation, analysis, and interpretation.

Finally, with the help of the data and information collected during the pre-fieldwork and fieldwork stages, mathematical, statistical, and cartographic techniques were applied for preparing graphs, charts, diagrams, models, and LULC, and geomorphological maps and cross-sections have been done to get the overall analysis of Indian Sundarbans. Algorithms and statistical analysis were done through computer GIS software like Qgis 3.14, Surfer 19, and Microsoft Excel 2016, respectively. Utmost care has been taken to avoid RMS errors during the preparation of the maps based on topographical maps and satellite imageries of certain intervals.

9.4 Background of the Study Area

During the preindependence, many British government officials carried out lots of brief or detailed works on the Sundarbans delta of undivided Bengal. Pergitter's (1765–1870) descriptions of the Sundarbans delta may be considered as the first written document of the Sundarbans. He may be regarded as the founder of the geographical researches on the Sundarbans delta. Fergusson (1863) did a memorable work on the geology of the Sundarbans. The other resourceful referred cited works that are assets in this regard are maps prepared by J. Rennell (1780), maps of W.E. Morison (1811–1814), and modified maps prepared by his brother Hugh Morison (1818) and survey records of Princep (1822–1823) and Dampier and Hodges (1828). The other important cited works are the descriptive study on the processes of present alluvium formation by Oldham (1870), the geological findings recorded in the "Calcutta Review" (1859), and W.W. Hunter's statistical account of 24-Parganas (1875). Besides, the resourceful works are the L.S.S. O'Malley's Gazetteer of the District of 24-Parganas (1914); Structure of the Gangetic Plains by Oldham (1917); A.C. Lahiri's Final Report on the Survey and Settlement Operations in the District of 24-Parganas, 1924–33 (1936); F.D. Ascoli's Revenue History of the Sundarbans, 1870–1920 (1921); *Journal of the Asiatic Society of Bengal*, the "Calcutta Review," Ptolemy's map of the second century A.D. and Heuen Tsiang (Yuan Chwang) descriptions in seventh century A.D. in the "Ancient Geography of India" (1871) (Cunningham, 1871); the Bengal Records and the periodical reports of the various departments of the Government of West Bengal; etc. have also been taken with keen concern. Chakraborti (1908) has an excellent note on the historical evolution of the geography of Sundarbans. One of the most unique classical works on the geography of the Ganges delta was carried out by Bagchi (1944). His work is still acknowledged without any pause by eminent researchers. Chatterjee (1946) dealt with another work of land utilization of the districts of 24-Parganas. A detailed study on the geology of the Bengal basin was carried out by Morgan, J.P. and McIntire (1959). Kar (1962) dealt with the pattern of urban growth in lower West

Bengal. The study on the geological pattern of the delta region was done by Sengupta (1972) and Neogi (1975). Bagchi and Mukherjee (1978) worked intensively on the historical development of the settlement and on the land use pattern of Sundarbans region.

9.4.1 *Locational Significance of Sundarbans in India*

Sundarbans in West Bengal is the deposited output of the Ganges-Padma and Brahmaputra-Meghna and their allied tributaries and distributaries in the Bengal basin (India and Bangladesh). The southward flowing distributaries of those major rivers entering the Bay of Bengal for thousands of years have left their load during their sluggish flow along the Bengal coast and helped in the formation of the vast floodplain which has been geomorphologically categorized into the (i) *moribund delta* with the oldest or older alluvium deposits where land building is practically ceased and drainage channels became functionally inoperative (Nadia and Murshidabad); (ii) *mature delta* with relatively younger than the moribund alluvium but older than the recent alluvial deposits where rivers still have well-defined channels and provide some impetus for irrigation and induce drainage disposal under normal conditions, for example, mostly the northern part of the 24-Parganas (S), Calcutta, roughly to the north of a line joining Kakdwip with Canning; and the (iii) *active delta* where dynamic functions of the hydrological system are more manifested (Bagchi, 1944). The channels and creeks are wide, deep, and well defined; land building is still in its current phase in which upland flow changes regime with marine incursions (Fig. 9.1).

The Ganges and Brahmaputra rivers currently contribute about a billion tons (1012 kg) of sediment annually to the Bengal basin, and over the Holocene these sediments have built a deltaic feature with a subaerial surface area of 110,000 km^2. The modern delta continues offshore as a prograding clinoform, extending 125 km across the continental shelf from the river mouths and about 250 km along shelf from the eastern coast of Bangladesh west to the "Swatch-of-No-Ground" submarine canyon, roughly at the border of India and Bangladesh. Thus, the total area of the combined subaerial and submarine delta is about 140,000 km^2, an area roughly equivalent to the size of Britain (Kuehl et al., 2005).

Climate

Indian Sundarbans in North and South 24-Parganas, experiences the tropical maritime type of climate due to its locational attributes. The Bay of Bengal dominates the climatic characteristics over the entire Sundarbans.

Three prominent seasons, i.e., (i) hot summer (March–May), (ii) humid summer (June–September), and (iii) cool dry winter (October–February), are mainly experienced in this deltaic part of West Bengal. Another significant season in this deltaic

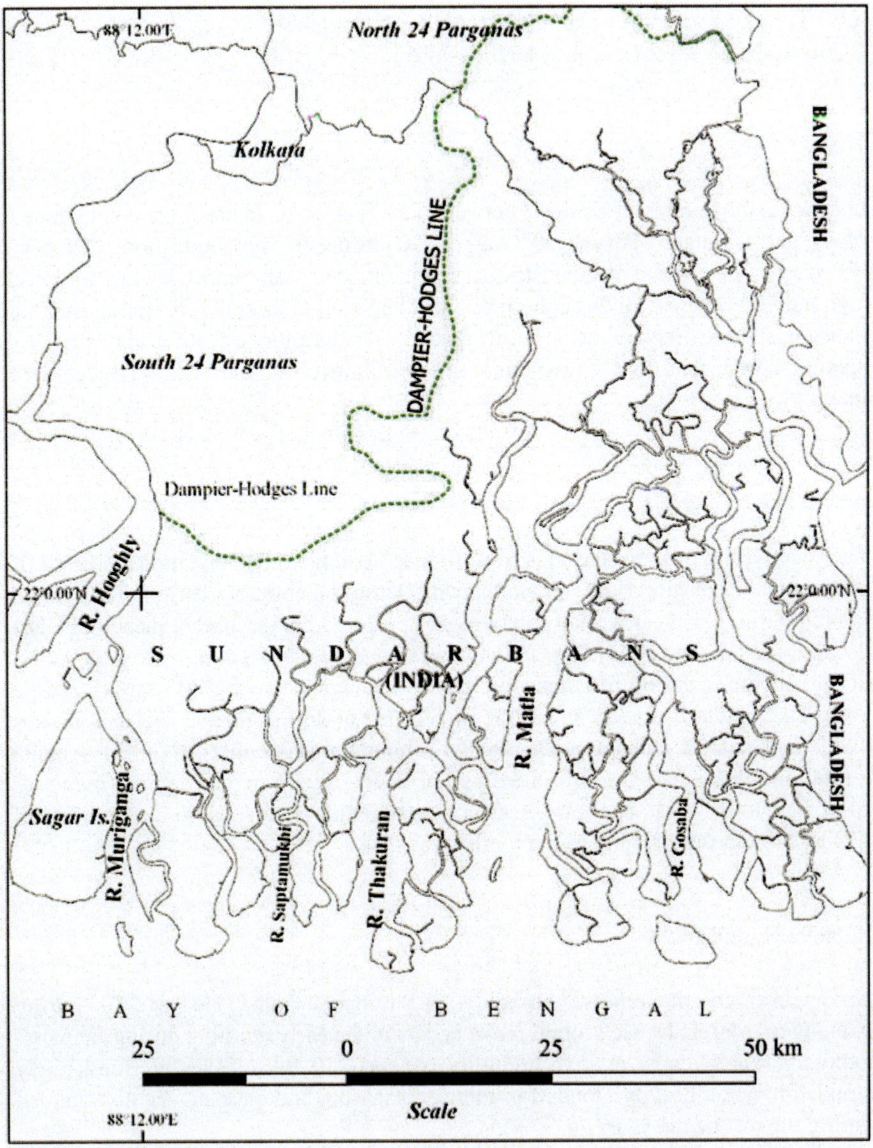

Fig. 9.1 Location of Sundarbans within Dampier-Hodge's Line in West Bengal, India

part of the Sundarbans is the "storm season," i.e., the periods April to May and October to December during which most of the cyclonic storms occur in the Bay of Bengal and Arabian Sea. Cyclonic storms mainly occur over the Bay of Bengal from April to May and from October to mid-December. May is the hottest month and January is the coldest. During the summer, the afternoon humidity is comparatively less. The cold season is about the middle of November to the end of February followed by the summer from March to May.

Temperature

In general, the temperature ranges between 28°C and 36°C in pre-monsoon (March–May), and it ranges between 10°C and 24°C during the post monsoon (October–February). The climate of the district is characterized by an oppressive hot summer, high humidity almost all throughout the year and well-distributed rainfall during the monsoons (June–September). The average temperature in the district varies from a maximum of around 38°C (May–June) to a minimum of around 13.5°C (December–January).

Rainfall

The southwest monsoon season is from June to September. October and the first half of November constitute the post-monsoon (October–February) season. The annual rainfall during this year was also very high, i.e., 26.55 cm (second highest in the last decade after the year 1995). The annual rainfall averages 1800 mm, more than 80.0% of which comes during the monsoon. On an average, there are 80 rainy days in a year. The heaviest rainfall in 24 hours recorded at any station in the district was 424.9 mm. (16.73 inches) at Diamond Harbour on September 20, 1900. Annual rainfall in the Sundarbans is in the range of 1640–2000 mm, and rainfall increases from west to the east. The rainy season prevails with an average annual rainfall of 150 to 240 cm during June to September.

Relative Humidity

On an average, the relative humidity (%) remains from 71% to 74% during November–March. Dense ground moist occurs in the early morning during January–February. The average relative humidity is nearly 80–84%. Skies are moderately clouded in May, heavily clouded in monsoon season, and clear or slightly clouded during the rest of the year.

Wind Direction and Wind Velocity

The mean wind velocities during the pre-monsoon, monsoon, and post-monsoon are 11.5 Kmph, 11.1 Kmph, and 6.65 Kmph, respectively. The SSW-SW wind direction during the monsoon (June–September) and pre–monsoon (late February–September) converts to NNE-NE during the post-monsoon periods, i.e., during October to February. Storms are common during spring (February–early March) and autumn (late September–early October) seasons. During spring and autumn, the wind velocity generally exceeds 100 kmph. Winds are generally stronger in Sundarbans and its surroundings. Nor'westers from March to May and the bay cyclones during the monsoon ravage the land every year.

9.5 Relief

Sundarbans delta is a gently sloping floodplain having an average height of 2–4 m above mean sea level near the Bay of Bengal (Fig. 9.2). Mukherjee (1978) prepared a relief map of the entire Rarh Bengal having a contour interval of 3 m. The cross profile prepared along the southernmost limit of the Indian Sundarbans depict the facts that Matla River has the widest mouth through which the high tide water during bore gush northward become violent during severe cyclonic storms, whereas the rest of the rivers' mouth have been decayed too much and are less deep than that of Matla River.

Taking cross-section lines from west to east latitudes starting from 21°00′00″N latitude toward north at an interval of 06′N based on WGS84 UTM zone 45N coordinate system, several profiles have been prepared. Analyzing the DEM of the Indian Sundarbans projected on WGS84 UTM 45N from west to east Profile 1 along A-B section line (extending from 88°00′ 00″ E to about 88°50′ 00″ E longitudes) along 21°00′ 00″ E latitude, it has been found that the HTL of ≤4.0 m range overtops the southwestern-southern onshore fringe of the Sundarbans, i.e., Lowe longshore sand, north Sagar sand, and south Sagar sand located at the south of the Sagar Island (Figs. 9.3 and 9.4). To the next, Profile 2 along C-D section line reveals the fact that 2/3 parts of the island of Namkhana along the Baratala or Muri Ganga River is less prone to be inundated if the tidal range remains below 3.5 m. But the range >4.0 m may inundate the eastern bank adjacent areas (particularly the Mousuni Island) of Namkhana along the Muri Ganga and can create tidal flood, like the situation particularly during cyclonic conditions. As the interlinking khals and channels

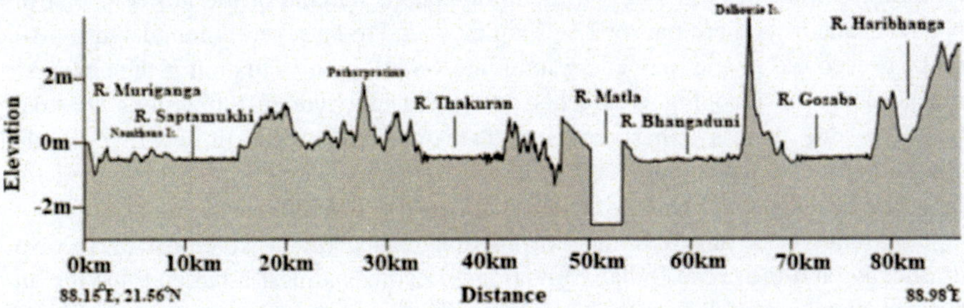

Fig. 9.2 Cross profile along the mouths of major tidal rivers of Indian Sundarbans along the Bay of Bengal (BOB)

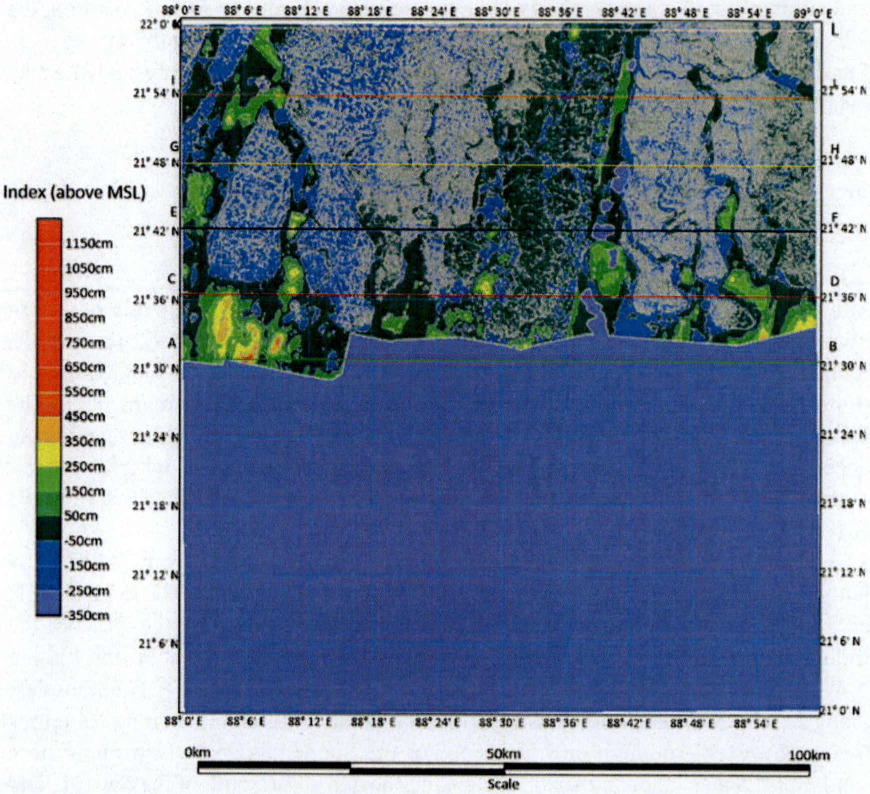

Fig. 9.3 West-East section lines (A–L) along the Indian Sundarbans. (Source: Based on C1_DEM_16B_2005-2014_v3_R-1_88E21N_f45q)

have been decaying at a rapid rate, this may lead to longtime saline water stagnation in lowland areas of Namkhana.

Profile 3 indicates the fact that the homogenous flatland of the northern part of Namkhana and southern part of Kakdwip may tend to be severely inundated if tidal range exceed 4.0 m and may cause longtime waterlogging situation if the embankments along the Hooghly is beached by increasing hydraulic pressures by tidal waves or due to less cohesiveness of the embankment during severe cyclonic conditions like Aila or Amphan. The situation may worsen if accompanied by lingering heavy shower with wind speed above 100 Kmph.

In Profiles 4, 5, and 6 further toward inland, the Matla River narrows down drastically, and the entire Indian Sundarbans becomes almost a flatland, having an average elevation of less than 2 m. Thus, the entire southeastern part of Indian Sundarbans is lower than the rest of this deltaic region in India.

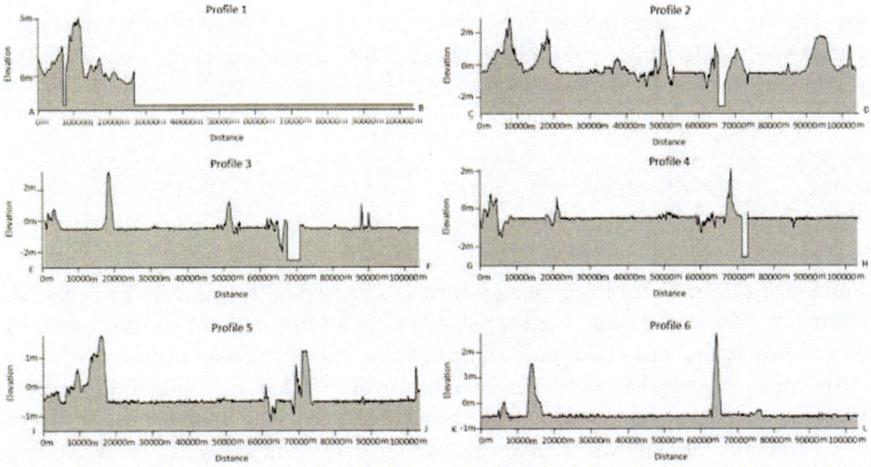

Fig. 9.4 Profiles (A–L) along the Indian Sundarbans showing the apparent relief features

Courtesy: U.S. Mandal, July 20, 2020

Fig. 9.5 Tidal water intrusion during bore tide at Chargheri, 24 Parganas (S) along Datta River, July 20, 2020. (Courtesy: US Mandal, July 20, 2020)

Thus, it can easily be presumed that the rise of a few centimeters in sea level in the future matters a lot to the entire fluvio-geomorphological environment in the present Sundarbans (Fig. 9.5). A little change in the drainage system of the active deltaic parts will affect the maturing or mature deltaic parts to north. The rise in river water level due to rapid decaying of the riverbeds and increasing trend of tidal bore range along with accelerating magnitude of the cyclonic storms will bring more complex fluvio-geomorphological environment in the entire Sundarbans.

9.6 Natural Calamities

Sundarbans from its very origin is associated with natural calamities like cyclones, storms, floods, earthquakes, bank erosion, etc. From the previous records and field experience, it has been observed that the number of the cyclones occurring, whether severe or less severe, have left their footprints in the mind of the residents living with the delta. As the studied interfluve region lies at the toe of the Bengal basin, so every 2–5 years interval, this area faces severe or less severe natural calamities. According to the Indian Meteorological Department, depression (the wind speed varies from 20–30 knots or 37.04–55.56 Kmph) is the common climatic phenomena prevailing almost throughout the year. Cyclonic storms of more than 34 knots (62.97 Kmph) also occur from the month of April to November. Sometimes the wind speed reaches to about 60 knots (111.12 Kmph). It occurs maximum in the month of May, June, July, October, and November. Records of the Indian Meteorological Department say that a maximum intense cyclonic storm prevailed over this region during October 26–27, 2008 and during May 22–25, 2009. Riverside villages along major tidal rivers experienced havoc bank erosion. "Aila," a severe cyclone, prevailed over this interfluve during the third and fourth week of May in 2009, and super cyclone Amphan during May 20–21, 2020, caused havoc bank erosion, and tidal floods ravaged the Sundarbans. The rainfall data of IMD supports the fact that the intensity of rainfall is also high during those storm-affected months. "According to data collected by the School of Oceanography, Jadavpur University, the surge reached the Sundarbans shores the night before Aila struck. Water level had climbed 0.5 m to 0.7 m above the normal mark 12 hours before the cyclone devastated the region." The low pressure formed by a cyclone pulls the seawater up. The swelling of the water leads to a surge toward the shore that precedes the storm system to the coast. Their combined effect could be tremendous. In Aila's case, this is exactly what had happened. "It was the flooding, rather than the cyclonic wind, that caused more damage," said Gautam Sen, Joint Director, School of Oceanography. Um Pun has also left its devastating footprints, and geomorphologists also experienced the power tidal magnitude (Fig. 9.5).

The heavy rain from June to September in association with minor to major cyclonic storm surges triggers the tidal bore to breach the earthen embankment. It ultimately increases the salinity of the soil and changes the pH value and other attributes of the soil in this deltaic region of Sundarbans.

9.7 Tidal Bores

Tides occur twice a day along this deltaic tract. Generally spring tides become more violent in the narrowing upper reaches of the tidal channels like Matla, Bidyadhari, etc. in Sundarbans. The tides in the Sundarbans are occasionally so strong that it gives rise to the phenomenon known as a bore (locally known as "Ban Daka" meaning "calling the flood"). The name is given to the head wave formed due to the unusual high tides checked by the narrowing of the river channel. "The obstructed influx," writes Hunter in *The Indian Empire*, "no longer able to spread itself out, rises into a wall of waters which rushes onward at a rate nearly double that of a storage coach. Rennell stated that the Hooghly bore ran from Hooghly Point to Hooghly town, a distance of about 70 miles, in four hours." The average range of the tidal bore fluctuates within 4–5m here during the last two decades.

A writer in the *Calcutta Review* of 1859 has provided an interesting account of this "Ban," a portion of which is as follows: "Numerous boats from the up-country provinces are lost every year from the crews being ignorant either of the existence of the bore, or from their not knowing the correct position to take up so as to meet it. Ships at anchor in Calcutta, though not exposed to the breaking portion of the wave, frequently part their cables when struck with the wave. Standing on the shore during the rapid rushing passage of the bore, it is curious sight to see the lower portion of the river, or that nearest the sea, six or eight feet higher than the upper portion of the river, the tide rising that number of feet in an instant." Later however such highly destructive waves were brought under control by the British Government by deepening the river channel to certain extent the deep water of which helped robbing the bore of its force, and still later the Tidal Semaphore stations set up by the Government of West Bengal in different places of the district have organized the transmission of information regarding numerous occurrences including that of "Ban" through their sounding serangs. The tidal bore locally termed as "Sanrasari Ban" in Bengal estuarine delta is an extreme case of shallow water effects (O'Malley, 1914). It occurs usually at low-water springs when a tide with a large range is funneled into a river or estuary with a steeply shelving bottom. Sometimes the front of the rising tide moves forward along the river as a bore, a sudden churning and tumbling wall of water advancing up a river. The creation of a bore requires a large rise of tide at the mouth of a river, some sandbars, or other restrictions at the entrance to impede the initial advance of the tide, and a shallow and gentle sloping riverbed.

The data of tides of the middle of eighteenth century reveals that the maximum range of tidal bore was 7.111 m during the Monsoons. The rising of the riverbeds of Saptamukhi, Thakuran, Matla, Bidyadhari, and Gosaba Rivers helps the saline tidal water overtop the banks in some places higher than the adjacent settlements along the banks here (Fig. 9.6). The wave dash actions become more devastating and accompanied by the monsoon rain inundate the area along the tidal rivers like riverside villages in Basanti and Gosaba Block along Matla, Hogol Nadi, Karatal Ganga, and Bidyadhari. However, the possibility of havoc bank erosion or breaching

Fig. 9.6 Settlement along Matla River during high tide (21°18′20.5″N, 88°40′57.8″E: December 29, 2018)

of embankment in dry seasons is comparatively less as the average range of tidal bores is lesser (below 5 m) during the post monsoons (November to March). The maximum range of the tidal bore is concentrated within the limit of 4–5 m in and around Hooghly River, and there is a negative trend of the expected tidal bore during the year 2010.

From this hydrological analysis, it is found that the maximum range of the tidal bore in Matla and Bidyadhari Rivers also remained within 4–5 m. In the year 2007, there was a steep fall of the maximum range in tidal bore from March to June 2007 and a rise in July 2009. The range of tidal bores was comparatively higher during the post and pre-monsoon and lower during the monsoon period in 2007–2008. There was 21.69 cm of annual rainfall during the year 2007, but this interfluve was least affected by the cyclone that occurred during the month of July in 2007 which mainly entered the western parts of Bangladesh. Whereas the range of tidal bore was maximum during (June-August), the monsoon period in 2009, and was affected by the prevailing low pressure just near the coast of West Bengal, the tidal range remained higher during (July–October), the monsoon period in 2010.

Thus, the higher range of tidal bores during the monsoon periods and sometimes during the pre-monsoon and post-monsoon accompanied by the cyclonic storm originated by the intense low pressure over the Bay of Bengal accelerates the damages caused either by storms or by bank tidal floods. The combined activities of the cyclones and tidal bores are important hydro-climatic phenomena in this interfluve of Sundarbans.

Tides and ebbs alternatively sculpt the islands of Sundarbans since its very origin. The discharge, velocity, and turbulence of the totalized rivers and khals always alter the configuration from small- to large-scale changes along with the changing sea level of the Bay of Bengal. Hence, “the rates of sea level changes for the stations are found to be –3.82, +0.89, +2.43, and +4.85 mm year^{-1}, respectively, connoting a significant positive relation between landward distances of the stations and the rates of sea level rise (Nandy et al.) and no relation exhibit between the erosion and accretion of its tidal island along Hooghly River.”

As per the records from different official maps, reports published from time to time by the British Government officials during British rule in India, the tidal mechanism was quite different in Sundarbans, and the present islands were also quite differently characterized by their fluvio-geomorphological, pedological, and hydro-morphological characters. Hence, lots of research works have revealed the facts that climate change and the changing sea level are also are becoming a significant factor in controlling the nature of the major tidal channels like Baratala or Muri Ganga, Saptamukhi, Jamira, Matla, Bidyadhari, Gosaba, and Thakuran rivers along with their intricate network of intertidal channels, creeks, and khals in Sundarbans.

According to L.D. Wright (1978), 1 cm of sea level change takes place with the change in 1 mb change in atmospheric pressure. In compliance with that, it has also been observed that the cyclonic wind circulation and the changing direction with varying velocity of occasionally forming cyclonic storms along the Bay of Bengal also trigger and accelerate the movement of tidal circulation within the islands through the interlinked creeks in Sundarbans. The nature of tidal range varies accordingly with the seasonal changes in atmospheric pressure also. The seasonal variations in the range of tidal bore can easily be noticed.

From the data obtained from Kolkata Port Trust, tidal bore range analysis reveals the fact that the mean monthly range of the tidal bore is >5 m during April to October since 2007 to 2019 (except in March 2015), and it never exceeded 5 m in the years 2013, 2018, and 2019 (Figs. 9.7a, 9.7b, 9.7c, 9.7d, 9.7e and 9.7f). During the last 13 years, the maximum and minimum ranges were noticed in August 2015, i.e., 5.38 m, and in July 2011, i.e., 4.07 m. Generally, the mean range of tidal bore remains low from the month of November to January almost every year. The mean tidal bore range steadily remained above 4.5 m from the years 2012 to 2017. But it lowered below 4.5 m later (except during monsoon period). The highest annual mean tidal bore range was 4.89 m in 2015 and lowest of 3.83 m during 2007. The total increase in annual mean range of tidal bore from 2007 to 2019 is +1.03. From this point of view, it may be concluded that the range of tidal bore has increased from 2007 to 2015 but decreased a little bit by 0.31 m from 2016 to 2019.

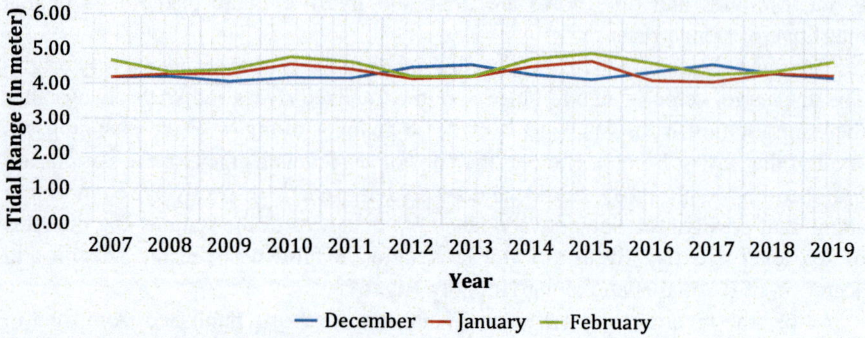

Fig. 9.7a Tidal bore during winter in Sundarbans

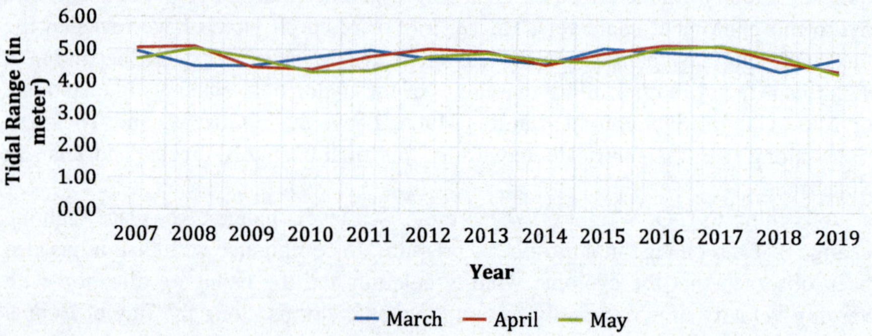

Fig. 9.7b Tidal bore during summer in Sundarbans

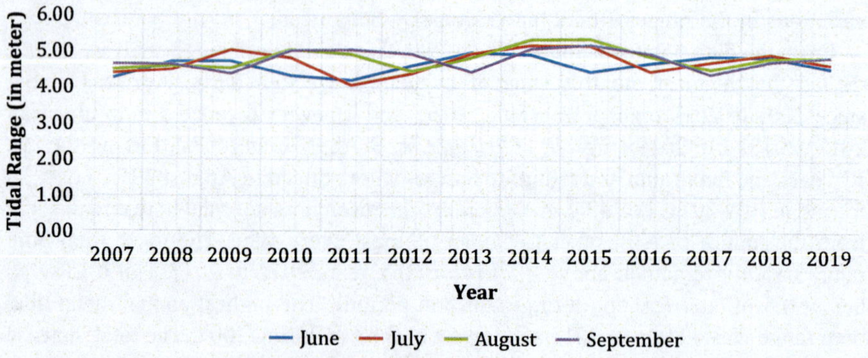

Fig. 9.7c Tidal bore during monsoon in Sundarbans

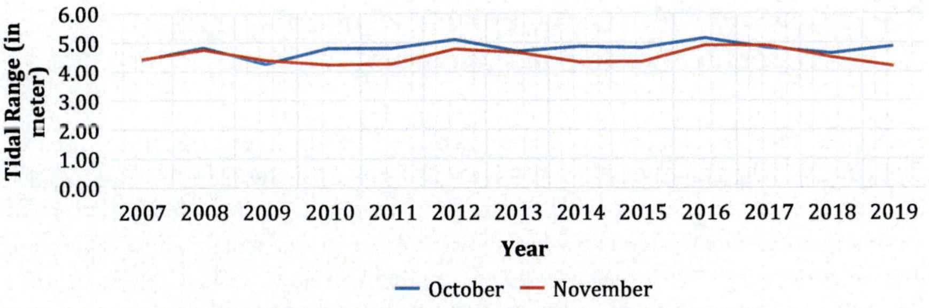

Fig. 9.7d Tidal bore during post-monsoon period in Sundarbans

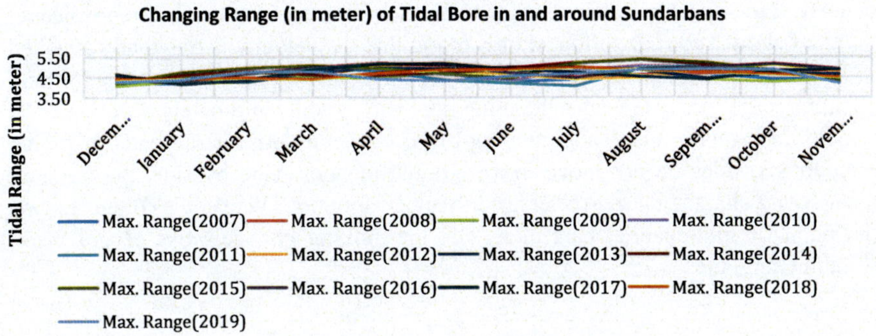

Fig. 9.7e Changing Tidal Bore Range in Sundarbans from 2007 to 2019

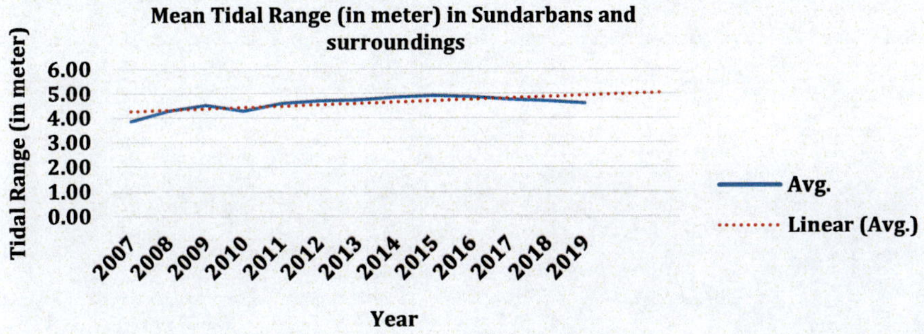

Fig. 9.7f Mean Tidal Bore Range in Sundarbans from 2007 to 2019

9.8 Impact of Tides on the Fluvio-Geomorphological Environment in Indian Sundarbans

Generally, the tidal range remains between 2m and 4m in the meso-tidal area of Sundarbans, whereas it is >4 m in micro-tidal zones of Sundarbans (Das, 2006). According to the River Research Institute of Government of West Bengal, the mean incoming tidal velocity and ebb velocity in Thakuran River remain about 48 cm/sec and 58 cm/sec. Further, it has also been recorded that almost the same pattern has been noticed in major rivers of Indian Sundarbans. But side by side, the upstream discharge by tides is also controlled and modified to some extent by the meandering and shifting courses of the tidal channels in Indian Sundarbans (Fig. 9.8). In this context, if the linear movement of the tidal track (limited within an average channel width of about 100m) is followed, it is found that the velocity is frequently checked in some tidal rivers particularly in their upper courses where heavy siltation and choking of the sources reverse incoming tidal waves which ultimately are diverted in a multidirectional route. Thus, moving toward the northern mature parts of Indian Sundarbans, tides do not move in a rhythmic manner. Side by side, the frequent meandering character also accelerates in vortex formations here eddies often reshape the point or mid-channel bars along the interlinking channels like Hogol Nadi, Karatal Gang, etc.

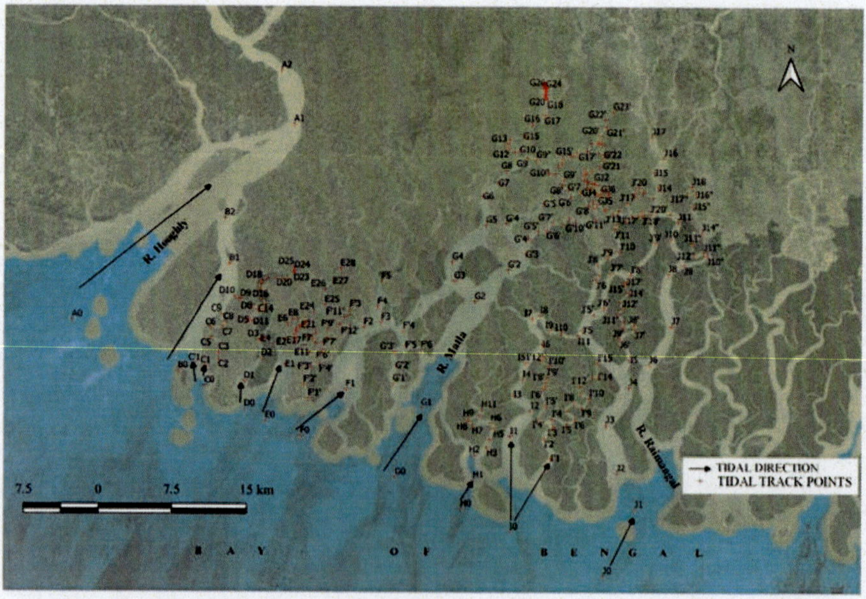

Fig. 9.8 Tidal track and the number of meandering points along the major tidal river in Indian Sundarbans (Source: Based on Google satellite)

It noteworthy to mention that the ratio between the velocity of the surface of the tidal currents and the wind force depends upon the stress of the wind and value of eddy viscosity (Sverdrup, 1942). The longest straight line free advancing of tidal waves can be noticed along Hooghly River having a fetch of about 25.92 km, and the shortest fetch can be noticed along Saptamukhi River as its mouth is only about 1.49 km wide with high sinuous course near its mouth. Next to Hooghly River, Matla River has the second widest mouth, i.e., about 9.91 km in Sundarbans delta (Gour, 2011) with a free south-north tidal fetch of about 19.72 km (Fig. 9.8). Gosaba River has the third highest onshore tidal fetch with a length of about 9.38 km, and its mouth is about 5.42 km wide. Haribhanga River also has free tidal fetch of about 9.26 km with mouth of about 5.98 km wide.

Hence, the longer the onshore tidal fetch and the wider the mouth, more impact of tidal bore will be there, and the effect of it will be experienced by the interior or mature part rather than estuarine part of Sundarbans delta. Besides this, the angle of deflection between the resulting tidal current and the prevailing wind could be 45° on the right (Nansen & Ekman, 1904). Although this many oceanographers like C.S. Durst in 1924 pointed out a 40° deflection (Fig. 9.8) from the present tidal flow direction and previous cyclonic records reveals the fact that blocks like Namkhana along Muri Ganga and Saptamukhi rivers, Patharpratima block along Thakuran River, and Basanti block along Matla River have always suffered severely, whenever any severe or super cyclonic storms accompanied by tidal bore with a wind velocity of more than 100 Kmph occurred.

9.9 Conclusion

Tides are common natural phenomenon and are the lifeline of Sundarbans since its very origin. Tidal water entering from the Bay of Bengal is not only the source of marine nutrition for survival of the aquatic flora and fauna in estuarine Sundarbans but is the natural distributer of the mangroves. The regular and uninterrupted scouring washing out of the riverbeds by the ebbs and tides not only maintains the hydro-morphological equilibrium state in Sundarbans but maintains the ecological property and enriches the biological diversity as well. It is the rapid reclaiming and anthropogenic abandoning of the upper courses of the rivers by embankments which are solely responsible for the increasing complexity in the natural distribution of tidal water. The more the deeper will be the tidal channels and interlinking khals, the lower will be the risk of tidal flooding during any natural calamity. As the channel morphology of the major tidal rivers of Sundarbans determines the nature of tidal activities in Sundarbans, the restriction of these tidal channels by anthropogenic activities may invite more devastating future. Existing islands in Sundarbans will also be modified, keeping abreast of the time with natural shifting phenomena of the major tidal rivers. Tidal gauge stations along major rivers in Sundarbans like Saptamukhi, Thakuran, Matla, Bidyadhari, and Gosaba should be set up to acquire sufficient data particularly during the cyclonic situations, and a trend analysis of spatiotemporal changes in tidal water discharge, velocity, dimensions, etc. must be taken into consideration before embanking of the bank erosion-prone and reclaimed riverside dwelling areas. Meteorological and sea level rising-related chronological data can unveil so many facts about the tidal dynamism and help in analyzing the changing fluvio-geomorphological environment of the interfluves which may open up new dimensions in mitigating and managing the upcoming awaiting disaster in Indian Sundarbans.

References

ABSTRACT, http://nopr.niscair.res.in/handle/123456789/13260,0975-1033(Online), http://nopr.niscair.res.in/handle/123456789/13260

Aitch, P., Bhandari, G., & Das, S. C. (2005). A study on Bank line migration of Bhagirathi-Hooghly River by GIS. In Proceedings of international conference on crisis management in water and environment (vol. 1, pp. 64–67), ICCMWE.

Amjad, A. S., Kasawani, I., & Kamaruzaman, J. (2007). Degradation of Indus Delta mangroves in Pakistan. *International Journal of Geology, 1*(3), 27–34.

Aucan, J., & Ridd, P. V. (2000). Tidal asymmetry in creeks surrounded by saltflats and mangroves with small swamp slopes. *Wetlands Ecology and Management, 8*, 223–232.

Bach, L. B., Stern, M. A., Crandall, J. D., Aldous, A. R., Stevens, C., & Elseroad, A. C. (2005). Restoring large-scale functional lake-fringe and floodplain wetlands at the Williamson River Delta, Oregon, USA, 2005, Abs. World Conference on Ecological Restoration Zaragoza, Spain, Europe 12–18 September 2005, p. 30

Bagchi, K. (1944). *The Ganges Delta,* (print. & pub.) D. Gangulee, B.A. Supt., Cal. Univ., p. 153

Bagchi, K. G., & Mukherjee, K. N. (1978). *Diagnostic Survey of Deltaic West Bengal* (p. 45). Calcutta University, Kolkata.

Banerjee, A. (1998). Environment, population and human settlements of Sundarban Delta, Concept Publishing Co. (p. 424). New Delhi-59.

Bardhan, R., & Banerjee, S. (2011). Mitigating disaster with salt tolerant paddy IR72046 in 'Ayala' affected coastal lands of West Bengal, India. http://delta.iwmi.org

Bhandari, G. (2005). Probable causes of failure of existing Embankments at Sundarbans and its remedial measures, *Embankments of Sundarbans and Related Issues*, Wksp. cum Semi., (Org.) Sundarban Affairs Department, Science & Technology Deptt. and I. & W. Dept., Govt. of W. B., Manikanchan, Kolkata, 21 July 2005, Com. Vol., pp. 62–76

Bhandari, G., Sarkar, S. S., & Rao, G. V. (1997). Erosion control with geosynthetics, WKSP. Geosynthetics technology, Bangalore, Geohorizon-State of Art in Geosynthetics Technology (pp. 151–169). Oxford IBH.

Bhattacharya, R. (2008). Ground water contamination in and its consequences: Case studies from Deltaic West Bengal. *Geographical Review of India, 70*(1), 105–109.

Biswas, S. K., & Agarwal, A. (1992). Tectonic evolution of the Bengal Foreland Basin since the Early Pliocene and its implication on the development of Bengal Fan, GSI Spl. Pub., pp. 5–19.

Biswas, S. K., & Neogi, B. B. (1988). In B. B. Bhattacharya (Ed.), *Frontiers in exploration geophysics for hydrocarbon prospect in eastern India basins* (pp. 68–88). Oxford and IBH Publishing Pvt. Ltd.

Chakraborti, M. (1908). *Notes on the geography of old Bengal* (p. 26).

Chakrabarty, S., & Bhandari, G. (2001). Improvement of bearing capacity using compacted PFA and Geo-Jute reinforcement over soft clay layer. Proceeding Indian Geotechnical Conference, 2001 (IGC-2001), Indian Geotechnical Society, Indore Chapter, Indore, India, 14–16 December, 2001, Vol. 1, pp. 243–246.

Chakraborty, S. C. (2005). The Sundarbans—Terrain, legends, gods & myths. *Geographical. Review of India, 67*(1), 1–11.

Chatterjee, S.P. (1946). Land utilization in the district of 24 Parganas, Bengal, (pub.) Cal. Geog. Soc., p. 342–408.

Currey, J. R., & Munasinghe, T. (1989). Timing of intraplate deformation, northeastern Indian Ocean. *Earth and Planetary Sciences Letters, 94*, 71–77.

Dasgupta, R., & Hazra, S. (2005). Fishing scenario in Sundarbans—A case study in coastal West Bengal. *Indian Journal of Landscape Systems and Ecological Studies*, *28*(2).

Das, G. K. (2006). *Sundarbans- environment and ecosystem* (p. 254). Sarat Book Distributors, M. De, Levant Books.

Das, G. K. (2008). Evidences of Some Erosional Features Along the River Bank of Sundarban, India. *Geographical Review of India, 70*(2), 139–145.

de Farias, S., & Gomes, E. T. A. (2005). The Brazilian coastline environment restoration - a case study at the northeast coast of Brazil. In *World conference on ecological restoration* (pp. 12–18) September 2005:65.

Dhar, S., & Garg, V. K. (2011). Implications of Climatic Changes for the Rural Population Residing in the Vicinity of Piyali River, Sundarbans, India available at http://delta.iwmi.org

Fergusson, J. (1863, April 1). *Delta of the Ganges* (vol. 19, pp. 28–32 & 321–354), Qtly. Jour., Geological Society, Longman, Brown, Green and Longmans

Ghosh, T., Bhandari, G., & Hazra, S. (2001). Assessment of land use/land cover dynamics and shoreline changes of Sagar Island through remote sensing. In *22nd Asian Conference on Remote Sensing (Singapore)* (Vol. 2, pp. 848–852).

Ghosh, T., Bhandari, G., & Hazra, S. (2002). Coastal Erosion and flooding in Ghoramara and adjoining islands of Sundarban deltaic System. In *Analysis and practice in water resource engineering for disaster mitigation* (Vol. 1, pp. 37–41). New Age.

Ghosh, T., Bhandari, G., & Hazra, S. (2003). Application of a Bio engineering technique to protect Ghoramara Island (Bay of Bengal) from severe erosion, Jour. *Coastal Conservation Sweden*, Opulus Press, Uppsala, EUCC 9, pp. 171–178.

Gour, J. (2011). Effects of Land Reclamation programmes in Matla-Bidyadhari interfluve of Sundarbans, West Bengal, *Practising Geographer (Jour. of the IGF)*, Winter 2010, Vol. 14(2).

Govt. of West Bengal: *District Human Development Report, South 24 Parganas,* October 2009, p. 318

Govt. of West Bengal: Report on predicted Bore tides for the years 2007–2019: Kolkata Port Trust available at www.kolkataporttrust.gov.in/

Hait, A. K., Das, H. K., Ghosh, S., Ray, A. K., Saha, A. K., & Chanda, S. (1996). New dates of Pleisto-Holocene suburb samples from South Bengal. *India, Earth Science, 23*(1–2), 79–82.

Hazra, S., & Baksi, A. (2003). *Environmental refugees from vanishing islands: Environment & human security* (pp. 219–227). Lancer Books.

Hazra, S., & Mukherjee, M. (2007). Coastal Disaster Management and Fisher Folk in Sunderban Wetlands, (pub.) Dept. of Fisheries, Govt. of W.B., p. 100–114.

Hazra, S., Ghosh, T., Dasgupta, R., & Sen, G. (2001). Sea level and associated changes in Sundarbans. *Science and Culture, 68*, 309–321.

Hazra, S., Samanta, K., Dasgupta, R., & Sen, G. (2003). A preparatory assessment of Vulnerability of the ecologically sensitive Sundarban Island System, West Bengal, in the perspective of climate Change, Wksp. Proc., Vulnerability Assessment & Adaptation due to Climate Change on Indian Water Resource, Coastal Zone and Human Health, pp. 66–82

Hong, P. N., Natarajan, I., Hoang, T. S., Nguyen, H. T., & Mai, S. T. (1996). Community participation in conservation, sustainable use and rehabilitation of mangroves in Southeast Asia, Proc. ECOTONE-V, UNESCO, MAB, MERD, Ho Chi Minh City, Vietnam, 8–12 January 1996, pp. 68–74.

Hunter, W. W. (1875). Preface, *Statistical Account of Bengal*, London, Vol. 1, p. 404

Islam, S. N., & Gnauck, A. (2007). Mangrove wetland ecosystems in Ganges-Brahmaputra delta in Bangladesh. *Frontiers of Earth Science in China, 2*(4), 439–448.

Kar, N. R. (1962). Urban hierarchy and central functions around Calcutta in lower West Bengal, India, and their significance, p. 253.

Khalil, S. (1999). Economic valuation of the mangrove ecosystem along the Karachi coastal area. In *The economic value of the environment: Cases from South Asia*. IUCN, The World Conservation Union. Jan. 1999: 11.

Kjerfve, B., Drude de Lacerda, L., Rezende, C. E., & Coelho Ovalle, A. R. (1999). Hydrological and hydro geochemical variations in mangrove ecosystems. In A. Yáñez-Arancibia & A. L. Lara-Domínquez (Eds.), *Ecosistemas de Manglar en América Tropical. Instituto de Ecología A. C. México* (pp. 71–82). UICN/ORMA, Costa Rica, NOAA/NMFS Silver Spring MD U.S.A.

Kuehl, S. A., et al. (2005). The Ganges-Brahmaputra delta. In *River Deltas—Concepts, models, and examples* (Vol. 83, pp. 413–434). SEPM.

Linh, L. H. et al. (2011). Improving rice tolerance of submergence and salinity stress to cope with climate change in coastal areas of Vietnamese Deltas, Abs. Int. conference Deltas under climate change: The challenges of adaptation, Hanoi, Vietnam, 2–4 March 2011, p. 26.

Linh, L.-H., Ham, L.-H., Toan, N.-T., Quang, V.-D., Huyen, L.-T. N., Cuc, L.-M., To-Phuc Tuong, Ismail, A. M., & Wassmann, R. (2001). Improving rice tolerance of submergence and salinity stress to cope with climate change in coastal areas of Vietnamese deltas. In *Int. Conf. Deltas under climate change: The challenges of adaptation*. Pap. No. 26, 2–4 Mar. 2011, Hanoi, Vietnam.

Loon, A. F. Van (2005). Water flow and tidal influence in a mangrove-delta system, Can Gio, Vietnam. M.Sc. thesis, Department Of Environmental Sciences, Subdepartment Water Resources, Wageningen University, The Netherlands, p. 159.

Mitra, A., Gangopadhyay, A., Dube, A., Schmidt, A. C. K., & Banerjee, K. (2009). Observed changes in water mass properties in the Indian Sundarbans (northwestern Bay of Bengal) during 1980–2007. *Current Science, 97*(10), 1445–1446.

Mitra, R., Bhattacharya, R. N., Hazra, S., & Santra, S. C. (2006). Mangrove conservation efforts and the local economy, a case study. *Economic & Political Weekly, 5041*(33), 3613–3616.

Morgan, J. P., & McIntire. (1959). Quaternary Geology of Bengal Basin, East Pakistan and India, (Bull.). *Geological Society of America, 70*, 319–342.

Mukherjee, K. N. (1969). Nature and Problems of Neo-reclamation in the Sundarban. *Geographical Review of India, 31*(4), 1–9.

Mukherjee, K. N. (1976). Harmonious solutions of the basic problems of Sundarban Reclamation. *Geographical Review of India, 38*(3), 311–315.

Mukherjee, A., & Hazra, S. (1997). Changing paradigm of petroleum exploration in Bengal Basin. *Indian Journal of Geology, 69*(7), 41–64.

Mukherjee, D., Banerjee, A., & Sen, G. K. (2006). Physico-chemical properties of water and fish availability at the Muriganga Estuary (adjoining Bakkhali Region of Western Indian Sundarban). *Environment & Ecology, 24*(2), 385–388.

Mukhopadhyay, P. S., Mandal, G. K., Sen, G. K., & Sinha, D. K. (1997). A simple model to study wave-surge interaction. *Mausam, 48*(2), 323–328, Govt. of India.

Nansen, F., & Ekman, V. W. (1904). In F. Nansen (Ed.), *"On dead water" in Norwegian North Polar Expedition 1893–1896* (pp. 1–150). Longmans, Green and Co.

Neogi. (1975). Quaternary geology of the coastal plain in W.B. and Orissa. *Indian Journal of Earth Sciences, 2*(1), 51–61.

Oldham, D. T. (1870). *Process of formation of the present alluvial surface of the Gangetic delta.* Proc. Asiatic Society of Bengal.

Oldham, R. D. (1917). Structure of the Gangetic plains. *Memoirs of the Geological Society of India, 3*(1), 16–18.

O'Malley, L. S. S. (1914). *24 Parganas*. The Bengal Secretariat Book Depot.

Pargitter, F. E. (1765–1870). The revenue history of the Sundarbans (Delta of the Ganges). *The Journal of the Royal Asiatic Society of Great Britain and Ireland, 1&2*, 1–4, 309–524.

Parua, P. K. (1990). Riverbank erosion: A potential threat to land resources, WKSP. Paper, Wetland Development, Eastern Regional Center, Jadavpur University, Kolkata, 13–16 1990.

Paul, A. K. (1987). Coastal Dune systems of Sundarbans. *Geographical Review of India, 50*(1), 14–23.

Paul, A. K., Bandyopadhyay, M. K., & Chowdhury, A. (1986). Morpho-ecological variations in the Sundarban Mudflats. *Geographical Review of India, 49*(1), 1–14.

Ratna, N. (2005). Restoration and management of mangrove forests through people's participation: Experiences from India. In *World conference on ecological restoration* (pp. 12–18) Sept. 2005:238.

Ridd, P. V., Wolanski, E., & Mazda, Y. (1990). Longitudinal diffusion in mangrove-fringed tidal creeks estuarine. *Coastal and Shelf Science, 31*, 541–554.

Rivera-Monroy, V. H., de Mutsert, K., Twilley, R. R., Castaneda-Moya, E., Romigh, M. M., & Davis, S. E. (2007). Patterns of nutrient exchange in a riverine mangrove forest in the Shark River Estuary, Florida, U.S.A. *Hydrobiological, 17*(2), 169–178.

Sarkar, R. K. et al. (2011). Extent of tolerance of multiple abiotic stresses in rice- Landraces adapted to coastal zones of the Ganges and Mahanadi Deltas, India, and Prospects for Use in Breeding. http://delta.iwmi.org

Sarkar, K. (2007, November 11). *Itihas Kotha Koy* (1st ed., p. 93). Bhagabati Press.

School of Oceanographic Studies: Jadavpur University: Report on Aila by School of Oceanography: available at www.savesundarban.org

Sen, G. K., Chakrabarti, D., Sanyal, P., Saha, R. & Mukherjee, A. D. (2002, December 5–6). Evaluation of water quality parameters leading to pollution hazard from east of Sagar Island, Proc. International Conference on Water Related Disaster (ICWRD 2002), Analysis and Practice in Water Resources Engineering for Disaster Mitigation, S.K. Banerjee, A. Roy & S. C. Das (Eds.), (pp. 209–215) New Age International Publisher, New Delhi.

Sengupta, S. (1972). Geological framework of the Bhagirathi-Hooghly Basin, Proc. Vol. Bhagirathi-Hooghly Basin, University of Calcutta Publication, pp. 1–8.

Shukla, P. (2007). Population spectrum in the light of urbanisation in Sagar District during 1991–2001. *Geographical Review of India, 69*(1), 33–38.

Sidik, A. S. (2008). The changes of mangrove ecosystem in Mahakam Delta, Indonesia: A complex social-environmental pattern of linkages in resources utilization, ABS. The South China Sea Conference, Thailand, p. 22.

Sil, A. K. (1990). Adaptation of the Migrant Oraon in their present habitat in Sundarbans. *Geographical Review of India, 52*(1), 89–90.

Sverdrup, H. U. (1942). *Oceanography for Meteorologists* (p. 125). Prentice-Hall, Inc.

W.M.O Technical Document, WMO/TD No. 84: Tropical Cyclone Programme Report No. TCP21: Tropical Cyclone Operational Plan For The Bay Of Bengal and The Arabian Sea, (ed.)

Wolanski, E., Jones, M., & Bunt, J. S. (1980). Hydrodynamics of a tidal creek–mangrove swamp system. *Australian Journal of Marine and Freshwater Research, 31*, 431–450.

Wright, L. D. (1978). River Deltas. In R. A. Davis (Ed.), *Coastal sedimentary environments*. Springer.

Yu, B. et al (2011). Impacts of climate change on agriculture and policy options for adaptation-The case of Vietnam. http://www.ifpri.org

Chapter 10
Appraisal of the Variation of TSS and Turbidity on Fish Production in Mahanadi Delta Region Post Fani and Phailin Cyclones

Jayashree Karmakar and Abira Dutta Roy

10.1 Introduction

Estuarine water system is one of the most sensitive and productive aquatic environments in the world (Lansing et al., 1998). The water quality here is impacted by continuous variations in rainfall, irradiance, wind, freshwater runoff, and water level changes (Mateus et al., 2008). Biochemical oxygen demand (BOD), chemical oxygen demand (COD), dissolved oxygen (DO), total dissolved solids (TDS), total solids (TS), electrical conductivity (EC), power of hydrogen (pH), and surface water temperature are nonoptical water quality parameters (KC et al., 2019). Turbidity and total suspended solids (TSS) impact the optical property of water. TSS and turbidity, although are essential for the development of aquatic ecosystems as they are consisted of various minerals, nutrients, and organic matter, can be fatal in higher quantities. TSS concentrations >1000 mg/l prevents light penetration and has an adverse effect on aquatic life (Bian et al., 2010; Kulkarni, 2011; Rossi et al., 2006). The turbidity and TSS content in different water bodies are affected by different natural factors (soil erosion, landslides, sediment transport, phytoplankton presence, storms, freshets, tidal flows, water currents) and some anthropogenic factors (different construction works, agricultural activities, and industrial activity like quarrying, mining and coal recovery, dredging, dredge material disposal, pile driving and pile removal, jetting/jet plowing, barrier removals, culvert replacements, and cofferdams) along the coast (Wilber & Clarke, 2001; Anchor Environmental, 2003; KC et al., 2019; Kulkarni, 2011). Tropical cyclones often prove to be a major factor in the disrupting the water quality, especially TSS and turbidity, and henceforth the aquatic ecosystem to an enormous extent. Research works by Gong and

J. Karmakar · A. D. Roy (✉)
Bankura Zilla Saradamani Mahila Mahavidyapith, Bankura, West Bengal, India

G. S. Bhunia et al. (eds.), *Anthropogeomorphology*, Geography of the Physical Environment, https://doi.org/10.1007/978-3-030-77572-8_10

Shen (2009), Lou et al. (2016), and Seers and Shears (2015) have extensively discussed about impacts of cyclone on coastal limnology. Studies on impact of varied TSS and turbidity on marine biodiversity have been carried out by Bhattacharya et al. (2014), Das et al. (2016), and Forbes and Cyrus (1992) especially in the context of cyclonic events. Johnson (2018) assessed that fish and invertebrates in particular are negatively impacted by high concentrations of TSS and turbidity. He identified high concentrations of suspended sediment and longer exposure times cause more severe impacts than exposure to lower concentrations and shorter exposure times. He summarized that fishes receive oxygen from the water rather than the air, and harmful effects included the clogged gills and reduced feeding ability, thus leading to emigration from the affected areas, disturbing important biological behaviors. There has been a paucity of assessment of relation of impacts of cyclone on TSS and turbidity and in turn on fish production in and around the Bay of Bengal. This paper as a consequence aims at identifying the impact of cyclonic storms on the water quality of the Mahanadi deltaic region with special focus on the turbidity and TSS content. It also tries to have a simplistic estimation of the impact of the altered turbidity and TSS on the fishery production.

Remote sensing had proved to be very beneficial to analyze the impact of the cyclones on the water quality of the seas. Though turbidity can be measured directly using a light turbidity meter, or visually using Secchi disc depth (Ritchie et al., 2003) and TSS can be measured by portable TSS meter, in the recent past, various studies have been led for the measurement of the variations of TSS concentration and turbidity, using different remote sensing data, and the results have been strongly validated. Peterson et al. (2018) have studied that due to complexities and inaccessibility of in situ data collection, remote sensing has become a significant way to assess cyclone-induced TSS and turbidity variations in estuarine waters. Ye et al. (2014) used MODIS-based algorithm, in the Pearl River estuary, to study sediment variation caused by Typhoon Vicente using two high-resolution bands at 645 nm and 555 nm to retrieve the TSS concentration. Wang and Lu (2010) and Rodrigues et al. (2017) also used MODIS for retrieving TSS in Yangtze River and reservoirs, respectively. Ritchie and Cooper (1988) used Landsat MSS to determinate water quality conditions; Ma and Dai (2005) and Shahzad et al. (2018) used Landsat-7 ETM+ to assess TSS and chlorophyll content in water; Wibisana et al. (2018), Yanti et al. (2016), KC et al. (2019), and Emiyati et al. (2017) mapped TSS with the help of Landsat-8 OLI; Zhou et al. (2006) and Baban (1995) used Landsat TM; Wang et.al., (2017a, b) used both TM and OLI; and Parwati and Purwanto (2017) used Landsat-5 TM, Landsat-7 ETM, and Landsat-8 sensors to determine TSS using different band ratioing techniques. Because of a better resolution compared to TM and ETM, Landsat-8 has been widely used in coastal management and especially has been regarded to be a prominent tool for monitoring of coastal sediment concentrations (Vanhellemont & Ruddick, 2014).

Mahanadi delta like most of the other tropical estuaries too faces the onslaught of cyclonic storms which effect the anthropogenic and physical condition of the delta region. Tropical depressions and cyclones have shown an increasing frequency over the recent years along the Bay of Bengal coast (Ghosh et al., 2019). Of all the east coast states of India, Odisha alone experiences 44% of the total cyclonic disturbances (Mishra et al., 2019). Odisha coast experiences 29% of the total cyclonic disturbances in comparison to other states like West Bengal (14%), Andhra Pradesh (13%), and Tamil Nadu (7%) (Memorandum, Fani, 2019). Between 2001 and 2015, eight high-intensity flooding events were reported (Ghosh et al., 2019). In October 1999, the state had experienced one of the most hazardous super cyclones of all times. During this cyclone, more than 8000 people were killed in Jagatsinghpur district (Ghosh et al., 2019). On 12 October 2013 and 3 May 2019, respectively, cyclonic storms Phailin (Padhy et al., 2015) and Fani (Memorandum, Fani, 2019) made their landfall in Odisha, creating havoc not only on the infrastructure and property but also on the environment (Hazra et al., 2019). The cyclone Phailin hit Odisha with heavy rainfall and strong winds with the speed of 200–220 km an hour. The storm surge rose about 3.5 m from normal and inundated large areas in the districts of Ganjam, Puri, Khurdha, and around Chilika lagoon (Padhy et al., 2015). The brunt of such extreme weather events had mostly affected the Mahanadi deltaic region (Ghosh et al., 2019). Several studies have been conducted on the Mahanadi delta like dealing with the risk of extreme environmental events (Ghosh et al., 2019), delving into the changes that have occurred in Mahanadi delta region, both in terms of biophysical and socioecological change (Hazra et al., 2019) and management of cyclone Phailin 2013 (Padhy et al., 2015).

10.2 Study Area

The Mahanadi is the eighth largest basin having a total catchment area of 139,681.51 km^2 (Central Water Commission, 2014) and covering more than 4% of the total geographical area of India (Patil et al., 2017). The Mahanadi River emerges from the Maikal (Amarkantak) hills in Dhamtari district of Chhattisgarh. It enters the state of Odisha below Baloda Bazar and crosses the Eastern Ghats to enter the delta plain near Naraj. The river is of 851 km in length (Mishra et al., 2019). The delta is drained by a network of three main rivers: Mahanadi, Brahmani, and Baitarini (Mahalik et al., 1996). All the major branches and subbranches (except the rivers Daya and Bhargovi which fall into Chilika Lake) finally fall in BOB and main branch at false point near Paradip. The delta covers an area of about 9000 km^2 (Maejima & Mahallk, 2001). The coastline, the expansion of which is Chilika in the south to Dhamra River in the north, is about 200 km in length (Ghosh et al., 2019). The Mahanadi annually delivers 15.74 million tonnes of sediment to the Bay of Bengal, and more than 80% of the sediment load is carried in the coarse silt fraction (Chakrapani and Subramaniam, 1990). Our study area is located between 85°9′29″E to 56°53′10″E and 20°42′11″N to 19°38′38″N. Figure 10.1 shows the location and extent of the Mahanadi delta. With respect to the geological time scale, the age of rocks is in the basin ranges between Archaean and Recent, covering the entire geological time scale of the Earth (Patil et al., 2017). In the upper Mahanadi region, the rocks are generally comprised of Pleistocene to Archean age and in the

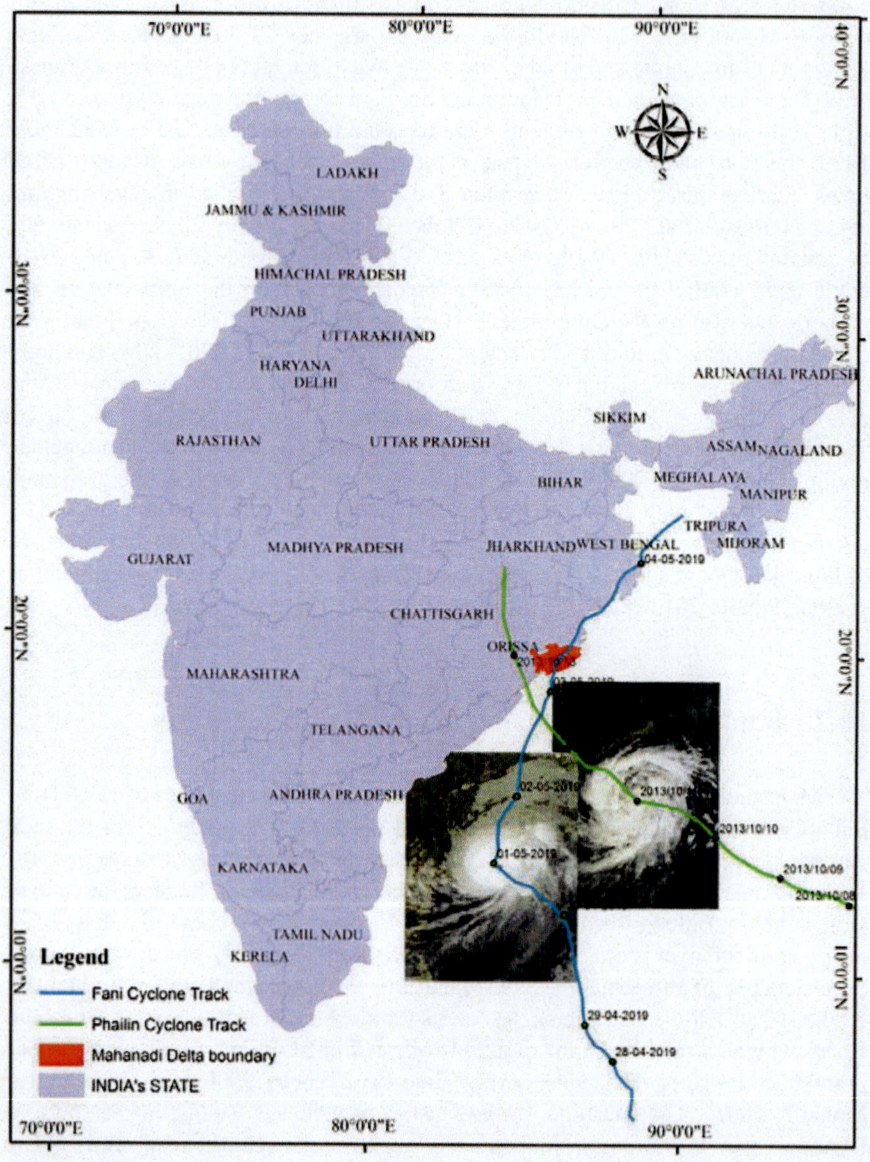

Fig. 10.1 Mahanadi delta (study area) with cyclonic tracks of Phailin and Fani along the Indian sea coast

lower Mahanadi basin belonging to alluvial formation of recent origin (Water Year Book, 2017). The main soil types found in the basin are red and yellow soils, mixed red and black soils, laterite soils, and deltaic soils. 41.95% of the area comes under fine-textured soil followed by covering 51.27% area under medium-textured soil

(Central Water Commission, 2014). The upper part of the delta is highly populated compared to the lower part of the delta. As per the 2011 Census, the total population is around 8.03 million (Hazra et al., 2016). Most portion of the basin is covered with agricultural land estimated at 54.27% of the total basin area. Out of this total agricultural area, 45.87% accounts for cropland area and fallow land is 7.88%. Under the forest area, deciduous forest comes under 28.73% and scrub forest covers 3.57%. Water bodies including reservoirs, lakes, rivers, etc. cover 4.45% of the basin area (Central Water Commission, 2014). The Mahanadi delta experiences a tropical with hot and humid monsoonal climate (Hazra et al., 2016). The average highest relative humidity in the basin is 82%, and the average lowest relative humidity is 31.6% (Central Water Commission, 2014). Most of the portion of the basin area receives rainfall from 1200–1400 mm (Water Year Book, 2017). December and January are the coldest months in this basin area with the minimum temperature of 12 °C. April and May are the hottest months where the maximum temperature ranges from 39 °C to 40 °C (Central Water Commission, 2014). Phailin crossed Odisha coast near Gopalpur at the evening of 12 October 2013. Due to this very severe cyclonic storm (VSCS), about 13.2 million people were affected (Padhy et al., 2015), and economic loss of about Rs. 14,373 crore was reported (Padhy et al., 2015). The cyclone "Fani" made landfall on 3 May 2019 between Satapada and Puri as an extremely severe cyclonic storm. The Khordha district recorded highest 187.8 mm rainfall, and 64 persons have lost their lives due to this cyclone (Memorandum, Fani, 2019).

10.3 Data Source and Methodology

This paper uses Landsat-8 OLI with a panchromatic band which provides moderate spatial resolution of 15 m. This dataset has been frequently used in some literatures to analyze TSS and turbidity in deltaic waters (Wang et al., 2017a, b; Chen et al., 2015; Gholizadeh and Melesse, 2017; Dorji et al., 2016; Yanti et al., 2016; Zhang et al., 2016). The absence of cloud-free images along the coast are rare, more so immediately pre- and post-cyclonic events, which hinder the process of accurate estimation of various water quality parameters. Nevertheless in this study, 11 scenes (with less than 10% cloud cover) of 2013 and 2019, covering Mahanadi River delta, were used to examine the changes of TSS and turbidity pre- and post Phailin and Fani cyclone, respectively. The details of the scenes used in the study are listed in Table 10.1.

Figure 10.2 shows the broad methodological flow chart according to which the analytical procedures were carried on. The cloud-free satellite images were downloaded from USGS Earth explorer (http://earthexplorer.usgs.gov/). Atmospheric correction, radiometric correction, and sun angle rectification of each of the bands of the images were done in Erdas 2014 to remove the scattering and absorption effects from the atmosphere to acquire the surface reflectance

Table 10.1 Retrieved images of Landsat-8 OLI

Cyclonic event	Path and row	Date of acquisition
Pre-Phailin	140_46	17/04/2013
	139_46	17/09/2013
	139_47	Data not available
Post Phailin	140_46	11/11/2013
	139_46	04/11/2013
	139_47	04/11/2013
Pre-Fani	140_46	18/04/2019
	139_46	11/04/2019
	139_47	11/04/2019
Post Fani	140_46	20/05/2019
	139_46	29/05/2019
	139_47	29/05/2019

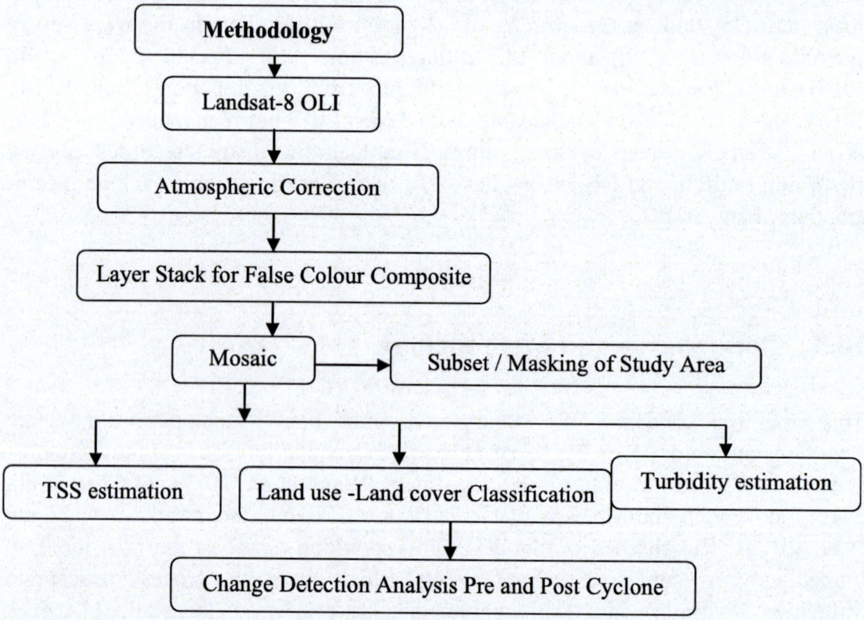

Fig. 10.2 Methodological flow chart

characterization (KC et al., 2019). Equation 10.1 was used for radiometric correction of each band of the Landsat-8 image.

$$p\lambda' = mpQcal + Ap \tag{10.1}$$

TOA reflectance with a correction for the sun angle is then

$$p\lambda' = \frac{p\lambda}{\cos(\theta sz)} = \frac{p\lambda}{\sin(\theta SE)} \tag{10.2}$$

where

$p\lambda'$= TOA planetary reflectance, without correction for solar angle
mp = band-specific multiplication rescaling factor from the metadata (REFLECTANCE_MULT_BAND_X where "X" is the band number)
Ap = band-specific additive rescaling factor from the metadata (REFLECTANCE_ADD_BAND_X where "X" is the band number)
$Qcal$ = quantized and calibrated standard product pixel values (DN)
SE = solar elevation

Layer stacking and false color composite generation were followed by mosaicking of the images. Based on the hydrogeological classification, the deltaic area was delineated, and the study area was masked from the satellite images. Twenty-four nautical miles off the coastline in the Bay of Bengal was considered within the study area. Twenty-four nautical miles is considered under the territorial sea and contiguous zones as per the Coastal Regulation Zone of UNCLOS (United Nations Convention on the Law of the Sea). Supervised classification of the images pre- and post cyclonic storms (Phailin and Fani) was carried out assisted by Google Earth temporal high-resolution images for accuracy assessment. A change detection analysis was carried out to identify the areas inundated as a result of the cyclonic storms.

10.3.1 Retrieval Method of Total Suspended Solid Concentration

Many Landsat-based models like empirical, semi-empirical, semi-analytical, or analytical algorithms have been used for calculating the TSS concentration (Chen et al., 2014; Montanher et al., 2014; Ramakrishnan et al., 2013; Wu et al., 2013; Zhang et al., 2014). In different water bodies, the application of semi-analytical and analytical models is higher than the empirical or semi-empirical methods for its more rigorous theoretical derivation (Wang et al., 2017a). But, because of the difficulties in getting accuracies on initialization parameters, there are limitations in the application of the models (Binding et al., 2012; Chen et al., 2015; Wu et al., 2013). Therefore, with the advantages of simplicity and sufficient accuracies, the empirical and semi-empirical methods are still used extensively to retrieve TSS concentration (Wu et al., 2013). This study used the semi-empirical method using relatively higher spatial resolution (30m) Landsat-8 imagery. Equations 10.1 and 10.2 developed by Wang et al. (2016, 2017a) were used to retrieve TSS concentration of Mahanadi River estuary during pre- and post Fani and Phailin period. The semi-empirical model has a higher accuracy compared to the previous models conducted by Montanher et al. (2014), Wu et al. (2013), and Zhang et al. (2014).

$$\frac{\log(\mathrm{R1})}{\log(\mathrm{R2})} = \mathrm{a} * (\log(\mathrm{TSS}))2 + \mathrm{b} * \log(\mathrm{TSS}) + \mathrm{c} \tag{10.3}$$

$$\log(\mathrm{TSS}) = \frac{-b \pm \sqrt{b^2 - 4a\left(c - \frac{\log(R1)}{\log(R2)}\right)}}{2a}, \left(\mathrm{b2} - 4\mathrm{a}\left(\mathrm{c} - \frac{\log(R1)}{\log(R2)} \geq 0\right)\right. \tag{10.4}$$

where R1 and R2 represent near infrared band and red band of OLI. Parameters a, b, and c are −0.3575, 1.1135, and 0.7162 (Landsat OLI sensor). The unit of suspended sediment concentration is in mg/l (Wang et al., 2017a, b, c). According to the above analysis, TSS concentration can be retrieved by using Eq. 10.4 in the form of a positive squared root if the reflectance of red band is lower than 0.032 (OLI sensor) and in the form of a negative squared root if the reflectance of red band is greater than 0.032 (OLI sensor) (Wang et al., 2017a). In the study for all the dates, the red band had a reflectance greater than 0.032 and hence negative square root was used.

10.3.2 Retrieval Method of Turbidity

Traditionally, turbidity is estimated visually using a Secchi disk or measured directly with nephelometry. The following functional model is the measuring procedure of turbidity in dry and wet season as per Gholizadeh and Melesse (2017):

Dry season

$$\mathrm{Turbidity} = a + (b * \mathrm{Blue}) + (c * \mathrm{Red}) + \left(d * \frac{\mathrm{Green}}{\mathrm{NIR}}\right) + \left(e * \frac{\mathrm{Red}}{\mathrm{Green}}\right) \tag{10.5}$$

Wet season

$$\mathrm{Turbidity} = a + \left(b * \frac{\mathrm{Green}}{\mathrm{Blue}}\right) + \left(e * \frac{\mathrm{Green}}{\mathrm{NIR}}\right) + \left(d * \frac{\mathrm{Red}}{\mathrm{NIR}}\right) + \left(e * \frac{\mathrm{NIR}}{\mathrm{Blue}}\right) \tag{10.6}$$

The value of "a," "b," "c," "d," and "e" are given in Table 10.2.

In this current study, Equation (10.6) was used because of previously occurring rainfall events during the course of approach of the cyclones. Moreover, Phailin

Table 10.2 The values of a, b, c, d, and e for Eqs. 10.5 and 10.6

Dry season	Turbidity (NTU)	= −0.24−53.34 × (blue) + 66.1 × (red) +1.48 × (green/NIR) + 0.08 × (red/green)
Wet season	Turbidity (NTU)	= −5.54+6.67 × (green/blue)−1.2 × (green/NIR)+3.21 × (red/NIR)−1.51 × (NIR/blue)

occurred in November, which is a post-monsoon month (wet season) in Indian subcontinent, and Fani occurred in May which is a monsoon month (wet season).

The maps of both TSS and turbidity were computed and comparative choroplething was done for ease in visual interpretation. Besides investigating the changes in TSS and turbidity, it also becomes imperative to look into the effects of limnological changes on fishery as the population residing in the Mahanadi deltas are heavily dependent on. Most studies by Paul et al. (2020), Paerl et al. (2001, 2018) have carried out primary survey and collected data in order to estimate the changes in the quantity of copepods, phytoplanktons, and various marine organisms in the Ganges estuary and coastal north Caroline estuary, respectively. This paper uses the secondary data on fish catch procured from the Annual Activity Report, 2013 and 2019, published by the Fisheries and Animal Resource Development Department, Government of Odisha, in order to understand the impact of the cyclones on the fishery.

10.4 Results and Discussions

The land use/land cover (LULC) maps were mainly generated to estimate the areas inundated as a result of storm surge post the occurrences of cyclone Phailin and Fani. The areal estimation showed that post Phailin an area of 805 km^2 was inundated which included areas east and northeast of Chilika Lake. Agricultural land, inland fisheries in the wetlands, and settlements such as Bhusandpur, Dochhian, Badaora, Paikarma, Bhudkera, Golara, Denai, and Chandumal were badly affected. The coastal area between the mouth of Devi River and Mahanadi River was also inundated. Barkuda, Naugaon, Polanga, and Ambiki were one of the worst-hit settlements. Figure 10.3 zooms into the inundated areas and shows the LULC pre- and post cyclones. In the case of Fani, the areas inundated were comparatively lesser than that of Phailin. The areal statistics generated showed that 263 km^2 were inundated along the coastline and in the northeast areas of the Chilika Lake.

The total suspended solids that have been calculated using raster calculator in ArcGIS showed that in the pre-Phailin phase the TSS concentration ranged from 297.1 4 milligram per liter to about 353.69 milligram per liter. The concentration of TSS is seen to be maximum near the Chilika Lake where Daya and Bhargavi river drain its sediment load. The TSS concentration is also seen to be high along the western half of the Mahanadi estuary because it belonged to the satellite image with Path 140 Row 46 which was mosaiced with the other scenes. The image having P140R46 numbered was acquired on 17 April 2013 unlike the other image P139R46 which was acquired in 17 September 2013. The image P140R46 as was captured during a hotter, pre-monsoon season experienced higher temperature, which lead to higher evaporation, less rainfall, and hence less influx of runoff. This could have increased the total suspended sediment concentration in that particular scene, resulting in the depiction of more TSS values of 344.27 to 353.69 mg/l. Higher TSS values are also found near Nehru bungalow and fishery harbor where the Devi

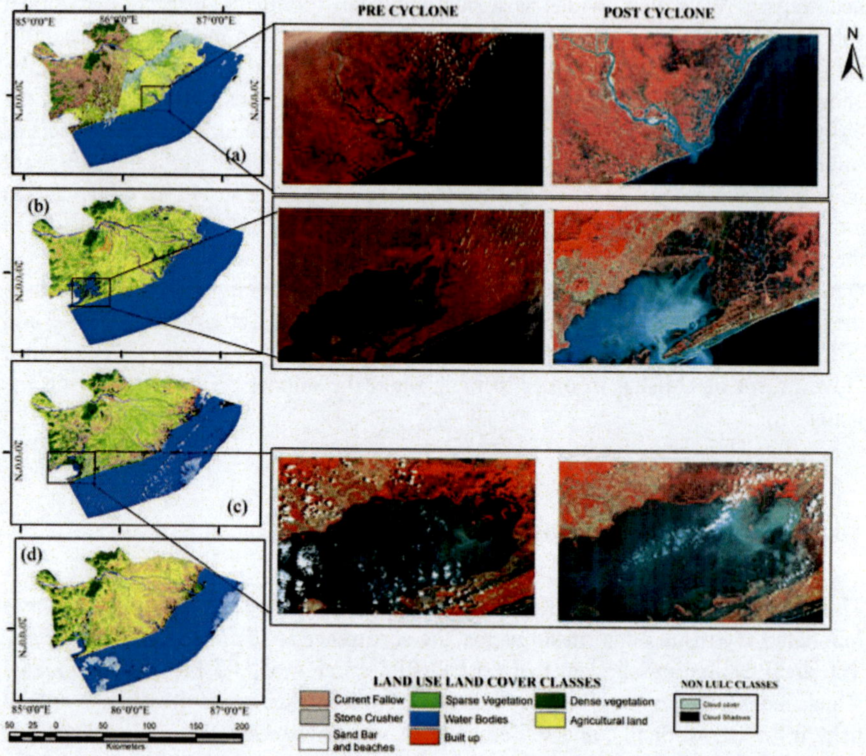

Fig. 10.3 (**a** and **b**) Pre-Phailin cyclone and post Phailin LULC condition, respectively, with insets highlighting mouth of Devi River and the Chilika lake along with inundated areas in FCC (false color composite. (**c** and **d**) LULC condition of pre- and post Fani, respectively, with insets highlighting parts of Chilika Lake

River meets the Bay of Bengal. Here the TSS concentration ranged from 325.4 to 334.84 milligram per liter under pre-Phailin condition. Toward the east of the Delta near Jammu Deep and Paradeep, too high concentration of TSS is observed. But after the Phailin struck, the concentration of TSS had increased by threefolds which could be attributed to the disturbance caused by the settled sediment load, contribution from coastline erosion, etc. The spatial variation of TSS has remained the same with highest TSS concentration in the Chilika Lakes, Devi river mouth, Paradeep and Jambu Deep, and near the mouth of the Mahanadi River; the concentration of TSS values has increased ranging from 779.32 to 848.98 mg/liter. Figure 10.4 shows the spatiotemporal variation of pre- and post-cyclonic condition in TSS concentration. The cyclone Fani left a minor impact on the TSS concentration of the Mahanadi estuary. The cause could be attributed to the angle in which the cyclone made landfall. Cyclones with tracks perpendicular to the coast are shown to produce the highest water levels and broadest inland and offshore extent (Ramos-Valle et al., 2020). Phailin cyclone's track was perpendicular to the coast of Odisha when it made

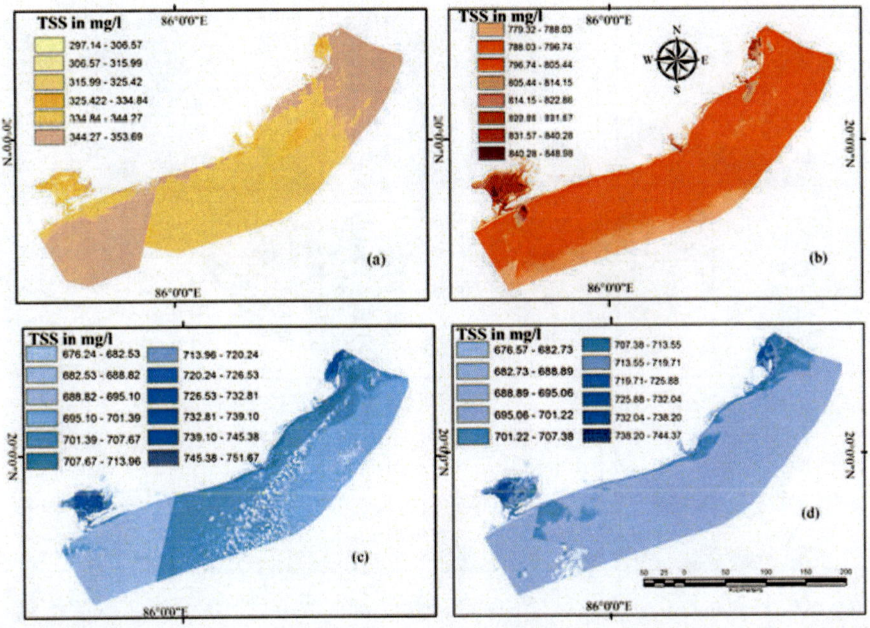

Fig. 10.4 TSS concentration in mg/liter in (**a**) pre-Phailin, (**b**) post Phailin, (**c**) pre-Fani, and (**d**) post Fani

its landfall, whereas cyclone Fani's track ran prarallel to the coast of Odisha as can be seen in Fig. 10.1. Thus, Fani created lesser disturbance to the TSS concentration.

It can be observed from Fig. 10.5 that during the pre-Phailin phase, the turbidity in the Chilika Lake varied between 1.631 and 4 NTU and similar values were seen at the mouth of Devi and Mahanadi rivers. The turbidity at 12 nautical miles off the coast ranged from 1.01 to 1.62 NTU, but this turbidity values spread over the eastern part of the delta along the coast near the mouth of Devi and Mahanadi rivers, showing a disturbance in the sedimentation. The Paradip and Jambu Dweep coast too had increased turbidity values post Phailin cyclone. Figure 10.5 (iv, v, and vi) shows the increased turbidity inside the river channels, coastline, lakes, and lagoons where seawater with under 0 to 1 NTU turbidity had increased its values ranging from 3.01 to 6.59 NTU turbidity in post Phailin situation.

Figure 10.6 shows a completely different picture pre- and post Fani. It is observable that there has been a dilution and decrease in turbidity post Fani cyclone. The spread of turbidity has also decreased as can be seen in Fig. 10.6 (c) and (d) along the coast from Puri to Paradeep. The pre Fani condition shows turbidity ranging from 3.01 to 5.00 NTU near the Chilika Lake, 2.01–3.00 NTU at the mouth of Devi River, and 1.631 to 4 near the mouth of Mahanadi River.

The values of turbidity has reduced and faded away in the post Fani condition as can be seen in Fig. 10.6 (iv, v, and vi) where the turbidity values ranged from 1.631 to 4 NTU. The wind speed in the case of Phailin was lesser than Fani. As per the

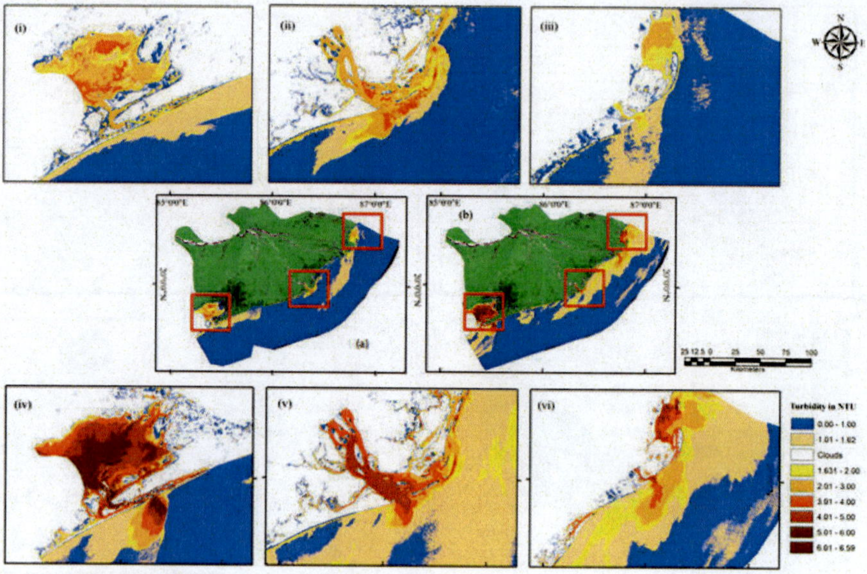

Fig. 10.5 Turbidity conditions pre- (**a**) and post Phailin (**b**) cyclone in the Mahanadi delta. (i) Pre Phailin-Chilika Lake, (ii) pre-Phailin-Diver river mouth, (iii) pre-Phailin-Mahanadi river mouth, (iv) post Phailin-Chilika Lake, (v) post Phailin-Diver river mouth, (vi) post Phailin-Mahanadi river mouth

India Meteorological Department (IMD), Fani was an extremely severe cyclonic storm (94 knots), while Phailin had a wind speed of over 64 knots and was a very severe cyclonic storm. According to IMDs preliminary report of Phailin and Fani, it was found that the average rainfall in the case of Phailin was 78 cm and that of Fani was 64 cm. Heavy downpour can also result in the turbulence of the water leading to increased turbidity in the case of Phailin cyclone. Additionally the IMDs preliminary report of Phailin and Fani also states that the storm surge in the case of Phailin was around 2–2.5 m, whereas in the case of Fani, it was 1.5 m, which also explains the increased TSS and turbidity conditions in the case of Phailin cyclone.

Though the TSS values could not be checked or validated from other sources, a publication by Samantray et al. (2009) shows similar turbidity values which were measured at the field. The turbidity values obtained by Samantaray at locations also present in this study are tabulated in Table 10.3. As can be seen in Figs. 10.5 and 10.6 in the index, the value 1.63 NTU is missing. It has been classified as cloud and background values during image processing and has been omitted.

Mahanadi delta occupies 200 km of the 480 km stretch of Odisha's coastline (Hazra et al., 2019) and over 1 lakh ha from a total of 4.181 lakh ha of brackish water resource of the state. According to the survey conducted by CMFRI (Central Marine Fisheries Research Institute) in 2011, about 605,514 persons along the coastal belt of Odisha were fishermen and a total of 114,238 were fishermen families. This population and their livelihoods were at stake because of the frequent cyclones occurring

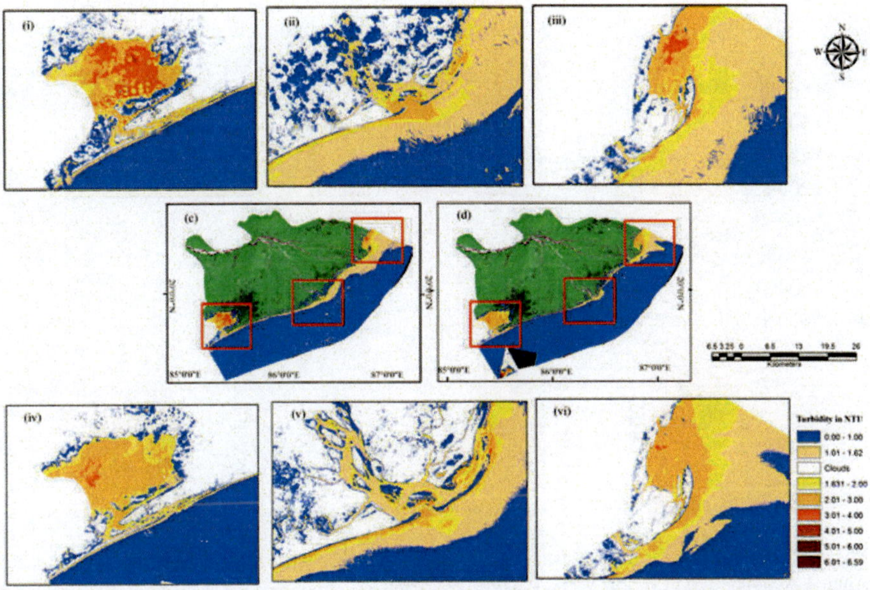

Fig. 10.6 Turbidity conditions pre- (**a**) and post Fani (**b**) cyclone in the Mahanadi delta. (i) Pre-Fani-Chilika Lake, (ii) pre-Fani-Diver river mouth, (iii) pre-Fani-Mahanadi river mouth, (iv) post Fani-Chilika Lake, (v) post Fani-Diver river mouth, (vi) post Fani-Mahanadi river mouth

Table 10.3 Turbidity in NTU according to Samantray et al. (2009)

Sampling locations	Post monsoon	Winter	Summer
Nehru Bunglow	7	5	5
Fishing harbor	5	7	8
Jambu Dweep	5	6	6
Musadia	4	5	6

in the delta. The Chilika Lake alone occupied an area of 906 km^2 in the summers and 1165 km^2 in monsoon, serving 132 fishermen villages with 122,339 population located around it. There were 18 landing centers where different fishes, prawns, and crabs were produced. The fisheries statistics in the Annual Activity Report of the Fisheries and Animal Resources Development Department of 2013–2014 and 2019–2020 is shown in Fig. 10.7 for the entire Odisha, specifically for the fish production in marine and brackish waters.

A decline in brackish and marine water fish production is evident from Fig. 10.7. With government efforts, funding, and technological interventions, fishing production has led to a rising trend in fish production except for the years 2012–2013, 2013–2014, and 2019 where there has been a decline from 30062 MT in 2011–2012 to 30007 MT in 2013–2014 and from 94033 MT in 2018–2019 to 91354 MT in 2019–2020 for brackish water fish production. Similar decline in not observed in the case of marine fish production during 2013–2014, but in 2019–2020, marine water fish production has fallen from 158321 MT in 2018–2019 to 122302

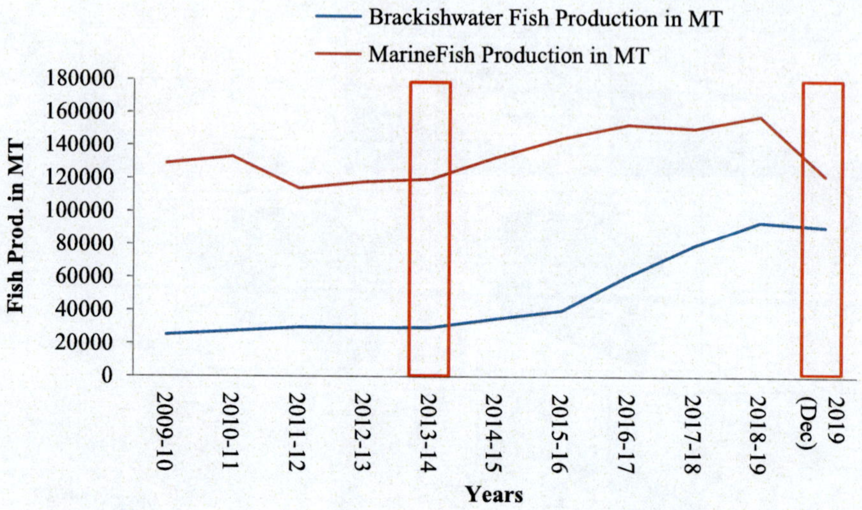

Fig. 10.7 Brackish and marine water fish production in Odisha from 2009 to 2020. (Source: Annual Activity Report of 2013–2014 and 2019–2020)

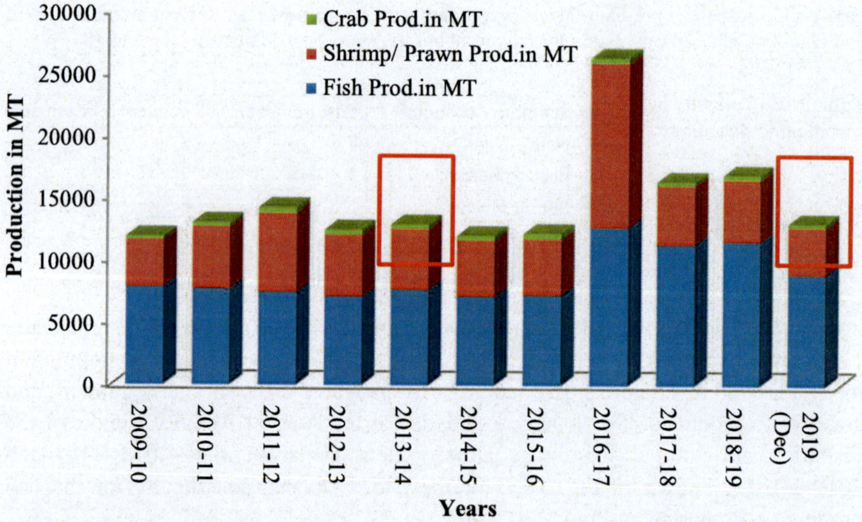

Fig. 10.8 Crab, shrimp/prawn, and fish production in Chilika Lake from 2009 to 2020. (Source: Annual Activity Report of 2013–2014 and 2019–2020, Fisheries and Animal Resources Development Department)

MT. Figure 10.8 shows the trends in crab, fish, and prawn production in one of the largest marine resource bases, the Chilika Lake. The data obtained from the Annual Activity Report of 2013–2014 and 2019–2020 of the Fisheries and Animal Resources Development Department was used to graphically plot Fig. 10.8, which

Table 10.4 Decadal fish production (in MT) trends in the coastal districts of Mahanadi delta

Year	Jagatsinghpur	Kendrapara	Puri
2009–2010	33,012	4798	31,431
2010–2011	35,656	6853	31,880
2011–2012	28,675	6339	31,000
2012–2013	32,971	4897	11,348
2013–2014	30,395	7474	30,938
2014–2015	36,632	7009	11,659
2015–2016	34,503	9059	37,979
2016–2017	40,284	9409	39,598
2017–2018	41,319	8566	38,906
2018–2019	42,469	9475	38,033
2019–2020	29,786	7382	27,534

showed the decline in the production of shrimp/prawn and crabs and an unaffected rise in production of other fishes in the Chilika Lake post Phailin. But the production of crab, prawns/shrimp, and other fishes all declined in the post Fani conditions.

The fish production fell from 11618.05 to 897.67 MT, shrimp/prawn production from 4951.37 to 3845.86 MT, and crab production from 331.9 to 260 MT between 2018–2019 and 2019–2020. Similar results have been observed in the coastal belts of the districts of Puri, Jagatsinghpur, and Kendrapara which comprise the coasts of the Mahanadi delta. Table 10.4 shows the year-wise and district-wise marine fish production over the decade, the data being obtained from the Annual Activity Report of 2013–2014 and 2019–2020.

It is evident from Table 10.4 too that there has been a fall in the production of marine fishes post cyclone Fani and Phailin in the three districts. Research works by Eddiwan et al. (2020) and Ray et al. (2010) have shown detrimental impacts of increased TSS and turbidity conditions. TSS levels are shown to have adverse effects on benthic communities (crabs, crustaceans, mollusks) when they exceed 390.0 mg/L (EPA, 1986). Mortality for eggs/larvae can occur at anywhere between 10 mg/L and 120 mg/L, depending on the duration of exposure (Wilber & Clarke, 2001). The loss of life, property, houses, equipment, and livelihood has henceforth threatened the very existence of the fishermen community dwelling in this region. The survey conducted by CMFRI (Central Marine Fisheries Research Institute) in 2016 saw a reduction in the number of fishermen. The number of fishermen families increased to 11,5228, but the fisherman population reduced to 51,7623 persons during 2011–2016. Hazra et al. (2019) showed through a perception study that environmental stress have been the major reason behind fishermen migrating to other sources of livelihood and also emigrating to nearby urban centers.

10.5 Conclusion

The study clearly emphasizes the importance of remote sensing in understanding the dynamics of limnology, especially the turbidity and TSS concentration changes as a result of cyclonic conditions in coastal regions along the Mahanadi delta which has been rarely researched upon. It has been clear that Phailin has been considered as a severe cyclonic storm in comparison to the extremely severe cyclonic storm Fani, but the direction of the track, the rainfall intensity, and the height of the storm surge had a major role to play in determining the spatial variability and concentration of turbidity and TSS post the cyclonic disturbance. Phailin had increased the TSS values from 297.149 to 353.69mg/l in the pre-phase to 780.86 to 848.98mg/l in the post cyclonic phase and tripled the turbidity values and spatial extensions to 6 NTU unlike that of Fani which did not make much of a difference to the TSS and turbidity concentrations of the coastal waters of the Mahanadi delta. The adverse effects of TSS and turbidity were noticeable on the fishery production. Fish kills are common post cyclonic events, but even in the long run, fish habitat is destroyed; food availability is hampered resulting in declined fish production. The statistics of annual fish production have shown the adverse effect as there was tremendous decline in brackish water fish production (from 30062 MT in 2011–2012 to 30007 MT in 2013–2014 and from 94033 MT in 2018–2019 to 91354 MT in 2019–2020), and in 2019–2020, marine water fish production has fallen from 158321 MT in 2018–2019 to 122302 MT. The fall in fish production, together with loss of equipment, was a hard economic blow to the fishing community. Due to floods following Phailin, 8198 boats and 31,058 fishing nets were damaged (UNDP, 2013). Post Fani, 90% of the fishing boats were damaged, which included 6390 traditional marine fishing boats. In the fisheries sector, 7240 nets, 279 fish ponds in an area of 66.92 Ha., 3 fishing harbors, 6 fish landing centers, 5 fish farms, building, and other infrastructures have been fully/partially damaged (UNICEF, 2019; Red Cross, 2019). Thus in order to improve the socioeconomic condition of the marginalized fishing community, it is essential to model impacts of cyclone on water quality and fish production together with disaster risk reduction methods and secure their livelihood through insurances and aids for procurement of fishing equipment as well as fast rectification of water quality of aquaculture ponds.

References

Anchor Environmental. Literature review of effects of resuspended sediments due to dredging operations. Prepared for Los Angeles Contaminated Sediments Task Force, Los Angeles, California. 140 (2003). https://www.coastal.ca.gov/sediment/Lit-ResuspendedSediments.pdf

Annual Activity Report 2013–14, Directories of Fisheries, Department of Fisheries and Animal Resource Development, Government of Odisha, https://odishafisheries.nic.in/?p=report&type=3&pg=52

Annual Activity Report 2019–20, Directories of Fisheries, Department of Fisheries and Animal Resource Development, Government of Odisha, https://odishafisheries.nic.in/?p=report&type=3&pg=52

Baban, S. M. J. (1995). The use of Landsat imagery to map fluvial sediment discharge into coastal waters. *Marine Geology, 123*(3–4), 263–270. https://doi.org/10.1016/0025-3227(95)00003-H

Bhattacharya, B. D., Bhattacharya, A., Rakshit, D., & Sarkar, S. K. (2014). Impact of the tropical cyclonic strom 'Aila' on the water quality characteristics and mesozooplankton community structure of Sundarban mangrove wetland, India. *Indian Journal of Geo-Marine Sciences, 43* (2), 216–223. https://www.researchgate.net/publication/285716558

Bian, C., Jiang, W., & Song, D. (2010). Terrigenous transportation to the Okinawa trough and the influence of typhoons on suspended sediment concentration. *Continental Shelf Research, 30,* 1189–1199. https://doi.org/10.1016/J.CSR.2010.03.008

Binding, C. E., Greenberg, T. A., & Bukata, R. P. (2012). An analysis of MODIS-derived algal and mineral turbidity in Lake Erie. *J. Great Lakes Res., 38*, 107–116. https://doi.org/10.1016/j.jglr.2011.12.003

Central Water Commission and National Remote Sensing Centre, Mahanadi Basin. (2014). https://indiawris.gov.in/downloads/Mahanadi%20Basin.pdf

Chakrapani, G. J., & Subramanian, V. (1990). Factors controlling sediment discharge in the Mahanadi River Basin, India. *Journal of Hydrology, 117*(1–4), 169–185. https://doi.org/10.1016/0022-1694(90)90091-B

Chen, Y., Kong, J., Sun, X., Du, D., & Cheng, P. (2014). Retrieval of sea surface suspended sediment concentration in Bohai Gulf Offshore area based on semi-analysis model. *Geographic Information Science, 30*, 33–36. https://doi.org/10.3969/j.issn.1672-0504.2014.03.008

Chen, J., Quan, W., Cui, T., & Song, Q. (2015). Estimation of total suspended matter concentration from MODIS data using a neural network model in the China eastern coastal zone. *Estuarine, Coastal and Shelf Science, 155*, 104–113. https://doi.org/10.1016/j.ecss.2015.01.018

Das, S., Giri, S., Das, I., Chanda, A., Akhand, A., Mukhopadhyay, A., Maity, S., & Hazra, S. (2016). Tide induced annual variability of selected physio-chemical characteristics in the northern Bay of Bengal (nBoB) with a Special emphasis on Tropical Cyclone-Phailin, 2013. *Indian Journal of Geo-Marine Sciences, 45*(8), 952–959. https://www.researchgate.net/publication/307593614

Dorji, P., Fearns, P., & Broomhall, M. (2016). A semi-analytic model for estimating total suspended sediment concentration in Turbid coastal waters of Northern Western Australia using MODIS-Aqua 250 m Data. *Remote Sensing, 556*(8), 1–23. https://doi.org/10.3390/rs8070556

Eddiwan, E., Sukendi, S., Siregar, Y. I., & Saam, Z. (2020). The effect of water quality variables on Vannamei Shrimp productivity (Litopenaeus Vannamei) in the mining area of the Sungai Pinang Village, Lingga Timur District, Lingga Regency, Riau Islands Province. *IOP Conference Series: Earth and Environmental Science, 430*(012039), 1–6. https://doi.org/10.1088/1755-1315/430/1/012039

Emiyati, Manoppo, A. K. S., & Budhiman, S. (2017). Estimation on the concentration of total suspended matter in Lombok Coastal using Landsat 8 OLI, Indonesia. *IOP Conference Series: Earth and Environmental Science, 54*(012073), 1–8. https://doi.org/10.1088/1755-1315/54/1/012073

EPA (Environmental Protection Agency). Quality Criteria for Water. EPA 440/5-86-001. (1986). https://www.epa.gov/sites/production/files/2018-10/documents/quality-criteria-water-1986.pdf

Forbes, A. T., & Cyrus, D. P. (1992). Impact of a major cyclone on a southeast African estuarine lake system. *Netherlands Journal of Sea Research, 30*, 265–272. https://doi.org/10.1016/0077-7579(92)90064-L

Gholizadeh, M. H., & Melesse, A. M. (2017). ISSN: 2469-4134). Study on spatiotemporal variability of water quality parameters in Florida Bay using remote sensing. *Journal of Remote Sensing & GIS, 6*(3), 1–11. https://doi.org/10.4172/2469-4134.1000207

Ghosh, A., Das, S., Ghosh, T., & Hazra, S. (2019). Risk of extreme events in delta environment: A case study of the Mahanadi delta. *Science of the Total Environment, 664*, 713–723. https://doi.org/10.1016/j.scitotenv.2019.01.390

Gong, W., & Shen, J. (2009). Response of sediment dynamics in the York Estuary, USA to tropical cyclone Isabel of 2003. *Estuarine, Coastal and Shelf Science, 84*, 61–74. https://doi.org/10.1016/j.ecss.2009.06.004

Hazra, S., Cazcarro, I., Arto, I., & Bhattacharya, R. (2016). *Biophysical and socioeconomic State of the Mahanadi Delta region of India from the perspectives of gender and spatial relations* (pp. 1–21). DECCMA. https://doi.org/10.13140/RG.2.2.15850.36800

Hazra, S., Das, S., Ghosh, A., Raju, P. V., & Patel, A. (2019). ISBN 978-3-030-23516-1 ISBN 978-3-030-23517-8 (eBook)). The Mahanadi Delta: A rapidly developing delta in India. In *Deltas in the Anthropocene* (pp. 53–77). Palgrave Macmillan. https://doi.org/10.1007/978-3-030-23517-8_3

Indian Red Cross society. Odisha FANI cyclone Assessment Report. (2019). https://www.indianredcross.org/notices/OdishaFaniAsessmentReport.pdf

Johnson, A. (2018). *The effects of turbidity and suspended sediments on ESA-listed species from projects occurring in the Greater Atlantic region* (Greater Atlantic Region Policy Series 18-02). NOAA Fisheries Greater Atlantic Regional Fisheries Office. www.greateratlantic.fisheries.noaa.gov/policyseries/. 106p

KC, A., Chalise, A., Parajuli, D., Dhital, N., Shrestha, S., & Kandel, T. (2019). ISSN : 2676-1416). Surface water quality assessment using remote sensing, GIS and artificial intelligence. *Technical Journal, 1*(1), 113–122. https://doi.org/10.3126/tj.v1i1.27709

Kulkarni, A. (2011). Water quality retrieval from Landsat TM imagery. *Procedia Computer Science, 6*, 475–480. https://doi.org/10.1016/j.procs.2011.08.088

Lansing, J. S., Lansing, P. S., & Erazo, J. S. (1998). The value of a river. *Journal of Political Ecology, 5*, 1–22. https://pdfs.semanticscholar.org/9a71/4ae96b909934b9001ddc9b3c68845a017806.pdf

Lou, S., Huang, W., Liu, S., Zhong, G., & Johnson, E. (2016). Hurricane impacts on turbidity and sediment in the Rookery Bay National Estuarine Research Reserve, Florida, USA. *International Journal of Sediment Research, 31*(4), 330–340. https://doi.org/10.1016/j.ijsrc.2016.06.006

Ma, R., & Dai, J. (2005). Investigation of chlorophyll-a and total suspended matter concentrations using Landsat ETM and field spectral measurement in Taihu Lake. *China. International Journal of Remote Sensing, 26*(13), 2779–2795. https://doi.org/10.1080/01431160512331326648

Maejima, W., & Mahallk, N. K. (2001). The role of prevalent wind on the geomorphic development of the Marine-Marginal Zone of the Mahanadi Delta, India. *Journal of Geosciences, Osaka City University, 44*(9), 173–180. ISSN: 0449-2560. https://core.ac.uk/download/pdf/35269191.pdf

Mahalik, N., Das, C., & Maejima, W. (1996). Geomorphology and evolution of the Mahanadi delta, India. *Journal of Geosciences, Osaka City University, 39*(6), 111–122. https://core.ac.uk/download/pdf/35269244.pdf

Mateus, M., Mateus, S., & Baretta, J. W. (2008). Basic concepts of Estuarine Ecology. In *Perspectives on Integrated Coastal Zone Management in South America (EDS.)* (pp. 2–14). IST PRESS. https://doi.org/10.13140/2.1.4497.0562

Memorandum, Extremely Severe Cyclonic Strom 'Fani',.Special Relief Commissioner, Revenue & Disasater Management Department, Government of Odisha. 1–77 (May 2019). https://srcodisha.nic.in/calamity/Memorandum_Cyclone%20FANI_3rd%20May%202019.pdf

Mishra, S. P., Sethi, B. K., & Barik, K. K. (2019). ISSN 0974-5904). Delta Partitioning, Geospatial Changes, Anastomosis of Mahanadi Tri-delta, India. *International Journal of Earth Sciences and Engineering, 12*(1), 21–39. https://doi.org/10.21276/ijee.2019.12.0103

Montanher, O. C., Novo, E. M. L. M., Barbosa, C. C. F., Renno, C. D., & Silva, T. S. F. (2014). Empirical models for estimating the suspended sediment concentration in Amazonian white water rivers using Landsat/TM. *International Journal of Applied Earth Observation and Geoinformation, 29*, 67–77. https://doi.org/10.1016/j.jag.2014.01.001

Padhy, G. K., Mishra, A., Das, S., Padhy, R. N., & Padhy, J. K. (2015). A review on Management of cyclone Phailin 2013 in Odisha. *Indian Journal of Forensic and Community Medicine, 2*(4), 182–192. https://doi.org/10.5958/2394-6776.2015.00001.6

Paerl, H. W., et al. (2001). Ecosystem impacts of three sequential hurricanes (Dennis, Floyd, and Irene) on the United States' largest lagoonal estuary, Pamlico Sound, NC. *Proceedings of the National Academy of Sciences of the United States of America, 98*(10), 5655–5660. https://doi.org/10.1073/pnas.101097398

Paerl, H., Crosswell, J., Dam, B., Hall, N. S., Rossignol, K., Osburn, C., Hounshell, A. G., Sloup, R. S., & Harding, L. W. (2018). Two decades of tropical cyclone impacts on North Carolina's estuarine carbon, nutrient and phytoplankton dynamics: Implications for biogeochemical cycling and water quality in a stormier world. *Biogeochemistry, 141*, 307–332.

Parwati, E., & Purwanto, A. D. (2017). P-ISSN 0216-6739; E-ISSN 2549-516X). Time series analysis of Total Suspended Solid (TSS) using Landsat data in Berau Coastal Area, Indonesia. *International Journal of Remote Sensing and Earth Sciences, 14*(1), 61–70. https://doi.org/10.30536/j.ijreses.2017.v14.a2676

Patil, S., Kulkarni, H., & Bhave, N. (2017, June). Groundwater in the Mahanadi River Basin. Advanced Center for Water Resources Development and Management (ACWADAM), Forum for Policy Dialogue on Water Conflicts in India, Pune 1–51. http://waterconflictforum.org/lib_docs/Groundwater-in-the-Mahanadi-basin.pdf

Paul, S., Karan, S., & Bhattacharya, B. D. (2020). Daily variability of copepods after successive tropical cyclones in the Ganges River estuary of India. *Estuarine, Coastal and Shelf Science, 246*, 107048. https://doi.org/10.1016/j.ecss.2020.107048

Peterson, K. T., Sagan, V., Sidike, P., Cox, A. L., & Martinez, M. (2018). Suspended sediment concentration estimation from Landsat imagery along the lower Missouri and Middle Mississippi Rivers using an extreme learning machine. *Remote Sensing, 10*(1503), 2–17. https://doi.org/10.3390/rs10101503

Ramakrishnan, D., Bharti, R., & Das, M. (2013). A technique for estimation of suspended sediment concentration in very high turbid coastal waters: An investigation from Gulf of Cambay. *India. Marine Geology, 346*, 256–261. https://doi.org/10.1016/j.margeo.2013.10.001

Ramos-Valle, A. N., Curchitser, E. N., & Bruyère, C. L. (2020). Impact of Tropical cyclone landfall angle on storm surge along the Mid-Atlantic Bight. *Journal of Geophysical Research: Atmospheres*, 1–19. https://doi.org/10.1029/2019JD031796

Ray, A. J., Lewis, B. L., Browdy, C. L., & Leffler, J. W. (2010). Suspended solids removal to improve shrimp (*Litopenaeus vannamei*) production and an evaluation of a plant-based feed in minimal-exchange, superintensive culture systems. *Aquaculture, 299*, 89–98. https://doi.org/10.1016/j.aquaculture.2009.11.021

Ritchie, J. C., & Cooper, C. M. (1988). Comparison of measured suspended sediment concentrations with suspended sediment concentrations estimated from Landsat MSS data. *International Journal of Remote Sensing, 9*(3), 379–387. https://doi.org/10.1080/01431168808954861

Ritchie, J. C., Zimba, P. V., & Everitt, J. H. (2003). Remote sensing techniques to assess water quality. *Photogrammetric Engineering and Remote Sensing, 69*(6), 695–704. https://doi.org/10.14358/PERS.69.6.695

Rodrigues, T., Mishra, D. R., Alcantara, E., Watanabe, F., Rotta, L., & Imai, N. N. (2017). Retrieving total suspended matter in tropical reservoirs within a Cascade system with widely differing optical properties. *IEEE Journal of Selected Topics in Applied Earth Observations and Remote Sensing, 10*(12), 5495–5512. https://doi.org/10.1109/JSTARS.2017.2745700

Rossi, L., Fankhauser, R., & Chevre, N. (2006). Water quality criteria for suspended solids (TSS) in urban wet- weather discharge. *Water Science & Technology, 54*(6-7), 355–362. https://doi.org/10.2166/wst.2006.623

Samantray, P., Mishra, B. K., Panda, C. R., & Rout, S. P. (2009). Assessment of water quality index in Mahanadi and Atharabanki Rivers and Taldanda Canal in Paradip Area, India. *Journal of Human Ecology, 26*(3), 153–161. https://doi.org/10.1080/09709274.2009.11906177

Seers, B. M., & Shears, N. T. (2015). Spatio-temporal patterns in coastal turbidity e Long-term trends and drivers of variation across an estuarine-open coast gradient. *Estuarine, Coastal and Shelf Science, 154*, 137–151. https://doi.org/10.1016/j.ecss.2014.12.018

Shahzad, M. I., Meraj, M., Nazeer, M., Zia, I., Inam, A., Mehmood, K., & Zafar, H. (2018). Empirical estimation of suspended solids concentration in the Indus Delta region using Landsat-7 ETM+ imagery. *Journal of Environmental Management, 209*, 254–261. https://doi.org/10.1016/j.jenvman.2017.12.070

UNDP. Assessment of DRM effectiveness responding to very severe cyclonic storm "Phailin" in Odisha., 1–114 (2013). http://www.ndmindia.nic.in/images/pdf/PhailinReport-UNDP-Odisha.pdf

UNICEF. India Cyclone Fani Situation Report. 1–11 (2019). https://www.humanitarianlibrary.org/sites/default/files/2019/05/UNICEF%20Odisha%20Cyclone%20Fani%20Situation%20Report%20No.%202%20-%202%20May%202019.pdf

Vanhellemont, Q., & Ruddick, K. (2014). Turbid wakes associated with offshore wind turbines observed with Landsat 8. *Remote Sensing of Environment, 145*, 105–115. https://doi.org/10.1016/j.rse.2014.01.009

Wang, J. J., & Lu, X. X. (2010). Estimation of suspended sediment concentrations using Terra MODIS: An example from the Lower Yangtze River, China. *Science of the Total Environment, 408*, 1131–1138. https://doi.org/10.1016/j.scitotenv.2009.11.057

Wang, C., Li, D., Wang, D., Chen, S., & Liu, W. (2016). A total suspended sediment retrieval model for multiple estuaries and coasts by Landsat imageries. In *2016 4th International Workshop on Earth Observation and Remote Sensing Applications, IEEE.* https://doi.org/10.1109/EORSA.2016.7552785

Wang, C., Chen, S., Li, D., Wang, D., Liu, W., & Yang, J. (2017a). A Landsat-based model for retrieving total suspended solids concentration of estuaries and coasts in China. *Geoscientific Model Development, 10*, 4347–4365. https://doi.org/10.5194/gmd-10-4347-2017

Wang, C., Li, D., Wang, D., & Chen, S. (2017b). ISSN 1819-3412). Detecting the temporal and spatial changes of suspended sediment concentration in Hanjiang River Estuary during the past 30 years using Landsat imageries. *Research Journal of Environmental Sciences, 11*(4), 143–155. https://doi.org/10.3923/rjes.2017.143.155

Wang, C., Li, W., Chen, S., Li, D., Wang, D., & Liu, J. (2017c). The spatial and temporal variation of total suspended solid concentration in Pearl River Estuary during 1987–2015 based on remote sensing. *Science of the Total Environment, 618*, 1–14. https://doi.org/10.1016/j.scitotenv.2017.09.196

Water Year Book. Central Water Commission. Mahanadi Basin, Hydrological Observation Circle Bhubaneswar. 1–13 (2017). http://cwc.gov.in/sites/default/files/admin/7A_MERO_Bhubaneshwar_Mahanadi_WYB_2016-17.pdf

Wibisana, H., Soekotjo, B. M., & Umboro, L. (2018). Preliminary study of total suspended solid distribution in coastal Ujung Pangkah Gresik based reflectance value of Landsat satellite imagery. *Indonesian Journal of Geography, 50*(1), 42–48. https://doi.org/10.22146/ijg.38967

Wilber, D. H., & Clarke, D. G. (2001). Biological effects of suspended sediments: A review of suspended sediment impacts on fish and shellfish with relation to dredging activities in estuaries. *North American Journal of Fisheries Management, 21*(4), 855–875. https://doi.org/10.1577/1548-8675(2001)021<0855:BEOSSA>2.0.CO;2

Wu, G., Cui, L., Duan, H., Fei, T., & Liu, Y. (2013). An approach for developing Landsat-5 TM-based retrieval models of suspended particulate matter concentration with the assistance of MODIS. *ISPRS Journal of Photogrammetry and Remote Sensing, 85*, 84–92. https://doi.org/10.1016/j.isprsjprs.2013.08.009

Yanti, A., Susilo, B., & Wicaksono, P. (2016). The application of Landsat 8 OLI for Total Suspended Solid (TSS) Mapping in Gajahmungkur Reservoir Wonogiri Regency 2016. *IOP Conference Series: Earth and Environmental Science, 47*(012028). https://doi.org/10.1088/1755-1315/47/1/012028

Ye, H., Chen, C., Tang, S., Tian, L., Sun, Z., Yang, C., & Liu, F. (2014). ISSN: 1463-4988 print/ 1539-4077 online). Remote sensing assessment of sediment variation in the Pearl River Estuary induced by Typhoon Vicente. *Aquatic Ecosystem Health and Management, 17*(3), 271–279. https://doi.org/10.1080/14634988.2014.944475

Zhang, M., Dong, Q., Cui, T., Xue, C., & Zhang, S. (2014). Suspended sediment monitoring and assessment for Yellow River estuary from Landsat TM and ETM+ imagery. *Remote Sensing of Environment, 146*, 136–147. https://doi.org/10.1016/j.rse.2013.09.033

Zhang, Y., Zhang, Y., Shi, K., Zha, Y., Zhou, Y., & Liu, M. (January 2016). A Landsat 8 OLI-Based, semianalytical model for estimating the total suspended matter concentration in the slightly turbid Xin'anjiang reservoir (China). *IEEE Journal of Selected Topics in Applied Earth Observations and Remote Sensing, 9*(1), 398–413. https://doi.org/10.1109/JSTARS.2015.2509469

Zhou, W., Wang, S., Zhou, Y., & Troy, A. (2006). Mapping the concentrations of total suspended matter in Lake Taihu, China, using Landsat-5 TM data. *International Journal of Remote Sensing, 27*(6), 1177–1191. https://doi.org/10.1080/01431160500353825

Chapter 11
Impact of Sand Mining on the Physical Health of the River and the Livelihood of the People: A Case Study of Umtyngngar River, Meghalaya

Baiaroihun War Shymbin and Gardinia Nongbri

11.1 Introduction

The genesis of human civilization sprouts out in the bank of rivers. When it comes to this planet's rivers, man has changed their nature by controlling floods, constructing large impoundments, overexploitation of living and non-living resources and using rivers for disposal of wastes. Among these, indiscriminate extraction of non-living resources like sand and gravel from the riverbed is one of the most disastrous activities as this threatens the very existence of the river ecosystem (Kondolf, 1993). Sand mining refers to the actual process of removal of sand or gravel from a place of their occurrence (Langer, 2003). As a resource, sand by definition is 'a loose, incoherent mass of mineral material which is a product of natural processes'. River sand is one of the world's most plentiful resources, and an estimated 40 billion metric tonnes of sand is extracted annually in the world over for different uses such as construction and glass manufacturing (Edwards, 2015). The durability of river-borne coarser clastics (e.g. sand and gravel) and their sorting by fluvial action make them the best suitable raw material/ingredient for building constructions (Kondolf, 2002), thus becoming a very important mineral for the expansion of society.

Sand is considered a 'minor mineral' under Indian law, unlike coal, diamonds or gold, which are classed as major minerals. The extraction of minor minerals is governed by the state rather than by federal laws. Illegal sand mining in India is something of an open secret. According to the Mines and Minerals Development and Regulation Act 1957, amended in May 2012, those who want to mine sand must

B. W. Shymbin (✉)
Department of Geography, Mawsynram Border Area College, Mawsynram, India

G. Nongbri
Department of Environmental Studies, North-Eastern Hill University, Shillong, India

G. S. Bhunia et al. (eds.), *Anthropogeomorphology*, Geography of the Physical Environment, https://doi.org/10.1007/978-3-030-77572-8_11

obtain a license from the state authorities and pay royalty on the amount they extract. The state government also has the power to make rules preventing illegal mining (Sugden, 2013). This Act remains unhelpful, especially in the tribal areas of India where sand has been and continues to be mined illegally every day from the river to meet the requirements of building construction.

Illegal sand mining in India is something of an open secret. Within 50 kilometres of any construction site in India, there is likely to be sand mining going on along river banks and coastal areas. Almost all of this mining is happening without any license because the demand is unbridled and regulatory consequences are minimal. It is a quick buck making industry with very little investment. All that is needed is a truck, labourers and a place to go and mine (Goenka, 2013).

Before the sand-dredging activity, the channel had remained in a basic balance between deposition and erosion on the whole (Weli et al., 2009). The actual dredging or scarping of sand and gravel during mining operations can alter stream channels and banks. Dredging or scarping usually involves enlargement or widening of the stream channel (Etnier, 1972; Boyd, 1976; Yorke, 1978), which creates uniform conditions of either deep or shallow reaches throughout the channel (Yorke, 1978). These physical effects can change the stream length, gradient, width and depth of the channel (Boyd, 1976). Channel deepening can also cause stream banks to become unstable and eroded (Bull & Scott, 1974). In-stream mining also introduces major environmental issues to a river system. With the removal of sediment, erosion can become widespread and cause loss of land and wooded areas. The loss of aquatic and riparian habitats can also result from in-stream mining (Sandecki, 1989).

Rivers are the most important life-supporting systems of nature. For centuries, humans have been enjoying the natural benefits provided by rivers without understanding much how the river ecosystem functions and maintains its vitality (Padmalal et al., 2008). Indiscriminate sand mining poses a great threat to the river affecting the health of a river because it leads to erosion of the river bank and damages biodiversity (Goenka, 2013). Indian Supreme Court in February 2011 observed that 'sand mining on either side of river in-stream or upstream is one side of the causes of environmental degradation and also a threat to biodiversity'.

Since time immemorial, sand has been exploited for its varied uses, and with the rapidly growing population and urbanization, there arises an unprecedented demand for sand to cater to the needs of the construction industry (Victor, 2013). This ever-increasing demand has led to many people making sand extraction and production as a means of livelihood. It creates employment opportunities and thereby income provision for people living near the rivers by being hired as sand diggers, loaders, transporters and suppliers (Salifu, 2016). The flipside of this activity is that there are consequences on the environment, that further have an impact, in one way or another, on other forms of livelihoods of the sand mining fringe communities (Victor, 2013). Therefore, the effects of sand mining on livelihoods could either be positive, negative or a combination of both (Akabzaa, 2009). Notwithstanding the outcomes and numerous challenges arising due to sand mining and its activities, the significant contributions to livelihood enhancement and economic development of many nations cannot be ignored (Salifu, 2016).

The rapid rise in urbanization and construction of large-scale infrastructure projects are driving increasing demands for construction materials globally. United Nations Environment Programme (UNEP) estimated that between 32 and 50 billion tonnes of sand and gravel are extracted globally each year with demand increasing, especially in developing countries (Schandl et al., 2016). Rivers are a major source of sand and gravel for numerous reasons: cities tend to be located near rivers so transport costs are low; river energy grinds rocks into gravels and sands, thus eliminating the cost of mining, grinding and sorting rocks; and the material produced by rivers tends to consist of resilient minerals of angular shape that are preferred for construction (whereas wind-blown deposits in deserts are rounder and less suitable). Here, we use "sand mining" as a generic term to embrace the extraction of riverine aggregates regardless of particle size. Sand mining activities are one of many recognized pressures affecting riverine ecosystems, where biodiversity is already in rapid decline (World Wildlife Fund, 2018). Increasingly, there are media reports about the negative environmental and social impacts of sand mining, and as calls grow for stronger regulation of mining (Schandl et al., 2016).

11.1.1 River Health

The term 'river health' is a useful and widely understood concept in recent years (Song et al., 2015). However, it is difficult to define it in precise scientific terms. There is a debate over the meaning of health when applied to ecosystem. The new emphasis on biota and ecosystem has led many people to embrace the term 'river health', which is often seen as being analogous to human health, giving a general sense of understanding (Rapport, 1989; Resh et al., 1995). Unfortunately, the meaning of 'river health' remains obscure. Clearly, it is not the same as human health because ecosystems have a life of their own without the human (Rapport, 1989).

River health is taken to mean the degree of similarity to an unimpacted river of the same type, particularly in terms of physical characteristics, its biological diversity and ecological functioning. Ecologically, river health can be defined as 'the ability of the aquatic ecosystem to support and maintain key ecological processes and a community of organisms with a species composition, diversity and functional organization as comparable as possible to that of undisturbed habitats within the region' (Schofield & Davies, 1996).

River health means physical, chemical and biological integrity, which refers to maintaining natural ecosystems' structures and functions (Norris & Thoms, 1999). Schofield indicated that river health refers to the similarity to the undisturbed (original) river of the same type on the biological integrity and ecological function (Clean Water Act (CWA), 1972).

11.2 Statement of the Problem

River health is one of the main problems faced for centuries especially for rivers flowing through urban areas. Environmentalists focus on river health mainly in terms of disposal of pollutants in the form of solid waste and chemical substances. Unfortunately, few consider indiscriminate sand mining as a factor in the deterioration of river health in rural areas. In the past few decades, demand for construction grade sand is increasing in many parts of the world due to rapid economic development and subsequent growth of building activities (Salifu, 2016). This has led to sand mining proving to be a lucrative form of livelihood for the people. However, with sand and gravel mining continuing to occur in or near rivers, questions about its physical and environmental effects have increased. Given the urgent need for economic development, the long-term impacts of indiscriminate sand mining with reference to river health and the livelihood of the people have rarely been studied. River sand mining has been practiced in the Umtyngngar River for many years, and therefore, the present study aims at exploring these two aspects of sand mining activities in Umtyngngar River of Meghalaya as a case study. It focuses on assessing the river health by estimating the extent of sand mining in the area and by taking into consideration only one aspect, i.e. bank erosion or change in the channel width of the river due to sand mining. Other impacts of sand mining on river health, which are equally important, are excluded from the scope of the study due to time and resource constraints. The positive and negative impacts of sand mining on the livelihood of people residing in the study area are also investigated. The research primarily seeks answers to the issue of sand mining as affecting the health of the river and the implications it has on the livelihood of the local people.

11.3 Objectives

- To detect the impact of sand mining on the river health of Umtyngngar River with reference to the channel of the river.
- To highlight the impact of sand mining on the livelihood of the people in the study area.

11.4 Study Area

The village of Umtyngngar is a medium-size village located in Khatarshnong Laitkroh Block of East Khasi Hills district, Meghalaya, with a total of 43 families residing in it. The village has a population of 245, of which 119 are males while 126 are females (Census of India, 2011). The physiography of Umtyngngar consists of undulating tract with moderately gentle sloping land to very steep sloping lands with small pockets of small to medium valley lands. There are high hills present, and the slope gradient is found to be more than 33% (Jena et al., 2016). The undulating topography of the area blended in a mosaic of rocky outcrop exposures where soil depth is very low with sandy soil texture has resulted in a low prevalence of natural vegetation. As a result of these factors, pine (*Pinus kesiya*) has become the dominant tree species across the landscape. The primary vegetation of the area can be seen only on a few scattered pockets along depressions having good moisture concentration. Some tree species include *Quercus* spp., *Castanopsis* spp., *Schima khasiana*, *Betula alnoides* and *Alnus nepalensis* (MBDA, 2014).

The Umtyngngar River is one of the major rivers in East Khasi Hills district of Meghalaya and flows from the east to the south-western part of Meghalaya, joins the Umiew River, and falls into the Mawphlang dam. The study area of Umtyngngar River is geographically located toward the southern part of Shillong, the capital city of Meghalaya. It has an elevation ranging between 1100 and 1700 m above sea level and a depth of 0.42–0.58 m. This river is most suitable to study the impact sand mining on the river health because the river is subjected to sand mining for many years for its high-grade sand that is exported to the city for building construction. This area is also well suited for the study in the perspective of sand mining because it is an area which consists of high-grade crystal sand, deposited after erosion and transportation from the upstream of the river. Sand mining is being practiced all along the Umtyngngar River, but for the study, the different sites of the study area have been selected where sand is mostly extracted from the river. The study area is confined to only a small area and an extension of 25°27′55.52″N to 25°27′36.42″N latitude and 91°49′27.99″E to 91°47′34.17″E longitude. The study area extended to about 3.28 kilometres in length with an average altitude of 1640 metres (Fig. 11.1).

11.5 Materials and Methods

This research work adopted both qualitative and quantitative approaches for studying the impact of sand mining on the river health as well as the livelihood of the people. The study was carried out by conducting an extensive field survey in the study area where sand mining is mostly practiced and also by using GIS tools and technology for achieving the results.

Primary data collection was carried on in the period from February to June 2015 by close field observation, ground truth observation and verification, interviews, focus group discussions with residents of Umtyngngar, the local leaders and workers around the area for obtaining unbiased ideas and accurate information. For the estimation of sand quantity extracted from the selected sites, methods by Padmalal et al. (2008) in their study on Environmental effects of river sand mining: a case from the river catchments of Vemband Lake, Southwest coast of India has been used. In the present study, the sand extracted from the selected sites has been counted by observing vehicles loaded with sand moving out from the respective sites during the sand extraction period.

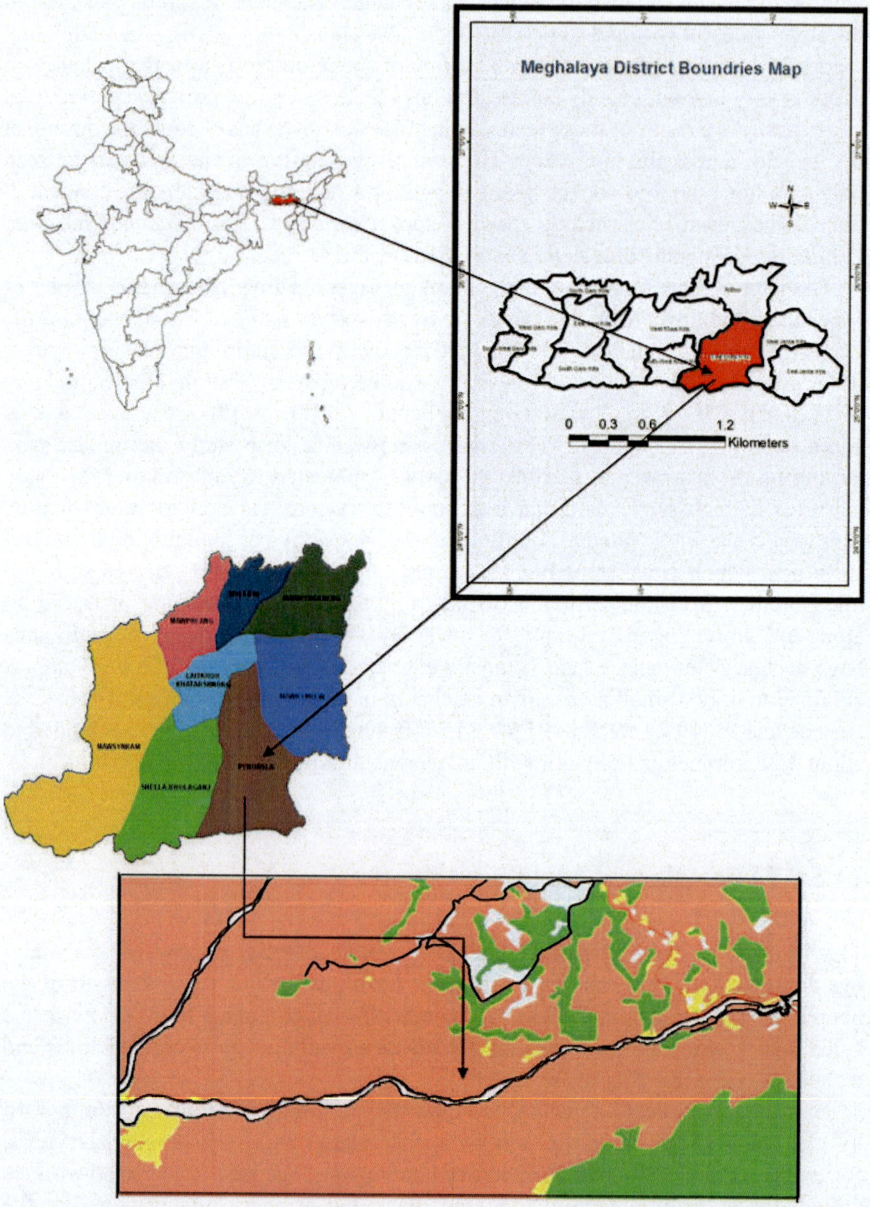

Fig. 11.1 Location map of the study area

The secondary data were attained with the aid of GIS tools, whereby the methods used by Shaji and Anilkuar (2014), in their study of the Socio-Environmental Impact of River Sand Mining: An Example from Neyyar River, Thiruvananthapuram District of Kerala, India, have been adopted. Where:

(a) The widening of river bank or the channel change detection due to sand mining is analysed by comparing the area represented in two Google Earth Imageries, i.e. the year 2002 and 2014 (representing a 12-year difference), respectively, by dividing the river channel into three sections.
(b) Using Google Earth Imagery as base maps, the river channels have been digitized by creating different data base or layers for different years which is differentiated by the colours of the line. After digitizing, the data has been saved as *.kmz*, which is then imported in ArcGIS software 9.3 versions. The two digitized maps of the years 2002 and 2014 are then overlayed, and using the Measure tool on the Tools toolbar of the software, a cross-section of the two overlayed river channels has been generated to measure the change in the channel and width of the river in metres. The changes detected from the cross-section of the final layout map are then analyzed and interpreted.

Assessment of the river health of the area is based and focused only on one of the parameters, namely, the physical health of the river wherein the bank erosion and channel shifting is taken into consideration in the present study. The impact of sand mining on the physical health of the river has been assessed by comparing between two maps of two different years by detecting the change in the channel and also the increase and decrease of the width of the river.

A simple measurement of the eroded land has been carried out using a measuring tape and also the depth and thalweg of the river were measured by using a ranging measuring rod. But the overall and correct measurement of the river to compare the past and the present condition of the river has been measured using Google earth imagery of the year 2002 and the year 2014 and through the import of the .kmz file generated from Google Earth in Arc GIS software 9.3 version for the overall measurement of the eroded land of the Umtyngngar River.

The river channel has been divided into left bank and right bank. The blue colour line shows the bank of 2002, and the red colour line shows the bank of 2014. When the green colour line moves beyond the red colour line in the right means that the bank has been eroded or shifted toward the right bank and if the green colour line moves beyond the left of the red colour line means that the river channel has been eroded or shifted toward the left. The measurement taken which mentioned the right and the left banks as well as the directions of the shift and how much the channel has been shifted (in metre) has been recorded in a Microsoft Excel spreadsheet for further analysis.

The following is a schematic presentation of the methodology adopted (Fig. 11.2).

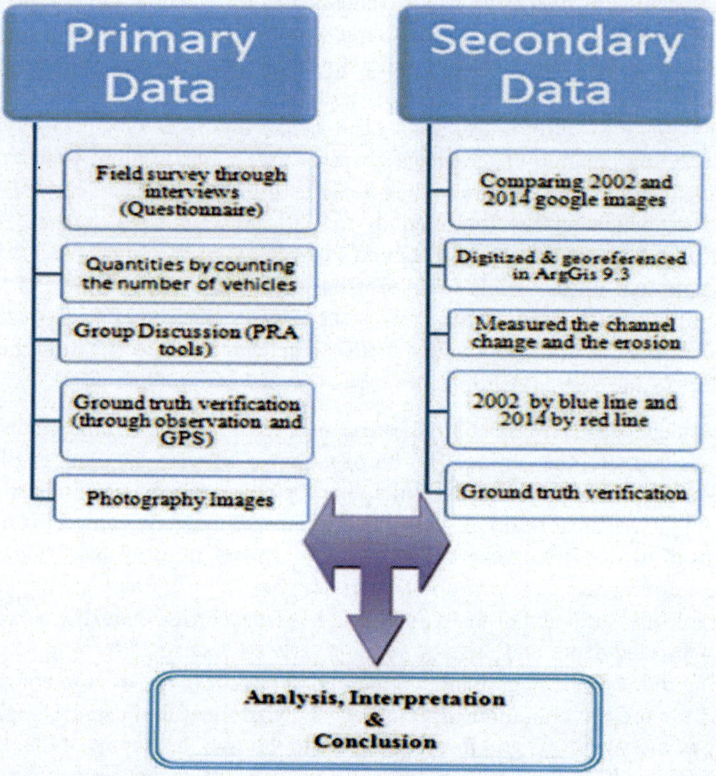

Fig. 11.2 Methodology flowchart

11.6 Results and Discussion

The primary and secondary data collected from the selected sites of Umtyngngar River are analysed to assess one aspect of the physical health of the river as well as the impact that sand mining has on the livelihood of the people in the study area. This includes records of the quantities of sand mining extracted from the study site, method of sand mining in the area and livelihood information by interviewing local people. The images and maps depicting the channel change and river bank erosion in the study area have been prepared using Google earth satellite imagery as well as ArcGIS software of 9.3 versions.

Umtyngngar River is a tributary of Mawphlang Dam. The selected site of study area is from the Umtyngngar Bridge (upstream) up to the estuary of Mawphlang Dam (downstream), which is a highly sand deposited site of the river. Mining of sand has been in practice in the Umtyngngar River for many years, and this has eventually led to a great impact on the overall river health, be it biologically, chemically or physically as well as the livelihood of the people residing in the study area. However, in the present study, focus has been given to the physical health as the river has shifted its channel and the river width has also increased to a great extent due to sand mining. The environmental stress in this fraction is also very high as sand is being mined at upper stream at an alarming rate. Sand mining has affected the water flow and destabilized the river banks leading to loss of land. The impact of sand mining in the study area has been highlighted as follows:

11.6.1 Types and Methods of Sand Mining

River sand is mined extensively from the drainage networks of the Umtyngngar River, depending on the physiographic characteristics, river orders and ecological significance. However, the intensity of mining is high in the alluvial reaches of the main channels. In addition to the mining of sand from active channels (in-stream mining), a substantial amount of sand is also being extracted from the overbank areas (floodplain mining) of the river as well. Generally, two types of in-stream mining are in practice in the study area—(i) pit excavation and (ii) bar skimming.

(i) Pit excavation is the extraction of sand and gravel from the riverbed or floodplain areas by uncontrolled digging in which draglines or dredges remove material from below the water table or directly from a stream channel (Fig. 11.3).
(ii) Bar skimming, on the other hand, is the controlled extraction of sand from the exposed sand bars (in-stream bars and point bars) in the channel environment. Usually, bar skimming would be done above the water table and within a minimum width buffer that separates the excavation site from the low flow channel and the adjacent active channel bank (Fig. 11.4).

Fig. 11.3 Wet pit mining (25°27′41″N to 91°49′37″E)

Fig. 11.4 Bar skimming method of sand mining (25°27′47″N to 91°49′40″E)

Fig. 11.5 Some tools used for sand mining

Of the two types of sand mining, pit excavation is the more widespread sand mining method adopted in the alluvial reaches of the rivers.

Extraction of sand from the Umtyngngar River has been practiced by using old and traditional methods or tools by the local people (Fig. 11.5). For bar skimming or flood plain mining, the people use a shovel to extract the sand and carry it in a basket from the bank of the river to the roadside for easy transport as there are no roads which lead to the river for easy transportation of sand. On the other hand, for wet pit mining, the people also use a long shovel which has a curved tin tie with a long wooden stick for easy extraction of the sand from the riverbed. The extracted sand is then placed in a wooden boat of 4/6 inch, and the sand is then carried by a basket into the roadside.

11.6.2 *Quantity of Sand Extraction*

As per the field survey that was carried out in the months of February to June 2015, an estimate of the quantum of sand mining in the study sites in different sections of the river has been attempted. The methods adopted to estimate the extent of sand mining are based on the approach of Padmalal et al. (2008). The estimation is based on the interview carried among the local labourers in the mining sites to obtain information about the sand extracted from these areas as per the number of trucks on a particular day. One person was hired to observe the mining sites to get an exact number of trucks where sand has been extracted from the area. It has been found in the study area that it has a number of both active and abandoned sand mines. However, the information on the quantities of sand extracted from the abandoned mines was obtained by interviewing elderly people who were engaged in sand mining activity in that area for a long period of time. This information was collected towards the end of the field data collection period (June 2015) as per the convenience of the responders. They proved to be a very useful resource as they had first-hand experience of the activity of sand mining that has been going on in the river for a long time.

11.6.3 *Quantity of Sand Extracted in Two Mining Sites of the Area*

The tables below describe the quantity of sand extracted from two mining sites in the study area as per the number of trucks on a particular day. Mining site 1 (25°27′41″N to 91°49′37″E) (Table 11.1) and mining site 2 (25°27′47″N to 91°49′40″E) has been selected for the study (Table 11.2).

Table 11.1 Sand extraction in mining site 1

Quantities extracted	**2–3 trailers a day (bolero pick up)**		
	Average of 2 trucks		
	In a week	10 trucks	20 tonnes
	In a month	40 trucks	80 tonnes
	In 8 months (Mining season: October–may)	320 trucks	640 tonnes

Table 11.2 Sand extraction in mining site-2

Quantities extracted	**1–2 trailers a day (bolero pick up)**		
	Average of 2 trucks		
	In a week	5 trucks	10 tonnes
	In a month	20 trucks	40 tonnes
	In 8 months (Mining season: October–may)	200 trucks	240 tonnes

11.7 Impact on River Health

The health of the river in the study area has been affected physically due to the sand mining taking place for so many years. The flow regime of the river has been changing, bank erosion is high in the mining sites, channel has been shifting, etc. The sand mining activities have affected the physical health of the river to a large extent.

Bank Erosion and Channel Shifting For assessing the impact of sand mining on the physical river health of the river, only one of the parameters was used in this study which is the bank erosion and channel shifting of the river. This has been assessed by comparing two digitized maps of two different years by detecting the change in the channel and width of the river.

In order to assess the physical health of the river, the study area has been divided into three sections for a clear and in-depth analysis which is as follows:

Section 1

In the first section of the river, which extends over 283 metres of the study area and as highlighted in Fig. 11.6 and Table 11.3, respectively, it shows that there has been a slight shift of the channel on both the right bank and left bank of the river. This section was an old mining site that has been abandoned for the last 3 years, and due to overexploitation of the sand from the river, most of the banks has been eroded. In the cross-section A1–A2, about 4.83 metres of land on the left bank and 5.24 metres of land from the right bank were eroded, which causes the channel of this section of the river to widen on both sides, respectively. Similarly in cross-sections B1–B2, C1–C2 and D1–D2, it is recorded that 0.24, 6.78 and 4.26 metres of land, respectively, have been eroded from the left bank of the river, while 6.23, 3.12 and 0.11 metres of the land, respectively, have been eroded on the right bank of the river. This has eventually resulted in the shifting and widening of the river channel on both the directions. Bar skimming method of sand extraction is the main factor that led to the high level of erosion.

Section 2

In the second section (Fig. 11.7) of the river, as seen in Fig. 11.6 and Table 11.4, about 0.14, 0.44 and 6.13 metres of land from the left bank as well as 2.24, 6.23 and 0.12 metres of land from the right bank of all the tree cross-sections, respectively, have been eroded leading to the widening and shifting of the river channel on both left and right direction (Table 11.4).

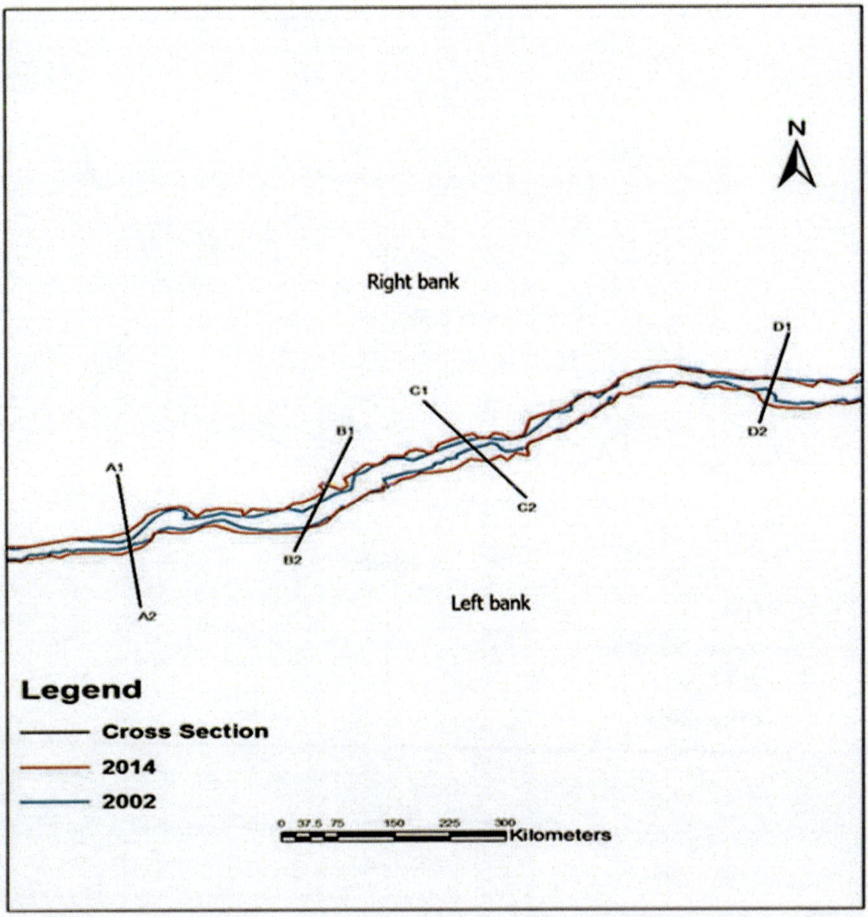

Fig. 11.6 Umtyngngar River: Section 1

Table 11.3 Channel shifting and erosion of the bank in the first section

Section	Left bank (in metres)	Direction	Right bank (in metres)	Direction
A1–A2	4.83	Left	5.24	Right
B1–B2	0.24	Left	6.23	Right
C1–C2	6.78	Left	3.12	Right
D1–D2	4.26	Left	0.11	Left

Section 3

As per Fig. 11.8 and Table 11.5, it can be observed that in the third section of the river, there has been a drastic shifting of the channel in the left bank of cross-section B1–B2 as well as C1–C2, respectively, whereby about 9.31 metres of land on the

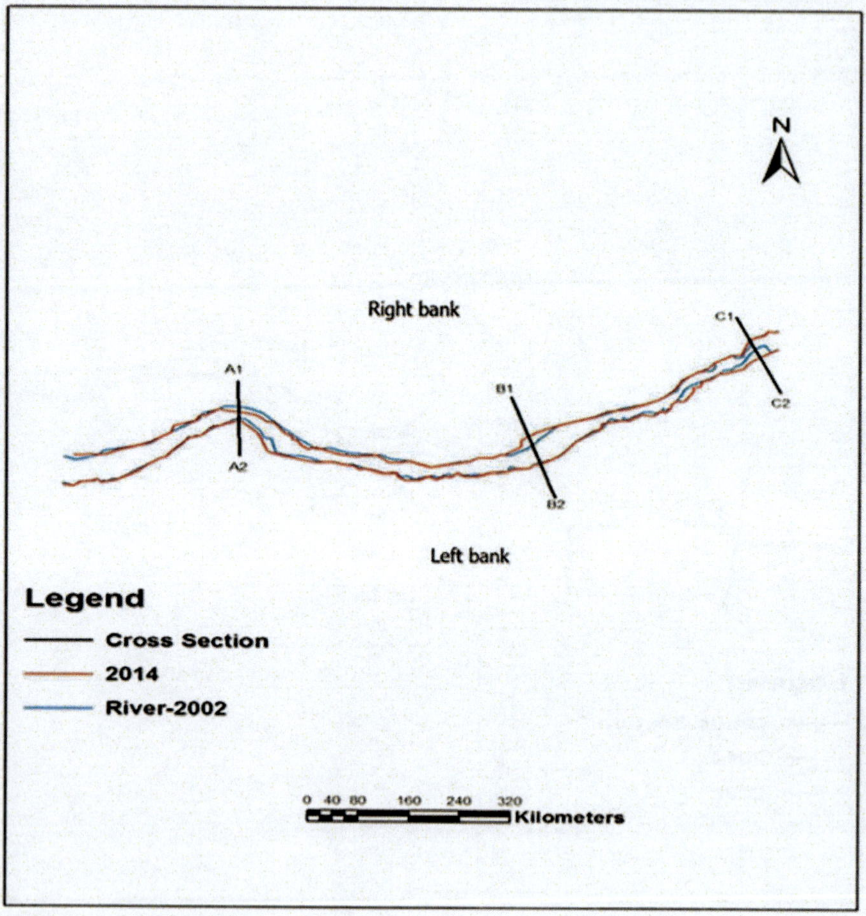

Fig. 11.7 Umtyngngar River: Section 2

Table 11.4 Channel shifting and erosion of the bank in the second section

Section	Left bank (in metres)	Direction	Right bank (in metres)	Direction
A1–A2	0.14	Left	2.24	Right
B1–B2	0.44	Left	6.23	Right
C1–C2	6.13	Left	0.12	Right

left bank of the former cross-section and 10.22 metres of the right bank of the later cross section have been eroded on both directions. As highlighted in the figure and table above, in all the four cross-sections, there has been a slight shifting of the river channel on both sides of the river. However, on average, the total land eroded is highest in this third section of the river as compared to the other two, and this is mainly because most of the mining sides are found in this section.

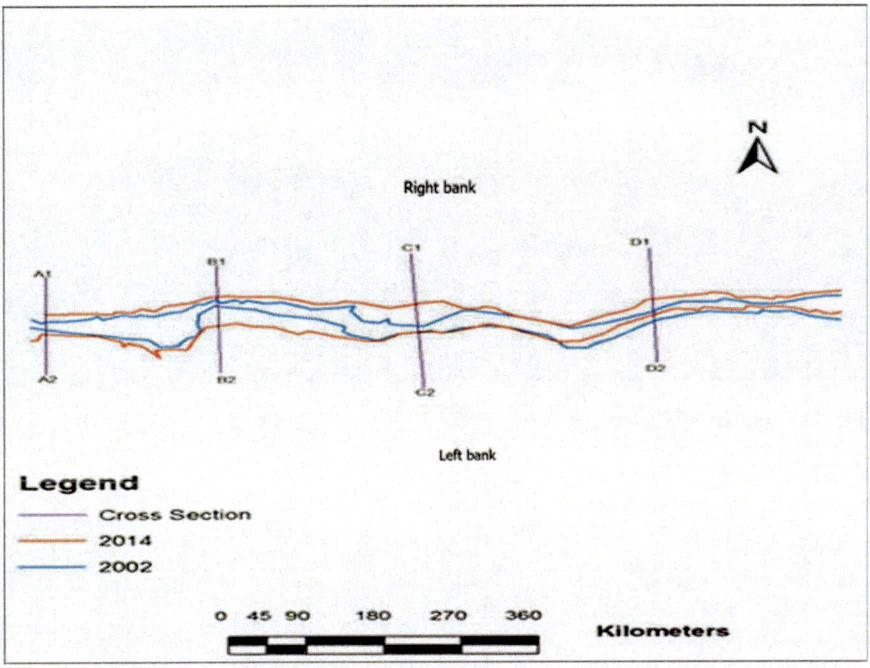

Fig. 11.8 Umtyngngar River: Section 3

Table 11.5 Channel shifting and erosion of the bank in the third section

Section	Left bank (in metres)	Direction	Right bank (in metres)	Direction
A1–A2	2.34	Left	3.32	Right
B1–B2	9.31	Left	0.33	Right
C1–C2	0.71	Left	10.22	Right
D1–D2	2.83	Right	4.11	Right

Threat to Mawphlang Dam As per the field observation and interviews with some of the local people, it was observed that during the rainy season, the colour of the water at Mawphlang dam (Figure 11.9.), which supplies water to the whole city under the Greater Shillong Water Supply Scheme, changed to a muddy brown colour. One of the major contributing factors to this phenomenon is the unscientific and indiscriminate sand mining and quarrying as all the sand and pebbles, compounded by the rainfall factor, flow from the extraction sites to the Umtyngngar River before heading to the Mawphlang dam. In June 2014, the Deputy Chief Minister, in charge of Public Health Engineering (PHE) Department, R.C. Laloo, said that the department had taken steps to control the sedimentation at Mawphlang dam, primarily through the construction of silt retention dam across Umtyngngar River. However, even with the construction of the silt retention dam, one cannot expect it to be a complete foolproof preventive measure until and unless sand mining regulations are put in place and enforced by the state government firmly (Shillong Times, 2014; The Hindu Business Line, 2013).

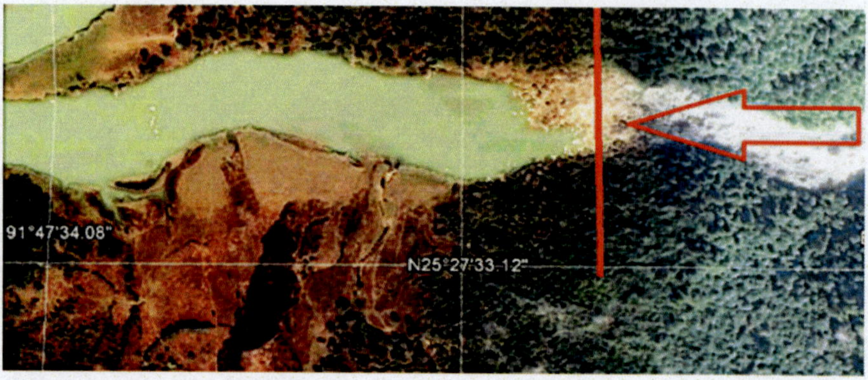

Fig. 11.9 Estuary of Mawphlang Dam

Table 11.6 Positive effects of sand mining on livelihoods

Positive effects of sand mining	Strongly agree	Agree	Disagree	Strongly disagree	Total %	Total no. of respondents
Sand mining has created more employment	50	10	–	–	100	60
Sand mining provides improvement in roads infrastructure	–	16	36	8	100	60
Sand mining brings higher income	16	42	2	0	100	60
Sand mining brings increased sales	21	34	5	–	100	60

11.8 Effects of Sand Mining on Livelihoods

Sand mining and its associated activities have impacted the social, economic and environmental aspects of the communities living in the mining areas. There are outcomes of sand mining activities that can be considered positive when beneficial and profitable results are attained and negative when there are adverse and destructive outcomes.

Positive Effects of Sand Mining on Livelihoods In this section, the positive effects of sand mining on the livelihood of the people in the study area are analysed. The analysis is based on questionnaires that focused on four main effects of sand mining, as mentioned in Table 11.6. Sand mining and its varied activities are increasingly becoming alternative avenues for employment for many people, especially those living in rural areas. This can be attributed to the fact that employment opportunities have become harder to come by for the majority of the population. As shown in Table 11.6, out of the 60 respondents sampled for the study, 50 of them (83%) strongly agree, and 10 (17%) of them agree that sand mining has created more employment avenues. None of the respondents disagreed with this positive effect. The respondents resonated with the fact that sand mining has opened up many job opportunities, both direct and indirect, for many people living in the sand mining fringe communities. These job opportunities come in the form of sand diggers, excavator operators, loaders and transporters. The indirect avenues are the additional services required by sand miners such as food vendors, water sellers and vehicle mechanics.

Another aspect of the positive effects of sand mining on livelihoods is the improvement in road infrastructure. According to the data collected, the majority of the respondents (73%) disagree and strongly disagree with this point, and only 27% of the respondents agree. The ones who disagree are of the view that road accessibility is already prevalent in the study area, but the roads are in a dilapidated condition with false promises of improvement by the local elected representative as well as sand mining owners and contractors. Some have pointed out that the bad condition of the roads in the form of potholes is a result of the frequent plying of the trucks used for transporting the sand. Contrary to this, 16 out of the 60 respondents agree that there is improvement in road infrastructure. This is attributed to the fact that these respondents belong to the section of the study area where road repair and construction have been carried out by some of the sand miners. This resulted in better road links to more remote hamlets of the study area. Sand miners construct these untarred access roads in order to access sand mining sites and also transport sand to intended destinations. The results from the table depict that majority of the respondents do not agree with this positive aspect that sand mining provides improvements in road infrastructure.

Income Level of Sand Miners The third positive effect of sand mining on livelihoods is that it generates higher income for the sand miners. For this aspect of the positive effects, Fig. 11.9 has been drawn up to illustrate the income level of sand miners before and after taking up sand mining as a livelihood. To understand the income level of the miners better, different categories of income in ascending order were used, as shown in Fig. 11.10. From the interview, it was found that before taking up sand mining, the income levels of the sand miners were distributed across the various income categories. From the figure, it can be seen that the majority of the miners received relatively lower monthly income. 32% of the sand miners fall below the ₹5000 income category and 42% come under the ₹5000–10,000 category with only 19% and 7% under the ₹10,000–₹15,000 and above ₹15,000 categories, respectively. This is because before sand mining started in Umtyngngar River, many of the sand miners were farmers and others had odd jobs, which did not generate much money to support themselves well. From Fig. 11.9, the bars depicting the income level after sand mining became their source of livelihood and income indicate that as high as 67% of the miners fall under ₹10,000–₹15,000 income category with 22% and 11% of them fall under ₹5000–10,000 and ₹15,000 categories, respectively. No one was earning less than ₹5000. This cements the fact that sand mining brings higher income to the communities with a majority of the respondents agreeing with this point, as seen in Table 11.6.

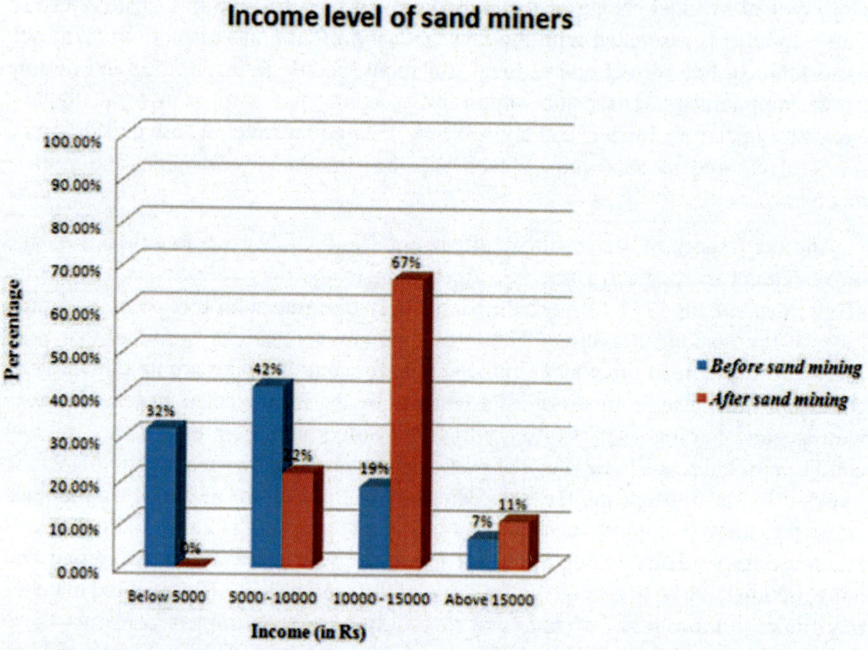

Fig. 11.10 Sand miners' income before and after sand mining

The fourth positive effect of sand mining on livelihoods considered for this study is sand mining fetches increased sales. This would include commercial sales and services within the fringe communities and even beyond. With the increase in sand mining across the study area through the years, there has been a stark increase in the influx of people to the area. As a result of this, there has been an increasing demand for goods and services by the sand miners. This in turn, resulted in people from the community to take up various forms of commercial activities to meet these demands and also serve as a source of livelihood for them. From the study, it was found that many of the responders (55%) agree and strongly agree the view that sand mining does indeed open up many indirect job opportunities in the form of commercial activities, thereby increasing sales. It was revealed that the reason for this is because as the sand mining activities progress throughout the day, sand miners buy several products like food items, cigarettes, gutkha, water and other general provisions from these shops. They also require the services of mechanics for their trucks.

Negative Effects of Sand Mining on Livelihoods Sand mining and its activities result in bank erosion or an increase in channel width which subsequently leads to environmental degradation such as removal of topsoil. As a result, agricultural land in these areas is slowly reduced. Despite the success of sand mining, farming is still one of the major sources of income for a large part of the population. For the first negative effect of sand mining on livelihoods, it was found that a majority of the respondents (46%) strongly agree that farming land has been slowly diminishing and

Table 11.7 Negative effects of sand mining on livelihoods

Negative effects of sand mining	Strongly agree	Agree	Disagree	Strongly disagree	Total %	Total no. of respondents
Lands for farming reduced/destroyed	46		14	–	100	60
Reduction in employment opportunities	41	8	10	1	100	60
Reduced fish population	52	7	1	–	100	60

destroyed after sand mining started. The 14% of the respondents disagree with the point as they believe that there is still plenty of land available for farming. During data collection in the field, a transect walk was carried out, and it was observed and informed by the village elders that many of the lands have been converted to sand mines. This paints a grim picture of choosing profit over environmental and livelihood sustainability. Additionally, the destruction of agricultural land due to sand mining activities can lead to many farmers losing their only source of livelihood, subsequently leading to amplified poverty.

The other negative impact regarding reduction in employment opportunities can be linked to the first negative impact. As discussed above, a large part of the population of the study area is also engaged in farming, and thereby loss of agricultural land to sand mining activities can result in lesser cultivable land available for the farmers. This issue can also create conflicts for acquiring the land between thc farmers and sand miners. From Table 11.7, it can be seen that 49% of the respondents agree that there has been a reduction in employment opportunities. The respondents attributed this solely due to the loss of agricultural land. Another flip side to the unemployment among those who have made sand mining as part of their livelihood is that due to the NGT ban of sand mining, there has been reduced employment in this line of work as well. This has had its consequence, and many young people are either migrating out looking for jobs or are planning to do so.

The third negative effect of sand mining on livelihoods is the reduction in fish population in the river. The major aspect of the general negative impacts of sand mining is ecological destruction caused to the river. The Umtyngngar River is a habitat for many life forms and supports the survival of different fishes and planktons. A river should remain unimpacted so as to maintain a healthy ecosystem. However, sand mining activities are directly responsible for the destruction of water bodies in every aspect. Water quality and aquatic life have been adversely affected by sand mining activities in the area. A majority of the respondents (52%) strongly agree that there has been a reduced fish population over the years. The discussion with village elders brought to light the fact that some people of the village earn their livelihood by selling fish caught from the river. However, due to these adverse impacts by sand mining on the fish population, the people have to migrate to other sources of livelihood. This can be supported by 99% of the respondents, including sand miners agreeing to this negative impact which subsequently affects one of the sources of livelihood of the people in the study area.

Therefore, in general, this particular research found out that sand mining has affected positively and negatively in many aspects, thereby expanding the much-needed knowledge base on how sand mining has played a significant role in influencing people's livelihoods.

11.9 Conclusion

Bulky scale mining of sand and gravel has led to irreparable damages to the land, water, biotic and social as well as human environments related to many ecosystems. The problem is serious in the case of the rivers in the northeast of India, especially in Meghalaya, where the rivers are comparatively minute with limited riverbed resources. The present research was undertaken to study this impact on river health by estimating the extent of sand mining in the area and by taking into consideration only one aspect, i.e. bank erosion or change in the channel width of the river due to sand mining. Bank erosion or increase in channel width is considered the most significant of all impacts as it directly affects the livelihood of people living close to the river as they face problems of loss of agricultural land and other resources. The health of the Umtyngngar River, especially the physical health, has been affected badly where the morphology of the river or the width of the river has changed drastically in most sections, whereby because of erosion, the channel of the river has shifted in both directions leading to loss of land resources. Mawphlang dam, which is the main source of drinking water which the entire town, is bearing the repercussions of indiscriminate mining in the study area as the Umtyngngar River converges with the Umiew River, which then flows to the dam and deposits sediments.

From the study, it was found that issues pertaining to environmental sustainability were a common grievance among the respondents. Despite the positive effects of sand mining as a source of livelihood, respondents practising other forms of livelihood voiced out their concern for livelihood security and a call for concerted efforts by all the stakeholders in the sand mining industry to find strategies to address the issues. It can be stated that there is a dire need to further understand the complex issues arising due to sand mining activities on the livelihoods of the people in the fringe communities.

There is a rising trend in the research community to apply various technologies and techniques to address various problems and to find solutions to them. Geographic Information Systems (GIS) and Remote Sensing are two such important tools that can aid in identifying the trends of the river channel and as a source of input data to determine river changes and bank erosions. The results of the present study indicate the usefulness and validity of these techniques. The information attained from such studies is crucial for understanding river health characteristics and also to aid in planning and management strategies which have remained poorly understood. Therefore, this knowledge gap can only be bridged by further academic research.

Acknowledgements The authors are thankful to the village headman and the local people, including sand miners, village elders, farmers and shopkeepers of the studied sites, for providing us with detailed information on sand mining activities. We sincerely thank Mr. Preciouswell Nongsiej, whose insight and expertise greatly enhanced the research. We are indebted to Mr. Jerry Buiting and Ms. C. Zohlupuii for assisting us in the field and providing their expertise with ArcGIS.

References

Akabzaa, T. (2009). *Mining in Ghana: Implications for national economic development and poverty reduction*. Researchgate.

Boyd, G. B. (1976). *Aggregate extraction in Yolo County: A study of impacts and management alternatives*. Woodward Clyde Consultants.

Bull, W. B., & Scott, K. M. (1974). Impact of mining gravel from urban streambeds in the southwestern United States. *Geology, 2*(4), 171–174.

Census of India. (2011). https://www.census2011.co.in/census/state/meghalaya.html. Last accessed 12 July 2020.

Edwards, B. (2015). The insatiable demand for sand. *Finance & Development, 52*, 4.

Environmental Protection Agency (EPA). (1972). Clean Water Act (CWA), United States of America.

Etnier, D. A. (1972). The effects of annual rechanneling on a stream fish population. *Transactions of the American Fisheries Society, 101*(2), 372–375.

Goenka, D. (2013). Conservation Action Trust, Mumbai.

Government of India Ministry of Mines. (2018). Mines and Minerals (Development and Regulation) Act, India.

Jena, R., Duraisami, V., Shanmugasundaram, R., Krishnan, R., Padua, S., Bandyopadhyay, S., Ramachandran, S., Ray, P., Deb Roy, P., Singh, S. K., & Ray, S. K. (2016). Characterization and classification of soils of Jirang Block in Meghalaya Plateau. *Agropedology, 26*(1), 47–57.

Kondolf, G. M. (1993). The reclamation concept in regulation of gravel mining in California. *Journal of Environmental Planning and Management, 36*(3), 395–406.

Kondolf, G. M. (2002). The reclamation concept in regulation of gravel mining in California. *Journal of Environmental Planning and Management, 36*(3), 395–406.

Langer, H. (2003). *A general overview of the technology of in-stream mining of sand and gravel resources, associated potential environmental impacts, and methods to control potential impacts*. Researchgate.

Meghalaya Basin Development Authority (MBDA). (2014). Study of traditional usage and availability of plants in Meghalaya. SENES Consultants India Pvt. Ltd. & ENDEV, Kolkata.

Norris, R. H., & Thoms, M. C. (1999). What is river health? *Freshwater Biology, 41*, 197–109.

Padmalal, D., Maya, K., Sreebha, S., & Sreeja, R. (2008). Environmental effects of river sand mining: A case from the river catchments of Vembanad lake, southwest coast of India. *Environmental Geology, 54*, 879–889.

Rapport, D. J. (1989). What constitute ecosystem health. *Perspectives in Biology and Medicine, 33*, 120–132.

Resh, V. H., Norris, R. H., & Barbour, M. T. (1995). Design and implementation of rapid assessment approaches for water resources monitoring using benthic macroinvertibrates. *Australian Journal of Ecology, 20*, 108–121.

Salifu, B. (2016). *Implications of sand mining on the environment and livelihoods in Brong Ahafo region* (Vol. 10). Department of Geography and Rural Development, Kwame Nkrumah University of Science and Technology.

Sandecki, M. (1989). Aggregate mining in river systems. *California Geologist, 42*(4), 88–94.

Schandl, H., Fischer-Kowalski, M., West, J., Giljum, S., Dittrich, M., Eisenmenger, N., & Fishman, T. (2016). *Global material flows and resource productivity*. Assessment report for the UNEP International Resource Panel. UNEP, Paris.

Schofield, N. J., & Davies, P. E. (1996). Measuring the health of our rivers. *Water (Australia), 23*, 39–43.

Shaji, J., & Anilkuar, R. (2014). Socio-environmental impact of river sand mining: An example from Neyyar River, Thiruvananthapuram District of Kerala, India. *IOSR Journal of Humanities and Social Science (IOSR-JHSS), 19*(1), 01–07.

Song, J., Cheng, D., Li, Q., He, X., Long, Y., & Zhang, B. (2015). An evaluation of river health for the Weihe River in Shaanxi Province, China. *Advances in Meteorology, 2015*. https://doi.org/10.1155/2015/476020

Sugden, J. (2013). Why India has a 'Sand Mafia'. *The Wall Street Journal*. https://www.wsj.com/articles/BL-IRTB-19756. Last accessed 20 Aug 2020.

The Hindu Business Line. (2013). https://www.thehindubusinessline.com/news/national/meghalaya-not-to-ban-sand-mining-but-regulate-it/article23110144.ece. Last accessed 8 November 2020.

The Shillong Times. (2014). https://theshillongtimes.com/2014/06/10/mawphlang-reservoir-sedimentation-in-check/. Last accessed 8 November 2020.

Victor, O. A. K. (2013). *An assessment of the environmental and socio-economic impacts of sand mining activities; a case study in the Anayeri catchment in the Kassena-Nankana East and Bolgatanga municipality*. University for Development Studies, Ghana.

Weli, V. E., Wizor, C. H., & Chukwu-Okeah, G. O. (2009). Impact of sand dredging on river morphology of catchment areas to the New Calabar river. *Journal of Nigerian Environmental Management, 5*(2), 67–72.

World Wildlife Fund: Living planet report. (2018). *Risk and resilience in a new era*. WWF International, Gland.

Yorke, T. H. (1978). *Impact assessment of water resource development activities: A dual-matrix approach*. Department of the Interior, Fish and Wildlife Service, Eastern Energy and Land Use Team, Washington, D.C. (1978) a new era. WWF International, Gland.

Chapter 12
Assessment of Water Quality and Landscape Dynamics in Some Selected Pit Lakes of Andal Block, Paschim Bardhaman, West Bengal, India: A Geospatial Appraisal

Debnath Palit, Saikat Mandal, Swarupa Das, Papia Nandy Palit, and Soumik Bhattacharya

12.1 Introduction

When open-cut mining procedure ceases, pit lakes establish and the residual pit is filled with ground-, surface- and rainwater. In order to minimize mining costs, mine pits, i.e. pit lakes, generally have higher depth to surface area ratios with comparatively level bottoms and shallow edges. As a feature of regional geomorphology and catchment connection, water quality differs between pit lakes. In peculiarly hot, dry atmospheric areas of India with comparatively few natural water sources, water in pit lakes has the ability to be effective for a range of applications. Their importance as recreational opportunities, fish culture, supply of water and the habitat structure depends mainly on their topography, protection and stability. Coal mining began even during British East India era in the Raniganj coalfield (RCF) area in 1774. Moreover, in the wake of mining closes, such exercises additionally leave a tradition of open mine pits. In addition to being ecologically endangered and vital aquatic environments, pit lake natural surroundings are in reality a wellspring of huge

D. Palit (✉)
Department of Botany, Durgapur Government College, Durgapur, West Bengal, India

S. Mandal
Department of Zoology, Raghunathpur College, Raghunathpur, West Bengal, India

S. Das · P. N. Palit
Department of Geography, Barjora College, Barjora, West Bengal, India

S. Bhattacharya
Aerospatial Solution Pvt. Ltd., Thane, Maharashtra, India

G. S. Bhunia et al. (eds.), *Anthropogeomorphology*, Geography of the Physical Environment, https://doi.org/10.1007/978-3-030-77572-8_12

biological assets for what's to come. Open pit lakes seem to be a component of many cities in India over the years, maintained by inhabitants as the primary source of potable water for various recreational uses. But nowadays, due to rapid urban growth, demand, improvements in the nature and quantities of solid waste and sanitation, as well as human error, this area is under attack from multiple directions. In these pit lakes, the two major severe problems are that of encroachment due to urbanization and sedimentation. The biodiversity of these pit lakes has been influenced by the continuous alteration of the land use pattern. Several broad pit lakes were turned into wasteland.

The concept of water quality relies mostly on the optimal utilization water. Different applications also include various water quality requirements, and also common procedures for documenting and evaluating water analysis findings (Khodapanah et al., 2009). GIS is a very useful tool for designing solutions to water supply issues in order to measure water quality, evaluate water resources and consider the local and/or regional scope of the natural ecosystem. Recognition of shift is the method of detecting changes in the status of an entity or occurrence by examining it at various times (Singh, 1989). The essential rule of utilizing remotely sensed information for the detection of adjustments is that shifts in the focal points can bring about changes in reflectance or surface investigation esteems that are recognizable from the progressions initiated by different factors, like variations in ambient conditions, angle of lighting and perception and moisture content of the soil. A significant area in global studies on environmental change is land use land cover (LULC) change. Inventory and monitoring of LULC changes are important aspects of increasing recognition of the mechanism of transition and modelling the impacts of climate change and related habitats at various scales (William et al., 1994). Earth observation data and sophisticated geospatial technology have supported the inventory, evaluation and monitoring approach to the framework of natural resources, including globally occurring wetlands (Davidson & Finlayson, 2007). Various methods to regulate and manage natural phenomena utilizing GIS have been suggested in the area of geoscience, and related literature on GIS applications has been extensively undertaken (Wang & Xie, 2018; Gajos & Sierka, 2012). But while GIS has been extensively used for the construction of mines in the mining industry, relatively little emphasis has been directed to the assessment of GIS-based approaches and applications related to mine preparation or environmental conservation (Choi et al., 2020), but Yucel et al. (2014) investigated the identification and visualization of shifts in acid mine lakes through GIS time series satellite image information that offered a range of information on the geographic scale of the land and variations in water sources on a yearly basis. The essential point of the current chapter is to analyse the water quality and landscape dynamics of the selected pit lakes of Andal Block under Raniganj coal mining area, West Bengal by using remote sensing and GIS techniques. The visual interpretation method is used to delineate the pit lake area and also the other land use in GIS environment.

12.2 Methodology

12.2.1 Description of the Study Area

The Raniganj coalfield is situated in West Bengal and partially in the eastern part of the Damodar Valley coalfield in the Jharkhand (23°37′44″N 87°06′54″E). Geographically, a large section of the Raniganj coalfield is wedged between the Ajoy and Damodar Rivers and minor parts fall north of the Ajoy River, south of the Damodar River and west of the Barakar River. The Raniganj coalfield is the country's cradle of coal mining. On the basis of geographic distribution, 12 pit lakes were selected for study in the Andal Block, Paschim Bardhaman, West Bengal, India (Table 12.1, Figs. 12.1 and 12.2).

12.2.2 Analysis of Physico-chemical Parameters

Water samples were collected from the studied pit lakes during 2016–2019 using standard protocols. The pH, conductivity and TDS were measured in the sites. The pH was estimated by using pH meter, and conductivity by conductivity meter. Dissolved oxygen was measured using Winkler's methods in situ. Total alkalinity, phosphate, nitrate, hardness, chloride and biological oxygen demand were determined using the protocol of the American Public Health Association manual (APHA, 2005) in laboratory.

Table 12.1 Brief profile of the study sites

Sl. no.	Pit lake name	Area in (ha)	Elevation in (m)
1	Jambad	1.85	92
2	Western Kajora	1.22	86
3	Atewal	9.79	92
4	Khadan Kali	1.55	83
5	Babuisol Shibmandir	6.97	101
6	Real Kajora	11.85	84
7	Chakrambati	8.74	85
8	Dhanderdihi 1	12.77	89
9	Dhanderdihi 2	6.91	85
10	Dhanderdihi 3	6.29	87
11	Babuisol Colony	1.84	86
12	Sankarpur	3.90	111

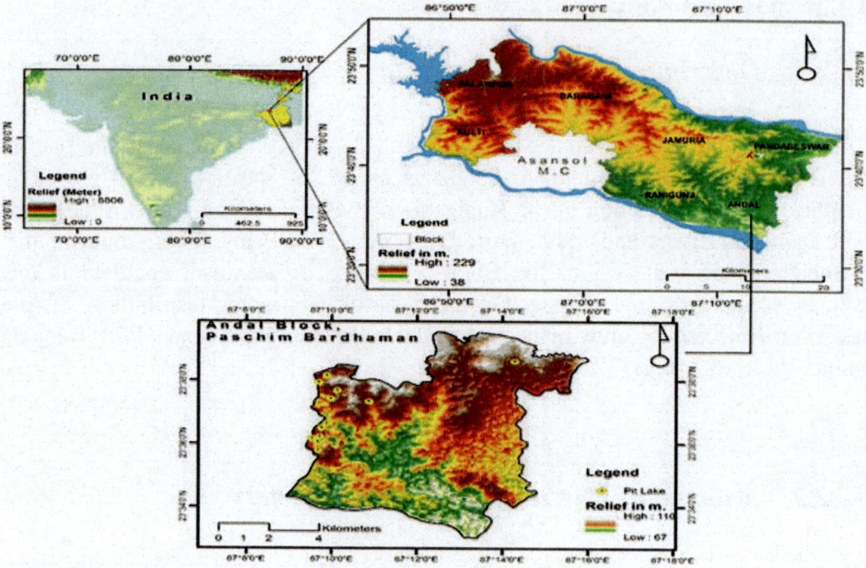

Fig. 12.1 Topographic details of the study area

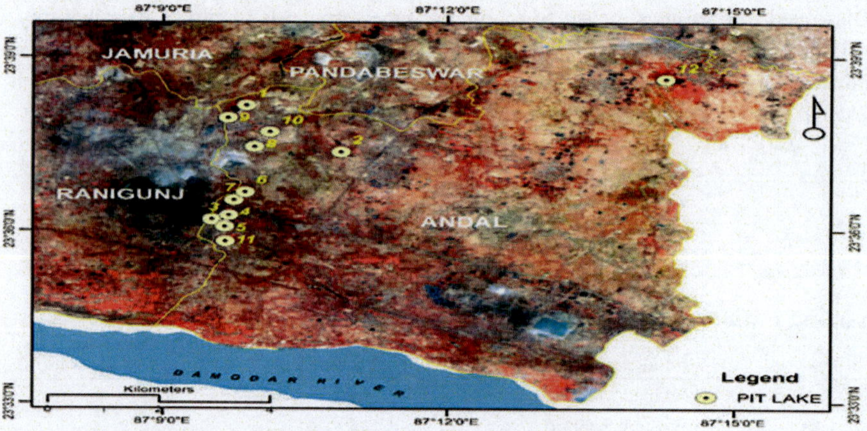

Fig. 12.2 Location of pit lakes (numbers indicate their location)

12.2.3 WQI Analysis

WQI sets a framework for providing a cumulatively generated, numerical term that specifies a specific standard of water quality (Miller et al., 1986). Nine parameters have been chosen to compute the WQI by using the quality standards endorsed by the BIS (2003) in the present work. The pit lakes' WQI was determined using Brown et al. (1970) process, and the equation is described beneath.

$$\mathrm{WQI} = \Sigma \mathrm{QnWn}/\Sigma \mathrm{Wn}$$

Calculation of quality rating (qn)

$$\mathrm{Qn} = 100[(\mathrm{Vn} - \mathrm{Vi})/(\mathrm{Vs} - \mathrm{Vi})]$$

Vn = Studied value, Vi = Ideal value.
[Vi is 0, except for pH (7) and DO (14.6 mg/l)],
Vs = Standard permissible limit

Calculation of unit weight (Wn)

$$\mathrm{Wn} = k/\mathrm{Vs}\ [K \text{ reflects proportionality constant}]$$

Calculation of k

$$k = [1/\Sigma\, 1/\mathrm{Vs} = 1, 2, \ldots n]$$

12.2.4 Method for Analysing the Evolution of Pit Lake

Satellite image data have large coverage and an increased volume of information. In addition, these are multi-temporal and usable on a wide scale (Schroeter & Glasser, 2011). To analyse temporal change of pit lakes in Andal Block region, Google Earth historical imagery has been used. In order to identify the change of area and perimeter of the pit lakes, Dhanderdihi 1 (Table 12.1) lakes have chosen to extract temporal information among the 12 lakes. This pit lake is very newly formed lake due to open-cast mine activity. Utilizing Q-GIS program, time series satellite images were processed, pit lake perimeter digitized and transformed to vector data format. Alterations in areal extent and periphery/perimeter were measured. Therefore, thematic maps of the pit lakes were created utilizing topographical maps.

12.2.5 Scheme for Analysis of Land Use/Land Cover

Landsat TM (1990) and OLI (2020) data were collected from Global Visualization Viewers (USGS-GLOVIS) of the US Geological Survey. Data from the 1990 Landsat TM (Thematic Mapper), bands 1-5 and 7 were used for land use mapping, band 6 was not used as a thermal band, whereas only bands 1-7 were being used from Landsat OLI (Operational Land Imager) data (Yuan & Bauer, 2007). To illustrate improvements in LULC over the span of three decades, regulated image recognition technique with maximum likelihood methodology was used. Among the most efficient regulated classification techniques used in the remote sensing image datasets with normal distribution in each input band is the maximum probability algorithm (MLC).

To establish the relationship between rising temperatures with respect to LULC change due to huge urbanization and open-cast coal mining activity of this region, land surface temperature (LST) has been calculated using Landsat images taken from USGS data archive. We extracted the thermal band (band 6) from Landsat 5 TM of 1990 and (band 10 and band 11) Landsat 8 OLI of 2020 because only the thermal band is designed to extract LST data.

The LST data from the Landsat thermal bands should pass across six different stages for recovery utilizing Q-GIS software. These measures have been pursued to derive ground surface temperature from Landsat imagery thermal bands (Ding & Shi, 2013). Any thermal electromagnetic energy is reflected by any object in the universe when it is at temperature greater than 0 (zero) K, which is defined when absolute 0 (zero). Using that same concept, signals that can be transformed into a sensor radiance are received by the thermal sensor (ETM+). The calculation of spectral radiance (Lk) was determined using the following formula (Landsat Project Science Office, 2002):

$$\text{L⍰} = \text{ML} * \text{QCAL} + \text{AL}$$

where, L⍰ indicates TOA spectral radiance, ML indicates radiance multiband (X), AL indicates radiance add band and QCAL means quantized and calibrated standard product pixel values.

$$\text{TOA spectral radiance} = \text{LMIN}\lambda + [(\text{LMAX}\lambda - \text{LMIN}\lambda)/(\text{QCALMAX} - \text{QCALMIN} \times \text{QCAL}]$$

where QCALMIN = 0, QCALMAX = 255, QCAL = DN of each pixel.

LMINλ AND LMAXλ are the spectral radiance of thermal band digital number from 0 to 255.

Adjustments to emissivity (e) have been extended to radiant temperatures due to the nature of the ground cover. The value of 0.95 is primarily for vegetated area and 0.92 for non-vegetated area (Nichol, 1994). Following Artis and Carnahan (1982), the adjusted reflectivity of surface temperature was determined:

$$T = \frac{\text{K}_2}{\ln(\text{K}1/L1 + 1)}$$

where T = At-satellite brightness temperature

$L\lambda$ = TOA spectral Radiance
K_1 and K_2 = constant band

For Landsat 8, OLI values of K1 for bands 10 and 11 are 774.8853 and 480.8883, respectively, and K2 for band 10 and 11 are 1321.0789 and 1201.1442, respectively.

For Landsat TM value of K1 for band 6 is 607.76 and K2 for band 6 is 1260.56, respectively.

The emissivity (*e*) of the ground surface can be obtained by

$$e = 0.004 * P_v + 0.986$$

12.3 Result and Discussion

Limnological characteristicsof the studied pit lakes were recorded and highlighted in terms of Mean ± SD values in Table 12.2. The importance of limnological characteristics for water quality assessment is widely accepted (Sunkad, 2008). Limnological parameters are very important in regulating the aquatic ecosystem and have a spatio-temporal variation. The pH has a promising role in shaping the aquatic biota since it regulates different biochemical functions (Jalal & Sanalkumar, 2013). The recommended pH level is 6.5–8.5 as per BIS (2003). In our study, the pH varied between 7.58 and 8.46 ± 0.28. Conductivity measures the electric current transmission capability of a solution (Lodh et al., 2014). Conductivity was found to be the lowest of 204.74 ± 46.78 ms/cm in the site K and highest of 391.68 ± 171.05 ms/cm in the site G. TDS represents different salts (mainly inorganic) and a small number of organic substances in the water. In our study, TDS ranged from 159.67 to 348.33 ± 54.31 ms/cm. The hardness of water is mainly due to the presence of calcium ion and magnesium ion (Salve & Hiware, 2006). The hardness of the sampling sites ranged from 132 to 271.89 ± 47.65 mg/l. The total alkalinity ranged between 22.22 and 34.00 ± 3.86 mg/l in the studied pit lakes, which lies within the limits. Dissolved oxygen is a significant contributor to the ecosystem's health in any aquatic bodies. Oxygen (O_2) can enter into the water from the air directly or through the photosynthetic activity of the aquatic plant species (Kotadiya & Acharya, 2014). It also serves as a potential indicator of pollution levels in the aquatic ecosystems (Gopalkrushna, 2011). The permissible limit of DO is 5 mg/l as per BIS (2003). In our study, DO ranged between 3.37 mg/l and 5.15 ± 0.35 mg/l. BOD represents the quantity of O_2 needed for aerobic microorganisms to degrade different organics present in water (Solanki & Pandit, 2004), and hence high BOD values indicate pollution (Patel et al., 1983). The mean BOD in this study ranged from 1.83 to 2.37 ± 1.64 mg/l. The concentrations of phosphate in different pit lakes ranged from 0.84 to 9.94 ± 2.43 mg/l. Excessive nitrate content can lead to different types of health-related issues (Kumar et al., 2014). The N content in the studied pit lakes ranged from 0.63 to 2.41 ± 0.60 mg/l. The presence of nitrate in the water bodies is mainly due to human activities. Chloride concentration in the studied pit lakes ranged from 22.54 to 63.34 ± 12.38 mg/l which is below the recommended value

Table 12.2 Mean values and standard deviations of water quality parameters obtained from studied pit lakes ~~in/~~during the period from 2016 to 2019

Sites	pH	TDS mg/l	CON μS/cm	TH mg/l	TA mg/l	DO mg/l	BOD mg/l	PHOS mg/l	NIT mg/l	Cl mg/l
A	8.03 ± 0.63	292.22 ± 83.17	297.39 ± 128.41	176.44 ± 74.59	33.11 ± 6.05	4.22 ± 0.48	1.83 ± 0.19	1.78 ± 1.67	1.11 ± 0.52	52.86 ± 2.21
B	8.44 ± 0.38	348.33 ± 89.11	375.37 ± 129.43	261.33 ± 41.40	23.11 ± 11.59	4.09 ± 0.44	1.97 ± 0.20	9.94 ± 3.03	2.40 ± 1.49	29.81 ± 8.07
C	7.84 ± 0.84	242.33 ± 68.15	276.33 ± 95.35	258.89 ± 78.70	26.44 ± 4.23	3.73 ± 0.08	1.89 ± 0.42	0.85 ± 0.78	1.28 ± 0.41	24.38 ± 3.66
D	7.97 ± 0.44	236.33 ± 72.53	339.96 ± 103.16	193.78 ± 17.97	23.78 ± 3.08	4.48 ± 0.69	2.10 ± 0.41	2.67 ± 1.54	1.29 ± 0.34	30.98 ± 12.40
E	7.63 ± 0.92	232.11 ± 35.86	298.55 ± 84.43	154.67 ± 34.08	24.44 ± 6.54	5.15 ± 0.30	2.15 ± 0.38	3.19 ± 1.19	0.72 ± 0.06	22.54 ± 7.31
F	7.89 ± 1.10	310.81 ± 91.25	218.32 ± 199.72	177.55 ± 89.73	22.66 ± 8.19	4.64 ± 0.43	2.23 ± 0.15	2.07 ± 1.39	2.41 ± 1.27	38.75 ± 31.35
G	8.46 ± 0.54	316.66 ± 88.54	391.68 ± 171.05	271.89 ± 48.24	34.00 ± 14.05	4.35 ± 0.67	2.09 ± 0.05	2.07 ± 1.55	2.06 ± 2.54	63.34 ± 6.13
H	7.85 ± 0.37	249.88 ± 13.42	248.33 ± 80.13	149.78 ± 110.50	28.00 ± 4.67	4.25 ± 0.32	2.19 ± 0.13	1.60 ± 1.43	1.14 ± 0.28	44.81 ± 10.68
I	8.10 ± 0.24	250.55 ± 6.05	310.95 ± 124.11	156.22 ± 51.04	24.22 ± 1.02	4.60 ± 0.36	2.37 ± 0.14	0.84 ± 0.50	1.41 ± 0.33	33.64 ± 10.10
J	7.86 ± 0.22	253.44 ± 2.99	250.18 ± 83.13	163.78 ± 84.50	25.11 ± 6.19	4.63 ± 0.23	2.35 ± 0.16	1.07 ± 0.49	0.63 ± 0.14	41.03 ± 2.40
K	8.21 ± 0.39	159.67 ± 17.10	204.74 ± 46.78	132.00 ± 43.67	22.22 ± 5.75	4.47 ± 0.47	2.22 ± 0.19	3.10 ± 078	0.95 ± 0.06	23.82 ± 3.18
L	7.58 ± 0.43	328.11 ± 161.87	356.23 ± 99.01	184.44 ± 44.58	25.56 ± 2.70	4.55 ± 0.54	2.12 ± 0.16	2.55 ± 2.38	1.26 ± 0.97	32.36 ± 20.29

A Jambad, *B* Western Kajora, *C* Atewal, *D* Khadan kali, *E* Babuisol, *F* Real Kajora, *G* Chakrambati, *H* Dhanderdihi 1, *I* Dhanderdihi 2, *J* Dhanderdihi 3, *K* Babuisol colony, *L* Sankarpur. Con, conductivity, *TDS* total dissolved solids, *DO* dissolved oxygen, *BOD* biological oxygen demand, *TH* total hardness, *TA* total alkalinity, *PHOS* phosphate, *N* nitrate, *Cl* chloride

Table 12.3 Calculation of water quality index of site A

Parameters	Vn	Vs	Wn	Qn	WnQn
pH	8.03	8.5	0.211265126	68.67	14.5068
TDS	292.22	500	0.003565893	58.444	0.2084
CON	297.39	300	0.005943155	99.13	0.5892
HARD	176.44	300	0.005943155	58.813	0.3495
TA	33.11	120	0.014857889	27.592	0.4099
DO	4.22	5	0.356589328	108.125	38.5562
BOD	1.83	5	0.356589328	36.6	13.0512
NIT	19.11	45	0.039621036	42.467	1.6826
CL	52.86	250	0.007131787	21.144	0.1508
	WQI = ΣWnQn/ΣWn = 69.40				

Table 12.4 Calculation of water quality index of site B

Parameters	Vn	Vs	Wn	Qn	WnQn
pH	8.44	8.5	0.211265126	96	20.2815
TDS	348.33	500	0.003565893	69.666	0.2484
CON	375.37	300	0.005943155	125.123	0.7436
HARD	261.33	300	0.005943155	87.110	0.5177
TA	23.11	120	0.014857889	19.258	0.2861
DO	4.09	5	0.356589328	109.479	39.0391
BOD	1.97	5	0.356589328	39.4	14.0496
NIT	20.31	45	0.039621036	45.133	1.7882
CL	29.81	250	0.007131787	11.924	0.0850
	WQI = ΣWnQn/ΣWn = 76.92				

Table 12.5 Calculation of water quality index of site C

Parameters	Vn	Vs	WN	Qn	WnQn
pH	7.84	8.5	0.211265126	56	11.8308
TDS	242.33	500	0.003565893	48.466	0.1728
CON	276.33	300	0.005943155	92.11	0.5474
HARD	258.89	300	0.005943155	86.297	0.5129
TA	26.44	120	0.014857889	22.033	0.3274
DO	3.73	5	0.356589328	113.2291667	40.3763
BOD	1.89	5	0.356589328	37.8	13.4791
NIT	1.28	45	0.039621036	2.844	0.1127
CL	24.38	250	0.007131787	9.752	0.0695
		WQI = ΣWnQn/ΣWn = 67.33			

(250 mg/l). High chloride content can lead to health-related problems (Sadat-Noori et al., 2014).

The values of WQI from all the four pit lakes are presented in Tables 12.3, 12.4, 12.5, 12.6, and 12.7. Our study highlights that the water quality of all the four

Table 12.6 Calculation of water quality index of site D

Parameters	Vn	Vs	WN	Qn	WnQn
pH	7.97	8.5	0.211265126	64.66666667	13.6618
TDS	236.33	500	0.003565893	47.266	0.1685
CON	339.96	300	0.005943155	113.32	0.6735
HARD	193.78	300	0.005943155	64.593	0.3839
TA	23.78	120	0.014857889	19.817	0.2944
DO	4.48	5	0.356589328	105.417	37.5905
BOD	2.1	5	0.356589328	42	14.9768
NIT	18.58	45	0.039621036	41.289	1.6359
CL	30.98	250	0.007131787	12.392	0.0884
		WQI = ΣWnQn/ΣWn = 69.37			

Table 12.7 Calculation of water quality index of site E

Parameters	Vn	Vs	WN	Qn	WnQn
pH	7.63	8.5	0.211265126	42	8.8731
TDS	232.11	500	0.003565893	46.422	0.1655
CON	298.55	300	0.005943155	99.517	0.5914
HARD	154.67	300	0.005943155	51.557	0.3064
TA	24.44	120	0.014857889	20.367	0.3026
DO	5.15	5	0.356589328	98.438	35.1018
BOD	2.15	5	0.356589328	43	15.3333
NIT	18.89	45	0.039621036	41.978	1.6632
CL	22.54	250	0.007131787	9.016	0.0643
		WQI = ΣWnQn/ΣWn = 62.31			

studied pit lakes was of poor quality (WQI > 50); hence, proper management of these pit lakes is recommended before any use (Table 12.7). Our earlier study about the WQI of the other pit lakes in the RCF also reflected a similar kind of findings (Palit et al., 2018). Ranges of WQI values from all the studied pit lakes were 69.40 (site A), 76.92 (site B), 67.33 (site C), 69.37 (site D), 62.31 (site E), 67.34 (site F), 75.75 (site G), 68.06 (site H), 71.19 (site I), 69.41 (site J), 73.19 (site K) and 62.35 (size L) (Tables 12.3, 12.4, 12.5, 12.6, 12.7, 12.8, 12.9, 12.10, 12.11, 12.12, 12.13, and 12.14). Assessment of WQI was done by so many researchers in various water bodies (Akumtoshi et al,. 2020), but our result is not the same, maybe due to the characteristics of pit lakes or maybe due to the absorption of pollutants by the aquatic vegetation in these pit lakes or could be due to improper management and poor post-mining strategies. Several earlier studies have demonstrated promising results in enhancing water quality through effective management practices and mitigation actions (Naubi et al., 2016). The involvement of certain human habitat in the adjacent areas of these pit lakes has a significant impact on the overall water quality. Anthropogenic sources such as wastewater discharges by residents living in catchment and fertilizer runoff have also contributed to water quality deterioration (Shah

Table 12.8 Calculation of water quality index of site F

Parameters	Vn	(Vs)	WN	Qn	WnQn
pH	7.89	8.5	0.211265126	59.33333333	12.5351
TDS	197.18	500	0.003565893	39.496	0.1408
CON	218.32	300	0.005943155	72.77333333	0.4325
HARD	177.55	300	0.005943155	59.183	0.3517
TA	22.66	120	0.014857889	18.883	0.2806
DO	4.64	5	0.356589328	103.75	36.9961
BOD	2.23	5	0.356589328	44.6	15.9038
NIT	7.81	45	0.039621036	17.356	0.6876
CL	38.75	250	0.007131787	15.500	0.1105
		WQI = ΣWnQn/ΣWn = 67.34			

Table 12.9 Calculation of water quality index of site G

Parameters	Vn	Vs	WN	Qn	WnQn
pH	8.46	8.5	0.211265126	97.33333333	20.5631
TDS	316.66	500	0.003565893	63.332	0.2258
CON	391.68	300	0.005943155	130.56	0.7759
HARD	271.89	300	0.005943155	90.630	0.5386
TA	34	120	0.014857889	28.333	0.4209
DO	4.35	5	0.356589328	106.7708333	38.0734
BOD	2.09	5	0.356589328	41.8	14.9054
NIT	2.06	45	0.039621036	4.578	0.1814
CL	63.34	250	0.007131787	25.336	0.1806
		WQI = ΣWnQn/ΣWn = 75.75			

Table 12.10 Calculation of water quality index of site H

Parameters	Vn	Vs	WN	Qn	WnQn
pH	7.85	8.5	0.211265126	56.667	11.97169047
TDS	249.88	500	0.003565893	49.976	0.178209083
CON	248.33	300	0.005943155	82.777	0.491954599
HARD	149.78	300	0.005943155	49.927	0.296721942
TA	28	120	0.014857889	23.333	0.346684069
DO	4.25	5	0.356589328	107.813	38.44478692
BOD	2.19	5	0.356589328	43.8	15.61861256
NIT	7.79	45	0.039621036	17.311	0.685884164
CL	44.81	250	0.007131787	17.924	0.127830142
		WQI = ΣWnQn/ΣWn = 68.06			

Table 12.11 Calculation of water quality index of site I

Parameters	Vn	Vs	WN	Qn	WnQn
pH	8.1	8.5	0.211265126	73.33333333	15.4928
TDS	236.33	500	0.003565893	47.266	0.1685
CON	339.96	300	0.005943155	113.32	0.6735
HARD	193.78	300	0.005943155	64.593	0.3839
TA	23.78	120	0.014857889	19.817	0.2944
DO	4.48	5	0.356589328	105.4166667	37.5904
BOD	2.1	5	0.356589328	42	14.9768
NIT	18.58	45	0.039621036	41.289	1.6359
CL	30.98	250	0.007131787	12.392	0.0884
		WQI = ΣWnQn/ΣWn = 71.19			

Table 12.12 Calculation of water quality index of site J

Parameters	Vn	Vs	WN	Qn	WnQn
pH	7.86	8.5	0.211265126	57.333	12.1125
TDS	253.44	500	0.003565893	50.688	0.18075
CON	250.18	300	0.005943155	83.393	0.4956
HARD	163.78	300	0.005943155	54.593	0.3245
TA	25.11	120	0.014857889	20.925	0.3109
DO	4.63	5	0.356589328	103.854	37.0332
BOD	2.35	5	0.356589328	47	16.7597
NIT	24.79	45	0.039621036	55.089	2.1827
CL	41.03	250	0.007131787	16.412	0.1170
		WQI = ΣWnQn/ΣWn = 69.41			

Table 12.13 Calculation of water quality index of site K

Parameters	Vn	Vs	WN	Qn	WnQn
pH	8.21	8.5	0.211265126	80.6667	17.0421
TDS	159.67	500	0.003565893	31.934	0.1139
CON	204.74	300	0.005943155	68.24666667	0.4056
HARD	132	300	0.005943155	44.000	0.2615
TA	22.22	120	0.014857889	18.517	0.2751
DO	4.47	5	0.356589328	105.5208333	37.6276
BOD	2.22	5	0.356589328	44.4	15.8326
NIT	19.11	45	0.039621036	42.467	1.6826
CL	23.82	250	0.007131787	9.528	0.0679
		WQI = ΣWnQn/ΣWn = 73.19			

Table 12.14 Calculation of water quality index of site L

Parameters	Vn	Vs	WN	Qn	WnQn
pH	7.58	8.5	0.211265126	38.66666667	8.1689
TDS	328.11	500	0.003565893	65.622	0.2340
CON	356.23	300	0.005943155	118.7433333	0.7057
HARD	184.44	300	0.005943155	61.480	0.3654
TA	25.56	120	0.014857889	21.300	0.3165
DO	4.55	5	0.356589328	104.6875	37.3304
BOD	2.12	5	0.356589328	42.4	15.1194
NIT	1.26	45	0.039621036	2.800	0.1109
CL	32.36	250	0.007131787	12.944	0.0923
		WQI = ΣWnQn/ΣWn = 62.35			

& Joshi, 2017). The submerged aquatic vegetation reflects the good quality of water that was reported by Dennison and Orth (1993).

12.3.1 *Genesis and Evolution of Pit Lake*

The whole method of surface mining causes dramatic shifts in the landscape. They lead to the formation in the mining areas of massive overburden dumps, vast voids and pit lake habitats. A pit lake is a lake, by design, created by the drainage of an unearthed mining pit. In their substantially greater relative depths, pit lakes differ physically from natural lakes. Groundwater is drained through open pits throughout active extraction, and dewatering systems are being used to regulate the inflow and subsequent rainfall. Whenever the pumps are turned off, pit lakes emerge, and post-mining surface/underground water runoff and precipitation continues to accumulate within the dormant pit. Plenty of examples of legacy sites need persistent care, although various advantageous end uses have been accomplished by several other pit lakes. However, the scarcity of a publicly available report on the development, distribution and consistency of pit lakes does not enable both mining companies and societies in this field to demonstrate the suitability of these water supplies. Using geospatial technology, we can easily find out the distribution and evolution of pit lake to gather the primary physical information regarding their location, surrounding landscape, perimeter and total evolution process of these lakes along with their surrounding landscape dynamics. Here in the Andal Block among the 12 pit lakes, we have analysed the genesis and evolution process using satellite imagery of Dhanderdihi 1 pit lakes (Fig. 12.3).

Here using Google Earth's historical imagery, the total evolution process from the primary stage to the ultimate stage was observed from 2007 to 2019.

Fig. 12.3 Evolution of a Pit Lake

12.3.2 Land Use/Land Cover Change Analysis

LULC applies to two different, interchangeably employed concepts (Rawat & Kumar, 2015; Butt et al., 2015). Land cover may be defined as the physical characteristics of the surface of the planet, including trees, water, soil, etc., made by human activities like dwellings, while land use refers to land used mostly by humans for ecosystems linked to economic activities (Rawat & Kumar, 2015). In terms of natural and socioeconomic growth across time and space, LULC trends rely on human use. In other terms, shifts in land use will have the potential to impact the coverage of land and conversely. LULC shifts that affect habitats and marine environments are very associated with ecological skills (Wright & Wimberly, 2013). Water quality can be impacted by LULC adjustments in a waterbody, contributing to higher surface runoff, decreased retention of groundwater and pollutant movement (Butt et al., 2015). Consequently, for the collection, planning, control and management of water supplies, LULC knowledge at the catchment level is relevant so that improvements in land use satisfy the growing need for people's desires and health without sacrificing the quality of water. Multiple research works on the change analysis of waterways have also been performed, which are critical in designing successful water quality security management practices (Gajbhiye & Sharma, 2013; Hu et al., 2012). Water resources management is essential because these are not only a hydrological unit (Singh et al., 2014) and plays a vital role in socio-ecological terms by offering local communities with physical, nutritional and social welfare and also life support services (Fernald et al., 2012). Our research is a leader in the pit lake ecology of the Raniganj coal mining region's Andal Block. The research has additionally proven that natural and human activities have contributed to enhancements in LULC affecting the assets provided via way of means of the riparian surrounding services (Chu et al., 2013; Khan et al., 2017). The key drivers of LULC transition, which have an impact on the

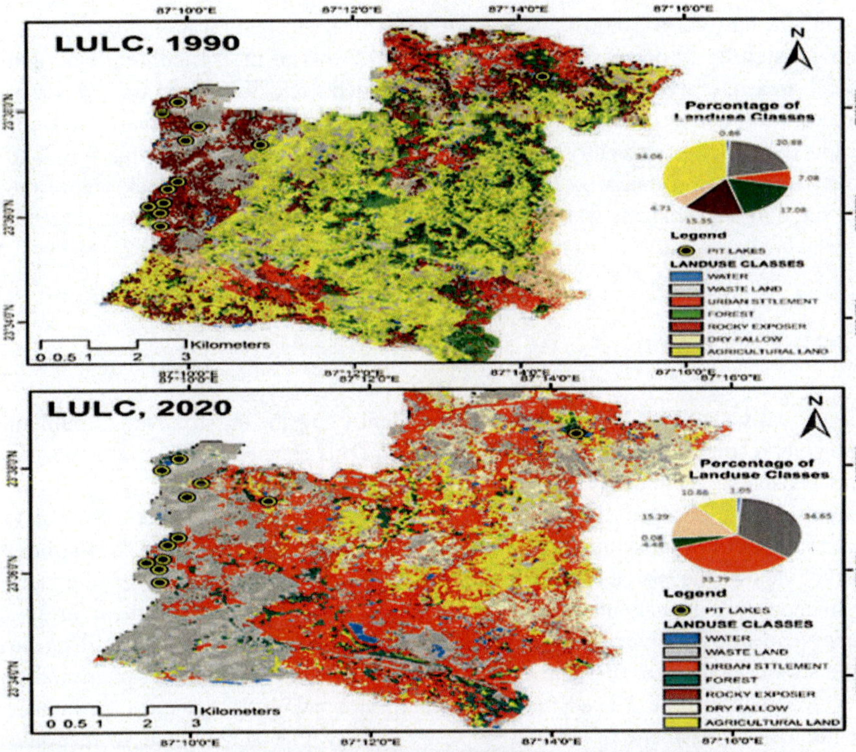

Fig. 12.4 Land use/land cover pattern in Raniganj coalfield area between 1990 and 2020

fineness of water sources, had been defined as anthropogenic activities together with irrigation, deforestation and urbanization (Olusola et al., 2018). The efficiency of the water supply is impaired by these practices uniquely or collectively (Li et al., 2014).

The LULC class has been divided into seven groups in this study. Water bodies, wastelands, urban areas, woodland, mountainous outcrops, dry fallow and farmland. The analysis of land use and land cover has been extensively studied over the past three decades (1990, 2000 and 2020) (Fig. 12.4). The reliability evaluation for classification of land use was evaluated. One-hundred seventy-five random points for 1990 and 196 random points for 2020 were gathered from Google Earth's imagery to verify the classified maps for this reason. The average precision (percent) for classified images is 86.12 for the year 1993 and 89.91 for 2020. With the assistance of the Kappa coefficient, the accuracy of the analysis was obtained, and the Kappa coefficient values were 0.86 and 0.90 for 1990 and 2020, respectively. The trend and spatial distribution, i.e. landscape patterns for the year between 1990 and 2020, have been conveniently studied by planning land use/land cover mapping. The built-up area has dramatically grown from 576.09 ha to 2750.04 ha (7.08–33.779%), wasteland 1699.11 ha to 2819.88 ha (20.88–34.65%) and dry fallow 382.95 ha to 1244.34 ha (4.71–15.29%). Agricultural land from 2771.64 ha to 867.78 ha (34.06–10.66%) and forest 1389.69 ha to 364.59 ha (17.08–4.48%) show a significant decrease from 1990 to 2020. Most of the agricultural land and forest area has been converted to urban settlement, and wasteland indicates that urbanization and opencast mining activity in this Andal region dominate the landscape dynamicity. Presently all those 12 pit lakes except Shankarpur pit lake (Sl. No. 12) are situated on the extreme western part of Andal block where the surrounding landscape is wasteland, built-up and some portion of agricultural land.

12.4 Conclusion

Geospatial technology plays a huge role in identifying the environmental problem and gives a solution for good decision making. In this chapter, using remote sensing and geospatial technology helps to find out the spatial changes during a long temporal scale in respect to pit lake genesis and its evolution. On the other hand, assessment of the surrounding landscape and water quality of pit lakes during three decades shows the spatial changes regarding existing landscape gives a broader view to take proper management plan for those pit lakes to conserve them in situ ecosystem and biodiversity also and take proper management plan to utilize the lake's water in the nearest urban area and agricultural land. The findings presented in this chapter illustrate a need further research, along with the pursuit of a human health risk assessment of criteria of water quality that surpass the recreational standards and conducting a correlation analysis between the quality of the sediment/surrounding soil and the quality of the water. Before the lakes are encouraged for use for domestic or agricultural purposes, greater studies in these proposed lines of study are recommended.

References

Akumtoshi, L. K. R., Singh, M. R., & Puro, N. (2020). Assessment of water quality status of Doyang River, Nagaland, India, using water quality index. *Applied Water Science, 10*, 46.

American Public Health Association (APHA). (2005). *Standard methods for the examination of water and wastewater*. APHA-AWWA.

Artis, D. A., & Carnahan, W. H. (1982). Survey of emissivity variability in thermography of urban areas. *Remote Sensing of Environment, 12*(4), 313–329.

Brown, R. M., McClelland, N. I., Deininger, R. A., & Tozer, R. G. (1970). A water quality index – Do we dare? *Water & Sewage Works, 117*, 339–343.

Bureau of Indian Standards, BIS 10500. (2003). Manak Bhavan, New Delhi, India.

Butt, A., Shabbir, R., Ahmad, S. S., & Aziz, N. (2015). Land use change mapping and analysis using remote sensing and GIS: A case study of Simly watershed, Islamabad, Pakistan. *Egyptian Journal of Remote Sensing and Space Science, 18*(2), 251–259.

Choi, Y., Baek, J., & Park, S. (2020). Review of GIS-based applications for mining: Planning, operation, and environmental management. *Applied Sciences, 10*, 2266.

Chu, H. J., Liu, C., & Wang, C. (2013). Identifying the relationships between water quality and land cover changes in the Tseng-Wen Reservoir Watershed of Taiwan. *International Journal of Environmental Research and Public Health, 10*, 478–489.

Davidson, N. C., & Finlayson, C. M. (2007). Earth observation of wetland inventory, assessment and monitoring. *Aquatic Conservation: Marine and Freshwater Ecosystems, 17*(3), 219–229.

Dennison, W. C., & Orth, R. J. (1993). Assessing water quality with submersed aquatic vegetation. *Bioscience, 43*, 86–94.

Ding, H., & Shi, W. (2013). Land-use/land-cover change and its influence on surface temperature: a case study in Beijing City. *International Journal of Remote Sensing, 34*(15), 5503–5517.

Fernald, A., Tidwell, V., Rivera, J., Rodríguez, S., Guldan, S., Steele, C., Ochoa, C., Hurd, B., Ortiz, M., Boykin, K., & Cibils, A. (2012). Modeling sustainability of water, environment, livelihood, and culture in traditional irrigation communities and their linked watersheds. *Sustainability, 4*(11), 2998–3022.

Gajbhiye, S., & Sharma, S. K. (2013). Land use and land cover change detection of Indra river watershed through remote sensing using multi-temporal satellite data. *International Journal of Geomatics and Geosciences, 3*(1), 89.

Gajos, M., & Sierka, E. (2012). GIS technology in environmental protection: Research directions based on literature review. *Polish Journal of Environmental Studies, 21*, 241–248.

Gopalkrushna, M. H. (2011). Determination of physico-chemical parameters of surface water samples in and around Akot city. *International Journal of Research in Chemistry and Environment, 1*(2), 183–187.

Hu, H. B., Liu, H. Y., Hao, J. F., & An, J. (2012). Analysis of land use change characteristics based on remote sensing and GIS in the Jiuxiang River watershed. *International Journal on Smart Sensing and Intelligent Systems, 5*(4), 811–823.

Jalal, F. N., & Sanalkumar, M. G. (2013). Water quality assessment of Pampa river in relation to pilgrimage season. *International Journal of Research in Chemistry and Environment, 3*(1), 341–347.

Khan, A., Khan, H. H., & Umar, R. (2017). Impact of land-use on groundwater quality: GIS-based study from an alluvial aquifer in the western Ganges basin. *Applied Water Science, 7*(8), 4593–4603.

Khodapanah, L., Sulaiman, W. N. A., & Khodapanah, D. N. (2009). Groundwater quality assessment for different purposes in Eshtehard District, Tehran, Iran. *European Journal of Scientific Research, 36*(4), 543–553.

Kotadiya Nikesh, G., & Acharya, C. A. (2014). An assessment of lake water quality index of Manipu lake of district Ahmedabad, Gujarat. *International Journal of Science and Research, 3* (4), 448–450.

Kumar, S. K., Logeshkumara, A., Magesh, N. S., Godson, P. S., & Chandrasekar, N. (2014). Hydrogeochemistry and application of water quality index (WQI) for groundwater quality assessment, Anna Nagar, part of Chennai City, Tamil Nadu, India. *Applied Water Science, 5*, 335–343.

Landsat Project Science Office. (2002). *Landsat 7 science data user's handbook*. http://ltpwww.gsfc.nasa.gov/IAS/handbook/handbook_toc.html. GoddardSpace Flight Center, NASA, Washington, DC.

Li, X., Nian, Y., Zhou, J., & Hu, X. (2014). Impact of land use change on water resource allocation in the middle reaches of the Heihe River Basin in North Western China. *Journal of Arid Land, 6* (3), 273–286.

Lodh, R., Paul, R., Karmakar, B., & Das, M. K. (2014). Physico chemical studies of water quality with special reference to ancient lakes Udaipur city, Tripura, India. *International Journal of Scientific and Research Publications, 4*(6), 1–9.

Miller, W. W., Joung, H. M., Mahannah, C. N., & Garrett, J. R. (1986). Identification of water quality differences Nevada through index application. *Journal of Environmental Quality, 15*(3), 265–272.

Naubi, I., Zardari, N. H., Shirazi, S. M., Ibrahim, N. F. B., & Baloo, L. (2016). Effectiveness of water quality index for monitoring Malaysian river water quality. *Polish Journal of Environmental Studies, 25*(1), 231–239.

Nichol, J. E. (1994). A GIS-based approach to microclimate monitoring in Singapore's high-rise housing estates. *Photogrammetric Engineering and Remote Sensing, 60*, 1225–1232.

Olusola, A., Onafeso, O., & Durowoju, O. S. (2018). Analysis of organic matter and carbonate mineral distribution in shallow water surface sediments. *Osun Geographical Review, 1*(1), 106–110.

Palit, D., Mondal, S., & Chattopadhyay, P. (2018). Analyzing water quality index of selected Pit-Lakes of Raniganj Coal Field Area, India. *Environment and Ecology, 36*(4A), 1167–1175.

Patel, S. G., Singh, D. D., & Harshey, D. K. (1983). Pamitae (Jabalpur) sewage polluted water body, as evidenced by chemical and biological indicators of pollution. *Journal of Environmental Biology, 4*, 437–449.

Rawat, J. S., & Kumar, M. (2015). Monitoring land use/cover change using remote sensing and GIS techniques: A case study of Hawalbagh block, district Almora, Uttarakhand, India. *The Egyptian Journal of Remote Sensing and Space Science, 18*(1), 77–84.

Sadat-Noori, S. M., Ebrahimi, K., & Liaghat, A. M. (2014). Groundwater quality assessment using the water quality index and GIS in Saveh-Nobaran aquifer, Iran. *Environment and Earth Science, 71*, 3827–3843.

Salve, B. S., & Hiware, C. J. (2006). Studies on water quality of Wanparkalpa reservoir, Nagapur, Near ParliVaijinath, Dist Beed, Marathwada region. *Journal of Aquatic Biology, 21*(2), 113–111.

Schroeter, L., & Glasser, C. (2011). Analyses and monitoring of lignite mining lakes in Eastern Germany with spectral signatures of Landsat TM satellite data. *International Journal of Coal Geology, 86*, 27–39.

Shah, K. A., & Joshi, G. S. (2017). Evaluation of water quality index for River Sabarmati. *Applied Water Science, Gujarat*. https://doi.org/10.1007/s13201-015-0318-7

Singh, A. (1989). Digital change detection technique using remotely sensed data. *International Journal of Remote Sensing, 10*(6), 989–1003.

Singh, P., Gupta, A., & Singh, M. (2014). Hydrological inferences from watershed analysis for water resource management using remote sensing and GIS techniques. *Egyptian Journal of Remote Sensing and Space Science, 17*(2), 111–121.

Solanki, H. A., & Pandit, B. R. (2004). Trophic status of lentic waters of ponds water of Vadodara, Gujarat, India. *International Journal of Bioscience Reporter, 4*, 191–198.

Sunkad, B. N. (2008). Water quality assessment of Kanabargi water body (Belgaum). *Environment and Ecology, 26*, 191–194.

Wang, X., & Xie, H. (2018). A review on applications of remote sensing and geographic information systems (GIS) in water resources and flood risk management. *Water, 10*, 608.

William, E. R., William, B. M., & Turner, B. L. (1994). Modelling land use and land cover as part of global environmental change. *Climate Change, 28*, 45–64.

Wright, C. K., & Wimberly, M. C. (2013). Recent land use change in the Western Corn Belt threatens grasslands and wetlands. *Proceedings of the National Academy of Sciences of the United States of America, 110*(10), 4134–4139.

Yuan, F., & Bauer, M. E. (2007). Comparison of impervious surface area and normalized difference vegetation index as indicators of surface urban heat island effects in Landsat imagery. *Remote Sensing of Environment, 106*, 375–386.

Yucel, D. S., Yucel, M. A., & Baba, A. (2014). Change detection and visualization of acid mine lakes using time series satellite image data in geographic information systems (GIS): Can (Canakkale) County, NW Turkey. *Environmental Earth Sciences, 72*, 4311–4323.

Chapter 13
Utility of Low-Cost Geospatial Tools for Monitoring of Water Resources for Their Conservation and Optimum Management: A Case Study of a Small River in Tripura

Sima Majumdar

13.1 Introduction

Water resources, their conservation and judicial use are imperative for sustainability of human kind. These resources are often prone to exhaustive stress and mismanagement. The scope of geospatial technologies has broadened with the ultimate goal of ensuring judicious resource utilization and sustainability. Majumder et al. (2019) have recently summarized the steps that can be taken to minimize water body-related hazards like floods. According to the study, systematic monitoring of floods both spatially and temporally at a large scale is very limited, and therefore, the following methods are proposed by Majumder et al. (2019):

1. Preparation of a master plan for flood management.
2. Development of embankment for physical flood protection.
3. Regulation of settlement.
4. Comprehensive coastal land management plan.
5. Improvement of flood hazard maps at the local level.

Along with this measure, monitoring of water resources and tracking the impact of anthropogenic factors are some of the real-time tools that can be harnessed for impact analysis studies for water resources.

Due to complex human-nature interaction, the anthropogenic impact leading to gross alteration in geomorphological signatures often requires meticulous analysis. The mechanism of these changes is hidden in the way the smaller alterations are manifested. These changes can be assessed through geospatial tools. In resource limited set-ups, utility of low-cost geospatial tool-based visual time series analysis

S. Majumdar (✉)
Department of Education, Government of Tripura, Agartala, Tripura, India

G. S. Bhunia et al. (eds.), *Anthropogeomorphology*, Geography of the Physical Environment, https://doi.org/10.1007/978-3-030-77572-8_13

Table 13.1 Some issues and challenges of rapid urbanization upon river and water resources

Sl no	Issues and challenges	Author	Year
1	Metal pollution in marine water	Virkanen, J.	1998
2	Long-term hydrologic impact	Bhaduri, B., Minner, M., Tatalovich, S., and Harbor, J.	2001
3	River water quality depletion	Ouyang, T., Zhu, Z., and Kuang, Y.	2006
4	Water resources constraint	Fang, Chuang-lin, Chao Bao, and Jin-chuan Huang.	2007
5	Variations of surface water quality	Wang, Junying et al.	2008
6	Bacterial pollution in coastal water bodies	Walters, S. P., Thebo, A. L., and Boehm, A. B.	2011
7	Ground water quality depletion	Liu, Zhenhuan et al.	2013
8	River water pollution	Zhao, H., Duan, X., Stewart, B., You, B., and Jiang, X.	2013
9	Urban water body challenges	Hughes, R. M., Dunham, S., Maas-Hebner, K. G., Yeakley, J. A., Harte, M., Molina, N., and Kaczynski, V. W.	2014
10	Urban flood risk	Chen, Yangbo et al.	2015

may have great significance to divulge these processes. In the meantime, there has been rapid urbanization in almost all places that have resulted in some irreversible impact on various biotic and abiotic factors of the environment. Among the major concern of rapid and unsustainable urbanization, there are transformations of agricultural landscapes, land fragmentation and land use change. Rapid urbanization has also caused severe alterations in other ecological parameters and even rainfall pattern (Song et al., 2014; Majumdar & Pan, 2019).

However, in this study, the focus is on the impact of rapid urbanization on river geomorphology and related aspects. Among major findings on the issues of river and water resources with respect to the impact of rapid urbanization, some are summarized below (Table 13.1):

Rapid urbanization and industrialization lead to pollution in water bodies (Virkanen, 1998) with long-term hydrological impact (Bhaduri et al., 2001) and river water quality depletion (Ouyang et al., 2006), etc., all of which cause constraints in water resources (Fang et al., 2007). Challenges like variations of surface water quality (Wang et al., 2008), bacterial pollution (Walters et al., 2011), groundwater quality depletion (Liu et al., 2013; Zhao et al., 2013), etc. can cause urban water body challenges (Hughes et al., 2014) as well. In the long run, all these factors' may cause large-scale risks such as urban flooding (Chen et al., 2015).

13.2 Materials and Methods

Limiting cost of research yet maintaining novelty and insight is highly anticipated in Geoinformatics research today. A coupled approach of RS-GIS applications with respect to satellite image analysis, detection of LU/LC changes and monitoring of digital elevation map (DEM) is described in the text. This study is novel and aids to acquire vital information on how human-nature interaction is shaping the local geomorphology of smaller rivers.

Hydrodynamic factors influenced by climatic condition, geology of the area as well as anthropogenic activity would impact riverbank erosion. In the long run, these factors lead to massive riverbank alterations. These long-term effects negatively impact the people of the area (Fig. 13.1).

Study Site In this chapter, the concept of coupled approach of low-cost geospatial tools will be applied to the Khowai River in Tripura. The Khowai drainage system extends for 387.89 km^2, and Khowai River is a sixth-order stream while Maharani Cherra (46.05 km^2) being the largest tributary of the system.

The study area is bounded by Atharamura hill range in the east and Barmura hill range in the west. Khowai River is the second longest river (166 km long) in Tripura, which flows for 78 km between Baramura and Atharamura hill ranges (Fig. 13.2). Geographically, the study area located in Khowai district lies between 23°45′N to 24° N latitude and 91°30′20″E to 91°44′10″E longitude covering an area of 352.12 sq. km.

Data Sources, Processing and Analysis In order to assimilate the overall characteristic of the studied river and to emphasize on the impact of rapid urbanization on the River Khowai, a multitude of tools and techniques were used.

Here the utility of low-cost geospatial tool-based visual time series analysis may demonstrate great significance to divulge the mechanism of geomorphological change of the studied river. A coupled approach of RS-GIS applications with respect

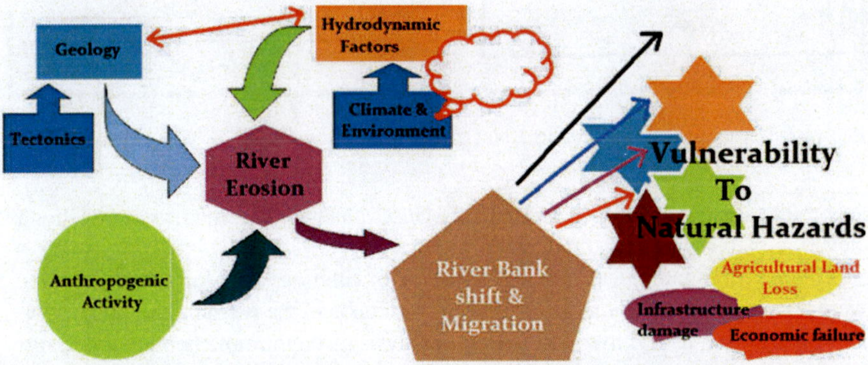

Fig. 13.1 Interplay of multiple factors in river geomorphic alteration

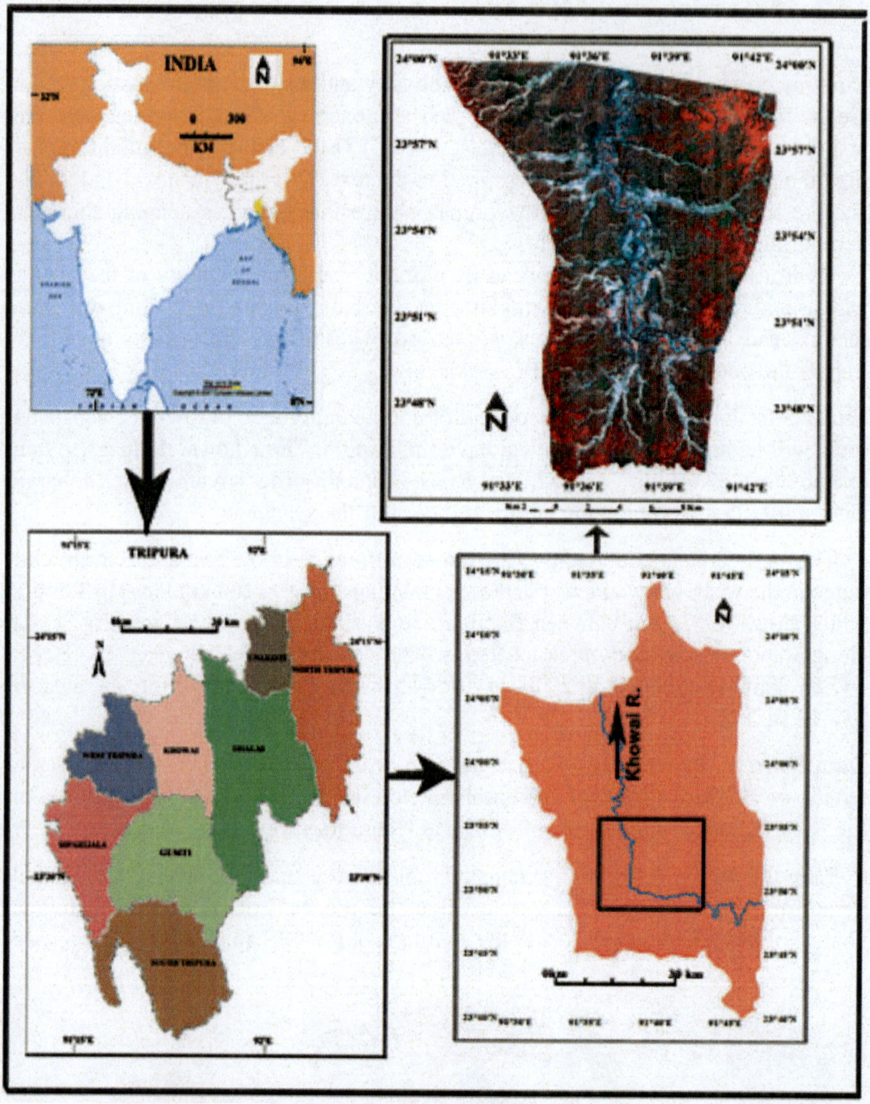

Fig. 13.2 Location map of study site; Khowai River. (Map prepared by Dr. Sima Majumdar)

to satellite image analysis, detection of LU/LC changes and monitoring of digital elevation map (DEM) is attempted.

The baseline geological analyses in terms of lithology, geology, tectonics, etc. were done to understand the nature of the riverbank and the surrounding of the River Khowai. Subsequently, low-cost RS-GIS analysis through implementing the spatio-temporal analysis was done meticulously. The proximity of brick kilns, morphological indicator such as river width and formation of sand bars were extensively studied. Finally, the analysis was build up on the observation of detection of LU/LC changes and monitoring of digital elevation map (DEM). The preliminary results are presented in this chapter.

13.3 Results and Discussion

The study area falls under Surma, Tipam, Dupitilla and Alluvium formations. Anticlinal hill range and synclinal valley with flood plain are the dominant physiographic divisions. The area experiences tropical monsoon climate. Average monthly monsoon rainfall is about 200 mm. Pedagogically, the area is characterized by red loamy soil and younger alluvial soil. The study area is thickly populated with 68,269 people, and population density is 106 persons/km^2 (2011 Census). Here, agriculture is the main activity.

On the River Khowai, the Chakmaghat barrage is established. According to the study published in American Journal of Environmental Sciences (2012), Muhammad opined that (Muhammad, 2012) the two dams over the Khowai River – one at Chakma Ghat (Figs. 13.3a and 13.3b) and the other at Kalyanpur in Tripura – are instrumental for the irrigation of about 25,000 acres of land for cultivation of Irri and Boro. In the dry season, its discharge is about 8.5 cumecs (cubic meters per second).

During subsequent years when the extent of urbanization increased around the River Khowai, a number of brick kilns were established. Interestingly, the brick kilns are situated at very close proximity to the riverbed (Fig. 13.4) (mean distance, 136.30 m; maximum distance, 300 m; minimum distance, 50 m).

Fig. 13.3a Chakma Ghat Barrage on Khowai River. (Satellite image from an altitude approx. 6000 ft.)

Fig. 13.3b Measuring distance of a brick kiln from Khowai riverbed

The brick kiln industry is a resource demanding enterprise in terms of natural resources. It requires sand of the riverbank in large quantity, and due to random utilization of riverbank material, the characteristics of the riverbank are altered significantly.

Industrialization is the need for a developing economy and for creating wealth. However, sustainable consumption should be practiced by the consumers. To do so, the general people should be made aware of the long-term consequences of random, unethical and unsustainable consumption of resources. The stress that builds up on the environment due to such practices may lead to catastrophic consequences. Various studies have indicated that severe environmental degradation may be linked to environmental as well as biotic disasters as the present emergence of infective diseases, the COVID-19 pandemic.

The analysis of satellite imagery, assessment of the Khowai riverbank with respect to the brick kilns & calculation of linear distance (in metres), is done by in-built tool in https://earthengine.google.com/.

Changes in the landscape mosaic are the result of complex interactions between physical, biological, economic, political and social factors. Comparative analysis showing status of territorial management by natural or anthropogenic changes is needed for better understanding of land use and land cover pattern change.

However, anthropogenic activity is the main cause of agricultural land loss and rapid changes in land use and land cover, and soil erosion is largely influenced by land use pattern. Riverbank erosion due to various factors is a cause of concern already (Majumdar & Das, 2014a).

For analysis of LU/LC, the golden standard is spatio-temporal analysis using a combination of remote sensing techniques, geographical information systems (GIS) and statistical analysis. These methods are potentially helpful for characterization studies of land cover, which integrates the resulting information to support the decision-making process. Another vital tool is digital processing of satellite images

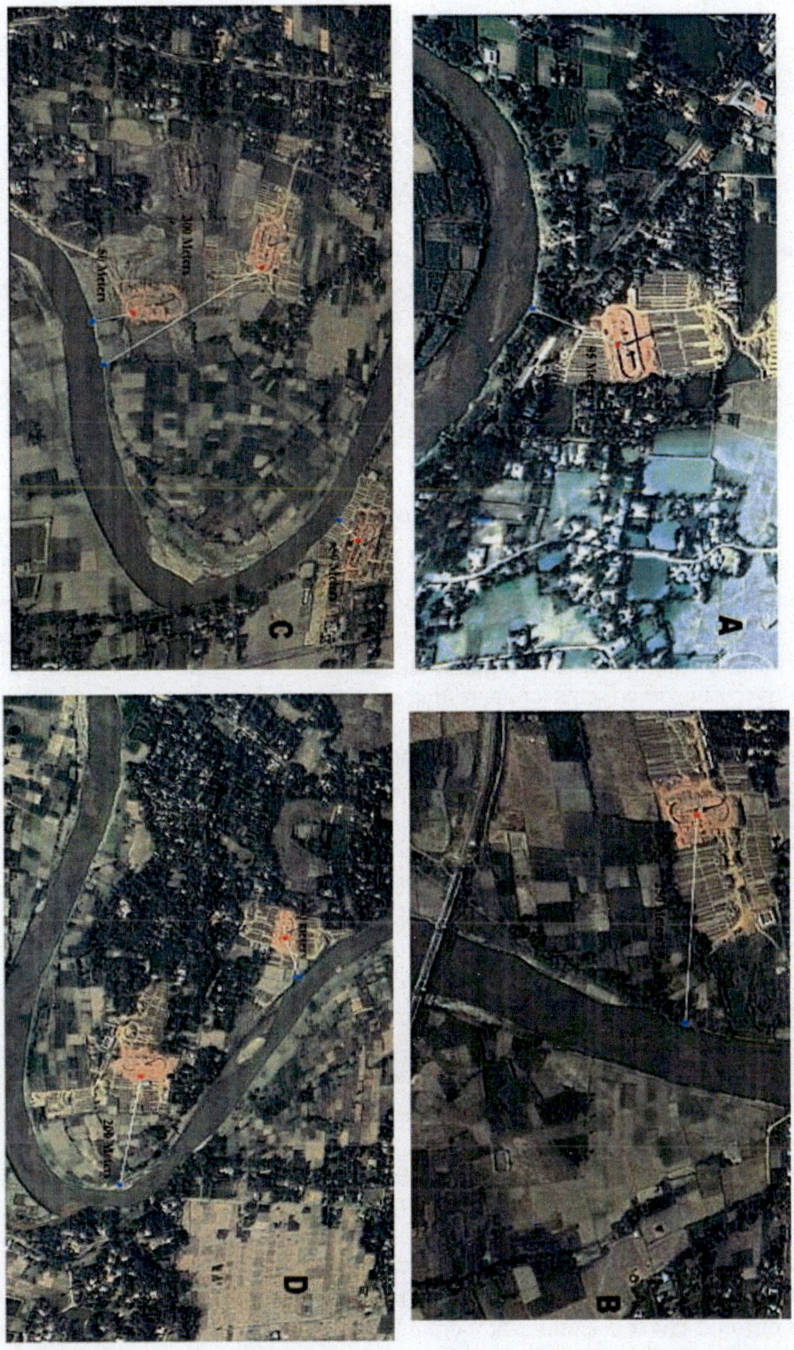

Fig. 13.4 Brick kilns on the Khowai riverbank (satellite image from an altitude approx. 6000 ft.). (Image analysed by Dr. Sima Majumdar)

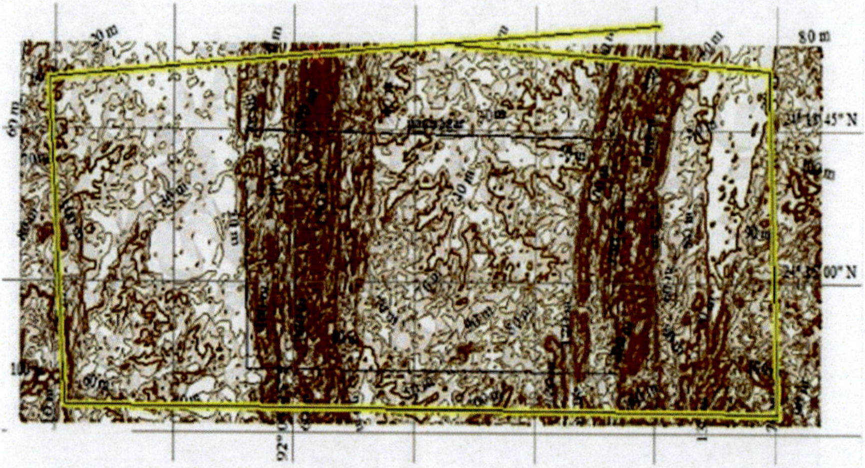

Fig. 13.5 DEM analysis of Khowai river region – Atharmura and Baramura hill ranges

that are classified depending on the response of different spectral signatures present in land cover. Therefore, in this work, RS-GIS coupled with satellite imagery analysis has been extensively used for analysis of LU/LC pattern in the study area. The river Khowai presents an interesting case study (Majumdar & Das, 2013).

In the context of riverbank erosion and migration, it was observed that the combined effects of changes in land use patterns and climate may have a stronger impact on water yield as compared to either of the two taken separately, and change in forest cover might be the primary cause of river hazards (Tarigan, 2016). The river Khowai is already under investigation by the author (Majumdar & Das, 2014b).

Also climate-related patterns are exacerbated by deforestation, and, in turn, the negative effects of deforestation are further aggravated by increasing climate variability and intensity of climate-related extreme events (Oviedo et al., 2016). Therefore, in this work, attempts to evaluate LU/LC as well as climatic factor such as rainfall or precipitation trend on the studied drainage systems have been carried.

From this literature review, it was understood that riverbank erosion, riverbank shift and migration lead to severe consequences on people. Integrated river management and ability to forecast the long-term effects may be able to mitigate vulnerability to fluvial hazards.

The Khowai river system flows in the plateau, and on two sides the two hilly ranges of Tripura, namely, the Atharmura ranges and the Baramura ranges, are situated. The dark contour (dark brown) on the above map indicates the higher elevation of the two hilly ranges (Fig. 13.5).

The LU/LC map of the region where the River Khowai flows indicates that the sign of forest cover is significantly diminished. As per norms of LU/LC maps, the dark brown pixels indicate geomorphic structures such as hilly ranges. The light brown to off-white pixels indicate anthropogenic activity such as habitation and

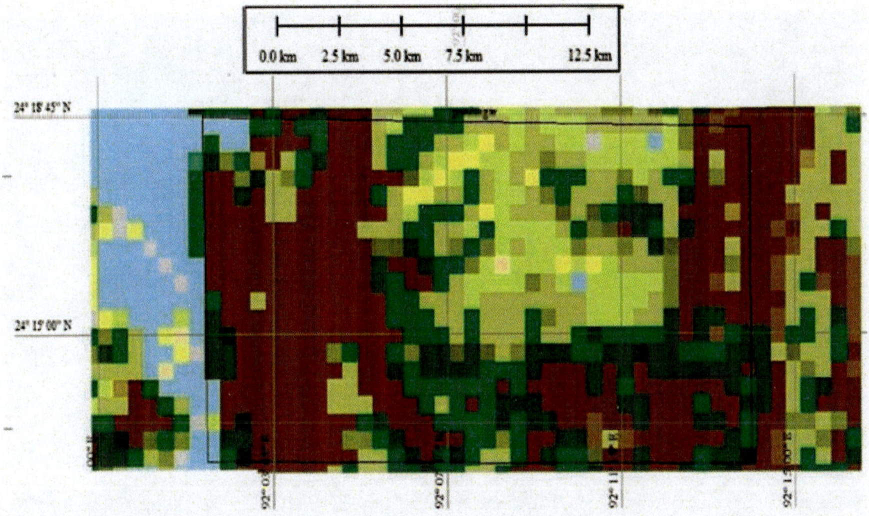

Fig. 13.6 LU/LC map of the studied region – Khowai River

some sort of urbanization. The green pixels indicate vegetation and tree cover. The intensity of the green colour is in proportion to the tree density or forest cover.

The LU/LC map of the region clearly indicates rapid urbanization and loss of vegetation around the Khowai river (Fig. 13.6).

The small-scale geographical time series analysis of Khowai River also points towards significant anthropogenic activity. The satellite imagery of the Khowai River captured at an approximate altitude of 6813 ft. of the GPS location 23°50′13.74″N and 91°39′6.28″E reveals that in 15 years time, there has been significant depletion of width of the Khowai River. This depletion of river width (Fig. 13.4) may be a result of significant change in river course due to geomorphic as well as anthropogenic impact. Bank material released from one site in the river course may have been deposited in another site leading to depletion of river width.

13.4 Conclusion

From the above study, the impact of anthropogenic activity in the form of rapid urbanization is evident. The Khowai River that is situated in the middle of the two hilly ranges, Atharmura and Baramura, is very significant for the population of Tripura (Majumdar & Pan, 2014). The major barrages are situated on the river that caters to various economic factors of the state. The Khowai riverbank also serves as a base for a large population in Tripura (Majumdar & Pan, 2015). Due to rapid urbanization, industrial setup in the form of bricks kiln is situated on the riverbank in good number. Due to large demand for resources, the brick kiln as well as other anthropogenic activity exerts great pressure on the riverbank. The impact of

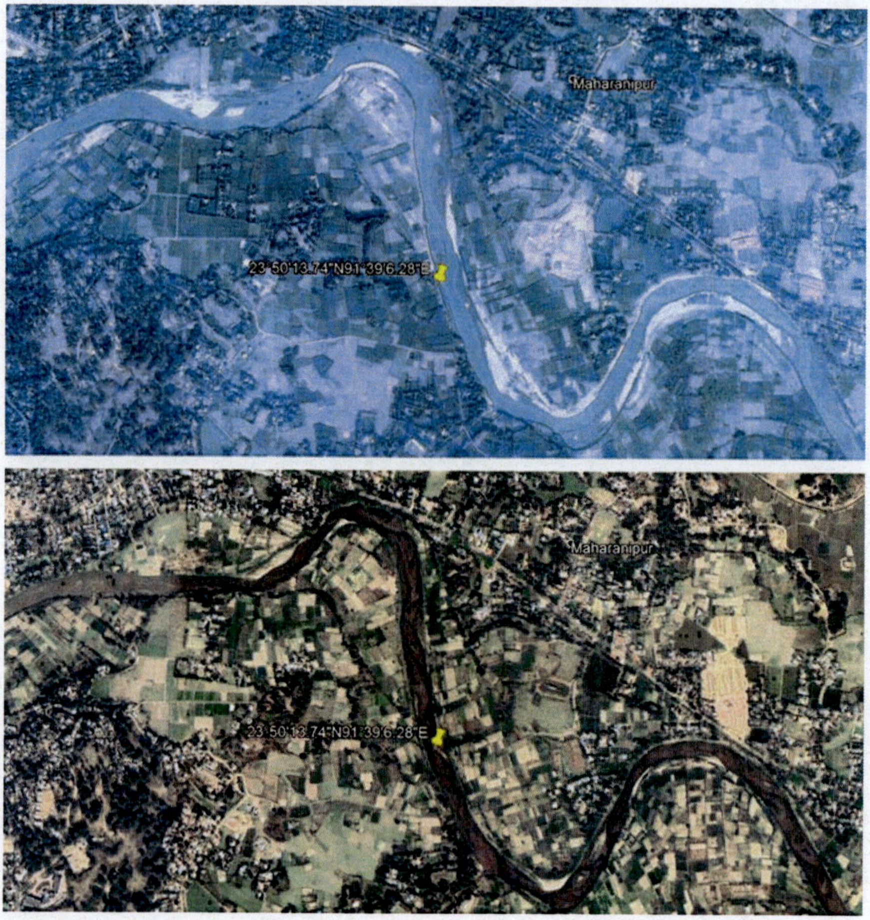

Fig. 13.7 Changes in river course in 15 years time series – Khowai River of Tripura

anthropogenic factors is clearly evident in the LU/LC map where depletion of forest cover and vegetation is apparent. Similarly, the morphometric changes on the Khowai River as evident from the satellite image time series indicate that the river is responding in a dynamic fashion. The stream morphology and its changes over time demands more studies to develop a full proof strategy for restoration of water resources in the context of the Khowai River system (Fig. 13.7).

One of the objectives of my study is optimum use of open-source GIS application that demands minimum dependence on technically demanding application. Therefore, to ensure this, I have used the LULC generated by evaluation copy of Global Mapper with default parameters. This may be a limitation in this study.

Research and development in limited resource setup is a challenge in itself. However, meaningful research demands dedication, technical skill as well as a clear understanding of the problem with an insight to solve it. The Khowai River is the major drainage system in Tripura. The Khowai River barrage serves to irrigation and related activities to a large number of people. Due to rapid urbanization, there have been consequences on the geomorphic characters of the river. The people residing in the close proximity to the river should be made aware of the impact of the deterioration of the river character. Government initiative, public organizations, NGOs and concerned bodies must come forward to work in synergy that would ultimately bring about sustainable development and responsible consumption of water resources.

References

Bhaduri, B., Minner, M., Tatalovich, S., & Harbor, J. (2001). Long-term hydrologic impact of urbanization: A tale of two models. *Journal of Water Resources Planning and Management, 127*(1), 13–19.

Chen, Y., Zhou, H., Zhang, H., Du, G., & Zhou, J. (2015). Urban flood risk warning under rapid urbanization. *Environmental Research, 139*, 3–10.

Fang, C. L., Bao, C., & Huang, J. C. (2007). Management implications to water resources constraint force on socio-economic system in rapid urbanization: A case study of the Hexi Corridor, NW China. *Water Resources Management, 21*(9), 1613–1633.

Hughes, R. M., Dunham, S., Maas-Hebner, K. G., Yeakley, J. A., Schreck, C., Harte, M., Molina, N., Shock, C. C., Kaczynski, V. W., & Schaeffer, J. (2014). A review of urban water body challenges and approaches: (1) rehabilitation and remediation. *Fisheries, 39*(1), 18–29.

Liu, Z., Wang, Y., Li, Z., & Peng, J. (2013). Impervious surface impact on water quality in the process of rapid urbanization in Shenzhen, China. *Environmental Earth Sciences, 68*(8), 2365–2373.

Majumder, Raja, et al. (2019). Assessment of flood hotspot at a village level using GIS-based spatial statistical techniques. *Arabian Journal of Geosciences, 12*(13), 1–12.

Majumdar, S., & Das, N. (2013). Temporal change of the Khowai River within Teliamura Sub-Division, Tripura and its impact. In S. Majumder (Ed.), *Contemporary issues on environmental development in India and adjacent countries, proceedings* (pp. 418–424).

Majumdar, S., & Das, N. (2014a). Bank erosion risk: A study on the Khowai River, Tripura. *Indian Journal of Landscape Systems and Ecological Studies, Kolkata, 37*(2), 83–91.

Majumdar, S., & Das, N. (2014b). Causes of bank line change of the Gumti River within Amarpur Sub-Division, Tripura. In D. K. Khan & S. Nath Bhadra (Eds.), *Proceedings of application of modern techniques for the management of contemporary environmental hazards and disasters* (pp. 215–221).

Majumdar, S., & Das, N. (2015). Bank erosion of the Gumti River between Barmura-Atharamura Hill ranges and its impact on socio-economic condition of the local people. In D. Sarkar & N. Gupta (Eds.), *Emerging environmental issues in today's society with special reference to North East India, proceedings* (pp. 78–89). Supriya Publication.

Majumdar, S., & Pan, N. D. (2014). Spatio-temporal shift of right bank of the Gumti River, Amarpur Town, Tripura and its impact. In *Landscape ecology and water management* (pp. 221–232). Springer.

Majumdar, S., & Pan, N. D. (2019). Combining open source GIS and meta-analysis to link rainfall trend and human activity: Case study on Gumti and Khowai drainage systems, Tripura, India. *Spatial Information Research, 28*(20), 1–12.

Miah Muhammad, A. (2012). Downstream ecocide from upstream water piracy. *American Journal of Environmental Sciences, 8.5*, 528.

Ouyang, T., Zhu, Z., & Kuang, Y. (2006). Assessing impact of urbanization on river water quality in the Pearl River Delta Economic Zone, China. *Environmental Monitoring and Assessment, 120*(1–3), 313–325.

Oviedo, A. F., Mitraud, S., McGrath, D. G., & Bursztyn, M. (2016). Implementing climate variability at the community level in the Amazon floodplain. *Environmental Science & Policy, 63*, 151–160.

Song, X., et al. (2014). Rapid urbanization and changes in spatiotemporal characteristics of precipitation in Beijing metropolitan area. *Journal of Geophysical Research-Atmospheres, 119*(19), 11–250.

Tarigan, S. D. (2016). Land cover change and its impact on flooding frequency of Batanghari Watershed, Jambi Province, Indonesia. *Procedia Environmental Sciences, 33*, 386–392.

Virkanen, J. (1998). Effect of urbanization on metal deposition in the bay of Töölönlahti, Southern Finland. *Marine Pollution Bulletin, 36*(9), 729–738.

Walters, S. P., Thebo, A. L., & Boehm, A. B. (2011). Impact of urbanization and agriculture on the occurrence of bacterial pathogens and stx genes in coastal water bodies of central California. *Water Research, 45*(4), 1752–1762.

Wang, J., Da, L., Song, K., & Li, B. L. (2008). Temporal variations of surface water quality in urban, suburban and rural areas during rapid urbanization in Shanghai, China. *Environmental Pollution, 152*(2), 387–393.

Zhao, H., Duan, X., Stewart, B., You, B., & Jiang, X. (2013). Spatial correlations between urbanization and river water pollution in the heavily polluted area of Taihu Lake Basin, China. *Journal of Geographical Sciences, 23*(4), 735–752.

Chapter 14
Modeling and Monitoring Soil Erosion by Water Using Remote Sensing Satellite Data and GIS

Suresh Kumar and Justin George Kalambukattu

14.1 Introduction

Soil is among the most fundamental natural resources because it determines our food as well as environmental security. As a valuable nonrenewable natural resource, the soil provides and regulates ecosystem services. The resources are under tremendous pressure and are being used unscientifically to meet the food requirements of a growing population. The soil is being adversely affected and is degrading rapidly by this improper utilization and unsustainable use. It has been estimated that nearly one third of the soil resources in the world have lost their production capacity (Blanco & Lal, 2010). Land degradation is a major global issue and is considered as the most important international agenda in the twenty-first century. Land degradation "refers to any reduction or loss in the biological or economic productive capacity of the land, quantitatively or qualitatively, through various processes such as wind or water erosion, salinization, acidification, contamination, compaction, soil carbon decline and depletion of soil nutrients." (UNCCD Secretariat, 2013). Physical, chemical, and biological processes are among those factors responsible for land degradation. The dominant physical process includes soil erosion by wind or water, resulting in decline in soil structure, soil compaction, and poor hydrological characteristics. The chemical process includes salinization, alkalization, Latinization, acidification, and heavy metal toxicity, and major biological degradation is related to a decline in soil organic matter as well as decline in the microbial properties of the soil.

Soil deterioration is the most critical aspect of land degradation, leading to irreversible changes that result in desertification. Soil degradation results in reduction of soil quality by anthropogenic activities and their interaction with natural environments. Soil degradation in general can be classified into (i) physical (soil

S. Kumar (✉) · J. G. Kalambukattu
Agriculture & Soils Department, Indian Institute of Remote Sensing, Dehradun, India
e-mail: suresh_kumar@iirs.gov.in; justin@iirs.gov.in

G. S. Bhunia et al. (eds.), *Anthropogeomorphology*, Geography of the Physical Environment, https://doi.org/10.1007/978-3-030-77572-8_14

erosion by water or wind), (ii) chemical (e.g., acidification, salinization), and (iii) biological (e.g., decline in soil organic matter) degradation. Among these, soil erosion is the most predominant of land degradation types globally. It is a worldwide phenomenon whose magnitude varies across countries, depending on land use/land cover, landform, climate, topography, and land management practices followed in place. Worldwide, approximately 24,000 million tons of fertile soil is estimated to be lost annually as a result of water erosion (FAO, 2011). Globally, the area of croplands estimated to be lost as a result of erosion processes stands at about 10 million hectares (mha) annually (Pimentel, 2006). Erosion has also been recognized as a key non-point source of environmental pollution, having adverse impacts on water quality as well as ecosystems. A recent estimate revealed that nearly 120.7 mha land of India that accounts for 36.7% of the total geographic area (including both arable and nonarable lands) experience various forms of degradation. Water erosion forms the major cause of land degradation, constituting 83 Mha (68.4%) (NAAS, 2010). The average annual erosion rate of the country has been assessed to be 15.35 tonnes ha^{-1} year^{-1} (Sharda & Ojasvi, 2016), amounting to an estimated nutrient loss of 5.37–8.4 million tonnes, decline in crop productivity, sedimentation of reservoir capacity, and loss of biodiversity. Reduction in crop productivity is the most negative impact of water erosion, having severe consequences on people's foods, livelihoods, and environmental security.

14.1.1 Soil Erosion Factors and Processes

Erosion, being a natural geologic process occurring throughout eons of time, is the wearing away, transportation, and deposition of earthen materials by various forces such as water, wind, or glaciers. These processes are vital in the creation of diverse landforms/landscapes, in soil formation and development, and nutrient cycling, feeding local aquatic ecosystems by supplying essential nutrients as well as cleansing the soil by removing useless materials. These processes are extremely important for different ecosystem processes, thus maintaining their quality as well as environmental sustainability. Recognizing these key roles, erosion becomes a matter of grave concern when natural erosion is accelerated by various causative agents aided by different anthropogenic activities, thus exceeding the rate of natural soil formation and disrupting the equilibrium.

Tracing the origin of soil erosion takes us to various weathering processes, leading to the formation of soil from rocks by means of disintegration and decomposition, thus forming the material to be eroded later. Disintegration involves the breaking down of rocks into smaller fragments without any change in chemical composition, whereas decomposition involves a change in chemical composition of fragments also caused by various physical, chemical, and biological forces. The type, extent, and rate of various weathering processes are determined by several factors, such as rock characteristics (mineralogical composition, texture, grain size, pattern of joints, cracks, and fissures), climate, and vegetation. The interplay of various weathering processes results in soil material that is carried away during different erosion events. The major mechanisms involved in soil erosion are categorized as detachment, transportation, and deposition. The detachment of soil from its original site is caused by various erosive agents, namely water, wind, gravity, and glaciers. The dominant erosive agent causing detachment varies depending on the geographic region as well as the climatic conditions prevalent in the area: wind is the predominant erosive agent in dry arid environments, whereas in the humid tropics and snowclad areas, water as well as snow/glaciers, respectively, becomes the major erosion-causing agent.

14.1.2 Factors Affecting Water Erosion

Soil erosion by water being the most dominant form of erosion, accounting for nearly 80% of all the soil loss by erosion, further discussions are limited to various aspects of water erosion only. Rainfall forms the major erosive factor in water erosion. The transportation of soil material from the site by moving water depends on various characteristics especially related to rainfall, such as soil characteristics, topography, hydrology, and ground cover.

Among the different rainfall characteristics, rainfall intensity as well as duration deserves special mention as their role in erosion processes is vital. Rainfall has the major function of breaking the soil aggregates by beating action on direct impact. The falling raindrops not only break the aggregates but also splash the soil particles, initiate creeping of particles on the surface, sort soil particles, and force fine-grained particles into soil voids, causing sealing of pores and resultant reduction in infiltration rates. Apart from intensity and duration, the mass, diameter, and velocity of the raindrops also have significant impact in determining soil erosion rates. These different aspects of rainfall are assimilated into a single factor widely described as the kinetic energy of rainfall, which gives a measure of the potential erosion-causing ability of rainfall, defined as rainfall erosivity. Rainfall erosivity can be estimated using techniques based on different facets of continuous rainfall observation using field-based instruments, long-term climate records, and satellite-derived rainfall products. A soils response to rainfall also depends on preceding meteorological conditions such as duration of dry spell and amount of last received rainfall.

Soil characteristics are crucial in influencing the ability of soil to resist or withstand erosion processes, widely denoted as soil erodibility. Erodibility indicates the resistance offered by the soil to both detachment and transportation: it depends mainly on various soil physical characteristics such as texture, aggregate stability, infiltration capacity, and organic matter content. These different characteristics have crucial roles in providing resistance toward easy disruption of soil aggregates, transportation of particles, and regulating the infiltration rates and thus the runoff/overland flow generations. The estimation of erodibility has no universal method, because so many factors are involved that it is practically difficult to measure. The prominently adopted methods mainly involve relationships using different, easily measurable soil characteristics such as textural composition, organic matter content, soil structure, and permeability status. In general, soils with high levels of organic matter, improved infiltration rates, as well as good soil structure, tend to offer higher resistance to erosion processes. Particles such as silt, very fine sand, and clay-textured soils are found to be more erosible in comparison to sand, sandy loam, and other coarse-textured soils.

The major topographic factors influencing water erosion are slope length and gradient, which assume greater importance in undulating as well as hilly terrains. In general, the risk of erosion disproportionately increases manifold with longer and steeper slopes in comparison to flatter terrains. Greater slope lengths result in increased soil erosion rates, mainly from the greater runoff accumulation (high runoff volume) in the area as well as amplified runoff velocity. Both these factors drastically increase the sediment-carrying capacity of runoff water, also termed degree of scouring. Similarly, the shape of slopes indicated by curvatures also affects erosion rates by altering flow characteristics as well as the chances of deposition. Uniform slopes tend to have greater flow lengths and possess higher chances of material deposition, whereas the convex slopes tend to have higher erosion rates because of the higher velocities associated with turbulent flow.

Vegetation in the form of both natural and cultivated crops as well as residues has a prominent and undisputable role in determining the erosion rates associated with rainfall events. Vegetation impacts erosion processes by numerous effects such as intercepting rainfall by the canopy; decreasing runoff velocity by physical obstruction; covering the soil surface by fallen litter/debris; improving soil strength with granulation as well as porosity by action of roots; insulating soil against higher temperatures; and drying soils by means of transpiration. Interception of rainfall significantly reduces its kinetic energy by absorbing the impact, thereby decreasing the beating action and detachment of soil particles at the surface. The physical obstruction-induced slowing down of runoff increases the time available for infiltration and thus alters the transportation process. The impact of vegetation on erosion processes depends on many characteristics such as crop type, canopy type, extent as well as quantity, canopy height from the ground, planting distance, root architecture, and residue management, which determines the extent of surface protection available during various periods in a year, corresponding to the amount of erosive rainfall events occurring during that period. Crops that keep the soil bare for major portions of the year, especially during highly erosive rainfall periods such as summer and monsoon (e.g., cereals, row crops) increase the erosion rates much more in comparison to crops providing full protective cover during a major portion of the year (e.g., legumes and other cover crops). Various crop management practices such as intercropping, contour farming, conservation tillage, contour strip cropping, staggered planting, vegetative barriers, and retention of crop residue on the surface can reduce the amount of erosion.

Other important factors with significant influence on water erosion are the tillage practices, depending on the time, direction, and depth of plowing, and the implements used. Increased tillage activities are critical in disturbing the soil structure, aggregates as well as vegetative/residue cover at or near the soil surface, thus drastically increasing the erosion risk. Tillage activities and other cultural operations performed in the along-the-slope direction (up–down field slopes) could accelerate the erosion process by creating waterways for surface runoff, whereas cultivation and tillage in the across-slope direction and contour farming practices can reduce erosion loss by limiting the amount of concentrated runoff. Reduced tillage or no-tillage practices are found to be very effective in the reduction of soil erosion losses.

14.1.3 Soil Erosion Types

Depending on the processes and mode of occurrence, water erosion has been categorized into different types, summed briefly next.

Splash Erosion Widely considered as the first stage in the erosion process, splash erosion occurs from the beating action or direct impact of falling raindrops on bare soil or on thin water surfaces. The high-energy impact disrupts and breaks the soil aggregates, releasing finer individual soil particles that are 'splashed' on the surface. The major rainfall characteristics determining the rates of splash erosion include raindrop mass, size, shape, velocity, and direction, as well as the intensity of rainfall. Other factors include the previous moisture status of the soil, canopy cover on the land, and residue retention.

Sheet Erosion The surface runoff/overland flow (when the rainfall exceeds infiltration rate) will remove soil particles released by splash erosion as well as detach particles from the surface in a uniform fashion in gentle sloping lands, which is termed sheet erosion. It is commonly referred to as sheet erosion owing to the fairly even or uniform removal of soil particles from the surface and goes unnoticed until a fairly large amount of productive topsoil is lost (resulting in overall lowering of land surface). Muddy runoff from land following a rainfall event is the primary indication of the sheet erosion process. Splash and sheet erosion are often combined and referred to as inter-rill erosion. Rainfall intensity, runoff velocity, soil particle characteristics, and slope are major factors determining the soil loss rates.

Rill Erosion Rill erosion involves the detachment as well as transportation of soil particles from concentrated runoff in small channels or rivulets (a few millimeters deep and wide), commonly referred as rills. Rills usually do not interfere with various field operations such as tillage and can be obliterated by ploughing as well as weathering processes. The concentrated flow in rills removes soil particles by scouring action and tends to increase its depth and width dimensions in the long run, if left unattended. The rills formed normally expand by migrating in the upslope direction corresponding to the retreat of headcut at the top portion of the channel.

Gully Erosion In this most advanced form of erosion, wherein soil particles are removed by concentrated flow in relatively permanent deeper and wider channels with steep side-slopes, generally with cross-sectional area greater than 1 m^2 (Poesen,

1993), these are referred to as gullies. Gullies in general are several meters wide as well as deep, causing physical obstruction of tillage operations as well as movement of people and livestock across the slope. If left unchecked, the gully becomes wider and deeper in the course of time, resulting in dissection of land initially, with subsequent regular networking leading to the formation of ravenous lands. The amount of runoff and sediment load transported via gullies is much higher in comparison to rill erosion.

Apart from the afore-described four major forms of water erosion, several other forms also exist: stream bank erosion (from prolonged undercutting, side cutting, and scouring of drainage channels or waterways); coastal or seashore erosion (soil and land removal by action of tidal waves); glacial erosion (from movement of glaciers); and mass wasting (from rainfall-aggravated landslides). These forms are beyond the scope of this chapter and hence are not discussed here.

14.2 Remote Sensing

Remote sensing (RS) is defined as "the science and art of obtaining information about an object, area, or phenomenon through the analysis of data acquired by a device that is not in contact with the object, area, or phenomenon under investigation" (Lillesand et al., 2004). Remote sensing enables us to observe and study nature in many different ways that would otherwise have been beyond human capability, covering great distances and at wavelengths ranging from visible to even invisible to human eyes. The entire remote sensing process can be categorized into these stages:

- Emission of electromagnetic radiation (EMR) under different wavelengths by an energy source such as the sun (passive remote sensing) or artificial sources such as satellites (active remote sensing).
- Transmission of EMR from source through atmosphere towards earth surface (target). During this transmission, the EMR interacts with various atmospheric components such as ozone, carbon dioxide, and water vapor. Clouds and dust aerosols, for example, via scattering as well as absorption mechanisms, are affected before being transmitted toward the surface of the Earth. These atmospheric interferences greatly alter the nature of the EMR reaching the surface. Of the entire electromagnetic spectrum, only the wavelength regions outside the main absorption bands of atmospheric constituents/gases (referred as atmospheric transmission windows) are useful for the purpose of remote sensing. RS types, namely optical, thermal, and microwave remote sensing, are generally defined based on the wavelengths used, which correspond to different atmospheric windows such as visible and reflected infrared region (between 0.4 and 2 μm, where optical sensors operate); thermal infrared region (two windows of about 3 and 5 μm, respectively, and a comparatively broad window (from 8 to 14 μm, where thermal sensors operate), and a microwave region (a more or less transparent region beyond 1 mm, where microwave sensors operate).

- Interaction of EMR passing through atmosphere with various features present on the Earth surface. Depending on the nature as well as condition of object and wavelength of the radiation, the incident energy may be absorbed, transmitted, and/or reflected by the features, at different proportions.
- The reflected or radiated energy from the targets will be received and recorded by various sensors.
- The sensor recorded data will be converted into a sensor output and transmitted to a ground receiving and processing station on Earth, and then processed and converted into an analogue and/or digital image.
- The generated image will be analyzed and interpreted by means of visual as well as digital image processing techniques to extract the object, or area, or to study the phenomena associated with the target feature.
- Finally, the information obtained from the image may be put into use for addressing such applications as pertaining to agriculture, forestry, land use, atmosphere, geology, hydrology, disaster management, navigation, and defense.

The platforms that provide RS images at varying coverage scales include ground-based, airborne (UAVs, drones, aircraft, etc.), as well as space-borne satellites. The various features as well as applicability of these different platform types are well explained by Uday Chatterjee. Among the two broad categories of satellites, namely, geostationary and sun synchronous satellites, sun synchronous satellites located at an altitude of about 700–800 km above the Earth's surface are mainly used for earth observation applications, such as IRS series, Cartosat, and Landsat, whereas the geostationary satellites at an altitude of nearly 36,000 km from the surface are widely used for applications requiring very frequent repetitive observations such as communication as well as meteorological applications (e.g., INSAT, GSAT). Irrespective of the source of illumination and altitude above the Earth's surface, sensors measure and store electromagnetic energy in the form of digital images having a regular grid format, whose single image element is referred to as a pixel or picture element. Ali (2010) has defined a digital image as "a 2-dimensional grouping of distinct image elements or pixels demonstrating the geographical distribution of various parameters such as electromagnetic radiation, temperature, or topographical elevation, etc." Each pixel contains a quantized measurement of energy in the form of a digital number, which correspond to variations in surface reflectance values or tonal variations. Such images or datasets are generated and stored separately for each measured wavelength range, designated as bands/channels/layers. The combination of such images is used for information retrieval with respect to the targeted application by employing various digital image processing/interpretation techniques. A detailed account of the various fundamental or essential steps in image processing has been provided by various authors (Sabins, 1978; Lillesand & Kiefer, 1994; Joseph & Jaganathan, 2018). The image characteristics are mainly determined by four different types of resolution, that is, spatial, spectral, radiometric, and temporal resolution, which correspond to ground sampling distance, number or width of bands in the spectrum in which data are collected, the slightest variations in energy that can be recorded by sensors, and the time frequency between two successive image acquisitions of the same area (Joseph & Jagananthan, 2018).

The choice of sensor and RS data should be based on intended purpose/end use, scale of mapping, processing capabilities available, cost-effectiveness, and area coverage as well as repeated availability. The chosen images will be interpreted based on visual as well as digital image interpretation schemes depending on the purpose as well as detailing required. The images will be classified as well: various features will be extracted ultimately.

A plethora of satellite data with a wide range of resolutions that could be chosen and utilized for diverse applications are available from different space agencies as well as commercial players across the world. A comprehensive list of earth observation satellites launched and operated by the Indian Space Research Organization (ISRO), widely utilized in diverse applications, including the Indian Remote Sensing (IRS) series of satellites (data ranging from 0.6 to nearly 200 m resolution), is available at *https://www.isro.gov.in/spacecraft/list-of-earth-observation-satellites*. Among the foreign satellite data most widely and frequently used by various user groups for different applications, the *Landsat* series and the recent *Sentinel* series need special mention. More detailed information regarding the Landsat and Sentinel series of satellite data can be obtained from *https://landsat.gsfc.nasa.gov/*, *https://www.usgs.gov/land-resources/nli/landsat* and *https://sentinel.esa.int/web/sentinel/missions/sentinel-2*. Apart from the various freely available satellite data just mentioned, many commercial agencies are providing us with very high resolution RS data having sub-meter resolutions. A comprehensive list of satellites and sensors along with their technical specifications has been hosted and regularly updated for worldwide users by ITC, Faculty of Geo-Information Science and Earth Observation, University of Twente (https://www.itc.nl/research/research-facilities/labs-resources/satellite-sensor-database/). A few very common freely available data and data product sources are listed in Table 14.1. This list is not a comprehensive one, as many freely available datasets are available worldwide from which the users can choose.

14.3 GIS Analysis

Making use of RS and other georeferenced data for meeting the objectives of our targeted application requires a computer-based system commonly referred to as a Geographic Information Systems (GIS). GIS is capable of handling georeferenced data by providing capabilities to input, manage (storage and retrieval), manipulate, and analyze as well as generate outputs (Aronoff, 1989). The most common and widely used GIS software for geospatial data handling and processing is the commercial licensed ArcGIS software. Apart from this a number of open-source, competent GIS software are available for data analysis with respect to various applications (Table 14.2).

Table 14.1 List of commonly available RS data sources

Bhuvan: Indian Geo-Platform of ISRO (*For Indian Satellite Data/Data products only*)	https://bhuvan.nrsc.gov.in/bhuvan_links.php#
USGS Earth Explorer	https://earthexplorer.usgs.gov/
Sentinels Scientific Data Hub (*ESA's Sentinel satellite data*)	https://scihub.copernicus.eu/dhus/
NOAA CLASS *(Free high quality atmospheric data sets)*	https://www.avl.class.noaa.gov/saa/products/welcome
JAXA's Global ALOS 3D World *(Global-scale elevation data)*	https://www.eorc.jaxa.jp/ALOS/en/aw3d30/
DigitalGlobe: Very high resolution satellite data *(free product samples)*	http://www.digitalglobe.com/products/satellite-imagery
Geo-Airbus: For SPOT, Pleiades, RapidEye and TerraSAR data	https://www.intelligence-airbusds.com/en/8262-sample-imagery

Table 14.2 List of commonly used open source GIS software

GIS software	Salient features	Website
GRASS	Used for geospatial data management and analysis with capabilities of image processing, maps/graphics production, spatial modelling as well as visualization	https://grass.osgeo.org/
QGIS	User-friendly open source GIS software that can run on different platforms (Windows, Linux, Unix, Mac, OSX, etc.). It can support vector, raster and various other database formats	http://www.qgis.org/
SAGA	GIS designed keeping geosciences application in mind, especially for hydrographic and terrain analysis	http://www.saga-gis.org/
PCRaster	Computer language mainly focused on the construction of iterative spatiotemporal environmental models. Encompasses raster-based analysis with strong hydrological modeling	https://pcraster.geo.uu.nl/
PostGIS	Adds support for geographic objects to the PostgreSQL object-relational database. Used mainly as backend spatial database for GIS	https://postgis.net/
gvSIG	Desktop tool designed to manage geographic information both in raster and vector data formats	http://www.gvsig.com/
ILWIS	PC based GIS & RS software acting as a complete package for image analysis/processing, digital mapping and spatial analysis	https://52north.org/software/software-projects/ilwis/
uDIG	Both a platform and geospatial application providing developers with an opportunity to create new derived applications	http://udig.refractions.net/

14.4 Soil Erosion Models

Soil erosion is the most widespread form of land degradation, occurring in all climatic conditions with varying intensity and severity. Soil erosion caused by water involves three major processes: soil particle detachment from the land surface, transportation of particles through surface runoff, and their deposition elsewhere on the landscape or in the reservoir/water bodies. It is initiated by raindrops falling onto a bare soil surface, causing splash erosion followed by sheet, rill, and gully erosion. Erosion modeling is the process of mathematically describing the detachment, transportation, and deposition of soil particles on land surfaces. In sheet erosion, a thin uniform layer of soil is removed from the surface whereas in rill erosion runoff water over the land surface flows along depression areas/special pathways, forming fine channels known as rills. Sheet and rill erosion processes are considered together in modeling soil erosion. Gully erosion is the advanced form of rill erosion, where by continuous sheet and rill erosion processes rills become widened and deepened by the incising action of runoff water. All the erosion models only simulate splash, sheet, and rill erosion to estimate soil erosion in the landscape.

Erosion models vary with regard to intricacy, processes involved, and data requirement for calibration as well as validation. In general, as such we cannot pinpoint a 'best' model for all landscapes. The model selection for simulating soil erosion depends on the planned use and the watershed/catchment characteristics. In general, models are categorized based on the physical processes involved and the data requirements in the model (Merritt et al., 2003). The models are primarily classified into three major groups: (i) empirical or statistical models, (ii) conceptual models, and (iii) physical process-based models.

14.4.1 *Empirical Models*

Empirical models are primarily defined based on significant factors generated via field measurements and observation, experimentation, and statistical techniques, thus relating different erosion-causing factors to soil loss. Empirical models are the simplest model and do not describe any processes. The data requirements and computational complexities of empirical models are usually less compared to conceptual and physics-based models. These simple models are referred as black box type models that relate sediment loss to either rainfall or runoff. These models have been developed for estimating sheet and rill erosion rates based on a large set of experimental data generated from agricultural plots. The first empirical model, the Universal Soil Loss Equation (USLE) developed by Wischmeier and Smith (1978), and its derivatives such as the Revised Universal Soil Loss Equation (Renard, 1994) and Modified Universal Soil Loss Equation (MUSLE) (Williams, 1975) are the most extensively used as well as accepted empirical models. These models are predominantly beneficial as a primary step in detecting sediment loss sources and soil erosion. They ignore the heterogeneity of catchment characteristics and provide long-term estimates of soil erosion and ignore the rainfall–runoff processes. The data requirements of these models are limited as compared to other models.

14.4.2 Conceptual Models

Conceptual models have a transitional role among empirical and process-based models (Beck, 1987). In general, conceptual models lump characteristic processes over the output simulated scale (Wheater et al., 1993). These models predict sediment yields primarily using the concept of unit hydrographs: conceptual (lumped) models accounting for spatially averaged catchment parameters and estimate runoff and sediment loss for the whole catchment region. These models include an overall depiction of processes operating in the catchment without any exact details regarding interactions among various processes because detailed catchment information is lacking (Sorooshian, 1991). The models generally integrate the fundamental transfer mechanisms of runoff generation and sediment transport representing flow paths in the catchment as a series of storages. Parameterization of these models is done by calibration against observed data related to runoff and sediment load (Abbott et al., 1986) and better represent reality. These models are known to offer an account of the quantitative and qualitative consequences associated with land use changes, without the need of a large quantity of temporally as well as spatially distributed input data (Merritt et al., 2003).

14.4.3 Physical Process-Based Models

Physically based models are centered on the knowledge regarding various fundamental processes involved in erosion and incorporate the law of conservation of energy and mass (Bennett, 1974). These models describe essential processes of soil erosion through mathematical expressions: these are fundamental physical equation-based models and their outputs describe the stream flow as well as sediment transport in the catchment. The equations regulating various processes in these models are generally derived at small scale, under definite physical settings (Beven, 1989). The equations constitute the synthesis of distinct soil erosion components, comprising the various complex interactions between different factors and their temporal as well as spatial variability. Physically based models are spatially distributed and are extensively used in the identification of critical source areas within a watershed with improved simulation accuracy in comparison to the other two types of models (empirical or conceptual), thus providing better understanding of the erosion and runoff generation processes. The models provide accurate simulation and prediction with respect to soil erosion and the associated sediment as well as nutrient loss through overland flow. They could simulate the effect of change in climate as well as land use/land cover on erosion/deposition processes under various landscapes, in detailed manner. Physical models could also reflect the influence of different physical features such as land use, topography, plant growth, geology as well as stream flow characteristics. These models are data intensive and termed as white box models as all the mechanism of processes are well defined through mathematical expressions. These models run these equations with uninterrupted spatial and temporal scale although the used input data are often point source, representing an entire grid cell in the watershed.

The widely used models are listed in Table 14.3. A brief description of these models under various categories follows.

Table 14.3 List of frequently used soil erosion models

Model	Scale	Time Scale	Input parameters	Output Parameters
Empirical				
USLE	Hillslope, plot	Annual	Land use/ land cover, soil, climate, topography	Soil erosion
MUSLE	Watershed	Event/ annual	Topography, climate, soil, land use/ land cover	Sediment loss
RUSLE	Watershed, landscape	Seasonal/ annual	Soil, climate, topography, land use/ land cover	Soil erosion
Conceptual				
MMF	Plot, hillslope	Annual	Climate, soil, topography, plant parameters, land use/land cover	Runoff, soil erosion
SWAT	Watershed/ catchment	Daily, seasonal, annual	Climate, soil, topography, management practices, land use/land cover	Sediment loss, runoff
APEX	Farm scale, watershed	Daily, monthly	Climate, soil, topography, management practices, land use/land cover	Runoff, sediment and nutrient loss
AGNPS	Watershed	Daily, annual	Climate, soil, topography, management practices, land use/land cover	Runoff, sediment and nutrient loss
Physical process				
WEPP	Hill slope, small watershed	Daily, monthly	Climate, soil, topography, crop types, land cover, management practices	Runoff, erosion rate, sediment loss
GeoWEPP	Small watershed	Daily, monthly	Climate, soil, topography, crop types, management practices, land cover, drainage network runoff	Runoff, sediment loss
ANSWERS	Small watershed	Daily, monthly	Climate, soil, topography, crop types, management practices, drainage network runoff	Sediment loss

14.4.4 Universal Soil Loss Equation (USLE)

The universal soil loss equation (USLE) is the most widely used model for estimation of soil erosion in the world (Wischmeier & Smith, 1978). The model development utilized soil loss rates estimated from 10,000 erosion plot-years of data in the USA, by means of field experiments conducted under both natural and simulated rainfalls. The erosion plots were maintained at a uniform size of 72.6 feet in length with a width of either 6.0 or 12.0 feet (0.01 or 0.02 acres) and 9% slope. The length dimension was chosen by keeping in mind the easiness in computation of runoff and resultant erosion on a unit area basis. This model provided a fundamental concept for developing other existing erosion and sediment transport models. This model was primarily devised for estimating annual erosion rates in small plots with an approximate area of nearly 40 m^2. Huge errors may be expected when the model is applied to individual rainfall events and much larger study areas (Kinnell, 1999, Sadeghi et al., 2007). The model can be written as:

$$A = RKLSCP$$

where A is the average annual soil loss (tons ha^{-1} $year^{-1}$); R is the rainfall erosivity; K is the soil erodibility factor; LS is the slope length and steepness factor; C is the cover-management factor; and P is the management practice factor.

The LS factor is computed as (Wischmeier & Smith, 1978):

$$LS = (1/22.13)^{m}.(65.4\ \sin^{2}B + 4.56\ \sin\ B + 0.0654)$$

where l indicates slope length (in meters) and the value of m is equivalent to 0.5 for areas with slope steeper than 5%, 0.4 for slopes between 3% and 4%, 0.3 for slopes between 1% and 3%, and 0.2 for slopes less than 1%; and B represents slope angle (in degrees).

Soil erodibility (K) factor is computed as:

$$K = 2.8 * 10^{-7} M^{1.14}\ (12 - a)\ 4.3 * 10^{-3} \times (b - 2) * 3.3 * 10 - 3(c - 3)$$

where K = soil erodibility factor (t ha^{-1} per unit R), M = particle size parameter (% silt + % very fine sand) $*$ (100 − % clay), a = organic matter content (%), b = soil structure code, and c = soil permeability class.

C and P values varies with land use/land cover type as well as management practices, respectively. R factor is calculated using long-term rainfall data of the past 20 years.

14.4.5 Revised Universal Soil Loss Equation (RUSLE)

USLE was updated and revised by incorporating a substantial amount of erosion research information generated after the unveiling of USDA-Agriculture Handbook 537 (Wischmeier & Smith, 1978). The applicability of USLE was extended to many land uses in addition to agriculture. The methodology was revised to compute factors of soil erosion described in the USLE model. The revised model was described in USDA-Agriculture Handbook 703 and was termed as "revised universal soil loss equation," abbreviated as RUSLE (Renard et al., 1997). RUSLE estimates loss of soil caused by inter-rill as well as rill erosion, from a hillslope, but does not assess erosion from gullies or stream channels. In the revised USLE, the annual R factor is estimated as a sum of the product of the storm kinetic energy times the maximum 30-min intensity for each storm occurring in an "n" year period/season. The K factor was improved by considering its intraannual variability. The LS factor equations were modified, improving their accuracy, and extended to incorporate hillslopes with much steeper gradients than those defined in USLE. The C factor values were determined employing a sub-factor approach, which includes input values depicting the key features of a cover-management system. As a result, RUSLE became more applicable to diverse field conditions providing more site-specific C values compared to USLE. The basic structure of USLE is retained in RUSLE, and it is the commonly adopted model to compute annual soil erosion in the landscape/watershed in GIS environment. Watershed-scale sediment yields are not estimated by RUSLE. In GIS it is used to estimate the soil of each pixel/grid cell, which can be described as:

$$\mathrm{A(x)} = R * K * LS(r) * C * P$$

where A(x) = average soil loss per year of a grid cell (t ha^{-1}year^{-1}) at a point x (geographic location of grid cell), R = rainfall intensity factor (mt ha^{-1} cm^{-1}), K = soil erodibility factor (t ha^{-1} per unit R), LS(x) = topographic (length-slope) factor at a grid cell (x) and is dimensionless, C and P are the land cover factor and soil conservation practices factor, respectively (both dimensionless).

Renard et al. (1997) defined slope steepness factor values for an area having slope lesser than 9% and steeper than 9%. The S-factor estimation equation is given by

$$S = 10.8 \times \sin \Theta + 0.03, \text{where slope gradient} < 9$$
$$S = 16.8 \times \sin \Theta - 0.05, \text{where slope gradient} \geq 9$$

where Θ is the gradient of slope in degrees.

The *LS* factor is calculated for each grid cell as

$$LS(\mathrm{x}) = (m+1)(A(\mathrm{x})/22.13)^m \\ sin\beta(\mathrm{x})/0.09)^n \tag{14.1}$$

where β(x) is land surface slope expressed in degrees, and *m* and *n* are constants equal to 0.6 and 1.3, respectively.

14.4.6 Modified Universal Soil Loss Equation (MUSLE)

The USLE model was modified by replacing the rainfall factor (R) with a runoff factor to determine the stream sediment yield corresponding to individual storms (Williams, 1975). The modification used the data collected from 18 small watersheds accounting for 778 storm-runoff events, with areas ranging from 15 to 1500 ha, and slopes extending from 0.9% to 5.9% as well as slope lengths varying from 78.64 to 173.74 m (Williams & Berndt, 1977; Haan et al., 1994). MUSLE simulates erosion and sediment yield by making use of storm-based runoff volumes and runoff peak flows. It uses variables corresponding to runoff instead of rainfall erosivity, which enables the estimation of sediment yields related to individual storm events. The model assumes the form of an equation:

$$Y = 11.8\ (Q * \mathrm{pr})^{0.56} * K * C * P * \mathrm{LS}$$

where Y is the sediment yield (metric tons); Q is the runoff volume (m^3); pr is the peak runoff rate (m^3s^{-1}); K is the soil erodibility factor; LS is slope length and steepness factor (terrain factor); C is crop management factor; and P is management practice factor.

The SCS curve number method and a modified version of the rational equation (USDA-SCS) are used for the estimation of runoff volume (Q) and peak runoff rates, respectively.

14.4.7 Revised Morgan, Morgan & Finney (MMF) Model

The Morgan, Morgan & Finney (MMF) model provides a stronger physical basis than the empirical models such as USLE and at the same time retains the simplicity and advantages of an empirical model. A revised MMF model was proposed by Morgan (2001) that takes into account plant canopy height and leaf drainage, and another component of soil particle detachment by runoff water was added. The model describes erosion processes in two parts: water phase and sediment phase. The water phase estimates the energy of the rainfall available to detach soil particles from the soil mass and volume of runoff. In the sediment phase, soil particle detachment rate caused by rainfall and runoff is estimated along with transport capacity of runoff. Erosion rate was predicted by comparing both the soil particle detachment and transport capacity, based on the lesser of both rates. The revised model provides an improved description of the water erosion processes operating on hillslopes of a small watershed/catchment. It is used to predict annual runoff and soil loss erosion and is briefly describe as follows.

Water Phase This phase includes prediction of detachment by rainfall and runoff. In this phase, kinetic energy of the direct throughfall (KE (DT) in J m^{-2}) and leaf drainage (KE) are estimated:

$$KE\,(DT) = DT\,(11.9 + 8.7\,\log\,I)$$
$$KE\,(LD) = \left(15.8 * PH^{0.5}\right) - 5.87$$
$$KE = KE\,(LD) + KE(LD)$$

Runoff estimated as :

$$Q = R\,\exp\,(-Rc/Ro)$$
$$Rc = 1000MS * BD * EHDC\,(Et/Eo)$$

Where, Rc-moisture storage capacity; Ro- mean rain per rain day; Ms- Soil moisture context at field capacity, BD-bulk density of soil; EHD- the effective hydrological depth of soil; Et/Eo- the ratio of actual to potential evapotranspiration.

Soil particle detachment by raindrop impact is computed as:

$$F = K * KE * 10^{-3}$$

K – the erodebility of soil

Soil particle detachment by runoff is computed as:

$$H = ZQ^{1.5}\,\sin\,S\,(1 - GC) * 10^{-3}$$
$$Z = 1/(0.5\,COH)$$

GC- percent ground cover; COH- value of cohesion of saturated soil; S- Slope angle

Transport capacity of runoff is computed as:

$$TC = CQ^2\,\sin S * 10^{-3}$$

The estimates of soil particle detachment by raindrop impact and by runoff are added together to find a total annual detachment rate: it is compared with the annual transport capacity, and the lower of the two values is defined as the annual soil erosion rate.

14.4.8 Soil & Water Assessment Tool (SWAT)

SWAT is a distributed, continuous time model that operates on a daily basis. It used to predict land use changes and impact of land management on surface runoff and sediment yield in meso- to macroscale basins or watersheds (Arnold et al., 1996). A watershed is divided into several micro watersheds and each micro watershed is composed of several hydrological response units (HRUs). A HRU is characterized by a unique combination of land use/land cover, soil, slope, and management type. The input parameter of rainfall and other climate parameters are required in a daily time step. The model uses the SCS curve member (SCS-CN) method to estimate surface runoff. It computes the peak runoff rate using a modified rational method. It predicts sediment yield using the Modified Universal Soil Loss Equation (MUSLE) (Williams & Berndt, 1977). It computes surface runoff and sediment yield for each HRU and routes it to estimate runoff water and sediment yield for watershed/ catchment and basins.

The surface runoff is calculated using the SCS-CN method as:

$$Q = (R - Ia)^2/(R - Ia + S)$$

where Q = daily surface runoff; R = daily rainfall; Ia = initial abstraction; and S = retention parameter.

Initial abstraction (Ia) includes surface storage, interception, and infiltration before runoff. Runoff takes place when R exceeds Ia, which is commonly estimated as 0.2S.

The retention parameter (S) is computed as:

$$S = 25.4\ (1000/CN - 10)$$

where S = retention factor (mm), and CN = curve number.

14.4.9 Water Erosion Prediction Project (WEPP) Model

The Water Erosion Prediction Project (WEPP) is a continuous physical process-based model that is used to simulated surface runoff and soil loss (Flanagan & Nearing, 1995). It is based on the fundamentals of infiltration theory, hydrology, soil physics, plant science, hydraulics, and erosion mechanics; the erosion component of the model is based on a steady-state continuity equation. The model comprises nine components: weather generation, irrigation, soils, hydrology, plant growth, winter process, residue decomposition, erosion, and deposition. This model is available in two versions, such as hillslope profile and small watershed. The hillslope version computes runoff and soil erosion along a slope (1–200 m in length) whereas the watershed version simulates a small watershed (up to 260 ha). The model is widely used for conducting soil erosion simulation research where large amounts of field and observed data are available regarding land use/land cover, plant growth, soil, topography, and soil hydrology as well as climate/weather data for soil erosion simulation. It serves as the best model for conducting very detailed simulation studies.

14.4.10 GeoWEPP

The GeoWEPP model was developed as a collaborative project of the USDA National Soil Erosion Research Laboratory and Purdue University, USA. The geospatial interface for WEPP (GeoWEPP) uses georeferenced information such as a digital elevation model (DEM), land use/land cover, and soil to derive model input parameters to predict sediment yield at watershed scale (Renschler, 2002). It provides spatially distributed soil erosion in the watershed. The WEPP model was originally interfaced with Arc View 3.2 extension, and currently GeoWEPP for ArcGIS 10.4.0 has been released (http://geowepp.geog.buffalo.edu/version/arcgis10-4/). In GeoWEPP, the TOPAZ algorithm uses DEM to generate channel network and then the watershed outlet as defined by user. Thereafter, the water boundary and sub-catchment areas contributing to the channels are delineated using TOPAZ. Land cover and soil type information stored as raster layers were used to derive soil and land cover characteristics for each grid cell of the hillslopes to run the model simulation. The WEPP model is a physical process-based model that continuously simulates surface runoff and sediment yield for hillslope profile and small watersheds.

14.4.11 Agricultural Policy/Environmental eXtender (APEX)

APEX is a continuous simulation model used to predict surface runoff, sediment yield, sediment-bound nutrients, soil carbon, and pesticides loss from the watershed. It simulates hydrological, erosion, and sediment processes in the watershed (Williams, 2008). APEX is an advanced improved version of the predecessor Environmental Policy Impact Climate (EPIC) model. The APEX model predicts daily surface runoff for daily rainfall using the modified NRCS Curve Number method (USDA-NRCS, 2004) and uses the Modified Universal Soil Loss Equation (MUSLE) and its three variants such as MUST and MUSI to simulate erosion and sediment yield. The APEX model also contains USLE, RUSLE, and RUSLE2 models to predict soil erosion. Sediment yield at the watershed outlet is estimated by applying a sediment delivery ratio. The APEX model integrates land use/land cover, soil type, and slope maps of the watershed to subdivide the watershed into homogeneous subareas in terms of land cover, soil type, slope, and management types for simulation. It simulates at daily time steps and can perform for long-term simulation to study impacts of different nutrient management practices, conservation practices, and other management practices/alternative cropping systems on surface runoff, sediments, and nutrient loss.

14.4.12 Agricultural Non-Point Source (AGNPS)

AGNPS is a single event non-point source pollution model developed to predict surface runoff, sediment, and nutrient load as water quality parameters from a catchment of a few to 30,000 hectares (Young et al., 1989). The model calculates upland erosion, overland runoff volume, pollutants [sediment and sediment-bound (nutrients of N, P, and carbon)] from point source inputs. It uses a grid cell representation of the catchment and simulates potential pollutants for each grid cell routed through cells to the catchment outlet. Runoff of each grid cell of the catchment is simulated using the USDA SCS curve number method (SCS, 1975). Soil erosion is estimated using the Revised Universal Soil Loss Equation (RUSLE), and sediment transport is simulated using a modified Einstein deposition equation and the Bagnold suspended sediment formula. Model input parameters are catchment morphology, land slope, shape factor, channel slope, soil erodibility, vegetation cover, management practice factors, and other soil parameters such as soil nutrients and pollutants. AGNPS was improved as Annualized Agricultural Non-Point Source Pollutants (AnnAGNPS) for continuous simulation to predict runoff sediment and nutrient transport for catchment/watersheds of varying sizes (Bingner & Theurer, 2003). It also used to access the impact of alternative management practices on runoff and sediment yield.

14.5 Satellite Remote Sensing in Soil Erosion Assessment

Soil erosion is a dynamic process requiring up-to-date information of vegetation cover and its spatial distribution. These data are increasingly being used to detect the areas susceptible to soil erosion as well as to map eroded areas. Soil erosion processes are characterized by sheet, rill, and gully erosion in the landscape. Sheet and rill erosion types are difficult to detect and map, whereas gully erosion of various severity can be easily mapped using remote sensing data. The recent availability of high-resolution multispectral data has improved the mapping of gully erosion to a great extent and also made it possible to map rill erosion to some extent in bare land conditions. Optical multispectral remote sensing data are commonly used to detect erosion features such as gully-eroded lands. Several studies have demonstrated use of remote sensing methods to identify eroded areas and monitor erosion processes at regional levels (Prince et al., 2009; Sepuru & Dube, 2018; Ilienko et al., 2019).

Several researchers have demonstrated the use of spectral indices to characterize soil surface reflectance to study soil erosion types in the landscape (Lobser & Cohen, 2007; Sayao et al., 2018).

Soil erosion models simulate splash, sheet, and rill erosion processes and are used to assess erosion in the landscape. None of the models simulates gully erosion processes. Soil erosion models are commonly employed to estimate soil erosion rate and soil erosion risk area in the landscape. These models require input parameters, which are generated using satellite remote sensing data. Remote sensing data are being widely used to generate thematic maps of land use/land cover, vegetation, geology, geomorphology, soil and terrain types. Applications of remote sensing data in generating model input parameters are discussed as follows.

14.5.1 Land Use/Land Cover

Remote sensing satellites offer a variety of data comprising spatial, temporal, and spectral resolutions and facilitate users/scientists/researchers for characterization, mapping, and temporal monitoring of land use/land cover. Remote sensing technology offers unique opportunities in collecting land use/land cover information in great detail. Today, several remote sensing data sources such as Landsat 8, Sentinel 1 &2, ASTER, MODIS, of coarse to fine spatial and temporal resolution are freely available, and can be used to prepare land use/land cover maps of the area (Thenkabail, 2015). Visual analysis and digital techniques such as supervised and unsupervised classification methods are used in mapping land use/land cover. Crop cover (C) and crop management practice (P) factors are defined in the model based on a land use/land cover map. Jain and Kothyari (2000) used such maps derived using remote sensing data for soil erosion and sediment yield modeling. In recent years, the availability of high-resolution satellite data has facilitated a large-scale inventory of land use/land cover, vegetation, soil, and terrain parameters. Remote sensing data with fine temporal and spatial resolution are useful in monitoring of field-scale soil coverage information. Panagos et al. (2012) used time series remote sensing data in soil erosion mapping using month-step rainfall erosivity data. Xue and Su (2017) used spectral indices in feature extraction techniques to map soil erosion.

Physically based process models require biophysical parameters such as canopy cover fraction, plant height, plant density, leaf area index, and biomass information of the land use types to simulate soil erosion. Availability of the time series data helps in deriving biophysical parameters (leaf area index, plant height, biomass), enabling dynamic soil erosion risk assessment (Vrieling et al., 2014; Alexandridis et al., 2015). Among various spectral indices, the normalized difference vegetation index (*NDVI*) is the index most used for estimating vegetation cover on soil surface (Rouse et al., 1974; Vrieling et al., 2014; Prabhakara et al., 2015). NDVI is used to derive the vegetation cover (C) factor as it is directly correlated to the C-factor (Durigon et al., 2014). Remote sensing-based C-factor provides an opportunity for improving the spatial and temporal soil erosion modeling (Borrelli et al., 2018). Many studies have used NDVI to calculate fractional vegetation cover (FVC) to estimate the C-factor model (Yang, 2014), although the estimated C-factor derived from remote sensing should be compared and established with C-factor values obtained from measured data (Oliveira et al., 2015).

14.5.2 Topographic Information

Topography largely governs soil erosion by water. The topographic factor is represented by the slope length (L) and slope steepness (S) factors in the erosion models. The L factor gives the impact of slope length whereas the S factor accounts for the effect of slope steepness on soil erosion. Empirical models such as USLE, MUSLE, and its revised version (RUSLE) define the topography factor as the LS factor whereas other erosion models use terrain slope as factor. The LS factor is usually estimated from a digital elevation model (DEM). The Digital Elevation Model (DEM) is a quantitative representation of the elevation of the land used to derive terrain parameters such as slope, aspect, drainage network, and terrain indices Wischmeier and Smith (1978) defined slope length (l) as the distance from the origin point of overland flow to the point where either the slope gradient decreases enough for deposition to start or runoff water is drained into a stream/channel. It is applicable to a simple slope such as plot or field length. Moore and Wilson (1992) proposed the unit stream power theory accounting for the influence of terrain shape and for complex topographic conditions. They replaced slope length (l) with the concept of the unit-contributing area (Desmet & Govers, 1996, 1997).

Earlier topographical maps (1:10,000, 1:50,000, or 1:250,000 scale) were used to extract contour or elevation data of terrain to calculate the LS factor. Now, several DEMs such as ASTER (30 m) (Advanced Spaceborne Thermal Emission and Reflection Radiometer), SRTM (30 and 90 m) (Shuttle Radar Topography Mission), and CartoDEM (10 and 20 m for the Indian region) are available to users to derive an LS factor map. Several GIS software provide functionality to process DEM in retrieving these terrain attributes. Several erosion models such as RUSLE, SWAT, AGNPS, and GeoWEPP have been interfaced with GIS. The accuracy of derived topographic (LS) factors depends on the spatial resolution of DEM (Mukherjee et al., 2013; Fu et al., 2014; Shan et al., 2019). A larger error in LS factor computation has been observed for steeper slopes but smaller for gentler slope (valley areas).

14.5.3 Soil Information

Soil information derived from soil maps is used to compute the soil erodibility (K) factor of each soil type. Soil maps are prepared by conventional methods of soil physiographic analysis using remote sensing data (Manchanda et al., 2002). Remote sensing data are visually interpreted to delineate a soil landscape unit similar in soil-forming factors. These soil–landscape units represent similar soils in the unit, and soil profiles are studied and characterized to prepare soil maps. The soils map is composed of soil map (soil–landscape) units. K values for each soil map unit are calculated, and a soil erodibility map is prepared by reclassifying soil map in GIS. Today, digital soil mapping is being used to map spatial variability of surface soil properties of soil texture and organic carbon contents. It has become more popular with the advances in statistical models and the availability of high-resolution DEMs as well as remote sensing data. Several researchers have demonstrated the use of the quantitative soil–landscape modeling method by integrating digital terrain analysis derived from DEMs and statistical models to predict soil properties and their mapping (Gessler et al., 1995; McKenzie et al., 2000). Kumar and Gupta (2016) developed a model using multiple linear regression model to derive soil erodibility map using terrain parameters. The ISRIC SoilGrid database provides the 1 km spatial resolution soil database of World Soil. It was used to calculate the K factor based on texture, organic matter, and coarse fragments obtained from the database, which is available to download (https://soilgrids.org/).

14.5.4 Rainfall Information

Rainfall erosivity (R) is a major factor in causing soil erosion by water. The rainfall erosivity (R) of rainfall events is calculated by estimating the rainfall energy of the event (E) and multiplying by the maximum 30-min rainfall intensity of the rainfall (I_{30}). Annual and seasonal or monthly rainfall erosivity is then estimated by summing the values of the E^*I_{30} over the period of time. Several meteorological stations are equipped with automatic weather station (AWS) to provide the weather data to calculate the rainfall erosivity (R) factor. Weather data from meteorological stations with 15-min rainfall are required to calculate R factor. Nearing et al. (2017) discussed the concept and their significant improvements in the calculations of rainfall erosivity in detail. Nearing et al. (2017) reviewed development of rainfall erosivity calculations based on an empirically based index. A Global Rainfall Erosivity Database (GloRED) with spatial resolution of 30 arc-seconds (~1 km) was developed using a Gaussian Process Regression (GPR) (Panagos et al., 2017).

14.6 Monitoring Soil Erosion with Remote Sensing

Monitoring of soil erosion at frequent time periods is required to assess land degradation in relation to studying the impact of anthropogenic activities and climate change. Remote sensing satellite data offer a unique opportunity to investigate for their potential and impact on monitoring soil erosion phenomena. Such data facilitate mapping, monitoring, quantifying, and analyzing soil erosion processes in detail. The increasing availability of remote sensing data and the constant improvement in change detection techniques have made possible assessing dynamics and monitoring erosion employing soil erosion models. Land use/land cover, crop growth, and vegetation cover information can be monitored over a period of time using multispectral satellite data. Multitemporal remote sensing data are used to study change detection and to quantify temporal change in land use/land cover. Land use/land cover change analysis is carried out following multi-date classification, image differencing, rationing, vegetation index differencing, and principal component analysis. In general, remote sensing data of two dates are used to classify a land use/land cover map, and post-classification comparisons are carried out to monitor soil erosion using erosion models. Jiu et al. (2019) analyzed change in land use/land cover to study the soil erosion risk pattern in the Three Gorges Reservoir Region (TGRR) in China.

Remote sensing data are also directly interpreted to study soil spectral reflectance to map eroded features (Price, 1993). Fadul et al. (1999) used multi-date aerial photographs along with Landsat data in mapping and monitoring gully erosion over a small area. Fulajtar (2001) used high spatial resolution SPOT PAN Image to detect soil erosion patterns. Several researchers used remote sensing-derived spectral data and spectral vegetation indices to monitor erosion (Dhakal et al., 2002; Lu et al., 2007). Karami et al. (2015) used Indian Remote Sensing (IRS–P6) multispectral data for mapping gully erosion features. Studies show use of vegetation indices (VIs) derived from satellite data for mapping soil erosion features (King et al., 2005; Vrieling, 2006). Very high spatial resolution images obtained from an unmanned aerial vehicle (UAV) are also being used for soil erosion monitoring at a small scale (d'Oleire-Oltmanns et al., 2012). Some studies have used photogrammetry methods to determine the volume of soil loss by concentrated flow erosion. Seitz et al. (2020) used photogrammetry techniques with terrestrial, airborne (UAV), single lens reflex (SLR) cameras, and a terrestrial laser scanner (TLS) to determine event-based initial erosion rates based on volume estimation, calculating the differences between multitemporal DSMs or point clouds. Moritani et al. (2010) demonstrated the use of digital photogrammetry method to measure the erosion rates under simulated rainfall conditions. Xu et al. (2019) used remote sensing data to estimate fractional vegetation cover, bare soil index, and slope parameters in monitoring soil erosion.

14.7 Case Study

14.7.1 Soil Erosion Risk Assessment in a Watershed Using the RUSLE Model

The study was carried out to assess soil erosion risk in a watershed based on revised universal soil loss equation (RUSLE) model use of various RS-derived information in combination with field-collected information in a geospatial environment. The study was carried out in a watershed located near Chamba adjacent to Tehri dam in

Tehri Garhwal district of Uttarakhand in the middle Himalayan region of India and lies between 30°16′30″ to 30°21′30″N latitude and 78°24′0″ to 78°29′30″E longitude. The elevation values within the watershed ranged from 556 to 2147 m above mean sea level, and the total geographic area is 52.49 km^2.

The watershed boundary was delimited with the help of an SOI toposheet and Cartosat DEM of 10 m spatial resolution employing digital terrain analysis procedures in GIS software. The boundary was used for extraction of study area from false color composite (FCC) images generated using IRS P6 LISS IV (5.8 m resolution) and Landsat 8 (30 m resolution). A physiographic map and land use/land cover maps at 1:10,000 scale were prepared by visual interpretation of multitemporal LISS IV images. A field survey was undertaken during which soil samples were collected from different physiographic units for further characterization as well as ground truthing with respect to land use/land cover (LULC) types in the area. The precise locations of soil sampling as well as field data collection were recorded with the help of a handheld GPS device. Ground-based observations corresponding to the distribution of land use/land cover types such as forests, croplands, scrub, as well as identification of different slope categories and landforms in the field, were vital for the accurate representation of field condition via satellite data interpretation and subsequent incorporation into the model. The soil information, a vital component in soil erosion modeling, was generated by undertaking extensive soil sampling in the study area. The collected soil samples were analyzed in the laboratory using different standard procedures to estimate soil textural composition, organic carbon content, and structure as well as permeability status, which are essential in the estimation of soil edibility factor.

The field verification details were incorporated and a LULC map was finalized that revealed that forests are the predominant land use type in the area, followed by agriculture. The different forest types constituted nearly 56.59% of the area, followed by agriculture (26.95%), scrubland (15.96%), and settlements (0.50). The soil samples collected from various physiographic units during field surveys were air dried and analyzed in the laboratory for estimation of different soil properties that are essential for the estimation of the soil erodibility factor of RUSLE model. Terrain analysis based on CartoDEM revealed that nearly 59% percentage of the total study area is exhibiting slope values greater than 25%, indicating the mountainous nature of the area. The aspect analysis revealed that the area had nearly equal proportions of north-facing (52.26%) and south-facing (47.74%) slopes.

For generating various RUSLE model parameters, different field-collected as well as RS-based geospatial datasets were used. Rainfall erosivity (R factor) was computed from the long-term (23 years) rainfall data collected from an observatory located in the immediate vicinity of the watershed. Soil texture (sand, silt, and clay percentages) and organic carbon content estimated from the field-collected soil samples based on a physiographic map were utilized for soil erodibility (K factor) computation. The slope length–steepness (LS) factor, which signifies the impact of topography on soil erosion rates, was computed by terrain analysis using slope and flow accumulation values on a pixel basis, for the entire area. The LS factor algorithm discussed in detail under the RUSLE model description section was employed for generating the spatial distribution of the slope length steepness factor in the entire watershed. The estimation of C and P factors from real field conditions is quite tedious and requires extensive field experimentation and long-term observations of numerous parameters. Different researchers have generated databases under different climatic, vegetation, and management conditions worldwide, which could be useful for ready reference and adoption in situations or areas where such field estimated values are not available. The cover management (C factor) and support practice factors were generated by assigning appropriate values from literature to the different LULC types identified and mapped in the area. All the generated RUSLE model parameters (R, K, LS, C, and P factors) were converted into raster format with the same projection system and resampled to 10 m grid size and for maintaining uniformity before model execution. Thus, the use of remote sensing and other geospatial technologies enabled generating spatial coverages of different erosion factors as well as soil erosion risk in the area (Fig. 14.1.).

The soil erosion risk assessment based on the RUSLE model revealed that nearly 40.74% of the area exhibited a soil erosion risk rate less than 8 t ha^{-1} $year^{-1}$, whereas 26% and 33.26% have erosion risk rates of 8 to 20 t ha^{-1} $year^{-1}$ and more than 20 t ha^{-1} $year^{-1}$, respectively. The average annual soil erosion rates were predicted to be highest in case of scrub land (53.51 t ha^{-1} $year^{-1}$) and least in case of dense mixed forest (1.2 t ha^{-1} $year^{-1}$), whereas different categories of agricultural land were predicted to have erosion rates in the range of 10–14 t ha^{-1} $year^{-1}$. Similar results were reported in the studies by Kumar and Kushwaha (2013), Justin and Kumar (2017), Gupta and Kumar (2017), and Singh and Kumar (2019) conducted in the Himalayan landscape employing the RUSLE model by integrating remote sensing and GIS.

Further, the micro watersheds constituting the entire the watershed were delineated using DEM employing terrain analysis procedures. Forty-five micro watersheds were delineated based on the catchment area of prominent streams in the area and crossed with soil erosion risk assessment map for estimating their average erosion risks and further prioritization (Fig. 14.2). Among all, 3 micro watersheds were categorized under those requiring high priority (more than 40 t ha^{-1} $year^{-1}$) and 16 under medium priority (20–40 t ha^{-1} $year^{-1}$) from a conservation planning point of view. These results could be helpful in planning and implementation of various management and conservation practices aimed at better utilization of natural resources, improved agricultural production, and environmental sustainability.

14.8 Simulating Climate Change Impact on Soil Erosion Using SWAT

Hydrologic models are in general used with a climate scenario generated for General Circulation Models (GCMs) to assess potential scenarios of climate change impact on the watershed hydrology and sediment yield/loss. In future climate scenarios, future simulated monthly rainfall data are used in the model to simulate soil loss,

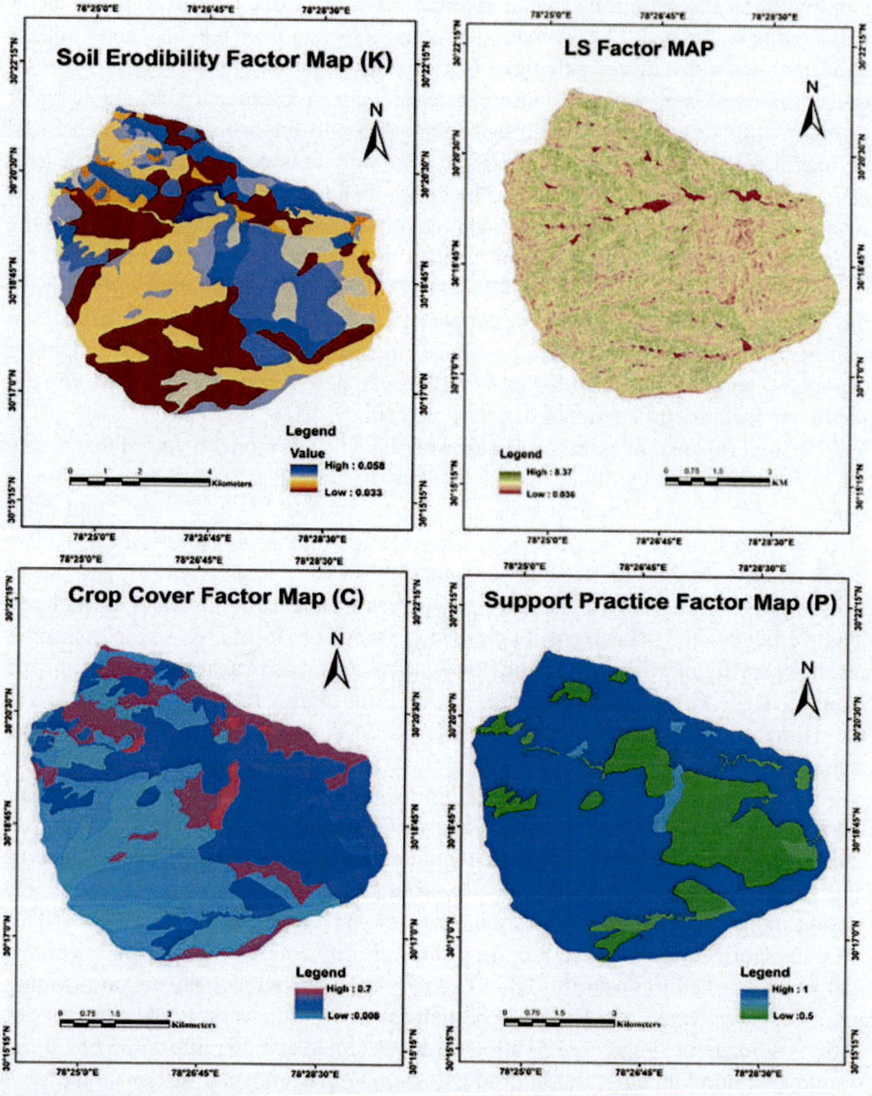

Fig. 14.1 Different RUSLE model parameters estimated for the study area

keeping other model parameters unchanged for the watershed. In the present study, a quantitative assessment of the annual soil loss with respect to the climate change was carried out using SWAT (Soil & Water Assessment Tool) model in a watershed of the Lesser-Himalayan region Utharakhand, India (Sooryamol et al., 2020). MarkSim Weather Generator Tool was used in the study to downscale future climate change scenario under RCP 4.5 and 8.5 scenario for the 2020s, 2050s, and 2080s at a regional scale. Model output showed increasing temperature and rainfall in the future scenario. The 0SWAT model was calibrated and validated for surface runoff and

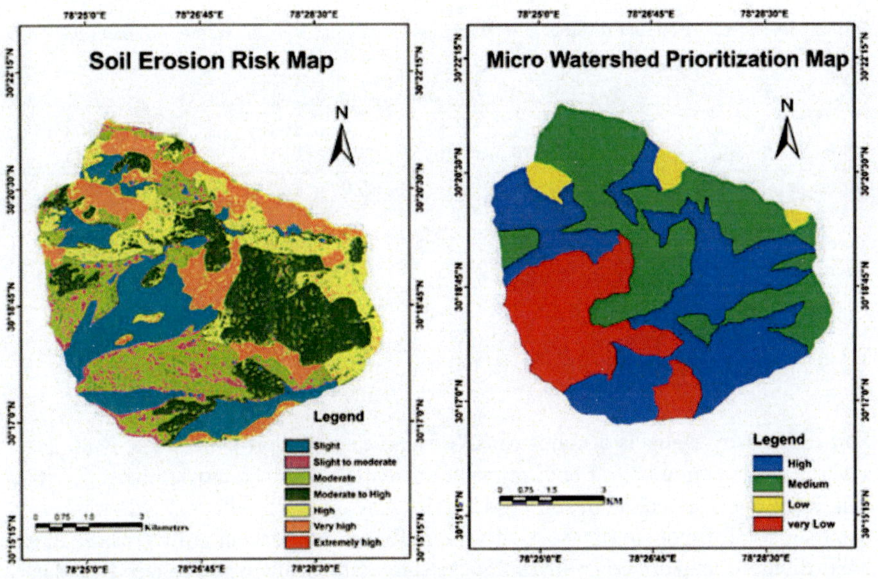

Fig. 14.2 Soil erosion risk and micro watershed prioritization map of the area

sediment yield for the watershed on measured data of daily runoff and sediment yield. The calibrated model parameters were used to simulate annual soil erosion scenario for the 2020s, 2050s, and 2080s for the watershed.

Analysis of rainfall and temperature under RCP 4.5 and RCP 8.5 showed an increasing trend in the future from the baseline period. The study revealed an increase in rainfall of 10.5% in the 2020s, 9.9% in 2050s, and 12.3% in the 2080s under the RCP 4.5 scenario. Under the RCP 8.5 scenario, expected increases in rainfall of 8.7%, 14.2%, and 10.1% were predicted in the 2020s, 2050s, and 2080s, respectively.

Currently, the average annual soil loss from the watershed was estimated as 29.3 ton ha^{-1} $year^{-1}$. Model simulation results showed an increase up to 34.2 ton ha^{-1} $year^{-1}$ in the 2020s, 33.9 ton ha^{-1} $year^{-1}$ in the 2050s, and 35 ton ha^{-1} $year^{-1}$ in 2080s under the RCP 4.5 scenario. Under RCP 8.5, soil erosion is expected to increase up to 33.3 ton ha^{-1} $year^{-1}$ in the 2020s, 35.9 ton ha^{-1} $year^{-1}$ in the 2050s, and 34 ton ha^{-1} $year^{-1}$ in 2080s. The average soil loss RCP 4.5 scenario increased up to 18.1% from the baseline during 2080, 15.5% during 2020, and 14.6% during the 2050s. Under the RCP 8.5 scenario, soil erosion increases 12.8% and further increases to 21%, then is further reduced to 15% during the 2020s, 2050s, and 2080s, respectively (Table 14.4).

Table 14.4 Average annual soil loss and the change in percentile of the watershed under RCP 4.5 and 8.5 scenarios

	Watershed Average soil erosion (ton ha^{-1} year^{-1})						
Climate Scenarios	Present	2020s	2050s	2080s	2020s (%)	2050s (%)	2080s (%)
Average (RCP 4.5 scenario)	29.3	34.2	33.9	35.0	15.5	14.6	18.1
Average (RCP 8.5 scenario)	29.3	33.3	35.9	34.0	12.8	20.9	14.9

14.9 Conclusions

Soil erosion by water is a major cause of land degradation worldwide, leading to several socioeconomic and environmental threats. Erosion removes the top fertile soil, resulting in a decline in soil productivity that adversely affects food security. It occurs in the form of splash sheet, rill, and gully erosion: except gully erosion, these are difficult to map directly with remote sensing data. Soil erosion caused by splash, sheet. and rill erosion together are estimated employing erosion models ranging from the empirical to physical process. Remote sensing data are used to derive model input parameters such as land use/land cover, soil, and terrain data in soil erosion estimation. These models are interfaced or integrated with GIS to provide spatial assessment of soil erosion rate and soil erosion severity areas. Increasing availability of remote sensing data from fine to coarse spatial resolution have facilitated soil erosion mapping and modeling from large- to small scale, meeting varying requirements of users. Temporal remote sensing data help to analyze change in land use/land cover and spectral characteristics of bare soils to model for monitoring soil erosion in the region. Information on soil erosion risk area with its severity levels is necessary for soil and water conservation planning in the watershed.

References

Abbott, M. B., Bathurst, J.C., Cunge, J.A., O'Connell, P.E., Rasmussen, J. (1986). An introduction to the European hydrological system—systeme hydrologique Europeen, SHEQ: 1. History and philosophy of a physically-based, distributed modeling system. *Journal of Hydrology, 87*, 45–59.

Alexandridis, T. K., Sotiropoulou, A. M., Bilas, G., Karapetsas, N., & Silleos, N. G. (2015). The effects of seasonality in estimating the C-factor of soil erosion studies. *Land Degradation & Development, 26*(6), 596–603.

Ali, K. A. (2010). *Remote sensing*. Laser Branch Department of Applied Sciences, University of Technology. Available at: http://www.uotechnology.edu.iq/appsciences/Laser/Lacture_-laser/thrid_class/Remote_Sensing/3-Remote_Sensing.pdf

Arnold, J. G., Williams, J. R., Srinivasan, R., & King, K. W. (1996). *SWAT: Soil and water assessment tool* (p. 190). User's Manual USDA Agriculture Research Service Grassland, Soil and Water Research Laboratory.

Aronoff, S. (1989). *Geographic information systems: A management perspective*. WDL Publication.

Beck, M. B. (1987). Water quality modelling: a review of uncertainty. *Water Resources Research, 23*(8), 1393–1442.

Bennett, J.P. (1974). Concepts of mathematical modeling of sediment yield. *Water Resources Research, 10*(3), 485–492.

Beven, K. (1989). Changing ideas in hydrology—the case of physically based models. *Journal of Hydrology, 105*, 157–172.

Bingner, R. L., & Theurer, F. D. (2003). AnnAGNPS technical processes documentation, version 3.3, USDA-ARS.

Blanco, H., & Lal, R. (2010). *Soil erosion and food security. Principles of soil conservation and management* (pp. 493–512). Springer.

Borrelli, P., Meusburger, K., Ballabio, C., Panagos, P., & Alewell, C. (2018). Objectoriented soil erosion modelling: A possible paradigm shift from potential to actual risk assessments in agricultural environments. *Land Degradation & Development, 29*(1270), e1281. https://doi.org/10.1002/ldr.2898

d'Oleire-Oltmanns, S., Marzolff, I., Peter, K. D., & Ries, J. B. (2012). Unmanned aerial vehicle (UAV) for monitoring soil erosion in Morocco. *Remote Sensing, 4*, 3390–3416.

Desmet, P. J. J., & Govers, G. (1996). A GIS procedure for automatically calculating the USLE LS factor on topographically complex landscape units. *J. Soil Water Conservation, 51*, 427–433.

Desmet, P., & Govers, G. (1997). Comment on 'Modelling topographic potential for erosion and deposition using GIS'. *Int. J. Geogr. Inf. Sci., 11*, 603–610.

Dhakal, A. S., Amada, T., Aniya, M., Sharma R. R. (2002). Detection of areas associated with flood and erosion caused by a heavy rainfall using multi temporal Landsat TM data. *Photogrammetric Engineering and Remote Sensing, 68*(3), 233–239.

Durigon, V. L., Carvalho, D. F., Antunes, M. A. H., Oliveira, P. T. S., & Fernandes, M. M. (2014). NDVI time series for monitoring RUSLE cover management factor in a tropical watershed. *International Journal of Remote Sensing, 35*, 441–453. https://doi.org/10.1080/01431161.2013.871081

FAO (2011). The state of the world's land and water resources for food and agriculture (SOLAW): Managing systems at risk. FAO, Rome and Earth scan, London. http://www.fao.org/docrep/015/i1688e/i1688e00.pdf.

Fadul, H. M., Salih, A. A., Imad-eldin, A. A., & Inanaga, S. (1999). Use of remote sensing to map gully erosion along the Atbara River, Sudan. *International Journal of Applied Earth Observation, 1*(3–4), 175–180.

Flanagan, D. C., & Nearing, M. A. (Eds.). (1995). *USDA-Water Erosion Prediction Project: Hillslope profile and watershed model documentation* (NSERL Report No. 10). USDA-ARS National Soil Erosion Research Laboratory.

Fu, S., Cao, L., Liu, B., Wu, Z., & Savabi, M. R. (2014). Effects of DEM grid size on predicting soil loss from small watersheds in China. *Environment and Earth Science, 73*, 2141–2151.

Fulajtar, E. (2001). Identification of severely eroded soils from remote sensing data tested in Risnovce, Slovakia. In D. E. Stott, R. H. Mohtar, & G. C. Steinhardt (Eds.), *Sustaining the Global Farm* (pp. 1075–1081). Purdue University.

Joseph, G., & Jeganathan, C. (2018). *Fundamentals of remote sensing*. Universities Press (India) Private Limited. ISBN 978-93-86235-46-6.

Gessler, P. E., Moore, I. D., McKenzie, N. J., & Ryan, P. J. (1995). Soil-Landscape modelling and spatial prediction of soil attributes. *Geographical Information Systems, 9(4)*, 421–432.

Gupta, S., & Kumar, S. (2017). Simulating climate change impact on soil erosion using RUSLE model – A case study in a watershed of mid-Himalayan landscape. *Journal of Earth System Science, 126*, 43.

Xu, H., Hu, X., Guan, H., Zhang, B., Wang, M., Chen, S., & Chen, M. (2019). A remote sensing based method to detect soil Erosion in forests. *Remote Sensing, 2019*(11), 513. https://doi.org/10.3390/rs11050513

Haan, C. T., Barfield, B. J., & Hayes, J. C. (1994). Design hydrology and sedimentology for small catchments. Academic Press 588 pp.

Ilienko, T., Tarariko, O., Syrotenko, O., & Kuchma T. (2019). Merging remote and in-situ land degradation indicators in soil erosion control system. In: Proceedings of the global Symposium on Soil Erosion. Rome, 190–195.

Jain, M. K., & Kothyari U. C. (2000). Estimation of soil erosion and sediment yield using GIS. *Hydrological Science Journal, 45(5)*, 771–786.

Jiu, J., Wu, H., & Li, S. (2019). The implication of land-use/land-cover change for the declining soil Erosion risk in the three Gorges reservoir region. *International Journal of Environmental Research and Public Health, 16*, 1856. https://doi.org/10.3390/ijerph16101856

Justin, G. K., & Kumar, S. (2017). Modelling soil erosion risk in a mountainous watershed of Mid-Himalaya by integrating RUSLE model with GIS. *Eurasian Journal of Soil Science, 6*(2), 92–105.

Karami, A., Khoorani, A., Noohegar, A., Shamsi, S. R. F., & Moosavi, V. (2015). Gully erosion mapping using object-based and pixel-based image classification methods. *Environmental and Engineering Geoscience, 27*(2), 101–110.

Kinnell, P. I. A. (1999). Discussion on The European soil erosion model (EUROSEM): a dynamic approach for predicting sediment transport from fields and small catchments. *Earth Surface Processes and Landforms, 24*, 563–565.

King, C., Baghdadi, N., Lecomte, V., & Cerdan, O. (2005). The application of remote sensing data to monitoring and modelling of soil erosion. *Catena, 62*, 79–93.

Kumar, S., & Gupta, S. (2016). Geospatial approach in mapping soil erodebility using CartoDEM – A case study in hilly watershed of lower Himalayan range. *Journal of Earth System Science, 125*, 1–10.

Kumar, S., & Kushwaha, S. P. S. (2013). Modelling soil Erosion risk based on RUSLE-3D using GIS in a Shivalik sub-watershed. *Journal of Earth System Science, 122*(2), 389–398.

Lillesand, T. M., Kiefer, R. W., & Chipman, J. W. (2004). *Remote sensing and image interpretation* (5th ed.). Wiley.

Lillesand, T. M., & Kiefer, R. W. (1994). *Remote sensing and image interpretation*. Wiley.

Lobser, S. E., & Cohen, W. B. (2007). MODIS tasselled cap: Land cover characteristics expressed through transformed MODIS data. *International Journal of Remote Sensing, 28*, 5079–5101.

Manchanda, M. L., Kudrat, M., & Tiwari, A. K. (2002). Soil survey and mapping using remote sensing. *Tropical Ecology, 43*(1), 61–74.

McKenzie, N. J., Jacquier, D. W., Ashton, L. J., & Cresswell, H. P. (2000). Estimation of soil properties using the Atlas of Australian soils. CISRO Land and Water, Technical report 11/100.

Merritt, W. S., Latcher, R. A., Jakeman, A. J. (2003). A review of erosion and sediment transport models. *Environmental Modelling & Software, 18*, 761–799.

Morgan, R. P. C. (2001). A simple approach to soil loss prediction: a revised Morgan–Morgan–Finney model, Catena, 44(4), 305–322.

Moore, I. D., & Wilson, J. P. (1992). Length-slope factors for the revised universal soil loss equation: Simplified method of estimation. *Journal of Soil and Water Conservation, 47*, 423–428.

Moritani, S., Yamamoto, T., Andry, H., Inoue, M., & Kaneuchi, T. (2010). Using digital photogrammetry to monitor soil erosion under conditions of simulated rainfall and wind. *Australian Journal of Soil Research, 48*(1), 36–42. https://doi.org/10.1071/SR09058

Mukherjee, S., et al. (2013). Evaluation of vertical accuracy of open source Digital Elevation Model (DEM). *International Journal of Applied Earth Observation and Geoinformation, 21*, 205–217.

NAAS. (2010). *Degraded and wastelands of India – Status of spatial distribution*. National Academy of Agricultural Sciences.

Nearing, M. A., Yin, S., Borrelli, P., & Polyakov, V. O. (2017). Rainfall erosivity: An historical review. *Catena, 157*, 357–362.

Oliveira, J. A., Dominguez, J. M. L., Nearing, M. A., & Oliveira, P. T. S. (2015). A GIS based procedure for automatically calculating soil loss from the universal soil loss equation: GISus-m. *Applied Engineering in Agriculture, 31*, 907e917. https://doi.org/10.13031/aea.31.11093

Panagos, P., Karydas, C. J., Gitas, I. Z., & Montanarella, L. (2012). Monthly soil erosion monitoring based on remotely sensed biophysical parameters: A case study in Strymonas river basin towards a functional pan-European service. *International Journal of Digital Earth, 5*(6), 461–487. https://doi.org/10.1080/17538947.2011.587897

Panagos, P. V., Liedekerke, M., Jones, A., Montanarella, L. (2012). European Soil Data Centre (ESDAC): response to European policy support and public data requirements. *Land Use Policy, 29(2)*, 329–338.

Price, K. P. (1993). Detection of soil erosion within pinyon-juniper woodlands using Thematic Mapper (TM) data. *Remote Sensing of Environment, 45*(3), 233–248.

Panagos, P., Borrelli, P., Meusburger, K., Yu, B., Klik, A., Lim, K. J., Yang, J. E., Ni, J., Miao, C., Chattopadhyay, N., Sadeghi, S. H., Hazbavi, Z., Zabihi, M., Larionov, G. A., Krasnov, S. F., Gorobets, A. V., Levi, Y., Erpul, Y. G., Birkel, C., Hoyos, N., Naipal, V., Oliveira, P. T. S., Bonilla, C. A., Meddi, M., Nel, W., Dashti, H., Boni, M., Diodato, N., Van Oost, K., Nearing, M. A., & Ballabio, C. (2017). Global rainfall erosivity assessment based on high-temporal resolution rainfall records. *Scientific Reports, 7*, 4175. https://doi.org/10.1038/s41598-017-04282-8

Pimentel, D. (2006). Soil erosion: a food and environmental threat. *Environment, Development and Sustainability, 8*, 119–137.

Poesen, J. (1993). Gully typology and gully control measures in the European loess belt. In S. Wicherek (Ed.), *Farmland erosion in temperate plains environment and hills* (pp. 221–239). Elsevier.

Prabhakara, K., Hively, W. D., & McCarty, G. W. (2015). Evaluating the relationship between biomass, percent groundcover and remote sensing indices across six winter cover crop fields in Maryland, United States. *International Journal of Applied Earth Observation and Geoinformation, 39*, 88–102.

Prince S.D., Becker-Reshef I., Rishmawi K. (2009). Detection and mapping of long-term land degradation using local net production scaling: Application to Zimbabwe. *Remote Sensing of Environment, 113*, 1046–1057.

Renard, K. G., Foster, G. R., Weesies, G. A., McCool, D. K., & Yoder, D. C. (1997). Predicting soil erosion by water: a guide to conservation planning with the Revised Universal Soil Loss Equation (RUSLE) (Vol. 703). Washington, DC: US Government Printing Office.

Renard, K. G., Foster, G. R., Yoder, D. C., & McCool, D. K. (1994). RUSLE revisited: status, questions, answers, and the future. *Journal of Soil and Water Conservation*, 213–220.

Renschler, C. S., Flanagan, D. C. Engel, B. A., & Frankenberger, J. R. (2002). GeoWEPP: The geospatial interface to the Water Erosion Prediction Project. ASAE Paper No. 022171. St. Joseph, Mich.: ASAE.

Rouse, J., Jr., Haas, R. H., Schell, J. A., & Deering, D. (1974). *Monitoring vegetation systems in the Great Plains with ERTS, NASA SP-351* (Third ERTS-1 Symposium) (Vol. 1, pp. 309–317). WNASA.

Sadeghi, S. H. R., Gholami, L., Khaledi Darvishan, A., & Saeidi, P. (2007). Conformity of MUSLE estimates and erosion plot data for storm-wise sediment yield estimation. *Terrestrial, Atmospheric and Oceanic Sciences, 18*(1), 117–128.

Sayao, V. M., Dematte, J. A. M., Bedin, L. G., Nanni, M. R., & Rizzo, R. (2018). Satellite land surface temperature and reflectance related with soil attributes. *Geoderma, 325*, 125–140.

Sabins, F. F., Jr. (1978). *Remote sensing: Principles and interpretation* (p. 1). W.H. Freeman and Co.

SCS (Soil Conservation Service) (1975). Urban hydrology for small watersheds. Technical release no. 55, Soil Conservation Service, United States Dept. of Agric., Washington DC, USA.

Seitz, S., Scholten, T., & Schmidt, K. (2020). Soil erosion monitoring at small scales: Using close range photogrammetry and laser scanning to evaluate initial sediment delivery, EGU General

Assembly 2020, Online, 4–8 May 2020, EGU2020–16685, https://doi.org/10.5194/egusphere-egu2020-16685.

Sepuru, T. K., & Dube, T. (2018). Understanding the spatial distribution of eroded areas in the former rural homelands of South Africa: Comparative evidence from two new non-commercial multispectral sensors. *International Journal of Applied Earth Observation and Geoinformation, 69*, 119–132.

Shan, L., Yang, X., & Zhu, Q. (2019). Effects of DEM resolutions on LS and hillslope erosion estimation in a burnt landscape. *Soil Research, 57*, 797.

Sharda, V. N., & Ojasvi, P. R. (2016). A revised soil erosion budget for India: Role of reservoir sedimentation and land-use protection measures. *Earth Surface Processes and Landforms, 41*, 2007–2023.

Singh, A. K. & Kumar, S. (2019). Modelling soil erosion and predicting sediment yield for sub-watershed prioritization using geospatial technique in North West Himalayan region. *Indian Journal of Soil Conservation* (Accepted).

Sooryamol R., Kumar, S., Mary, R. F, & Annu, D. (2020). Calibrating SWAT model in simulating climate change impact on sediment loss – Case study in a watershed of lesser-Himalayan landscape. *Earth Systems and Environment* (Communicated).

Sorooshian, S. (1991). Parameter estimation, model identification, and model validation: Conceptual-type models. Recent Adv. Model. Hydrol. Syst., 443–467.

Thenkabail, P. S. (2015). Land Resources Monitoring, Modeling, and Mapping with Remote Sensing, Taylor and Francis Inc. CRC Press. ISBN: SBN 9781482217957 - CAT# K22130.

UNCCD secretariat (2013). Role of parliamentarians in the implementation process of the UN Convention to Combat Desertification. A guide to parliamentary action.

USDA-NRCS. (2004). Chapter 10: Estimation of direct runoff from storm rainfall. In *Part 630: Hydrology: NRCS National Engineering Handbook.* USDA National Resources Conservation Service. Available at: http://www.wsi.nrcs.usda.gov/products/W2Q/H&H/tech_refs/eng_Hbk/chap.html. Accessed 16 June 2008

Vrieling, A. (2006). Satellite remote sensing for water erosion assessment: a review. *Catena, 65*, 2–18.

Vrieling, A., Hoedjes, J. C., & van der Velde, M. (2014). Towards large-scale monitoring of soil erosion in Africa: Accounting for the dynamics of rainfall erosivity. *Global and Planetary Change, 115*, 33–43.

Wheater, H. S., Jakeman, A. J., & Beven, K. J. (1993). Progress and directions in rainfall-runoff modelling. In: Jakeman, A.J., Beck, M.B., McAleer, M.J. (Eds.), Modelling Change in Environmental Systems. *John Wiley and Sons, Chichester*, 101–132.

Williams, J. R. (1975). Sediment routings for for agricultural watersheds. *Wat. Resour Bull., 11*, 965–975.

Williams, J. R., & Berndt, H. D. (1977). Sediment yields prediction based on watershed hydrology. *Transactions of ASAE, 20*(6), 1100–1104.

Williams, J. R. (2008). Agricultural policy/environmental eXtender model: Theoretical documentation version 0604 (Draft). BREC Report # 2008–17. Texas AgriLIFE Research, Texas A&M University, Blackland Research and Extension Center, Temple, TX.

Wischmeier, W. H., & Smith, D. D. (1978). Predicting Rainfall Erosion Losses-A Guide to Conservation Planning. In: Agriculture Handbook 537. US Government Print Office, Washington, DC.

Xue, J., & Su, B. (2017). Significant remote sensing vegetation indices: A review of developments and applications. *Journal of Sensors, 2017*, 1–17.

Yang, X. H. (2014). Deriving RUSLE cover factor from time-series fractional vegetation cover for hillslope erosion modelling in New South Wales. *Soil Research, 52*, 253–261.

Young, R. A., Onstad, C. A., Bossch, D. D., & Anderson, W. P. (1989). AGNP S: A non-point source pollution model for evaluating agricultural watersheds. *Journal of Soil and Water Conservation, 44*(2), 168–173.

Chapter 15
Soil Erosion Assessment Using RUSLE Model in the Randigad Catchment in Northern India

Kishor Chandra Kandpal and Neelam Rawat

15.1 Introduction

Soil erosion is a major problem in India, especially in mountain regions. The world's 80% of agricultural land is influenced by soil erosion (Ritchie & Crouch, 2003). The global per capita food supply is currently declining because of this erosion-induced loss of productivity and population growth. It is a far-flung trouble that threatens the sustainability of production in agriculture and causes land and reservoir storage capacity to be reduced. In certain parts of the world, on-site impacts of increased soil degradation are also combined with extreme off-site impacts correlated with increased sediment mobilisation and distribution to the rivers. These off-site effects include water contamination, river sedimentation, aquatic habitat destruction and increased water treatment costs. Soil erosion assessment helps to define erosion-prone zone for preventing the soil erosion. Knowing the spatial distribution of the eroded areas through soil would help government and stakeholders to plan suitable precaution measures and plan sustainable soil erosion prevention methods. Governmental practices that prevent soil erosion can only be implemented if the spatial distributions of soil erosion are known.

In the previous decades, many models have been developed such as physical and empirical-based models for soil erosion estimation. The physical-based model includes WEPP (Foster & Lane, 1987), SWAT (Arnold et al., 2012), ANSWERS (Beasley et al., 1982), CREAMS (Knisel, 1980), EUROSEM (Morgan et al., 1998), PESERA (Cerdan & Team, 2003) and TREX (Velleux et al., 2008). These physical-based models require a large number of input parameters and field variables, and these models do not perform better than empirical models (Nearin et al., 1989). The empirically-based model includes USLE (Wischmeier & Smith, 1965), RUSLE, MUSLE, E30 (Honda, 1996) and SEDD (Ferro & Porto, 2000). The

K. C. Kandpal (✉) · N. Rawat
Uttarakhand Space Application Centre, Dehradun, India

G. S. Bhunia et al. (eds.), *Anthropogeomorphology*, Geography of the Physical Environment, https://doi.org/10.1007/978-3-030-77572-8_15

empirical models are widely used for predicting soil erosion and planning soil conservation (Millward & Mersey, 1999; Lal, 2001; Fernandez et al., 2003; Mbugua, 2009; Chen et al., 2011; Owusu, 2012; Rahaman et al., 2015; Zhu, 2015; Ganasri & Ramesh, 2016; Kamuju, 2016).

The Universal Soil Loss Equation (USLE), developed by Wischmeier and Smith in 1965, is one of the most widely used empirical models for the assessment of sheet and rill erosions (Jain et al., 2001). Agriculture Handbook 703 (Renard, 1997) along with the RUSLE is a guide to conservation planning. Mainly, USLE was developed primarily for estimation of soil erosion in croplands or topography that was gently sloping. USLE is still being used in a significant number of soil loss estimation studies with its revised (RUSLE) and updated (MUSLE) versions (Wischmeier & Smith, 1978; Van Remortel et al., 2001; Pandey et al., 2009). The RUSLE model is an empirical model that is extensively used to estimate soil loss, identification of soil risk-prone areas, management and conservation purposes. There are various complexities in the soil erosion models but RULSE is one of the most widely applied model, which is easy to use and also requires less time and data than other soil erosion models.

The remote sensing and GIS techniques are useful tools for estimating soil erosion in larger extents at a reasonable cost. When these techniques employ along with the RUSLE model, it makes accurate estimation of soil erosion (Moore & Wilson, 1992). In the present study, the RUSLE model was used in combination with remote sensing and GIS techniques to determine the spatial distribution of soil erosion-prone sites in the Randigad catchment at the micro-watershed level. We also focused mainly on each micro-watershed across the entire catchment for soil erosion evaluation in this study.

15.2 Study Area

Randigad catchment falls in Pauri district of Uttarakhand, which lies between 30° 06′ 00.0″ N and 30° 09′ 00.0″ N, 78° 42′ 00.0″ E to 78° 48′ 00.0″ E, covering an area of 24.377 km^2 (Fig. 15.1). The catchment maximum and minimum elevations are 1249 m and 2114 m above the mean sea level, respectively. The tributaries Ujyari gad, Kuan gad, Magrau gad, Nalai gad and Tamlak gadhera join the mainstream Randigad. The soil in the catchment comprises of calcareous loamy, loamy and sandy skeletal, and the catchment mainly predominated by agriculture land and forest covers all around the studied area. The main crops grown are wheat, rise, mustered, etc. The annual rainfall of the catchment is more than 1068 mm, but majority of the area faces irrigation problem due to scarcity of water in the area, and the agricultural lands are at varying height from the main source of water (Randigad).

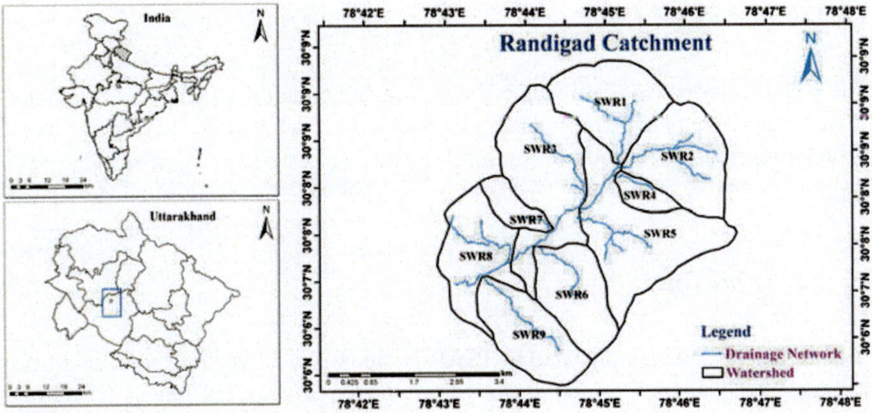

Fig. 15.1 Location map of the study area

15.3 Materials and Methods

The RUSLE model needs various thematic layers such as Digital Elevation Model (DEM), satellite images, soil information, and rainfall data. The concise description of each thematic layer is explained in the accompanying sections.

15.3.1 Digital Elevation Model (DEM)

The Cartosat-1 Digital Elevation Model (DEM) with 30 m resolution was used to generate slope. Slope and flow accumulation are the basis for preparation of LS (Slope Length Steepness) factor.

15.3.2 Rainfall Data

Rainfall data was downloaded from Global Weather data for SWAT. In order to prepare R (rainfall–runoff erosivity) factor, data from five weather stations (Edachota, Chinbo, BawasanaTalla, Balodi, Ali) had been collected for the past 5 years, and mean annual rainfall was calculated. The result of each station was entered in Kriging (Spatial Analyst) interpolation tool to interpolate the whole catchment.

15.3.3 LISS IV Data

The NDVI (Normalized Difference Vegetation Index) and LULC (Land Use/Land Cover) map were prepared using LISS IV data. NDVI is the basis for preparation of C factor (crop management).

15.3.4 Soil Data

A polygonised soil map prepared by ICAR-National Bureau of soil survey and land use planning was used to derive soil texture covered in the catchment. There are three varieties of soil textural classes identified from 'Randigad' catchment. These are calcareous loamy, loamy skeletal to sandy skeletal. The maximum portion of the catchment covered with calcareous loamy and a minimum area of soils is covered with loamy skeletal.

15.4 Methodology

In the present work, the RUSLE model is integrated with remote sensing and GIS techniques for estimating the soil erosion over the Randigad watershed. The RUSLE model utilizes an equation with different parameters for predicting average annual soil loss (Eq. 15.1). Figure 15.2 represents the overall process of the study.

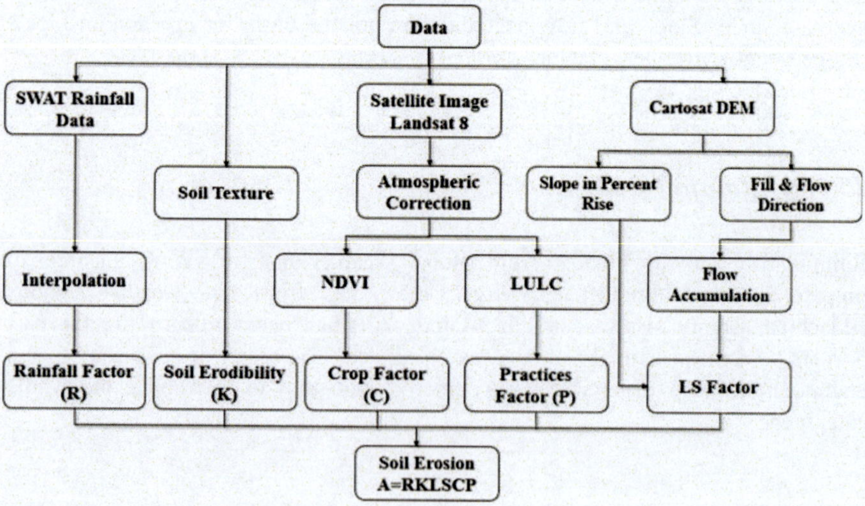

Fig. 15.2 Flow chart of methodology

$$A = R * K * LS * C * P \tag{15.1}$$

where

A = Computed annual soil loss per unit area [ton/ha/year]
R = Rainfall runoff erosivity factor
K = Soil erodibility factor
LS = Slope length steepness factor
C = Crop management factor
P = Conservation practices

15.4.1 Calculation of RUSLE Factor

Rainfall Erosivity Factor (R)

In the RUSLE model, rainfall erosivity is an important and integral component in soil erosion estimation. Due to rainfall, the detaching strength of the soil surface is directly related to soil depletion and also to the contribution of runoff along the ground. Low-slope areas should have a low degree of erosivity (R values), which means that flatter areas will pound higher water on the surface, thus protecting soil particles from being eroded by raindrops. There are many other methods of finding the rainfall erosivity; here, it is derived by the equation given by regression analysis (Hurni, 1985).

$$R = -8.12 + 0.562 * P \tag{15.2}$$

where P is mean annual rainfall in mm.

Rainfall data from five weather stations were used to compute the mean annual rainfall during the past five years (2010 to 2014). The result of each station was entered in Kriging (Spatial Analyst) interpolation tool to interpolate the result over the whole catchment (Fig. 15.3d).

Soil Erodibilty Factor (K)

Soil erodibility depends on soil and/or geological characteristics, such as parent material, texture, composition and quality of organic matter, porosity, catena and also more (Schwab et al., 1993). Soils become low erodible with decrease in silt percentage, regardless of comparable increases in the sand or clay proportion (Mhangara et al., 2012). Soil erodibility factor reflects soil or surface material susceptibility to erosion, while sediment transportability includes amount and rate of runoff at a specific rainfall input, as measured under standard condition. The higher value of K means that the higher the erodibility.

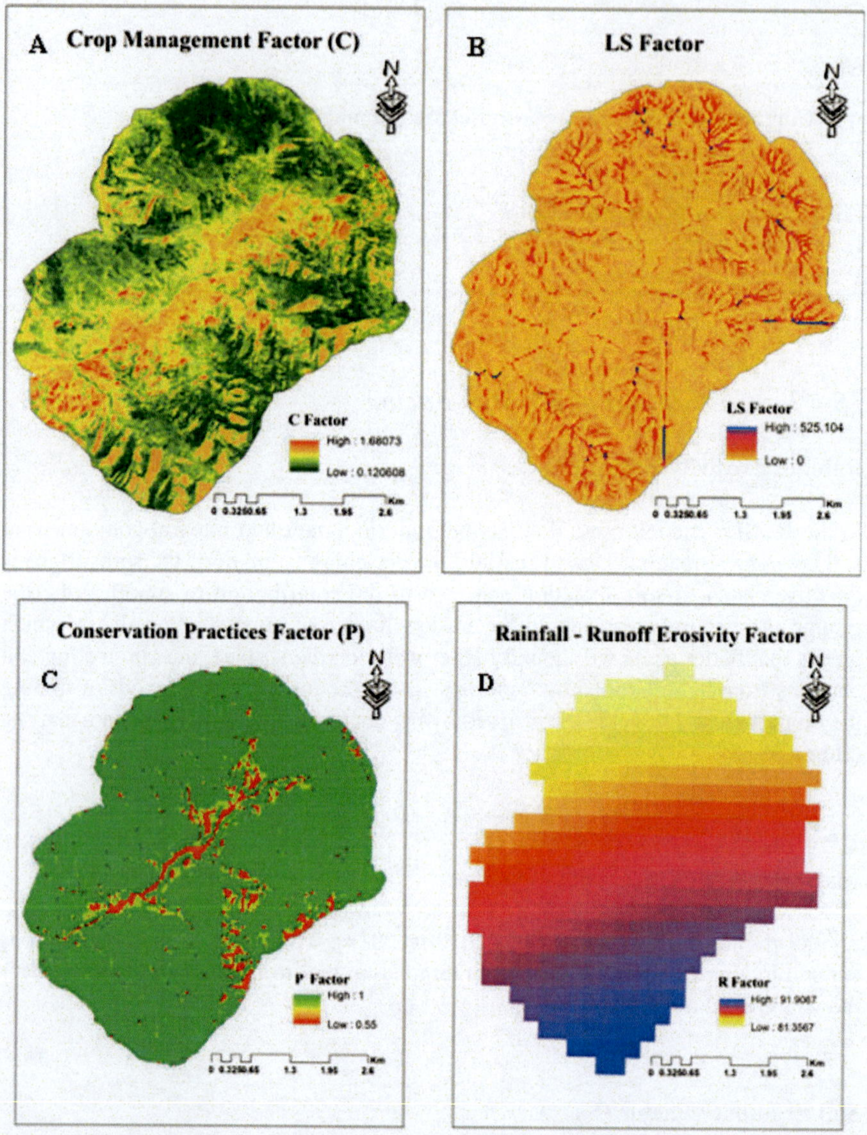

Fig. 15.3 Spatial distribution of variables in RUSLE equation for soil erosion estimation (**a**) crop management factor (**b**) slope length factor, LS (**c**) conservation practices factor (**d**) rainfall-runoff erosivity factor

The K-factor of the soil will be 0.3 because the whole catchment area was found to be composed of sandy loam. Due to high erosion rates and reduced runoff, sandy soils have low K values, and the sediment eroded from those soils is not easily transported.

15.4.2 *Slope Length and Steepness Factor (LS)*

Slope and steepness are one of the key factors for the rate of soil erosion, combining influence of slope length and slope steepness as well. The longer the length of the slope, the greater the volume of accumulated runoff, because the higher the land slope, the velocities of the runoff will be steeper and will contribute to the erosion. In the present study, the method suggested by Pelton et al. (2012) has been followed to find out the LS factor using digital elevation model (Fig. 15.3b), which can be expressed as:

$$\text{Power}\left(\text{FA} * \frac{[\text{CR}]}{22.1}, 0.40\right) * \text{Power}\left(\text{Sin}\left(\text{SD} * \frac{0.0017545}{0.09}, 1.4\right) * 1.4\right) \quad (15.3)$$

where

FA = Flow accumulation
CR = Resolution of DEM in m
SD = Slope raster in degree

Crop Management Factor, C

The crop management factor (C) is used to assess the relative effectiveness of soil and crop management systems in avoiding soil loss. It is generally related to the percentage of cover of the vegetation and is described as the soil loss ratio from specific crops to the similar loss from tilled, bare test plots. Higher values for the C factor suggest that the corresponding land cover type results in more soil erosion than the other types of land cover (Fig. 15.3a).

In the present study, the method suggested by Karaburun (2010) has been followed to find out the C factor using Normalized Difference Vegetation Index (NDVI), which can be expressed as:

$$\text{C factor} = 1.02 - 1.21 * \text{NDVI} \quad (15.4)$$

Conservation Practices, P

Conservation practice (P) factor represents the impact of activities that will minimize the water runoff quantity and intensity and thereby reduce the amount of erosion. The most important aiding activities are contour tillage, contour strip-cropping and terrace systems. The P-factor was calculated using the slope layer of the catchment at Randigad (Fig. 15.3c). The value of P varies from 0 to 1, where 0 reflects the resistance to manmade erosion and 1 no manmade resistance.

15.5 Result and Discussion

The RUSLE model, which is integrated with remote sensing and GIS techniques, was used in this current study to estimate annual soil loss for the catchment of Randigad, Pauri Garhwal and Uttarakhand. The average annual soil loss of 18 tonnes/ha/year for the whole catchment was estimated, and the soil loss map shows the minimum and maximum values, i.e. 0 and greater than 150 tonnes/ha/year, respectively (Fig. 15.4). In general, if the estimated value is high, this indicates the higher sediment yield rate, whereas the lower value shows the lower sediment yield rate.

The Watershed Randigad was classified into four risk groups for soil erosion, i.e. no erosion, low, moderate and high according to the severity of soil loss. Table 15.1 classes represent the proportion and area of risk over soil erosion.

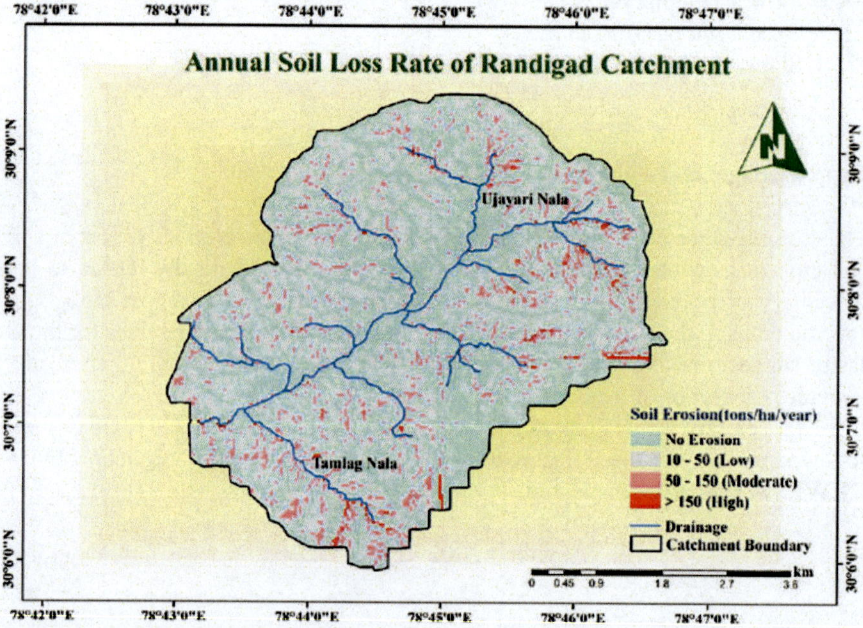

Fig. 15.4 Spatial distribution of erosion risk categories

Table 15.1 Soil erosion risk categories with annual soil loss

Soil erosion severity classes	Area in (Km^2)	Annual soil erosion rate (t/ha/year)	Area in (%)
Nil	11.427	0–10	48.23
Low	10.458	10–50	44.14
Moderate	1.648	50–150	6.95
High	0.157	>150	0.66

Distribution of risk levels and soil severity zones indicates that there is no soil loss at 48.23% of the watershed, 44.14% is low, 6.95% and 0.664% is moderate and high of the watershed. It is observed that large part of area comes under lower erosion category, which could be found in almost all areas, while moderate erosion exists in only 6.95% of the study area where barren land exists with mild slope. It also found that few sections of the region under study have a higher soil loss value, which may be due to the steep slope (Table 15.1).

15.5.1 *Prioritisation of Micro-Watersheds*

Arc SWAT toolwas used to determine the micro-watershed of all the tributaries joining the Randigad. A quantitative assessment of average soil loss on each micro-watershed was calculated using RUSLE model with a view to know the spatial distribution of soil erosion in each micro-watershed so that check dams can be created to stop further erosion. Table 15.2 and Fig. 15.5 indicate the distribution of nine micro-watersheds of Randigad catchment according to soil erosion intensity. It is observed that maximum area of higher erosion class comes under SWR4 micro-watershed, which may be due to high value of LS factor and steeper slope.

15.5.2 *Assessment of Soil Erosion Risk with LULC*

Agriculture crop land, agriculture fellow land, oak dense forest, pine open forest and barren land are the main LULC class type in the Randigad catchment (Tables 15.3 and 15.4). The total amount of contribution of these five classes towards soil erosion is 41.829% from these five LULC class types. *Cedrus deodara*, *Cupressus torulosa*,

Table 15.2 Micro-watershed wise soil erosion

	Soil erosion category area (km^2)			
Micro-watershed code	Nil	Low	Moderate	High
SWR1	1.47	1.28	0.15	0.0100
SWR2	1.41	1.35	0.30	0.0300
SWR3	0.22	0.28	0.06	0.0020
SWR4	0.22	0.28	0.06	0.0390
SWR5	0.28	0.35	0.03	0.0018
SWR6	2.23	2.17	0.20	0.0093
SWR7	1.16	1.25	0.15	0.0090
SWR8	1.08	0.99	0.17	0.0180
SWR9	0.93	1.15	0.34	0.0300
Total	9.00	9.10	1.46	0.1490

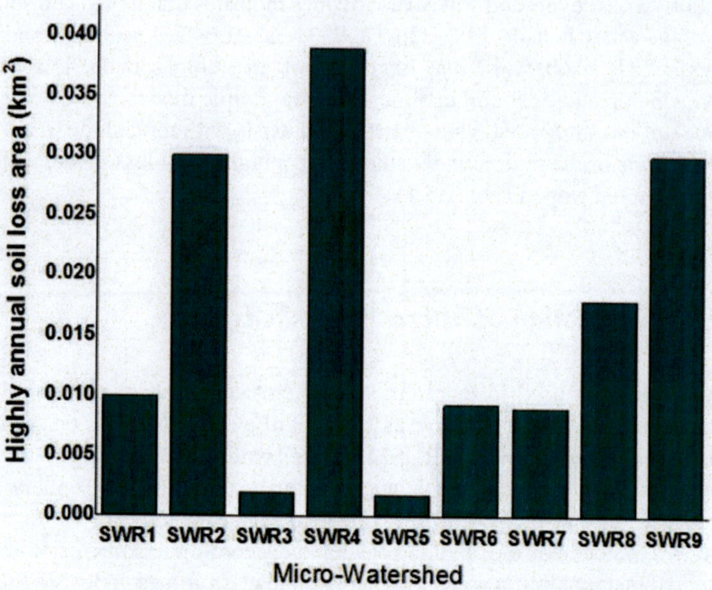

Fig. 15.5 Area of annual soil loss rate (tonnes/ha/year) in high category (>150)

Table 15.3 Assessment of soil erosion with LULC

LULC class	Area (km^2)	Mean soil loss (tonnes/ha/year)	Soil loss %
Agriculture fellow land	3.162	17.52	8.422
Agriculture crop land	3.838	12.50	6.009
Barren land	2.758	20.05	9.638
Cedrus deodar forest	0.109	3.79	1.822
Cupressus torulosa forest	0.086	3.51	1.687
Shrub	1.006	23.44	11.268
Settlement	0.764	12.77	6.139
Pine dense forest	2.485	17.56	8.441
Pine open forest	3.612	16.17	7.773
Mixed forest	2.568	22.23	10.686
Oak dense forest	3.514	20.73	9.965
Oak open forest	0.875	7.56	3.634
Water body	0.037	30.18	14.508

oak open forests, settlement and water body have covered the minimum area of the catchment. Shrub land and water body had the highest erosion risk in the 12 LULC class types. Water body had highest erosion risk due to it has high LS value, while the shrub land is generally located on topographically steeper slope and has LS value comparatively high (Fig. 15.6).

Table 15.4 Assessment of soil erosion risk and associated LULC distribution

	Soil erosion risk area (km^2)				
LULC class	Nil (0–10)	Low (10–50)	Moderate (50–150)	High (>150)	Total %
Agriculture fellow land	1.510	1.260	0.179	0.019	12.75
Agriculture crop land	1.930	1.478	0.285	0.018	15.47
Barren land	0.981	1.425	0.200	0.007	11.12
Deodar forest	0.086	0.008	0.000	0.000	0.44
Surai forest	0.022	0.043	0.014	0.000	0.35
Shrub	0.837	0.502	0.102	0.006	4.06
Settlement	0.575	0.279	0.028	0.000	3.09
Pine dense forest	1.172	1.069	0.164	0.020	10.11
Pine open forest	1.449	1.717	0.163	0.011	14.56
Mixed forest	2.059	0.977	0.196	0.031	10.36
Oak dense forest	1.585	1.246	0.314	0.037	14.16
Oak open forest	0.297	0.355	0.060	0.004	3.53
Water body	0.006	0.007	0.003	0.002	0.16
Total	**50.43**	**41.78**	**6.91**	**0.65**	**100.00**

15.6 Conclusion

Remote sensing and GIS-based Revised Universal Soil Loss Equation (RUSLE) methodology were used to estimate annual soil loss and identify the spatial distribution of soil erosion-prone areas in the Randigad catchment, Pauri Garhwal. Many areas in the catchment fall between low and moderate risk for erosion, but the moderate and high erosion areas lie in northwest and southeast areas of the catchment. The erosion severity map showed that about 7.61% of the area comes under moderate and high erosion category. The high soil erosion rate was in the southeast part of the catchment. The Uttarakhand government is planning to construct a small dam on the southeast part of the catchment. There is a high erosion rate at the suggested dam site, so ignoring this occurrence will impact the region's agriculture and land use (Fig. 15.7). To minimize soil loss, it is important to adopt effective soil conservation practices and to adjust land-use practices within some areas of the catchment. The present research would be useful in many ways for the policy-makers before planning for area sustainability, e.g. constructing a dam and land use, and it would be a key factor in determining the loss in the areas.

For Uttarakhand, soil erosion remains a key challenge for agriculture. So proper management is required for this valuable resource to sustain long-term agricultural productivity. Soil conservation practices should be taught to the farmers so that they can use this as tools to prevent soil degradation and build organic matter and adopt practices which include rotation of crops seasonally, farming through cross-slope, crop mulching, reduction tillage and covering of crop.

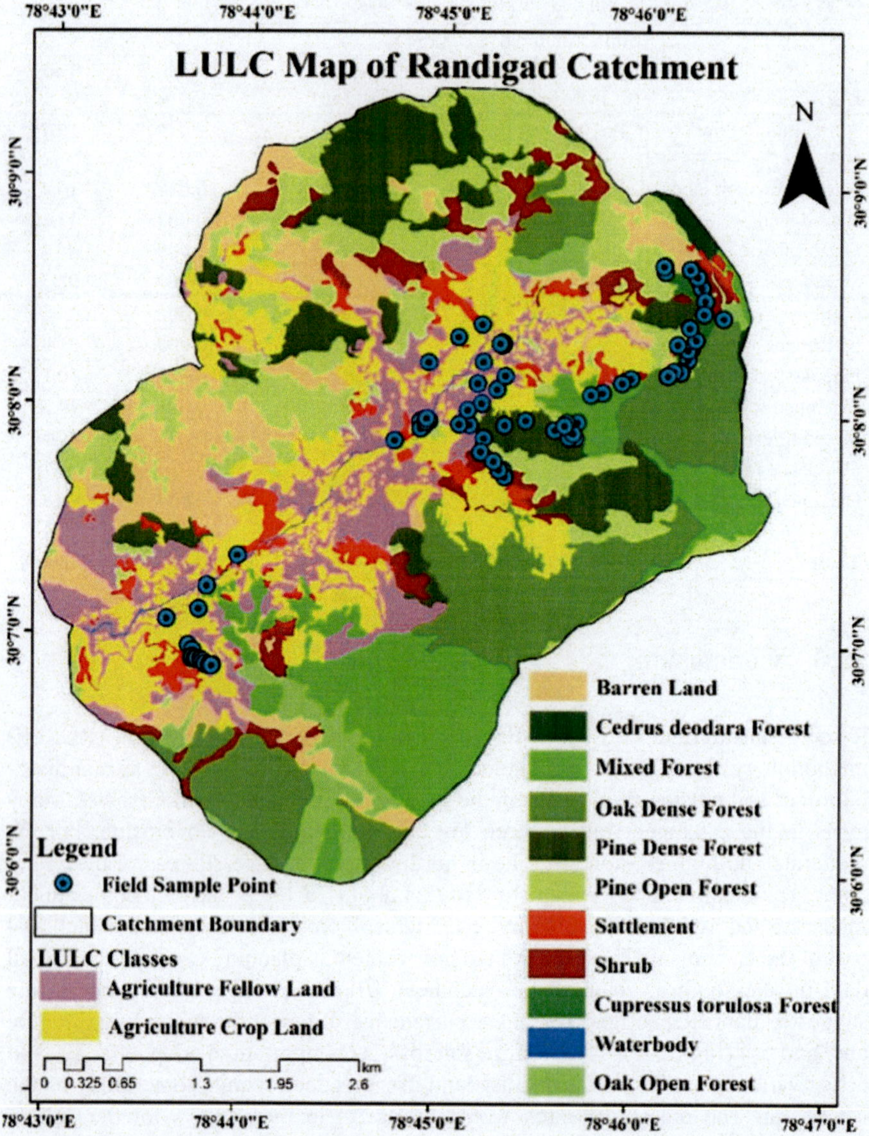

Fig. 15.6 Land use land cover map of Randigad catchment

Fig. 15.7 (**a**) Gully erosion control with wire net (loose stone) (**b**) soil erosion along with agriculture land (**c**) soil erosion along with roadside area (**d**) gully erosion in grassland. (Source: during field visit)

References

Arnold, J. G., Moriasi, D. N., Gassman, P. W., Abbaspour, K. C., White, M. J., Srinivasan, R., Santhi, C., Harmel, R. D., van Griensven, A., Van Liew, M. W., Kannan, N., & Jha, M. K. (2012). SWAT: Model use, calibration, and validation. *Transactions of the ASABE, 55*(4), 1491–1508.

Beasley, D. B., Huggins, L. F., & Monke, E. J. (1982). Modeling sediment yields from agricultural watersheds. *Journal of Soil and Water Conservation, 37*(2), 113–117.

Cerdan, O., & Team, H. (2003). *Long-term soil erosion plot data to evaluate the PESERA (Pan-European Soil Erosion Risk Assessment) approach*. EAEJA, 439.

Chen, T., Niu, R. Q., Li, P. X., Zhang, L. P., & Du, B. (2011). Regional soil erosion risk mapping using RUSLE, GIS, and remote sensing: A case study in Miyun Watershed. *North China. Environmental Earth Sciences, 63*(3), 533–541.

Fernandez, C., Wu, J. Q., McCool, D. Q., & Stockle, C. O. (2003). Estimating water erosion and sediment yield with GIS, RUSLE and SEDD. *Journal of Soil and Water Conservation, 58*(3), 128–136.

Ferro, V., & Porto, P. (2000). Sediment delivery distributed (SEDD) model. *Journal of Hydrologic Engineering, 5*(4), 411–422.

Foster, G. R., & Lane, L. J. (1987). User requirements: USDA, water erosion prediction project (WEPP) Draft 6.3. NSERL report (USA).

Ganasri, B. P., & Ramesh, H. (2016). Assessment of soil erosion by RUSLE model using remote sensing and GIS-A case study of Nethravathi Basin. *Geoscience Frontiers, 7*(6), 953–961.

Honda, K. (1996). Remote sensing and GIS technologies for denudation estimation in a Siwalik watershed of Nepal. In *Proceedings of the International Seminar on Water Induced Disaster (ISWID), Kathmandu.*

Hurni, H. (1985). Erosion-productivity-conservation systems in Ethiopia.

Jain, S. K., Kumar, S., & Varghese, J. (2001). Estimation of soil erosion for a Himalayan watershed using GIS technique. *Water Resources Management, 15*(1), 41–54.

Kamuju, N. (2016). Spatial identification and classification of soil erosion prone zones using remote sensing and GIS integrated 'RUSLE' model and 'SATEEC' GIS system. *International Journal of Engineering Science and Research Technology*, 2277–9655. https://doi.org/10.5281/zenodo.163095

Karaburun, A. (2010). Estimation of C factor for soil erosion modeling using NDVI in Buyukcekmece watershed. *Ozean Journal of Applied Sciences, 3*(1), 77–85.

Knisel, W. G. (1980). *CREAMS: A field scale model for chemicals, runoff, and erosion from agricultural management systems (No. 26).* Department of Agriculture, Science and Education Administration.

Lal, R. A. T. T. A. N. (2001). Soil degradation by erosion. *Land Degradation & Development, 12* (6), 519–539.

Mbugua, W. (2009). *Using GIS techniques to determine Rusle's "R" and "LS" factors for Kapingazi River Catchment* (pp. 9–11). Master of Science Research Project Report submitted to the Department of Geomatic Engineering and Geospatial Information Systems, Jomo Kenyatta University of Agriculture and Technology.

Mhangara, P., Kakembo, V., & Lim, K. J. (2012). Soil erosion risk assessment of the Keiskamma catchment, South Africa using GIS and remote sensing. *Environmental Earth Sciences, 65(7)*, 2087–2102.

Millward, A. A., & Mersey, J. E. (1999). Adapting the RUSLE to model soil erosion potential in a mountainous tropical watershed. *Catena, 38*(2), 109–129.

Moore, I. D., & Wilson, J. P. (1992). Length-slope factors for the revised universal soil loss equation: Simplified method of estimation. *Journal of Soil and Water Conservation, 47*(5), 423–428.

Morgan, R. P. C., Quinton, J. N., Smith, R. E., Govers, G., Poesen, J. W. A., Auerswald, K., & Styczen, M. E. (1998). The European Soil Erosion Model (EUROSEM): A dynamic approach for predicting sediment transport from fields and small catchments. *Earth Surface Processes and Landforms, 23*(6), 527–544.

Nearing, M. A., Foster, G. R., Lane, L. J., & Finkner, S. C. (1989). A process-based soil erosion model for USDA-Water Erosion Prediction Project technology. *Transactions of ASAE, 32*(5), 1587–1593.

Owusu, G. (2012). A GIS-based estimation of soil loss in the Densu Basin in Ghana. *West Africa Journal of Applied Ecology, 20*(2), 41–51.

Pandey, A., Chowdary, V. M., & Mal, B. C. (2009). Sediment yield modelling of an agricultural watershed using MUSLE, remote sensing and GIS. *Paddy and Water Environment, 7*(2), 105–113.

Pelton, J., Frazier, E., & Pickilingis, E. (2012). Calculating slope length factor (LS) in the revised Universal Soil Loss Equation (RUSLE).

Rahaman, S. A., Aruchamy, S., Jegankumar, R., & Ajeez, S. A. (2015). Estimation of annual average soil loss, based on RUSLE model in Kallar watershed, Bhavani basin, Tamil Nadu, India. ISPRS Annals of the Photogrammetry, Remote Sensing and Spatial Information Sciences, 2(2), 207. Simplified method of estimation. *Journal of Soil and Water Conservation, 47*, 423–428.

Renard, K. G. (1997). *Predicting soil erosion by water: A guide to conservation planning with the Revised Universal Soil Loss Equation (RUSLE).* United States Government Printing.

Ritchie, J. B., & Crouch, G. I. (2003). *The competitive destination: A sustainable tourism perspective*. Cabi.

Schwab, G. O., Frevert, R. K., Edminster, T. W., & Barnes, K. K. (1993). Soil and Water Conservation Engineering. John Willey & Sons. Inc., New York, NY USA, 4.

Van Remortel, R. D., Hamilton, M. E., & Hickey, R. J. (2001). Estimating the LS factor for RUSLE through iterative slope length processing of digital elevation data within Arclnfo grid. *Cartography, 30(1)*, 27–35.

Velleux, M. L., England, J. F., Jr., & Julien, P. Y. (2008). TREX: Spatially distributed model to assess watershed contaminant transport and fate. *Science of the Total Environment, 404*(1), 113–128.

Wischmeier, W. H., & Smith, D. D. (1965). *Predicting rainfall-erosion losses from cropland east of the Rocky Mountains: Guide for selection of practices for soil and water conservation* (*No. 282*). Agricultural Research Service, US Department of Agriculture.

Wischmeier, W. H., & Smith, D. D. (1978). *Predicting rainfall erosion losses: A guide to conservation planning (No. 537)*. Department of Agriculture, Science and Education Administration.

Zhu, M. (2015). Soil erosion assessment using USLE in the GIS environment: A case study in the Danjiangkou Reservoir Region, China. *Environmental Earth Sciences, 73*(12), 7899–7908.

Chapter 16
Role of Geospatial Techniques in Soil Erosion Modelling in South Koel Basin, Jharkhand, India

Nusrat Rafique, Durdanah Mattoo, Tahir Hussain Muntazari, and Shoib B. Wani

16.1 Introduction

United Nations Convention to Combat Desertification (UNCCD) defines land degradation as "any reduction or loss in the economic or biological productivity of the land brought about by anthropogenic activities, accelerated by natural processes and thereafter magnified by the impact of biodiversity losses and climate change climate" (UNCCD, 1994). Universal estimation of land deterioration displays an increase in the highly deteriorated areas from 15% in 1991 to 25% by 2011, and it is anticipated that if land degradation continues to occur at the present rate over the next 25 years, it would lessen the universal food availability by 12% (IFPRI, 2012). The whole land debasement process has expanded during the past century with a reckoning loss of 24 million tons of arable topsoil from the agricultural lands across the globe (FAO, 2011). Therefore, soil disintegration is one of the predominant reasons for land corruption, which includes a ceaseless cycle of the expulsion of soil particles from land surfaces by spillover, subsequently causing degradation of soil and unfavorably influencing the efficiency of every biological system including farming, forests, and woodlands environments (Lal & Stewart 1990; Pimentel et al., 1995). Aridness, extreme meteorological scenes over a long period, and scarcity are considered as innate causes of earth crumbling, whereas rampant or irresponsible land use by

N. Rafique (✉) · D. Mattoo
Department of Geography and Regional Development, University of Kashmir, Srinagar, Jammu and Kashmir, India

T. H. Muntazari
Department of Civil Engineering, National Institute of Technology, Srinagar, Jammu and Kashmir, India

S. B. Wani
Civil Engineering Department, B.S. Abdur Rahman University, Chennai, Tamil Nadu, India

G. S. Bhunia et al. (eds.), *Anthropogeomorphology*, Geography of the Physical Environment, https://doi.org/10.1007/978-3-030-77572-8_16

humans such as shifting cultivation, desertification, over-farming, and over-grazing as well are socioeconomic drivers (Eckert et al., 2015). As per the FAO, 2002, one of the leading environments reforming anthropogenic activities is agriculture and also at the all-inclusive scale, agriculture leads as human land use, and considerably more land is under agronomic control than is under woodlands. More than 80% of the land of the Republic of extreme Africa (more than 1002 m hectare) is employed for agronomic uses, and 11% of the space may be used for cropping (DEAT, 2006). Earth deterioration is increasing in severity and extent throughout the world, as 30% of woodlands, 20% of all developed spaces, and 10% of pastures are showing deterioration (Bai et al., 2008). Several hectares of land annually are being deteriorated in all environmental condition regions. It is calculated that about 2600 lakh individuals are affected with land debasement and geologic processes in an additional 100 nations, impacting one third of the globe's land face (Adams & Eswaran, 2000). The decrease in ground character caused by anthropogenic actions also has been seen as a serious world affair for as much as the recent century and is likely to stay significantly on the global calendar within the twenty-first century (Eswaran et al., 2001). In more than 100 countries, the livelihoods of more than 900 million persons are exactly and negatively influenced through land debasement (UN, 1994), and this is seen in almost all the agricultural zones of the world. The Global Assessment of Soil Degradation (GLASOD) estimated that 25% of global land and 13% of the soil in the Pacific and Asia region have deteriorated. The deterioration in China is almost 2.67 million km^2, which is about 28% of the country (FAO, 2005); climate is one of the causes in China, but anthropogenic causes are prevailing (FAO, 2005).

According to Indian Space Research Organisations, analysis shows that of the total geographic area of India, 29.32%, that is, 96.40 mha of the space of India, is enduring the means of soil debasement all through 2011–2013, whereas during all of 2003–2005 the area experiencing soil debasement was 94.53 mha. Investigations show that more than 24% (2011–2013) and 23.64% (2003–2005) of soil debasement of the total TGA is shared by eight states/union territories, and the remaining 1% (individually) of desertification/land degradation is contributed by other states/union/territories. However, the analysis shows that of the total area of the sole states of Jharkhand-Rajasthan-Delhi-Gujarat and states contribute 50% to desertification/land degradation, whereas states with 10% space beneath debasement/land deterioration are the Kerala-Assam-Mizoram- Haryana-Bihar-Uttar Pradesh geographic region and Arunachal Pradesh.

16.1.1 Case Study

This is the collaborative work of Reshma Parveen and Uday Kumar, from the Geology Department of the University of Ranchi, Ranchi, India.

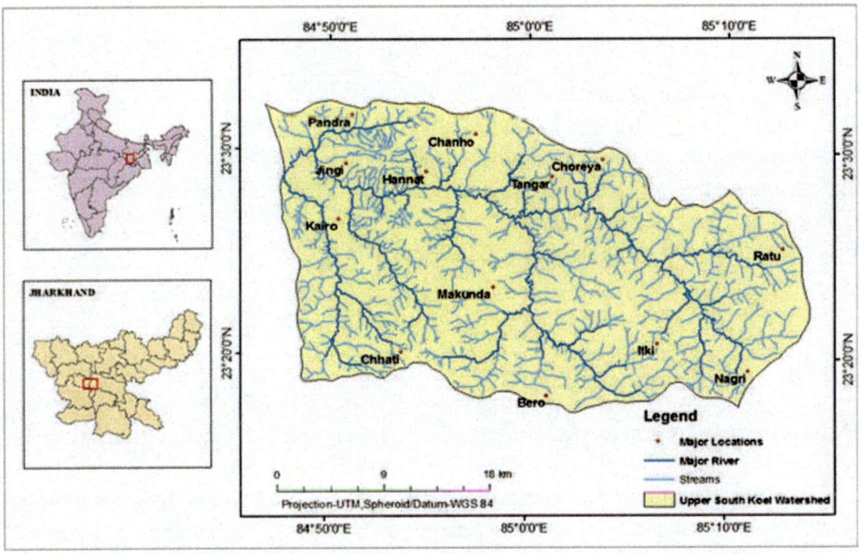

Fig. 16.1 Location of study area

16.2 Geographic Location

The research space is the portion of South Koel basin of area of 942.4 km^2 (Fig. 16.1). The area is at 23°17′16″N and 23°32′16″N and 84°14′15″E and 85°46′51″E. The area under study falls in the jurisdiction of two districts of Jharkhand via Lohardaga and Ranchi. The drainage system of the area is formed by South Koel with major tributaries such as the Kandani and Saphi Rivers. The climate is characterized by cold to mild winters and hot and humid summers. The area experiences an average rainfall of 1380 mm with mean maximum and mean minimum temperature of 36 °C and 7 °C accordingly. The elevation height of the study region ranges from 640 m to 925 m asl. About 84% of the area of the research space is under agriculture.

16.3 Data Sets and Methodology

16.3.1 Data Sets

USLE (Wischmeier & Smith, 1965) was used to ascertain the soil loss with few alterations in its criterion assessment to ensemble surroundings of India. The universal soil loss equation is as follows:

$$A' = R * K * LS * C * P \tag{16.1}$$

where A' is average annual soil loss (ton/ha/year)
R is rainfall–runoff erosivity (in MJ mm/ha/h/year)
K is soil erodibility factor (in ton/MJ/mm)
LS is slope and length of slope factor
C is cropping management factor
P is supporting conservation practice factor

16.3.2 Data Preparation

Rainfall Erosivity Factor (*R*)

The R factor determines the impact of precipitation on soil loss and requires definite accounts for its computation (Wischmeier & Smith, 1978). Several workers in various parts of the globe (Grimm et al., 2003; Shamshad et al., 2008) have attempted to estimate the erosivity factor using precipitation knowledge with great duration of spells using the rainfall information of the various stations located in the study space. Amid mean yearlong precipitation and computed EI_{30} values, a relationship was established for varied belts of the Indian sub-continent and iso-erodent maps from yearlong and opportune EI_{30} values (Singh, 1981). The borrowed formula is as follows:

$$R = 79 + 0.363R_N \tag{16.2}$$

where R_N = average annual rainfall (mm)

Five years of mean yearlong information is used for computation of yearlong R factor. Because the information was not uniform, therefore interpolation was done in the raster calculator of Arc 10.2 GIS software, and a rainfall distribution map was calculated that was used as input for the computation of the R factor map.

16.3.3 Erodibility Factor (K)

K is defined as the speed at which different soils erode, alternatively depending upon factors such as precipitation, slope, and yield cover. The K issue reflects the speed of ground loss per R index. The land design of the space used was taken from the soil survey done by the NBSS & LUP. Parameters such as sand, silt, clay, and organic matter units were taken into consideration. During the study, K of the study space may be characterized through the link amid soil texture category and organic matter content projected by Schwab et al. (1981). The soil erodibility issue (K) is shown in Table 16.1.

Table 16.1 K factor values

Textural class	Organic matter content (%)		
	0.5	2	4
Fine sand	0.16	0.14	0.1
Very fine sand	0.42	0.36	0.28
Loamy sand	0.12	0.1	0.08
Loamy very fine sand	0.44	0.38	0.3
Sandy loam	0.27	0.24	0.19
Very fine sandy loam	0.47	0.41	0.33
Silt loam	0.48	0.42	0.33
Clay loam	0.28	0.25	0.21
Silt clay loam	0.37	0.32	0.26
Silty clay	0.25	0.23	0.19

Source: Schwab et al. (1981)

16.3.4 LS Factor

The LS factor in the soil loss equation accounts for the result of two parameters such as the length and steepness of slope on rill and sheet erosion. The gradient of slope length and steepness of the slope is measured from the digital elevation model (DEM).

Several relationships are seen for the estimation of LS issue. The most effective suited relationship for unification with GIS is the equation projected by Moore and Wilson (1992) and Moore and Burch (1986), as given next:

$$\mathrm{LS} = (\mathrm{A}/22.13)^{0.6}(\sin\mathrm{B}/0.089)^{1.3} \tag{16.3}$$

where *A* is the upslope contributing factor and *B* is the slope angle.

Magnitude of erosion is affected by steepness of slope, the interlinkage of angle, and slope length. The effects of both length of slope and steepness of slope need to be taken into account together (Edwards, 1987).

By the use of a Digital Elevation Model in the Geographic Information System, the steepness of slope (S) and slope length (L) are also calculated efficiently. The efficiency of estimation relies on the resolving power of the DEM.ASTER DEM of 30 m resolution as used. For LS, a map calculation raster calculator was used. The expression is as follows:

$$\begin{aligned}\mathrm{LS} = {} & \mathrm{Pow}\,([\text{flow adjustment}]\,\text{resolving power}/22.13, 0.6) \\ & \times \mathrm{Pow}\,(\sin\,([\text{slope of DEM}] \times 0.01745/0 - 0896, 1.3)\end{aligned} \tag{16.3a}$$

The LS factor must be filled initially, which may be done by using the spatial analyst tool in Arc GIS. Initially the DEM must be filled and so flow direction has to be determined. Using flow direction as an input in flow accumulation is derived in Arc GIS. The slope for every grid was found by using the slope operations in Arc GIS.

16.3.5 C Factor Map

Land use forms the base for the identification of C factors, thus lessening the issue to soil disintegration weakness, the USLE being the significant factor as these factors stand for the surroundings that could be effectively transformed to diminish disintegration. The conservation factor is essentially the flora spread rate, which is characterized by the proportion of land suffering from explicit harvests to the proportionate misfortune from ploughed, uncovered plots. Information about the land use arrangement within the space is required for dependable conservation issue values. Here, the normalized differential vegetation index (Landsat TM) was used as per the following equation:

$$\text{NDVI} = \text{NIR} - \text{RED}/\text{NIR} + \text{RED} \tag{16.4}$$

NIR is the near-infrared reflection of the EM gamut and IR in the visible band. Normalized differential vegetation index values ranges between −1.0 and + 1.0 where green vegetation is represented by higher values and common surface materials by low values. NDVI values of 0 represent bare lands and negative NDVI values represent water bodies (Lillesand et al., 2004; Sader & Winne, 1992).

The 30 m resolution Landsat TM data were taken into account for normalized differential vegetative index (NDVI) image generation. The following accustomed equation is used to bring about the conservation issue design after the generation of NDVI from NVDI values (Knijff et al., 2000; Sammons et al., 2004)

$$C = \ e^{(-\alpha((\text{NDVI})/(\beta - \text{NDVI})))} \tag{16.5}$$

where α and β are a unit-less framework that confirms a form of the curve about the NVDI and also the cover factor.

Knijff et al. (2000) established that good results are given by the scaling approach compared to linear relationships. In the end, the values of 2 and 1 were handpicked for the constraints α and β, respectively and it was seen that this gives better results as per relevant literature. *C* issue varies between 0 and 1: some picture elements with negative values were assigned 0 and pixels with values greater than 1 were assigned value 1.

16.3.6 Practice Management Factor Map

The values of control method aspects as per the methods of cultivation and also the slope of the area are given in Table 16.2 (Shin, 1999). The P factor represents practices such as contouring agriculture terracing. The conservation practice factor ranges from 0 to 1 where 0 represents erosional resistance facility, mostly manmade, and no manmade resistant facilities are represented by 1.

Table 16.2 P factor values as per slope and cultivation practice

Slope (%)	Contouring	Strip cropping	Terracing
0.0–7.0	0.55	0.27	0.1
7.0–11.3	0.6	0.30	0.12
11.3–17.6	0.8	0.40	0.16
17.6–26.8	0.9	0.45	0.18
>26.8	1	0.5	0.2

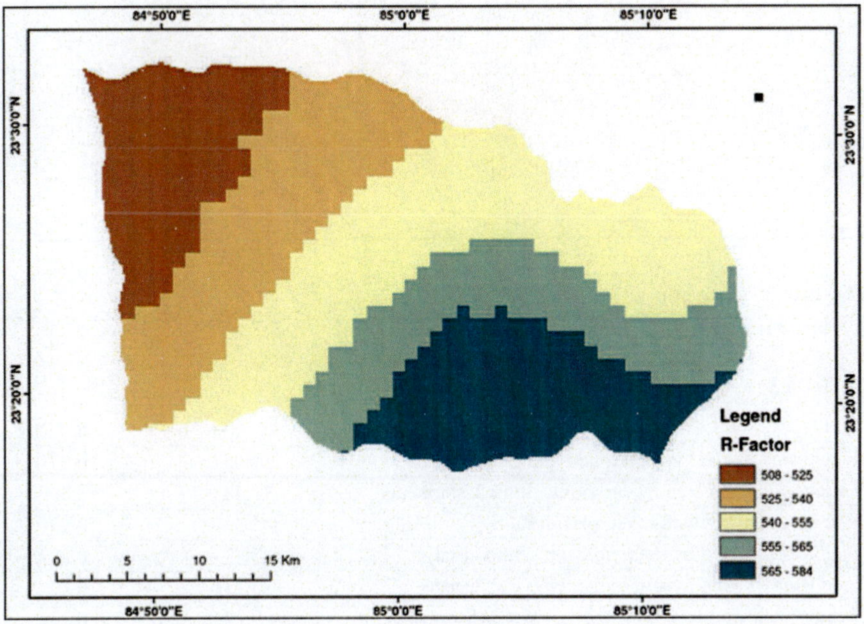

Fig. 16.2 R factor map

16.4 Results and Discussion

16.4.1 R Factor Map

The normal yearly precipitation information of 19 encompassing precipitation check stations was utilized to obtain the precipitation circulation guide of the whole watershed. The normal yearly precipitation in the investigation zone was from 880 mm to 1480 mm.

This issue map (Fig. 16.2.) was generated from a mean yearly precipitation map with R factor varying from 508 to 584 MJ ha/mm/h/year.

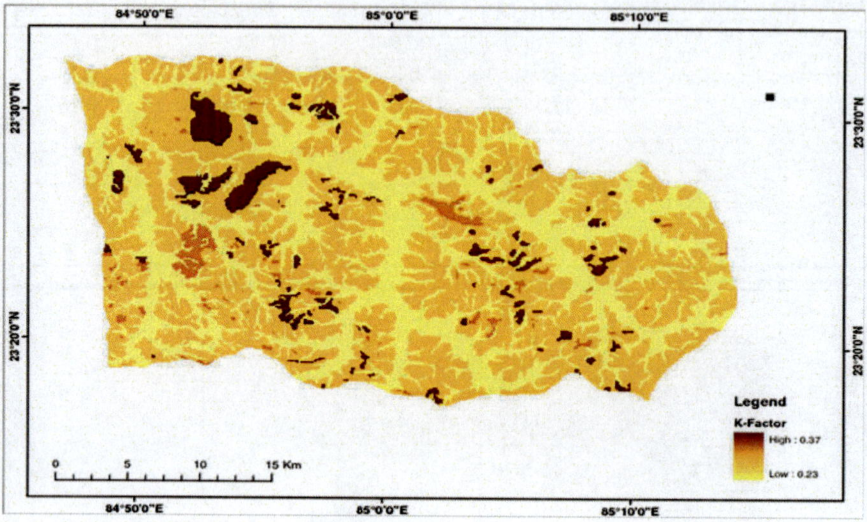

Fig. 16.3 K factor map

Table 16.3 K factor values

Soil types	Area (km^2)	K factor
Fine loamy mixed hyperthermic typic Dystrustepts	34.05	0.27
Loamy skeletal mixed hyperthermic, lithic Ustorthents	11.43	0.26
Fine mixed hyperthermic ultic Haplustalfs	385.39	0.23
Fine loamy mixed hyperthermic typic Rhodustalfs	487.77	0.28
Fine loamy mixed hyperthermic typic Haplustepts	13.23	0.37
Loamy skeletal mixed hyperthermic typic Ustorthents	9.27	0.27

16.4.2 K *Factor Map*

The soil erodibility issue map arranged the soil map of the study spacesupported completely different soil textures (Fig. 16.3). The obtained erodibility issue is given in Table 16.3.Various factors stimulating the potential of soil to disintegrate are its soil arrangement, soil balance, permeability, and organic matter.

16.4.3 LS *Factor Map*

Erosion is considerably affected by slope length and steepness. The two maps. the flow accumulation map and the slope map, were prepared with the use of the ASTER Digital Elevation Model.

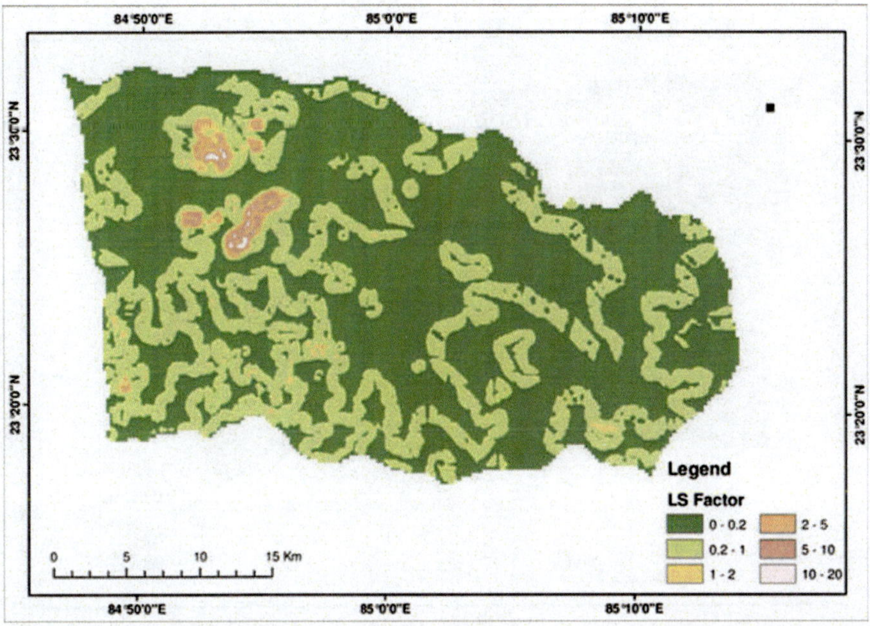

Fig. 16.4 LS factor map

Using the slope function from ARC GIS, slope was calculated. Using the two derived maps such as flow accumulation map and slope map, the topographic map was generated (Fig. 16.4). Slope length and slope steepness issue vary from 0 to 20 but the greater part of the study space is in zero to two ranges with some greater values occurring in the undulating areas.

16.4.4 C & P Issue Map

An NDVI Index image was extracted from Landsat data for C factor map generation (Fig. 16.5) and the C value ranged between 0.058 and 1. Woodland areas C factor ranges from 0.05 to 0.2 whereas waterbody values approached 1 and conservation issue values ranged from 0.3 to –0.6 for agricultural areas.

There are no foremost conservation drills to be followed except the bunds in cultivated agricultural areas, thus the P factor value was assigned as 0.28 whereas for all other land uses the value was assigned as 1 (Rao, 1981).

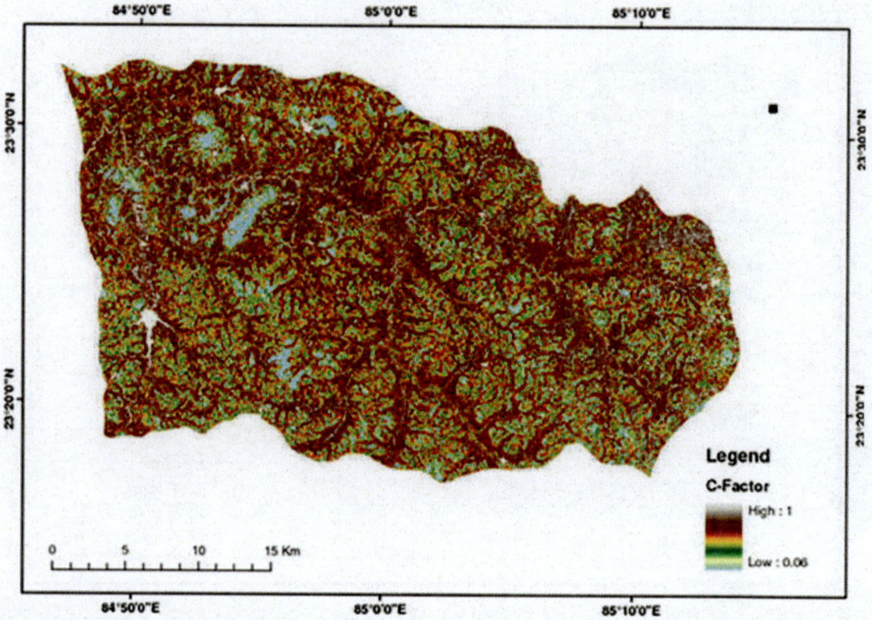

Fig. 16.5 Conservation practice (C) map

16.4.5 Soil Loss Map

The aim of the undertaken research was to check the erosion loss in the study space. Parameters used are in either numerical format or spatial format. All the factors were integrated with the GIS environment (Eq. 16.1). The erosion maps achieved were categorized into slight (0–5 t/h/year), moderate (5–10 t/h/year), high (10–20 t/h/year), very high (20–40 t/h/year), severe (40–80 t/h/year), and very severe (>80 t/h/year) as per the protocols of Singh et al. (1992) (Fig. 16.6). Table 16.4 shows the geographic allocation of various categories. The annual erosion for the whole study area is 12.20 ton/h/year, while the bare areas show more than 80 ton/h/year of annual loss. Most of the space falls in the category of slight to moderate, and 18.2% of the space is in the high to very severe category.

16.5 Conclusion

The creation of a directory is a very difficult task. Various thematic layers such as rainfall erosivity maps, soil erodibility maps, slope length and steepness maps, and crop management and practice maps generated were overlaid in the GIS settings as per the RUSLE equation to generate the map of erosion for the study space. All these

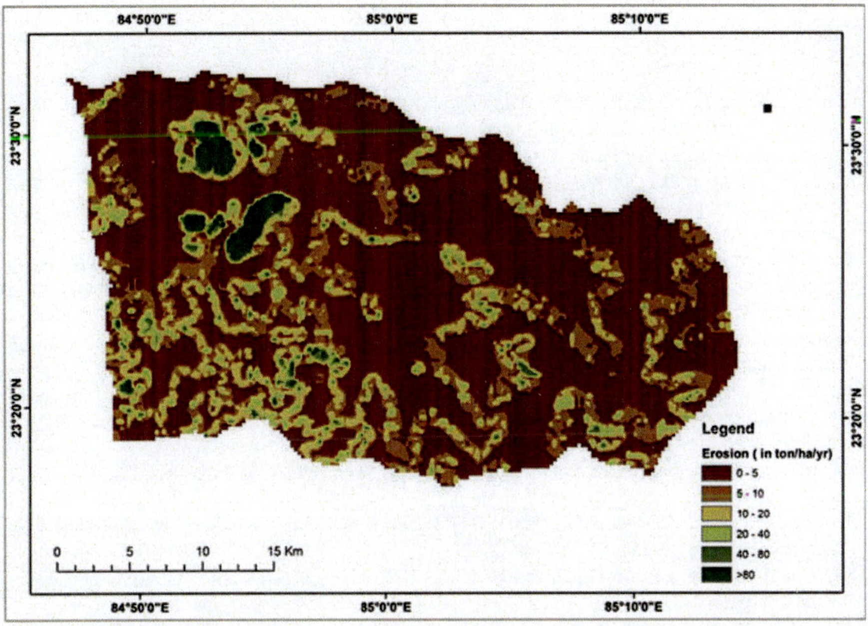

Fig. 16.6 Soil erosion map

Table 16.4 Areal extent of various soil loss zones

Sample no.	Soil loss zone	Range (t/h/year)	Area (km^2)	Area (%)
1	Slight	0–5	609.72	64.7
2	Moderate	05–10	161.15	17.1
3	High	10–20	94.67	10.05
4	Very high	20–40	43.8	4.65
5	Severe	40–80	15.12	1.6
6	Very severe	>80	17.94	1.9

factors were integrated in the GIS environment according to the RUSLE equation to obtain the results. An approximate estimation of mean soil loss was in dire straits in the selected space. The area of interest was divided into slight (64.70%), moderate (17.10%), high (10.05%), very high (4.65%), severe (1.60%), and very severe (1.90%) soil loss potential zones. It can be inked for resembling and anticipating soil loss as well as for environmental monitoring and water resources management. The said map will serve as a database for the implementation of conservation mitigation in the study area.

References

Adams, C. R., & Eswaran, H. (2000). Global Land Resources in the context of food and environmental security. In *Advances in Land Resources management for the 20th century. Soil Conservation Society of India, New Delhi* (pp. 35–50).

Bai, Z. G., Dent, D. L., Olsson, L., & Schaepman, M. E. (2008). *Global assessment of land degradation and improvement: 1. Identification by remote sensing* (No. 5). ISRIC World Soil Information.

DEAT (Department of Environmental Affairs and Tourism). (2006). *South Africa environment outlook. A report on the state of the environment.* Department of Environmental Affair and Tourism, Pretoria. 371pp.

Desertification and land degradation: Atlas of India, space applications centre, Indian space research organisation: Ahmadabad, India, 2016.

Eckert, S., Husler, F., Liniger, H., & Hodel, E. (2015). Trend analysis of MODIS NDVI time series for detecting land degradation and regeneration in Mongolia. *Journal of Arid Environments, 113*, 16–28.

Edwards, K. (1987). *Runoff and soil loss studies in New South Wales*. Soil Conservation Service of NSW.

Eswaran, H., Lal, R., & Reich, P. F. (2001). Land degradation: An overview. In *Responses to land degradation* (pp. 20–35). CRC Press LLC.

FAO (Food and Agriculture Organization of the UN). (2002). *FAO-STAT Statistics Database*. Rome, Italy.

FAO (2005). Global Forest Resources Assessment 2005 – progress towards sustainable forest management. FAO Forestry Paper No. 147. Rome. http://www.fao.org/docrep/008/a0400e/a0400e00.htm

FAO. (2011). *The state of the world's land and water resources for food and agriculture (SOLAW) – Managing systems at risk.* Food and Agriculture Organization of the United Nations.

Grimm, M., Jones, R. J., Rusco, E., & Montanarella, L. (2003). Soil erosion risk in Italy: A revised USLE approach. *European Soil Bureau Research Report, 11*, 23.

IFPRI. (2012). 2011 Global food policy report. International Food Policy Research Institute, Washington, DC. USA.. Available at: http://www.ifpri.org/cdmref/p15738coll2/id/126897/filename/127108.pdf [access date: 05.08.2016]

Lal, R., & Stewart, B. A. (1990). Advances in Soil Science. Vol. 11, Soil Degradation. Springer Verlag, New York, USA. 345p.

Lillesand, T. M., Kiefer, R. W., & Chipman, J. W. (2004). *Remote sensing and image interpretation* (5th ed.). Wiley.

Moore, I. D., & Wilson, J. P. (1992). Length-slope factors for the revised universal soil loss equation: Simplified method of estimation. *Journal of Soil and Water Conservation, 47*(5), 423–428.

Moore, I. D., & Burch, G. J. (1986). Modelling erosion and deposition: Topographic effects. *Transactions of the ASAE, 29*(6), 1624–1630.

Pimentel, D., Harvey, C., Resosudarmo, P., Sinclair, K., Kurz, D., McNair, M., Crist, S., Shpritz, L., Fitton, L., Saffouri, R. Blair, R. (1995). Environmental and economic costs of soil erosion and conservation benefits. *Science 267*(5201), 1117–1123.

Rao, Y. P. (1981). Evaluation of cropping management factor in Universal Soil Loss Equation under natural rainfall conditions of Kharagpur, India.

Sader, S. A., & Winne, J. C. (1992). RGB-NDVI colour composites for visualizing forest change. *International Journal of Remote Sensing, 13*, 3055–3067.

Schwab, G. O., Frevert, R. K., Edminster, T. W., & Barnes, K. K. (1981). *Soil water conservation engineering* (3rd ed.). Wiley.

Singh, G. (1981). Soil loss prediction research in India. Dynamics. *International Journal of Remote Sensing, 13*(16), 3055–3067.

Singh, G., Babu, R., Narain, P., Bhushan, L. S., & Abrol, I. P. (1992). Soil erosion rates in India. *Journal of Soil and Water Conservation, 47*(1), 97–99.

Shamshad, A., Azhari, M. N., Isa, M. A., Hussin, W. W., & Parida, B. P. (2008). Development of an appropriate procedure for estimation of RUSLE EI30 index and preparation of erosivity maps for Pulau Penang in Peninsular Malaysia. *Catena, 72*(3), 423–432.

Shin, G. J. (1999). *The analysis of soil erosion analysis in watershed using GIS*. Department of Civil Engineering, Gang-won National University, Gangwon-do, South Korea, Ph. D. dissertation.

UNCCD. (1994). *Elaboration of an International convention to combat desertification in countries experiencing serious drought and/or desertification, particularly in Africa*. United Nations, General Assembly. A/AC.241/27 12 September 1994, Article 2. Available at: http://www.unccd.int/Lists/SiteDocumentLibrary/conventionText/conv-eng.pdf

United Nations. (1994). Earth summit convention on desertification. *Proceedings of the United Nations Conference on Environment and Development (UNCED),* Rio De Janeiro, Brazil, Department of Public Information, United Nations, New York, USA.

Wischmeier, W. H., & Smith, D. D. (1965). *Predicting rainfall-erosion losses from cropland east of the rocky mountains: Guide for selection of practices for soil and water conservation*. U.S. Department of Agriculture, Agricultural Research Service, Issue 282 of Agriculture Handbook. Washington DC, USA. p. 47.

Wischmeier, W. H., & Smith, D. D. (1978). *Predicting rain- fall-Erosion losses: A guide to conservation planning, agricultural handbook no. 537*. Agricultural Research Service—US Department of Agriculture.

Chapter 17
A Multiple Criteria-Based Approach in Monitoring Soil Water Stress in Sonitpur District, Assam

Kaushik Kharghoria and Karishma Dutta

17.1 Introduction

A crop is a floral product: in a habitual sense it act as a life preserver for entire living organisms. It is flourished and reaped substantially for benefit and maintenance. Crops are mainly accumulated from the wild or may be from other sources (Ratna Raju et al., 2016). Subsistence, plantation, dairy, cooperative, and collective are some of the farming activities across the globe. However, the tiny liquid water droplets are crucial in shaping our agricultural economy. Topography creates a restraint on hydrological processes (Testi et al., 2008). Topography has notable consequences on the various characteristics of soil because it regulates the runoff. Moreover, its direction influences vegetative cover (Dyer, 2009). Bisrat and Berhanu (2018) determined the water-storing sites of Ethiopia with the Topographic Wetness Index (TWI) and Normalized Difference Vegetation Index (NDVI) along with satellite data and concluded it is a cost-effective, noncomplicated potential GIS-based technique for representing water work. Soil is the permeable, biologically agile layer that has grown in the Earth's crust (Laurent et al., 2017). Ayele et al. (2019) in their research examined the properties of soil around the watersheds of Rift Valley basins, its effect on physiochemical properties, and tried to establish a relationship between topographic features and soil characteristics. The physical and chemical properties of soil along with topography influences the denitrification process (Li et al., 2018). Vegetative cover works as a shield layer on the surface because it holds the soil deeply and protects it from erosion. However, when Yang et al. (2018) examined the connection between the vegetative cover of Horqin Sandy Land of China with soil water content, they found a negative relationship between them. Moreover, the authors portrayed a clear-cut reason for the outcome that varied

K. Kharghoria (✉) · K. Dutta
Department of Geography, School of Human and Environmental Sciences, North Eastern Hill University, Shillong, Meghalaya, India

G. S. Bhunia et al. (eds.), *Anthropogeomorphology*, Geography of the Physical Environment, https://doi.org/10.1007/978-3-030-77572-8_17

vegetation groups have different water consumption capacities. Some of our countries and states are facing a deficiency of rainwater in correspondence to such conditions: irrigation facilities assist those insufficiencies (González-Dugo et al., 2005). An analysis investigated the agricultural drought and predicted the future rainfall of Tegal City, Indonesia using the Standardized Precipitation Index (SPI). Water deficit in crop roots and an excess amount of sunlight along with poorer quality of soil cover leads the crop to its wilting point (Alghory & Yazar, 2019). A comparison of agricultural indices such as the Crop Water Stress Index (CWSI) and the Water Deficit Index (WDI) by Tanriverdi et al. (2017) pointed out better indices for irrigation management. Erdem et al. (2006) investigated the potentiality of the CWSI for irrigation scheduling in sunflowers and determined threshold values for the desired index that was ultimately proven as a significant tool in observing and evaluating water stress. Our India is an agrarian country where a large percent of the people are engaged in agricultural sectors: rice and wheat are regarded as the staple crops (Dangwal et al., 2015). The conventional methods for investigating the water stress of crops are based mainly on in situ soil water measurements (Taghvaeian et al., 2012). Recent advances in technology gave rise to several other sophisticated and less time consuming methods. Among these, the infrared thermometer is used for computing plant water stress. CROPWAT 8.0 is a computer instruction recruit to compute water requirements in crop (Ihuoma & Madramootoo, 2017). Li et al. (2010) in their research tried to quantify crop water stress with the help of canopy temperature and the balance of energy measurement above wheat and maize of North China Plain during winters. El-Shirbeny & Abutaleb (2017) conducted research to evaluate crop water stress employing the remote sensing techniques of Sentinel-1 and Radar. Saseendran et al. (2015) identified the associations between various factors, that is, soil and plant water availability, evapotranspiration of crop and biomass, measurement of the grain yield, and corn canopy cover: keeping all these factors as the base, they investigated crop water stress through the RZWQM2 model.

The Sonitpur district of Assam is one of the historical abode places. Located in the foothills of Himalaya, it receives a sufficient amount of precipitation during monsoonal months, and in the pre-monsoonal months it shows some drought conditions for more than a quarter period of time, which minimizes the moisture content of the soil. Many factors affect the distribution of precipitation in a particular region. So, here our study focuses on a few components that affect the physical platform: elevation, slope, texture, and topography. To some extent the results of uneven distribution of these elements are noticeable. Here, our study upholds the crop water stress as a part and parcel of unwelcome consequences in some areas of Sonitpur. Although much other momentous work has been done, this particular work emphasized the unexplored portion, giving more concentration on soil moisture content of Sonitpur district that were not yet considered by previous works. The chief objectives undertaken follow:

(a) To compute the annual rainfall peculiarity of Sonitpur district using the Rainfall Anomaly Index (RAI)

(b) To estimate the duration of drought from 2014 to 2018 using the Standardized Precipitation Index (SPI)
(c) To determine the topographic wetness index (TWI) from a Digital Elevation Model (DEM) for monitoring soil moisture
(d) To investigate the relationship among different variables, that is, topographic wetness, elevation, slope, and soil texture
(e) To examine the crop water stress of Sonitpur from 2014 to 2018

17.2 Materials and Methods

17.2.1 Study Area

The Sonitpur district is located (Fig. 17.1) between 26°°30′N and 27°°2′N and 92°°17′E to 93°°47 E (Srivastava et al., 2002). The district is a land of culture and ancient heritage composed of the pleasant park Chitralekha udyan, familiarly known as the 'Cole Park,' Agnigarh (the fire in name itself), 'Padum Pukhuri,' an attractive lake with an island, the Bhairab temple where the goddess Durga is worshipped, Mahabhairav temple, the long concrete bridge over Brahmaputra River 'Kolia Bhomora Setu' (bridge), and many more features to include (Lama & Bordoloi, 2017). The district includes a total area of 5324 km^2, holding second position in the area proceeding after Karbi-Anglong, surrounded by Arunachal Pradesh in its north, Udalguri, Darrang in its west, Nagaon, Morigaon, and Golaghat districts in its south,

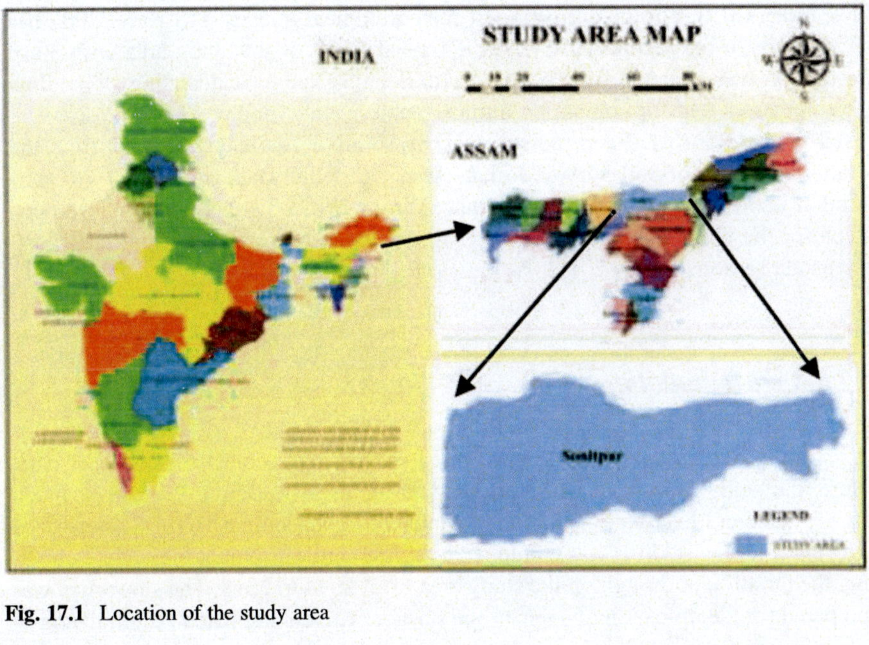

Fig. 17.1 Location of the study area

Lakhimpur in the east (Dutta et al., 2010). The district in its true sense means 'the city of blood.' The main drainage system of the district is controlled by the mighty River Brahmaputra and its tributaries: Jia-Bharali, Gabharu, Ghiladhari, Bargang, and Beliri. The climatic framework of the district indicates a subtropical and humid type with a high amount of rainfall (Sonitpur District Inventory of Agriculture 2015–2016). The average annual rainfall of the district is 2393 mm. The summers are warmest, with 29 °C average temperature. Sonitpur is also the abode of tea gardens; even Asia's largest Monabari tea estate, is located in this district (Dutta et al., 2010).

17.2.2 Database

Data as a general concept refers to some existing information that is represented or coded in suitable form for further processing. To proceed in any work, datasets have an important role. Here, to accomplish the study, the following satellite products and remote sensing software have been enrolled. Data may be primary sources of data (which are ground-based information that is collected directly); and secondary sources of data (already existing information that is collected by others) (Prada-Ramallal et al., 2018). This particular study is based on secondary sources. For the preparation of the Rainfall Anomaly Index (RAI), monthly precipitation data were acquired for the years 2014–2018 from Rainfall Statistics of India 2014, 2015, 2016, 2017, 2018. The mean air temperature of Sonitpur District was obtained from NASA Prediction of Worldwide Energy Resources for calculating Standardized Precipitation Index (SPI). For the derivation of meteorological indices, MS-Excel 2007 has been used. In the second turn, ASTER Global DEM of the study area with 30 m resolution was obtained from NASA Earth Search Data for calculating slope, flow accumulation, and for computing a topographic wetness index. Here, the software used is ArcGIS 10.1. Even the soil map of Sonitpur has been collected from the National Bureau of Soil Survey, Jorhat, Assam. Satellite Data LANDSAT 8 (Operational Land Imager) of Sonitpur acquired from USGS Earth Explorer have been used for the preparation of Crop Water Stress Index Maps for both pre- and post-monsoon seasons (2014–2018).

17.2.3 Methodology

A multiple criteria analysis was performed scheduling secondary sources of data. Here, the study proceeded with various data inputs. The Rainfall Anomaly Index (RAI) was recruited for computation of annual rainfall peculiarity. The Standardized Precipitation Index (SPI) was employed as a drought determinant index for estimating the duration of drought in the study area from 2014 to 2018. The slope map was prepared for identification of angle of the surface. The direction and accumulation of flow were computed on Digital Terrain Model (DTM) ground for obtaining topographic wetness. In the long run, the wetness of the topography was enrolled for predicting soil moisture and to observe the relationship with other variables of slope and soil texture. Here, the soil map was collected for examining the variation in soil texture in different areas of Sonitpur district. The relationships between topographic wetness and elevation were evaluated on the ground of regression analysis with 35 scattered plots. For investigating crop water stress in pre- and post-monsoon seasons, an index of irrigation scheduling called the crop water stress index (CWSI) was used for each of the years from 2014 to 2018. To make the study an explicit one, 30 reference points (pixel values of CWSI) were taken for investigating the seasonal shift between crop water stress values. A regression analysis with 35 points was again arranged between CWSI and TWI to verify the water stress level in respect to topographic wetness.

17.3 Analysis

17.3.1 Rainfall Variability Analysis Using Different Indices

Sufficient amounts of indices were evolved by various researchers for representation of spatial and temporal fluctuations of rainfall (Joshi et al., 2014). Up to a limited extent our present study on Sonitpur district was based on two precipitation indices: RAI (Rainfall Anomaly Index) and SPI (Standardized Precipitation Index).

Computation of Rainfall Anomaly Index (RAI)

The Rainfall Anomaly Index (RAI) was first obtained by Rooy (1965). This specific index was used for drought monitoring. Here, in this study it considers both positive and negative anomalies. The rainfall data are put in order from highest to lowest (Salehnia et al., 2017). The maximum 10 values are averaged to doorstep the positive anomaly again; the minimum 10 values are averaged to doorstep the negative anomaly (Salehnia et al., 2017). It can mathematically be equated as:

For positive anomaly,

$$\mathbf{RAI} = 3\left[\frac{\mathbf{N} - \overline{\mathbf{N}}}{\overline{\mathbf{M}} - \overline{\mathbf{N}}}\right] \tag{17.1}$$

For negative anomaly,

$$\mathbf{RAI} = 3\left[\frac{\mathbf{N} - \overline{\mathbf{N}}}{\overline{\mathbf{X}} - \overline{\mathbf{N}}}\right] \tag{17.2}$$

where N denotes the monthly, seasonal. and yearly rainfall (in mm); $\overline{\text{N}}$ denotes average monthly, seasonal, and yearly rainfall of historical series (in mm); $\overline{\text{M}}$denotes the 10 highest average monthly/seasonal/yearly precipitation of historical series; and $\overline{X}$ denotes the 10 lowest average monthly/seasonal/yearly precipitations of historical series.

Estimation of Standardized Precipitation Index (SPI)

The Standardized Precipitation Index (SPI) was established by McKee et al. (1993). This particular index was calculated for the first time for 3-, 6-, 12-, 24-, and 48-month timescales (McKee et al., 1993). The SPI is a quantification index that evaluates precipitation for a given time (Shah et al., 2015). Moreover, it is broadly used to assess drought conditions and vegetation production (Meroni et al., 2017).

Here, in the present study, the standardized precipitation was computed for all the existing months between 2014 and 2018 (Akinsanola & Ogunjobi, 2014). The values of this index can be determined: if the SPI value becomes negative (−1 or below), it indicates a dry/extremely drier condition, whereas if it continues to be positive (1 or more) it reflects a wet/extremely wet condition (Subash & Ram Mohan, 2011).

$$\mathbf{SPI} = \frac{\mathbf{x} - \overline{\mathbf{x}}}{\boldsymbol{\sigma}} \tag{17.3}$$

where X is the actual value of each variable (air temperature and rainfall); $\bar{X}$ denotes a long-term mean value; and σ is the standard deviation.

17.3.2 Quantification of Topographic Wetness Using DEM

The topographic wetness index generated from the DEM attraction model and Shuttle Radar Topography Mission were compared by Hojati and Mokarran (2016). In the present study, manipulation of the topographic wetness index considers some parameters for monitoring soil moisture content such as flow direction and flow accumulation slope analysis (Beven & Kirkby, 1979). The flow direction of groundwater carried for 16 water wells in Yenagoa, Nigeria, was investigated conventionally with the triangulation method (Oborie & Nwankwoala, 2017). A fresh procedure evolved a grid DEM for the computation of flow directions and representation of upslope by Tarboton (1997). For the extraction of drainage networks, the method of flow accumulation is necessary (Do et al., 2011).

In our study, the overall computation took place in ArcGIS 10.1. In the first turn, DEM identification was done with a projected coordinate system. The direction of flow was calculated on the ground of a Digital Terrain Model (DTM). After the computation of flow direction, another step proceeds for determination of accumulated flow. The areas represented with valleys specifies a high value of accumulation while the areas with watersheds denotes 0 value. Accordingly, to the plan slope analysis was performed and tangent of slope was calculated. The last procedure continues with computation of TWI by using the following formula suggested by Beven and Kirkby (1979):

$$\mathbf{TWI} = \mathbf{ln}\ (\boldsymbol{\alpha}/\mathbf{tan}\boldsymbol{\beta}) \tag{17.4}$$

where α = local upslope contributing area per unit length of contour; $\tan\beta$ = topographic slope of the cell.

17.3.3 Determination of Crop Water Stress

The crop water stress index is one of the most sophisticated techniques for detection of canopy temperature (Veysi et al., 2017). A study planned by Prueger et al. (2019) to assess the significance of the crop water stress index (CWSI) by adopting manifold canopy temperature sensors; the author also inquired the dairy trend in the index of irrigated vineyard of California. This particular index is derived from air temperature and canopy temperature, which is basically raised as an application to investigate plant water status (Alvino & Marino, 2017). However, it can be determined as:

$$\mathbf{CWSI} = [(\mathbf{T_c} - \mathbf{T_a}) - (\mathbf{T_{nws}} - \mathbf{T_a})]/\left[(\mathbf{T_{dry}} - \mathbf{T_a}) - (\mathbf{T_{nws}} - \mathbf{T_a})\right] \tag{17.5}$$

In this equation, the variable T_c is the canopy temperature in (°C), T_a is the air temperature in (°C), T_{nws} is the non-water stressed canopy temperature (in °C), and T_{dry} is the water-stressed canopy temperature (in °C) (Jackson et al., 1981). The values of the crop water stress index range between 0 and 1, where 0 indicates no stress and a value of 1 indicates maximum stress (Yuan et al., 2004).

17.4 Results and Discussions

17.4.1 Examination of Annual Rainfall Peculiarity Using Rainfall Anomaly Index (RAI)

In Sonitpur district, this specific Rainfall Anomaly Index (RAI) was computed (Table 17.1.) after observation of rainfall data from 2014 to 2018, taking averages of monthly totals to estimate the intensity of wet and dry months. Here, the threshold for positive anomalies +4 or above (humid months) and negative anomalies −4 or below (dry months) (Freitas et al., 2019; Araújo et al., 2009).

So, after calculating the RAI (using (Eq. 17.1) for positive anomalies and (Eq. 17.2) for negative anomalies) it becomes easier to figure out the wettest and driest months for the years under consideration. In 2014 the rainy months continue

Table 17.1 Computed RAI values from 2014 to 2018

2014		2015		2016		2017		2018	
Month	RAI	Month	RAI	Month	RAI	Month	RAI	Month	RAI
Jan	−8.95	Jan	−8.21	Jan	−7.32	Jan	−8.78	Jan	−9.78
Feb	−7.02	Feb	−8.35	Feb	−10.25	Feb	−5.48	Feb	−9.25
Mar	−8.39	Mar	−8.28	Mar	−7.66	Mar	−6.96	Mar	−5.29
Apr	−7.88	Apr	4.82	Apr	13.09	Apr	1.96	Apr	−0.97
May	21.22	May	10.38	May	21.40	May	5.61	May	6.48
June	25.84	June	28.88	June	4.42	June	27.50	June	7.04
July	20.14	July	13.04	July	17.58	July	23.41	July	24.06
Aug	6.37	Aug	25.24	Aug	4.75	Aug	11.34	Aug	23.22
Sep	24.39	Sep	−1.38	Sep	6.58	Sep	1.19	Sep	15.58
Oct	−8.36	Oct	−6.45	Oct	−3.89	Oct	−3.28	Oct	−7.57
Nov	−8.96	Nov	−7.29	Nov	−11.65	Nov	−8.44	Nov	−7.53
Dec	−8.95	Dec	−5.70	Dec	−11.44	Dec	−8.90	Dec	−8.07

for only 5 months with positive RAI starting in May and ending in September. The rest of the months with negative RAI were January, February, March, April, October, November, and December, signifying the driest months of the year where the maximum negative anomaly was noticeable in February: −7.02 (rainfall received 29.5). The condition in 2015 depicts a quite changeable scenario. In this year, the rainy months started in April and ended up in August. Here, the driest months with negative RAI were January, February, March, September, October, November, and December, where the maximum positive anomaly was noticeable in the month of June 28.88 (rainfall received, 482.3 mm). In the year 2016, the negative anomalies were observable in the months from January to March, and October to December. Here, the lowest anomaly was spotted in November with −11.65 (rainfall received, 0.1 mm only). Consequently, the humid months continued from April to September (time about 6 months) with maximum humidity in May 21.40 (rainfall received, 295.8 mm). The year of 2017 was observable with 6 months wettest and 6 driest months where the months of April and September show a positive anomaly of 1.96 (rainfall received, 196.4 mm) and 1.19 (rainfall received, 187.6 mm) respectively. It was seen that the month of June carried a positive anomaly of 27.50 (amount of rainfall received, 490.6 mm). In 2018, the driest months were discernible for 7 months, from January to April and October to December. In it the maximum negative anomaly was noticeable in January, 9.78 (rainfall received, 0.0 mm) and the month with highest positive anomaly was perceived in July, 24.06 (rainfall received, 343.2 mm) (Fig. 17.2).

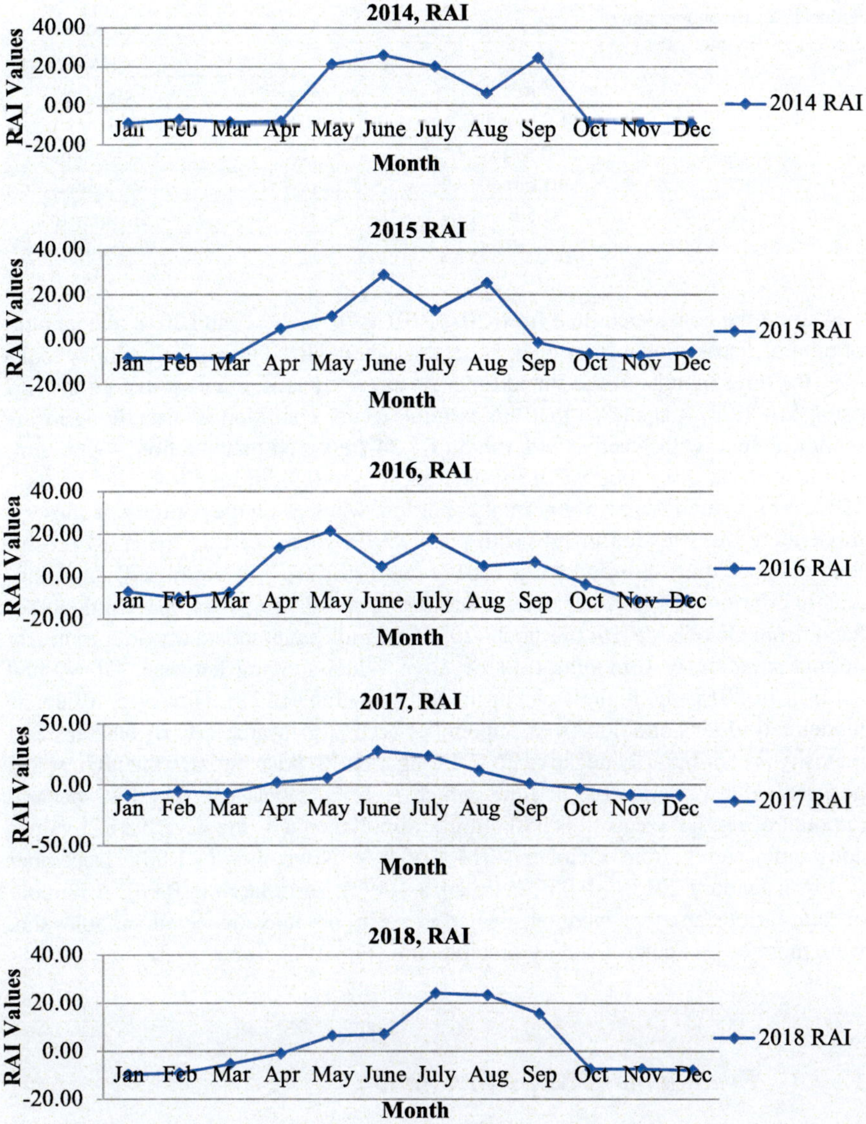

Fig. 17.2 Representation of annual rainfall peculiarity (2014–2018) using the Rainfall Anomaly Index (RAI)

17.4.2 Assessment of Drought Duration Using Standardized Precipitation Index (SPI)

For assessing the drought event of Sonitpur district, the categorization of SPI value resembles Table 17.2.

Table 17.2 Categorization of SPI values (Pramudya et al., 2019)

SPI values	Categories
2.0+	Extremely wet
1.5 to 1.99	Very wet
1.0 to 1.49	Moderately wet
−0.99 to 0.99	Near normal
−1.0 to −1.49	Moderately dry
−1.5 to −1.49	Severely dry
−2 or less	Extremely dry

During the considered time from 2014, 2015, 2016, 2017, and 2018, the amount of rainfall distribution is dissimilar. Here, this Standardized Precipitation Index looks over the drier months of the stated time. At the study area, after estimating the SPI using Eq. 17.3, it is shown that the extreme wetter condition is discernible in the month of June with positive SPI value of 2.34 (received precipitation, 490.6 mm) during the year 2017, but in 2018 it decreases up to 0.33 in June (received rainfall, 195.2 mm), which is a nearly normal condition, whereas on the contrary, it shows a moderately wet situation in July with positive SPI values of 1.33 (received precipitation, 343.2 mm). From October (2017) to April (2018), it portrays 7 long moderately drier months (SPI negative values between −0.95 and −1.00). On the other hand, from October (2016) to March (2017) are indicated moderately drier to nearly normal situation of 6 months (SPI negative values ranging between −0. 99 and −1.00). In 2015, the highest pickup of SPI is in June (2.28). However, it depicts moderately drier condition from September (2015) to March (2016) whereas SPI intensity is continuously negative up to 7 long months. Also, in 2014 the highest SPI positive value is noticeable in June, which is 1.52, which is a very wet scenario (amount of rainfall received is 370.3 mm). Simultaneously, the negative SPI values constantly varies from October 2014 (−0.94); November (−1.00); December (−1.00); January 2015 (−0.95); February (−0.97); and March (−0.96). A number of multifarious analyses were terminated to verify the interconnected variables that were more or less determined by precipitation (Fig. 17.3, Table 17.3).

17.4.3 Evaluation of Slope Inclination

Slope is the perpendicular digress of a line, basically indicating the steepness, rise, or fall of a surface (Wang et al., 2019), as different objects behave distinctly in different wavelengths. Similarly, each soil texture has contrasting characteristics, because from the very beginning of years before topography is significant in predicting soil texture and distribution of flora (Moeslund et al., 2013). In the current study, the gradient of the slope (Fig. 17.4.) is recognized by observing the slope inclination in degrees: that is, <2° denotes very gentle; 2°–3° indicates gentle slope; 3°–5° is considered as a moderate slope; and the last two angles as moderately steep slope (5°–14°), and >53° (as steep slope). The linear areas in the north of Biswanath,

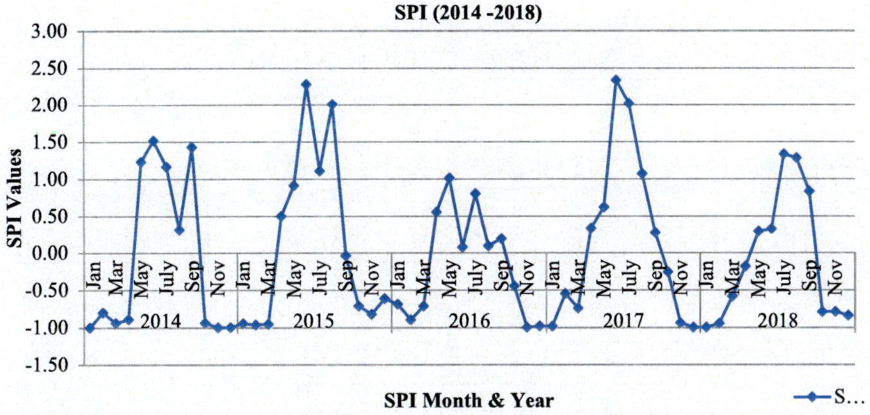

Fig. 17.3 Representation of drought duration (2014–2018) using the Standardized Precipitation Index (SPI)

Chariduar, and Dhekiajuli possess a steep slope (>53°); some moderately steep slopes can be noticed in Gohpur, the middle part of Biswanath, and Helem. Even so, a condition with moderate terrain is generally absent. Furthermore, the sub-divisions of Dhekiajuli, Chariduar, Tezpur, Na-Duar, and the southeastern part of Gohpur are characterized with a gentle slope having an elevation of 2°–3°. The southeastern part of Sonitpur where the mighty River Brahmaputra flows possesses a perfectly gentle slope with elevation less than 2°.

17.4.4 Investigation of Relationship Among Topographic Wetness, Slope, and Soil Texture

The nature of the terrain is an essential variable in determining soil moisture (Mandal, 2015). The results obtained using (Eq. 17.4) elucidate that increase in topographic wetness value leads to gentle slope angle whereas, decrease in topographic value portrays opposite of it. The attainment of greater TWI values affirms a large amount of soil moisture whereas smaller TWI values denotes a lesser amount of soil moisture. A topography with a steep slope gradient cannot be good in soil water percolation. The areas of Biswanath, Helem, and northern parts of Dhekiajuli, Chariduar possess a topographic wetness value of 3–5 m. In the study area, the predominance of moderate elevation is more prevalent with 7–9 m of terrain wetness values. Moreover, a sufficient amount of topographic wetness value 9–13 m and 13–21 m are recognized in the lower section of Dhekiajuli, Tezpur, Na-Duar, Biswanath, and Helem. These particular areas were specified with a moderately gentle to gentle gradient. A linear gentle slant in the south–north direction is perceptible in Na-Duar where the tributary Jia-Bharali flows vertically. The areas where gentle or moderately gentle slopes are the principal controlling variable have

Table 17.3 Evaluated values of standardized precipitation index (API) (2014–2018)

Year	Month	SPI	Year	Month	SPI	Year	Month	SPI	Year	Month	SPI	Year	Month	SPI
2014	Jan	−1.00	2015	Jan	−0.95	2016	Jan	−0.69	2017	Jan	−0.99	2018	Jan	−1.00
	Feb	−0.80		Feb	−0.97		Feb	−0.90		Feb	−0.55		Feb	−0.95
	Mar	−0.94		Mar	−0.96		Mar	−0.71		Mar	−0.74		Mar	−0.58
	Apr	−0.89		Apr	0.50		Apr	0.55		Apr	0.34		Apr	−0.18
	May	1.23		May	0.91		May	1.01		May	0.62		May	0.29
	June	1.52		June	2.28		June	0.08		June	2.34		June	0.33
	July	1.17		July	1.11		July	0.80		July	2.02		July	1.33
	Aug	0.32		Aug	2.01		Aug	0.10		Aug	1.07		Aug	1.29
	Sep	1.43		Sep	−0.04		Sep	0.20		Sep	0.28		Sep	0.83
	Oct	−0.94		Oct	−0.71		Oct	−0.44		Oct	−0.25		Oct	−0.79
	Nov	−1.00		Nov	−0.82		Nov	−1.00		Nov	−0.94		Nov	−0.79
	Dec	−1.00		Dec	−0.61		Dec	−0.98		Dec	−1.00		Dec	−0.84

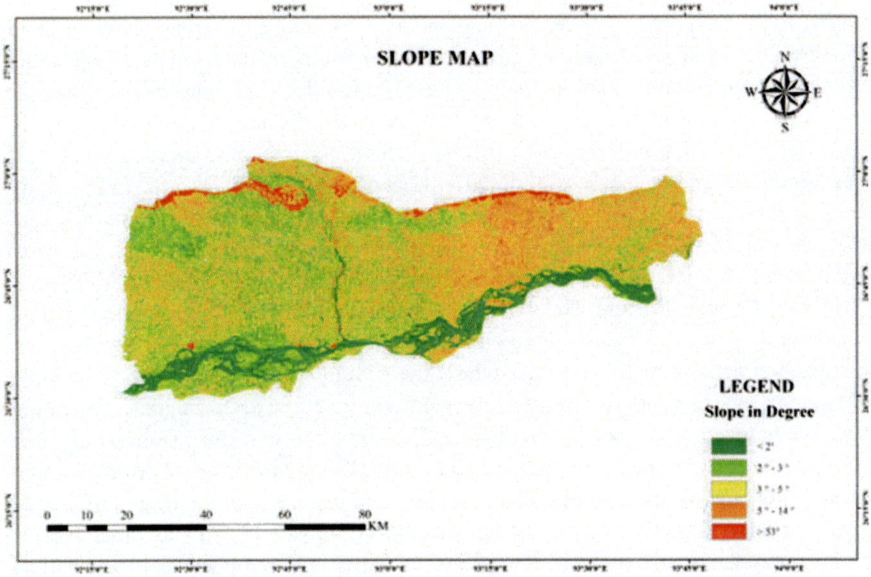

Fig. 17.4 Slope map of the study area

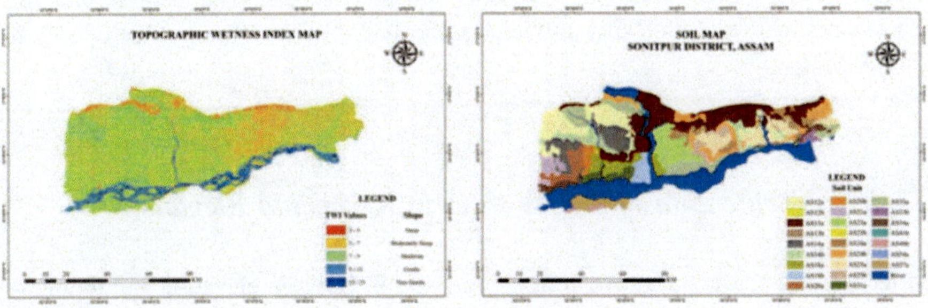

Fig. 17.5 Topographic wetness index map with a soil texture map of Sonitpur district

adequate water percolation. The soil texture of the present study area varies from coarse loamy to fine silt. Basically, the coarse loamy texture is more usual. This type of soil texture holds a small amount of moisture in soils with a greater variance of sand particles. The central part of Biswanath is composed of silty textured soil characterized with a wetness value of 3–6 m, whereas the upper part of Biswanath and Na-Duar, the southeastern parts of Chariduar, are more prominent to coarse loamy texture having a sandy appearance. The northwestern part of Dhekiajuli represents severe erosion with a loamy surface texture. The central area of Na-Duar is marked with a silty surface texture that holds a wetness value of 5–7 m (Fig. 17.5.).

The nature of the terrain is an essential variable in determining soil moisture (Mandal, 2015). The results obtained using Eq. 17.4 elucidate that an increase in topographic wetness value leads to a gentle slope angle whereas a decrease in topographic value portrays the opposite. The attainment of a greater TWI value affirms a large amount of soil moisture whereas a smaller TWI value denotes a lesser amount of soil moisture. Topography with a steep slope gradient cannot be good for soil water percolation. The areas of Biswanath, Helem, and northern parts of Dhekiajuli, Chariduar possess a topographic wetness value of 3–5 m. In the study area, the predominance of moderate elevation is more prevalent with 7–9 m of terrain wetness values. Moreover, a sufficient amount of topographic wetness values at 9–13 m and 13–21 m are recognized in the lower section of Dhekiajuli, Tezpur, Na-Duar, Biswanath, and Helem. These particular areas were specified with a moderately gentle to gentle gradient. A linear gentle slant in the south–north direction is perceptible in Na-Duar where the tributary Jia-Bharali flows vertically. The areas where gentle or moderately gentle slopes are the principal controlling variable have adequate water percolation. The soil texture of the present study area differs from coarse loamy to fine silt. Basically, the coarse loamy texture is more usual. This type of soil texture holds little moisture in soils with a greater variance of sand particles. The central part of Biswanath, composed of silty textured soil, is characterized with a wetness value of 3–6 m whereas the upper part of Biswanath and Na-Duar, southeastern parts of Chariduar, are more prominent with a coarse loamy texture having a sandy appearance. The northwestern part of Dhekiajuli represents severe erosion with a loamy surface texture. The central area of Na-Duar is marked with a silty surface texture that holds a wetness value of 5–7 m (Table 17.4.).

17.4.5 Verification of TWI Values in Respect to Elevation

In the present study a negative correlation (−0.28) between elevation and topography wetness is seen. Both the variables were indirectly proportional to each other with regression value of 0.0775. As the elevation increases, the wetness of the topography decreases. The highest elevated value noticeable was 295 m (where visible wetness value is 6.19), followed by 193 m (wetness value. 5.83) and 186 m (wetness value. 7.88). In contrast to these, the areas with low elevation depict higher wetness in topography. Accordingly, the areas having elevation of 47 m show wetness of 8.88 followed by 50 m (wetness value, 10.98–13.08). So, the areas with high altitude resist less moisture, leading to maximum evaporation, while the areas having less altitude can retain moisture with less evaporation (Table 17.5, Fig. 17.6).

Table 17.4 Soil classification and their characteristics (National Bureau of Soil Survey, Jorhat, Assam)

Soil unit	Taxonomy	Drainage	Particle size	Surface texture	Slope
AS12a	Loamy skeletal	Well	Loamy skeletal	Loamy	Gently
AS12b	Coarse loamy	Well	Coarse loamy	Loamy	Moderate
AS13a	Coarse-loamy	Well	Coarse loamy	Loamy	Gently
AS13b	Coarse-silty	Well	Coarse loamy	Loamy	Gently
AS14a	Coarse-silty	Well	Coarse loamy	Loamy	Gently
AS14b	Fine-loamy	Imperfectly	Fine	Loamy	Gently
AS18a	Fine-loamy	Well	Fine	Loamy	Gently
AS18b	Coarse-loamy	Mod. well	Coarse loamy	Loamy	Gently
AS20a	Fine-loamy	Well	Fine	Loamy	Very gently
As20b	Coarse-loamy	Imperfectly	Coarse loamy	Loamy	Very gently
AS21a	Fine-loamy	Poorly	Fine	Loamy	Gently
AS23a	Fine-silty	Poorly	Fine	Loamy	Very gently
AS23b	Coarse-loamy	Imperfectly	Sandy	Sandy	Nearly level
AS24a	Fine-loamy	Mod. well	Fine	Loamy	Very gently
AS24b	Fine-silty	Well	Fine	Silty	
AS25a	Coarse-loamy	Well	Coarse loamy	Loamy	Gently
AS25b	Coarse-loamy	Imperfectly	Coarse loamy	Loamy	Very gently
AS31a	Fine-silty	Imperfectly	Fine-silty	Silty	Nearly level
AS33a	Typic Udipsamments	Mod. well	Sandy	Loamy	Level
AS33b	Coarse-loamy	Well	Coarse loamy	Loamy	Nearly level
AS34a	Coarse-loamy	Well	Coarse loamy	Loamy	Very gently
AS41b	Clayey	Well	Clayey	Clayey	
AS48b	Coarse-loamy	Mod. well	Coarse loamy	Loamy	Gently
AS54a	Coarse-silty	Well	Coarse loamy	Loamy	Nearly level
AS57a	Fine-loamy	Well	Fine	Loamy	Nearly level

17.4.6 Examination of Crop Water Stress

In this part, using Eq. 17.5 for investigating crop water stress, it discloses that in 2014 (pre-monsoon) the area coverage under moderate plant response to water stress (CWSI value between 0.42 and 0.50) consists of 2078 km^2; at the same time, very high plant response is perceived by 543 km^2; and very low plant response to water stress is received by 483 km^2. In the same year, during post-monsoon the picture is very low: a response is observed by only 107 km^2 while very high and moderately

Table 17.5 Topographic wetness values in accordance to elevation

Elevation	TWI	Elevation	TWI	Elevation	TWI	Elevation	TWI
174	8.4	66	6.37	57	8.62	56	7.33
186	7.88	78	8.1	52	8.36	50	12.85
135	6.19	61	6.07	50	13.08	65	7.2
132	8.42	59	6.98	58	7.96	68	11.89
193	5.83	62	9.29	78	7.79	73	5.53
297	6.19	83	6.5	50	10.98	72	10.98
105	6.64	69	7.04	50	6.15	61	6.24
92	7.68	97	6.87	55	7.63	84	6.88
81	8.23	55	5.91	47	8.88		

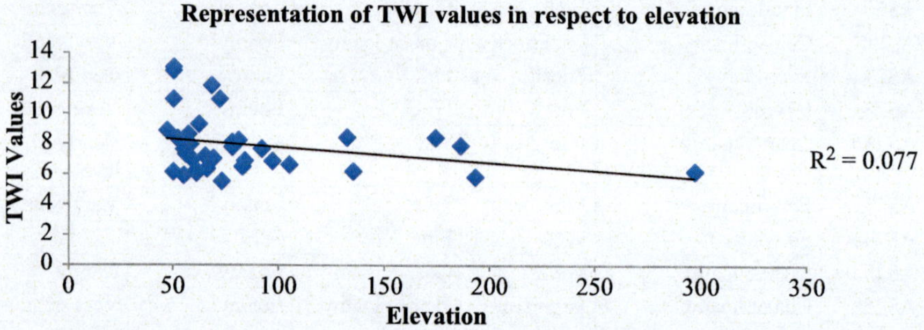

Fig. 17.6 Representation of TWI values in respect to elevation

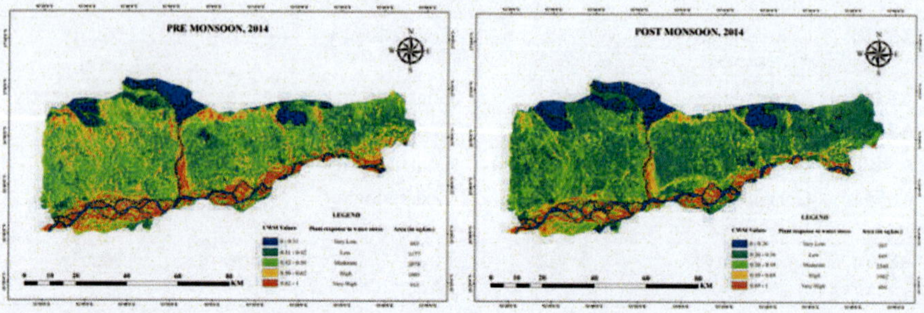

Fig. 17.7. Pre- and post-monsoonal map of Crop Water Stress Index (2014)

high plant response to water stress is perceived by about 404 and 1982 km^2, respectively. The northern parts of Dhekiajuli, Chariduar, and Biswanath are all non-water stress areas (Fig. 17.7).

In 2015 (pre-monsoon) the areas under low plant response to water stress are maximum (2193 km^2; CWSI values ranged between 0.27 and 0.38, which is visible in the middle portions of Gohpur, Helem, Dhekiajuli, and Na-Duar. Areas having very high plant response to stress cover 442 km^2, observed in the southern parts of

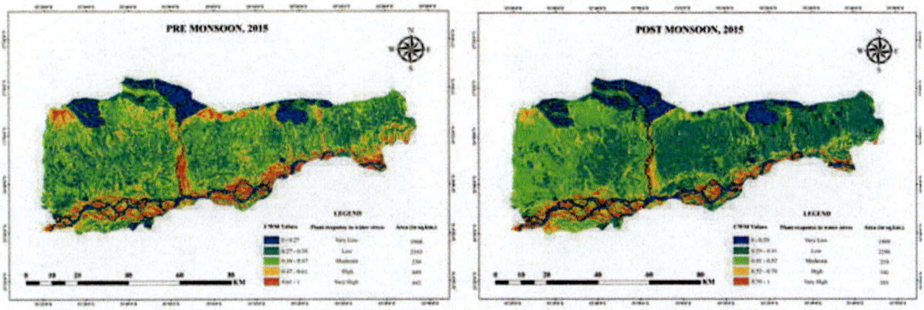

Fig. 17.8 Pre- and post-monsoonal map of Crop Water Stress Index (2015)

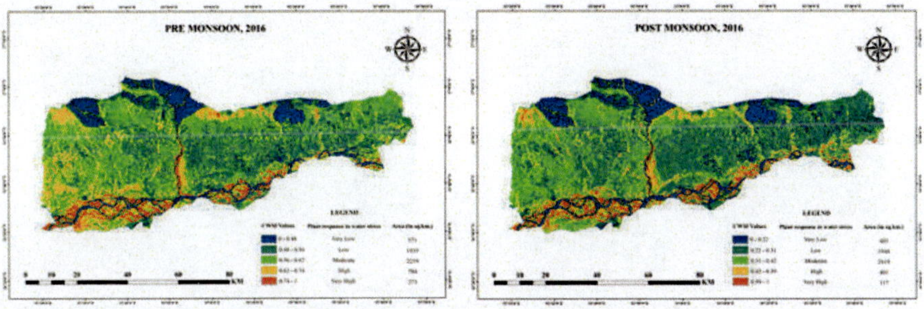

Fig. 17.9 Pre- and post-monsoonal map of Crop Water Stress Index (2016)

Dhekiajuli, Tezpur, Na-Duar, Biswanath, Helem; and the northern corner of Dhekiajuli and Na-Duar. In the post monsoon season, the area coverage under very low and moderately low plant response are maximum, at 1999 and 2296 km^2, respectively (Fig. 17.8.). Here, most of the parts of Helem, Gohpur, Biswanath, Tezpur, and the southern and northern parts of Chariduar, are in this category (CWSI varies between 0 and 0.41).

In the year 2016 (pre-monsoon), areas covered under moderate plant response to water stress are maximum (2219 km^2) and very high water stress is faced by only 273 km^2 (where CWSI ranged between 0.74 and 1). Moreover, the northwestern and southern parts of Dhekiajuli and northern parts of Biswanath featured a high plant response to water stress (784 km^2). On the other hand, in post-monsoon, the middle parts of Biswanath, Helemand Gohpur, are dominated by low plant response for about 1940 km^2 with CWSI value 0.22–0.31; areas under very high water stress are minimum, about 117 km^2 (Fig. 17.9).

During 2017, in pre-monsoon the maximum areas are under moderate water stress (2431 km^2) and the area under minimum plant water stress is only 170 km^2 (CWSI values ranged between 0 and 0.47). Basically, some patches of Dhekiajuli, Tezpur, Biswanath, and Helem are under high water stress (Fig. 17.10) for about 1608 km^2 (values ranged between 0.65 and 0.76). Now in post-monsoon the areas under very low response to water stress are only 18 km^2, and the values ranged between 0 and 0.27.

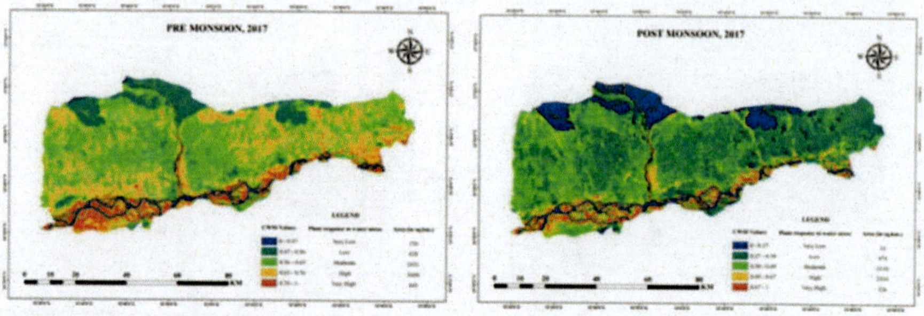

Fig. 17.10 Pre- and post-monsoonal map of crop water stress index (2017)

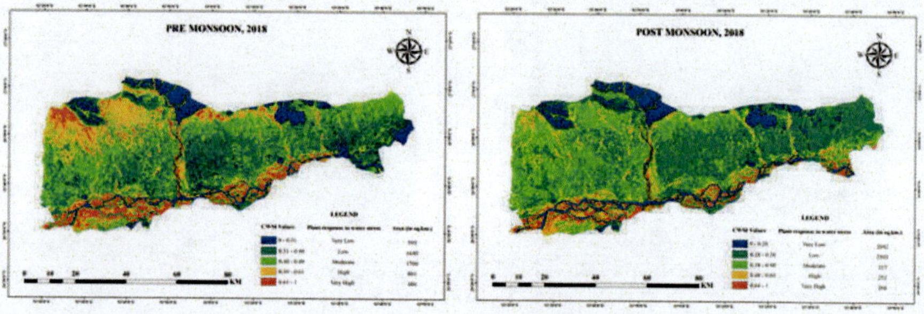

Fig. 17.11 Pre- and post-monsoonal map of crop water stress index (2018)

According to the layout depicted in 2018 (pre-monsoon) the areas under very high plant response to water stress are minimum (486 km^2), and some portion of this category is noticeable in the northern corner of Dhekiajuli. With it, the northern part of Dhekiajuli and Chariduar also indicates high plant response to water stress (values ranged between 0.49 and 0.61). The eastern section of Gohpur, southern part of Helem, northeastern part of Biswanath, Chariduar shows very low plant response to water stress (595 km^2) with CWSI values 0 and 0.31. Concurrently, in post-monsoon portions of Dhekiajuli, Chariduar, and Tezpur were characterized with moderate plant response to water stress (217 km^2), which is about 1700 km^2 during pre-monsoon. Moreover, the area coverage under very low and low plant response are 2042 (values, 0–0.28) and 2503 (values 0.28–0.38) km^2, respectively. The lowest area coverage, 252 km^2, had a high plant response with CWSI values between 0.48 and 0.64 (Fig. 17.11.)

17.4.7 Examining Seasonal Crop Water Stress Values with 30 Reference Points

As a last step toward the analysis, here about 30 points were determined for examining the seasonal (pre- and post-monsoon) variations in crop water stress.

According to the observed points of 2014, it can be stated that the fluctuations of CWS (crop water stress) were greater on reference point 20 (0.22 in March and 0.75 in November), 21 (March 0.48 and November 0.12), and 28 (March 0.69 and November 0.15). These variations were extremely less on point number 29 (March 0.72 and November 0.69) and 30 (March 0.44 and November 0.40). The values are supposed to be uniform between point 6 and point 19. In point number 20, its pre-monsoon crop water stress value seems to be less than that post-monsoon. During 2015, the greater variation in crop water stress was observed in reference number 29 (pre-monsoon 0.19 and post-monsoon 0.83), followed by point 22 (pre-monsoon 0.11 and 0.47 during post-monsoon). In point 21 the variation was 0.67 in March and 0.37 in November. The variance in stress values was small on points numbered 6, 8, 9, 11, 12, 15, and 19. The seasonal crop water stress varied more in 2016; the 16 observed points with greater fluctuations were 1, 4, 5, 9, 10, 11, 12, 14, 15, 16, 18, 19, 20, 21, 22, and 23. Here, the point with very little variance was 28. In the year 2017, very little disparity was noticeable in reference number 19 (March 0.48 and November 0.41), and 28 (March 0.89 and November 0.84). Here, the rest of the points showed a sizeable variance. In 2018, the points 4, 12, 14, 19, 22, and 29 showed scanty variance between the seasons although the points 23, 25, 27, and 30 depict greater fluctuation (Fig. 17.12.).

17.4.8 Verification of Topographic Wetness with Seasonal Crop Water Stress Through Scatter Plots

In this category, the relationship of topographic wetness with crop water stress (pre- and post-monsoon) was verified through regression analysis. The results obtained after the investigation were an increase in topographic wetness, leading to decrease in crop water stress. As per the graphs represented for the different years 2014, 2015, 2016, 2017, and 2018, these results imply that the points with maximum wetness show less water stress whereas the points with minimum wetness indicate more water stress in the crop. In the year 2014, a positive relationship (0.26) is observed in pre-monsoon whereas in post-monsoon a negative relationship (−0.33) is seen. A negative correlation of topographic wetness with both seasons existed in 2015 (March, −0.15; November, −0.21). A similar scenario was monitored in 2017 also where a positive (0.41) correlation existed in March and a negative (−0.10) correlation post-monsoon. The other 2 years, 2016 and 2018, determine a negative correlation in both seasons with topographic moisture in 2018, March (−0.08) and November (−0.05), and in 2016, March (−0.09) and November (−0.06). Sometimes the value of r^2 becomes small: here the variables that act as an intervening factor were the rainfall differences between two seasons (pre- and post-monsoon), variations in crop water stress (Fig. 17.13), and structure of the topography (as different topographic patterns can hold a differential degree of moisture).

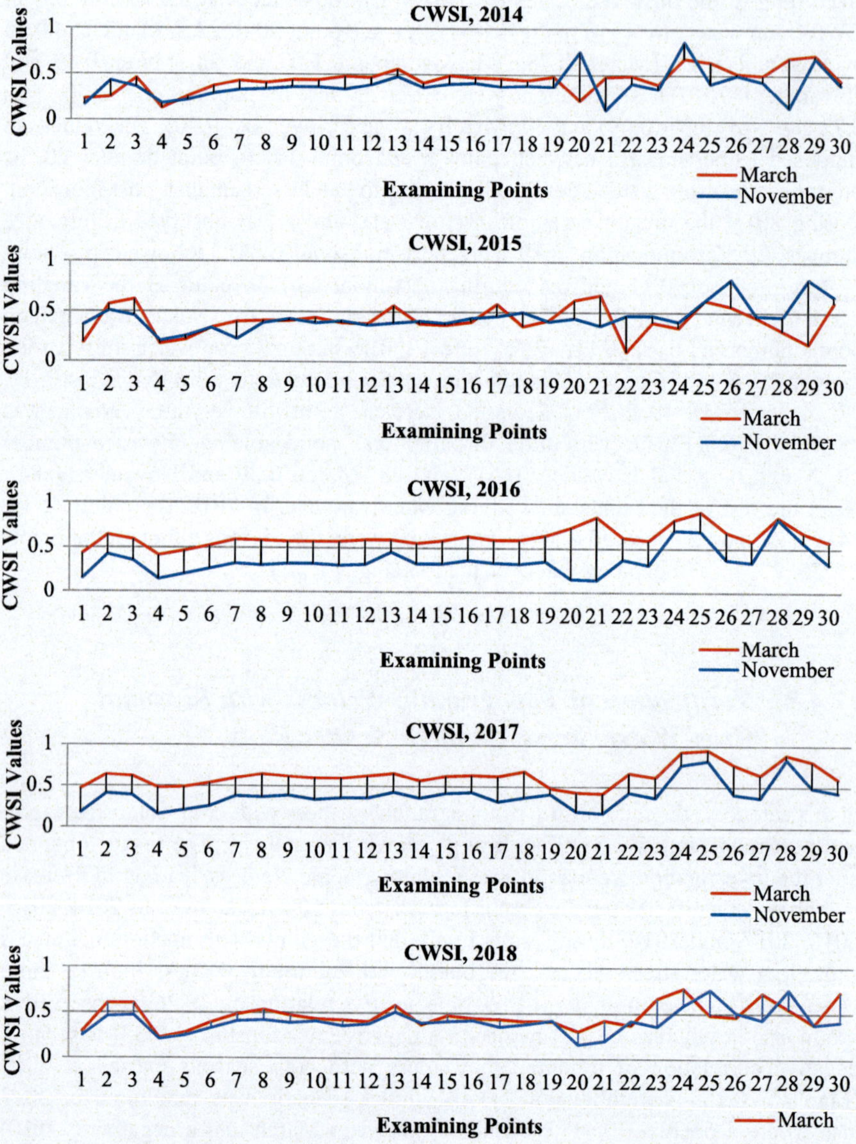

Fig. 17.12 Comparative representation of seasonal crop water stress (2014–2018)

17.5 Conclusion

This study displayed a clear and precise picture of geographic features in the Sonitpur district of Assam. An area with shortage of water for about 5 months or more is simply determined as a drought condition. After the utilization of two

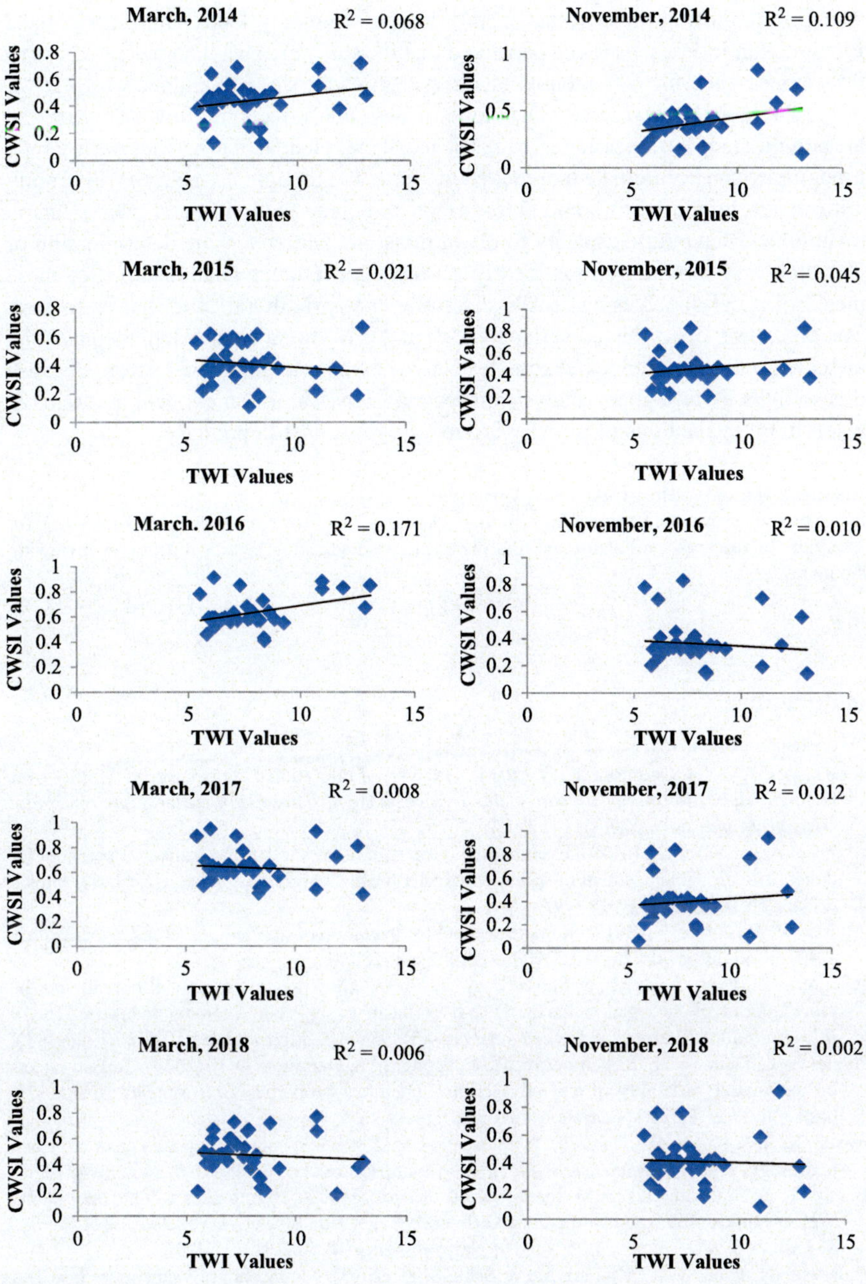

Fig. 17.13 Validation of topographic moisture with seasonal crop water stress (2014–2018)

drought-monitoring indices, that is, the Rainfall Anomaly Index and Standardized Precipitation Index, it has been perceived that the study area was characterized with a drier appearance for 6–7 months of interval in each of the examined years 2014, 2015, 2016, 2017, and 2018. The slope of any area affects the flow of water: the areas with steeper slopes failed to resist moisture whereas the patches with gentle topography can accumulate more moisture. Even the topography behaves differently with manifold soil texture and slope. Here, the areas with silty and coarse loamy textured soil have little capacity to retain moisture. Moreover, the determination of elevation and terrain wetness clearly portrays an indirect relationship. So, these meteorological variables can easily determine the agricultural condition of an area. The prediction of correlated variables TWI and CWSI to a great extent displays that an increase in topographic moisture leads to a decrease in crop water stress. Because this study is a preliminary analysis on secondary bases, for an eminent and reliable result more evaluations of primary ground are of utmost importance.

Acknowledgments This work is fully organized with some valuable support. The authors would like to express gratitude toward Juri Baruah, Assistant Professor of Dibrugarh University for assisting the research; and also thank the reviewers for dedicating significant time in reviewing the manuscript.

References

Akinsanola, A. A., & Ogunjobi, K. O. (2014). Analysis of rainfall and temperature variability over Nigeria. *Global Journal of Human Social Science: B Geography, Geo-Sciences, Environmental Disaster Management, 14*(3), 1–17.

Alghory, A., & Yazar, A. (2019). Evaluation of crop water stress index and leaf water potential for deficit irrigation management of sprinkler irrigated water. *Irrigation Science, 37*, 61–77. https://doi.org/10.1007/s00271-018-0603-y

Alvino, A., & Marino, S. (2017). Remote sensing for irrigation of horticultural crops. *Horticulture, 3*(40), 1–36. https://doi.org/10.3390/horticulturae3020040

Araújo, L. E., Neto, J. M. M., & Sousa, F. A. S. (2009). Classification of annual rainfall and the rainy quarter of the year in the Paraíba river basin using Rain Anomaly Index (RAI). *An Interdisciplinary Journal of Applied Science, 4*(3), 93–110. https://doi.org/10.4136/1980-993X

Ayele, G. T., Demissie, S. S., Jemberrie, M. A., Jeong, J., & Hamilton, D. P. (2019). Terrain effects on the spatial variability of soil physical and chemical properties. *Soil Systems, 4*(1), 1–21. https://doi.org/10.3390/soilsystems4010001

Beven, K. J., & Kirkby, M. J. (1979). A physically based, variable contributing area model of basin hydrology. *Hydrology Science B, 24*(1), 43–69. https://doi.org/10.1080/02626667909491834

Bisrat, E., & Berhanu, B. (2018). Identification of surface water storing sites using Topographic Wetness Index (TWI) and Normalized Difference Vegetation Index (NDVI). *Journal of Natural Resources and Development, 8*, 91–100. https://doi.org/10.5027/jnrd.v8i0.09

Dangwal, N., Patel, N. R., Kumari, M., & Saha, S. K. (2015). Monitoring of water stress in wheat using multispectral indices derived from Landsat – TM. *Geocarto International, 31*(6), 682–693. https://doi.org/10.1080/10106049.2015.1073369

Do, H.-T., Limet, S., & Melin, E. (2011). Parallel computing flow accumulation in large digital elevation models. *Procedia Computer Sciences, 00*, 1–10.

Dutta, J., Chetia, M., Sarmah Baruah, J. P., & Misra, A. K. (2010). Assessment of drinking water quality in different small tea gardens of Sonitpur District (Assam), India. *Archives of Applied Science Research, 2*(5), 226–238.

Dyer, J. M. (2009). Assessing topographic patterns in moisture use and stress using a water balance approach. *Landscape Ecology, 24*, 391–403.

El-Shirbeny, A., & Abutaleb, K. (2017). Sentinel-1 Radar Data assessment to estimate crop water stress. *World Journal of Engineering and Technology, 5*, 47–55. https://doi.org/10.4236/wjet.2017.52B006

Erdem, T., Erdem, Y., Orta, A. H., & Okursoy, H. (2006). Use of a crop water stress index for scheduling the irrigation of sunflower (Helianthus annuus L.). *Turkish Journal of Agriculture and Forestry, 30*, 11–20.

Freitas, M. A. S., Silveira, P. B. M., & Freitas, G. B. (2019). A resilient drought risk management approach in the semi-arid North East Brazil. *International Journal of Current Research, 11*(9), 6968–6974. https://doi.org/10.24941/ijcr.36577.09-2019

González-Dugo, M. P., Moran, M. S., Mateos, L., & Bryant, R. (2005). Canopy temperature variability as an indicator of crop water stress severity. *Irrigation Sciences, 24*(4), 233–240. https://doi.org/10.1007/s00271-005-0023-7

Hojati, M., & Mokarram, M. (2016). Determination of a topographic wetness index using high resolution Digital Elevation Models. *European Journal of Geography, 7*(4), 41–52.

Ihuoma, S. O., & Madramootoo, C. A. (2017). Recent advances in crop water stress detection. *Computers and Electronics in Agriculture, 141*, 267–275. https://doi.org/10.1016/j.compag.2017.07.026

Jackson, R. D., Idso, S. B., Reginato, R. J., & Pinter, P. J. (1981). Canopy temperature as a crop water stress indicator. *Water Resources Research, 17*(4), 1133–1138. https://doi.org/10.1029/wR017i004p01133

Joshi, S., Kumar, K., Joshi, V., & Pande, B. (2014). Rainfall variability and indices of extreme rainfall-analysis and perception study for two stations over Central Himalaya, India. *Natural Hazards, 72*, 361–374. https://doi.org/10.1007/s11069-013-1012-4

Lama, M., & Bordoloi, R. (2017). Potato productivity and its determinants: A case study in Sonitpur District, Assam. *Social Change and Development, XIV*, 70–81.

Laurent, F., Poccard-Chapuids, R., Plassin, S., & Martinez, G. P. (2017). Soil texture derived from topography in North-Eastern Amazonia. *Journal of Maps, 13*(2), 109–115. https://doi.org/10.1080/17445647.2016.1266524

Li, L., Nielsen, D. C., Yu, Q., Ma, L., & Ahuja, L. R. (2010). Evaluating the crop water stress index and its correlation with latent heat and CO_2 fluxes over winter wheat and maize in the North China plain. *Agricultural Water Management, 97*, 1146–1155. https://doi.org/10.1016/j.agwat.2008.09.015

Li, X., McCarty, G. W., Lang, M., Ducey, T., Hunt, P., & Miller, J. (2018). Topographic and physicochemical controls on soil denitrification in prior converted croplands located on the Delmarva Peninsula, USA. *Geoderma, 309*, 41–49. https://doi.org/10.1016/j.geoderma.2017.09.003

Mandal, S. (2015). Upslope contributing area, topographic wetness and land sliding: A case study of the shivkhola watershed, Darjiling Himalaya. *International Research Journal of Earth Sciences, 3*(7), 23–29.

McKee, T. B., Doesken, N. J., & Kleist, J. (1993). *The relationship of drought frequency and duration to time scale*. In Proceedings of the Eight Conference on Applied Climatology, Anaheim, California, 17–22 January 1993. Boston: American Meteorological Society, pp. 179–184.

Meroni, M., Rembold, F., Fasbender, D., & Vrieling, A. (2017). Evaluation of the Standardized Precipitation Index as an early predictor of seasonal vegetation production anomalies in the Sahel. *Remote Sensing Letters, 8*(4), 301–310. https://doi.org/10.1080/2150704X.2016.1264020

Moeslund, J. E., Arge, L., Bocher, P. K., Dalgaard, T., Odgaard, M. V., Nygaard, B., & Svenning, J.-C. (2013). Topographically controlled soil moisture is the primary driver of local vegetation patterns across a lowland region. *Ecosphere, 4*(7), 1–26. https://doi.org/10.1890/ES13-00134.1

Oborie, E., & Nwankwoala, H. O. (2017). Determination of groundwater flow direction in Yenagoa, Bayelsa state, Nigeria. *Journal of Scientific Achievements, 2*(9), 23–27.

Prada-Ramallal, G., Roque, F., Teresa Herdeiro, M., Takkouche, B., & Figueiras, A. (2018). Primary versus secondary source of data in observational studies and heterogeneity in meta-analyses of drug effects: A survey of major medical journals. *BMC Medical Research Methodology, 18*(97), 1–14. https://doi.org/10.1186/s12874-018-0561-3

Pramudya, Y., Onishi, T., Senge, M., Hiramatsu, K., Prasetyo, M. R. N., & Komariah, K. (2019). Evaluation of recent drought conditions by standardized precipitation index and potential evapotranspiration over Indonesia. *Paddy and Water Environment, 17*, 331–338. https://doi.org/10.1007/s10333-019-00728-z

Prueger, J. H., Parry, C. K., Kustas, W. P., Alfieri, J. G., Alsina, M. M., Nieto, H., Wilson, T. G., Hipps, L. E., Anderson, M. C., Hatfield, J. L., Gao, F., Mckee, L. G., McElrone, A., Agam, N., & Los, S. A. (2019). Crop water stress index of an irrigated vineyard in the central valley of California. *Irrigation Science, 37*, 297–313. https://doi.org/10.1007/s00271-018-0598-4

Ratna Raju, C., Yella Reddy, K., Satyanarayana, T. V., & Yogitha, P. (2016). Estimation of crop water requirement using CROPWAT software in Appapuram channel command under Krishna western delta. *International Journal of Agricultural Sciences, 8*(31), 1644–1649.

Rooy, M. P. V. (1965). A Rainfall Anomaly Index (RAI) independent of time and space. *Notos, 14*, 43–48.

Salehnia, N., Alizadeh, A., Sanaeinejad, H., Bannayan, M., Zarrin, A., & Hoogenboom, G. (2017). Estimation of meteorological drought indices based on AgMERRA precipitation data and station observed precipitation data. *Journal of Arid Land, 9*(6), 797–809. https://doi.org/10.1007/s40333-017-0070-y

Saseendran, S. A., Trout, T. J., Ahuja, L. R., Ma, L., McMaster, G. S., Nielsen, D. C., Andales, A. A., Chávez, J. L., & Ham, J. (2015). Quantifying crop water stress factors from soil water measurements in a limited irrigation experiment. *Agricultural Systems, 137*, 191–205. https://doi.org/10.1016/j.agsy.2014.11.005

Shah, R., Bharadiya, N., & Manekar, V. (2015). Drought index computation using Standardized Precipitation Index (SPI) method for Surat District, Gujarat. *Aquatic Procedia, 4*, 1243–1249. https://doi.org/10.1016/j.aqpro.2015.02.162

Srivastava, S., Singh, T. P., Singh, H., Kushwaha, S. P. S., & Roy, P. S. (2002). Assessment of large scale deforestation in Sonitpur district of Assam. *Current Sciences, 82*(12), 1479–1484.

Subash, N., & Ram Mohan, H. S. (2011). Trend detection in rainfall and evaluation of Standardized Precipitation Index as a drought assessment index for rice-wheat productivity over IGR in India. *International Journal of Climatology, 31*, 1694–1709.

Taghvaeian, S., Chávez, J. L., & Hansen, N. C. (2012). Infrared Thermometry to estimate crop water stress index and water use of irrigated maize in Northeastern Colorado. *Remote Sensing, 4*, 3619–3637. https://doi.org/10.3390/rs4113619

Tanriverdi, C., Atilgan, A., Degirmenci, H., & Akyuz, A. (2017). Comparison of crop water stress index (CWSI) and water deficit index (WDI) by using remote sensing (RS). *Polish Academy of Sciences*, 879–894. https://doi.org/10.14597/infraeco.2017.3.1.068

Tarboton, D. G. (1997). A new method for the determination of flow directions and upslope areas in grid digital elevation models. *Water Resources Research, 33*(2), 309–319.

Testi, L., Golhamer, D. A., Iniesta, F., & Salinas, M. (2008). Crop Water Stress Index is a sensitive water stress indicator in Pistachio trees. *Irrigation Science, 26*, 395–405. https://doi.org/10.1007/s00271-008-0104-5

Veysi, S., Ali Naseri, A., Hamzeh, S., & Bartholomeus, H. (2017). A satellite based crop water stress index for irrigation scheduling in sugarcane fields. *Agricultural Water Management, 189*, 70–86. https://doi.org/10.1016/j.agwat.2017.04.016

Wang, S., Zhang, Y., Wu, X., Tian, J., & Jia, H. (2019). Stability of step-shaped Dump slope and Reinforcement optimization analysis of Anti-slide piles. *Journal of Engineering Science and Technology Review, 12*(1), 110–116. https://doi.org/10.25103/jestr.121.13

Yang, T., Ala, M., Zhang, Y., Wu, J., Wang, A., & Guan, D. (2018). Characteristics of soil moisture under different vegetation coverage in Horqin Sandy Land, northern China. *PLoS One, 13*(6), 1–15. https://doi.org/10.1371/journal.pone.0198805

Yuan, G., Luo, Y., Sun, X., & Tang, D. (2004). Evaluation of a crop water stress index for detecting water stress in winter wheat in the North China Plain. *Agricultural Water Management, 64*, 29–40. https://doi.org/10.1016/S0378-3774(03)00193-8

Chapter 18
GIS-Supported Database Towards Transforming Monocropped Areas for Yearlong Cultivation in Paschim Medinipur, West Bengal, India

Pinakesh Das and Subhabrata Panda

18.1 Introduction

Generation of precision database, comprising areal extent of agricultural lands and water bodies, is very important for planning irrigation based on rainwater harvesting for increasing cropping intensity from 100% to 300% (i.e. monocropping to three crops in a year) in a drought-prone area of Paschim Medinipur District in southern part of West Bengal State in India towards increasing food production for achieving Sustainable Development Goal (SDG) No. 2.4 of Zero Hunger of the United Nations by 2030 with single Target Indicator 2.4.1 of proportion of agricultural area under productive and sustainable agriculture (UN SDG, 2020). This can be ensured by planning of irrigation from rainwater harvesting based on hydrological balance approach for improving cropping intensity and, thus, maintaining ecosystem and strengthening capacity for adaptation to climate change, drought, flood and progressively improving land and soil quality.

From the past decades to present times, research results have revealed that based on hydrological balance approach, conservation of rainwater can sufficiently satisfy crop requirements of two more crops during post-monsoon (i.e. *rabi* or winter crops) and pre-monsoon (summer or *pre-kharif*) other than winter rice (i.e. *Aman* or *Kharif*) growing during monsoon months (June to October). In that way, conserving

P. Das
Department of Soil and Water Conservation, Bidhan Chandra Krishi Viswavidyalaya, Mohanpur, Nadia, West Bengal, India

S. Panda (✉)
Department of Soil and Water Conservation, Bidhan Chandra Krishi Viswavidyalaya, Mohanpur, Nadia, West Bengal, India

AICRP on Agroforestry, Bidhan Chandra Krishi Viswavidyalaya, West Bengal, India

G. S. Bhunia et al. (eds.), *Anthropogeomorphology*, Geography of the Physical Environment, https://doi.org/10.1007/978-3-030-77572-8_18

rainwater is potential enough in transforming a monocropped *kharif* rice-growing area (i.e. with 100% cropping intensity) to a yearlong cultivable tract with another two non-monsoon crops through provisioning supply of irrigation for increasing cropping intensity to 300% for that area in southern West Bengal (Panda & De, 1989; Panda et al., 1990; Saren, 2019).

From a field survey, it has been ascertained that harvesting of rainwater in village-level ponds on one-fifth of agricultural area can sufficiently meet up such crop water requirements for raising three crops in that area where Paschim Medinipur District, the present study area, is located (Panda, 2005; Saren, 2019). Now, it has become conventional to excavate rainwater harvesting ponds of about 2.44 m (i.e. 8 ft) depth from surface with areal extent of one-fifth (i.e. 20%) of the cultivable lands.

Such database of cultivable lands and available water bodies forms one of the four core components of Spatial Decision Support System (SDSS) which is the basis of Web GIS-based SDSS to enhance cropping intensity. It is unanimously accepted that GIS is the key software in the database management (Rinner, 2003; Raju, 2004; Bailey, 2005; Choi et al., 2005a, b; Sugumaran & Sugumaran, 2007; Huisman & deBy, 2009) and dialog management functions of SDSS (Piarsa et al., 2012; Peshwa, 2015; Kumar & Babu, 2016; Bui & Pham, 2016; Robin et al., 2019). With the use of Google Map application programming interfaces (APIs) and online Google Internet Services (GIServices) database (here, cultivable lands and water bodies) in the study area are accessible from remote geodata repositories (Keenan & Jankowski, 2019; Lan et al., 2020).

Data required in spatial system is a combination of general data, here, cultivable land and data related to decision-making, here, areas under water bodies as percent of total cultivable lands in the district. The Landsat 8 OLI/TIRS Level 1 satellite images of the study area are required to detect those general data with the help of ERDAS imagine 2015 and ArcGIS 10.8 platform. Such Database Generation will be helpful for Block and District Administrators (i.e. Service providers) and end users, i.e. Farmers (or, Land Holders of Agricultural Plots). Such studies will also be helpful for service providers for assuring irrigation water to uplift cropping intensity from 100% to 300% either from the existing water bodies, or with the required renovation works, if any, or more excavations of new tanks required for that purpose up to one-fifth of total cultivable lands for attaining the objective of transforming a monocropped area to a yearlong cultivable tract only through the utilisation of conserved rainwater.

SDSS applications, in general, involve transformation of large data sets to much smaller output, suited for both clients (i.e. service providers) and server (i.e. base of programming platform) for creation of model. Along with mapping development, latitude- and longitude-based spatial data in combination with those regular data of cultivable lands and water bodies, as analysed in the present studies, can form the Decision Support Database for generation of Web GIS-based SDSS.

18.2 Materials and Methods

18.2.1 Study Area

Paschim Medinipur District (22°57′06.4″N Lat., 87°21′41.0″E Long. to 21°46′06.4″N Lat., 87°24′06.9″E Long. and 22°23′00.0″N Lat., 87°18′15.0″E Long. to 22°23′41.7″N Lat., 87°39′05.7″E Long.) is the selected study area in the southwestern part of West Bengal State in India (Fig. 18.1).

Paschim Medinipur district consists of three administrative subdivisions of Ghatal, Kharagpur and Medinipur Sadar with geographical areas of about 953.09 km^2, 2913.17 km^2 and 2441.50 km^2 respectively (Govt. West Bengal, 2014), with distinct geomorphological differentiations of Chotanagpur Flanks with hills, mounds and rolling lands in the western most portion, the *Rahr* Plain with lateritic uplands in the middle and alluvial plain in the east, having hilly region in the northwest, low basin in south eastern areas and dense dry deciduous forest cover in the western side. Vast expanse of well-cultivated lands of the district is contrasted with the gently undulating lands with ridges of thick cover of *sal* (*Shorea robusta*) trees on the western side (Govt. of West Bengal, 2011).

18.2.2 Data Sources

In the present study, both the primary and secondary data were collected from various sources. Primary data were extracted from Landsat 8 OLI (Operational Land Imager)/ TIRS (Thermal Infrared Sensor) Level 1 satellite images of the study area for the years of 2014 and 2019, and those images were downloaded from USGS earth explorer (Table 18.1) to estimate the areas under different types of lands in Paschim Medinipur district. SRTM (Shuttle Radar Topography Mission) 1 Arc–Second Global satellite images of that district were also downloaded from USGS earth explorer (USGS earth explorer, 2020) to find out the land elevation status in that study area. The secondary data were collected from 2011 Census of India data (Government of India, 2011) of that district and district official website (EGIYE BANGLA, 2020). Additionally, satellite imageries for two different years of studies of 2014 and 2019 were collected for LULC (Land Use Land Cover) information from Google earth pro (Google Earth, 2020) along with District Census Handbook (Government of India, 1971) and Agricultural Contingency Plan for 2011 of Paschim Medinipur district (Government of India, 2011).

18.2.3 Details of Band Structure of the Experiment

Landsat 8 OLI/TIRS Level 1 satellite image consists of 11 spectral bands (Table 18.2) with different wavelengths, out of which Bands 1–7 and 9 were in

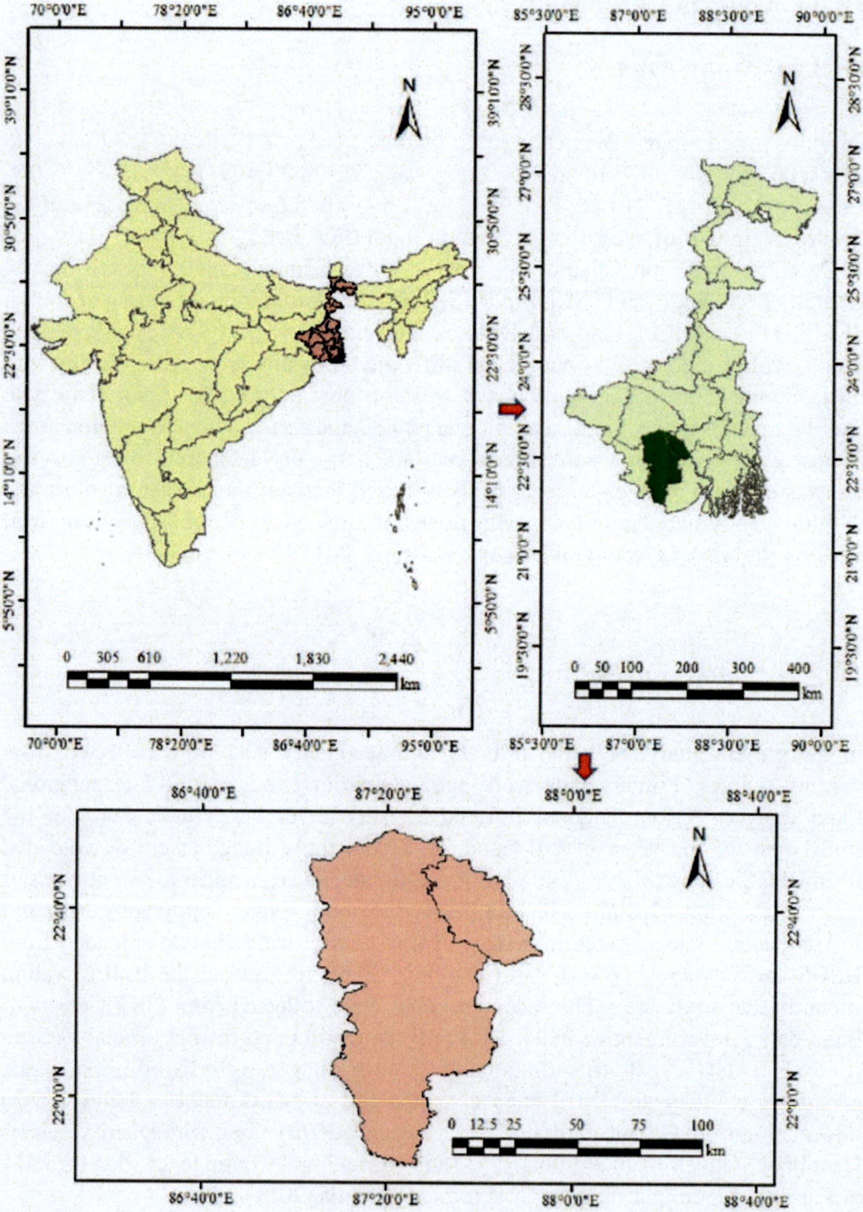

Fig. 18.1 Location map of the study area: Paschim Medinipur District, West Bengal, India

Table 18.1 Details of Landsat data and SRTM data[a]

Date of image	Satellite/sensor	Reference system/Path/Row	Year	Cloud cover (%)
28-03-2014	Landsat 8 OLI/TIRS Level 1	WRS-2 139/44,139/45	2014	<10
27-04-2019	Landsat 8 OLI/TIRS Level 1	WRS-2 139/44,139/45	2019	<10
23-09-2014	SRTM 1 Arc–Second Global	WRS-2 Coordinates: 21,87 22,87	2014	–

[a]Downloaded from USGS Earth Explorer (2020)

Table 18.2 Band structure of Landsat 8 OLI/TIRS Level 1 imagery[a]

Bands	Wavelength (micrometres)	Resolution (meters)
Band 1–coastal aerosol	0.43 – 0.45	30
Band 2–blue	0.45 – 0.51	30
Band 3–green	0.53 – 0.59	30
Band 4–red	0.64 – 0.67	30
Band 5–near infrared (NIR)	0.85 – 0.88	30
Band 6– short-wave infrared (SWIR) 1	1.57 – 1.65	30
Band 7–SWIR 2	2.11 – 2.29	30
Band 8–panchromatic	0.50 – 0.68	15
Band 9–cirrus	1.36 – 1.38	30
Band 10–thermal infrared (TIRS) 1	10.6 – 11.19	100
Band 11–thermal infrared (TIRS) 2	11.50 – 12.51	100

[a]Sources USGS (2020)

30-meter resolution, band 8 (panchromatic) in 15-meter resolution and bands 10 and 11 thermal sensors were useful for surface temperature analysis, as collected from 100-meter resolution (USGS, 2020). In this experiment, bands (1–7 and 9) in 30-meter resolution were selected for creating a composite image and obtaining a better spectral reflectance.

18.2.4 Satellite Data and Image Processing Operation

The satellite imageries of 2014 and 2019 were geometrically corrected in the Datum WGS84 (World Geodetic System 1984) and projected coordinate system UTM (Universal Transverse Mercator) zone N45 (Bhunia et al., 2012). Some image improved techniques were applied to fix atmospheric error correction to obtain improved image quality. The different bands of Landsat 8 images (coastal aerosol, blue, green, red, NIR, SWIR 1, SWIR 2 and Cirrus) having 30 m resolution were selected for layer stacking to get composite images, followed by mosaicking of different scenes to select geometry-based seamline generation option to make a single image covering the study area, i.e. Paschim Medinipur district, using ERDAS IMAGINE software (Dolui et al., 2014).

18.2.5 Creation of AOI (Area of Interest) Layer and Subsets

Extraction of the study area was done for creation of AOI layers in vector format for utilising the help of ERDAS imagine software, followed by the process of sub-setting to extract the study area from mosaicking image (Dolui et al., 2014).

18.2.6 Classification of Images

Supervised classification method, based on maximum livelihood (parametric rule) algorithm, was followed to classify the pre-processed images on training sites for both the years under study (Basu & Saha, 2018). Then the signatures of infinite numbers, based on their DN (digital number) values and spectral information of individual classes or categories, were collected and followed by merging of those signatures classified in each class. For LULC classification of study area, various LULC features like dense forest, arable land, water body, lateritic land, agriculture land, sand dune, etc. were selected. Then areas (in ha) of all individual classes were calculated from Landsat 8 imagery of Paschim Medinipur district.

The NDVI (Normalized Difference Vegetation Index) is most widely used to detect clear view of healthy vegetation of forest cover and other vegetations in the study area from satellite imageries (Myneni et al., 1995). Before LULC classification, NDVI values were assessed for both the years of 2014 and 2019 satellite images. The NDVI images could also be classified and estimated from visible (RED $\equiv 0.64 - 0.67\ \mu\text{m}$) and near-infrared (i.e. NIR $\equiv 0.85 - 0.88\ \mu\text{m}$) by using following formula (Dolui et al., 2014).

$$\text{NDVI} = (\text{NIR} - \text{RED})/(\text{NIR} + \text{RED}) \tag{18.1}$$

18.2.7 Digital Elevation Model (DEM)

From the shuttle radar topography mission (SRTM) of NASA's Jet Propulsion Laboratory (JPL) (NASA, 2020), SRTM 1 Arc-Second Global (30 meters) digital elevation model of the present study area was collected from USGS Earth explorer (Table 18.1) for assessing land variability status of Paschim Medinipur district and each subdivision with the help of ArcGIS 10.8 platform (USGS EROS Archive, 2020; Bhunia et al., 2012).

After completion of image classification (Fig. 18.2), accuracy assessments of both the classified maps were performed with the help of ERDAS Imagine software accuracy assessment tool by creating 50 random points of pixels from each resulting map and checking their labels against classes determined from reference data

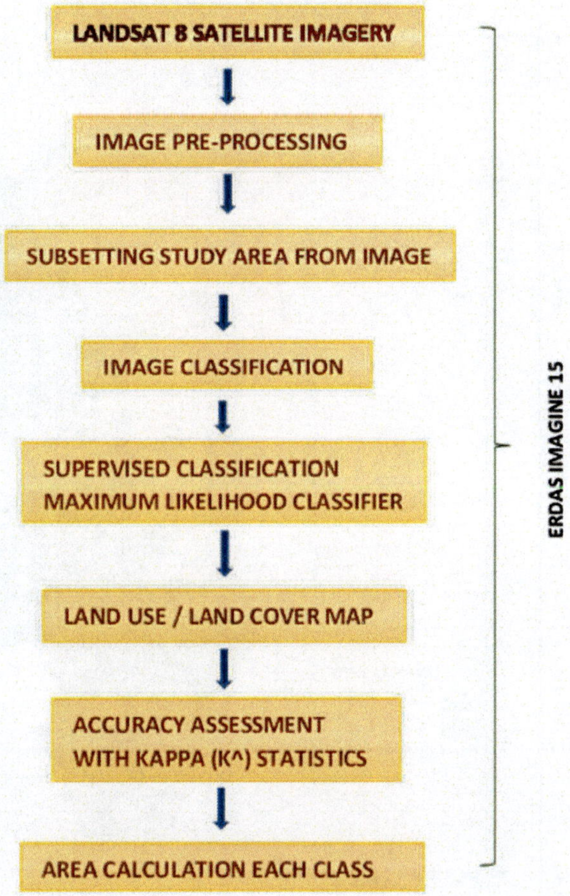

Fig. 18.2 Flow chart on techniques of land use land cover (LULC) classification

(Bhunia et al., 2012). Finally, accuracy of Paschim Medinipur district maps for 2014 and 2019 was checked on the basis of Kappa coefficient (Congalton & Green, 1999; Hudson & Ramm, 1987).

18.3 Results and Discussion

The DEM (digital elevation model) analysis (Fig. 18.3) shows the general land slope in Paschim Medinipur is northwestern to southeastern direction. Most of the low lands are found in eastern sides, whereas the western portion is highly elevated (Fig. 18.3). That is also evident from DEM maps of all three subdivisions of Ghatal, Kharagpur and Medinipur Sadar (Figs. 18.4, 18.5, and 18.6).

In Ghatal and Medinipur Sadar subdivisions (Figs. 18.4 and 18.6), most of the low lands are found towards southeastern side. In Kharagpur, low lands are mostly

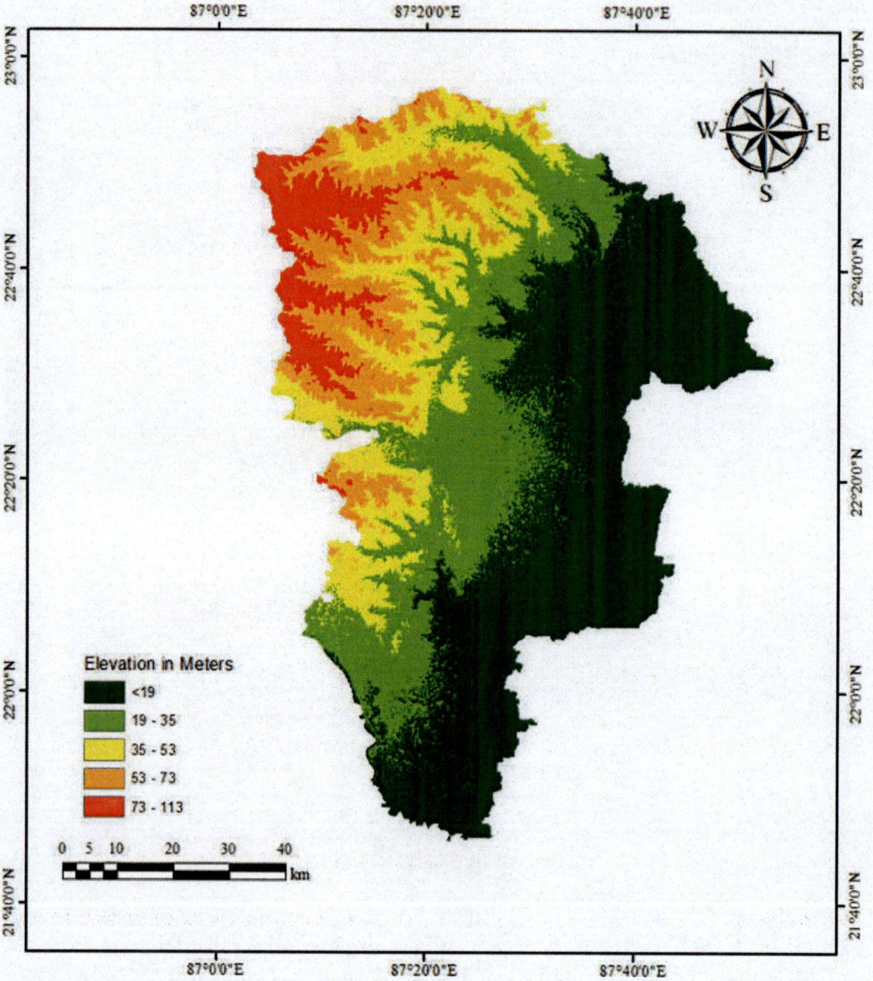

Fig. 18.3 DEM map of the District: Paschim Medinipur (2014)

found in eastern side as in the district as a whole (Figs. 18.5 and 18.3). Most of the low and medium lands are more than two third of total geographical areas in Ghatal and Kharagpur subdivisions, whereas those are about one-half of the total geographical area of the Medinipur Sadar subdivision (Fig. 18.6). High lands including upland agricultural areas constitute about 6% of the total geographical area of the district (Table 18.3).

Such DEM analysis of the physiography of the subdivisions and district would form a base work for further detailed works on creating GIS-based Spatial Decision Support Database for on field web-based application towards selection of suitable sites for excavation of rainwater harvesting tanks.

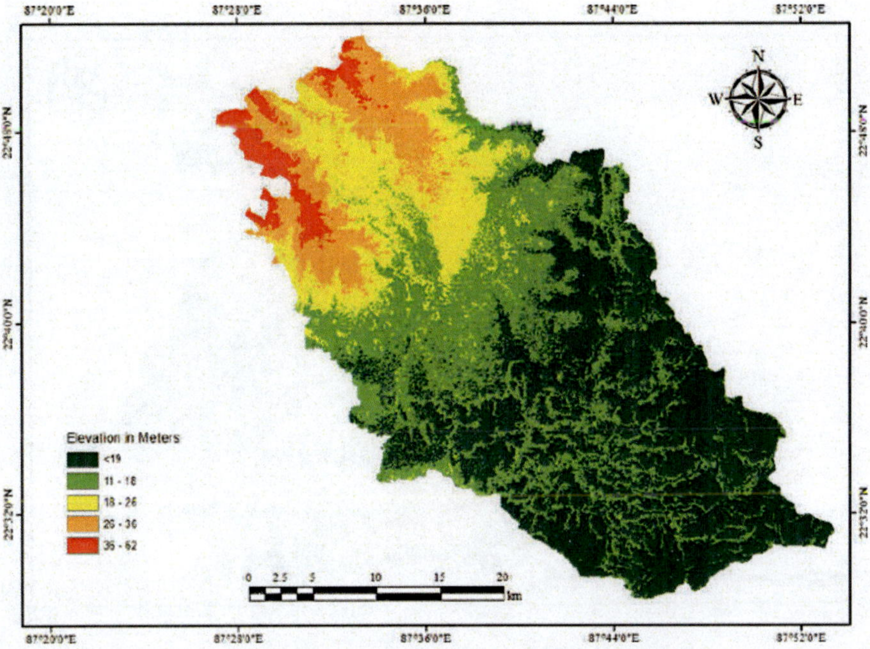

Fig. 18.4 DEM map of the subdivision: Ghatal (2014)

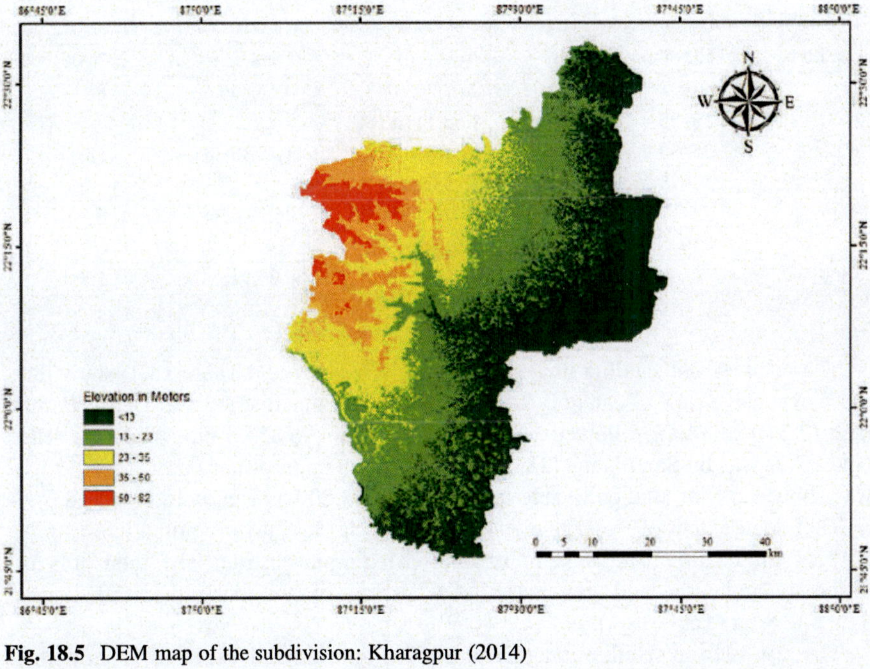

Fig. 18.5 DEM map of the subdivision: Kharagpur (2014)

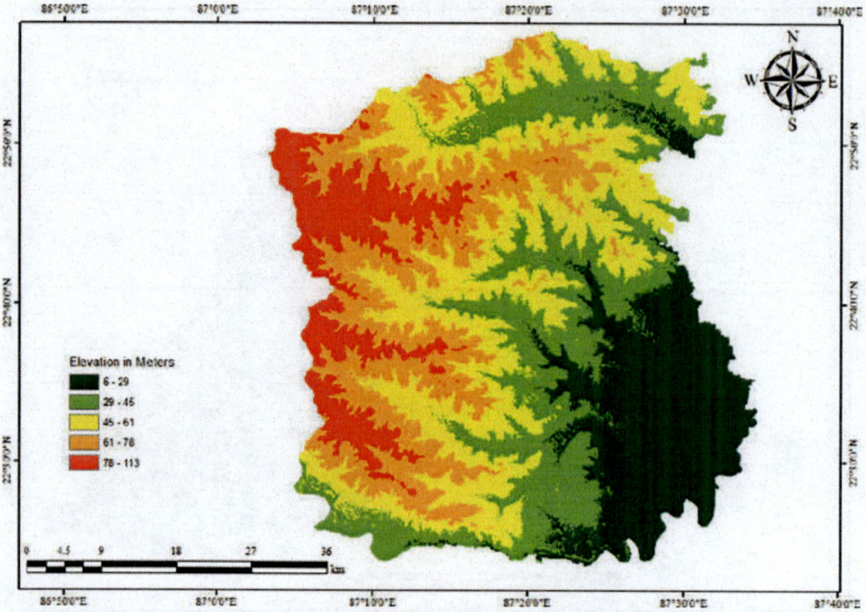

Fig. 18.6 DEM map of the subdivision: Medinipur Sadar (2014)

Table 18.3 DEM (digital elevation model) with areas under different land types in Paschim Medinipur District

Land type	Elevation in meters	Area (ha)	Land area (ha)	% of total
Low	<19	2,61,205.078	4,16,346.60	66
	19–35	1,55,141.518		
Medium	35–53	1,01,600.878	1,78,136.85	28
	53–73	76,535.968		
High	73–113	36,292.558	36,292.56	6
	<19	2,61,205.078		
Total			6,30,776	100

The results from satellite imagery of 2014 (Fig. 18.7 and Table 18.4) show that most expansive LULC category of Paschim Medinipur district was Dense Forest land (2,24,122.3 ha), followed by cultivated area (1,96,855.8 ha), while the least land cover was by Sand Dune (2832.6 ha). The area under water body (11591.3 ha) was about 3.6% of total cultivable land of 3,62,043.80 ha (Tables 18.5 and 18.6).

That areal extent of water bodies in 2014 was far less than required one-fifth or 20% of total cultivable lands in Paschim Medinipur district. The total area of different land use land cover (LULC) classes of the district in 2014 was 6,29,532.8 ha (Table 18.4).

The results from satellite imagery of 2019 (Fig. 18.8 and Table 18.4) show that most expansive LULC category of Paschim Medinipur district was cultivated area

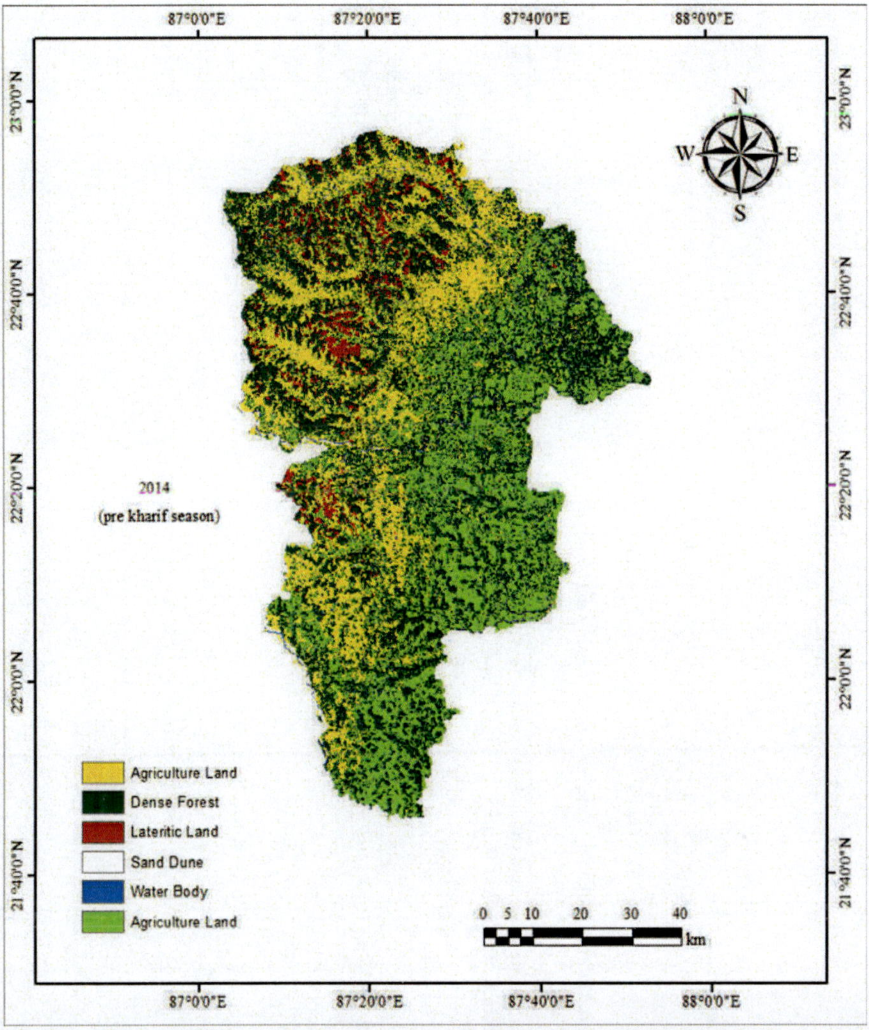

Fig. 18.7 LULC map of Paschim Medinipur district in the year 2014 (*prekharif* season)

Table 18.4 Area under different land classification classes in Paschim Medinipur district (Years: 2014 and 2019)

	2014	2019
Type of land	Area (ha)	Area (ha)
Lateritic land	28,942.8	72,924.8
Dense forest	2,24,122.3	1,56,133.0
Agriculture fallow	1,65,188.0	1,00,231.0
Sand dune	2832.6	3571.5
Agriculture land/crop land	1,96,855.8	2,84,519.0
Water body	11,591.3	12,153.6
Total	6,29,532.8	6,29,532.9

Table 18.5 Classification of land in Paschim Medinipur district based on remote sensing data analysis (Years: 2014 and 2019)

Subdivision	Year: 2014			Year: 2019		
	Area under cultivation (ha)	Cultivable wasteland (including pasture and groves) (ha)	Total cultivable land (ha)	Area under cultivation (ha)	Cultivable wasteland (ha)	Total cultivable land (ha)
(1)	(2)	(3)	(4) = (2+3)	(5)	(6)	(7) = (5+6)
Ghatal	38,715.83	22,674.64	61,390.47	49,282.15	7185.47	56,467.62
Kharagpur	1,27,151.27	51,843.55	1,78,994.82	1,29,898.12	79,202.84	2,09,100.96
Medinipur Sadar	30,988.70	90,669.82	1,21,658.52	1,05,338.74	13,842.68	1,19,181.42
District: Paschim Medinipur	1,96,856	1,65,188	3,62,043.8	2,84,519	1,00,231	3,84,750.0

Table 18.6 Classification of cultivable land and water bodies in Paschim Medinipur district based on remote sensing data analysis (Years: 2014 and 2019)

			Area under water bodies					
					% of cultivable land			Extra required area for irrigation of three crops from rainwater harvesting (ha)
	Total cultivable land (ha)		Area (ha)		Existing		Extra requirement to fulfil one-fifth of cultivated land	
Subdivision	2014	2019	2014	2019	2014	2019	2019	
Ghatal	61,390.47	5,64,67.62	3142.60	2288.97	5.1	4.1	15.9	9006.59
Kharagpur	1,78,994.82	2,09,100.96	4639.12	7828.02	2.6	3.7	16.3	33,999.82
Medinipur Sadar	1,21,658.52	1,19,181.42	3809.59	2036.61	3.1	1.7	18.3	21,798.28
District: Paschim Medinipur	3,62,043.80	3,84,750.00	11,591.31	12,153.6	3.6	3.2	16.8	64,804.69

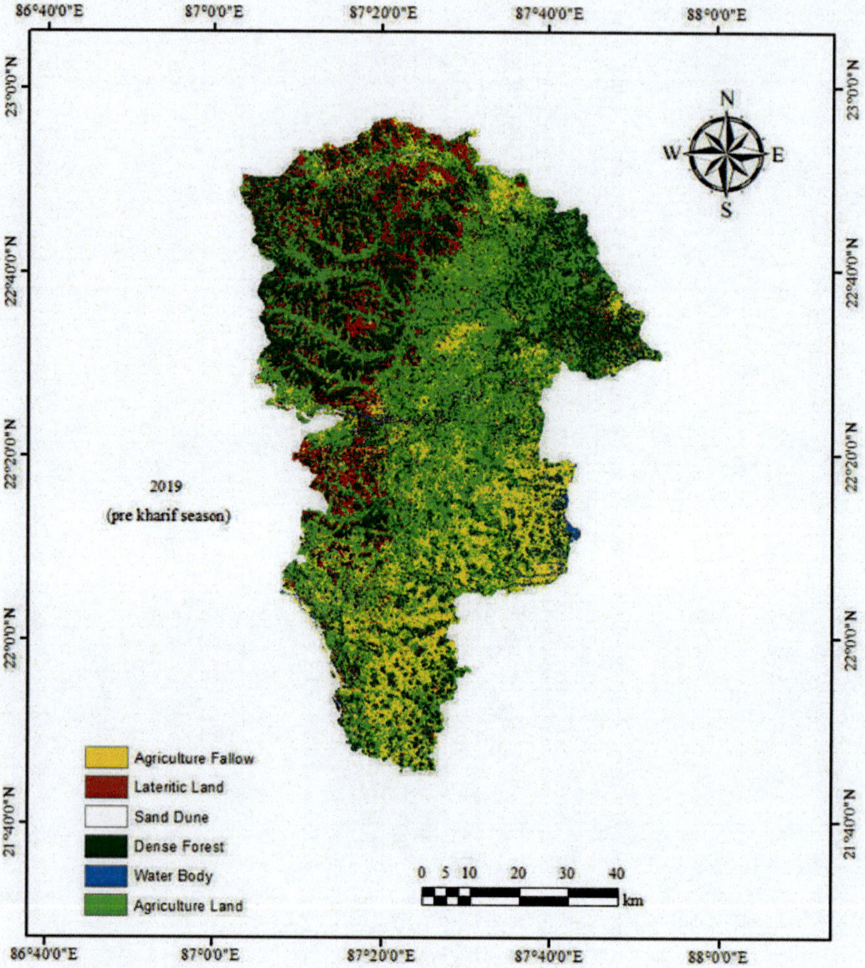

Fig. 18.8 LULC map of Paschim Medinipur district in the year 2019 (*prekharif* season)

(2,84,519.0 ha), followed by dense forest land (1,56,133.0 ha), while least land cover was by sand dune (3571.5 ha). The area under water body (12,153.6 ha) was about 3.2% of total cultivable land of 3,84,750.0 ha (Tables 18.5 and 18.6). That areal extent of water bodies in 2019 was far less than required one-fifth or 20% of total cultivable land. Total area of different land use land cover (LULC) classes of Paschim Medinipur district in 2019 was 6,29,532.9 ha (Table 18.4).

For assessing the scopes for rainwater harvesting and proportionate presence of water bodies in comparison with total cultivable areas, remote sensing imageries were also analysed separately for three subdivisions of Ghatal, Kharagpur and Medinipur Sadar in Medinipur district (Figs. 18.9, 18.10, 18.11, 18.12, 18.13, and 18.14; Tables 18.5 and 18.6).

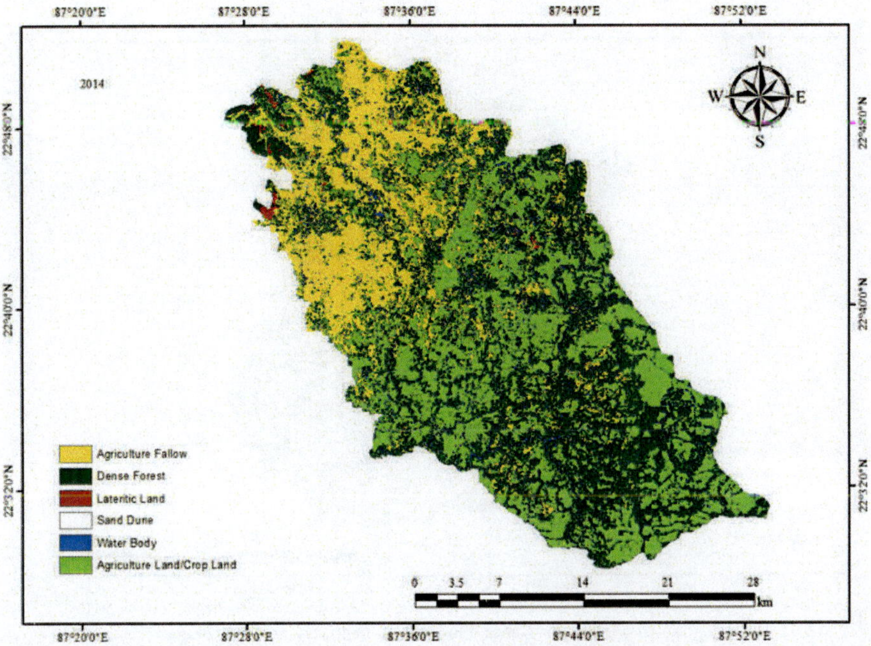

Fig. 18.9 LULC map of Ghatal subdivision in the year 2014 (*prekharif* season)

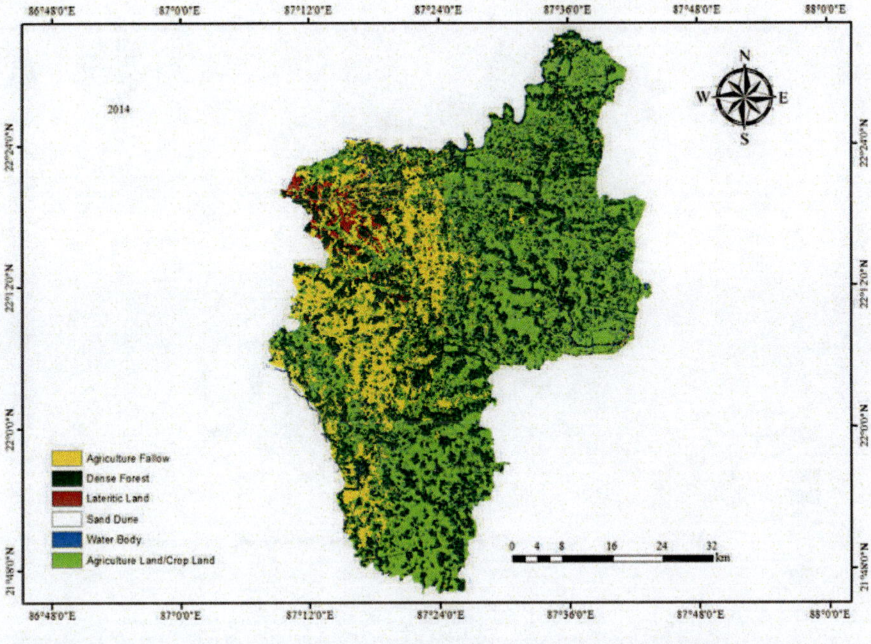

Fig. 18.10 LULC map of Kharagpur subdivision in the year 2014 (*prekharif* season)

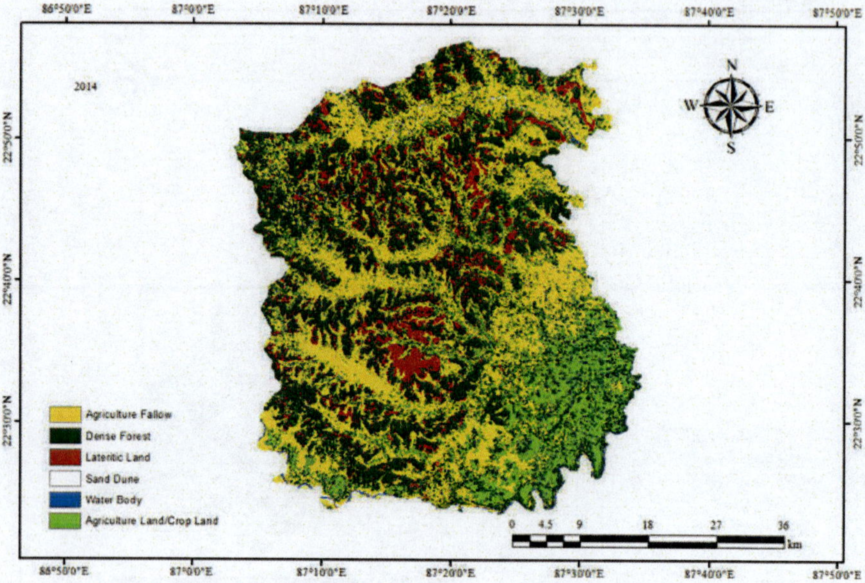

Fig. 18.11 LULC map of Medinipur Sadar subdivision in the year 2014 (*prekharif* season)

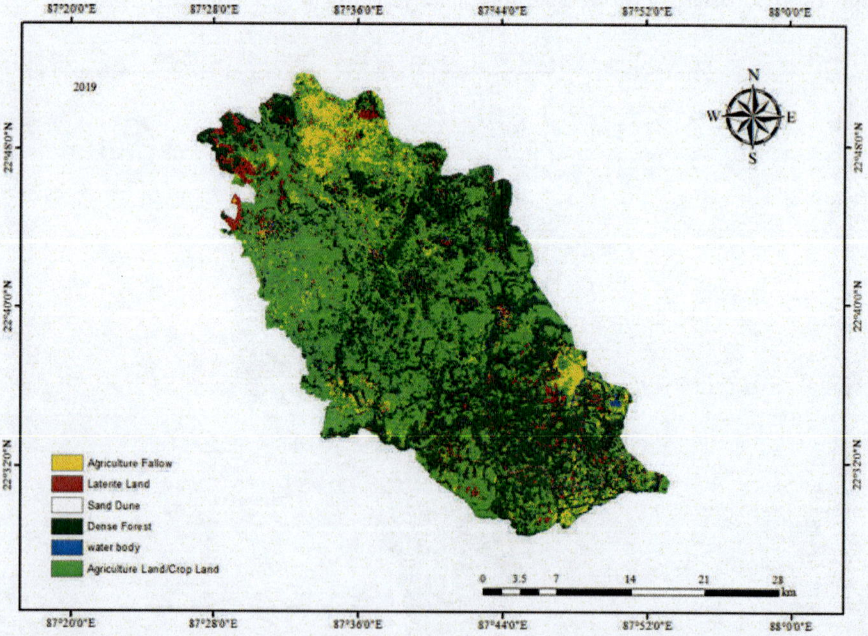

Fig. 18.12 LULC map of Ghatal subdivision in the year 2019 (*prekharif* season)

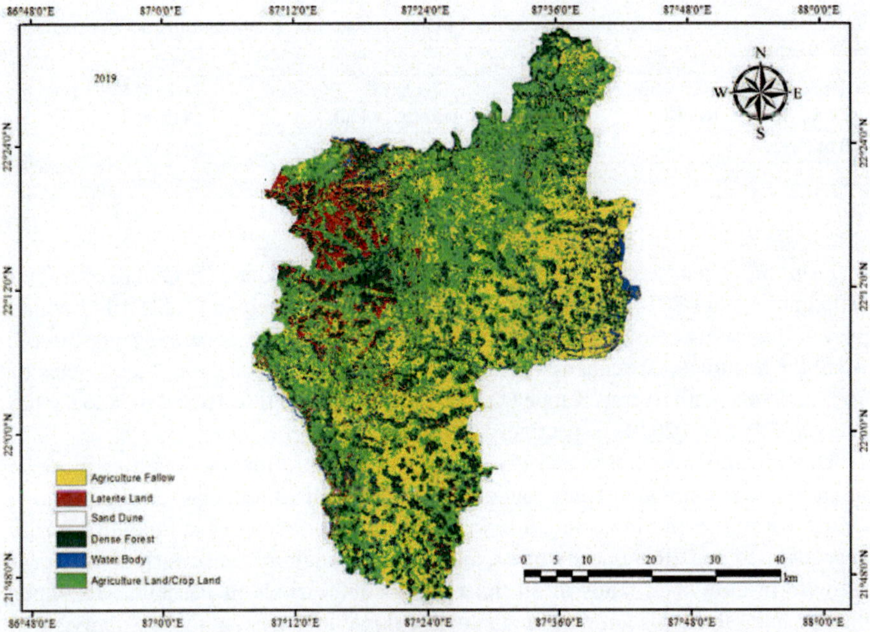

Fig. 18.13 LULC map of Kharagpur subdivision in the year 2019 (*prekharif* season)

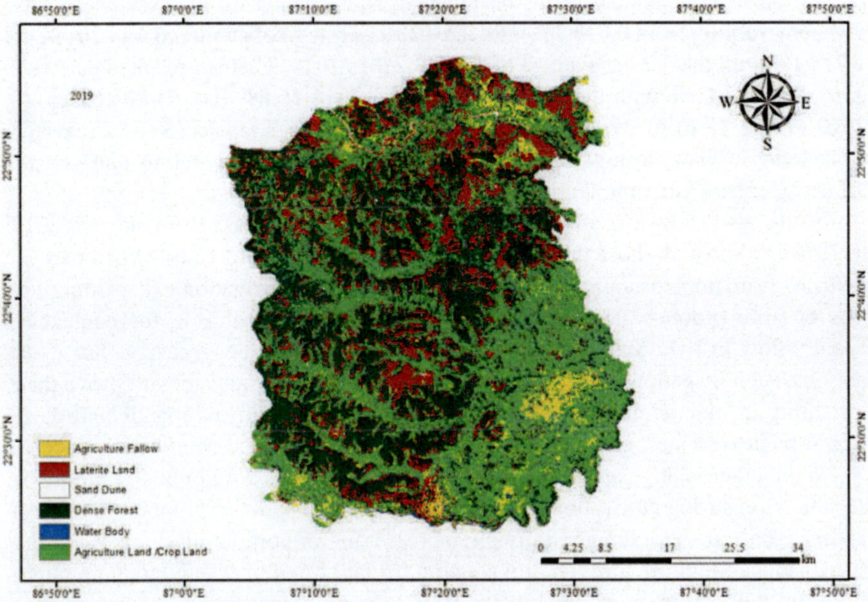

Fig. 18.14 LULC map of Medinipur Sadar subdivision in the year 2019 (*prekharif* season)

Table 18.7 Accuracy assessment of classified images of Paschim Medinipur district (Years: 2014 and 2019)

Classified images Paschim Medinipur district (for the years)	Overall classification accuracy (%)	Overall Kappa ($K^{\wedge}$) Statistics
2014	76	0.6716
2019	84	0.7826

From the discussions, it is revealed that total areas under LULC classes in two different years of 2014 and 2019 have been found mostly equal (Table 18.4) which proves that analysis of remote sensing imageries has been accurately performed. Also, the accuracy assessment of classified images shows the overall accuracies of 76% and 84% with overall Kappa ($K^{\wedge}$) statistics values of 0.6716 and 0.7826 for the years of 2014 and 2019, respectively (Table 18.7).

Though cultivable wastelands do not carry guided information regarding increase in areas under cultivable lands for imageries taken for LULC classification in *pre-kharif* season (i.e. pre-monsoon: first week of March to first week of June) in both the years of 2014 and 2019 under studies, analysis of agriculture area clearly shows areal increase in cultivated lands in all those three subdivisions in Paschim Medinipur district and, thus, also increases in total cultivable lands in those subdivisions (Tables 18.5 and 18.6).

The analysis also reveals a very grim picture of the presence of water bodies in proportionate comparisons with the total cultivable lands in all those three subdivisions varying from 2.6–5.1% with the district average of about 3.6% in 2014 and 1.7–4.1% with the district average of about 3.2% in 2019. That has resulted in a wide gap of 15.9–18.3% with the district average of about 16.8% (i.e. 64,804.69 ha) in 2019 (Table 18.6) of extra requirements of water bodies needed to be excavated considering soil hydrological properties through coordinated public and private efforts to ensure yearlong cultivation.

Results also show that areas under cultivation are in increase from the year 2014 to 2019 in almost all those three subdivisions (Tables 18.5 and 18.6), which may be revealed from non-sustainable practice of provisioning of irrigation either from river lifts or from groundwater extraction. Both are non-sustainable in the context of healthy flow in river and adequate potentiality of groundwater recharge. For those reasons, such irrigation water resources from river and groundwater will prove their instability in maintaining irrigational capacity for such cultivable lands and those would be proved inefficient in future unwanted environmental conditions.

On the other hand, excavations of required areal extent of water bodies would be capable of ensuring cultivation in three crop seasons (i.e. 300% cropping intensity) with a goal to zero hunger through increase in proportion of productive and sustainable agriculture through provisioning irrigation in increasing cropping intensity from monocropping condition (i.e. 100% cropping intensity) through rainwater harvesting based on hydrological balance approach, thus, most possibly ensuring Indicator 2.4.1 under Target 2.4 within SDG 2 of Zero Hunger of the UN by 2030 (UN SDG, 2020).

18.4 Conclusions

From the results of analysis of satellite imageries of Paschim Medinipur district, the following conclusions may be drawn:

1. Analysis of satellite imageries taken in pre-*kharif* season (i.e. pre-monsoon: first week of March to first week of June) has yielded highly accurate data.
2. There are extremely inadequate areas under water bodies in the district.
3. Analysis of remotely sensed data strongly suggests an opportunity for excavations of about 64,804.69 ha more water bodies on the basis of soil hydrological properties of the localities through coordinated public and private efforts to ensure irrigation for yearlong cultivation in Paschim Medinipur district.
4. Increase in agricultural areas proves increase in provisioning irrigation. It is speculated that totally river- and/or groundwater-dependent present mode of irrigation practice may be non-sustainable in future unwanted environmental conditions.
5. All those findings with the application of DEM analysis have potential scopes for creation of GIS-based Spatial Decision Support Database for solution of on-field problems of rainwater harvesting based on hydrological balance approach through web application in increasing sustainable agricultural areas for food security in line with SDG 2 (Target 2.4 and Indicator 2.4.1) of the UN by 2030.

References

Bailey, D. T. (2005). *Development of an optimal spatial decision-making system using approximate reasoning*. Ph.D. thesis submitted to the Queensland University of Technology, Australia.

Basu, T., & Saha, S. (2018). The analysis of land use land cover changes using geo-informatics and its relation to changing population scenarios in Barasat Municipality in North Twenty-Four Parganas, West Bengal. *International Journal of Humanities and Social Science Invention, 6*(8), 01–13.

Bhunia, G., Samanta, S., Pal, D., & Pal, B. (2012). Assessment of groundwater potential zone in Paschim Medinipur District, West Bengal–A meso-scale study using GIS and remote sensing approach. *Journal of Environment and Earth Science, 2*, 41–59.

Bui, T. Q., & Pham, H. M. (2016). Web based GIS for spatial pattern detection: application to malaria incidence in Vietnam. *Springer Plus, 5*, 1014. https://doi.org/10.1186/s40064-016-2518-5

Choi, J.-Y., Engel, B. A., Theller, L., & Harbor, J. (2005a). Utilizing web-based GIS and SDSS for hydrological land use change impact assessment. *Transactions of the ASAE (American Society of Agricultural Engineers), 48*(2), 815–822.

Choi, J.-Y., Engel, B. A., & Farnsworth, R. L. (2005b). Web-based GIS and spatial decision support system for watershed management. *IWA publishing: Journal of Hydroinformatics, 7*(3), 165–174.

Congalton, R. G., & Green, K. (1999). *Assessing the accuracy of remotely sensed data: Principles and practices* (p. 137). Lewis Publishers.

Dolui, G., Das, S., & Satpathy, S. (2014). An application of remote sensing and GIS to analyze urban expansion and land use land cover change of Midnapore Municipality, WB, India. *International Research Journal of Earth Sciences, 2*(5), 8–20.

EGIYE BANGLA: (The official portal of the District Paschim Medinipur, Government of West Bengal, https://www.paschimmedinipur.gov.in/maps. Last accessed 20 June (2020).

Google Earth: Google Earth Pro App. Downloaded from: google.com/intl/en_in/earth/versions/. Last accessed 25 June (2020).

Government of India: District Census Handbook, Midnapore, 1971. Directorate of Census Operations, West Bengal 1971, Series.22, Part. X-A, West Bengal (1971).

Government of India: Census of India. Office of the Registrar General and Census Commissioner, Ministry of Home Affairs, Government of India, India (2011). http://censusindia.gov.in. Last accessed 25 June 2020.

Government of India: Ministry of Agriculture & Farmers Welfare, Department of Agriculture, Cooperation & Farmers Welfare. Agricultural Contingency Plan for 2011 of Paschim Medinipur district (2011). http://agricoop.nic.in/agriculturecontingency/paschim-medinipur Last accessed 25 June 2020.

Govt. of West Bengal: District Human Development Report of Paschim Medinipur. Development and Planning Department, Government of West Bengal (2011). http://wbplan.gov.in/HumanDev/DHDR/DHDR_Paschim Medinipur. Last accessed 25 June 2020.

Govt. West Bengal: District Statistical Handbook of Paschim Medinipur. Bureau of Applied Economics and Statistics, Department of Statistics and Programme Implementation, Government of West Bengal. Table 2.2 (2014). http://www.wbpspm.gov.in/publications/District%20Statistical%20Handbook. Last accessed 25 June 2020.

Hudson, W., & Ramm, C. (1987). Correct formula of the Kappa coefficient of agreement. *Photogrammetric Engineering and Remote Sensing, 53*(4), 421–422.

Huisman, O., deBy, R. A. (Eds.). (2009). *Principles of Geographic Information Systems*. (ITC Educational Textbook Series; 1) (pp. 342–430). The International Institute for Geo-Information Science and Earth Observation (ITC), Hengelosestraat 99, P.O. Box 6, 7500 AA Enschede, The Netherlands (Chapter 6).

Keenan, P. B., & Jankowski, P. (2019). Spatial decision support systems: Three decades on. *Decision Support Systems, 116*(January), 64–76. https://doi.org/10.1016/j.dss.2018.10.010

Kumar, K. K., & Babu, D. B. S. (2016). A web GIS based decision support system for Agriculture Crop Monitoring System-A case study from Part of Medak District. *J Remote Sensing & GIS, 5* (4), 1–19. https://doi.org/10.4172/2469-4134.1000177

Lan, Y., Tang, W., Dye, S., & Delmelle, E. (2020). A web-based spatial decision support system for monitoring the risk of water contamination in private wells. *Annals of GIS, 26*(3), 293–309.

Myneni, R. B., Hall, F. B., Sellers, P. J., & Marshak, A. L. (1995). The interpretation of spectral vegetation indices. *IEEE Transactions on Geoscience and Remote Sensing, 33*(2), 481–486.

NASA. (2020). Jet Propulsion Laboratory, California Institute of Technology. https://www.jpl.nasa.gov. Last accessed 15 July 2020.

Panda, S. (2005). Systems approach for rainwater harvesting with emphasis on quality of surface water in the coastal region of Eastern India. In *International Conference on Crisis Management in Water and Environment (ICCMWE-2005), 15–16 July 2005*. Organised by: Indian Association of Hydrologists, Roorkee and West Bengal Regional Centre, at: Auditorium of Science City, J B S Haldane Avenue, Kolkata – 700 046, India. Proceedings Vol. 1: 92–96. https://www.researchgate.net/publication/340236315_Systems_approach_for_rainwater_harvesting_with_emphasis_on_quality_of_surface_water_in_the_coastal_region_of_Eastern_India

Panda, S., & De, P. (1989). Hydrological balance approach for planning irrigation in coastal areas of Contai, Midnapore, West Bengal. In *Proceedings All India Seminar on "Role of Hydrology in Efficient Management of Irrigation System"*, organised by Indian Association of Hydrologists, West Bengal Regional Centre. At the Auditorium, Birla Industrial & Technological Museum, 19 A, Gurusaday Road, Calcutta – 700 019, West Bengal, India, 19–20th May. https://www.researchgate.net/publication/340102421_Hydrological_balance_approach_for_planning_irrigation_in_coastal_areas_of_Contai_Midnapore_West_Bengal_Panda_S_De_P_1989

Panda, S., Roy, G. B., & Ghosh, R. K. (1990). Detection of agroclimatic feasibility for transforming an apparently water deficit monocropped area to a yearlong cultivable tract in Contai,

Midnapore, West Bengal. *Indian Journal of Landscape Systems and Ecological Studies, 13*(2), 174–175. https://doi.org/10.13140/RG.2.2.14316.41600. https://www.researchgate.net/publication/328275198_Detection_of_agroclimatic_feasibility for ransforming_an_apparently_water_deficit_monocropped_area_to_a_year_long_cultivable_tract_in_Contai_Midnapore_W_Bengal

Peshwa, P. (2015). *Spatial decision support system for infrastructure management – A case study of electrical utility structure of Jodhpur city*. Ph.D. thesis submitted to the Maharshi Dayanand Saraswati University, Ajmer, India.

Piarsa, I. N., Kompiang, A. A., Sudana, O., Wahyu, G., & Gunadi, M. (2012). Web-based GIS by using Spatial Decision Support System (SDSS) concept for searching commercial marketplace – Using Google MAP API. *International Journal of Computer Applications, 50*(7, July 2012), 1–5.

Raju, P. L. N. (2004). Fundamentals of Geographical Information System. In M. V. K. Sivakumar, P. S. Roy, K. Harmsen, & S. K. Saha (Eds.), *Satellite remote sensing and GIS applications in agricultural meteorology: Proceedings of the training workshop 7 – 11 July 2003, Dehradun, India* (pp. 103–120). World Meteorological Organisation.

Rinner, C. (2003). Web-based spatial decision support: Status and research directions. *Journal of Geographic Information and Decision Analysis, 7*(1), 14–31.

Robin, T. A., Khan, M. A., Kabir, N., Rahaman, S. T., Karim, A., Mannan, I. I., George, J., & Rashid, I. (2019). Using partial analysis and GIS to improve planning and resource allocation in a rural district of Bangladesh. *BMJ Global Health, 4*, e000832. https://doi.org/10.1136/bmjgh-2018-000832

Saren, S. (2019). *Mouza wise water management planning in Red and Laterite zone of West Bengal.* M.Sc. (Agriculture) in Soil and Water Conservation thesis submitted to Bidhan Chandra Krishi Viswavidyalaya, West Bengal, India, Unpublished.

Sugumaran, V., & Sugumaran, R. (2007). Web-based spatial decision support systems (WebSDSS): Evolution, architecture, examples and challenges. *Communications of the Association for Information Systems, 19*(40), 844–875.

UN SDG: Sustainable Development Goals Target 2.4. In: SDG 2: End hunger, achieve food security and improved nutrition and promote sustainable agriculture. Division for Sustainable Development Goals, Department of Economic and Social Affairs, United Nations Secretariat Building, 405 East 42nd Street, New York, NY 10017, USA (2020). https://sdgs.un.org/goals/goal2. Last accessed 29 Sept 2020.

USGS: What are the band designations for the Landsat satellites? https://www.usgs.gov/faqs/what-are-band-designations-landsat-satellites?qt-news_science_products=0#qt-news_science_products. Last accessed 6 July (2020).

USGS earth explorer: https://earthexplorer.usgs.gov/. Last accessed 15 July (2020).

USGS EROS Archive – Digital Elevation – Shuttle Radar Topography Mission (SRTM) Non-Void Filled, https://www.usgs.gov/centers/eros/science/usgs-eros-archive-digital-elevation-shuttle-radar-topography-mission-srtm-non?qt-science_center_objects=0#qt-science_center_objects. Last accessed 15 July (2020).

Chapter 19
Monitoring of Meteorological Drought Based on Rainfall Departure Using Remotely Sensed CHIRPS Precipitation Product over Tamil Nadu, India

Samykannu Venkadesh, Sellaperumal Pazhanivelan, and Kancheti Mrunalini

19.1 Introduction

Tamil Nadu situated in the southern peninsular of India, characterized by various climatic patterns with major hazards such as flood and drought, which leads to socio-economic losses. Northeast monsoon from October to December is the significant rainy season for the area accounting for nearly 48 (440.4 mm) percent of the annual rainfall. Coastal and interior districts receive about 50–60% of the annual rainfall, respectively. Drought is a creeping natural disaster and recurring event for all climatic regimes, leading to water scarcity in the environmental system, resulting in significant socio-economic impacts. According to the American Meteorological Society (1997), drought can be classified into four physical classes, i.e. meteorological (deficit in precipitation), hydrological (deficit in groundwater), agricultural (deficit in soil moisture) and socio-economic (supply and demand of goods and social response) droughts (Ghulam et al., 2007). Wang et al. (2018) state that in upcoming decades, droughts would further expand and intensify due to climate change. According to Narasimhan and Srinivasan (2005) below, normal rainfall is the root cause for metrological drought. Meteorological droughts are the cause of all other categories of droughts; consequently, it is necessary to develop the

S. Venkadesh (✉)
Agro Climate Research Centre, Tamil Nadu Agricultural University, Coimbatore, Tamil Nadu, India

S. Pazhanivelan
Department of Remote Sensing and Geographical Information System, Tamil Nadu Agricultural University, Coimbatore, Tamil Nadu, India

K. Mrunalini
ICAR-Indian Institute of Pulses Research, Kanpur, Uttar Pradesh, India

G. S. Bhunia et al. (eds.), *Anthropogeomorphology*, Geography of the Physical Environment, https://doi.org/10.1007/978-3-030-77572-8_19

more effective method to monitor drought in order to mitigate drought hazards. However, monitoring, predicting or tracking meteorological droughts and their spatial variability are highly challenging (Dai, 2011). Several meteorological drought indicators have been developed to identify drought severity for agricultural management purpose, such as the Percentage of Normal Precipitation (NDMC, 2018), Percentage of Precipitation Anomaly (Zhang et al., 2009), Deciles Index (Gibbs & Maher, 1967), Palmer Drought Severity Index (PDSI, Palmer, 1965) and Standardized Precipitation Index (SPI, McKee et al., 1993). Precipitation is the main input parameter to detect drought in most of the regions (Wilhite, 2000). Conventional rainfall data have been derived from in situ rain gauge networks, but weather station is sparsely and unevenly distributed in some regions (Tang et al., 2016; Wang et al., 2017). Alternatively, satellite-based precipitation products have been developed in the last decades and can monitor rainfall continuously over the globe in near real time (Kaptue et al., 2015). Therefore, it is essential to evaluate how reliable the remotely sensed rainfall data against the rain gauge network to choose the most suitable product for various applications and the confidence level also necessary for analysis. Indeed, satellite performance is an alternative to monitor droughts by observing climatic conditions (Liu et al., 2015; Naumann et al., 2012; Wardlow et al., 2011). PERSIANN (Remotely Sensed Information using Artificial Neural Networks), TRMM (Tropical Rainfall Measurement Mission), TMPA (Multi-Satellite Precipitation Analysis), CMORPH (Climate Prediction Centre morphing technique) and IMERG (Integrated Multi-Satellite Retrievals for Global precipitation measurement) are the commonly used precipitation products (Joyce et al., 2004; Huffman et al., 2007, 2010, 2012) with the high spatial and temporal resolution to compensate of observed rain gauge networks.

As reported by Funk et al., 2014, CHIRPS (Climate Hazards Group Infrared Precipitation with Stations) satellite precipitation product provides long-term data records of more than 30+ years since 1981 and presenting great potential for drought monitoring. Prakash (2019) evaluated the four multi-satellite precipitation products, namely, CHIRPS, MSWEP, SM2RAIN and TMPA, for the period of 1998–2015 at monthly scale across India; their results showed that CHIRPS had a good correlation coefficient with in situ measurement of spatial distribution over the study area. Many studies carried out in different parts of the Indian subcontinent found an increase in droughts frequency under climate change (Ahmed et al. (2018); Bisht et al. (2019). Tamil Nadu experienced one of the worst droughts in the year 2016 and declared all 32 districts drought-stricken as the worst drought in 144 years hit due to the failure of monsoon rain (Gazette, 2017). This imposed a severe threat to both agriculture and hydrology, resulting in huge economic loss. It is also noticed that the frequency of drought years has increased in recent decades. The India Meteorological Department (IMD) defined meteorological drought as excess (more than +20%), normal (+19% to −19%), deficient (−20 to −59%) and scanty (−60% or below), based on percentage rainfall departure from the climatological mean of that area. The agricultural drought vulnerability at the disaggregated level of Andhra Pradesh in India using crop-generic index was studied by Murthy et al. (2015); their results indicated that the vulnerable group of Mandals represents 57% of agricultural area and 67% of the rainfed crop area.

The objective of this study is to evaluate the performance of precipitation product against rain gauge data at an individual monthly scale over Tamil Nadu for monitoring meteorological drought at block level from October to December between the periods of 2015 and 2017. In this study, the latest version of CHIRPS v2.0 dataset is compared with 385 rain gauge stations, and a derived satellite product has been applied for the meteorological drought, which is related to the deviation of rainfall from normal conditions over a period of time.

19.2 Brief Description of the Study Area

The state of Tamil Nadu is located in the southernmost part of Indian peninsular region. The state covers an area of 13 million hectares and has a coastline of 1076 km which is about 15% of the coastline of India. Tamil Nadu state lies between 08° 05′ and 13° 35′ northern latitudes and 76° 15′ to 80° 20′ eastern longitudes (Fig. 19.1) and having 34 administrative districts with 385 blocks. The state is divided into seven agro-climatic zones, and the rain shadow region of Western Ghats is located in western part and is deprived of rain during the south-west monsoon season, which is the assured monsoon for most of India. Tamil Nadu is categorized under tropical monsoon climate. An average annual rainfall of the state receives 920.9 mm of which 440.4 mm was received during the northeast monsoon season (GoT, 2015). The maximum annual average temperature is around 33 °C and rises to 45°C in the

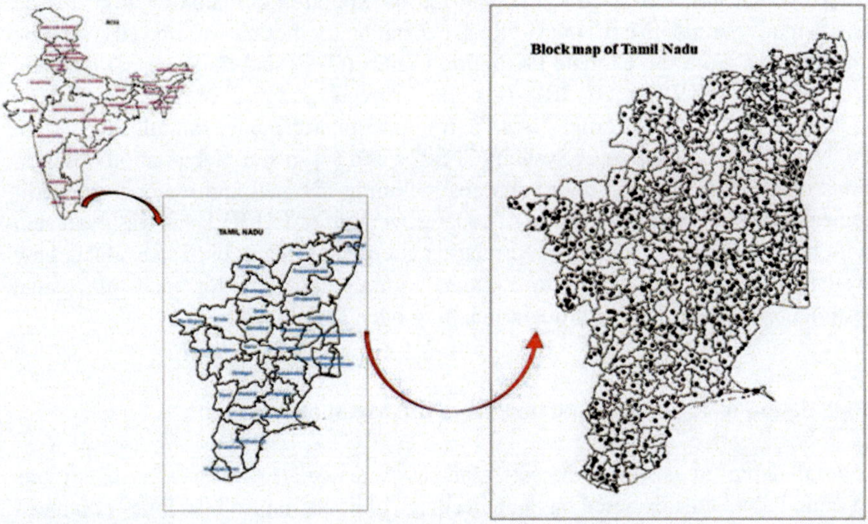

Fig. 19.1 Spatial distribution of rain-gauge stations with study area

summer. The minimum annual average temperature is 24 °C and drops to about 10 °C during the winter in the hilly areas.

19.2.1 Datasets and Methodology

The daily rainfall data were obtained for 381 rain gauge observation stations from Tamil Nadu Agriculture Weather Network, which is established by Tamil Nadu Agricultural University in collaboration with Department of Agriculture, Tamil Nadu (TAWN; available online at http://tawn.tnau.ac.in/General/HomePublicUI.aspx). Data is collected from October to December for the period 2015−2017. Most rain gauge data sets contained outliers that have unusually large or small values when compared, and the quality of all data sets was first corrected, and obvious outliers were removed and from the daily rainfall, the aggregated monthly rainfall data was used for validation with CHIRPS product.

CHIRPS-Based Rainfall Data

Climate Hazards Group Infrared Precipitation with Stations (CHIRPS) is a new quasi-global precipitation product of 50° S–50° N and 180° E–180° W with high temporal at daily, pentadal, monthly precipitation totals and seasonal scale and spatial resolution developed by the US Geological Survey Earth Resources Observation and Science Center in association with the Santa Barbara Climate Hazards Group at the University of California, available from 1981 to near real-time period of record which has been used by (i) pentadal precipitation climatology at grid scale (six pentads per month); (ii) quasi-global geostationary thermal infrared (IR) satellite observations from the Climate Prediction Center (CPC) and the National Climatic Data Center (NCDC) (B1 IR); (iii) the Tropical Rainfall Measuring Mission (TRMM) 3B42 product from NASA; (iv) atmospheric model rainfall fields from the NOAA Climate Forecast System (CFSv2); and (v) in situ precipitation observations obtained from a variety of sources including national and regional meteorological services (Funk et al., 2014).The latest version of CHIRPS v.2 was evaluated with rain gauge stations and the correlation coefficient metric used to evaluate how well rain-gauge data applied to compute rainfall departure for each rain gauge stations for meteorological drought mapping over Tamil Nadu.

Calculation of Rainfall Departure from the Mean or Median

Rainfall departure is simple indicator and easy to calculate local weather conditions of rainfall deficiencies based on the percentage of deviation from its long-term mean monthly rainfall. The negative and positive values represent the deficit and excess rainfall at a given station. The monthly drought condition was determined using the

Table 19.1 IMD classification of rainfall deviation

Deviation from normal rainfall (%)	Category
60 and more	Large excess
20 to 59	Excess
+19 to 19	Normal
−59 to −20	Deficient
−60 to more	Large deficient

Table 19.2 Correlation coefficient between CHIRPS and rain gauge observations

Month/year	2015	2016	2017
October	0.49	0.66	0.75
November	0.82	0.87	0.81
December	0.87	0.83	0.78

criteria suggested by India meteorological department (1976), and it is expressed in percentage. It is calculated using the following formula:

$$\text{Rfdev} = [(\text{Rfi} - \text{RFn})/\text{RFn}] \times 100$$

where Rfi is the current rainfall for a comparable period (in mm) and RFn is the normal rainfall (at least 30 years on average) for the same period (in mm). Here, the monthly rainfall data from CHIRPS satellite was used after carefully evaluated with 385 rain gauge stations. Based on rainfall deviations, six categories were applied to monitor drought across Tamil Nadu for the individual months of NEM season from 2015 to 2017, and the rainfall deviation classification of IMD is given in Table 19.1.

19.3 Results and Discussion

19.3.1 Comparison Between CHIRPS and Rain Gauge Observations

First step, we evaluated the performance of the CHIRPS precipitation product with rain gauge observation at individual monthly scale during October to December for three northeast monsoon seasons from 2015 to 2017 over Tamil Nadu. Correlation coefficient comparison of CHIRPS and rain gauge observations were done and found out that there is better correlation between CHIRPS and rain gauge observations with value of 0.87.

These values indicate that the monthly CHIRPS rainfall data captures the significance of rainfall well. From Table 19.2, the results showed there is good agreement of CHIRPS with precipitation and the correlation ranging from 0.49 to 0.87.

Table 19.3 RMSE for CHIRPS and rain gauge observations for the north east monsoon season

Month/year	2015	2016	2017
October	32.20	12.95	36.92
November	83.26	19.83	54.76
December	28.88	11.86	8.65

The high correlations also indicate that, we can very well use CHIRPS data for assessment of drought features over Tamil Nadu. Among these study period, the highest CC (0.87) is during December 2015 and November 2016, while the lowest CC (0.49) correlations show during October 2015. During all northeast monsoon season, value range from 0.49 to 0.75 in October, from 0.82 to 0.81 in November, and from 0.87 to 0.78 in December indicates reliability of CHIRPS product. It is to be noted that spatial distributions of correlation coefficient of CHIRPS show higher among the multi-satellite precipitation products across India (Prakash et al., 2019), and also the scatter plots have shown a high correlation coefficient of greater than 0.5 which means that the trends in the observed rain gauge data are well captured in the satellite. In CHIRPS rainfall product in 2015, which was the extremely wettest season, correlation values ranged from 0.49 to 0.87, whereas in 2016, values ranged from 0.66 to 0.83, which was drought year, and in 2017, the values ranged from 0.75 to 0.78 which was a normal year. The most perfect correlation is shown during November and December months of 2015, 2016 and 2016, respectively. The Root Mean Square Error (RMSE) varies from 8.65 to 83.26 mm/month, and the higher error was concentrated on the month of November, 2015 and 2016 (mainly lower than 8.65 in December, 2017). The variations of RMSE may be due to the moderate amount of rainfall during the study periods. As is shown in Table 19.3, RMSE for monthly precipitation ranged from 28.45 to 34.67, 71.04 to 91.62 and 28.88 to 33.57, respectively, during October, November and December of 2015–2017 as compared with TAWN rain gauge network. During November 2017, the RMSE showed less error with relatively close to observed rainfall with high correlation (0.78) for the same month.

19.3.2 Spatial Distribution of Rainfall Departure Based on CHIRPS Rainfall

From these evaluation results, CHIRPS rainfall data can be extended to examine the spatial drought severity conditions and was compared at block level for the same period. The spatial distribution maps depicting rainfall departure values were categorized into five classes, i.e. no drought, mild, moderate, severe and extreme drought based on the intensity of IMD classes to demarcate the spatial analysis of each block over Tamil Nadu. As seen in Figs. 19.2, 19.3, and 19.4, the spatial distribution of rainfall departure, Tamil Nadu had suffered from severe to mild drought during

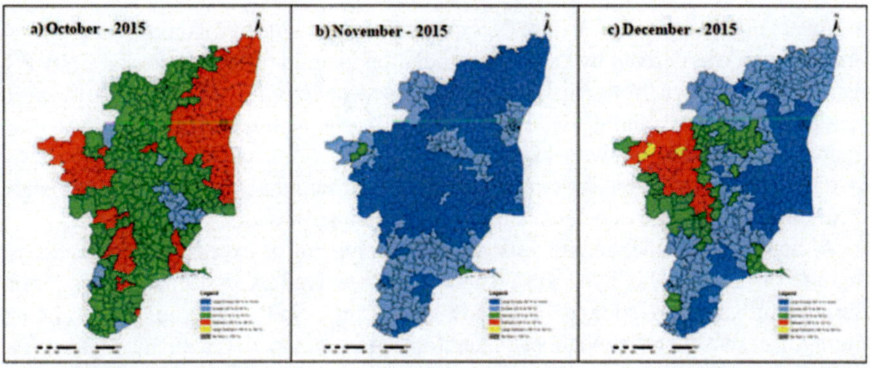

Fig. 19.2 (a–c) Spatial distribution of rainfall departure during October, November and December in Tamil Nadu during 2015

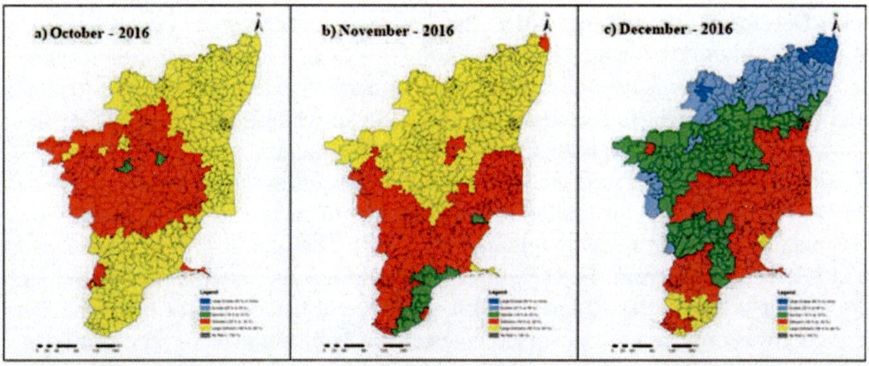

Fig. 19.3 (a–c) Spatial distribution of rainfall departure during October, November and December in Tamil Nadu during 2016

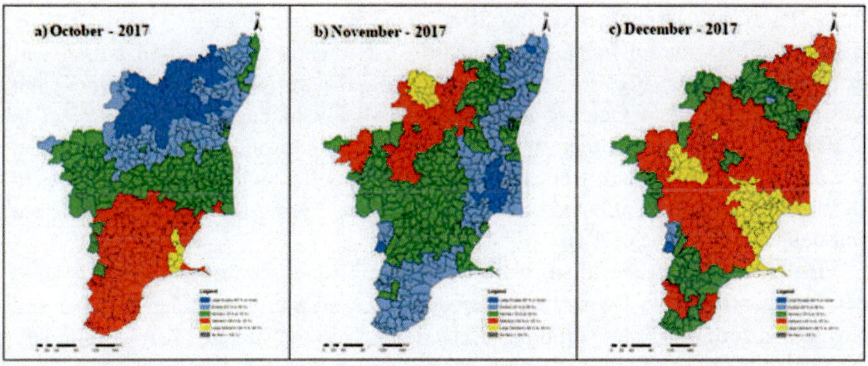

Fig. 19.4 (a–c) Spatial distribution of rainfall departure during October, November and December in Tamil Nadu during 2017

northeast monsoon season from 2015 to 2017. In this study, spatial distribution of drought map was derived by kriging interpolation method from point-based CHIRPS data. It can be seen from Fig. 19.2a–c that wettest season occurred in 2015 with spatial differences; rainfall departure-based drought monitoring results show that major parts of the blocks were experienced deficient to normal drought conditions in October and followed by November and December months it did not show any sign of dryness except few deficient category in December months.

Although November month was observed dominant of excess and large excess amount of rainfall, this is because Tamil Nadu received excess amount of rainfall during northeast monsoon period. However, Tamil Nadu received good rainfall during the month of November. Therefore, this method is realistic to identify drought in time and space. The deficient (moderate drought) areas are located in the northeastern and few parts on western regions of Tamil Nadu, especially during October and December months, respectively. This can be understood clearly, 53% of normal deviation and the unusual and sudden heavy rainfall during the November and December months of 2015 (http://www.imdchennai.gov.in/nemweb.pdf). According to the 2016 which is shown in Fig. 19.3a–c, it is seen that, from October to November, there is overall severity that was more visible between large deficient and deficient drought impacted across Tamil Nadu, while the month of December showed less deficient conditions in the eastern and southern parts.

However, the state faced one of the worst monsoon droughts in 150 years, and 27% of the total area has suffered from extreme to moderate drought conditions (source: IWMI-South Asia drought monitor 2017). The year 2017 was normal year, and it can be seen from the Fig. 19.4a–c that during the months of October and November detected less deficient and large deficient category. But most deficient category was observed in December, whereas less large deficient of rainfall affected areas also located in southeast regions experienced.

Overall, the rainfall departure index using CHIRPS data perform well in capturing the spatial distribution and used for monitoring of the monthly precipitation. In particular, the drought intensity distribution for different blocks for the same period from October to December during 2015–2017 has been shown in Fig. 19.5a–c. Based on IMD rainfall departure categories, it is seen that in 2015 which was extremely wet year (Fig. 19.5a), 225 and 34 blocks were affected under deficient and large deficient in October followed by only 49 blocks detected deficient in December. In Fig. 19.5b, there were 229 and 155 blocks observed in October, and 202 and 170 blocks in November were under large deficient to deficient, whereas in December the drought-affected blocks decreased to 153 and 17 under large deficient and deficient in the year 2016.

Finally, in Fig. 19.5c, there were 245 and 57 blocks under deficient to large deficient conditions in December, whereas the rest of the months, i.e. October and November, found with less drought-affected blocks in the year 2017. As a summary of the results obtained to monitor drought for severe (deficient) and (large deficient) extreme conditions, there has been a highest number of spatial variations and blocks that were detected in October and November months of 2016 due to the failure of

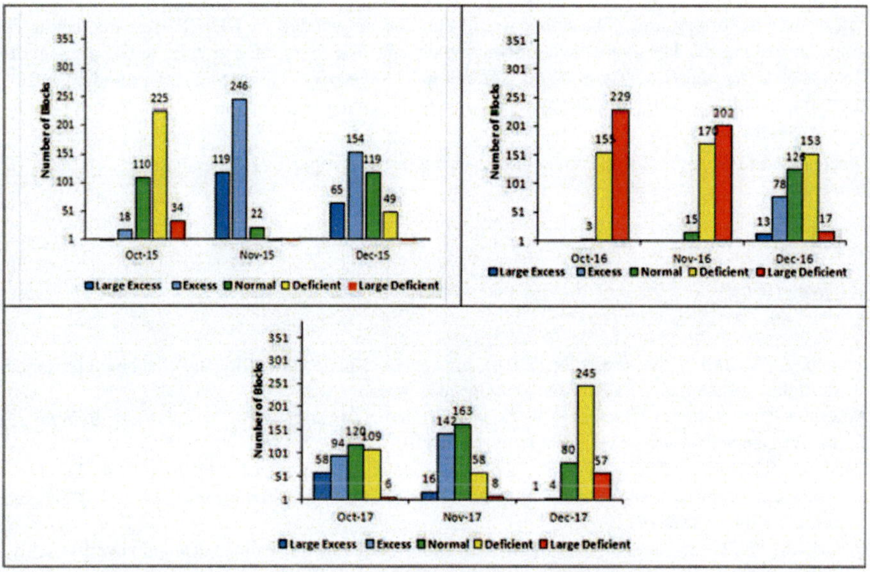

Fig. 19.5 (**a–c**) Drought intensity distribution for different blocks (Nos) during 2015–2017

monsoon rainfall, while considering 2015 and 2017, there were higher spatial variations in northeastern and southern part of Tamil Nadu.

19.4 Conclusion

This study examined the performance of CHIRPS-based precipitation product and observed rain gauge data for spatial pattern of reliable meteorological drought monitoring over Tamil Nadu during northeast monsoon season (October–November) from 2015 to 2017 using IMD-based rainfall departure index especially with precipitation data. The result revealed that based on the monthly CHIRPS precipitation data, there is a good linear relationship with correlation coefficient greater than 0.5, and it led us to calculate rainfall departure for drought intensity. On the temporal distribution, October to November months are more prone to extreme and severe drought conditions due to failure of monsoon rainfall and identify more than 50% of the blocks are found to be extreme to severe drought conditions, while CHIRPS can detect drought events, such as the worst drought of 2016, and also provide valuable information for decision-making. The monthly spatial distribution of meteorological drought occurrence can be detected using CHIRPS rainfall data for drought monitoring variations.

Acknowledgement The first author is thankful to the University Grants Commission: The National Fellowship for Scheduled Caste (UGC: NFSC) funded by Ministry of Social Justice and Empowerment Government of India for providing financial assistance for carrying out this study as part of the PhD programme. The author would like to acknowledge the CHIRPS research communities for making the data available to international users. The first is also grateful to Department of Remote Sensing and GIS, Tamil Nadu Agricultural University, Coimbatore, for providing infrastructural facilities to carry out this study.

Disclosure Statement No potential conflict of interest was reported by the authors.

References

Ahmed, K., Shahid, S., & Nawaz, N. (2018). Impacts of climate variability and change on seasonal drought characteristics of Pakistan. *Atmospheric Research, 214*, 364–374.

American Meteorological Society. (1997). Meteorological drought – Policy statement. *Bulletin of the American Mathematical Society, 78*, 847–849.

Bisht, D. S., Sridhar, V., Mishra, A., Chatterjee, C., & Raghuwanshi, N. S. (2019). Drought characterization over India under projected climate scenario. *International Journal of Climatology, 39*(4), 1889–1911.

Dai, A. (2011). Characteristics and trends in various forms of the Palmer drought severity index during 1900–2008. *Journal of Geophysical Research, 116*(D12), D12115.

Gazette, Tamil Nadu Government. 2017. Tamil Nadu Government Gazette.

Government of Tamil Nadu. (2015). *Season and crop reports Tamil Nadu 2015*. Department of Economics and Statistics, Government of Tamil Nadu.

Ghulam, A., Qin, Q., & Zhan, Z. (2007). Designing of the perpendicular drought index. *Environmental Geology, 52*, 1045–1052. https://doi.org/10.1007/s00254-006-s0544-2

Gibbs, W. J., & Maher, J. V. (1967). *rainfall deciles as drought indicators*. Melbourne.

Funk, C. C., Peterson, P. J., Landsfeld, M. F., Pedreros, D. H., Verdin, J. P., Rowland, J. D., Romero, B. E., Husak, G. J., Michaelsen, J. C., & Verdin, A. P. (2014). A quasi-global precipitation time series for drought monitoring. *US Geological Survey Data Series, 832*(4), 1–12.

Hufman, G., Bolvin, D., Nelkin, E., Wolf, D., Adler, R., Gu, G., Hong, Y., Bowman, K., & Stocker, E. (2007). The TRMM multi-satellite precipitation analysis (TMPA): Quasi-global, multiyear, combined-sensor precipitation estimates at fine scales. *Journal of Hydrometeorology, 8*(1), 38–55.

Huffman, G. J., Adler, R. F., Bolvin, D. T., & Nelkin, E. J. (2010). The TRMM multi-satellite precipitation analysis (TMPA). In *Satellite rainfall applications for surface hydrology* (pp. 3–22). Springer.

Huffman, G. J., Bolvin, D. T., Braithwaite, D., Hsu, K., Joyce, R., Kidd, C., Sorooshian, S., Xie, P., & Yoo, S. H. (2012). *Developing the integrated multi-satellite retrievals for GPM (IMERG)* (p. 6921). EGUGA.

India Meteorological Department. (1976). *Hundred Years of Weather Service, 1875–1975*. India Meteorological Department.

Joyce, R. J., Janowiak, J. E., Arkin, P. A., & Xie, P. (2004). CMORPH: A method that produces global precipitation estimates from passive microwave and infrared data at high spatial and temporal resolution. *Journal of Hydrometeorology, 5*(3), 487–503.

Kaptue, A. T., Hanan, N. P., Prihodko, L., & Ramirez, J. A. (2015). Spatial and temporal characteristics of rainfall in Africa: Summary statistics for temporal downscaling. *Water Resources Research, 51*, 2668–2679.

Liu, J., Duan, Z., Jiang, J., & Zhu, A. X. (2015). Evaluation of three satellite precipitation products TRMM 3B42, CMORPH, and PERSIANN over a subtropical watershed in China. *Advances in Meteorology, 2015*, 151239.

McKee, T. B., Doesken, N. J., & Kleist, J. (1993). The relationship of drought frequency and duration to time scales. In *Eighth conference on applied climatology* (pp. 179–184). American Meteorological Society.

Murthy, C. S., Laxman, B., & Sai, M. S. (2015). Geospatial analysis of agricultural drought vulnerability using a composite index based on exposure, sensitivity and adaptive capacity. *International Journal of Disaster Risk Reduction, 12*, 163–171.

Narasimhan, B., & Srinivasan, R. (2005). Development and evaluation of soil moisture deficit index (SMDI) and evapotranspiration deficit index (ETDI) for agricultural drought monitoring. *Agricultural and Forest Meteorology, 133*(1–4), 69–88.

Naumann, G., Barbosa, P., Carrao, H., Singleton, A., & Vogt, J. (2012). Monitoring drought conditions and their uncertainties in Africa using TRMM data. *Journal of Applied Meteorology Climatology, 51*, 1867–1874.

NDMC. (2018). *Putting science in your hands annual report* (pp. 6–7). National Drought Mitigation Center.

Palmer, W. C. (1965). *Meteorological drought.* US Department of Commerce.

Prakash, S. (2019). Performance assessment of CHIRPS, MSWEP, SM2RAIN-CCI, and TMPA precipitation products across India. *Journal of hydrology, 571*, 50–59.

Tang, G., Zeng, Z., Long, D., Guo, X., Yong, B., Zhang, W., & Hong, Y. (2016). Statistical and hydrological comparisons between TRMM and GPM level-3 products over a midlatitude basin: Is day-1 IMERG a good successor for TMPA 3B42V7? *Journal of Hydrometeorology, 17*(1), 121–137.

Wang, Z., Zhong, R., Lai, C., & Chen, J. (2017). Evaluation of the GPM IMERG satellite-based precipitation products and the hydrological utility. *Atmospheric Research, 196*, 151–163.

Wang, Z., Zhong, R., Lai, C., Zeng, Z., Lian, Y., & Bai, X. (2018). Climate change enhances the severity and variability of drought in the Pearl River Basin in South China in the 21st century. *Agricultural and Forest Meteorology, 249*, 149–162.

Wardlow, B., Anderson, M. C., & Verdin, J. (2011). *Remote sensing of drought: Innovative monitoring approaches* (p. 484). CRC Press.

Wilhite, D. (2000). Drought preparedness in the U.S. In J. V. Vogt & F. Somma (Eds.), *Drought and drought mitigation in Europe* (pp. 119–132). Kluwer.

Zhang, Z., Gong, D., Hu, M., Guo, D., He, X., & Lei, Y. (2009). Anomalous winter temperature and precipitation events in southern China. *Journal of Geographical Sciences, 19*(4), 471–488.

Chapter 20
Fluvial Characteristics of Dwarka-Mayurakshi Plain Causing Flood in Kandi Block and Impacts of Manmade Embankments on Riparian Environment

Swati Mollah

20.1 Introduction

Forming the apex of the Ganga delta, Murshidabad district stretches between 23°43′–24°50′ N and 89°49′–88°46′ E in central West Bengal. Its area is 5324 km^2, and the total population of the district is 7,103,807 (DOC-GoI, 2011). The district occupies the region between the Padma—which roughly constitutes its eastern boundary—and the Bhagirathi, the northernmost distributary of the Padma and several floodplains of the rivers joining the Bhagirathi on its west bank—Bansloi, Pagla, Dwaraka, Mayurakshi and Babla.

Flooding is almost an annual event in Murshidabad district with the major ones occurring in 1870, 1884, 1904, 1931–1934, 1956, 1959, 1978, 1986, 1998, 2000 and 2006, 2007 and 2010 (IWD-GoWB, 2007, 2008, 2009, 2016). But different parts of the district are subject to different degrees of flooding. Floods of the district are caused mainly by spillage from the Padma, spillage from the west-bank tributaries of the Bhagirathi (mainly due to tropical storms, sometimes aggravated by sudden water release from reservoirs) and water-logging due to embankments along the feeder canal constructed from the Farakka barrage for resuscitating the Bhagirathi. In recent times, deterioration of drainage due to human activities has also been held responsible for sustained flooding of southern part of the district.

Construction of embankments has its long history in the Lower Gangetic Plain as well as in the district. In the ninetieth century, landlords constructed these embankments to protect their property and agricultural lands from floods (Bhattacharya, 1998, 1999). During the Zamindari period, these embankments were mainly in random nature. But with the growing needs of defence from flood irrigation and

S. Mollah (✉)
Dumkal College, Murshidabad, West Bengal, India

G. S. Bhunia et al. (eds.), *Anthropogeomorphology*, Geography of the Physical Environment, https://doi.org/10.1007/978-3-030-77572-8_20

waterways department, the Government of West Bengal has made extensive attempts to control flood by constructing embankments in the entire Mayurakshi-Dwarka Plain region.

The Kandi block, situated in the Mayurakshi-Dwarka Plain of the district lying to the west of the Bhagirathi, gets flooded by a host of the scouring rivers which rise from the Chhotonagpur Plateau. Earlier these rivers used to traverse over a series of basins replete with *beels* and low lands to detain the courses of the rivers, and they acted as flood cushions; they would release stream-flows considerably late, when the flood-flow in the Bhagirathi receded. The rivers over this region would follow the natural sequence of flooding that prevailed—the Pagla, the Bansloi and the Ajay brought flood-flow first to be followed by the Mayurakshi-Babla system, and the Brahmani-Dwarka came last of all. Now, the British ruler Lord Cornwalis vested native landlords with power to rule their properties locally through the historic 'Permanent Settlement'. Obviously the native Zamindars wanted their tenants to cultivate more lands and thereby the revenue to be pushed up. Hence followed reclamation of lands, and embankments and encroachments intensified the direct flow down the channels across the Rarh region. The vast areas of land, which are used to absorb the floods of these flashy rivers, got segmented with unscientifically aligned and designed circuit embankments.

There was post-independence population influx into this border district in addition to natural growth. To provide for accommodation to this increased population, a drastic human encroachment on the detention basins and rivers across the area followed which evidently put the district's riverine ecosystem to an unprecedented stress. This upset the flood dynamics altogether over the region. Therefore, an un-wholesome extension of cultivation of the reclaimed lands, range lands in particular, has ensued since the independence. Recent spurt in the paddy cultivation has only added to the wrong ongoing, further worsening the riparian environment.

20.2 Study Area

Kandi block is located in the Mayurakshi-Dwarka Plain in the western part of Murshidabad district. The river Dwarka drains the western part of the block, and the river Babla forms the eastern boundary with Bahrampur block (Fig. 20.1). The Hijal Beel covers a good part of the block. The Mayurakshi and Dwarka meet together and their combined flow, i.e. Babla flows into this beel. The total area of the block is 227.48 km^2 with total population 220145 (DOC-GoI, 2011). Of the total population of the district, 3.29% live in this block.

The percentage of area likely to be flooded and the flood-prone population of Kandi block are 75.6 and 92.43, respectively (IWD-GoWB, 2009). The frequency of the occurrence of flood in the last two decades is 8. The central part of the block, i.e. Hijal Beel, and its surrounding areas and the areas along the river Dwarka were severely affected by the flood.

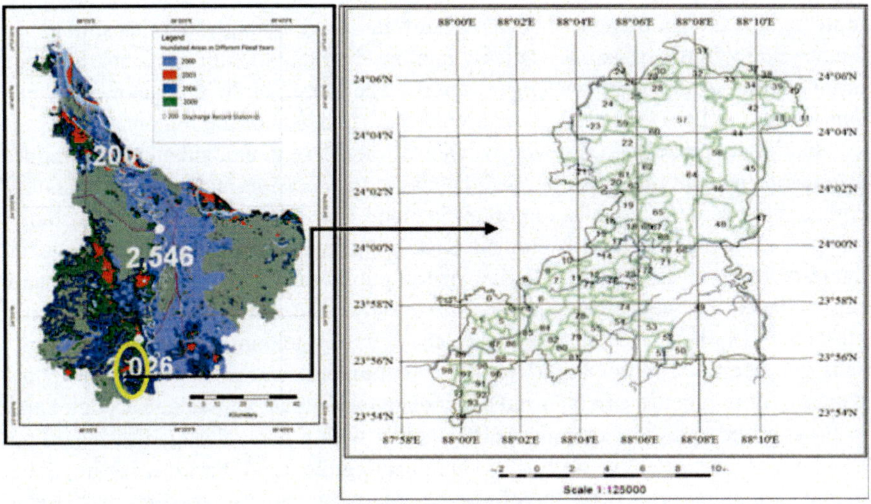

Fig. 20.1 Location of Kandi block in Murshidabad district

20.3 Background Problem

Murshidanad district has about 530 km long embankment along the different rivers; most of the embankments are 4.26 m. to 6.40 m. high and were built during the Zamindari period. Almost every 2–3 years, as flood level increases the height of embankment is elevated without considering its ecological impacts on fluvial environment. Most of the embankments were constructed during the 1950s and 1960s when reliable hydrologic data were not available; later on, they were found to offer inadequate protection. There are many deficiencies found in embankments, including their negative impacts on fluvial environment. On the one hand, the magnitude of flood of different return periods has not been taken under consideration during designing height of embankments, and on the other hand, its impacts on local ecosystem were overlooked. Frequently, embankments are overtopped during rainy season. Hydraulic gradients adopted are not based on actual tests. Drainage sluices provided in the embankments are inadequate resulting in drainage congestion behind the embankments. Embankments are not aligned through a suitable foundation, resulting in seepage through the seat of the embankment. Embankments are constructed close to the riverbanks to protect as many villages and towns as possible and thus costly anti-erosion works has to be undertaken subsequently to protect them. Spurs constructed to protect the embankments have adverse effect on the opposite bank. This is a very common feature in the district. It messes up proper maintenance, and embankments become susceptible to breaches during floods. These breaches result in unexpected flood, causing huge loss, which may not be if there is no embankment. Another aspect is that entire length of the river is not embanked; thus, quantum of floodwater reaching the unembanked reaches will be more and the time of travel will be less. Rainwater and the floodwater that spill in the countryside will not drain back quickly into the river due to embankments resulting in inundation of the areas for longer periods. Therefore, all the embankments were constructed without considering the fluvial characteristics of the region.

Flood management goes beyond technical means (e.g. embankments, spurs); it also recons perception of risk by the inhabitants and decision-makers when it counts community response equally important in flood risk assessment along with those components which are usually studied. Risk is a primary factor in many of political interests, often more important to the general public than other considerations, and certain facets of the perceived risk are strongly related to the demand for risk mitigation (Sjôberg, 1998, 2001). So, the decision-making process necessarily calls for people's participation for efficient implementation of flood protection policies. More practice-oriented and experience-based local knowledge accumulated in flood-prone areas' needs must be reconciled with a general expertise regarding flood issues. The effective ways of communicating the flood hazard message to the public, which are crucial for the efficacy of the participatory approach, have been still rather poorly explored.

The upper reach of the Mayurakshi has been embanked on its right bank—from Bhandirbad to Khatanga and Khatanga to Dhanagram. On the left bank, the embankments are marginal and meant for foiling the barrage pond spreading (Fig. 20.2). Having a very negative impact on the natural drainage regime, these structures are only too vulnerable to failure. The embankment on the left bank of the river stretches

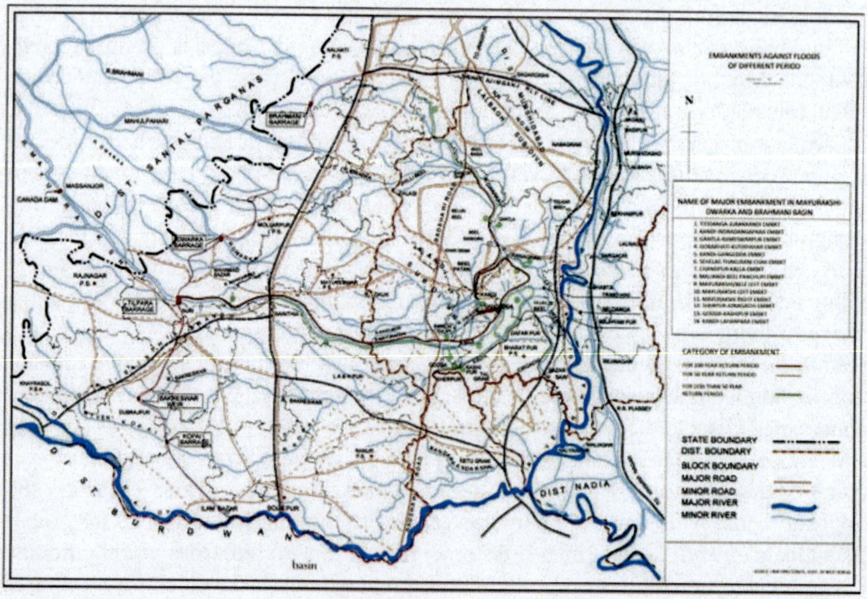

Fig. 20.2 Details of manmade embankments in the Mayurakshi-Dwarka Plain

from Khoirkuri to Sainthia (Birbhum). On the other hand, there is an embankment on its right bank, running from Sainthia to Sougram on the outskirts of Murshidabad. Again, on the left bank, an earthen bund runs from Hatinagar to Ulkinda and another from Ulkinda to Dwar. The extension of it runs from Ramnagar to Mondalpur Circuit embankment in Murshidabad via Rajarampur. In Murshidabad, on the right bank of the river, there run several short-length bunds, e.g. from Wdewar to Vestor, Vestor to Chaitpur, Rajhat to Shiuli, Chaitpur to Sundarpur, Maliandi to Panchthupi, etc. There occurs a bifurcation of the channel close to the village Goda to the right of the main channel. It carries about 40% of the flood flow and joins the river Kuye at Talgram. Following the bifurcation, the name of this reach of the river is the Bele, where the river Chuator falls into it. Abund, a part of Kandi- Gangedda Circuit embankment, runs along the left bank of the Bele from Gangedda to Golaermukh, on the right bank, one which forms a part of Rajarampur-Mondalpur Circuit embankment, from Singerbagan to Pirprikuri. Details of these embankments are shown in Fig. 20.3. Most of these embankments were ineffective in coping with the disastrous floods of 1997, 1998 and 2000 (Rudra, 2000) and thus have stimulated considerable government interest in finding a permanent solution to the flood problem of the area.

20.4 Objectives

The objectives of this paper are:

- To explore the causes of frequent occurrence of flood in Kandi block
- To identify the positive and negative impacts of the construction of embankment in their locality
- To quantify the responses of the inhabitants to the construction of embankment for flood management

20.5 Materials and Methods

To explore the causes behind the frequent occurrence of flood in the study area, LISS III satellite images and SRTM data are used. For perception study, 382 flood victims of 8 hamlets selected from the most flood affected villages of Kandi block are selected. The selected villages are Ranagram (14), Dakshin Lakshminarayanpur (15), Parbatipur (16), Uttar Lakshminarayanpur (30), Uday Chandpur (31), Durgapur (32), Hijal (41) and Benipur (42). These villages are considered based on their locations beside the embankments and frequency of occurrence of flood. A structured survey schedule is used to collect data from the field, and the overall response rate was 91.48%. The respondents have rated their perceptions by

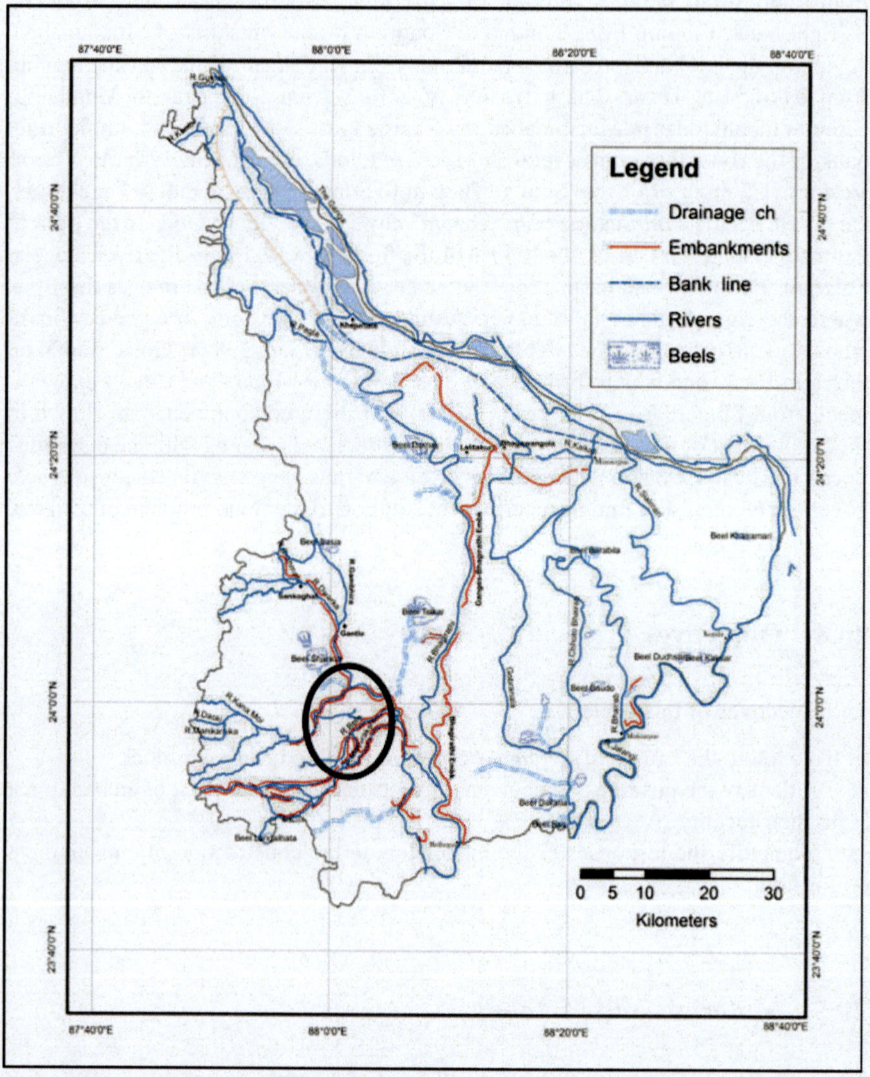

Fig. 20.3 Embankments along rivers of Murshidabad district; Source: IWD-GoWB, 2007 (Prepared from unpublished map)

answering the question in 5-point Likert scale—0 indicating least and 4 indicating most. They have been asked to scale their perception regarding the significance of construction of embankment as the key flood management approach. The assessment of perceptions is done with the help of RIDIT Analysis.

20.5.1 RIDIT Analysis

RIDIT analysis was first proposed by I. Bross (1978) and has been applied to the study of behavioural studies. If there are m items and n ordered categories listed from the most favoured to the least favoured in the scale, then, RIDIT analysis is as follows (Chien-Ho, 2007):

1. Compute ridits for the reference data set (Ref. Table 20.1).
2. Compute ridits and mean ridits for comparison data sets. Since there are m Likert scale items in this illustration, there will be m comparison data sets.

Table 20.1 RIDITS for the reference data sets

	Options	5	4	3	2	1	
1	OPI	96	114	78	62	32	382
2	OP2	73	77	106	89	37	382
3	OP3	42	58	85	93	104	382
4	OP4	95	85	66	49	87	382
5	OP5	42	56	77	97	110	382
6	OP6	31	43	98	122	88	382
7	OP7	48	92	86	74	82	382
8	OP8	109	107	91	44	31	382
9	OP9	40	51	78	74	139	382
10	OP10	80	78	76	73	75	382
11	OP11	77	87	79	76	63	382
12	OP12	120	93	65	58	46	382
13	OP13	111	105	95	37	34	382
14	OP14	95	128	105	33	21	382
15	OP15	90	101	70	65	56	382
16	OP16	96	104	99	53	30	382
17	OP17	35	33	80	109	125	382
18	OP18	100	97	93	52	40	382
19	OP19	103	89	78	69	43	382
20	OP20	66	88	109	67	52	382
21	OP21	118	122	72	41	29	382
22	OP22	121	90	74	54	43	382
	F	1788	1898	1860	1491	1367	8404
	1/2 F	894	949	930	745.5	683.5	
	Fj	894	2737	4616	6291.5	7720.5	
	Rj	0.10638	0.32568	0.54926	0.7486	0.91867	

Note: 1: Strongly disagree; 2: disagree; 3: agree; 4: strongly agree; 5: very strongly agree
Source: Field survey, 2018–2020

(i) Compute ridit value *rij* for each category of scale items.

$$r_{ij} = \frac{R_j \times \pi_{ij}}{\pi_i}, \text{where } i = 1, \ldots\ldots\ldots, m.$$

where π_{ij} is the frequency of category j for the ith scale item and π_i is the sum of frequencies for scale item i across all categories, i.e.

$$\pi_i = \sum_{k-1}^{n} \pi_{ik}$$

(ii) Compute mean riditρ_i for each Likert scale item.

$$\rho_i = \sum_{k=1}^{n} r_{ik}$$

(iii) Compute confidence interval for $\rho_{i.}$

$$\rho_i \pm \frac{1}{\sqrt{3\pi_i}}$$

(iv) Test the following hypothesis using Krushkal-Wallis Statistics W.

$$\begin{cases} H_0 : \forall i, \rho_i = 0.5 \\ H_a : \exists i, \rho_i \neq 0.5 \end{cases}$$

$$W = 12 \sum_{i=1}^{m} \pi_i (\rho_i - 0.5)^2$$

20.6 Results and Discussions

20.6.1 Causes of Flood in the Study Area

There are several prominent and permanently marshy depressions of various shapes and sizes in the study area. Some of them are Hijal Beel, Patan Beel, Belun Beel, Langolhata Beel, Karul Beel, etc. They are mostly abandoned beds of the rivers.

Fig. 20.4 Schematic diagram of Mayurakshi-Dwarka Plain; Source: IWD-GoWB, 2008

Among them Hijal Beel is quiet extensive one. The Hijal Beel, covering about 130 km^2, presents a very low topographic aspect. This part of the block is lower in height than the other parts, thus forming a saucer-shaped depression, and it is easily inundated. All these beels have great capacity to retain excess flood water during rainy season in the study area. But with time, due to jacketing of all rivers in the plain, they have lost their water retention capacity. Recently, the Mayurakshi and Hijal Beel have lost their carrying capacity and storage capacity significantly. The estimated volume of sand deposition of the river Mayurakshi is about 0.786×10 m^3 over the last 54 years. The ability to carry discharges of all the rivers collectively has descended to about 0.786×10^6 m^3. The storage capacity of the Hijal Beel area has got reduced to about 1.90×10^6 to 1.56×10^6 m^3 over the last five decades (Mukhopadhyay & Pal, 2011). Due to its loss of carrying capacity of these rivers and beels, they sometimes result in reverse flow from the points of confluence and inundate the surrounding low-lying area more frequently than earlier times (Fig. 20.4) (Plate 20.1).

Figure 20.5 shows the slope pattern of the northern part of the Dwarka-Mayurakshi Plain, where Kandi block is located, having the maximum slope as 24.3% and the average slope as 1.6% (along AB line). Below that, the average slope reduces to merely 0.6% in the southern part of the Dwarka-Mayurakshi Plain (along CD line). This change of slope in the Dwarka-Mayurakshi Plain from north to south

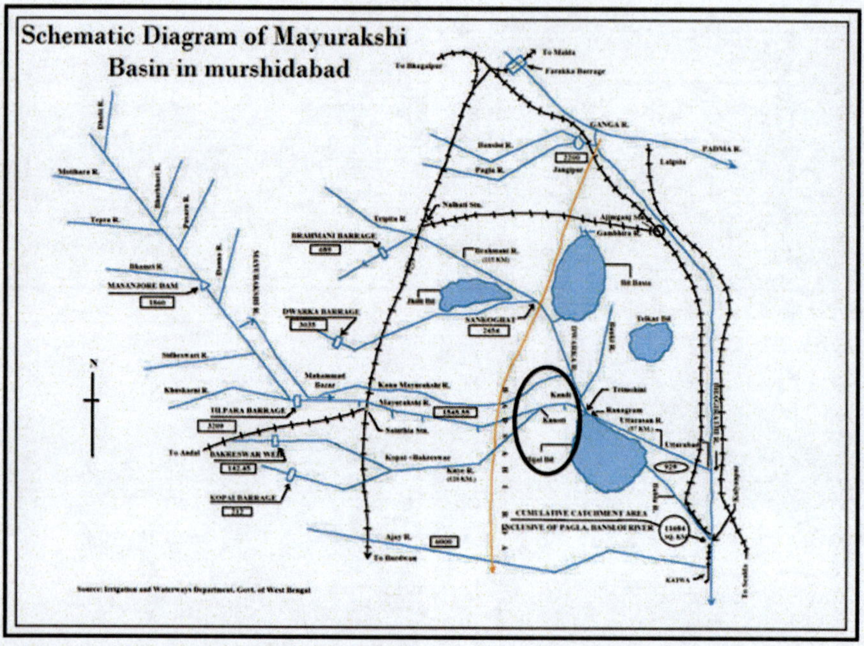

Plate 20.1 Field photos showing flood and embankments in the study area Mohllah (2013)

causes rapid flow of huge quantum of transboundary water from the upstream area during monsoon.

From Fig. 20.6, it is very clear that the topographical feature of the study area controls the occurrence of flood. The whole block is like a saucer, where the middle part forms the depression, inducing flood. Figure 20.7 shows detailed features of the drainage network of the study area, and it is found that there are two confluence points which play an important role in the inundation pattern of the study area. The single channel coming from the Hijal Beel flowing some distance divides into two channels—one in the northern part called river Uttarasan meets the river Bhagirathi near Uttarabad village. Another channel in the south called river Babla meets the river Bhagirathi near Kalyanpur village. Due to its loss of carrying capacity, the river Bhagirathi results in reverse flow from these two points of confluence and inundates the surrounding low-lying area.

20.6.2 *Perception Regarding Impacts of Embankment*

The above discussion reveals that the occurrence of flood in the study area is due to morphological and fluvial features of the area and its surrounding area. But unfortunately without considering all these aspects, only structural measures have been

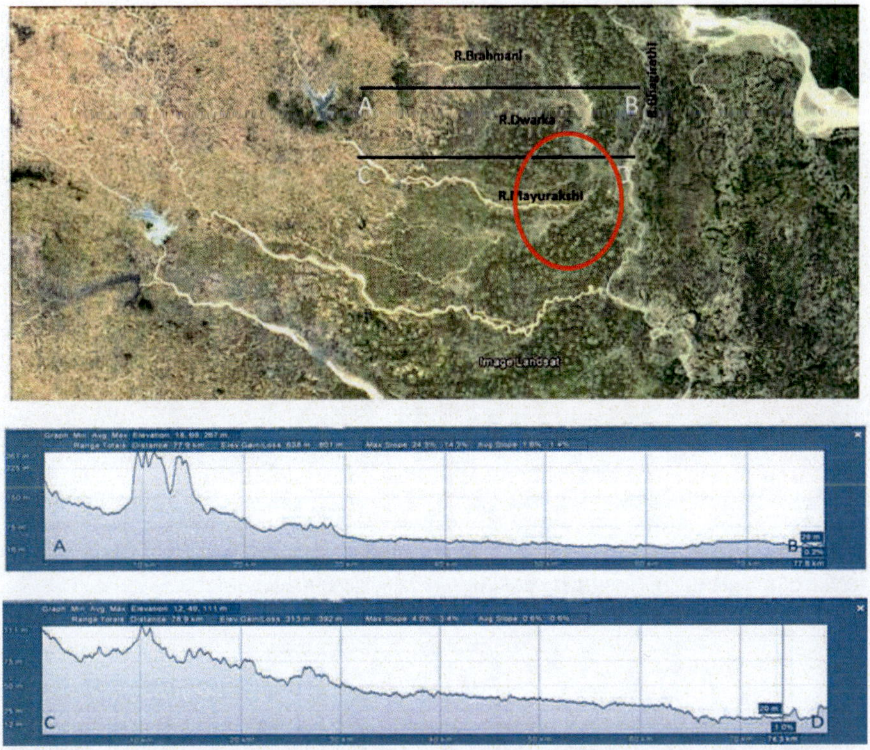

Fig. 20.5 Slope pattern in the Mayurakshi-Dwarka Plain; **ab**: max slope,,24.3; average slope, 1.6; **cd**, max slope, 4.0; average slope, 0.6; Source: Google Earth, 2012

taken to tackle the problem which proves futile in every event. Embanking one part of a river augment the other part more susceptible to flood and in this way every year losses pile up due to flood. But interestingly, the inhabitants of the study area having experience of flood perceive the problem differently than the local government. They are quite aware of the roles of embankment in their locality. Twenty-two impacts of embankments were identified in this study, and the respondents were asked to scale them according to their perceptions (Table 20.1). These options were as follows:

OP1: Increase frequency of flood in lower reach of Mayurakshi-Dwarka Plain.
OP2: Reduce property and road damage.
OP3: Helps in prosperity of the locality.
OP4: Creation of pool causing micro topographic change at the point of embankment breaching.
OP5: Totally stop flood damage..
OP6: Reduce flood frequency.
OP7: Possible to supply water through sluice gate for crop production in time of necessity (+).

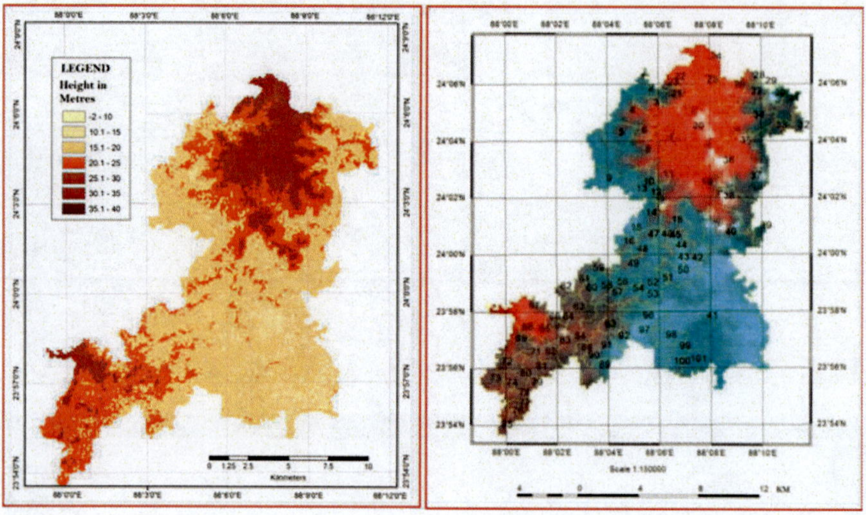

Fig. 20.6 Pattern of elevation (left) and inundation (right) of Kandi block; Source: Prepared from SRTM, 2010 (elevation) and IRS-1D WiFs image, 28.09.2000 (inundation)

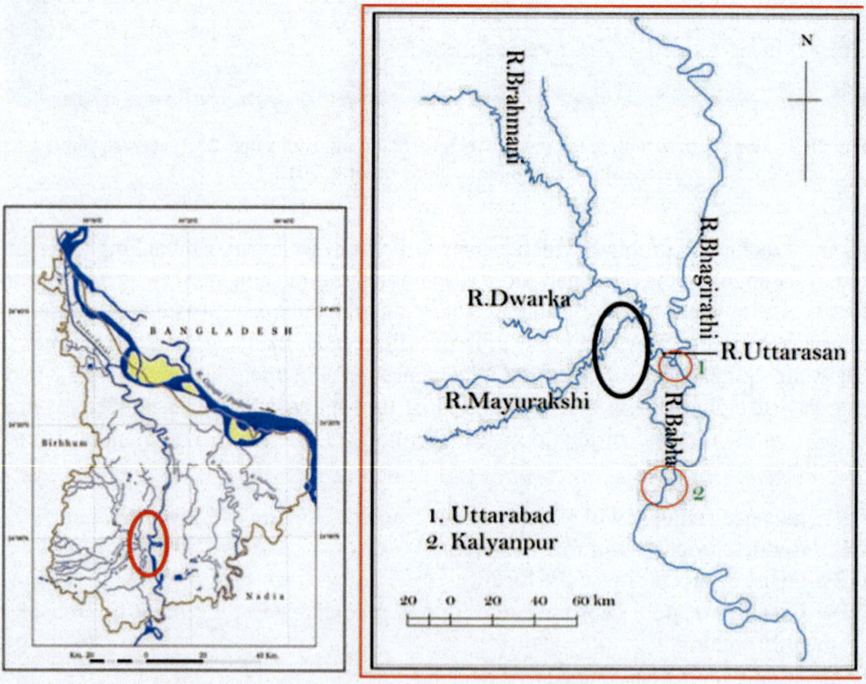

Fig. 20.7 Drainage congestion in Kandi block Source LISS III Data, 2010

Table 20.2 RIDITS for the comparison data sets

	Opinions	5	4	3	2	1	Pi	Rank
1	OPI	0.02673	0.09719	0.11215	0.12151	0.07696	0.0869	*8*
2	OP2	0.02033	0.06565	0.15241	0.17442	0.08898	0.1004	*13*
	OP3	0.01170	0.04945	0.12222	0.18226	0.25011	0.1231	*18*
4	OP4	0.02646	0.07247	0.09490	0.09603	0.20923	0.0998	*12*
5	OP5	0.01170	0.04774	0.11072	0.19010	0.26454	0.1250	*19*
6	OP6	0.00863	0.03666	0.14091	0.23909	0.21163	0.1274	*20*
7	OP7	0.01337	0.07844	0.12366	0.14502	0.19720	0.1115	*17*
8	OP8	0.03035	0.09122	0.13085	0.08623	0.07455	0.0826	*4*
9	OP9	0.01114	0.04348	0.11215	0.14502	0.33428	0.1292	*21*
10	OP10	0.02228	0.06650	0.10928	0.14306	0.18037	0.1043	*16*
11	OP11	0.02144	0.07417	0.11359	0.14894	0.15151	0.1019	*15*
12	OP12	0.03342	0.07929	0.09346	0.11367	0.11063	0.0861	*6*
13	OP13	0.03091	0.08952	0.13660	0.07251	0.08177	0.0823	*3*
14	OP14	0.02646	0.10913	0.15098	0.06467	0.05050	0.0803	*2*
15	OP15	0.02506	0.08611	0.10065	0.12738	0.13467	0.0948	*11*
16	OP16	0.02673	0.08867	0.14235	0.1039	0.07215	0.0868	*7*
17	OP17	0.00975	0.02813	0.11503	0.2136	0.30061	0.1334	*22*
18	OP18	0.02785	0.0827	0.13372	0.1019	0.0962	0.0885	*9*
19	OP19	0.02868	0.07588	0.11215	0.1352	0.10341	0.0911	*10*
20	OP20	0.01838	0.07503	0.15673	0.1313	0.12505	0.1013	*14*
21	OP21	0.03286	0.10401	0.10353	0.0804	0.06974	0.0781	*1*
22	OP22	0.0337	0.07673	0.1064	0.1058	0.10341	0.0852	*5*

Krushkal-Wallis W= 149.54

Source: Table 20.1

OP8: Increase siltation in Hijal Beel.
OP9: No sand deposition in crop fields.
OP10: Facilitate fish cultivation.
OP11: Reduce riverbank erosion.
OP12: Increase in amount of sediment load in the rivers.
OP13: Decrease soil fertility.
OP14: Unexpected loss of crop from sudden flood.
OP15: Loss of soil nutrient in floodplains.
OP16: Increase the speed of river flow.
OP17: Loss of fertile land due to construction of the embankment.
OP18: Loss of riparian wetlands.
OP19: Sand incursion over the agricultural field during breaching of embankments.
OP20: Drainage congestion due to malfunctioning of sluices.
OP21: The embankment-related flood generally comes fast, without warning, and hence is more destructive.
OP22: Hampers natural evolution of floodplains,

The results of the Ridit analysis (Table 20.2) reveal interesting facts. Respondents strongly agree the fact that, breach of embankment-related flood generally comes fast, without warning, and hence is more destructive. As a result this option ranks first in terms of perception of victims about the impact of embankment on the occurrence of flood. The other two negative impacts of embankments such as decrease of soil fertility and unexpected loss of crops rank second and third, respectively. The respondents also feel construction of embankment is not beneficial for flood management; instead, they result in a number of negative impacts on the fluvial environment such as increase in siltation of Hijal Beel, hampering of natural evolution of floodplains, increase in amount of sediment load in the rivers, increase in speed of river flow, increase in frequency of flood occurrence in lower reach of Mayurakshi-Dwarka Plain, etc. Out of all significant impacts of embankment, majority of them are related to negative effects on riverine environment. Almost all respondents feel that the occurrence of flood in Kandi block is due to its unique geomorphological features. Thus, construction of embankments cannot be the ultimate solution to the problem. This calls for rethinking of flood management in the Kandi block of Murshidabad district.

20.7 Conclusion

From the above discussion, it is clear that the topographic features of the study area and its surroundings are exclusively responsible for the frequent occurrence of flood. There are many factors of fluvial morphology to be considered here to manage the flood problem of study area. But without considering all these impacts, embankments are constructed along almost all the rivers draining the area. Local people with their experience of flood uncover the negative impacts of embankments on floodplains. Many times, the embankments cause unexpected flood in areas having no flood history in the past. Thus, embankments in the study area instead of protecting life and properties cause sufferings during the rainy season. This calls for new judgement focusing on the impact of structural measure of flood management.

References

Bhattacharya, K. (1998). Applied geomorphological study in a controlled tropical river-the case of the Damodar between Panchet reservoir and Falta, The University of Burdwan, Burdwan.

Bhattacharya, K. (1999). Floods, flood hazards and hazard reduction measures: a model - the case in the Lower Damodar River. *Indian Journal of Landscape Systems and Ecological Studies, 22*(1), 57–68.

Bross, I. D. J. (1978). Ridit analysis. *American Journal of Epidemiology, 107*, 263–264.

Chien-Ho, W. (2007). On the application of grey relational analysis and RIDIT analysis to likert scale surveys. *International Mathematical Forum, 2*(14), 675–687.

DCO. (2011). *Directorate of Census Operations: District census handbook* (Series-26, part-XII-A). Directorate of Census Operations.

IWD. (2007). *Irrigation and Waterways Department: A note on flood 2000 of West Bengal between 18th to 21st September 2000*. Public Relations-Cum-Statistical cell, Irrigation & Waterways Department, Government of West Bengal.

IWD. (2008). *Irrigation and Waterways Department: Flood damage report of Murshidabad, September–October 2000*. Office of the District Magistrate, Murshidabad, Irrigation and Waterways Department, Government of West Bengal.

IWD. (2009). *Irrigation and Waterways Department Murshidabad: Sechpatra (in Bengali), September–October issue*. Irrigation and Waterways Department, Government of West Bengal.

IWD. (2016). *Irrigation and Waterways Department Murshidabad: Sechpatra (in Bengali), September–October issue*. Irrigation and Waterways Department, Government of West Bengal.

Mohllah, S. (2013). Flood Hazard in Murshidabad district of West Bengal: an environmental appraisal. Ph.D. Thesis, University of Calcutta, Kolkata, India.

Mukhopadhyay, S., & Pal, S. (2011). Trend of flood at riverine Bengal Basin of Kandi block of Murshidabad district: A hydrogeomorphological overview. *Indian Journal of Geography and Environment, 12*, 9–18.

Rudra, K. (2000). *The flood of September 2000, 23*. Jayasree Press.

Sjoberg, L. (1998). Risk perception: Experts and the public. *European Psychologist, 3*, 1–12.

Sjoberg, L. (2001). Political decisions and public risk perception. *Reliability Engineering and System Safety, 72*, 115–123.

Chapter 21
Object-Based Mapping and Modelling of Sundarban Mangrove Forests in India

Sushobhan Majumdar, Uday Chatterjee, Bappaditya Koley, Gouri Sankar Bhunia, and Pravat Kumar Shit

21.1 Introduction

Mangrove forests are mostly developed in tropical and subtropical regions around the globe (Ghosh et al., 2019). They are supplied with numerous goods and services to the society in addition to benefitting both coastal communities' livelihoods and marine ecosystems (Giri et al., 2011; Wanga et al., 2019). Mangrove forests are one of the major barriers to obstruct tropical cyclones and save the coastal areas, and it also protects the coastal areas by dominating the intertidal areas of low tropical and subtropical coastlines (Lugo & Snedekar, 1974; Kathiresan & Bingham, 2001). The mangrove forests protect as a buffer from soil erosion, floods and natural threats to coastal populations (Ghosh et al., 2019; Ismail et al., 2012). Mangrove habitats and the species have significant ecological and economic value in India. It is one of the major ecological resources and also the highly productive ecological communities, and this is one of the integral parts of ecosystem functions (Ewel et al., 1998; Saenger, 2003; Hogarth, 2007; Nagelkerken et al., 2008). Climate changes and accelerated global warming have been blamed for the decline in mangrove forests

S. Majumdar (✉)
Department of Geography, Jadavpur University, Kolkata, India

U. Chatterjee
Department of Geography, Bhatter College, Dantan (Vidyasagar University), Paschim Medinipur, West Bengal, India

B. Koley
Department of Geography, BakimSardar College, Paraganas, West Bengal, India

G. S. Bhunia
Remote Sensing and GIS Expert TPF Gentisa Euroestudios S L, Jammu, India

P. K. Shit
PG Department of Geography, Raja N. L. Khan Women's College (Autonomous), Midnapore, West Bengal, India

G. S. Bhunia et al. (eds.), *Anthropogeomorphology*, Geography of the Physical Environment, https://doi.org/10.1007/978-3-030-77572-8_21

(Eslami-Andargoli et al., 2009). Sea level rise (SLR) is one of the important climate-related factors which affects mangrove forests and coastal wetlands in the long term (Kassakian et al., 2017). In this case, the mapping of mangrove forests is really needed. Forest mapping and forest research are very much needed to take decisions and formulate plans.

Remote sensing (RS) and geographical information system (GIS) have served as sustainable tools in studies of mangrove forests (Blasco et al., 2001; Kumar et al., 2013; Wang et al., 2019). Since the 1970s, remote sensing and GIS tools have been used for forest research (Peterson et al., 1987; Ardo, 1992; Congalton et al., 1993; Gemmell, 1995; Martin et al., 1998; Kilpeläinen & Tokola, 1999; Pax-Lenney et al., 2001; Tokola et al., 2001). Some of the research has been carried out on the ecology of the forest area (Ekstrand, 1994; He et al., 1998; Lucas & Curran, 1999; Coops & Culvenor, 2000).

Maximum forest-oriented studies have been carried out with the help of Landsat satellite images because it is moderate-to-high resolution and open access (Gemmell, 1995; Trotter et al., 1997; Salvador & Pons, 1998; Kilpeläinen & Tokola, 1999; Hyyppä et al., 2000; Bebi et al., 2001; Gao & Zhang, 2009; Lu et al., 2012; Kumar & Acharya, 2016; Alam et al., 2019).

Anderson et al. (1976) used Landsat thematic mapper (TM) images to find out the forest cover for larger regions. Fiorella and Ripple (1993) have also used Landsat TM images for the mapping of forest successional stages and to evaluate spectral differences between old-grown and mature forests in the central Cascade Range of Oregon.

Coastal areas of the Bengal delta were formed within the large basin during the last 11,000 years which extends offshore in the Bay of Bengal as a clinoform (Kuehl et al., 2005; Mikhailov & Dotsenko, 2007).

Sundarban, World's largest mangrove forest, covers 10,000 sq.km area of the lower Ganga-Brahmaputra-Meghnasubaerial delta plain; among these, only 40 per cent of the areas is located in India. Meaning 'beautiful forest' in the Bengali language, the Sundarbans tidal delta contains the most extensive and biologically productive halophytic mangrove forest on earth and is home to several endangered plant and animal species (including the Royal Bengal) (Rogers & Goodbred, 2014).

Sundarbans mangrove provides valuable ecosystem services with socio-economic and environmental functions to support coastal livelihoods (Raha et al., 2012; Getzner & Islam, 2013; Rahman et al., 2018). These forests provide a large variety of wood and non-wood forest products, and it also protects the coastal areas from several natural threats such as cyclones, storm surge, tsunamis, coastal erosion, etc. It is an area with huge biological diversities because of the presence of mammals, amphibians, birds, etc. (FAO, 2007). Their unique root systems create a great deal of physical roughness, thus capturing and storing vast quantities of sediment from upland and oceanic origin (Raha et al., 2012). But the extent and diversity of these forests are declining globally at a rapid rate, and much of the remaining forests are in degraded condition (Duke et al., 2007).

The area of the mangrove forest has decreased by 40 per cent over the past 300 years from harvesting of wood for timber and fuel (Islam et al., 1997).

Sundarbans tidal delta and mangrove forest appears to be morphologically stable; it is commonly cited as a deltaic ecosystem particularly vulnerable to anthropogenic activities and climate-related environmental change (Giri et al., 2007; Rogers & Goodbred, 2014). For example, deforestation in the Sundarbans has been considerably reduced since its classification as a preserved forest and tiger reserve, though illegal harvesting of mangroves for timber still occurs (Rogers & Goodbred, 2014). And yet, despite the value and vulnerability of the mangroves of Bengal delta, very little data exist on the spatio-temporal dynamics of these mangrove lands and the impacts of anthropogenic and natural disturbances on this ecosystem. In fact, recent reviews could find no published reports of ecosystem service loss of mangroves due to land use or climate change (Bouillon et al., 2008; Laffoley & Grimsditch, 2009).

Mangrove in the Sundarban deltas of West Bengal has been depleting very rapidly. So, in this paper, an attempt has been made to find out the spatio-temporal changes of Sundarban mangrove forests in India. Side by side, an attempt has also been made to find out the reasons behind this deforestation.

21.2 Study Area

The Sundarbans are located at the active delta created by the Ganges, Brahmaputra and Meghna Rivers in the Bay of Bengal. For the administrative purposes, it has been divided into two countries, i.e. India and Bangladesh. In 1987 (India) and 1997 (Bangladesh), the Sundarbans were classified as a protected UNESCO World Heritage Site. In India, Sundarbans are situated over two districts in West Bengal, i.e. North 24 Parganas and South 24 Parganas. Various areas, namely, Hasnabad, Baharu, Bakkhali I, Bali, Basanti, Canning, Koikhali I, Pakhirala, Sagar Islands, Jhingakhali, Bakkhali II, Burirdabri, Jharkhali, Koikhali II, Lothian Islands, Netidhopani, Sajnekhali and Sudhanyakhali areas are under Indian parts of Sundarbans. The Indian Sundarbans (between 21°13′N–22°40′N latitude and 88°03′E–89°07′E longitude) is bordered by Bangladesh in the east, the Hooghly River (a continuation of the Ganges River) in the west, the Dampier and Hodges line in the north and the Bay of Bengal in the south (Raha et al., 2012). The total area of Indian Sundarbans is 4260 sq. km (1640 mile). The name of the Sundarbans came from the name of ***SUNDRI*** trees. The study area is a low-lying floodplain with mean elevation of about 2m above mean sea level (MSL) (Spalding et al., 2010; Ghosh et al., 2015; Payo et al., 2016). The study area belongs to tropical monsoonal climate with annual rainfall varying from 1500 to 2000 mm and mean annual minimum and maximum temperatures 18–30 °C, respectively (Banerjee, 2002; Ghosh et al., 2015; Payo et al., 2016). Tidal amplitude within the estuaries is between 3.5 and 4 m, with seasonal variation between 1 and 6 m (Ghosh et al., 2015). Figure 21.1 shows location map of the study area.

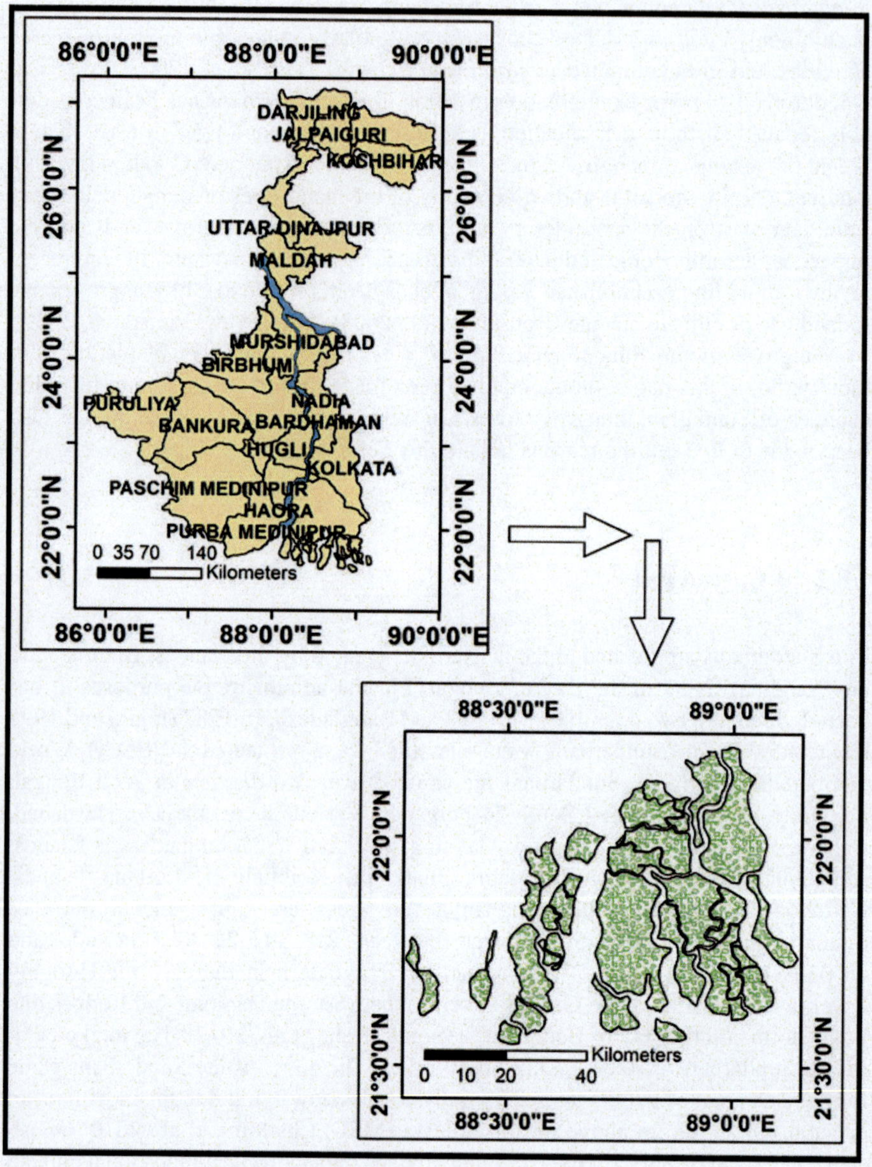

Fig. 21.1 Location map of Indian Sundarban

21.3 Data and Methods

In this study, mainly primary data and secondary data have been used. Secondary data has been collected from the various sources like district gazettes, Census data, district statistical handbook, etc. Among various earth observation (EO) programmes NASA's Landsat missions are known for free dissemination of extensive data. With longest (since 1972) continual global coverage at moderate-to-high resolution, the Landsat data was commonly used for vegetation cover change detection (e.g. Helmer et al., 2000; Lu & Weng, 2004; Gumma et al., 2011; Jia et al., 2014; Ghosh et al., 2015; Alam et al., 2019).

21.3.1 Data Collection

Various types of secondary data have been used. Secondary data has been collected from the various sources like Census of India, District Statistical Handbook, etc. A multi-temporal approach, based on remote sensing data, was used to detect changes in the mangrove-covered area over a span of over 30 years. Satellite data such as Landsat allowed monitoring land cover changes from the 1970s (Lillesand et al., 2004). For the present study, four multispectral cloud-free satellite images (Landsat) have been downloaded from the free available US Geological Survey (USGS:https://earthexplorer.usgs.gov/) in the selected study area.

These orthorectified images (with UTM projection and WGS 84 datum) consist of Landsat-5 Thematic Mapper (TM) data (Path 138, Row 44, dated 14.11.1990 and Path 138, Row 44 2010), Landsat-5 Enhanced Thematic Mapper Plus (ETM+) data (Path 137, Row 44, Dated 17.11.2000) and Landsat-8 Operational Land Imager (OLI) data (Path 138, Row 44, dated 08.03.2020) which have been used for vegetation cover change analysis. Google Earth has been used forgeo-referencing. Detailed information of the Landsat images are listed in Table 21.1. To find out the reasons behind the depletion of mangrove forests, extensive field survey has been done.

21.3.2 Data Processing

The whole image processing procedure was conducted using GIS platform (ArcGIS). Relative radiometric correction method has been used to reduce the influence of inconsistencies of remotely sensed data that may affect the ability to quantitatively analyse as well as interpret images (Song et al., 2001; Al-doski et al.,

Table 21.1 Characteristics of the satellite data used in the present study

Satellite	Date of acquisition	Sensor	Spatial resolution	Projection
Landsat 8	08.03.2020	OLI-TIRS	30m	WGS 84 UTM 45N
Landsat 5	21.01.2010	TM	30m	
Landsat 5	17.11.2000	ETM+	30m	
Landsat 5	14.11.1990	TM	30m	

Source: https://earthexplorer.usgs.gov/

2013). Firstly, converting digital numbers (DNs) has satellite reflectance by using standard calibration values to remove temporal differences in sensor calibration and in environmental factors between image acquisitions. Satellite images have been taken from different times; thus, the atmospheric conditions were different. As requirements for change detection analysis, it is necessary to standardize the effect of atmosphere. The dark object subtraction/histogram minimum method for atmospheric correction has been used. Ground control points (GCPs) have been used to improve the planimetric accuracy of the Landsat data from real-time satellite images of Google Earth.

21.3.3 Techniques

For the purposes of analysis, mainly geospatial techniques have been used. For the visual purposes, few cartographical techniques have been used. To analyse the spatio-temporal changes of the vegetation cover, few geo-statistical techniques have also been used.

21.3.4 Normalized Difference Vegetation Index

A wide range of spectral vegetation indices are generally used in the assessment of vegetation density and canopy closure. Normalized difference vegetation index (NDVI) has been used most extensively for the distribution of mangrove forests and its spatio-temporal changes. It is a numerical measurement that has been used for the mapping of mangrove forests. The main principal of NDVI is that the rates of reflection differ for the Band 4 (NIR) and Band 3 (red) of an image, and thereby these differences can produce an image which characterizes the status of green plants and is least influenced by topography (Datta & Deb, 2012). The NDVI has been calculated by the following equation:

$$\mathbf{NDVI} = \frac{\mathbf{NIR} - \mathbf{RED}}{\mathbf{NIR} + \mathbf{RED}} \tag{21.1}$$

where

'RED' band indicates visible red reflectance.
'NIR' band indicates near-infrared band of the Landsat images.

NDVI is mainly an equation between NIR and RED band which signifies the concentration of chlorophyll in plant leaf. The wavelength range of NIR band ranges between 750 and 1300 nm and RED band is 600 and 700 nm. The value of NDVI ranges between −1 and +1. The values of NDVI lower than 0.1 indicate the presence of barren areas, sandy or snowy surfaces. Moderate values of NDVI indicate the presence of the shrub and grassland, i.e. 0.2–0.3. The value between 0.6 and 0.8 indicates the presence of dense vegetation cover, while the values closest to 0 indicate the presence of water bodies.

21.3.5 Object-Based Classification

According to Baatz and Schape (2000), it is region-merging technique. Mainly spectral and spatial information for classification of the satellite images have been used for the object-based classification and object-oriented classification. Comparatively, it is new approach that has been developed in the recent decades. By the object-based classification, the researcher can easily identify similar pixels or pixels having same digital number values. There are mainly two processes in case of object-based classification: first to identify the objects and next to classify them.

21.4 Data Analysis and Results

Mangrove forest is one of the major resources in the Sundarban areas. From the field survey, it has been found that mangrove resources have been depleting very rapidly in the Indian Sundarban areas. For the purposes of analysis, NDVI techniques have been used. Four satellite images have been used for 30 years' time span, and the time interval is 10 years. By applying NDVI indices, it has been found that the value of the NDVI ranges from +0.66 to −0.62. The ranges of NDVI values are high in the southern parts of the Sundarban areas. In 2000, image showed that the values of NDVI range from +0.56 to −0.53. In this year, the average value of NDVI indices has been decreased. In 2010, the values of NDVI have also been decreased, which range between +0.55 and −0.37. But in 2020, the indices of NDVI have been hugely decreased which range between +0.37 and −0.38. The average NDVI indices have also been decreased. Rapid depletion of mangrove resources will give a serious threat to local land resources and also the forest ecosystem. Figure 21.2 shows NDVI map of Sundarban mangrove forests in India.

Table 21.2 shows detailed information about the mangrove forests in Sundarban area. In 1990, by applying NDVI indices, it has been found that the NDVI values are high in the southeastern parts of the Sundarban area which gradually decrease northwards. Vegetation is denser in the southern parts of the Sundarban area than any other parts. In 1990, very high index value ranges from 0.35 to 0.70 which can be seen in the southeastern parts of the Sundarbans. It is followed by the ranges, i.e. 0.00 to 0.35, which can be found in the northeastern and central portion of the Sundarban area which is Netidhopani and Sajnekhali area. It is followed by the ranges between −0.35 to 0.00 and −0.70 to −0.35 which can be found in the upper portion and eastern parts of the Indian Sundarban area. In this time period, the lower ranges of NDVI values can be found in the Hasnabad and Baharu areas.

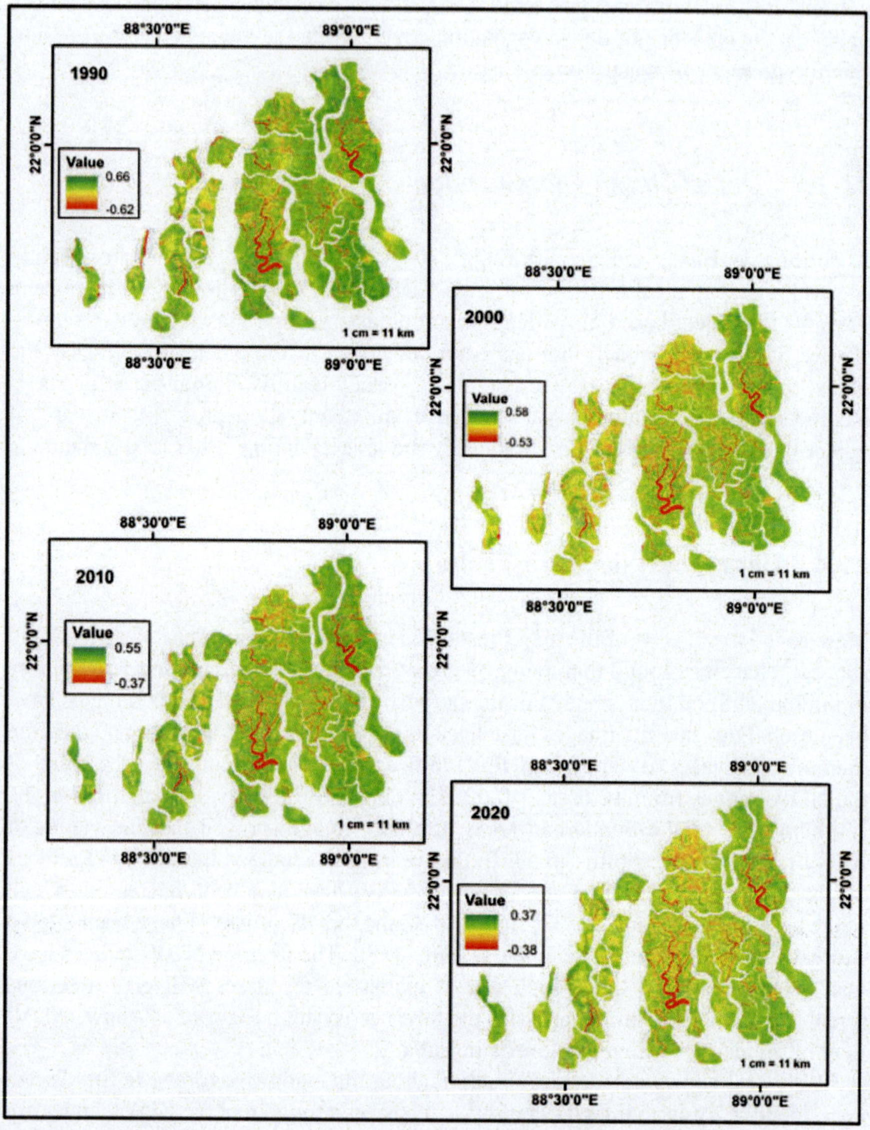

Fig. 21.2 NDVI map of Indian Sundarban mangrove forest (1990, 2000, 2010, 2000)

In 2000, high values of NDVI range from 0.35 to 0.70, which can be found in the extreme southeastern parts of the Sundarban area. It is followed by the range, i.e. 0.00 to 0.35, which can be found between Bakhhali and Sudhanyakhali areas of the South 24 Parganas. Lower ranges of NDVI, i.e. −0.35 to 0.00 and 0.70 to −0.35, can be found in the northern parts of the Indian Sundarban area, i.e. Canning, Hasnabad and Baharu area.

Table 21.2 Detailed information of vegetation

1990		2000		2010		2020	
Ranges of NDVI	Area (%)	Ranges of NDVI	Area (%)	Ranges of NDVI	Area (%)	Ranges of NDVI	Area (%)
−0.70 to −0.35	17.01	−0.60 to −0.30	22.47	−0.60 to −0.30	26.34	−0.40 to −0.20	28.26
−0.35 to 0.00	23.34	−0.30 to 0.00	26.21	−0.30 to 0.00	29.32	−0.20 to 0.00	33.32
0.00 to 0.35	27.54	0.00 to 0.30	23.22	0.00 to 0.30	18.57	0.00 to 0.20	17.41
0.35 to 0.70	32.11	0.30 to 0.60	28.1	0.30 to 0.60	25.77	0.20 to 0.40	21.01

Table 21.3 Change in NDVI indices

Ranges	1990–2000	2000–2010	2010–2020
Lower index value	5.46	3.87	1.92
Medium index value	2.87	3.11	4
High index value	−4.32	−4.65	−1.16
Very high index value	−4.01	−2.33	−4.76

In 2010, the high NDVI values range from 0.35 to 0.70. The values of NDVI are high in the southern and central portion of the Sundarban area. NDVI values and mangrove areas are high in the Burirdabri, Jharkhali area. Other areas experience relatively lower ranges of NDVI values than the other areas.

In 2020, NDVI ranges between 0.40 and −0.20. The values of NDVI are very high in the southern parts of Sundarban delta. Percentage of vegetation is quite high (0.00–0.20) in the central and eastern parts of delta. NDVI values are low in the northern part of the Sundarban delta. Table 21.2 shows vegetation characteristics in Sundarban area.

Table 21.3 shows changes in the NDVI values of the Sundarban delta. It indicates that the percentage of dense vegetation has been decreasing very rapidly which creates adverse impact on the local land resources (Fig. 21.3).

By applying object-based classification, it has been found that mangrove forests in the Sundarban delta areas are decreasing very rapidly which create adverse impact in the local ecosystems. Figure 21.4 describes the spatio-temporal changes of the vegetation cover from 1990 to 2020 (30 years). From 1990, drastic changes of vegetation can be seen in 2020. Various areas in Sundarban areas have been experiencing rapid loss of vegetation cover. Mangrove forests in the areas near Jharkhali, Bakkhali, Baharu and Canning areas have rapidly decreased which is mainly because of the population pressure and urban development in that area. In some areas like Canning, Bakkhali, etc., mangrove forests have been destructed because of the illegal tourism practices. In some areas like Jharkhali, Sagar islands, mangrove forests have decreased mainly because of the unplanned developments.

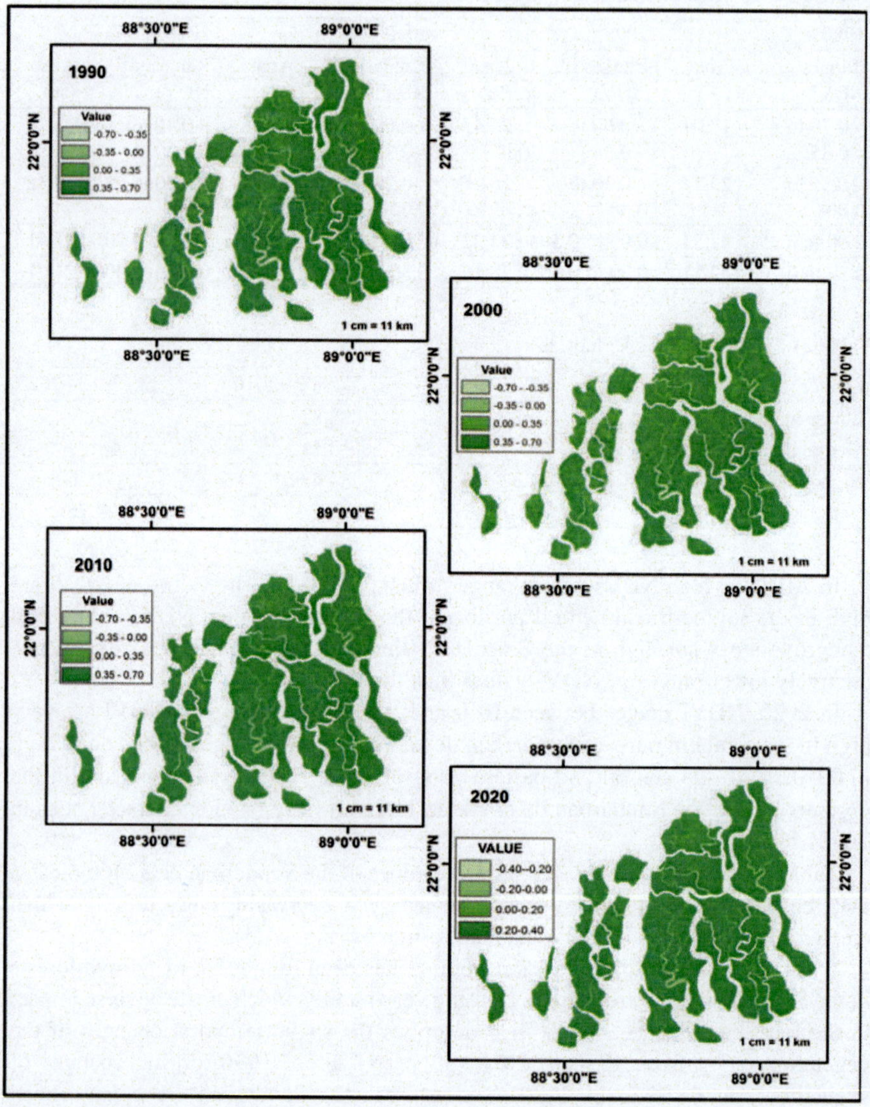

Fig. 21.3 Vegetation characteristics of Indian Sundarban mangrove forest

21.5 Discussion

From the above analysis, it has been found that mangrove forests in the Indian Sundarbans have been declining very rapidly. The depletion of the forest has been increasing in recent decades as it has been found during this study. After the field visit, few reasons have been observed, which are as follows:

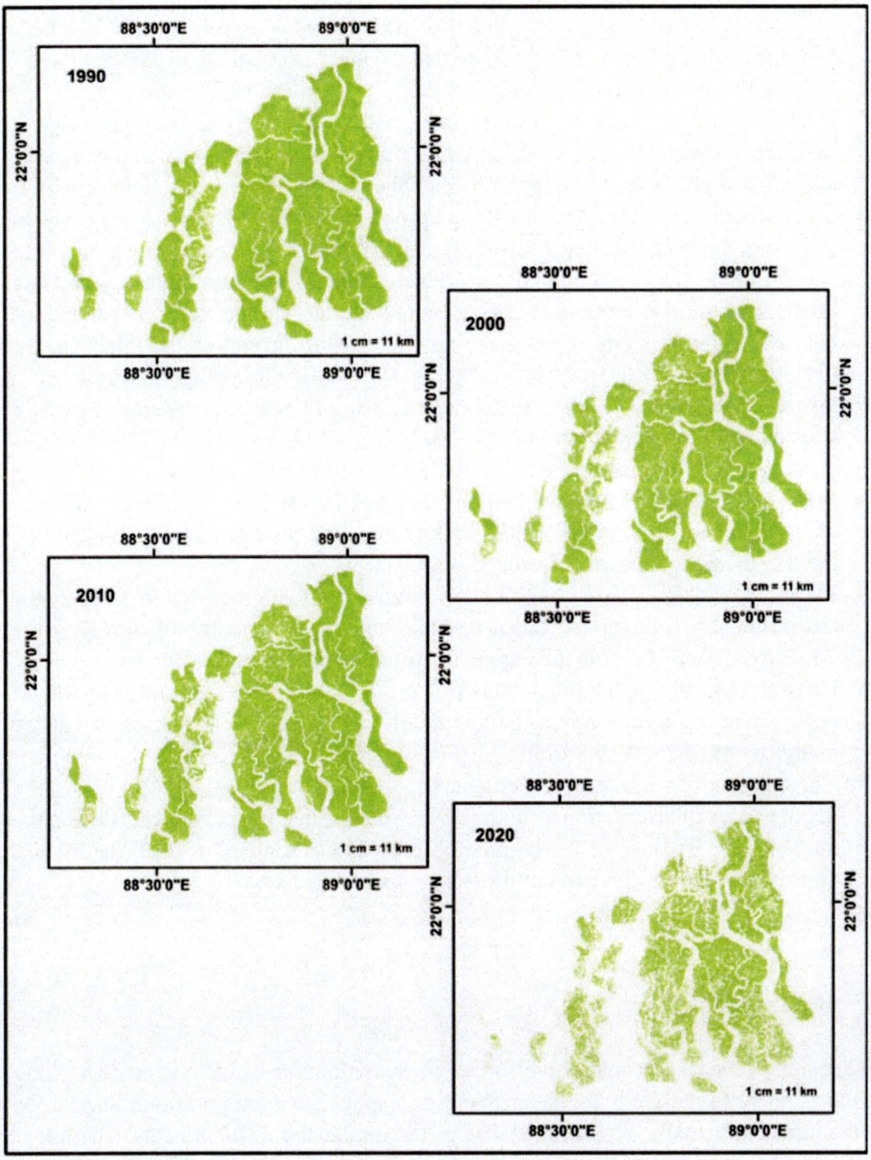

Fig. 21.4 Object-based classification of Indian Sundarban mangrove forest

1. Sea level rise is one of the major reasons in the coastal parts of India. Rate of sea level rise is high in the eastern parts of India than the western parts. According to SMRC (2003), the rate of sea level rise is high in the western parts, which is 4.0 mm year^{-1}, and in the eastern parts, it is 7.8 mm year^{-1}.

2. Increase in water level due to global warming is also a problem that has been faced by the mangrove forests in recent decades. According to Eriscson et al. (2006), rise of sea level in the Bay of Bengal is highest, i.e. 10 mm $year^{-1}$.
3. Deforestation because of various man-made reasons is also a problem in Sundarban areas. Because of the construction of human settlements, unscientific agricultural practices, mangrove forests have been declining very much rapidly.
4. Land subsidence rate and its associated problems i.e. land degradation are one of the major problems in the coastal areas of India. In the coastal areas near the Bengal delta, the mean annual subsidence rate is 15–50 mm (Stanley & Hait, 2000; Mikhailov & Dotsenko, 2007). Bengal delta is sinking at an enormous rate because of the removal of oil, gas and water from the soil of the inland deltas (Syvitski et al., 2009). Various reasons such as the trapping of sediment in upstream reservoirs, sinking in floodplain, rising of sea level etc. are the major reasons behind this lowering of sea level (Kummu & Varis, 2007; Bhattacharya et al., 2012; Unnikrishnan et al., 2014; Chowdhurya & Behera, 2015). Ghosh et al. (2017) noticed a positive trend of sea level rise during 1993–2010 in the Bay of Bengal. Han et al. (2010) and Unnikrishnan and Shankar (2007) showed that the rise of sea level is high in the Bay of Bengal.
5. Lack of policies is also a problem to protect the mangrove trees. Though the Government has sanctioned various policies to protect the mangrove forests in the Sundarban areas, but it is inadequate to solve these types of problems.
6. Unawareness among the inhabitants of the Sundarban areas regarding the advantages of the mangrove forests is completely unknown. So, they used to cut the mangrove tress for their various purposes.
7. Construction of the human settlements in the forest buffer areas is also one of the major causes of destruction of mangrove resources that has been found during the field survey. Because of the population growth and scarcity of land, the inhabitants of these areas have to cut those mangrove forests.

21.6 Conclusion

This study reveals that mangrove forests in Sundarban areas have been decreasing very rapidly. At present time, mangrove forests in the northern portion of Sundarban area have been totally depleted because of the unplanned developments. The intensity of mangrove forests has also decreased in the southern parts of Sundarban areas because of the illegal practices. Government officials and policy-makers should take various planning strategies to mitigate this problem. This type of study will throw a new light for the mapping of the forest areas.

References

Alam, A., Sultan Bhat, M., & Maheen, M. (2019). Using Landsat satellite data for assessing the land use and land cover change in Kashmir valley. *GeoJournal*. https://doi.org/10.1007/s10708-019-10037-x

Anderson, J. R., Hardy, E. E., Roach, J. T., & Witmer, R. E. (1976). *A land use and land cover classification system for use with remote sensor data* (Professional Paper 964) (p. 28). US Geological Survey.

Ardo, J. (1992). Volume quantification of coniferous forest compartments using spectral radiance recorded by Landsat Thematic Mapper. *International Journal of Remote Sensing, 13*(9), 1779–1786.

Banerjee, L. K. (2002). Sundarbans. In N. P. Shing & K. P. Shing (Eds.), *Floristic Diversity and Conservation Strategies in India, vol. V* (Botanical Survey of India) (pp. 2801–2829). Ministry of Environment and Forests.

Baatz, M., Schape, A. (2000). Multiresolution segmentation: an optimization approach for high quality multi-scale image segmentation. In: Strbl J, Blaschke T (eds) Angewandte geographische informationsverarbeitung. Wichmann, Heidelberg, pp 12–23.

Bebi, P., Kienast, F., & Schönenberger, W. (2001). Assessing structures in mountain forests as a basis for investigating the forests' dynamics and protective function. *Forest Ecology and Management, 145*(1), 3–14.

Bhattacharya, A. K., Bhattacharya, A., Sarkar, S. K., & Chatterjee, M. (2012). Bengal Basin, Sediment Sink. In L. Bengtsson, R. W. Herschy, & R. W. Fairbridge (Eds.), *Encyclopedia of Lakes and Reservoirs* (Encyclopedia of Earth Sciences Series). Springer. https://doi.org/10.1007/978-1-4020-4410-6_194

Blasco, F., Aizpuru, M., & Gers, C. (2001). Depletion of the mangroves of continental Asia. *Wetlands Ecology and Management, 9*, 255–266.

Bouillon, S., Borges, A. V., Castañeda-Moya, E., Diele, K., & Dittman, T. (2008). Mangrove production and carbon sinks: A revision of global budget estimates. *Global Biogeochemical Cycles, 22*, GB2013. https://doi.org/10.1029/2007GB003052

Chowdhurya, P., & Behera, M. R. (2015). A study on regional sea level variation along the Indian coast. *Procedia Engineering, 116*, 1078–1084.

Congalton, R. G., Green, K., & Teply, J. (1993). Mapping old growth forests on national forest and park lands in the Pacific Northwest from remotely sensed data. *Photogrammetric Engineering and Remote Sensing, 59*(4), 529–535.

Coops, N., & Culvenor, D. (2000). Utilizing local variance of simulated high spatial resolution imagery to predict spatial pattern of forest stands. *Remote Sensing of Environment, 71*(3), 248–260.

Datta, D., & Deb, S. (2012). Analysis of coastal land use/land cover changes in the Indian Sundarbans using remotely sensed data. *Geo-spatial Information Science, 15*(4), 241–250.

Duke, N., Meynecke, J., Dittmann, S., Ellison, A. M., & Anger, K. (2007). A world without mangroves. *Science, 317*(5834), 41–42.

Ekstrand, S. (1994). Landsat TM based forest damage assessment Correction for topographic effects. *Photogrammetric Engineering and Remote Sensing, 62*(2), 151–161.

Eriscson, J. P., Vörösmarty, C. J., Dingman, S. L., Ward, L. G., Meybeck, M. (2006). Effective sea-level rise and deltas: Causes of change and human dimension implications, *Global Planet Change, 50*, 63–82.

Eslami-Andargoli, L., Dale, P., Sipe, N., & Chaseling, J. (2009). Mangrove expansion and rainfall patterns in Moreton Bay, southeast Queensland, Australia. *Estuarine, Coastal and Shelf Science, 85*(2), 292–298.

Ewel, K. C., Twilley, R. R., & Ong, J. E. (1998). Different kinds of mangrove forests provide different goods and services. *Global Ecology and Biogeography Letters, 7*, 83–94.

Fiorella, M., & Ripple, W. (1993). Analysis of conifer forest regeneration using Landsat Thematic Mapper data. *Photogrammetric Engineering and Remote Sensing, 59*(9), 1383–1388.

FAO. (2007). *The world's mangroves 1980–2005* (FAO Forestry Paper, 153). Food, Agriculture Organization of the United Nations.

Gao, Y., & Zhang, W. (2009). LULC classification and topographic correction of Landsat-7 ETM? Imagery in the Yangjia River Watershed: The influence of DEM resolution. *Sensors, 9*(3), 1980–1995. https://doi.org/10.3390/s90301980

Gemmell, F. M. (1995). Effects of forest cover, terrain, and scale on timber volume estimation with thematic mapper data in a rocky mountain site. *Remote Sensing of Environment, 51*(2), 291–305.

Getzner, M., & Islam, M. S. (2013). Natural resources, livelihoods, and reserve management: a case study from Sundarbans mangrove forests, Bangladesh. *WIT Transactions on Ecology and the Environment, 8*(1), 75–87.

Ghosh, A., Schmidt, S., Fickert, T., & Nüsser, M. (2015). The Indian Sundarban Mangrove Forests: history, utilization, conservation strategies and local perception. *Diversity, 7*(2), 149–169.

Ghosh, M. K., Kumar, L., & Langat, P. K. (2019). Geospatial modelling of the inundation levels in the Sundarbans mangrove forests due to the impact of sea level rise and identification of affected species and regions, Geomatics. *Natural Hazards and Risk, 10*(1), 1028–1046. https://doi.org/10.1080/19475705.2018.1564373

Ghosh, S., Hazra, S., Nandya, S., Mondal, P. P., Wathama, T., & Kushwahaa, S. P. S. (2017). Trends of sea level in the Bay of Bengal using altimetry and other complementary techniques. *Journal of Spatial Science*. https://doi.org/10.1080/14498596.2017.1348309

Gumma, M. K., Thenkabail, P. S., Hideto, F. Nelson, A. Dheeravath, V. Busia, D., & Rala, A. (2011). Mapping irrigated areas of ghana using fusion of 30 m and 250 m resolution remote sensing data. *Remote Sensing, 3*, 816–835.

Giri, C., Zhu, Z., Tieszen, L. L., Singh, A., Gillette, S., & Kelmelis, J. A. (2007). Mangrove forest distributions and dynamics (1975–2005) of the tsunami-affected region of Asia. *Journal of Biogeography, 35*, 519–528. https://doi.org/10.1111/j.1365-2699.2007.01806.x.

Giri, C., Ochieng, E., Tieszen, L. L., Zhu, Z., Singh, A., Loveland, T., Masek, J., & Duke, N. (2011). Status and distribution of mangrove forests of the world using earth observation satellite data. *Global Ecology and Biogeography, 20*, 154–159.

Han, X., Feng, L., Hu, C., & Kramer, P. (2010). Hurricane-Induced changes in the Everglades National Park mangrove forest: Landsat observations between 1985 and 2017. *Journal of Geophysical Research: Biogeosciences, 123*, 3470–3488.

He, H. S., Mladenoff, D. J., & Radeloff, V. C. (1998). Integration of GIS data and classified satellite imagery for regional forest assessment and landscape modeling. *Ecological Applications, 8*, 1072–1083.

Helmer, E. H., Cohen, W. B., & Brown, S. (2000). Mapping montane tropical forest successional stage and land use with multi-date Landsat imagery. *International Journal of Remote Sensing 21,*, 2163–2183.

Hogarth, P. J. (2007). *The biology of Mangroves and Seagrasses* (p. 273). Oxford University Press.

Hyyppä, J., H. Hyyppä, M. Inkinen, M. Engdahl, S. Linko & Y. H. Zhu. (2000). Accuracy comparison of various remote sensing data sources in the retrieval of forest stand attributes. *Forest Ecology and Management, 128*, 109–120.

Ismail, H., Wahab, A. A., & Alias, N. (2012). Determination of mangrove forest performance in reducing tsunami run-up using physical models. *Natural Hazards, 63*(2), 939–963.

Islam, Md. Alam, S. & Maudood E. (1997). Remote sensing for change detection in the Sundarbands, Bangladesh, *Geocarto International, 12*(3), 91–100, https://doi.org/10.1080/10106049709354601.

Jia, M., Wang, Z., Li, L., Song, K., Ren, C., Liu, B., & Mao, D. (2014). Mapping China's mangroves based on an object-oriented classification of Landsat imagery. *Wetlands*, 34–41. https://doi.org/10.1007/s13157-013-0449-2.

Al-doski, J., Shattri, B. M., & Helmi Zulhaidi Mohd Shafri. (2013). NDVI differencing and post-classification to detect vegetation changes in Halabja City, Iraq. *Journal of Applied Geology and Geophysics, 1*(2), 01–10.

Kassakian, J., Jones, A., Martinich, J., & Hudgens, D. (2017). Managing for no net loss of ecological services: an approach for quantifying loss of coastal wetlands due to sea level rise. *Environmental Management, 59*(5), 736–751.

Kathiresan, K., & Bingham, B. L. (2001). Biology of mangroves and mangrove ecosystems. *Advances in Marine Biology, 40*, 81–251.

Kilpeläinen, P., & Tokola, T. (1999). Gain to be achieved from stand delineation in LANDSAT TM image-based estimates of stand volume. *Forest Ecology and Management, 124*, 105–112.

Kuehl, S. A., Allison, M. A., Goodbred, S. L., & Kudrass, H. (2005). The Ganges– Brahmaputra Delta. In L. Gosian & J. Bhattacharya (Eds.), *River deltas—Concepts, models, and examples* (Vol. 83, pp. 413–434). SEPM Special Publication.

Kumar, S., Panigrahy, P., Kumar, J. S., & Parihar. (2013). Classification of floristic composition of mangrove forests using hyperspectral data: case study of Bhitarkanika National Park, India *J. Coast. Conserv., 17*, 121–132.

Kumar, R., & Acharya, P. (2016). Flood hazard and risk assessment of 2014 floods in Kashmir valley: A space based multisensor approach. *Natural Hazards, 84*(1), 437–464. https://doi.org/10.1007/s11069-016-2428-4

Kummu, M., & Varis, O. (2007). Corrigendum to 'Sediment-related impacts due to upstream reservoir trapping, the Lower Mekong River. *Geomorpholgy, 85*, 275–293. https://doi.org/10.1016/j.geomorph.2006.03.024

Laffoley, D. A., & Grimsditch, G. (2009). *The management of natural coastal carbon sinks* (p. 53). IUCN.

Lillesand, T. M., Kiefer, R. W., & Chipman, J. W. (2004). *Remote sensing and image interpretation* (p. 704). Wiley.

Lu, D., & Weng, Q., (2004). Spectral mixture analysis of the urban landscapes in Indianapolis with Landsat ETM + imagery. *Photogrammetric Engineering and Remote Sensing, 70*, 1053–1062.

Lu, D., Hetrick, S., Moran, E., & Li, G. (2012). Application of time series landsat images to examining land-use/landcover dynamic change. *Photogrammetric Engineering and Remote Sensing, 78*(7), 747–755.

Lucas, N. S., & Curran, P. J. (1999). Forest ecosystem simulation modeling, the role of remote sensing. *Progress in Physical Geography, 23*(3), 391–423.

Lugo, A. E., & Snedekar, S. C. (1974). The ecology of mangroves. *Annual Review of Ecology and Systematics, 5*, 39–64.

Martin, M. E., Newman, S. D., Aber, J. D., & Congalton, R. G. (1998). Determining forest species using high spectral resolution remote sensing data. *Remote Sensing of Environment, 65*, 249–254.

Mikhailov, V., & Dotsenko, M. (2007). Processes of delta formation in the mouth area of the Ganges and Brahmaputra rivers. *Water Resources, 34*. 385–400. https://doi.org/10.1134/S0097807807040033.

Nagelkerken, I., Blaber, S. J. M., Bouillon, S., Green, P., Haywood, M., Kirton, L. G., Meynecke, J. O., Pawlik, J., Penrose, H. M., & Sasekumar, A. (2008). The habitat function of mangroves for terrestrial and marine fauna: A review. *Aquatic Botany, 89*, 155–185.

Pax-Lenney, M., Woodcock, C. E., Macomber, S. A., & Song, C. (2001). Forest mapping with a generalized classifier and Landsat TM data. *Remote Sensing of Environment, 77*, 241–250.

Payo, A., Mukhopadhyay, A., Hazra, S., Ghosh, T., Ghosh, S., Brown, S., Nicholls, R. J., Bricheno, L., Wolf, J., Kay, S., et al. (2016). Projected changes in area of the Sundarban mangrove forest in Bangladesh due to SLR by 2100. *Climatic Change, 139*(2), 279–291.

Peterson, D. L., Spanner, M. A., Running, S. W., & Teuber, K. B. (1987). Relationship of thematic mapper simulator data to leaf area index of temperate coniferous forest. *Remote Sensing of Environment, 22*, 323–341.

Raha, A., Das, S., Banerjee, K., & Mitra, A. (2012). Climate change impacts on Indian Sundarbans: A time series analysis (1924–2008). *Biodiversity and Conservation*. https://doi.org/10.1007/s10531-012-0260-z

Rahman, M., Jiang, Y., & Irvine, K. (2018). Assessing wetland services for improved development decision-making: A case study of mangroves in coastal Bangladesh. *Wetlands Ecology and Management, 26*, 563–580. https://doi.org/10.1007/s11273-018-9592-0

Rogers, K. G., & Goodbred, S. L. (2014). Chapter 18: The Sundarbans and Bengal delta: The world's largest tidal Mangrove and Delta system. In V. Kale (Ed.), *Landscapes and Landforms of India* (World Geomorphological Landscapes) (pp. 181–187). Springer. https://doi.org/10.1007/978-94-017-8029-2_18

Saenger, P. (2003). *Mangrove ecology, silviculture and conservation*. https://doi.org/10.1007/978-94-015-9962-7

Salvador, R., & Pons, X. (1998). On the reliability of Landsat TM for estimating forest variables by regression techniques, a methodological analysis. *IEEE Transactions on Geoscience and Remote Sensing, 36*(6), 1888–1897.

SMRC (2003). The vulnerability assessment of the SAARC coastal region due to sea level rise: Bangladesh Case, SMRC-No. 3, SMRC Publication, Dhaka, Bangladesh.

Song, C., Woodcock, C. E., Seto, K. C., Lenney, M. P., & Macombe, S. A. (2001). Classification and change detection using Landsat TM data: When and how to correct atmospheric effects? *Remote Sensing of Environment, 75*, 230–244.

Spalding, M., Kainuma, M., & Collins, L. (2010). *World atlas of mangroves*. Routledge.

Stanley, D. J., & Hait, A. K. (2000). Holocene depositional patterns, Neotectonics and Sundarban mangroves in the western Ganges–Brahmaputra delta. *J Coastal Res 16*, 26–39.

Syvitski, J. P. M., Kettner, A. J., Overeem, I., Hutton, E. W. H., Hannon, M. T., Brakenridge, G. R., Day, J., Vörösmarty, C., Saito, Y., Giosan, L., & Nicholls, R.J. (2009). "Sinking deltas due to human activities". *Nature Geoscience, 2*(10), 681–686.

Tokola, T., Sarkeala, J., & Linden, M. V. D. (2001). Use of topographic correction in Landsat TM-based forest interpretation in Nepal. *International Journal of Remote Sensing, 22*(4), 551–563.

Trotter, C. M., Dymond, J. R., & Goulding, C. J. (1997). Estimation of timber volume in a coniferous plantation forest using Landsat TM. *International Journal of Remote Sensing, 18*(10), 2209–2223.

Unnikrishnan, A., & Shankar, D.. (2007). Are sea-level-rise trends along the coasts of the north Indian Ocean consistent with global estimates?. *Global and Planetary Change. 57*, 301–307. https://doi.org/10.1016/j.gloplacha.2006.11.029.

Unnikrishnan, M. K., Veeresh, V., Yogendra, N., & Mudgal, P. P. (2014). Antidiabetic, antihyperlipidemic and antioxidant effects of the flavonoids. https://doi.org/10.1016/b978-0-12-398456-2.00013-x

Wanga, L., Jiab, M., Yina, D., & Tianc, J. (2019). A review of remote sensing for mangrove forests: 1956–2018. *Remote Sensing of Environment, 231*, 111223. https://doi.org/10.1016/j.rse.2019.111223

Wang, B., Wan, P., Qiu, Y., Su, Q., Guo, R., Wang, F., Sun, X., & Wu. (2019). Evaluating the performance of sentinel-2, landsat 8 and pléiades-1 in mapping mangrove extent and species. *Remote Sens., 10*, 1468–1476.

Chapter 22
Monitoring Land Use and Land Cover Analysis of the Barak Basin Using Geospatial Techniques

Wajahat Annayat, Kumar Ashwini, and Briti Sundar Sil

22.1 Introduction

Human beings need land for their survival and development (Xie et al., 2020). Land is becoming a scarce resource due to human-induced environmental changes such as expansion of agriculture, logging of forests, road hydropower and urbanization and setup of new mining industries (Ellwanger et al., 2020). Both natural and anthropogenic factors and their utilization by humans with space and time are responsible for the change in pattern of LULC of a particular region (Ganaie et al., 2020). From global to local scale, it has been widely accepted that LULC plays an important role on the functioning of the ecosystem. LULC of every continent has been comprehensively modified since the development of agriculture (Saunders et al., 1991). The nature of this modification is both dynamic and nonlinear which are activated by various factors such as social, political, economic and environmental factors which lead to the change in the climate (Reid et al., 2000; Webb & Honda, 2007; Miyamoto et al., 2014; Hassan et al., 2016). The process of LULC is not simple, as because there are structural complexities between various types of LULC changes which are changing both in space and time (Lambin et al., 2003).

All over the world, various researchers have investigated the analysis of changes in the LULC. Monitoring of LULC should be conducted on regular basis as because it causes various irreversible influences on the environment, particularly causing urban micro-climate warming (Coppin et al., 2004; Hahs et al., 2009; Niyogi et al., 2010; Wajahat & Sundar, 2018; Policelli et al., 2018). Satellite images of different spectral, spatial and temporal resolution have been used to carry out a LULC mapping successfully (Friedl et al., 2010; Solaimani et al., 2010; Aredehey et al., 2018). Monitoring and planning of LULC changes in the arid and semiarid regions, and the use of satellite images has been less used (Jia et al., 2014; Aredehey et al.,

W. Annayat (✉) · K. Ashwini · B. S. Sil
Department of Civil Engineering, National Institute of Technology, Silchar, Assam, India

G. S. Bhunia et al. (eds.), *Anthropogeomorphology*, Geography of the Physical Environment, https://doi.org/10.1007/978-3-030-77572-8_22

2018; Teka et al., 2018). In order to utilize the natural resources with proper planning and management, the study of LULC is mandatory. Traditional methods for collecting data are not sufficient to study the multi-complex environmental studies (Deng et al., 2009), since a lot of difficulties persists in order to present the multidisciplinary data set. In order to detect the changes in the LULC, accurate and timely data of various land utilizations of an area are required (Bastawesy et al., 2008). Hence, we require technologies like remote sensing (RS) and geographical information systems (GIS) to present these data sets. These technologies provide data at those places where field investigations are rare and difficult to attain.

Series of 30 years of data has been used to study the LULC changes in the Barak River which are based on RS and GIS technology. In the literature, authors found that in the Barak River till date, no study has been carried out which is based on the temporal and spatial channel changes in the LULC. So in order to fulfil this gap and to understand the nature and causes of LULC change in the Barak Basin, it is further planned to carry out a study through which the proper dynamics of the LULC changes of the Barak River can be understood. In this work, we have presented the changes in the LULC with the help of Landsat data by using a classification known as hybrid classification which consists of both supervised classification and unsupervised classification. The main aim of the present work is to apply RS and GIS techniques in order to understand temporal and spatial channel changes of LULC with a long period of time (1988–2018). In order to fulfil the main aim of the study, the following objectives are framed to carry out this study in the Barak River which is outlined below:

- To delineate the various LULC patterns
- To analyse the various changes in LULC categories from 1988 to 2018

The outcome of the study on the monitoring land use and land cover analysis of the Barak Basin using geospatial techniques is subdivided into seven sections. Following this introduction represents a synthesis of the relevant literature on river LULC analysis. Section 22.2 deals with the description of the study area highlighting the location and climatic peculiarities. Section 22.3 illustrates the materials and methods and data used and discusses the methodologies. Sections 22.4 narrate the results and discussion of the spatio-temporal variability of LULC. Section 22.5 summarizes the results and provides the major conclusions from the present study. The final section consists of reference section which is included at the end.

22.2 Study Area

The total length of Barak River is 900 km long and is flowing through the various states such as Manipur, Nagaland, Mizoram and Assam in India and into the Bay of Bengal via Bangladesh. In Bangladesh, this river is known by a name called as Surma River. From 900 km, 524 km is in India, and the rest is in Bangladesh. In the northeast India, Barak River is the second largest river which is flowing parallel to

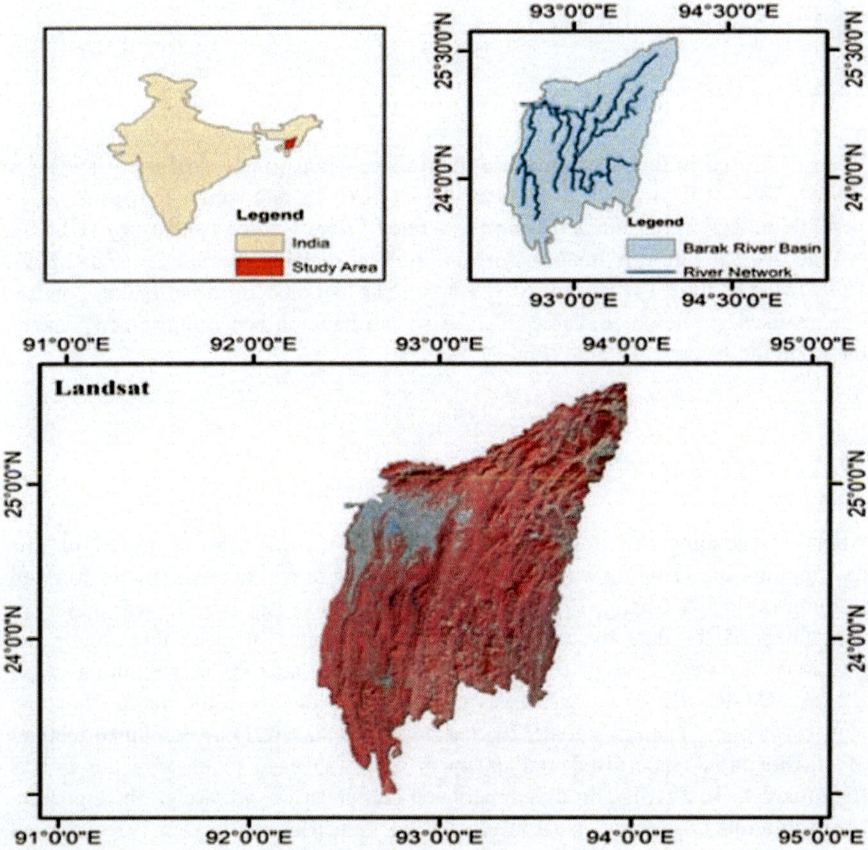

Fig. 22.1 Location of the study area

the river Brahmaputra. Barak River starts from the Manipur hills of the Japvo Mountain .The Barak Basin lies between 90°50′E to 95°0′E and 22°43′N to 25°57′N. It was found that the annual rainfall varies from 2500 to 4000 mm out of which 80% of the annual rainfall has occurred in the monsoon season (April to October) (Annayat & Sil, 2020a). The mean annual T_{min} and T_{max} ranged from 17.74 to 19.617 °C and 25.147 to 29.652 °C, respectively (Annayat & Sil, 2020b). In this study, analysis of LULC has been conducted for the whole of the Barak Basin located in India covering a distance of approximately 524 km. The location map of the study area is shown in Fig. 22.1.

22.3 Materials and Methods

22.3.1 Data Used

Basic data used in this study include multi-date Landsat images of the years 1988 (MSS), 1998 (MSS), 2008 (TM) and 2018 (OLI). In this study, four-time series satellite images were obtained from the United States Geological Survey (USGS). All the five satellite raw images were attained on separate years, i.e. 1988 (MSS), 1998 (MSS), 2008 (TM) and 2018 (OLI). The detailed information of Landsat images used is shown in Table 22.1. Also we have shown satellite raw images based on their capturing time (Fig. 22.2).

22.3.2 Methodology

After downloading the Landsat data, Landsat 1 (MSS) was subjected to line destripping (line stripping was found) as scanning error was removed by the filtering techniques (Pan & Chang, 1992). The resolution of Landsat 1(MSS) was 60 m; we georeferenced the data by using various control points in order to improve the positional accuracy and to resample the MSS data into 30 m resolution. Also ERDAS IMAGINE 14 software was used to carry out the radiometric error and haze corrections (Hazarika et al., 2015; Annayat & Sil, 2021). A combined method of classification (supervised and unsupervised) was used to classify the LULC (Hazarika et al., 2015). Through supervised classification, we assign the signature file of different classes by visual interpretation. Accordingly, the data is reclassified into eight classes, namely, water bodies, vegetation and dense vegetation, agricultural land, built-up, grassland, wasteland and riverbed (Aboelnour & Engel, 2018). We assess the classified data in order to find its accuracy by using a matrix known as confusion matrix (Abdelkareem et al., 2017). We compare the values of error in each class, which were classified one-to-one value in the ground truth data. A total of 240 ground truth points were used for the accuracy assessment. Same number of columns and rows were possessed by the table which is equal to the number of classes. The present chart of the methodology shows the schematic flow diagram of the whole work (Fig. 22.3).

A significant step in the classification of images is the accuracy assessment. From the error matrix, some of the processes of accuracy assessment are kappa coefficient

Table 22.1 Dataset utilized in the study

Data type	Sensor	Path/row	Scale	Acquisition date	Source
Landsat 1	MSS	136/43	60 m	05.01.1988	
Landsat 2	MSS	136/43	30 m	09.12.1998	http://earthexplorer.usgs.gov
Landsat 5	TM	136/43	30 m	04.02.2008	
Landsat 8	OLI	136/43	30m	20.12.2018	

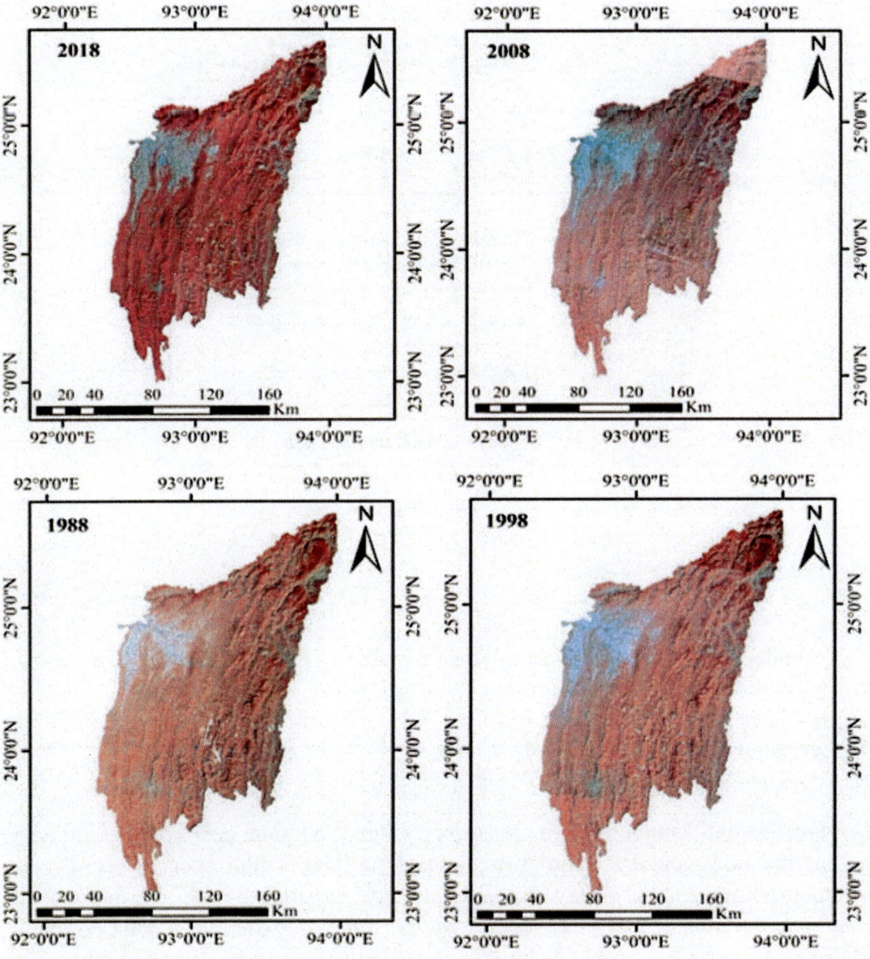

Fig. 22.2 Satellite raw images based on their capturing time

(K^), overall accuracy, producer's accuracy and user accuracy (Zhang et al., 2016). Then the kappa coefficient (K^) is then calculated using the following equation (Abdelkareem et al., 2017).

$$K = \frac{N\sum_{1=1}^{r} Xii - \sum_{i=1}^{r} X_{i+}X_{+i}}{N^2 - \sum_{i=1}^{r} X_{i+}X_{+i}} \tag{22.1}$$

where

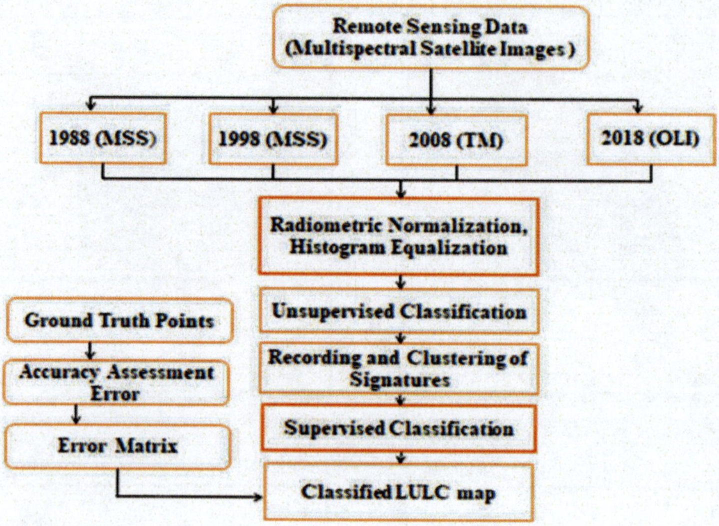

Fig. 22.3 Flow chart for methodology

r = "Number of rows/columns in confusion matrix"
Xii = "Number of observation in row i and column i"
X_{i+} = "Total number of row i"
X_{+i} = "Total number of column i"
N = "Number of observations"

Percentage of samples which are correctly classified in an error matrix is known as classification accuracy. In order to calculate the classification accuracy, we divide the total number of samples which are correctly classified by the total number of reference samples. It can be expressed by the following equation (Dada & Hahn, 2020):

$$\text{Overall accuracy} = \frac{''\text{Number of Sampling classes classified correctly}''}{''\text{Number of reference sampling classes}''} \quad (22.2)$$

In order to derive the mapping accuracy of LULC classes, both producers accuracy and user accuracy are calculated using the Eqs. 22.3 and 22.4 (Story & Congalton, 1986) and (Congalton & Green, 2019).

$$\text{Producers accuracy} = \frac{a_{ii}}{\sum_{i-1}^{n} a_{i+}} \quad (22.3)$$

$$\text{User accuracy} = \frac{a_{ii}}{\sum_{i-1}^{n} a_{+i}} \quad (22.4)$$

where

a_{ii} = "Number of samples correctly classified"
a_{i+} = "Column total for class i"
a_{+i}= "Row total for class i"

22.4 Results and Discussion

22.4.1 *Land Use Classification*

In order to analyse the LULC changes at different times, it gives an idea about the probable driving factors which are responsible for the level of changes and its consequences. Also respective management techniques are utilized in order to prevent the further loss of LULC in a particular area. The analysis of the entire land use and land cover of the study area in 1988, 1998, 2008 and 2018, respectively, reflected different proportions of changes that occurred in water bodies, vegetation and dense vegetation, agricultural land, built-up, grassland, wasteland and riverbed. Table 22.2 and Fig. 22.4 give the statistical results of LULC changes. Evaluation of LULC was done between 1988 and 2018. Changes were evaluated from the differences between 1988 and 1998, 1998 and 2008, 2008 and 2018 and 1988 and 2018. Eight major LULC types such as water bodies, vegetation, dense vegetation, agricultural land, built-up, grassland, wasteland and riverbed were registered and considered for the study area. In order, the abundance of different categories of LULC till 1988 are represented as dense vegetation (76.59)> vegetation (14.04)> agricultural land (5.23)> water (1.88)>urban and built-up area (1.49)> grassland (0.486)> wasteland (0.1)> riverbed (0.087), while as for the year 2018, the situation was somehow different and can be represented as dense vegetation (74.02)> vegetation (15.37)> agricultural land (5.69)> urban and built-up area (2.59)> water (1.44) > grassland (0.55)> riverbed (0.20)> and wasteland (0.10). During the recent years, significant momentum has been gained by the studies of the LULC on the environment and the society. LULC changes that are derived from the period of 30 years

Table 22.2 Land use/land cover changes of the Barak River Basin from 1988 to 2018

Area coverage				
LULC Class	1988	1998	2008	2018
Water	458.14	392.85	518.53	350.58
Vegetation	3407.12	3848.27	3934.42	3730.67
Dense vegetation	18584.95	18120.98	17811.47	17960.69
Agricultural	1270.63	1298.41	1254.86	1382.41
Urban/built-up	361.68	432.24	562.81	628.48
Grassland	118.99	108.40	131.24	134.61
Wasteland	40.17	44.69	14.20	25.73
Riverbed	21.16	17.02	35.31	49.66
Total	24262.87	24262.87	24262.87	24262.87

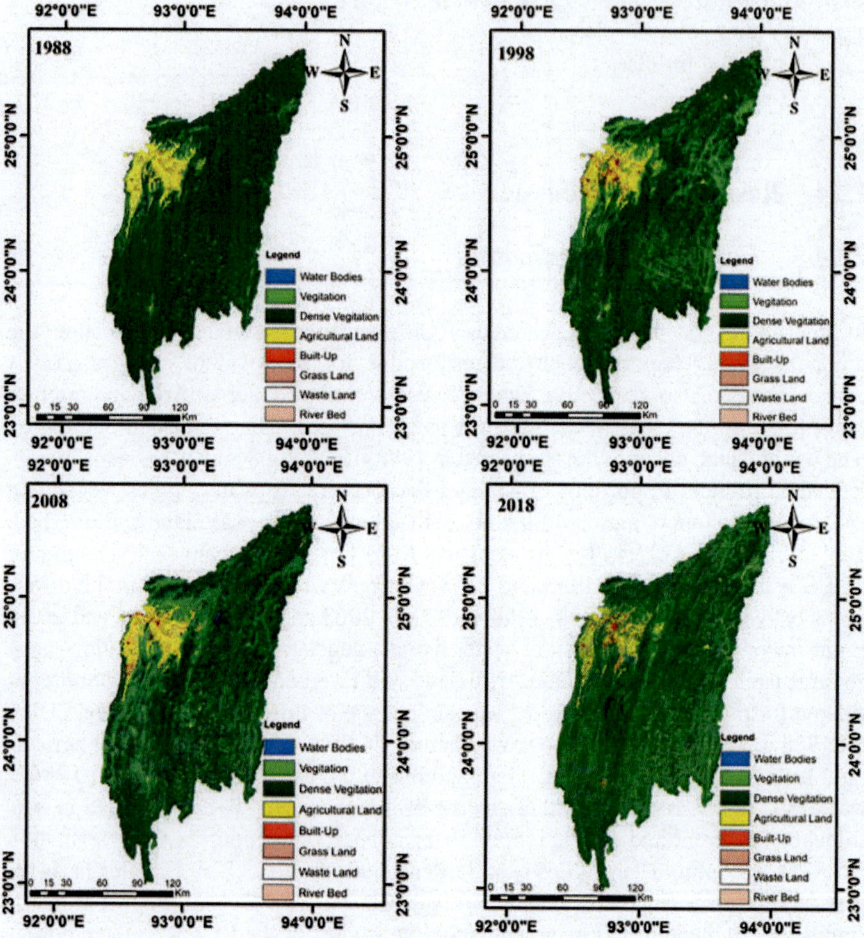

Fig. 22.4 Land use/land cover changes for three decades (1988–2018) in the Barak River Basin

(1988−2018) have been shown in Table 22.3 and Fig. 22.4. It is evident from Table 22.3 that the LULC changes were of highest amount in riverbed and urban from 1988 to 2018. Dominant land use that increased progressively over the study period was urban, grassland and agricultural land. For example, urban increased by 19.50% for 1988−1998, 30.20% for 1998−2008 and 73.76% for 1988−2018, while agricultural land increased by 2.18% for 1988−1998, 10.16% for 2008−2018 and 8.08% for 1998−2018. In addition, grassland increased by 21.06% and 13.12% from 1998 to 2008 and 1988 to 2018, respectively. In contrast, water decreased by 14.2% for 1988−1998 and 32.39% for 2008−2018 and 23.48% for 1988−2018. Dense vegetation showed a decreasing trend by 2.49% for 1988−1998, 1.70% for 1998−2008 and 3.48% for 1988−2018. Wasteland showed a decreasing trend of 35.92% for 1988−2018. Rapid population growth, migration of rural people to

Table 22.3 Change of LULC class through change matrix from 1988 to 2018

Class	1988–1998		1998–2008		2008–2018		1988–2018	
	Change in area in km^2	% Change*	Change in area in km^2	% Change*	Change in area in km^2	% Change*	Change in area in km^2	% Change*
Water	−65.29	−14.24	125.68	31.99	−167.95	−32.39	−107.56	−23.48
Vegetation	441.15	12.94	86.15	2.23	−203.75	−5.17	323.55	9.5
Dense vegetation	−463.97	−2.49	−309.51	−1.7	149.22	0.83	−624.26	−3.48
Agricultural	27.78	2.18	−43.55	−3.35	127.55	10.16	111.78	8.08
Urban/built-up	70.56	19.5	130.57	30.2	65.67	11.66	266.8	73.76
Grassland	−10.59	−8.89	22.84	21.06	3.37	2.57	15.62	13.12
Wasteland	4.52	11.27	−30.49	−68.22	11.53	81.24	−14.44	−35.92
Riverbed	−4.14	−19.57	18.29	107.44	14.35	40.63	28.5	134.62

*Cover change between periods is calculated as $\left(\frac{B_{Final\ Year} - B_{Intial\ Year}}{B_{Intial\ Year}}\right) * 100$, where B = area of LULC

Fig. 22.5 Field photo of LULC of the Barak River at various locations: (**a**) Fulertal (**b**) Anapurna (**c**) Badarpur

urban, unawareness of biophysical limitations and poverty are some of the probable reasons behind the changes in the LULC. In addition to industrial, agricultural and other activities, humans also affect climate through changes in land use (activities taking place on land, like growing food, cutting trees, or building cities) and land cover (the physical characteristics of the land surface, including grain crops, trees or concrete). Some field photographs (Fig. 22.5) show the LULC of the Barak River. From Table 22.3, it is clear that the change in area under urban and built-up was 19.5% during 1988–1998, while this change in area was 73.76% during 1988–2018. It is clear that the area under urban and built-up has increased; besides, this erosion of bank tries to capture the area which is nearby the settlement.

22.4.2 Overall Classification Accuracy and Kappa Statistics

It is very difficult to assess the classification accuracy as it is associated with lot of problems (Foody, 2002). Overall accuracy is one of the basic accuracy measures, and it is obtained by dividing the properly classified pixels by the total number of pixels (Banko, 1998). A statistical valid assessment for the quality of classification is provided by kappa coefficient, and it is a measure of overall agreement of matrix. All the diagonal elements are taken into consideration in case of kappa coefficient (Rosenfield & Fitzpatrick-Lins, 1986). The results of the hybrid classification of the LULC of the Barak River for the Landsat 1 (1988; MSS), Landsat 2 (1998; MSS), Landsat 5 (2008;TM) and Landsat 8 (2018;OLI) showed overall classification accuracy of 93.75%, 92.96%, 95.42% and 97.50% with overall kappa statistics of 0.9283, 0.9183, 0.9476 and 0.9714, respectively (Table 22.4). Some other studies have observed similar range of results such as in Umabdalla Natural Reserved Forest, South Kordofan, Sudan by Abdelkareem et al. (2017) and in Lodhran District of Pakistan by Hussain et al. (2019). As per (Pontius Jr (2000), the value of kappa coefficient greater than 0.5 is considered as satisfactory for modelling of LULC changes. However, Landis and Koch (1977) set a range of kappa coefficient as

Table 22.4 Overall classification and kappa statistics of the Barak River Basin using hybrid classification

	Barak River			
Accuracy	1988	1998	2008	2018
Overall classification accuracy (%)	93.75	92.96	95.42	97.50
Kappa statistics (K)	0.9283	0.9183	0.9476	0.9714

follows: a value of kappa coefficient greater than 0.7 is considered as excellent, values between 0.6 and 0.79 are considered as substantial and values of 0.59 or less indicate moderate or poor agreement. The overall accuracy and kappa coefficient of the present study are within the excellent range.

Concerning the overall producer and user accuracy for the classified imagery classes from 1988 to 2018, the result revealed excellent user accuracy for nearly all the classes in all years (Table 22.4). From all the classified images, water and agricultural land are the best classified with almost 100% accuracy as because of distinctive separation from the rest of the classes. For water, the best accuracy was obtained in the year 1998 where both producer and user accuracy are 100%. Similarly for agricultural land, the best accuracy was obtained in the year 2018 where both producer and user accuracy are 100%. Finally from Table 22.5, overall classification accuracies such as producer accuracy and user accuracy suggested that this classification is consistent and satisfactory for further analysis.

22.5 Conclusion

The Barak River is the second largest river in the northeast India. Main incentive of this chapter is to analyse the changes in the LULC of the Barak River using the remote sensing data and GIS technology. Four different Landsat images were used in order to under the dynamics of the LULC changes. The findings of the present study reveal that the LULC changes were significant from 1988 to 2018. It was found that water, dense vegetation and wastelands area have decreased by 23.48%, 3.48% and 35.92%, respectively, in 2018. However, agricultural, grassland, urban and riverbed have increased by 8.08%, 13.12%, 73.76 and 134.62%, respectively, which clearly indicate that there is a major impact of humans and their developmental activities on the LULC changes. The overall accuracy such as producer accuracy, user accuracy and kappa coefficient of the present study are within the excellent range.

Table 22.5 Summary of producer and user accuracy assessment for supervised classification of land use land cover in the Barak River during the years (1988, 1998, 2008 and 2018)

Year	1988		1998		2008		2018	
Class name	Producers accuracy	Users accuracy	Producers accuracy	Users accuracy	Producers accuracy	Users accuracy	Producers accuracy	Users accuracy
Water	96.77%	100.00%	100.00%	100.00%	93.55%	96.67%	100.00%	90.00%
Vegetation	93.33%	93.33%	90.32%	93.33%	100.00%	86.67%	100.00%	100.00%
Dense vegetation	93.33%	93.33%	96.55%	93.33%	88.24%	100.00%	90.91%	100.00%
Agricultural	93.75%	100.00%	100.00%	93.33%	100.00%	96.67%	100.00%	100.00%
Urban/built-up	90.00%	90.00%	79.41%	90.00%	93.55%	96.67%	100.00%	95.00%
Grassland	100.00%	90.00%	89.66%	86.67%	96.77%	100.00%	95.24%	100.00%
Wasteland	96.43%	90.00%	100.00%	93.33%	96.55%	93.33%	100.00%	95.00%
Riverbed	81.25%	92.86%	75.00%	100.00%	96.55%	93.33%	95.24%	100.00%

Acknowledgements 'We are thankful to USGS, for providing us data, and other support for completion of this work'.

Disclosure Statement 'The authors reported no potential conflict of interest'.

References

Abdelkareem, O. E. A., Elamin, H. M. A., Eltahir, M. E. S., Adam, H. E., Elhaja, M. E., Rahamtalla, A. M., Babatunde, O., & Elmar, C. (2017). Accuracy assessment of land use land cover in Umabdalla Natural Reserved Forest, South Kordofan, Sudan. *International Journal of Agricultural and Environmental Sciences, 3*(1), 5–9.

Aboelnour, M., & Engel, B. A. (2018). Application of remote sensing techniques and geographic information systems to analyze land surface temperature in response to land use/land cover change in Greater Cairo Region, Egypt. *Journal of Geographic Information System, 10*(1), 57–88.

Annayat, W., & Sil, B. S. (2020a). Assessing channel morphology and prediction of centerline channel migration of the Barak River using geospatial techniques. *Bulletin of Engineering Geology and the Environment*, 1–23.

Annayat, W., & Sil, B. S. (2020b). Changes in morphometric Meander parameters and prediction of Meander channel migration for the alluvial part of the Barak River. *Journal of the Geological Society of India, 96*(3), 279–291.

Annayat, W., & Sil, B. S. (2021). Predicting Meander migration of the Barak river by empirical and time sequence methods. *Water Security and Sustainability: Proceedings of Down To Earth, 2019*, 113.

Aredehey, G., Mezgebu, A., & Girma, A. (2018). Land-usc land-cover classification analysis of Giba catchment using hyper temporal MODIS NDVI satellite images. *International Journal of Remote Sensing, 39*(3), 810–821.

Banko, G. (1998). *A review of assessing the accuracy of classifications of remotely sensed data and of methods including remote sensing data in forest inventory*.

Bastawesy, M. A., Khalaf, F. I., & Arafat, S. M. (2008). The use of remote sensing and GIS for the estimation of water loss from Tushka lakes, southwestern desert, Egypt. *Journal of African Earth Sciences, 52*(3), 73–80.

Congalton, R. G., & Green, K. (2019). *Assessing the accuracy of remotely sensed data: Principles and practices*. CRC press.

Coppin, P., Jonckheere, I., Nackaerts, K., Muys, B., & Lambin, E. (2004). Review ArticleDigital change detection methods in ecosystem monitoring: A review. *International Journal of Remote Sensing, 25*(9), 1565–1596.

Dada, E., & Hahn, M. (2020). Application of satellite remote sensing to observe and analyse temporal changes of cocoa plantation in Ondo State, Nigeria. *GeoJournal*, 1–16.

Deng, J. S., Wang, K., Hong, Y., & Qi, J. G. (2009). Spatio-temporal dynamics and evolution of land use change and landscape pattern in response to rapid urbanization. *Landscape and Urban Planning, 92*(3–4), 187–198.

Ellwanger, J. H., Kulmann-Leal, B., Kaminski, V. L., Valverde-Villegas, J., Veiga, A. N. A. B. G., Spilki, F. R., Fearnside, P. M., Caesar, L., Giatti, L. L., & Wallau, G. L. (2020). Beyond diversity loss and climate change: Impacts of Amazon deforestation on infectious diseases and public health. *Anais da Academia Brasileira de Ciências, 92*(1).

Foody, G. M. (2002). Status of land cover classification accuracy assessment. *Remote Sensing of Environment, 80*(1), 185–201.

Friedl, M. A., Sulla-Menashe, D., Tan, B., Schneider, A., Ramankutty, N., Sibley, A., & Huang, X. (2010). MODIS Collection 5 global land cover: Algorithm refinements and characterization of new datasets. *Remote Sensing of Environment, 114*(1), 168–182.

Ganaie, T. A., Jamal, S., & Ahmad, W. S. (2020). Changing land use/land cover patterns and growing human population in Wular catchment of Kashmir Valley, India. *GeoJournal*, 1–18.

Hahs, A. K., McDonnell, M. J., McCarthy, M. A., Vesk, P. A., Corlett, R. T., Norton, B. A., Clemants, S. E., Duncan, R. P., Thompson, K., & Schwartz, M. W. (2009). A global synthesis of plant extinction rates in urban areas. *Ecology Letters, 12*(11), 1165–1173.

Hassan, Z., Shabbir, R., Ahmad, S. S., Malik, A. H., Aziz, N., Butt, A., & Erum, S. (2016). Dynamics of land use and land cover change (LULCC) using geospatial techniques: A case study of Islamabad Pakistan. *Springerplus, 5*(1), 812.

Hazarika, N., Das, A. K., & Borah, S. B. (2015). Assessing land-use changes driven by river dynamics in chronically flood affected Upper Brahmaputra plains, India, using RS-GIS techniques. *The Egyptian Journal of Remote Sensing and Space Science, 18*(1), 107–118.

Hussain, S., Mubeen, M., Ahmad, A., Akram, W., Hammad, H. M., Ali, M., Masood, N., Amin, A., Farid, H. U., & Sultana, S. R. (2019). Using GIS tools to detect the land use/land cover changes during forty years in Lodhran District of Pakistan. *Environmental Science and Pollution Research*, 1–17.

Jia, K., Liang, S., Wei, X., Yao, Y., Su, Y., Jiang, B., & Wang, X. (2014). Land cover classification of Landsat data with phenological features extracted from time series MODIS NDVI data. *Remote Sensing, 6*(11), 11518–11532.

Lambin, E. F., Geist, H. J., & Lepers, E. (2003). Dynamics of land-use and land-cover change in tropical regions. *Annual Review of Environment and Resources, 28*(1), 205–241.

Landis, J. R., & Koch, G. G. (1977). An application of hierarchical kappa-type statistics in the assessment of majority agreement among multiple observers. *Biometrics*, 363–374.

Miyamoto, M., Parid, M. M., Aini, Z. N., & Michinaka, T. (2014). Proximate and underlying causes of forest cover change in Peninsular Malaysia. *Forest Policy and Economics, 44*, 18–25.

Niyogi, D., Kishtawal, C., Tripathi, S., & Govindaraju, R. S. (2010). Observational evidence that agricultural intensification and land use change may be reducing the Indian summer monsoon rainfall. *Water Resources Research, 46*(3).

Pan, J.-J., & Chang, C.-I. (1992). Destriping of Landsat MSS images by filtering techniques. *Photogrammetric Engineering and Remote Sensing, 58*, 1417.

Policelli, F., Hubbard, A., Jung, H. C., Zaitchik, B., & Ichoku, C. (2018). Lake Chad total surface water area as derived from land surface temperature and radar remote sensing data. *Remote Sensing, 10*(2), 252.

Pontius, R. G., Jr. (2000). Comparison of categorical maps. *Photogrammetric Engineering and Remote Sensing, 66*, 1011–1016.

Reid, R. S., Kruska, R. L., Muthui, N., Taye, A., Wotton, S., Wilson, C. J., & Mulatu, W. (2000). Land-use and land-cover dynamics in response to changes in climatic, biological and socio-political forces: the case of southwestern Ethiopia. *Landscape Ecology, 15*(4), 339–355.

Rosenfield, G. H., & Fitzpatrick-Lins, K. (1986). A coefficient of agreement as a measure of thematic classification accuracy. *Photogrammetric Engineering and Remote Sensing, 52*(2), 223–227.

Saunders, D. A., Hobbs, R. J., & Margules, C. R. (1991). Biological consequences of ecosystem fragmentation: A review. *Conservation Biology, 5*(1), 18–32.

Solaimani, K., Arekhi, M., Tamartash, R., & Miryaghobzadeh, M. (2010). Land use/cover change detection based on remote sensing data (A case study; Neka Basin). *Agriculture and Biology Journal of North America, 1*(6), 1148–1157.

Story, M., & Congalton, R. G. (1986). Accuracy assessment: A user's perspective. *Photogrammetric Engineering and Remote Sensing, 52*(3), 397–399.

Teka, H., Madakadze, C. I., Botai, J. O., Hassen, A., Angassa, A., & Mesfin, Y. (2018). Evaluation of land use land cover changes using remote sensing Landsat images and pastoralists perceptions on range cover changes in Borana rangelands, Southern Ethiopia. *International Journal of Biodiversity and Conservation, 10*(1), 1–11.

Wajahat, A., & Sundar, S. B. (2018). Estimation and analysis of possible flood for the Silchar city-a case study. *Disaster Advances, 11*(1).

Webb, E. L., & Honda, K. (2007). Biophysical and policy drivers of landscape change in a central Vietnamese district. *Environmental Conservation*, 164–172.

Xie, H., Zhang, Y., Wu, Z., & Lv, T. (2020). A bibliometric analysis on land degradation: Current status, development, and future directions. *Land, 9*(1), 28.

Zhang, Z., Liu, S., Wei, J., Xu, J., Guo, W., Bao, W., & Jiang, Z. (2016). Mass change of glaciers in Muztag Ata–Kongur Tagh, Eastern Pamir, China from 1971/76 to 2013/14 as derived from remote sensing data. *PLoS One, 11*(1), e0147327.

Chapter 23
Assessing the Efficiency of Classification Techniques Between SVM and ML for Detecting Land Transformation in Bhawal Sal Forest

Rowshon Ara Toma, Md Fazla Rabby, Rezaul Roni, and Md Shahedur Rashid

23.1 Introduction

Forests play a fundamental part in preserving the Earth's ecological stability and environmental fitness due to their advancing part within the worldwide carbon cycle, water asset quality, and undertaking capacity. Bangladesh is a relatively smaller country with an area of 1,47,570 sq. km in Southeast Asia and a fundamentally high pace of population growth rate of 1.01% (United Nations -World Population Prospects, 2020). The livable land is fixed in Bangladesh. However, it is consistently gaining overweight of populace thickness around 1000 for every square kilometer. Therefore, the land actual attributes are overfocused on condition and the scenes are being changed quickly. Fragmentation of land is going on in a powerful manner, and along these lines the agriculture and essentially the forestry land is contracting consistently. A similar pattern is likewise forcing on low land and shallow water bodies, particularly closer to the urban and semiurban area. All the more fundamentally from the viewpoint of Bangladesh, such unfavorable impacts of land cover change and deforestation are considerably more articulated.

The area of worldwide forest cover is a little more than 4 billion hectares, 31% of is absolute land zone (FAO, 2010) and the all-out carbon stockpiling of worldwide backwoods environment is assessed to be 638 Gt for 2005, which is more than the measure of carbon in the whole air (FAO, 2006). The pace of carbon aggregation and arrival of a forest ecosystem because of forest growth and deforestation activities play a role in terrestrial carbon flux. Therefore, changes in forest cover are currently a matter of worldwide concern. Understanding forest designs, changes, and intuitive

R. A. Toma (✉) · M. F. Rabby · R. Roni · M. S. Rashid
Department of Geography and Environment, Jahangirnagar University, Dhaka, Bangladesh
e-mail: georoni31@juniv.edu; m.s.rashid@juniv.edu

G. S. Bhunia et al. (eds.), *Anthropogeomorphology*, Geography of the Physical Environment, https://doi.org/10.1007/978-3-030-77572-8_23

among human activities and herbal phenomenon are important for appropriate forest management and decision development. Nowadays, information from satellites is exceptionally appropriate and valuable for forest cover change detection research. Recognizing forest conditions as well as observing the changes of diverse forest structural and biophysical factors can permit to exact information of forest ecosystem services. Checking of woodland spread change is one of the principal utilizations of far-off detecting-based change discovery.

Different satellite image classification methods have turned out to be suitable widely for detecting land use land cover (LULC) modifications after free availability of optical multispectral images for the developing countries (i.e., Bangladesh). Repetitively updated LULC information is necessary to a number of socioeconomic and environmental applications, which include urban and regional planning, natural resources protection and management, etc. (Homer et al., 2007; Lu & Weng, 2007; Jensen, 2009). Remote sensing imagery, masking a giant geographical region with high temporal frequency, provides an interesting possibility for starting LULC data through the image interpretation and classification technique. In spite of the fact that LULC data perform a large role for monitoring and grasping of both physical and human environment, our understanding of LULC and its dynamics specifically in creating countries like Bangladesh nonetheless lie in the back due to (1) susceptible authorities' aid for mapping corporations and research institutions, (2) costly computer program and equipment, (3) inadequate finances distributions for facts purchases, and (4) conflict to adjustments especially with the aid of the traditionalist in the discipline of mapping. Effective satellite-derived information with suitable classification methods is vital for forest change detection. Here, we find out about the maximum likelihood (ML) classification technique, which is broadly used, and the support vector machine (SVM), both being utilized as they are very superior classification techniques.

The ML method is fast, easy to apply, and enables a clear interpretation of the results (Bolstad & Lillesand, 1991). This algorithm can get a spectral image of every LULC class through difference and covariance measurements of the arrangement of training sites identified within the image and figures the likelihood of having a place with each class as per the spectral signature; this technique has been proven in works like those of Hassan et al., 2016; Tahir et al., 2013 with satisfactory outcomes. ML is a method used for assurance of viewed-type conveyance as the greatest for a given measurement (Scott & Symons, 1971). It expects that the information for every land category in every band is generally dispensed and calculates the possibility that a given pixel belongs to a unique type and each pixel is assigned to the category that has the absolute best probability. The algorithm used by the maximum likelihood classification tool is based on two principles: (1) the cells in each class sample in the multidimensional space being normally distributed and (2) Bayes' theorem of decision making. ML is based on the Bayesian probability formula.

SVM is a team of supervised classification algorithms primarily based on statistical principle, which is used for classification and regression issues (Vapnik, 1995). It works via mapping information to a high-dimensional feature space with the goal that information focuses can be arranged, even when the information is no longer in any other case linearly separable. Yu et al. (2011) have utilized the SVM algorithm for the computerized lithological classification in a section of northwestern India and the usage of ASTER imagery. Many researchers in this discipline have discovered that a greater degree of accuracy can be performed by using SVM than different techniques of classification like ML, neural community (NN), etc. (Foody & Mathur, 2004; Pal & Mather, 2005; Oommen et al., 2008; Naguib et al., 2009).

Szuster et al. (2011) examined the performance of the SVM classification technique in the tropical coastal zone. They also compared this technique with the MLC and artificial neural network techniques and concluded that SVM is a better classifier for challenges like separating human infrastructures, for instance, buildings from rocks and sandy beaches, as they possess similar spectral signatures.

The aim of this study is to assess the efficiency of the land classification techniques between SVM and ML for detecting the land transformation in the Bhawal Sal Forest region between 2000 and 2019. Subsequently, the following objectives were set (1) to select a suitable land classification technique applicable for the study area in determining the more reliable and accurate spatial distribution of the land features and (2) to quantify and assess the land transformation through the selected land classification technique between 2000 and 2019 for the rapidly changing forested area.

23.2 Materials and Methods

23.2.1 Study Area

The chosen study area is located in the Bhawal Sal Forest, Gazipur District, Dhaka Division of Bangladesh, approximately 32.74 km from north of Dhaka, the capital city. The area is found at latitudes between 24°2′30″ N to 24°9′30″ N and longitudes between 90°21′30″ E to 90°26′00″ E in Gazipur, Bangladesh. The area is about approximately 38.16 sq. km. The study area (Fig. 23.1) is located beside Pirozali and Kuarchala, the two main localities that are linked by commercial and administrative activities through the Dhaka–Mymensingh highway. The forest of the study area is categorized as tropical moist deciduous forest (Rahman et al., 2010). Sal (Shorearobusta) is the main species of this natural forest and the only Sal forest in Bangladesh as the climate is suitable for growing trees. The climate of this area is tropical. The average rainfall is about 205 cm. The average temperature gradually rises from February and reaches its maximum in April (around 35 °C) (Salam & Pramanik, 2017). This forest contains large kinds of floral compositions. Moreover, different types of mammals, reptile, aves, and amphibian depend on these kinds of trees as their habitat as well as niche.

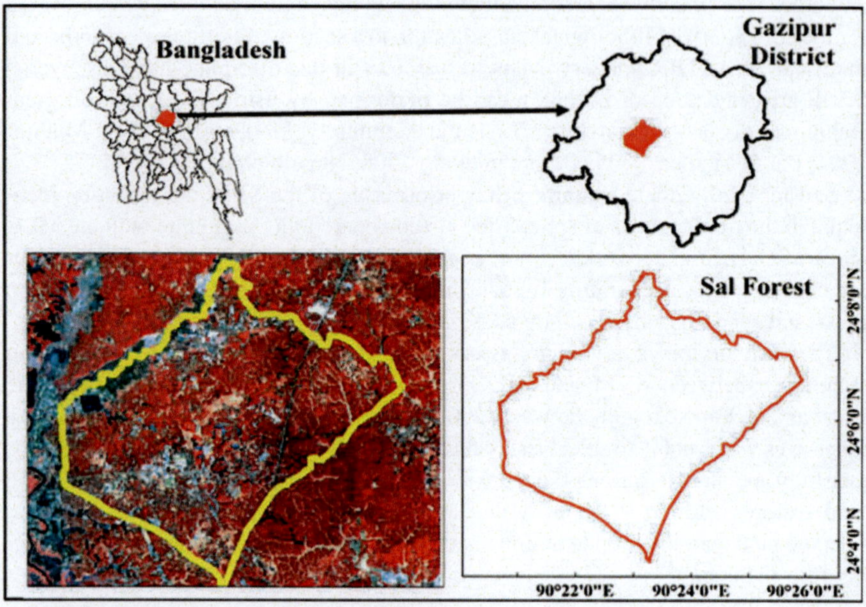

Fig. 23.1 Location map of the study area

Table 23.1 Details of Landsat images used in the study

Sl. no	Sensor name	Row/ path	Scene	Acquisition date	Resolution	Cloud cover
1	Landsat 5	43/ 137	LT51370432000019BKT00	19 January 2000	60 m	0
2	Landsat 8	43/ 137	LC81370432019023LGN00	23 January 2019	30 m	0

23.2.2 Imageries

The research needs the following Landsat images to set a temporary standardized thematic nomenclature for the analysis of the different land cover. Table 23.1 describes the Landsat satellite images details. The images were taken from the United States Geological Survey (USGS) official site (www.earthexplorer.usgs.gov) as it is freely available and the spatial resolution is in a considerable range to perform the analysis.

23.2.3 Image Processing

An ERDAS Imaging 14 software was utilized for image processing. Layer stacking option of this software was used to change over three bands (5, 4, and 3 for Landsat

Table 23.2 Description of identified covers and land uses

Class no.	Class	Description
1	Dense vegetation	Consisting of stunted trees or bushes
2	Light vegetation	Includes (scrub and grassland) and vegetation of low or scarce density
3	Water bodies	Water bodies, lakes, rivers, and seasonal water bodies
4	Built-up	Includes settlement and built-up areas
5	Bare soil	Under-construction land with bare soil exposed

8 and 4, 3, and 2 for Landsat TM) into a multispectral bundle layer. The study area was divided by utilizing vector layer mask. The subset image was then reprojected to UTM (Universal Transverse Mercator) zone 45 N and resampled to 30 m spatial resolution since two diverse scenes for various time frames have been utilized for the study reason, which has diverse sensor and distinct resolution.

For processing the images, spectral signatures were made from 20 training sites based on the identification of similar areas in various land covers, consolidating the information on the zone for an appropriate choice of the regions of interest (ROI). To distinguish the distinctive land spread classes, some visual interpretation was utilized. During the identification of the training sites, the class distinctness of the spectral signatures was confirmed for the five land use classes based on visual translation portrayed in Table 23.2.

23.2.4 Classification

To get the distinctive land use classes, two supervised classification techniques were used. First, ML method is fast, easy to apply, and enables a clear interpretation of the results (Bolstad & Lillesand, 1991). This algorithm can get a spectral image of every LULC class through difference and covariance measurements of the arrangement of training sites identified within the image and figures the likelihood of having a place with each class as per the spectral signature; this technique has been proven in works like those of Hassan et al., 2016; Tahir et al., 2013 with satisfactory outcomes.

The SVMs algorithm was the second technique applied (Fig. 23.2). This programmed learning algorithm trains linear and nonlinear learning capacities by changing the original information into an alternate space with a capacity (kernel) to acquire the hyperplane, which maximizes the edge of partition between at least two classes to be grouped (Vapnik, 1995). Presently, the SVM technique is among the most solid techniques; accordingly, it is utilized in numerous works (Mountrakis et al., 2011; Xie et al., 2017) with palatable outcomes. For the classification of images, the radial basis function (RBF) for not-directly detachable information was utilized.

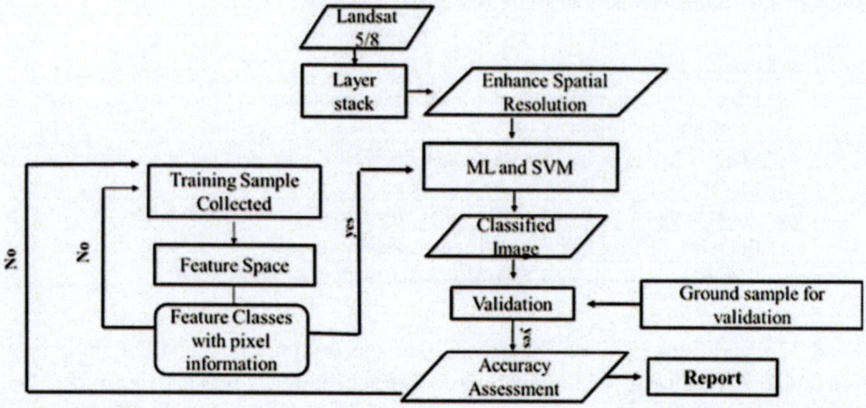

Fig. 23.2 Workflow for ML and SVM techniques

23.2.5 Accuracy Assessment

Error matrix is taken into account jointly of the basic methods for the estimation of thematic map accuracy (Foody, 2002). It is determined by getting an example from a specific class of a classified map so that the particular class is approved from the field (Congalton, 1991).

Totally 1020 points have been taken for the assessment of accuracy for ML classifications. According to the reference data, 150, 50, 578, 74, and 170 points are taken for dense vegetation, light vegetation, bare soil, built-up, and water bodies, respectively, while, according to the classified data, 125 points for dense vegetation, 62 points for light vegetation, 577 points for bare soil, 64 points for built-up, and 194 points for water bodies are taken. On the other hand, totally 1022 points have been taken for the assessment of the accuracy for the SVM classifications. According to reference data, 150, 50, 578, 74, and 170 points have been taken for dense vegetation, light vegetation, bare soil, built-up, and water bodies, respectively, while for classified data 149 points have been taken for dense vegetation, 44 points for light vegetation, 585 points for bare soil, 74 points for built-up, and 170 points for water bodies.

Producer's, user's, and overall accuracies are determined where the overall accuracy is gained for every class and is avowed as producer's and user's accuracy. Producer's accuracy is calculated by dividing the amount of sampling points that are exact in one class by the absolute number (Mondal et al., 2012). For the calculation of user's accuracy, classified units that are correct during a class are isolated by the all-out number of units that are as of now ordered in that specific class. The accuracy evaluation approved as the one with higher precision was utilized to speak the land cover mapping for the years 2000 and 2019.

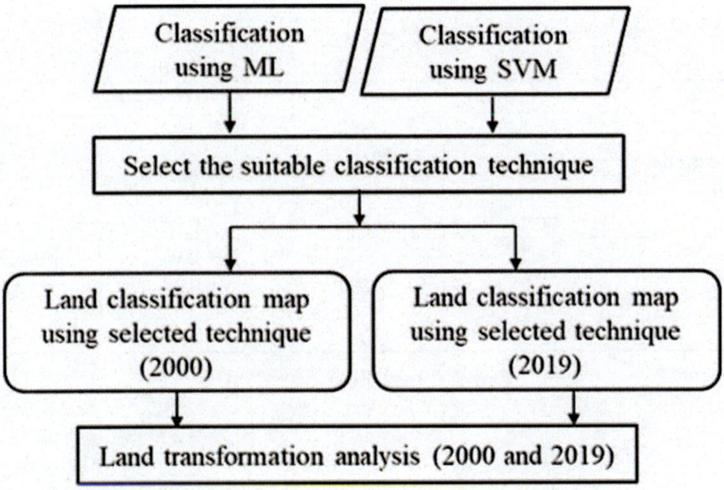

Fig. 23.3 Workflow of land transformation analysis

23.2.6 *Land Transformation Analysis*

The examination of LULC was directed through the cross-tabulation matrix to acquire losses, gains, and exchanges between the various covers and land uses, and from that point onward, the huge advances investigation was led by the strategy proposed by Pontius et al., 2004 and shown in Fig. 23.3.

Such significant transitions on each cover and land use were got by contrasting the increases and genuine losses against the additions as well as expected losses haphazardly, isolated by the additions and anticipated losses.

23.3 Results and Discussion

23.3.1 *Land Cover Classification Assessment*

Dense vegetation areas using ML and SVM techniques are 1.65 ha and 2.38 ha, respectively. There has been a difference of 0.73 ha area (17.16%) between the two methods. Light vegetation areas are 0.7 ha and 0.31 ha using ML and SVM, which are 16.31% and 7.31%, respectively, of the total area. Bare soil area covers 1.69 ha and 1.46 ha, which constitutes 39.51% and 34.10%, respectively. Built-up and water bodies as per both the methods encompass 0.14 ha and 0.04 ha and 0.1 ha and 0.08 ha, respectively. The details of the area distribution generated by both the ML and SVM methods are presented in Table 23.3 and correspondingly shown in Fig. 23.4.

Figure 23.4 shows the mapping of land covers obtained from classified Landsat 8 image and also represents the classification result by applying the ML and SVM

Table 23.3 Areal statistics of ML and SVM

	ML		SVM	
Class name	Area (ha)	Area (%)	Area (ha)	Area (%)
Dense vegetation	1.65	38.62	2.38	55.78
Light vegetation	0.7	16.31	0.31	7.31
Bare soil	1.69	39.51	1.46	34.10
Built-up	0.14	3.21	0.04	1.03
Water bodies	0.1	2.35	0.08	1.78
Total	4.28	100	4.28	100

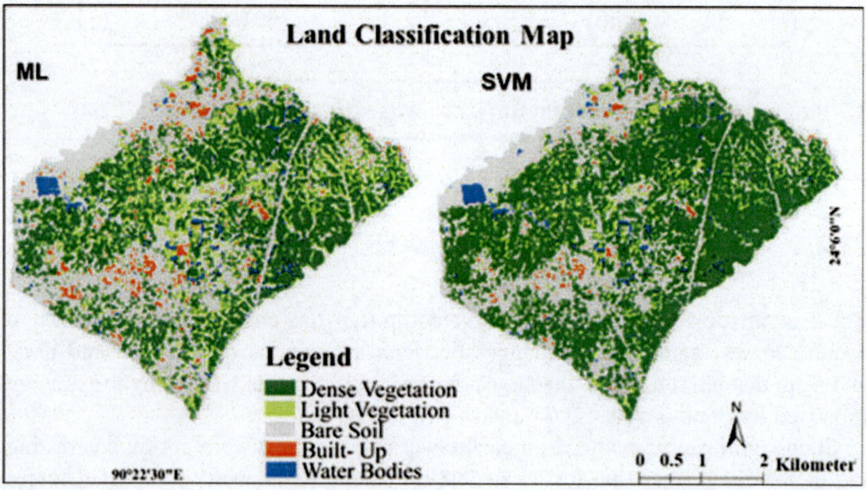

Fig. 23.4 Land cover map using ML and SVM classification techniques in Bhawal Sal Forest in 2019

techniques. The final land cover map presented in Fig. 23.4 shows the major classes like dense vegetation, light vegetation, bare soil, built-up, and water bodies and their transformations. It is noticeable that for both cases light vegetation is found far too abundant and encroaches markedly toward the forest areas, which is likely to be an ambiguous case. In other words, it is likely that misclassification occurs between dense vegetation and light vegetation in ML and SVM classifications (Table 23.4). These results are due to the similarities of spectral properties between dense vegetation and light vegetation.

23.3.2 *Comparison of the Two Classification Algorithms*

Table 23.5 shows the results from the process of classified images validation through the cross-tabulation matrix and the parameters obtained: kappa index, general

Table 23.4 Cross-tabulation matrix of ML and SVM

	Reference data							
	Techniques	Class name	Dense vegetation	Light vegetation	Bare soil	Built-up	Water bodies	Total
Classified data	ML	Dense vegetation	**125**	0	20	0	5	**150**
	SVM		**149**	0	0	0	1	**150**
	ML	Light vegetation	0	**50**	0	0	0	**50**
	SVM		0	**44**	0	0	6	**50**
	ML	Bare soil	0	12	**549**	3	14	**578**
	SVM		0	0	**575**	3	0	**578**
	ML	Built-up	0	0	8	**61**	5	**74**
	SVM		0	0	3	**71**	0	**74**
	ML	Water bodies	0	0	0	0	**170**	**170**
	SVM		0	0	7	0	**163**	**170**
	ML	Total	**125**	**62**	**577**	**64**	**194**	**1022**
	SVM		**149**	**44**	**585**	**74**	**170**	**1022**

Table 23.5 User's, producer's, and overall Accuracies of ML and SVM

	ML		SVM	
Class name	User's accuracy (%)	Producer's accuracy (%)	User's accuracy (%)	Producer's accuracy (%)
Dense vegetation	83.33%	100%	99.33%	100%
Light vegetation	100%	80.00%	100%	100%
Bare soil	95.00%	95.00%	98.46%	98.29%
Built-up	82.61%	95.00%	95.95%	95.95%
Water bodies	100%	85.00%	95.88%	95.88%
Overall accuracy	**93.44%**		**98.04%**	
Kappa statistics	0.97		0.89	

Table 23.6 Areas of land cover classes for the periods 2000 and 2019 according to the SVM method

		2000		2019
Class name	Area (ha)	Area (%)	Area (ha)	Area (%)
Dense vegetation	1.802	42.15	2.384	55.78
Light vegetation	1.192	27.89	0.312	7.31
Bare soil	1.197	27.99	1.457	34.10
Built-up	0.002	0.04	0.044	1.03
Water bodies	0.082	1.92	0.076	1.78
Total	4.28	100	4.28	100

accuracy, producer's accuracy, and user's accuracy. Validation statistics show better results when using the SVM classification method, from which a general accuracy of 98.04% is recorded unlike the ML, which indicates 93.44% accuracy.

This study found an overall accuracy of 91.11% by following the ML classification technique, but differences between user's accuracy and producer's accuracy in different classes is noticeable. In case of dense vegetation, the user's accuracy is low whereas the producer's accuracy is high. On the contrary, the result is reversed in the case of light vegetation. Here, the user's accuracy is high, but the producer's accuracy is pretty low. This difference is also observed in the water body areas. Only in the bare soil category is the user's accuracy and producer's accuracy the same.

Table 23.6 shows the changes in area for each class of cover occurring during 2000 and 2019. The bare soil and built-up have increased 7.1% from the overall analyzed area with 4.28 ha. At the same time, the vegetation cover has lost 6.95% of area in 19 years.

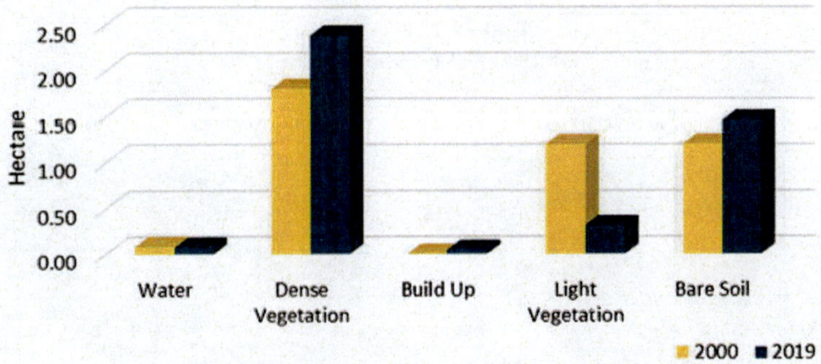

Fig. 23.5 Graphical presentation of land cover classes for the periods 2000 and 2019

The land cover transformation for 2000 and 2019 is represented in Table 5.1. It can be observed that light vegetation and water body areas decreased between the years, whereas dense vegetation, bare soil, and built-up area increased from 2000 to 2019. Although dense vegetation was found to have the largest area within the study region in both of the years, the rate of its change was observed to be unsystematic.

Figure 23.5 represents the data of Table 23.6 and visualizes the change over time. The dense vegetation, built-up, and bare land have increased from 2000, where the decrease of water body and light vegetation is prominent (Fig. 23.5).

The aforementioned graph reflects the overall changes of the study area and the result of this research. If we look at the Sal forest area in 2000, then it was found in the northeast and parts of the study area. During that time, the total area was 4.275 ha and dense vegetation covered 1.81 ha, which is 42.16% of the total area. But in 2019, 2.38 ha was been found to be 'dense vegetation,' which is 55.79% of the total area. So, it is clear that over the years (2000–2019) dense vegetation area has increased by a huge percentage.

On the other hand, a larger portion of the light vegetation transformed into dense vegetation and other categories in 2019, which reduced the overall area of light vegetation. In 2009, there was 1.192 ha of light vegetation, which reduced to 0.312 ha in 2019. It is also noteworthy that light vegetation has decreased in good rate (from 27.89% to 7.31%).

The amount of bare soil and built-up areas also increased over time, but the rate of change is very low. Overall, 6% and 1% growth has been detected for bare soil and built-up areas, respectively, from 2000 to 2019. Apart from that, 0.006 ha water bodies have disappeared over this time frame (Fig. 23.6).

The statistics in Table 23.7 shows that the dense vegetation areas have increased significantly, covering approximately 56% of the total area for the year 2019. A large fraction of light vegetation and dense vegetation has changed into bare lands. High expansion of agricultural land during last the 19 years has led to a significant increase in the bare soil areas.

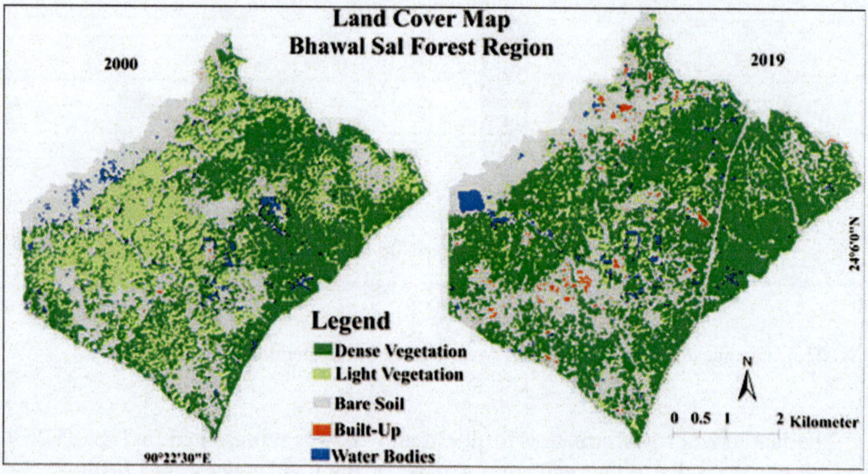

Fig. 23.6 Land cover map in Bhawal Sal Forest (2000 and 2019)

Table 23.7 Significant transition analysis (percent) for the periods 2000 and 2019

Changes from	Dense vegetation	Light vegetation	Bare soil	Built-up	Water
Dense vegetation	**31.5**	1.42	8.84	0.18	0.32
Light vegetation	17.81	**3.03**	6.58	0.19	0.05
Bare soil	7.15	1.89	**17.75**	0.48	0.94
Built-up	0.01	0.01	0.02	**0.01**	0.1
Water bodies	0.39	0.11	0.85	0.04	**0.36**

Data in bold represent unchanged fractions of each class

The transition analysis for each cover and land use is summarized in Table 23.7 and Fig. 23.7. Significant transitions for the analyzed periods are registered particularly in the vegetation cover change to built-up. During a 19-year period, 15.42% of vegetation cover transformed into bare soil and 0.37 into built-up.

Figure 23.8 illustrates that the overall dense vegetation area has increased over the years while there is a little loss in this category. Some other land cover classes have replaced some portion of the previously existed bare land and built-up areas, while a vast new area has converted into dense vegetation area from other classes. Gains in dense vegetation and bare land area are evident from 2000 to 2019.

Apart from that, light vegetation land cover area is decreasing at a larger scale in all the years. The differences (in terms of gains and losses) in other land cover types are almost the same or not influencing. Therefore, an increase in dense vegetation and bare land area and a decrease in light vegetation cover type are clear from this kind of analysis.

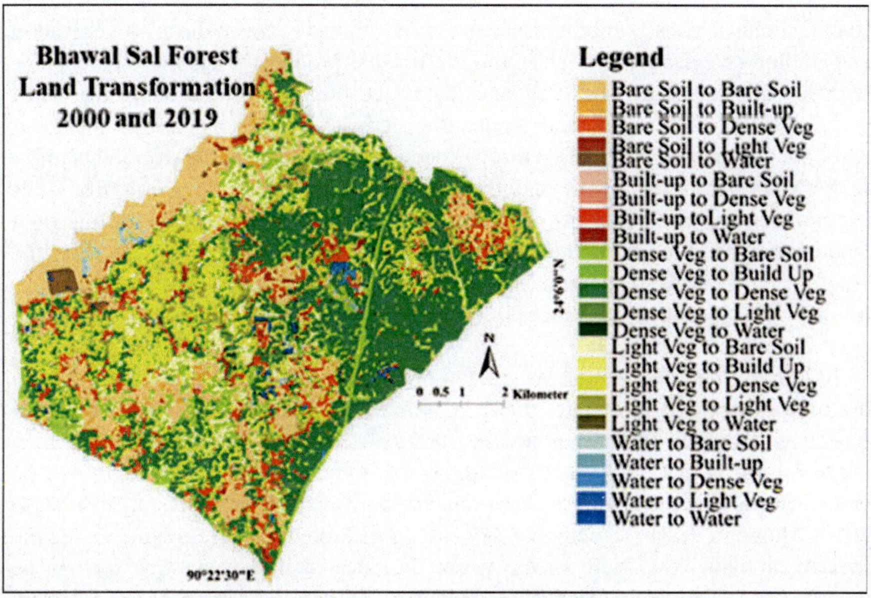

Fig. 23.7 Land transition map in Bhawal Sal Forest (2000–2019)

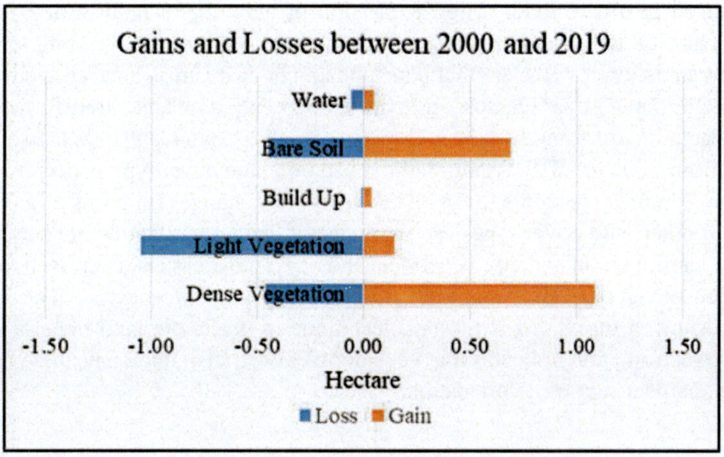

Fig. 23.8 Gains and losses of different categories between 2000 and 2019

23.4 Discussion

In this research, the performance of the ML and SVM algorithms was evaluated to categorize five land cover classes, the where accuracy was assessed by confusion matrix for the Bhawal Sal Forest in Bangladesh. The SVM method for ordering land cover was discovered to be more exact than the other methods. When choosing training fields, class separability issues were recognized between the thick vegetation and the light vegetation, which is reflected in the producer's accuracy percentage with lower esteems in ML. The best thing we noticed was that using the SVM method the user's accuracy and producer's accuracy were so much high for dense and light vegetation land types. The efficiency test revealed that the overall accuracy of SVM is better than ML, and the principal reason that came out from this examination is that SVM performs characterization by relegating pixel to the right land cover type by utilizing the hyperplane and maximum margin. This leads to the assignment of pixels to the right land covers, and consequently misclassification is limited. The classification process consequently influences the change detection assessment.

In light of the effectiveness evaluation of two calculations, the SVM calculation performed best and afterward SVM is utilized to recognize the change as per objective. The land cover changes for 2000 and 2019 are given in Table 23.6. It tends to be seen that light vegetation and water body areas diminished between the years, whereas dense vegetation, bare soil, and built-up areas expanded from 2000 to 2019. Although dense vegetation was found to include the biggest region inside the investigation locale in both of the years, the pace of its change was seen to be unsystematic. The dense vegetation, built-up, and bare land have expanded from 2000, where the abatement of water body and light vegetation is noticeable. Land transition analysis reveals that a huge part of light vegetation and dense vegetation has changed into bare lands. High expansion of agricultural land during the last 19 years has led to a significant increase in the bare soil areas. According to losses and gains statistics, it was clear that bare land and built-up areas have converted into some another land cover classes, and on the other side new areas transformed into dense vegetation from another class. Gains in dense vegetation and bare land area are evident from 2000 to 2019. Again, light vegetation land cover type is decrementing in large voluminous percentages in all the years. The changes (in terms of gains and losses) in other land cover types are proximately the equipollent or not impacting. Ergo, an incrimination in dense vegetation and bare land areas and decrement in light vegetation cover type are quite pellucid from this kind of analysis. The overall scenario showed that SVM is the best technique to detect the land transformation and the transformation indicates that vegetation cover transit into bare soil and built-up areas that is a sign of deforestation.

23.5 Conclusion

Two approaches to classifying land cover in Bhawal Sal Forests were compared using Landsat 5 and 8 data. ML and SVM classification techniques have been applied to distinguish the exact method for land use classification for 2000 and 2019. Consequences of classifications from both the strategies have been affirmed via land cover map for the year 2019. The level of accuracy has additionally been estimated through various proven parameters. It leads to the conclusion that the accuracy level is higher in SVM algorithm over the ML technique. Postclassification shows that the area covered by dense vegetation expanded during the period, yet the light vegetation is diminished. Management and control of forest department are one of the major reasons behind the expansion of dense vegetation cover. On the other side, light vegetation not only changed into dense vegetation but also transformed into built-up and bare soil areas. The expansion in populace and the overexploitation of forest assets can be seen as lost Sal forest. The geographical distribution (land cover) of the Sal forest for various targets can be improved with the advancement in satellite data and classification algorithms. In further studies, high-resolution satellite imageries like KOMPSAT-3 and ISS-Theia can be used that can improve the accuracy of the land cover mapping with more details. The study successively may help Sal forests' protection and management planning efforts.

References

Bolstad, P., & Lillesand, T. M. (1991). Rapid maximum likelihood classification. *Photogrammetric Engineering and Remote Sensing, 571*, 67–74.

Congalton, R. G. (1991). A review of assessing the accuracy of classifications of remotely sensed data. *Remote Sensing of Environment, 37*, 35–46.

Food and agriculture organization of the United Nations. (2006). Global Forest Resources Assessment 2005. Rome. http://www.fao.org/3/a-a0400e.pdf

Food and agriculture organization of the United Nations. (2010). Global Forest Resources Assessment 2010. Rome. http://www.fao.org/3/i1757e/i1757e.pdf

Foody, G. M. (2002). Status of land cover classification accuracy assessment. *Remote Sensing of Environment, 80*, 185–201. https://doi.org/10.1016/S0034-4257(01)00295-4

Foody, G. M., & Mathur, A. (2004). A relative evaluation of multiclass image classification by support vector machines. *IEEE Transactions on Geoscience and Remote Sensing, 42*(6), 1335–1343. https://doi.org/10.1109/2004.827257

Hassan, Z., Shabbir, R., Ahmad, S. S., Malik, A. H., Aziz, N., Butt, A., & Erum, S. (2016). Dynamics of land use and land cover change (LULCC) using geospatial techniques: A case study of Islamabad Pakistan. *Springerplus, 5*, 1–11. https://doi.org/10.1186/s40064-016-2414-z

Homer, C., Dewitz, J., Fry, J., Coan, M., Hossain, N., Larson, C., Herold, N., McKerrow, A., VanDriel, J. N., & Wickham, J. (2007). Completion of the 2001 National Land cover database for the conterminous United States. *Photogrammetric Engineering and Remote Sensing, 73*, 337–341.

Jensen, J. R. (2009). *Remote sensing of the environment: An earth resource perspective 2/e*. Pearson Education India.

Lu, D., & Weng, Q. (2007). Survey of image classification methods and techniques for improving classification performance. *International Journal of Remote Sensing, 28*, 823–870. https://doi.org/10.1080/01431160600746456

Mondal, A., Kundu, S., Chandniha, S. K., Shukla, R., & Mishra, P. K. (2012). Comparison of support vector machine and maximum likelihood classification technique using satellite imagery. *International Journal of Remote Sensing, 1*, 116–123.

Mountrakis, G., Im, J., & Ogole, C. (2011). Support vector machines in remote sensing: A review. *ISPRS Journal of Photogrammetry and Remote Sensing, 66*(3), 247–259. https://doi.org/10.1016/j.isprsjprs.2010.11.001

Naguib, A. M., Farag, M. A., Yahia, M. A., Ramadan, H. H., & Abd Elwahab, M. S. (2009). Comparative study between support vector machines and neural networks for lithological discrimination using hyperspectral data. *Egyptian Journal of Remote Sensing and Space Science, 12*, 27–42. https://doi.org/10.26833/ijeg.298951

Oommen, T., Misra, D., Twarakavi, N. K. C., Prakash, A., Sahoo, B., & Bandopadhyay, S. (2008). An objective analysis of support vector machine based classification for remote sensing. *Mathematical Geosciences, 40*, 409–422. https://doi.org/10.1007/s11004-008-9156-6

Pal, M., & Mather, P. M. (2005). Support vector classification in remote sensing. *International Journal of Remote Sensing, 26*, 1007–1011. https://doi.org/10.1080/01431160512331314083

Pontius, R. G., Shusas, E., & McEachern, M. (2004). Detecting important categorical land changes while accounting for persistence. *Agriculture, Ecosystems and Environment, 101*, 251–268. https://doi.org/10.1016/j.agee.2003.09.008

Rahman, M. M., Rahman, M. M., Alam, M. (2010). Disappearing forest tree species diversity in tropical moist deciduous forest and its implications: a case study in the Madhupur tract of central Bangladesh. 26, 161–170.

Salam, M. A., & Pramanik, M. A. T. (2017). Forest cover change analysis using remote sensing techniques in Madhupur Sal Forest of Bangladesh. *Journal of Environmental Science and Natural Resources, 10*(2), 73–78. ISSN 1999-7361.

Scott, A. J., & Symons, M. J. (1971). Clustering methods based on likelihood ratio criteria. *Biometrics, 27*(2), 387–397. https://doi.org/10.1016/S0169-7161(82)02012-4

Szuster, B. W., Chen, Q., & Borger, M. (2011). A comparison of classification techniques to support land cover and land use analysis in tropical coastal zones. *Applied Geography, 31*, 525–532.

Tahir, M., Iman, E., & Hussain, T. (2013). Evaluation of land use/land cover changes in Mekelle City, Ethiopia using remote sensing and GIS. *Computational Ecology and Software, 3*, 9–16.

Vapnik, V. N. (1995). *The nature of statistical learning theory*. Springer-Verlag.

Xie, L., Li, G., Xiao, M., Peng, L., & Chen, Q. (2017). Hyperspectral image classification using discrete space model and support vector machines. *Geoscience and Remote Sensing, 14*, 374–378.

Yu, L., Porwal, A., Holden, E. J., & Dentith, M. C. (2011). Towards automatic lithological classification from remote sensing data using support vector machines. *Computers & Geosciences, 45*, 229. Retrieved from https://doi.org/10.1016/j.cageo.2011.11.019

Chapter 24
Early Human Habitation and Environmental Adaptation in Central Tanzania in East Africa: An Archaeological and Geospatial Investigation

Krishna Rao Sadasivuni, N. Kasongi, and E. L. Temu

24.1 Introduction

Among very few African nations, which have potential archaeological and historical evidence, Tanzania is globally appreciated for early human origins as evident from the discovery of *Homo erectus*, the earliest human ancestor from Olduvai Gorge (OVG). It is known that the whole of Tanzania region, both mainland and islands, is fertile for archaeological sites dating to various time periods ranging from Oldowan up to historical Bagamoyo through Acheulian, Middle Stone Age (MSA) flake and blade tools, Late Stone Age (LSA) microliths, Neolithic/New Stone Age ground stone industry and Iron Age material culture. It is worth referring to prehistoric and historical cultural evidence that is found spread all over Zanzibar, Pemba and Mafia islands. The authors intended to focus on Central Tanzania because some of the areas in this region were recently explored. Moreover, Central Tanzania is potential for Stone Age, Iron Age cultures and rock art. Our recent investigations in the Dodoma Region of Central Tanzania (Krishna Rao, 2018; Ryano et al., 2020) have revealed significant cultural, geo-chronological information supported by archaeological and geospatial exercises.

K. R. Sadasivuni (✉)
Department of History and Archaeology, University of Dodoma, Dodoma, Tanzania

N. Kasongi
Department of Geography and Environmental Studies, University of Dodoma, Dodoma, Tanzania

E. L. Temu
Ministry of Culture, Dodoma, Tanzania

G. S. Bhunia et al. (eds.), *Anthropogeomorphology*, Geography of the Physical Environment, https://doi.org/10.1007/978-3-030-77572-8_24

Such spatio-temporal cultural inferences reflect on the then existed environmental adaptations for human habitation, and thus the need to focus on the determinants of human habitation and environmental adaptation. Review of literature relating to historical and geographical investigation will illuminate some insights on general understanding of human settlements and ecological adaptations. This kind of assertion reminds us of human geography and ecological models of ethnography (Chang, 1972). Multilinear evolutionary ideas and neo-evolutionary processual and post-processual approaches have therefore become part of scholarly interpretations. Of late, structuring structure theory has drawn the attention of many archaeologists for interconnecting people, space and objects pertaining to an archaeological site in their pragmatic archaeological explanations.

In this context, it is supportive to review on the archaeological references of evidence in the study areas in order to assess various geographical landscapes and environmental situations prevalent for the nurturing of people and understanding of their settlement patterns in the past. It is known that the major landmass of Tanzania comprises archaeological sites belonging to various cultural phases of different periods of time. A cursory look at the distribution of sites with different cultural phases as reported by several local and international scholars reveals some general ideas on human adaptations in different regions. Moreover this background research review on archaeological discoveries spanning from Early Stone Age or Early Palaeolithic up to historical evidence, which reflect the habitat, environmental situations and human adaptations in different landscapes of Tanzania as described below, will help understand the study region.

24.1.1 Sites with Evidence of Heavy-Duty Tool Complex

Generally heavy-duty tool component belongs to Early Stone Age (ESA). Archaeological sites of this period are spread into the western stream of the Great Rift in the northern highlands and a strip of southern highlands where the eastern strip of the Great Rift runs parallel to the plains and also the Bahi valley. In the northern Rift Valley region, the hominid fossil evidence was recorded from Olduvai Gorge (OVG), Laetoli, Peninj and Ndutu, which have yielded both human and cultural evidences (Leakey et al., 1972). Among these sites, OVG, which is familiar as 'Cradle of humankind', yielded complete record of the earliest human origins. It is a known fact that the northern part of Tanzania in general has the record of biological and cultural evidences. Stratigraphically seven major sections including Masek, Ndutu and Naisiusiu Beds were classified. Temporally these cultural beds cover a time span of about 2.1 million to 15,000 years ago. Former rivers, lakes and streams have contributed to sustainable hominid habitation.

Hominid fossils such as *Australopithecus boisei* (*Zinjanthropus* – see Appendix photo-1) and *Homo habilis* (see Appendix photo-2), along with Oldowan/Olduwan tools, were found in Bed I while the developed Oldowan (crude core tools) and Acheulian tools (advanced hand axes) were found in association with *Homo erectus* (Ergaster) in Bed II. Such kinds of cultural material embedded in the stream-laid detrital sediment were also found in Bed III and Bed IV. Fossil remains of *Homo erectus* (see Appendix photo-3) and other species of *Homo* were found together in the strata. While the Masek Beds are formed by stream-laid detrital sediment and Aeolian tuffs, the Ndutu and Naisiusiu beds are formed by the Aeolian sediments. This implies that the open grasslands that dominate the savannah seem to have offered sustainable habitat for early humans in the OVG area. It was suggested that the region experienced fluctuations in aridity, with dry periods dominated by grasslands and wet periods marked by lengthy woodland cover. The landscape of early humans, according to Magill et al. (2012), varied quickly back and forth between closed woodland and open grassland about five to six times during a period of 200,000 years. All of a sudden these alterations occurred with each transition indicated by over hundreds to few thousand years. The raw material used for tool manufacturing at OVG includes three major rock types such as igneous, metamorphic and sedimentary rocks.

The discovery of hominid fossils such as *Australopithecus afarensis* and archaic *Homo sapiens* or the Ngaloba skull (White, 1977; Day et al., 1980) encouraged Ndessokia (1990), who successfully found fauna at Ndolanya and Olpiro Beds in the Laetoli area dated to Ca 3.0–2.5 ma (see Appendix photo-4). This study helped in archaeological correlation, taphonomy and palaeo-environmental reconstruction. Accordingly, the upper Laetoli and Ndolanya Beds have been found deposited during arid environment. At Peninj on the east of OVG, palaeo-anthropological investigations yielded 27 palaeontological and 8 archaeological sites. Acheulian sites at Peninj like OVG are fluviatile fan depositional environments, while Oldowan cultural deposit was found mounted in distal alluvial fan or lacustrine floodplain.

The late Acheulian site, Isimila in the southern region, located little away from the eastern Rift, has yielded refined hand axes (see Appendix photo-5) (Howell et al., 1962). The research investigation on environment and activity at this site by Hansen and Charles (1971) is very prominent. The site bed has cut into the Pleistocene fill of a narrow valley. Its relief of the area at present is related to mid-tertiary and early Pleistocene tectonic activity, which caused fault and tilting of former eroded surface. Pre-Cambrian igneous and meta-sedimentary rocks constitute upland masses. Colluvial and alluvial deposits of quaternary period, representing fluviatile environment, mantle the crystalline rocks in the landscape of lowlands and valleys. Possibly seasonal flooding during the wet season drained the whole valley, causing the growth of woody shrubs and grasses. Yearly rainfall of 750–850 mm was reported during the summer season. Therefore the environment appears to be favourable to large carnivores and human population as well. Especially the habitat of Isimila after rainy season would have become more resourceful for the survival of population. The availability of water, raw material for tool manufacture and plant and animal food must have become a niche hive for Acheulian people to flourish in the Isimila site as evident from the occupational floor with implements.

24.1.2 Sites with Evidence of Flake and Blade Tool Complex

The implements such as scrapers, points and simple blades made on flakes and blades, which were detached from stone pebbles or cores jointly, constitute the characteristic trait of Middle Stone Age (MSA). The techniques like Levallois and retouched techniques are generally associated with the MSA tool preparation. Nevertheless, the heavy-duty tool matrix which includes core scrapers, core-axes and picks is inevitably associated with the early stage of MSA when more closed vegetation and woodland savannah, coastal evergreen forest and thicket environments were found. Production of light-duty tools in place of heavy-duty tools was dominant in the Holocene open country. The MSA tools like Levallois and discoidal cores, flakes with faceted striking platforms, retouched artefacts like scrapers and a few small choppers (Leakey et al., 1972: 332) were recorded from Ndutu Beds (Hay, 1976: 146) dating to late Middle to Early Upper Pleistocene of Olduvai sequence. These beds, which are dated to 12,000 years BP, seem to have been deposited when the climate was semi-arid. The open grassland environment might have favoured modern human adaptation. From the streambeds of sandstone and claystone of Ngaloba Beds of Laetoli zone two (Day et al., 1980), similar MSA artefacts made on lumps of lava and a skull were uncovered. These deposits are chronologically contemporaneous to Ndutu Beds.

Further from the two interesting MSA sites are Nasera and Mumba Hohle rock shelters. Nasera, which is dated between 23,000 and 26,000 years BP (Leakey, 1936; Mehlman, 1977), is represented by alluvial sediments. Here more concentration of MSA tools made on quartz and chert was noticed. The artefacts include prepared cores, retouched points and convex and straightedge side scrapers. The other MSA site is Mumba Hohle rock shelter (Kohl- Larsen, 1943; Mehlman, 1979) located in an inselberg of gneiss on the northern side of the saline lake Eyasi. Here former lakeshore was identified. Below this shore a stratigraphic section comprising loamy sand layer of Bed V has yielded large size crescents, backed knives, points, crescents, convex scrapers, Levalloisian flakes and cores, etc., resembling MSA material culture of the Howiesons poort industry of South Africa. The Bed VI at the depth of 6 m has also yielded similar tools like that of the previous Bed V, which can be comparable to the tool kit of Nasera. It is remarkable to note that the remains of archaic *Homo sapiens*, along with fossil mammalian bones and Levallois artefacts and heavy-duty tools typical of 'Sangoan', were ported from the sandstone bank on the northern edge of the Lake Eyasi (Mehlman, 1987). The fauna (Mehlman, 1989) in this region consists of modern East African grassland ungulates. The whole environment seems to have provided a potential habitat for MSA population.

The open air MSA site in the Northern Tanzania is located in the flood plain of Loiyangalani River in the Serengeti National Park which has yielded Levallois cores, flakes, borers and burins (Bower & Mabulla, 2013). Further it revealed that around 40–50 ka the *Homo sapiens* population strategically spread into the western margin of the Serengeti plains and adjacent areas.

Isimila, a deciduous woodland in the southern highland region, yielded flake tools reminding Sangoan cultural phase, and the tools that were collected from the top of the Upper Isimila Formation of the old valley floor, which is located in a swamp grassland depression, dated to the late Middle to early Upper Pleistocene (Howell et al., 1962: 279). From the coastal zone of forest/savannah landscape scrapers, Levallois flakes and cores resembling MSA were found in Tendanguru, 60 km northwest of Lindi towards the dinosaurs' sites (Isaac, 1974: 255).

24.1.3 Microlithic Tool Complex of Mesolithic Phase

This cultural phase is also known as Late Stone Age (LSA). Geometric blades such as lunates, points, triangles, etc., are called microlithic tools. Apart from the spatial distribution of MSA tools at the northeast corner of the Kilwa Island, woodland areas of the Mwemkuru River bank and Mchinga woodland, LSA blades and microlithic artefacts from excavation have been recovered from Kilwa island, Mnangole, Kitere and Mnaida sites (Kwekason, 2010: 138). A good number of blades and microliths were found associated with MSA-derived Levallois tools in the eroded gullies of Mbwemkuru valley of Mtwara region. Actually the earliest Late Stone Age (LSA) evidence dated to 3000 BC is reported from Zanzibar (Chami, 1996).

Majority of the rock paintings dated to LSA are reported from Kondoa District, Iramba area and Singida district of Central Tanzania (Leakey, 1936; Fosbrooke, 1950; Masao, 1976, 1979; Odner, 1971; Anati, 1980). Again early and later microlithic artefacts have been identified from Mumba Hohle (Mehlman, 1979) and Nasera (Mehlman, 1977) rock shelters and other popular rock art sites. Further Serengeti, Lake Natron, Manyara, Engaruka areas; Lake Eyasi and Mwanza rock shelters have yielded similar Mode-5 (LSA) industries (Soper & Ciolden, 1969). From rock shelters and open-air sites in Iringa and Mbeya areas on south and Rift Valley in northeastern Tanzania (Kwekason, 2010: 140) have yielded LSA microlithic tools.

24.1.4 Sites Rich in Ground Stone Tool Complex of Neolithic Period

Ground stone axes, chisels, adzes, celts and shouldered tools constitute Neolithic culture, which represents a major shift in the mode of subsistence from hunting-gathering to food production. Agricultural economy was initiated by the domestication of animals and plants. Tools like grindstones and pottery were introduced for food processing. Permanent settlements and villages were established. The sequence of great cultural changes during Holocene time is therefore termed as 'Neolithic

Revolution'. There are two cultural phases within Neolithic, Pastoral Neolithic (PN) and Neolithic, because domestication of animals began earlier than plant domestication in Africa unlike other parts of the world.

In the northern highlands, Ngorongoro Crater, West Kilimanjaro and Serengeti Plains (Bower & Gogan-Porter, 1981) Pastoral Neolithic sites were reported. The cultural material evidence from West Kilimanjaro and Serengeti dated to 2000 BC (Mturi, 1986) includes pottery, stone bowls, grindstones, lithics and bone remains of domesticated animals. The burial mounds (stone cairns) dated to 900 BC were reported from Ngorongoro Crater. The PN communities of highlands seem to have possessed Narosura pottery type while in Serengeti Akira type may have been practised. In addition to the above, Neolithic evidence, along with MSA flake and blade tools, was reported from the University of Dodoma (UDOM) campus in the Dodoma Region (Krishna Rao, 2018).

24.1.5 Cultural Evidence from Iron Age Sites

Generally the initiation of the iron component is evolved in the name of Early Iron Working (EIW). The efficiency of metal iron has drawn the attention of people who began to use iron in agricultural and defence activities and decoration. Various theories such as diffusion, independent invention and trade and contact on the appearance of iron in East Africa, in general, and Tanzania, in particular, have been floated. It is, however, convincing that the Bantu speakers introduced iron in the region. Several sites of iron working were reported from Tanzania. The new technology called iron smelting and iron working has gradually replaced the dominant use of stone, wood and bone tools. Consequently iron-using farming communities spread to various parts of Tanzania by about the first millennium AD. Depending on the intensity in the use of iron for subsistence it was classified into Early Iron Working (EIW) and Late Iron Working (LIW) (Chami, 1994). From Buhaya the evidence of iron for the first time was reported in the inter-lacustrine region, which was dated to around 500 BC (Soper, 1967).

Various traits of EIW cultural evidence dated between 100 and 1000 AD include iron objects, smelting furnaces (Schmidt & Childs, 1995), pottery with bevelled rims, dimple bases and elaborate decoration, domestic cattle, sheep and goats, cultivated cereals such as millet, sorghum and African yams. Contrastingly Later Iron Working (LIW) phase is post-dating to 1000 AD. Advancement in pottery styles with different techniques was represented in their association with the previous Kwale and Lelesu wares. The nature of evidence indicating mixed economy was supported by irrigation terraced platforms with stonewalls, stone-lined furrows and enclosures, and burials such as stone cairns and circles (Sassoon, 1966, 1967; Sutton, 1978) of Engaruka ruins in Northern Tanzania. Surplus food production through irrigation accelerated by efficient iron technology might have supported human adaptation for livelihood.

24.2 Archaeology of Historical Period

Several historical archaeological sites in the mainland and coast were reported. A historical Bagamoyo town, which had played a dominant role in slave trade and coastal trade, is located 70 km north of Dar es Salaam and about 40 km east of Zanzibar. Bagamoyo town is located on an elevated marine terrace called 'Mtoni' (Chittick, 1962), and its history started in the mid-nineteenth century AD. The earliest occupation was dated to late eighteenth century (Brown, 1970:71), and subsequently the town became the main gateway to the interior. There are two important sites, Kaole and Caravan Serai, located in Bagamoyo. The oldest evidence is the Kaole ruins, which include coral stone houses of a civilization on the Central Tanzania coast, dated back to twelfth century AD. These houses that were identified by Chami (1998) as Swahili culture were built with coral stone and lime. Carvan Serai is the other site, which was identified as post-Swahili tradition and was dated to sixteenth to seventeenth century AD. Mbuamaji (Ombori & Mabulla, 2013) in Dar es Salaam city is another site yielding lithics, beads, bones, shells, daubs, pottery, metal objects, and also Swahili architecture. The materials coral, rag and lime were used in carving architecture.

The whole review of the distribution of archaeological sites and related material evidence described above reveals various possibilities of human survival in diverse landscapes and environmental conditions in different periods of time. A brief review on the varied environmental circumstances, which played a dominant role in human survival through adaptations for admissible time periods, explains the strength in environmental resources. Complexity or any other unexpected barrier in the environment creates havoc in human survival bringing pressure on populations to either migrate to other areas or extinct totally. The human adaptations have been briefed as follows.

24.3 Environment and Human Adaptation from the Earliest Time till Historical Period

Complete history of human evolution covering the period from about 2 million to 15,000 years ago was drawn from the evidence of archaeological sites such as OVG, Masek, Ndutu and Naisiusiu located in Northern Tanzania. Natural water resources like former rivers, lakes and streams may have satisfied the basic needs of hominids. The remains of *Homo erectus* and other species of *Homo* were reported from the stream-laid detrital sediment of OVG Bed IV. The stream-laid detrital sediments, Aeolian tuffs and sediments, which caused the formation of Masek, Ndutu and Naisiusiu Beds, indicate open grassland landscapes dominating savannah. Early human habitation might have been supported by such grasslands for quite a long time as evident from the experiences of fluctuations in aridity with grassland dry periods and woodland wet periods. The three major rock types, igneous, metamorphic and sedimentary rocks, which are locally available, could be used as raw materials in tools preparation for foraging. At Peninj, an Acheulian site was found in fluviatile fan depositional environments, while Oldowan cultural debris is visible in either alluvial fan or lacustrine floodplain, which is confirmation of water resources for survival of humans.

In the southern highlands the relief of the area covering Isimila site was subjected to tectonic faulting and tilting. Uplands are characteristic of pre-Cambrian igneous and meta-sedimentary rocks, while lowlands and valleys comprised crystalline rocks, which are mantled by quaternary colluvial and alluvial deposits. It represents fluviatile environment generally. The valley seems to have been drained by seasonal streams, causing the growth of shrub jungle and grass extensively. The rainfall in summer season was recorded to be about 750–850 mm, and hence the overall environmental conditions appear to have supported large carnivores and also Acheulian people. Drinking water source, raw material for tool manufacture and plant and animal food contributed a lot to hunting and gathering activities of people. Evidence of the living floor with implements justified this fact. Further faunal evidence reveals that the prevailing sedimentological environment might have attracted even herbivores. That means people after rainy season appear to have stayed temporarily when there were more chances for hunting and gathering activities.

As far as MSA is concerned, the evidence at Ndutu Beds dating to late Middle to early Upper Pleistocene (approx. 120,000 BP) shows that there was semi-arid climate in the region. It was the grassland open habitat that must have supported the modern human populations. Typical MSA tools made on lava rock along with a skull were recovered from the stream deposits of sandstone and claystone from the beds of another site, Ngaloba, which is contemporaneous to Ndutu. The evidence of archaic *Homo sapiens* remains, fossil mammalian fauna and other MSA artefacts, along with a few heavy-duty core scrapers, choppers and axes typical of 'Sangoan' of early period from the sandstone bed on the northern edge of the Lake Eyasi around later Middle Pleistocene, constitutes potential habitat of MSA population. The associated faunal remains such as zebra, buffalo, etc. (Mehlman, 1989), were identified as East African grassland ungulates. An interesting site, Tendanguru on the coast, has yielded MSA artefacts like high-backed scrapers, Levallois flakes and cores. Again the coastal sites Kilwa and Lindi have become woodland habitat for the spread of MSA objects.

With regards to the LSA, similar woodland environments were noticed from Kilwa, Mnangole, Kitere and Mnaida sites with the evidence of blades and microlithic artefacts. Surprisingly, Central Tanzania, which includes Iramba, Singida and Kondoa rock art sites, was dated to LSA. The painted rock shelters of Mumba Hohle and Nasera rock shelters have also yielded early and later microlithic assemblages belonging to LSA. The Nasera landscape, east of Serengeti National Park, is covered by the biomes of grassland savannah. This landscape is represented by the driest zone in Northern Tanzania with an annual rainfall of 500 mm. Long rainy season followed by short rains occurs in the area like the bimodal pattern. Short grasses, shrubs and *Acacia* trees are the types of vegetation grown in the area. According to Sinclair (2005) Northern Tanzania during Pleistocene is believed to be similar to that of the present. The use of locally available raw material like quartz and quartzite rocks in tool making was an added advantage.

While LSA population is believed to have drawn food from specialized hunting, Neolithic people had introduced domestication of animals and plants, and agriculture for their subsistence. In fact within Neolithic there were two groups, Pastoral Neolithic and Neolithic. The former depended on herding while the latter depended on agriculture. The Neolithic groups seem to have continued their herding, gathering and agriculture activities in different areas under favourable environmental conditions. In the next phase, Early Iron Working communities (EIW) had given more emphasis to agricultural crops like cereals, millet, sorghum and African yam. The development of efficiency in agriculture production among EIW communities was derived from the efficiency of iron that was introduced by Bantu. Further efficiency in agricultural production through the introduction of irrigation techniques as evident from Engaruka ruins was reported. Gradually, the concept of burial in their belief system, as seen from their megalithic structures like stone cairns and circles (Sassoon, 1966), occupied a place of prominence.

Finally, the importance of historical archaeology is marked by the initiation of maritime trading between Tanzania coastal communities and other interior interlacustrine communities. Subsequently, increased trading activities on the coast and contacts with the Arabs from the Middle East, socio-economic relationships, marriages and coral stone houses expanded in the region. The great commercial centre, Bagamoyo with the townships of Kaole identified with Swahili culture and Caravan Serai representing post-Swahili tradition, flourished with Swahili civilization.

The above sequences of prehistoric and historical evidences, which were reviewed on the existence of geographical landscape, climate and possible human adaptation under different socio-economic, political and environmental conditions, seem to have based on various scientific approaches including ethnographic enquiry but with very limited application of geospatial techniques particularly in Central Tanzania. It is our main intention to concentrate on the application of geospatial techniques in the Central Tanzania region with specific focus on case studies in the Dodoma Region. Ethnographic information that was collected from a few localities is to just understand the present socio-economic situation in the region. That is why the following objective is formulated.

24.4 Objective

The main objective of this chapter is to identify the geographical locations of prehistoric and historic archaeological sites and related environmental conditions which may have favoured survival of human populations for different successive periods of time through adaptations.

24.5 Materials and Methods

In order to achieve the above objective, archaeological sites with reference to human adaptations and settlement patterns were examined using geospatial techniques and ethnographic documentation. Data relating to elevation were acquired and terrain analysis was undertaken.

Intrusive and non-intrusive methods were adopted for data collection and analysis. Intrusive methods include section scraping and test excavation for typo-technological classification of cultural material and geo-chronological assessments of specific sample archaeological sites in the Dodoma Region, Central Tanzania. For this, systematic and standard procedures were adopted with the help of scientific equipment like GPS, compass, rangefinders, digital camera, sieve set, Bunsel chart and other digging tools. Non-intrusive methods including aerial photos, satellite imagery and GPS coordinates were used to reconstruct geographical mapping pertaining to various location aspects and settlement patterns. Specifically data relating to satellite images representing climate, which includes temperature and rainfall for the past several decades, were collected to reconstruct past environments. Further the use of GPS coordinates helped in reconstructing terrain, which includes elevation, slope and slope aspect, in the settlement location analysis. These methods corroborate the idea of subsistence and adaptation to survival of early human populations within the existing ecological setting in space and time.

24.5.1 Acquisition of Elevation Data and Terrain Analysis

The elevation data were derived from Shuttle Radar Topography Mission (SRTM) Digital Elevation Model (DEM) data set. The DEM had a spatial resolution of 30 m. The mask function in raster package was used to extract the part of DEM data set that fits the extent of study area. The DEM was then re-projected from Geographical Coordinate System (WGS 84) to Projected Coordinate System (UTM zone 35 and 36), respectively. The slope raster was computed from DEM using mathematical Eq. 24.1 by using terrain function in r software. The aspect raster was derived from DEM by using terrain function in r software. The extract function found in raster package in r was used to extract the values of elevation, slope and aspect for individual study sites.

$$\begin{aligned} &\text{Degree of Slope} = \theta \\ &\text{Tangent } \theta = \frac{\text{Rise}}{\text{Run}} \end{aligned} \tag{24.1}$$

24.6 Statistical Analysis

24.6.1 The Mann–Kendall Test

The study used the rank-based non-parametric Mann–Kendall method on the long-term climate data of the study area to detect statistically significant trends. The trend package in r software was used to carry out the Mann–Kendal trend test. The analysis involved the following hypotheses:

Ho: There has been no trend in climate parameters (temperature and precipitation) over time.

HI: There has been a trend (increasing or decreasing) over time.

The mathematical equations (Eqs. 24.2–24.4) for calculating Mann–Kendall statistics, variance of Mann–Kendal statistics and standardized test statistics are as follows:

Mann–Kendal statistics (S):

$$\mathrm{S} = \sum_{k-1}^{n-1} \sum_{j=k+1}^{n} \mathrm{sign}\left(x_j - x_k\right) \tag{24.2}$$

where

$$\begin{aligned} \mathrm{Sign(xj - xk)} &= 1 \text{ if } \mathrm{xj - xk} > 0 \\ &= 0 \text{ if } \mathrm{xj - xk} = 0 \\ &= -1 \text{ if } \mathrm{xj - xk} < 0 \end{aligned}$$

where n is the length of the sample, k and j are from k=1, 2... n−1 and j= k+1... n. If n is bigger than 8, statistic S approximates to normal distribution. The mean of S is 0, and the variance of S can be acquired as follows:

Variance of Mann–Kendal statistics:

$$VAR(S) = \frac{1}{18}\left[n(n-1)(2n+5) - \sum_{p=1}^{g} t_p\left(t_p - 1\right)\left(2t_p + 5\right)\right] \tag{24.3}$$

Standardized test statistics:

$$Z = \frac{S-1}{\left[VAR(S)^{½}\right]} \quad \text{If } S < 0$$
$$= 0 \text{ if } S = 0 \tag{24.4}$$
$$= \frac{S+1}{\left[VAR(S)^{½}\right]} \quad \text{If } S < 0$$

If Z>0, it indicates an increasing trend, and vice versa. Given a confidence level α, the sequential data would be supposed to experience statistically significant trend if | Z|>Z (1-α/2), where Z (1-α/2) is the corresponding value of P=α/2 following the standard normal distribution. In this study, 0.05 and 0.01 confidence levels were used.

24.7 Results and Discussion

Rock art in Tanzania in general and Central Tanzania in specific:

Broadly speaking, Eastern Africa is reported to have different environmental conditions stretching from glaciated mountains to semi-arid deserts and mangrove-fringed ocean shores (Adams et al., 1996). Tanzania in East Africa has a rich database for not only archaeological material culture and other anthropological and palaeontological evidences of different time periods but also extensive number of rock art paintings in caves and rock shelters in different regions such as Lake Victoria, Lake Eyasi, Central Tanzania, Southern Tanzania and Mtwara and Lindi. Art styles in these rock shelters include hunter-gatherer, pastoralist and white Bantu images. The earliest East African rock paintings (Phillipson, 2002: 127) were known to have been depicted probably by microlithic culture practised by LSA people around the time span between 10,000 and 2000 years ago. Lewis-Williams (1984) opined that the painting scenes of dancing figures, phosphenes, nasal haemorrhage, attenuation of human figures, animals surrounded by hallucinations and therianthropos are all related to hunting-gathering trans-performances. Scholars such as Leakey (1936, 1983), Fosbrooke (1950), Tanner (1953), Odner (1971), Chaplin (1974), Masao (1976, 1979) and Anati (1986) have identified several rock art sites located in various regions of Tanzania. They are related to varieties of images with varied themes such as hunting, gathering, pastoral and other foraging activities, wild and domesticated animals, dancing and magico-religious activities.

24.7.1 Rock Art in Central Tanzania

Among all these painted rock shelters Central Tanzania is popular for rock paintings. Kondoa and Singida in Central Tanzania are more prominent among all the sites.

The region with hill valleys, aquatic resources and animal and edible plant resources must have provided an adequate niche hive for the livelihood of people in the past. Cushitic and Bantu speakers from outside also tend to settle down in the region. The rock art sites are spread in Kondoa District, some localities of Dodoma Rural district, Iramba and Singida rural and urban districts. Kondoa is the richest in rock art. Kolo, Kinyasi, Kandusi, Pahi, Chungai, Chora and Cheke etc (Masao, 1979; Anati, 1980) are some of the principal sites evident with dense concentration of human and wild animal figures, plant and domestic cattle images and other traps and symbols. Singida group of rock shelters has also yielded good number of rock art sites. Some potential rock art sites have been identified from Kisana, Kisiriri and Kinaliya villages of Iramba district. Apart from the above rock shelters, certain hills and mountains are found to have connections with ceremonies and rituals. Even now some painted caves and rock shelters are being worshipped (Krishna Rao & Temu, 2018).

24.7.2 Sites, Environments and Ethnographic Present

Kondoa and Singida districts are prominent in rock art, especially of hunter-gatherer and pastoralist thematic expressions in the Central Tanzania region. This region since ancient past has been occupied by foragers, pastoralists and later on agriculturists.

Kondoa The rock art sites of Kondoa are located on the slopes of the Maasai escarpment that slides from the western side of the Great Rift Valley. There are caves and rock shelters with paintings of elongated people, animals, hunting scenes and abstract markings. Like the chronological sequences of Eastern and Southern Africa, these line drawings are also dated to more than 2000 years ago. An interesting aspect is that these rock paintings are directly or indirectly associated with the traditions of communities living near the foothills of escarpment where caves and rock shelters are situated. The Kondoa paintings are popularly known as Kondoa Irangi paintings. There are two predominant communities, *Warangi* and *Wasi,* living in this district. The *Warangi* are the Bantu speakers who are identified by iron smelting technology. Their oral history states that they first settled around Haubi area (Lane et al., 2001; Kessy, 2005) and then spread to other areas, and so formed the largest ethnic group in Kondoa. The *Warangi* occupies two-thirds of the present total population of Kondoa. The other group, *Wasi,* is one of the three groups of Southern Cushitic speakers who live in various parts of the same district. They claim that they entered the Kondoa locality well in advance before the entry of *Warangi* community (Ehret, 1974, 1976; Clark, 1976; Bower, 1973; Smith, 1992). The *Wasi* seem to have brought with them the knowledge of cattle keeping, cattle breeding, irrigation technology, field agriculture and rainmaking. This knowledge gave them the ability to own and control the land. Different from ceremonies of *Warangi,* the *Wasi* hold rituals in caves, rock shelters, mountains and lake areas. Few scholars opine that the indigenous click-speakers of the region and the immigrant *Warangi* adopted some of these traits from the Southern Cushitic speakers (Kimambo, 1968; Sutton, 1968).

Distribution of Kondoa Sites There are about 16 sites such as Kisese 1, Kisese 2, Cheke, Chungai, Pahi, Kandusi, Kinyasi, Kolo, Iitololo and Masange. Kolo or Mongomi WaKolo (see Appendix: Kolo paintings 13 and 14) and Pahi sites are worth mentioning because they are geographically distributed in a pattern and accessible to the village settlements (Temu, 2016). On Kolo hills, three rock shelters, BI, BII and BIII, are located in three elevations. The first shelter (1757 m) has yielded 59 figures, which include humans, animals, plants and phosphenes (symbols), and are painted in red and white. The second rock shelter has shown 33 red paintings, which are humans, animals and plants. Similar paintings of total number of 18 were found in the third shelter located at lower elevation (1694 m).

Pahi cluster of rock shelters near Lusangi village (see Appendix photos 13–15) are located about 43 km from Kondoa town. There are three rock shelters identified here. *Warangi* settlements can be seen near this Pahi Lusangi village. The drawings were found painted in red, black and white. In the first rock shelter are 36 drawings belonging to humans, animals, plants, traps and phosphenes painted in red and white colours. Perhaps the two colours red and white belong to two different antiquities. In the second shelter, around 84 figures were found painted in red, black and white, indicating that different people might have drawn the paintings in different cultural periods. The depictions include humans with activities, wild and domesticated animals, symbols and traps. In the third rock shelter some 47 human figures painted in black and white colours were found.

Environmental Setting Prehistoric Kondoa seems to have extended natural resources such as hilly terrain, rocky outcrops, flora, monsoon and drainage, and animal movement to form into a habitable region for hunting, gathering, domestication of animals and plants, etc. The availability of rocks and soils also played a significant role in the preparation of tools and colouring materials. The vegetation cover of Kondoa is mostly woodland. The most common types of trees in this area include *Brachystegia* sp., *Pterocarpusangloensis* sp., *Baobab* and others. In the valley *Acacia kirkii, A. tortillis and Delenixalata* sp. are predominant. On the wide and flat valley of the Bubu River Valley there is a thick and evergreen bush comprising the *Lianas, Papyrus* and *Mbuga* types of vegetation (Masao, 1979). The ridge crests with their granite outcroppings and thin layer of soil with pellet gravel do not support much more than a handful of thorny shrubs of pseudo-*Posopis, Combretum, Burthia, Grewia* and *Bussiasp* (Newman, 1970).

The main geological systems observed in Kondoa District consist of a complicated group known as the basement system rocks (Masao, 1979), a highly metamorphosized sedimentary series. In some places, veins of pegmatite intrude into the basement system rocks. Much quartz and feldspars are seen between Gubali and Kolo, for instance, near the rock art site of Choka on the way to Haubi. The granitic rocks of the area are themselves to some extent regarded as products of metamorphism rather than basement systems (Aitken, 1950; Masao, 1976, 1979).

Further it is evident that there are major fault lines associated with the main rift valley systems. The Kondoa range of hills forms part of the eastern escarpment, which is associated with many huge inselbergs of granite or syenite boulders outcropping in small rocky ranges. In relation to the geology of the area, the paintings are very young. The rocks are of the pre-Cambrian era, dated to 500 million years ago (TNR No. 29) while the earliest rock paintings in Kondoa are dated to the Late Stone Age, between 30,000 and 2000 years ago (Masao, 1979; Inskeep, 1962; Leakey, 1983).

Present Situation/Ethnography Kondoa District has been inhabited by three major groups: hunter-gatherers (Sandawe and Hadza); pastoralists (Iraqw, Burenge, Wasi, Aragwa, Datoga and Maasai); and the semi-agriculturalists like Bantu language-speakers (Warangi). The Warangi (their name is known from Swahili name of Warangi and their Kirangi language) are the largest ethnic group in Kondoa (Bwasiri, 2016). At present, agriculture is the main economic activity for the people of Kondoa. The main types of crops planted include sunflower, onion, maize, beans and millet. Warangi have practised rituals for more than a century. There is no clear time when this practice initiated. Available evidence suggests that it is more than two generations old. In the past, elders met and discussed matters relating to their communities. Appeasement of their ancestors was through rituals, which are practised every year. Rain-making rituals were conducted every October/November before the beginning of the rainy season. Community members who had enough food to feed the circumcised youth conducted circumcision ceremonies. Other rituals like traditional healing practices took place at any time of the year. There were also rituals which were performed by elders to protect their communities from diseases and enemies. The rock art sites, especially Mongomiwa Kolo (Bwasiri, 2011) have been linked to these rituals. The main food of Warangi is Ugali made of sorghum, millet or maize flour. Beans, milk (masusu), slippery vegetables (mlenda/kirumbu) and meat were relished. Traditionally the Warangi used to dress in kaniki (black cloth), but due to Islamilization they have now adopted Swahili attires called Kanzu and hijab. The system of naming was related to the season, for example, Mwasi (sunshine), Mbula (rainy season), Nyange, Ndundya, Kichilo (heavy rainfall) and Salala (little rainfall).

Singida

Distribution of Sites Iramba and Mkalama districts of Singida Region have provided substantive rock art evidence. In the Iramba district, painted caves and rock shelters were identified at Kisana, Kisiriri and Kinaliya (see Appendix: Kyaga rock painting 15) villages. From the first site, Wangu Mountain, two caves have yielded four long wooden drums, each around 6 feet long and 2 feet wide. This site also yielded a few sacrificial pots. They are reported to have been related to rituals conducted by some ethnic groups in the past. From the second site, Ng'ombe rock shelter, domestic cattle cow and goat were found painted in red. From the Kinaliya site, around 24 figures of wild animals such as elephant, giraffe, antelope, zebra, buffalo and rabbit were evident with red and white colour paintings. Another cave called Mikiki cave in the same village has yielded six drums similar to those of Wangu Mountain.

Environmental Setting The dense concentration of rock paintings in the Singida group of rock shelters provides lot of insights with regard to socio-economic character and the palaeo-environmental support to the Holocene populations. The area is bordered to the north by Shinyanga Region, to the northeast by Manyara Region, to the east by Dodoma Region, to the southeast by Iringa Region, to the southwest by Mbeya Region and to the west by Tabora Region.

Singida Region, which abounds in rock art paintings, has a total area of 49,438 km^2, out of which water bodies of Lake Eyasi, Kitangiri, Singidani, Kindai and Balengida cover 95.5 km^2. The remaining 49,342.5 km^2 is land area. Climate-wise the region forms part of the semi-arid central zone of Tanzania, which experiences low rainfall and short rainy seasons which are often uneven, with fairly widespread drought for one year out of four years. Total rainfall ranges from 500 mm to 800 mm per annum, with high geographical, seasonal and annual variations. There are two seasons, short rainy season during the months of December to March which sometimes extends to April, and the long dry season from April to November.

The mean annual rainfall ranges from 600 mm to 800 mm over large areas of Iramba and Singida districts. On the eastern side of Manyoni district near the Bahi Swamp and the Rift Valley depression of Mgori and Shelui divisions lies the drier area in the region where the mean annual rainfall is less than 550 mm. Winds follow at a monsoonal pattern being northeasterly during the months of November to March and southeasterly for the rest of the year (dry season). In May to October, the winds are usually dry and contribute to the semi-arid climate of the region. The fact that maximum wind velocities coincide with the period of greatest water deficiency underlines the climatic impact of these winds on moisture losses and hence desertification.

Ethnography The tribes now inhabiting this region are the *Nyaturu* a tribe known to have reached the Singida district from the east about 200 years ago; the Iambi and *Irambab* arrived from the south about 150–200 years ago, and the *Isanzu* came southwards from Sukuma country only about 100 years ago (Hunter, 1953). According to Masao (1976), ethnographic evidence and oral traditions attest to use of rock shelters and caves for such purposes as home or camp bases, workshop areas, sanctuaries for magico-religious functions, refugee camps in times of warfare or hideouts for game. Indeed some of the contemporary local groups still use these caves for some of these functions including rain making; and there is no reason to doubt that they were used for more or less similar functions by the ancestors or the precursors of the present tribes. *Iramba* settlements are spread at the foot of the Wangu Mountains. They cultivate maize, sweet potato, onion, sunflower and sugarcane. On enquiry the local *Wanyiramba* community explained that human skin was used for the preparation of wooden drum blow covers. Though such drums were used during the time of rain making, nowadays they are no more in use. Yet, the local community continues to hold rituals of spiritual nature believed to be associated with the welfare of the clan at these painted shelters since the time of their ancestors, and in a way people protect the sites. That is how there exists a strong bondage between people and Wangu Mountains.

The geographical landscape and environment including lacustrine setting and monsoons would have influenced the development of woodland forest and animal movement. The availability of semi-precious rock types like quartz for manufacturing blade tools might have led the LSA populations initially to specialized hunting. Later on pastoralism, domestication of animals and plants and agriculture could have sequentially developed one after another for quite a long time in these regions as attested by rock art paintings in Central Tanzania, in general, and Kondoa and Singida Regions, in particular. Even now the local communities like Warangi, Wasi and Iramba conduct magico-religious activities and rituals connected with the socio-economic pursuits of development. Thus the Central Tanzania landscape seems to have offered sufficient quantity of natural resources to the ancestors of the Bantu speakers and Cushitic-speaking communities. Central Tanzania, particularly during Holocene, must have possessed substantial natural resources, a geographical and historical phenomenon for attracting different populations to habitat for a long time in the region.

Although the above environmental situation, which includes landscape, hilly terrain, monsoon and drainage in Kondoa and Singida districts, provides some feedback to understand the interaction of rock art communities with the then existed environment for leading sustainable livelihood during LSA period dated to 6000 years ago (Leakey, 1983), our inferences on terrain analysis for the above rock art areas shall add to the landscape aspects of the hilly tract that may have offered a pertinent habitat to the thriving ancient communities in the region. The terrain analysis for the above two areas is given in Table 24.1.

This table reveals that the hilly Kondoa terrain has the highest elevation, slope and aspect of 1484 m 4.82476° and 286.699°, respectively, compared to that of Singida locality which has only 1260 m, 4.41632° and 108.435°, respectively. Since Pahi locality is a part of Kondoa its topography need not be elaborated here. The intensive rains recorded at 460 mm in Kondoa and 544.4 mm in Singida contribute to facilitate water storage, growth of forest resources and animal movement. Unlike during historical time, prehistoric communities particularly during early Holocene

Table 24.1 Terrain analysis for Kondoa and Singida

Name	Annual average rainfall (mm)	Annual average temperature (degrees)	Elevation (meters)	Slope (degrees)	Aspect (degrees)
Singida	544.4	27.4	1260	4.41632	108.435
Kondoa	460	26.6	1484	4.82476	286.699
Pahi (Kondoa)	310.2	27.2	1112	2.9287	215.565

time appear to have mostly preferred natural rock shelters and caves for living. Such places were found to protect people from sun, wind and rain. Hilly terrain that have considerable slope and direction of 4.8° and 287° respectively at Kondoa and 4.4° and 108.4° at Singida may have shaped their livelihood comfortable to harvest the available forest resources through specialized hunting and gathering and agricultural activities. Apart from routine subsistence activities, religious and other aesthetic performances were conducted. The evidence of potential rock art paintings in the two areas and Kondoa in particular is the proof for such aesthetic dynamism of prehistoric communities. That is the reason why Kondoa became popular for rock art in Tanzania and occupied a prominent place in the list of World Heritage Sites. Such rock art centres need to be preserved and conserved to enhance tourism in Central Tanzania.

24.8 Case Studies: Dodoma Region

Dodoma is the capital city of Tanzania, and the region is spread in a sprawling area of 2576 km^2. The total population is 410,956. The area is a thicket landscape with hillocks and dry streams and valleys exposing red soil as a consequence of erosion of top humus. Seasonal rains convert the whole landscape verdant. The rainfall in Dodoma Municipality is below 500 mm and is poorly distributed. Because of such semi-arid condition and dependence of rain-fed agriculture, the area suffers from scanty crop production. Deforestation in different forms such as cutting trees for charcoal, firewood and building materials has resulted in soil erosion, scarcity of firewood and building materials and drying up of water sources. Geologically, Dodoma Urban and nearby areas are formed by unfoliated granites and foliated gneisses of various types. The unfoliated homogenous rock tends to be more resistant and forms watersheds and inselbergs, while valleys are mostly associated with areas containing foliated rocks (Quennell, 1955:2).

There are few village settlements such as Makulu, Ng'ong'ona, Ntyuka and Kikuyu inhabited by indigenous communities like Gogo/Wagogo. There is a CHSS foothill site, which was inhabited by historical *Wambambali* people approximately around 1300 AD (Ryano et al., 2020). The villagers grow agricultural crops like maize, groundnut, cassava and sunflower by using traditional hand hoe, plough and bush knife. Some families grow vegetables like tomato, cabbage and chilly with the help of tank water and dugout wells. Extensive agriculture does not yield good harvest because of sand soils and scarcity of rain in the area.

Most parts of the Dodoma Urban area in Central Tanzania are devoid of any archaeological activity. Therefore, this research was carried out on an archaeological survey (Krishna Rao, 2018) and rescue archaeology (Ryano et al., 2020) in the backdrop of Dodoma town including the University of Dodoma premises. The study area is shown in Fig. 24.1.

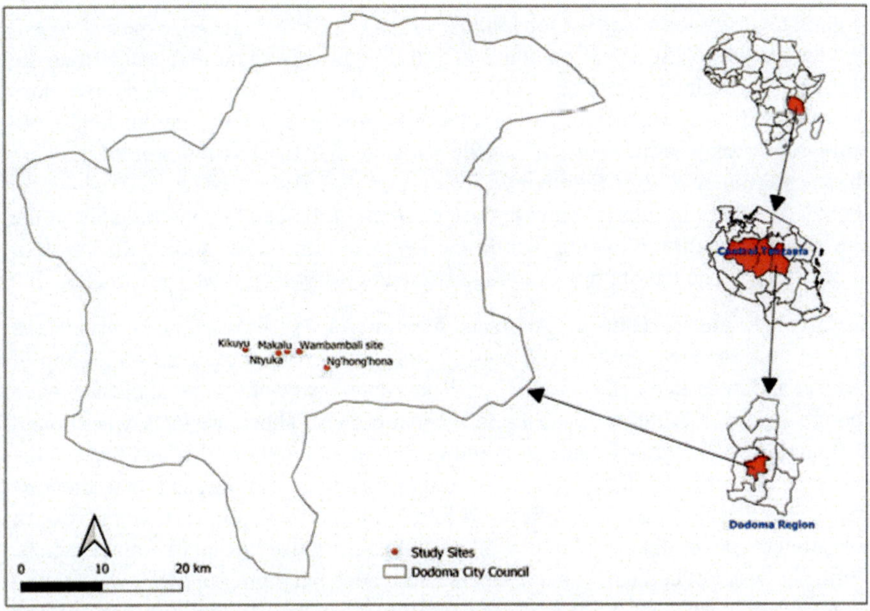

Fig. 24.1 Study area: Dodoma Region in Central Tanzania (Tanzania in Africa)

24.8.1 Cultural Evidence: Flake and Blade Tools, Microliths and Ground Stone Industry

The authors identified as many as a dozen archaeological sites in and around the University of Dodoma campus (see Map). One of the sites is a foothill site. Microlithic artefacts, hammer stones, flakes, cores and pottery were found separately or in an assorted condition. Some sample material collections and surface finds including unfinished tools, artefacts and potsherds of various colours and shapes and designs have been recorded (Krishna Rao, 2018). It must be noted that the quantity of unfinished finds mentioned in the tables is based on the physical counting of the finds within a limited archaeological site survey. Also the pottery with various shapes and designs on different types such as black ware, red ware, grey ware, etc., were counted. These sites are described as follows.

Makulu Site (MKL) Originally Makulu site (36° N 0807885, east 9312662 and elevation 1155 m) was identified in the year 2011 when one of the authors led a team of students of archaeology of the University of Dodoma to this place as part of field practicum. The preliminary survey in the area has yielded substantive evidence, which includes microliths, quartz flakes, pottery, hammer stones and grindstones. Lot of vegetation is found grown in the northern end of the site. At present some farmers cultivate agricultural crops and vegetable farms with the help of a water tank situated near the Makulu village. The landscape of the site consists of plains gently sloping towards north, and the top humus, which is mostly removed, exposes red soil and laterite gravel. Seasonal small hill streams that flow from north to south drain the site. Frequent rains seem to have caused intensive erosion on down slope to expose bedrock, which is weathered. At the time of section scraping during our fieldwork, stratigraphy of a section of soil profile showed that the top red soil of 120 cm thickness was devoid of humus by erosion and was laid on brown soil of 40 cm thickness, which in turn is found laid on bedrock at the end. At another spot in the site, red soil of 20 cm thickness is found deposited on brown soil of thickness of about 45 cm. The laterite gravel is found spread on weathered bedrock finally.

Cultural Evidence (Photo) Artefacts like microliths, flakes, cores, grindstone pieces, round stone hammers and pottery pieces were found exposed on the top of red soil and in certain cases brown soil. Flakes and cores were found in clusters even though fewer microlithic blade tools were reported. These artefacts were found scattered on pellet gravel patches on red or brown soil. Quartz is found used in the manufacture of microlithic tools. Nevertheless, it is also noticed that the raw materials chert and quartzite were used in microlithic tools preparation and also in the production of other implements like scrapers, grindstones and hammer stones. Further it was noticed that some scrapers and blades belonging to MSA were found to reveal cultural continuity from the end of MSA up to Neolithic period. The following are some details of the material culture of different periods.

Middle Stone Age (MSA) The flake and blade tools constitute MSA evidence. Sizable Levallois cores were found used for making the tools. The implements include side scrapers, round scrapers and blades. The tools might have been used to scraping the skin and cutting the flesh of the animals after being hunted.

Late Stone Age (LSA) It is also called Mesolithic. Small tiny tools called microliths such as geometric and non-geometric artefacts comprising backed blades and simple points characterize the LSA cultural phase. After the end of Pleistocene, the Holocene environment offered modern subsistence activities and techniques such as specialized hunting to exploit new environmental resources. Since global change brought changes in the landscape, new raw materials like semi-precious stones were inevitably used in the manufacture of tiny tools through reduction technology. Quartz microliths are predominant in the Makulu site. The artefacts like flakes and other debitage were found spread in small assemblages. Among finished forms of microliths, backed blades like lunates and other simple blades were found exposed on the eroded surface. Lunates may have been used in cutting and skinning animal flesh.

Neolithic Ground Stone Industry The time span to advance from one cultural phase to another during Holocene epoch is not as much longer as the time interval between various Palaeolithic phases of Pleistocene epoch. This variation could be attributed to cultural advancement and attitude of people towards sedentary life and socialization. That is the reason how people had shifted from foraging activities such as hunting, fishing and plant collection to pastoralism and food production. After few millennia years, people started domestication of animals and plants. Tanzania has experienced both Pastoral Neolithic and Food-Producing Neolithic. Ground and polished tools, grindstones and hammer stones/pounding stones, hand hoe, digging stick and pottery became the constituents of food producing, processing and storage. The pottery patches are found spread on the surface of the red soil where top humus (brown soil) is eroded. The red soil is found laid down bedrock, which is weathered. Makulu site has yielded evidence of grindstones, hammer stones and pottery in clusters resembling individual settlement patches in the locality. Three settlement localities are identified. The first two settlements have access for interaction while the third group of settlements is located at about 200 m away from the previous settlements.

Ngo'ng'ona (Ng'hong'hona) Site It is a foothill site (36° N 0812674, E 9310568) located at about half a kilometre north of the village Ng'ong'ona. The topography of the site comprises thorny bushes with red soil exposures and dry hill streams. It seems that the seasonal rain eroded the top brown soil. In some places of the site bedrock is exposed due to erosion of both humus and red soil. Pottery fragments are found scattered on the surface of the red soil. Thick black ware, red ware and grey ware pottery were found in the assemblage. Few pottery rims were found to have horizontal line designs on the external surface. A flake scraper resembling MSA and a Neolithic polished axe made on chert were found on the surface. On the laterite gravel patches, few quartz flakes and cores were noticed. Tools like round hammer stones were recorded from the surface. Chisel edged tools like polished axe were popular in the Neolithic period when they were used in agricultural activities like digging, weeding, etc. The distal end of the polished axe opposite to the medial working edge is usually set in a wooden handle in order to use it as a digging stick, a multi-functional tool. Grinding and polishing technique is usually adopted after the use of initially stone hammer in the preparation of a smooth edge tool.

Ntyuka Site The site (36° M 0806819, E 9312435 and elevation 1220 m) is located adjacent to Makulu site. Like Makulu and Ng'ong'ona villages, Ntyuka village is also occupied by the Gogo community. Gogo is more traditional and even their mortuary practices are very close to the ancient megaliths like stone circles. Apart from pastoralism, the community nowadays practices agriculture and cultivate crops like maize, beans, millet and vegetables.

Physiographically, there is not much change from that of the Makulu site except that the top humus is mostly eroded. Artefacts like flake and blade tools were found scantily scattered on top of the laterite exposures. Microlithic-backed blades such as lunates and trapezes made on glassy quartz were found in the assemblage. There is a chisel edged tool ($6 \times 4 \times 0.8$ cm) in the excavated collection. This tool may have been used either directly or as a chisel by setting it in a wooden shaft or a bamboo stick. Such chisel-edged tools are useful to remove soft materials like pulp of the vegetable gourd during the preparation of various storage/drinking water containers.

Stratigraphy and Geomorphology Geomorphological understanding of the archaeological site is of much significance in the assessment of spatio-temporal relationship of the spread of artefacts in different surfacial and sub-surfacial soil contexts. Moreover the nature of distribution of artefacts or other archaeological material remains including faunal and vegetative evidence in their geomorphological contexts cannot be understood without landscape and geomorphological studies. Study of soil sections or profile is therefore important. Stratigraphic exposure in the soil profile in the absence of excavation shall throw some knowledge in the evaluation of the habitat or living floor of concerned ancient population with which the spread of tools or objects of routine life activities would have been connected.

A few narrow seasonal streams and red bed exposure with intermittent alterations of grey soil and laterite gravel indicate the general topography of the Makulu site. Unexpected rains might have caused the cutting of the streambed and bedrock exposure. A bushy environment with thorny shrub jungle and cactus plants is spread all over. In a few spots the top red soil is totally washed off. Red colour appears to have been caused by the oxidation of elements like hematite. As a result, ferrous and ferric oxides are formed to give red colour appearance. A section facing north reveals the following stratigraphy. The top grey soil (60 cm thick) was found laid on laterite (22 cm), which was laid on again grey soil (42 cm). This third layer was deposited on laterite (36 cm). Finally the fourth layer laterite was deposited on bedrock. Quartz, quartzite, chert and greenstone are the dominant raw materials, which may have been used to prepare different tools. According to geological expertise, the microlithic blades are mostly found made on quartz. Levallois flakes and cores are found manufactured on either quartzite or chert. Prof. Vasudev, Department of Earth Sciences and Groundwater Engineering, University of Trinidad and Tobago, Saint Augustine Campus, West Indies, remarked on the geomorphological features of the Makulu site as follows. He explained that the study site is an undulating area being carved by the surface processes, especially rain and wind. From an elevated place, when a traverse was made along the course of a stream, layering was observed in the soil. There are five layers as described in Table 24.2.

The artefacts were found concentrated on the top lateritic soil (second layer from top). Further investigation revealed that they were concentrated on the top because the soil from that area was removed by surface processes like rain-wash and wind and they were in fact impregnated in the laterite. The presence of artefacts in the laterite obviously suggests that both the lateritic soil and the artefacts are not in situ. They were transported from some other place and deposited in the area, suggesting that the laterite is secondary and not primary. Moreover the artefacts were not worn out; they were with fresh facades and edges, suggesting that they were transported to this area from a place very close to the deposition of the present site.

Table 24.2 Soil layers and characteristics of Ntyuka site

S. no.	Layer and description
1	Top greyish white soil with bushes and grass grown over
2	Brownish red laterite soil with artefacts over and inside
3	Greyish white soil
4	Brownish red laterite soil
5	Bedrock

Kikuyu Site (36° M 0802697, E 9312959 and elevation 1214 m): It is a foothill site located half a kilometre northeast of Kikuyu village. On the east of the site are hills. The landmass of the area is approximately 5 km^2. Cactus plants and other thorny bushes were grown all over. Two such small dried hill streams orienting northeast–southwest were also found in the site. A soil deposit of about 4 m thickness is exposed in the periphery of the site. The depositional environment seems to be favourable for fossilization process as seen from one fossil insect (see photo). The undulating surface exposes soft humus where the Kikuyu farmers grow maize during rainy season. Below the humus, thin layer of brown soil with gravel pellets is spread throughout. The gravel layer is finally laid on the bottom layer, bedrock that is weathered. Black, brown and red pottery fragments were found spread on the brown soil layer. Sometimes the pottery evidence is found mixed with microlithic artefacts made of quartz and chert. Other objects include grindstone and stone pounders (see photo). The evidence of one big size grindstone (30 cm diameter and 30 cm height) and a semi-circular pounder with 12 cm diameter was interesting.

CHSS Foothill Site/*Wambambali* Site (36° M 0809368, E 9312626 and elevation 1253 m): The site is located near the water tank. It is known to have been inhabited by *Wambambali* community long ago, and it appears that they abandoned the site due to frequent drought in the region. Pottery pieces, grindstones and stone pounders resembling Pastoral Neolithic were noticed here. Surprisingly, few burnt daub pieces were found in the site, giving the impression that the walls of the huts might have been reinforced with mud and daub for strength. There is also a different explanation; that is, the houses might have been damaged in fire by enemy attack. Anyway, the *Wambambali* finally exited from the region. It is imperative that there was congenial environment for human adaptation prior to the arrival of *Wagogo* community.

24.9 Ethnography

Interviews with Focus Group Discussion (FGD) participants on *Wambambali* culture have provided some useful information to explain some cross-cultural periodization. Accordingly, it was *Wambambali* who might have first occupied Dodoma; *Wanyanzaga* came next and after them the *Wagogo,* who are currently settled in Dodoma. Extensive excavations for geo-chronological and chronometric dating are necessary to reconstruct cultural and chronological significances of both historic *Wambambali* and prehistoric cultures.

At present the stray occurrence of material evidences offers some important cues for future work. The evidence from surface and test pits includes pottery and burnt daub, and they appear to have been used in house construction (Ryano et al., 2020). The supporting reference for the existence of the *Wambambali* before the arrival of Bantu as explained by Mnyampala (1995) is thought provoking. The evidence seems to be similar to that of Pastoral Neolithic (PN) people. What is important is that the ethnographic interview reveals about the existence of Stone Age people around

70 thousand years ago before the arrival of the *Wambambali*. It looks apparent to reconstruct cultural sequence as evident from Middle Stone Age (MSA) flake and blade tools, grindstones related to PN, pottery and microlithic matrix in the research area. The PN people could be seen as none other than the ancestors of *Wambambali*. Perhaps before the introduction of burnt mud bricks, burnt daub could have been introduced in the construction of thatched houses. Changing environmental conditions, drought situation and deteriorating subsistence availability and grazing pastures might have forced the *Wambambali* to exit from the Dodoma Region.

24.10 Case Studies Overview

The archaeological material evidence of the sites such as Makulu, Ng'ong'ona and Ntyuka reveals that there are three characteristic assemblages. According to the typo-technological analysis, one group belongs to the Middle Stone Age as evident from the flake tools and Levallois cores and flakes. The second characteristic assemblage is related to LSA as seen from microlithic quartz blades while the third category consists of polished axe, grindstones, hammer stones and innumerable number of potsherds that resemble Neolithic culture. Specifically in the MSA cluster, scrapers made on flakes and blades were found. In addition to that, small Levallois cores were also found to be associated with the assemblage. These cores, which display regular and thin flake scars, indicate the trend towards shifting from large flake tool production to microlithic tool preparation. This development may have been influenced by slow changing environments at the end of the Pleistocene in the region.

The onset of Holocene immediately after the Pleistocene brought about several changes in the subsistence and survival strategies of modern man. Specialized hunting and other foraging activities along with incipient agriculture were practised. Geometric and other simple tool production had become the workshop of economic strategies. This is the reason perhaps microlithic blade tools made on chert and quartz were found particularly in the sites of Makulu and Ntyuka.

With regards to Neolithic tool assemblage, polished axe, round stone hammers, grindstones and a large number of potsherds were found in the sites. Makulu site is potential for pottery evidence. Some flake tools and microlithic blade tools were also found mixed with the pottery evidence in a few instances. Perhaps Pastoral Neolithic people during tough environmental conditions and relative ecological variability depended on microlithic tools including backed blades (Goldstein & Shaffer, 2017; Ambrose, 2002) as evident from Makulu material matrix. Whether the association of microlithic tools with Neolithic evidence represents any Neolithic tradition (Chami, 2000) or not is yet to be cleared. There is also the possibility to understand the association of microliths of Later Stone Age lithic technology and ceramic vessels probably along with more dependence on domestic stock as Pastoral Neolithic in East Africa (Bower et al., 1977).

Interestingly pottery is found scattered as settlement patches in certain spots of the Makulu site. The pottery designs exhibit patterns of incision and sigmas in horizontal line, single-line and double-line punctuation. Incision patterns were similar to the Narosura style typical of Nile Valley Neolithic sites (Robertson, 1991: 146 & 149). It was argued elsewhere that Bantu speakers should have been among the Negroid populations existing in the Rift Valley region and East African coast during the Neolithic (Chami & Kwekason, 2003: 78). It is also possible that during the spread of Negroid or Nilotic Bantu (Gramly & Rightmire, 1973: 578) from Kenyan Rift to the Rift Valley of Northern Tanzania and further East, Bantu population could have spread to some parts of Central Tanzania too.

It is possible that the Bantu speakers were in the region before the coming of iron technology and bevelled and fluted pottery as well (Chami & Kwekason, 2003: 78). This suggests that prior to these innovations the farming techniques, domestication of animals like goat/sheep, chicken, and cat, and pottery making had been popular on the East African coast and also in South Africa. Similar situations seem to have occurred in Central Tanzania including the Dodoma Region. The geomorphology and soil sectional profile studies revealed the stratigraphy of the archaeological sites. The presence of MSA, LSA and Neolithic tool matrix especially in the Makulu and the adjacent Ntyuka sites suggests that they belong to secondary laterites.

24.11 Results of Application of Geospatial Techniques in the Case Study Areas

Palaeoclimate and surrounding environs are mostly derived from rainfall, temperature, hilly terrain, elevation, slope, soil type and vegetation. The geospatial data from satellite images for the study of climate and terrain variables in the Dodoma Urban for the past 60 years (1958–2018) were collected for interpretation of our case studies, which confirmed what has been already observed. The climate data (1958–2018) showed that the rainfall initially began with high intensity around 1958 and gradually declined for about two decades. Then there was little increase for about another two decades from 1980 till 2000. Again the rainfall showed drastically decreasing trend from the year 2000 until the present (Fig. 24.2: climate graph–rainfall trend). Similarly, Mann–Kendal trend test also revealed a statistically significant declining rainfall trend (Mann–Kendal statistic (s) $= -14463$, p-value < 0.05). This corraborates the findings by Shemsanga et al. (2016) that rainfall in the study area had generally decreased during the past 91 years (1924–2015) because of climate change.

Contextually it is worth referring to the ancient environments. During ancient time sufficient rainfall with small periodical fluctuations for almost every two decades unlike the present seemed to be favourable for human habitation and survival. There were major fluctuations even during the prehistoric past like pluvials altered by interpluvials (wet phases versus dry phases contemporaneous to

Scandinavian glacial phases). For instance, Kagera region in Tanzania had also experienced intensive monsoons when animal and human species developed adaptation to such environments. After the arrival of Holocene, the entire globe was found to be prone to drastic changes, which inevitably exerted influence on the rest of the species. Accordingly, LSA or microlithic tools and later on ground stone industry had their appearance as a result of new adaptations. Since then, new subsistence activities like food production, storage and permanent settlements, intensive religious activities, art and architecture became prominent among the people.

Apart from rainfall, frequent oscillations in temperature (Figs. 24.3 and 24.4: Temperature max. and min. trends) in Dodoma Region were also reported to have minimal and maximal effects on biotic communities and vegetation as well. For instance, the temperature which was initially less than 17° in the year 1958 was shown to have further lessened by 0.5° after one and a half decades (up to the year 1973 approximately) which is a very normal condition. Thereafter it began increasing and peaked after two decades (up to 1997), when after that it showed downward trend until further 5 years (up to 2005). Again, upward trend started from 2005 until the present. In summary, the data analysis for the entire period of 60 years (1958–2018) showed a significant increasing trend for both minimum temperature (Mann–Kendal statistic (s) = 39117, p-value <0.05) and maximum temperature (Mann–Kendal statistic (s) = 51316, p-value <0.05). Previous studies in the study area have also shown the similar increasing trends in both maximum and minimum temperatures (Mayaya et al., 2015; Shemsanga et al., 2017).

In summary, the recorded maximum temperatures show that within a span of 60 years there have been gradual temperature raise in the first four decades (beginning from 28.5° in 1958) when it moved towards peaks (>32° in 2020). At the same time the minimum temperatures that began with 16.3° in 1958 continued its downward trend with maximum decline after one and a half decades. Then it again picked up an increasing trend of maximum 17.1° after two decades (around the year 1998). Thus from an analysis of the above data, it could be inferred that increasing temperature and decreasing rainfall brought several changes in the landscape environments, elevation, vegetation and soil and might have become a threat to the prehistoric human habitation and settlement in course of time. Even historical *Wambambali* civilization, which left its marks of evidence in the form of grindstones, pounders, pottery and burnt mud daub pieces, was found unnoticed in the region around 1300 AD (Ryano et al., 2020).

The following are the line graphs (Figs. 24.2, 24.3 & 24.4) of climate variables:

Secondly, the topographical terrain is one of the principal features of human settlement (Zhu et al., 2019). During prehistoric and historic periods most of the human settlements seem to have confined to foothill sites, slope, plateau and hill valley because such places have accessible physical environments congenial for livelihood. The terrain covers three metrics, namely, elevation (Fig. 24.5), slope (Fig. 24.6) and slope aspect (Fig. 24.7). The terrain analysis in the case study areas revealed that the elevation ranged between 1174 m and 1252 m (Fig. 24.5b). Makulu, Ntyuka, Ngo'ng'ona (Ng'hong'hona) and Kikuyu sites were located at

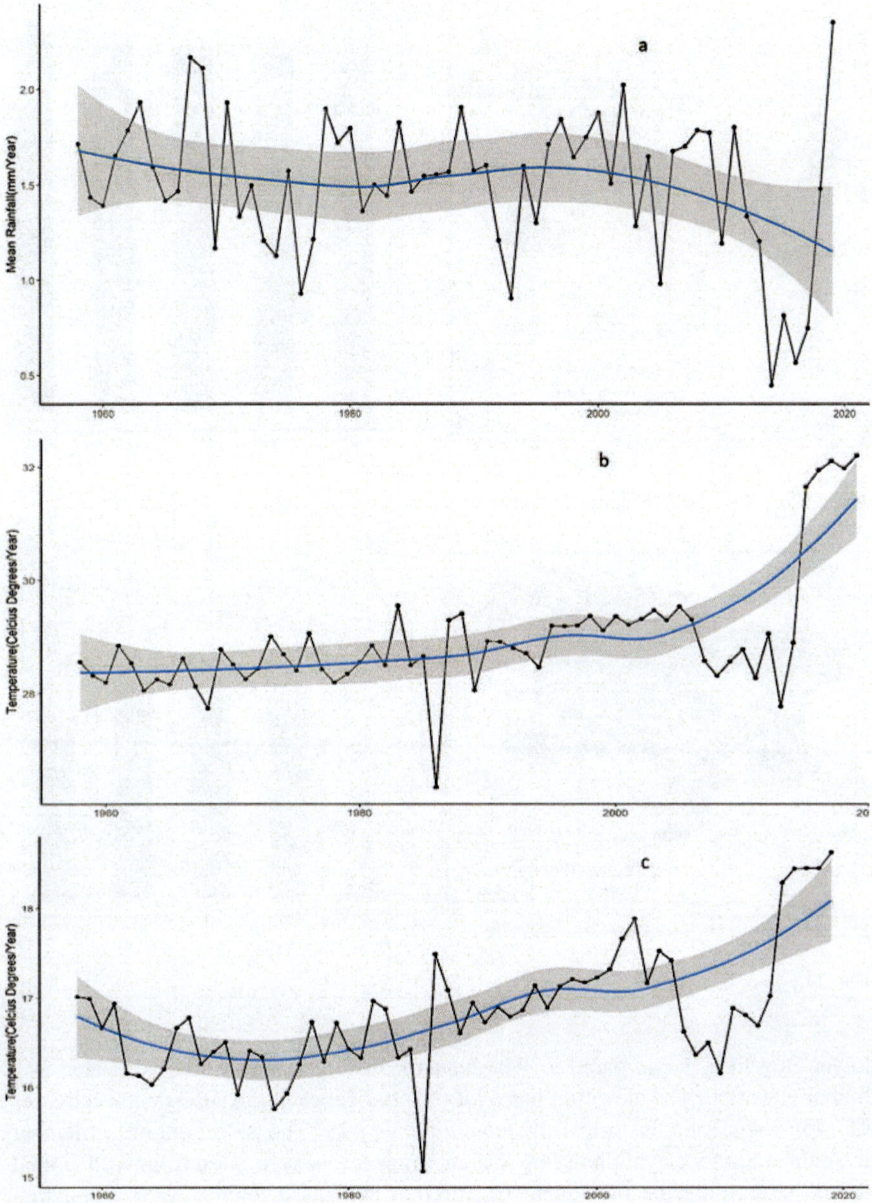

Figs. 24.2, 24.3 & 24.4 Climate graphs (1958–2018): (**a**: rainfall trend; **b** and **c**: temperature max. and min. trends)

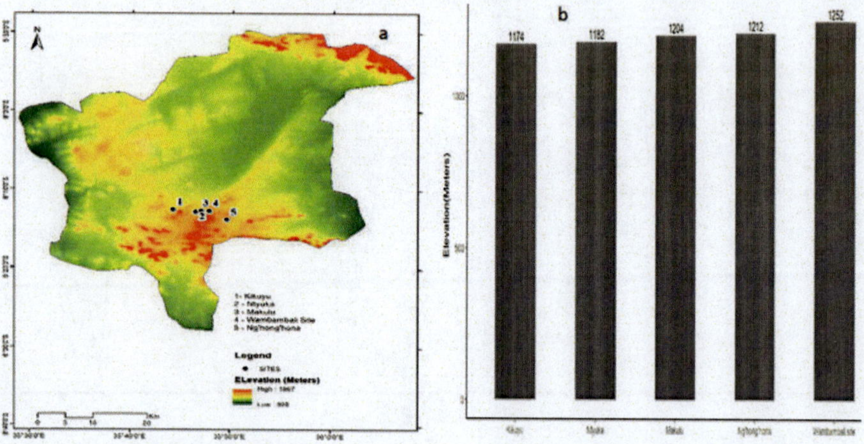

Fig. 24.5 a & b (elevation)

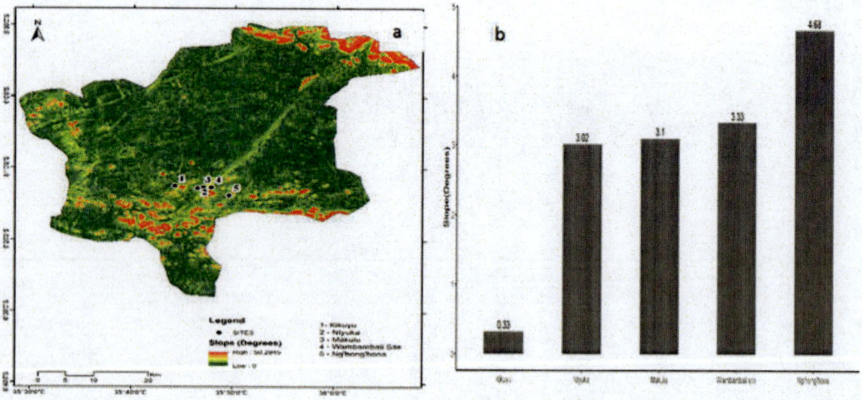

Fig. 24.6 a & b (slope)

lower elevation as compared to *Wambambali* historical site, which is located at a higher elevation. The elevation has a lot of significance in the subsistence activities of Stone Age and historical *Wambambali* people. The selection of settlement location at higher elevation during ancient times is to stay secured from wild animal attack, overseeing the movement of different animals, avoiding water stagnation during rain, etc. In fact, the surface water flow due to monsoonal rains from the nearby hills with a higher elevation might have removed the top humus in Makulu and Ntyuka sites, exposing the secondary laterites with the Stone Age artefacts. A similar study by Märker (2012) in Lake Manyara area also supports the above fact.

Further, the sites were found to have low slope values in the range of 0.33°–4.7° as shown in Fig. 24.6a and bar graph (Fig. 24.6b). This suggests that all sites were situated in relatively flat surfaces, offering an abundant scope for exploiting natural resources mainly for hunting, gathering, agriculture and pastoral activities. Previous

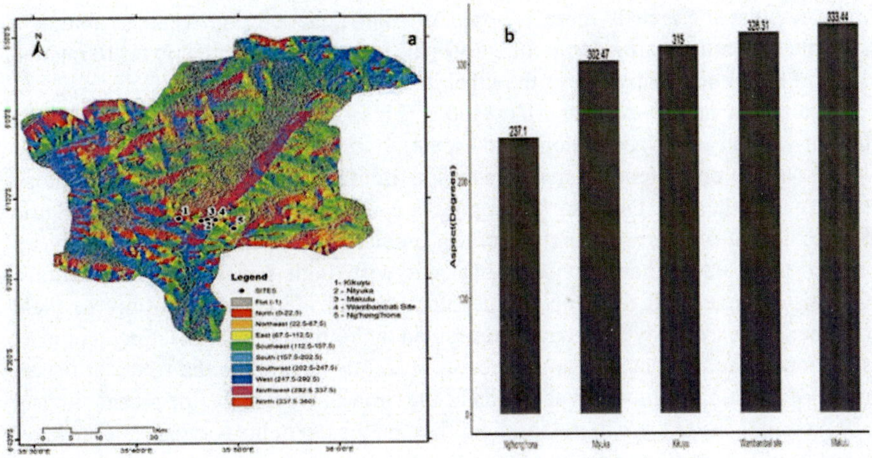

Fig. 24.7 a & b (aspect)

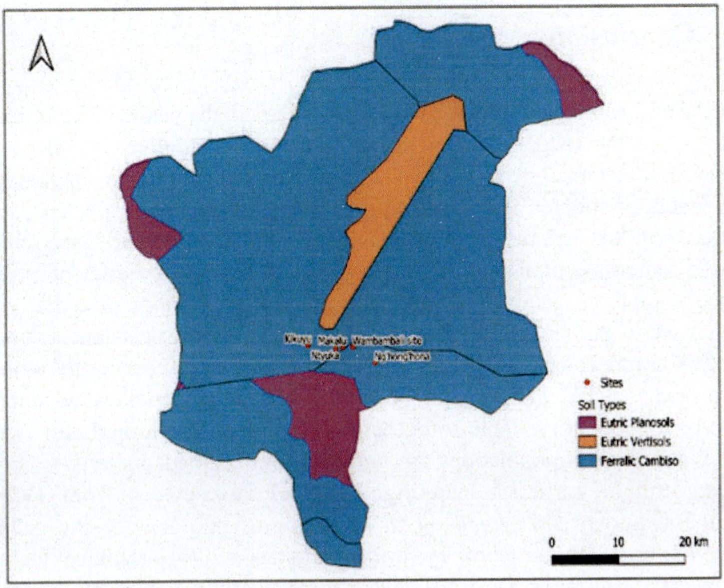

Fig. 24.8 Map showing the main soil types

studies on early human settlements have also shown that the slope below 5° was considered suitable and preferred for human settlements (Xi et al., 2018; Zhu et al., 2019).

In case of the aspect of slope (Fig. 24.7a and b), it is inclined towards west and northwest (237.1°–333.4°). Since the area is in the Southern Hemisphere, this indicates that the area is more exposed to solar radiation and receives abundant sunlight becoming warmer consequently. This makes the area suitable for living and support livelihoods activities of the inhabitants (Xi et al., 2018).

The following are maps and bar graphs (Figs. 24.5, 24.6, and 24.7) relating to terrain variables (elevation, slope and aspect):

Just as the promoters of environment like rainfall, temperature, elevation, slope and aspect, the soil types in the region are represented by three types: Eutric Planosols, Eutric Vertisols and Ferralic Cambisols. Tropical 'brown earth' has a higher base status than Luvisols, but otherwise both have similar limitations. Cambisols have high resilience to degradation and moderate sensitivity to yield decline. The case study sites were represented by Ferralic Cambisols. Such compact situations were evident as one proceeded to soil depths unlike the frequent occurrence of erosion by rain-wash at the top of the surface. Obviously, the archaeological artefacts in the sites were found to be exposed on the top of the compact laterite layer because of the erosion of the top humus layer.

24.12 Conclusion

Initially the archaeological studies of several scholars in the whole of Tanzania were reviewed to fix the status of different environmental conditions prevalent during prehistoric and historical periods and also to understand the human adaptations and cultural practices adopted in the prevailing environments. They are analysed and explained from the exhaustive archaeological and ethnographic data related to northern, southern, central and coastal regions of Tanzania. Aerial photos, GPS coordinates, archaeological survey and stratigraphy and ethnographic data correlation were used to explain the human biological and cultural aspects and adaptations. Especially Central Tanzania, with its rock art potential and other fertile Stone Age and historical material ruins including animal remains and pollen evidence, was found to be a good and reliable source of knowledge on the struggle for existence and survival of human populations through environmental adaptations.

Starting from the earliest human origins and culture as evident from OVG up to historical Bagamoyo and *Wambambali* civilizations, one can see the panoramic influence of varied environments in various altitudes and topographical features of the landmass of Tanzania. Such differences were clearly indicated by the available raw materials such as stone, bone, wood and iron; and environments such as grassland savannah, shrub jungle, rainforests and woodlands, riverine and lacustrine have offered aquatic, animal and edible plant resources to develop hunting, gathering, fishing, pastoralism, agriculture and trading activities for human survival. Such access to biome is primarily regulated by natural phenomena such as climate, terrain and soil. Prehistoric and historical communities nurtured their life ways successfully in the adaptable environments until they met with sudden changes in the natural environments causing droughts and other subsequent socio-economic and political problems. That is why palaeoclimatic fluctuations in Central Tanzania in general and Dodoma Region, in particular have led to several changes in biome which in turn impaled on the very survival of humanity. Such situations occurred quite frequently in prehistoric and historical periods. It is evident from cultural discontinuity in the archaeological areas. For instance, the historical *Wambambali* communities approximately dated (not carbon dated yet) to 1300 AD have left their habitat unnoticeably. The emergence of famine conditions as a result of catastrophic situation in climate might have caused their exit by abandoning their settlements forcibly. It is revealed from our studies on the application of geospatial techniques.

Tourism development in these archaeological site areas of Tanzania also needs immediate attention as that of other listed World Heritage Sites. Changes in the aspirations of tourist tendency are also being evaluated from time to time. For instance, eco-tourism which includes natural environment, wildlife, indigenous folk dances etc., is currently more appreciative. Natural environment, which consists of habitat of ancient and present communities, of late has become one of the features of tourist attractions. Whether it is eco-tourism or cultural tourism protection of prehistoric and historical archaeological sites in their original forms of landscape is an immediate task of heritage managers through conservation and preservation. For instance, fossil bones of *Zinjanthropus boisei* hominid in the OVG are left in the open air exposed to the wind and rain. Also some of the rock art sites like Kondoa and Singida groups of rock shelters of Central Tanzania also have such threats. Stone Age sites around the Dodoma Region face similar problems. Such threats need to be taken care of on an emergency basis before the precious cultural proofs of ancient and historical periods are prone to be destructed.

Appendix (Site Photos)

Top first row: 1. *Zinjanthropus* (OVG) 2. *H. habilis* site (OVG) 3. *H. erectus* site (OVG)

Second row: 4. Laetoli foot prints (preserved beneath the stone structure, and so copy preserved in the museum) 5. Isimila ESA (late Acheulian) 6. Makulu Stone Age materials

Third row: 7. Ntyuka Chisel edged blade 8. Kikuyu blades and pounders 9. Kikuyu stone pounders.

Fourth row: 10. Wambambali site (UDOM) grindstone 11. Wambambali burnt daub 12. Gogo settlement near Makulu

Fifth row: 13 & 14. Kolo Kondoa rock paintings & 15. Singida: Kinaliya-Kyaga rock paintings.

References

Adams, W. M., Goudie, A. S., & Orme, A. R. (Eds.). (1996). *The physical geography of Africa*. Oxford University Press.

Aitken, W. (1950). *Geomorphology of parts of the Kondoa District* (Tanganyika Notes and Records, No. 29). Dar es Salaam.

Ambrose, S. H. (2002). Somethings remembered: Origins and early microlithic tools industries in sub-Saharan Africa. In R. G. Elsion & S. L. Kuhn (Eds.), *Thinking Small: Global Perspectives on microlithictoolsization* (Archaeological Papers of the American Anthropological Association) (Vol. 2, pp. 9–29).

Anati, A. (1980). The state of research in rock art: the rock art of Tanzania and East African sequence. *In Bottllttino del Centro camuno di studPreistorical 23. Pp*, 15–68.

Anati, E. (1986). 'The rock art of Tanzania and the East African sequence'. Bulletino del Centro Camuno di Studi Preistorici 23: 15–68.

Bower, J., & Mabulla, A. Z. (2013). Setting in: Evidence of territorial exclusion in the late Middle Stone Age in Northern Tanzania. *Studies in African Past, 11*, 1–24.

Bower, J. R. F. (1973). Serenora: excavations at a stone bowl site in Serengeti National Park Tanzania. *Azania, 8*, 71–104.

Bower, J.R.F., Charles M Nelson, Albert F Waibel & Simiyu Wandibba. (1977). The University of Massachusetts Later Stone Age/Pastoral Neolithic Comparative study in central Kenya'. Azania 12, pp. 119-46.

Bower, J. R. F., & Gogan-Porter. (1981). *Prehistoric cultures of the Serengeti National Park* (Tanzania Papers in Anthropology. No. 3). IOWA State University.

Brown, W. T. (1970). Bagamoyo: 'An Historical Introduction'. *Tannzania Notes and Records, 71*, 69–84.

Bwasiri, E. J. (2016). *The Rock Art of Kondoa District, Tanzania: Approaches toward understanding authorship and meaning*. PhD thesis, University of the Witwatersrand, Johannesburg.

Bwasiri, E. J. (2011). The implications of the management of Indigenous living heritage: the case study of the MongomiWaKolo rock paintings world heritage site, central Tanzania. *South African Archaeological Bulletin, 66*(193), 60–66.

Chami, F. A., & Kwekason, A. (2003). Neolithic pottery traditions from the islands, the coast and the interior of East Africa. *African Archaeological Review, 20*(2), 65–80.

Chami, F. A. (1994). The Tanzania coast in the First Millennium A.D. *Studies in Archaeology* 7. Societas Archaeologica Upasliensis.

Chami, F. A. (1996). The excavation of Kiwangwa Late stone Age site. In G. Pwiti & R. Soper (Eds.), *Aspects of African archaeology papers of 10th congress of pan African association of prehistory and related studies* (pp. 307–316). University of Zimbabwe.

Chami, F.A. (1998). 'A Review of Swahili Archaeology'. African Archaeological Review, 5 (3) 199–218.

Chami, F. A. (2000). Further archaeological research on Mafia island. *Azania, 35*, 208–214.

Chang, K. C. (1972). Settlement Patterns in Archaeology. *An Addision-Wesley Module in Anthropology. Module, 24*, 1–26.

Chaplin, J. H. (1974). The prehistoric rock art of the Lake Victoria Region. *Azania, IX*, 1–50.

Chittick, N. (1962). Recent discoveries in Tanganyika. *Acts du VI Congress PanafricandePrehistoireet de l'etude du Quaternaire, Turvuren: Musee Royal d l'Afrique Centrale, 3*, 215–223.

Clark, J. D. (1976). The domestication process in sub-Saharan Africa with reference to Ethiopia. In E. S. Higgs (Ed.), *Origine de l'Élévageet de la Domestication* (pp. 56–115). Union International des Science Pre- et Proto-historique.

Culwick, A. T. (1931). Some Rock-Paintings in Central Tanganyika. *Journal of Royal Anthropological Institute of Great Britain and Ireland., 61*, 443–453.

Day, M. H., Leakey, M. D., & Magori, C. (1980). A Ne Hominid fossil skull (L.H. 18) from the Ngaloba Beds, Laetoli, northern Tanzania. *Nature, 284*, 55–56.

Ehret, C. (1974). *Ethiopians and East Africans: The problem of contacts*. East African Publishing House.

Ehret, C. (1976). Cushitic prehistory. In M. L. Bender (Ed.), *The non-semitic languages of Ethiopia* (pp. 85–96). Michigan State University Press.

Fosbrooke, H. A. (1950). The age and meaning of paintings. *Tanganyika Notes and Records, 29*, 11–14.

Goldstein, S. T., & Shaffer, C. S. (2017). Experimental and archaeological investigations of the backed microlith function among Mid-to-Late Holocene herders in southwestern Kenya. *Archaeological and Anthropological Sciences, 9*(8), 1767–1788.

Gramly, R. M., & Rightmire, G. P. (1973). A fragmentary cranium and dated Later Stone Age assemblage from Lukenya Hill, Kenya. *MAN, 8*(4), 571–579.

Hansen, C. L., & Charles, M. K. (1971). 'Environment and activity patterning at IsimilaKorongo, Iringa district, Tanzania: A preliminary report. *Amarican Anthropologist, 73*, 1201–1211.

Hay, R. L. (1976). *Geology of the Olduvai Gorge*. University of California Press.

Howell, F. C., Cole, G. H., & Kleindienst, M. R. (1962). Isimila: An Acheulian occupation site in the Iringa Highlands Province, Tanganyika. In G. Mortelmans & J. Nenquin (Eds.), *Actes du IV e Congress Panafrican de Prehistoireetl'Etude du Quarternaire* (Vol. 40, pp. 45–60). AnnalesMus'ee Royal de l' Afrique Central. Serie in 8 Sciences Humaines.

Hunter, G. (1953). Hidden drums in Singida district. *Tanzania Notes and Records, 34*, 28–32.

Inskeep, R. (1962). The Age of the Kondoa Rock Paintings in the hight of the recent Excavations at Kisese II Rock shelter. In G. Mortelmans & J. Nenquin (Eds.), *Actes du Ive Congress Panafrican de Prehistore et del 'Etude du Questionnaire, Liopoldville 1959.* Annals du Musee Royal de l' AfriqueCenttraleSerie 1N- 80 (Science Humains) 40: Section II: 249–256.

Isaac, G. (1974). Stone age remains on Kilwa Island. In N. Chittick (Ed.), *Kilwa: An Islamic trading city on the East African coast* (Vol. II, pp. 254–256). Nairobi.

Kessy, E. T. (2005). *The relationship between the Later Stone and Iron Age Culture of Central Tanzania.* Unpublished PhD thesis, Simon Fraser University, Vancouver.

Kimambo, I. N. (1968). Eastern Bantu peoples. In B. A. Ogot (Ed.), *Zamani: A survey of East African history* (pp. 195–209). Longman.

Kohl- Larsen, L. (1943). *Auf den Spuren des Varmenschen*. Vols. 1 and 2. Strecker and Schroder.

Krishna Rao, S. and Temu, E. L. (2018). Tanzania rock art heritage and ceremonies of indigenous people: A new phenomenon of tourism attraction. In M. N. Amutabi (Ed.), *Politics of development in Africa.* IIC proceedings volume, Nairobi, Kenya. ISBN 978-9966-116-79-6.

Krishna Rao, S. (2018). Archaeology and Heritage around University of Dodoma Campus in Tanzania. *Journal of African Interdisciplinary Studies* (JAIS): ISSN 2523-6725 (online) November 2018 Vol. 2 (11), Nairobi, Kenya.

Kwekason, A. P. (2010). *Holocene archaeology of the Southern coast of Tanzania.* In *E & D Vision Publishing*.

Lane, P., Mapunda, B. B., & Eriksson, M. (2001). Soil erosion, iron smelting and human settlement in the Haubi Basin, north-central Tanzania. *Antiquity, 75*(290), 803–804.

Leakey, L. S. B. (1936). *Stone age Africa.* Oxford University Press.

Leakey, M. (1983). *Africa's vanishing art: The rock paintings of Tanzania*. Hamish Hamilton/ Rainbird.

Leakey, M. D., Hay, R. L., Thurber, D. L., Protsch, R., & Berg, R. (1972). Stratigraphy, archaeology, and age of the Ndutu Beds and Naisiusiu Beds, Olduvai Gorge. *World Archaeology, 3*, 328–341.

Lewis-Willams, D. J. (1984). The empiricist impasse in South African rock art studies. *South African Archaeological Bulletin, 39*, 58–66.

Magill, C., Ashley, G., & Freeman, K. (2012). Landscape variability and early human environments in Africa. *Proceedings National Academy Sciences, USA, 110*, 1167–1174.

Märker, M. (2012). Terrain Analysis and Stochastic modelling for Archaeological site prediction and landscape reconstruction in the Lake Manyara area, Northern Tanzania. *Acta Agraria Debreceniensis, 49*, 47–53. https://doi.org/10.34101/actaagrar/49/2478

Masao, F. (1976). 'Some common aspects of the rock paintings of Kondoa and Singida'. *Tanzania Notes and Records*. No. 77 and 78. Dar es Salaam.

Masao, F. (1979). *The later stone age and rock paintings of Central Tanzania*. Weisbaden.

Mayaya, H. K., Opata, G., & Kipkorir, E. C. (2015). *Understanding climate change and manifestation of its driven impacts in the semi arid areas of Dodoma region. Tanzania, 43*(4), 364–376.

Mehlman, M. J. (1977). Excavations at Nasera Rock, Tanzania. *Azania, 12*(1), 111–118.

Mehlman, M. J. (1979). Mumba-Hohle revisited: The relevance of a forgotten excavation to some current issues in East African Prehistory. *World Archaeology, 2*(1), 81–94.

Mehlman, M. J. (1987). Provenience, age and associations of archaic *Homo sapiens* crania from Lake Eyasi, Tanzania. *Journal of Archaeological Science, 14*, 133–162.

Mehlman, M. J. (1989). *Late Quaternary archaeological sequences in Northern Tanzania*. Ph.D thesis, University of Illinois, Urbana.

Mnyampala, M. E. (1995). *The Gogo: History, customs, and traditions*. Translated, Introduced, and Edited by Gregory H. Maddox. Routledge.

Mturi, A. (1986). The pastoral neolithic of West Kilimanjaro. *Azania, XXI*, 53–63.

Ndessokia, P.N.S. (1990). The mammalian fauna and archaeology of the Ndolanya and Olpiro Beds, Laetoli, Tanzania. PhD dissertation, University of California, Berkeley.

Newman, J. (1970). The Ecological Basis for Subsistence change Among the Sandawe of Tanzania. National Academy of Science, Washington D.C.

Odner, K. (1971). An archaeological survey of Iramba, Tanzania. *Azania, 6*, 151–198.

Ombori, T. L., & Mabulla, A. Z. P. (2013). The archaeology of Mbuamaji: An early iron working site in Dar es Salaam City, Tanzania. *Studies in African Past, 11*, 90–113.

Phillipson, D. W. (2002). *African Archaeology*. Cambridge University Press.

Quennell, A. M. (1955). Proposed Mkonze dam site, Dodoma: summary report on geological conditions. *Geological Survey of Tanganyika, Report* No. AMQ/5.

Robertson, R. (1991). The Late Neolithic ceremonies from Shaqadud cave. In A. E. Marks & A. Mohammed- Ali (Eds.), *The late prehistory of the Eastern Sahel*. Southern Methodist University Press.

Ryano, K. P., Mwakipesile, A., Krishna Rao, S., Kasongi, N., Temu, E., Ngowi, E., & Kilonzo, R. (2020). Rescue archaeology at open-air sites around the University of Dodoma, Central Tanzania. *South African Archaeological Bulletin, 75*(212).

Sassoon, H. (1966). *Guide to Kunduchi*., Dar es Salaam: Antiquity department.

Sassoon, H. (1967). 'New views on Engaruka, Northern Tanzania: Excavations carried out for Tanzania Government in 1964 & 1966. *Journal of African History, 8*, 201–217.

Schmidt, P. R., & Childs, S. T. (1995). Ancient African iron production. *American Scientist, 83*, 524–533.

Shemsanga C, Muzuka ANN, Martz L, Komakech H, Omambia AN (2016). Statistics in Climate Variability, Dry spells, and Implications for Local Livelihoods in Semiarid Regions of Tanzania: the way forward. In: Chen W-Y, Seiner J, Suzuki T, Lackner M (eds) Handbook of climate change mitigation and adaptation. Springer Science and Business Media, New York.

Shemsanga, C., Muzuka, A. N. N., Martz, L., Komakech, H., & Omambia, A. N. (2017). Statistics in climate variability, dry spells, and implications for local livelihoods in semiarid regions of Tanzania: The way forward. In W.-Y. Chen, T. Suzuki, & M. Lackner (Eds.), *Handbook of climate change mitigation and adaptation* (pp. 801–848). Springer. https://doi.org/10.1007/978-3-319-14409-2_66

Sinclair, A. R. E. (2005). Serengeti past and present. In A. R. E. Sinclair & P. Arcese (Eds.), *Serengeti II: Dynamics, management and conservation of an ecosystem* (pp. 3–31). University of Chicago Press.

Smith, A. B. (1992). *Pastoralism in Africa: Origins and development ecology*. Hurst and Company.

Soper, R. (1967). Iron Age sites in North-Eastern Tanzania. *Azania, 2*, 19–38.

Soper, R. & Ciolden, B. (1969). 'Archaeological Survey of Mwanza Region, Tanzania'. Azania 4: 15–79.

Sutton, J. E. G. (1978). Engaruka and its waters. *Azania, XIII*, 37–70.

Sutton, J. E. G. (1968). The settlement of East Africa. In B. A. Ogot (Ed.), *Zamani: A survey of East African history* (pp. 70–98). Longman.

Tanner, R. E. S. (1953). A series of rock paintings near Mwanza. *Tanganyika Notes and Records, 34*, 62–67.

Temu, E. L. (2016). *The role of local communities in conservation of archaeological and historical heritage in Central and Coastal Tanzania*. PhD thesis, University of Dodoma, Tanzania.
White, T. D. (1977). New Fossil hominids from Laetoli, Tanzania. *American Journal of Physical Anthropology, 46*, 197–230.
Xi, C. b., Qian, T. l., Chi, Y., Chen, J., & Wang, J. c. (2018). Relationship between settlements and topographical factors: An example from Sichuan Province, China. *Journal of Mountain Science, 15*(9), 2043–2054. https://doi.org/10.1007/s11629-018-4863-z
Zhu, M., Dong, J., & Gao, Y. (2019). The research on temporal-spatial distribution and morphological characteristics of ancient settlements in the Songhua River Basin. *Sustainability (Switzerland), 11*(3). https://doi.org/10.3390/su11030932

Chapter 25
Anthropogenic Interventions and Urban Hydro-Geomorphic Hazards in Kolkata, India

Anwesha Haldar, Nasira Khatun, Anusree Dutta, and Lakshminarayan Satpati

25.1 Introduction

Rapid urbanisation has been taking place within the overpopulated Asian cities of the Global South, and in their surrounding fringe areas that are continually being reclaimed from open spaces, vegetated lands, water bodies and wetlands, and thereby in most of the cases degrading the natural environment including geo-hydrological systems. The character of climate too, in the region, has been observed as changing with frequent occurrence of heavy rainfall in short periods (Kalnay and Cai, 2003). Often growth of urban areas is responsible for altering the natural as well as artificial drainage systems (Sen, 2011). Price Simon et al. (2011) opined that human beings are active geological and geomorphic agents and had been the dominant factor in landscape evolution through settlement and widespread industrialisation and urbanisation. Urbanisation has major implications for the recent global development goals – economic, social, environmental, etc. With reference to geomorphic significance, the onset of the Industrial Revolution in the eighteenth century or the Anthropocene times are most challenging to the natural landscape and increasing urbanisation require the levelling of relief, by the excavation of heights, filling of depressions and slope terracing (A.G. Brown et al., 2017;

A. Haldar
Department of Geography, East Calcutta Girls' College, West Bengal State University, Kolkata, India

N. Khatun
Department of Geography, University of Calcutta, Kolkata, India

A. Dutta
Purwanchal Vidyamandir, Kolkata, West Bengal, India

L. Satpati (✉)
Department of Geography & UGC-HRDC, University of Calcutta, Kolkata, West Bengal, India

G. S. Bhunia et al. (eds.), *Anthropogeomorphology*, Geography of the Physical Environment, https://doi.org/10.1007/978-3-030-77572-8_25

Brandolini et al., 2020). The anthropogenic sedimentological record and human impact on the fluvial regime, therefore, provide a marker on which the Anthropocene is characterised (Downs and Piegay, 2019). Even the most rugged terrains have been drastically altered by the socio-economic and habitability drivers to unplanned urbanization in the Anthropocene times (Fan et al., 2016, Burak et al., 2017). According to Estoque and Murayama (2015), the knowledge about intensity and spatial pattern of urban land changes (ULCs), that is, changes from non-built-up to built-up lands, is important to a wide range of issues, from understanding of human–environment interactions and provisioning of urban ecosystem services, to land use policy development for landscape and urban planning towards sustainable urbanisation. Minnig et al. (2018) stated that a better understanding of changes in groundwater recharge in urban areas is required to move towards a sustainable water management. It was concluded that this expansion had placed great stress on forest, land and water ecosystems; and the rapidly growing population had to be accommodated in relatively risky land surfaces associated with potential geomorphic hazards.

The problems in the cities are manifold and interlinked among each other, and any attempt to find issue-oriented solution for one item fails because of not adopting a holistic approach to solve the problems together under a broad environmental management planning for the city. Implementation of any adaptive and integrated policy needs participation of all sections of the community and society. During the monsoon months, many South Asian cities located in floodplains for availability of reliable water supply experience the problem of urban flooding because of inadequacy and congestion of drainage networks. Urban flooding is the inundation of land or property in a built-up environment, particularly in more densely populated areas, caused by excessive rainfall and below-adequate capacity of drainage systems (Eldho et al., 2018). The increasing trend of urban flooding is a universal phenomenon and poses a great challenge to urban planners throughout the world. Problems associated with urban floods range from relatively localised incidents to major incidents, resulting in cities being inundated for a few hours to several days. The impacts too can also be widespread such as temporary relocation of people, damage to civic amenities, deterioration of water quality and risk of epidemics.

Urban areas are centres of economic activities with vital infrastructure which needs to be protected and kept functional at all hours. In most of the cities, damage to vital infrastructure has a bearing not only locally but could even have global implications as these are usually densely populated areas; and thus, people living in vulnerable areas, both rich and poor, suffer due to flooding. Cities have impervious and concretised surfaces that prevent effective infiltration of rainfall into soil. This causes large runoff and high flood levels, leading to inundation and significant flood damages (Ebisemiju, 1989). Highly populated urban areas carry high economic values, and when subject to flooding, result in disaster that can slacken urban development for years. China and Bangladesh are often worst hit by waterlogging that has been increasing due to rapid urban growth, which poses a serious threat to livelihood and property (Tawhid, 2004, Youpeng et al., 2010, Ning et al., 2017, Haque et al., 2018). In India, big cities like Kolkata, Chennai, Delhi, Mumbai and Ahmedabad witness urban floods of various intensities during the monsoon (Rafiq et al., 2016). In Chennai, urban flooding takes place due to a number of issues relating to land use change, for example, decreased natural surface, loss of water bodies, encroachment upon the rivers or streams and other drainage channels, uncontrolled expansion of built-up, etc. (Gupta & Nair, 2010). Urban flood mitigation is one of the most critical issues for both water administration and city management agency, in which urban flood modelling is vital and necessary (Liu et al., 2020). Cities in India are increasingly at risk from natural hazards and the effects of climate change. Increasing populations, enhanced per capita water use, climate change and allocations for water conservation are potential threats to adequate water availability. Cities are especially vulnerable to water shortages because of their high concentration of population and the high demand for reliable water supplies to maintain their intense pace of human and economic activity (Mukherjee et al., 2018). The exceedingly high and unprecedented anthropogenic activities that have led to a series and variety of urban hydro-geomorphic hazards are mostly in the fast-growing cities. The urban geomorphic hazards of lowland cities of the humid tropics include coastal and river bank erosion, channel degradation and siltation, waterlogging and urban flooding, water contamination, land subsidence and resultant structural collapse or hazards from precarious construction of geo-forms (Haldar & Satpati, 2018). Of all the major urban hydro-geomorphic hazards, the problem of changing slope patterns causing urban flooding is most critical in the city of Kolkata.

In Kolkata, urban development on the low-lying periphery has produced uneven geographies of risk, coupled with suburban or rurban settlements surrounded by overcrowded slums with poor infrastructure and lack of basic services (Rumbach, 2017). In the last 300 years, as the city expanded and as the Kolkata Urban Agglomeration took shape, the anthropogenic changes due to urbanisation have changed the original topography of Kolkata. The metropolitan city has experienced numerous transformations of its physical and cultural landscapes from its initiation (Satpati & Haldar, 2020). The Kolkata Metropolitan Area (KMA) is one of the most densely populated areas in the world, where extreme precipitation events result in frequent waterlogging and flood inundation conditions which can adversely affect the city. Human-induced land use changes like agriculture and urban development occur commonly at the cost of natural wetlands, which require immediate attention and sustained efforts from urban planners and concerned governments (Malik et al., 2020). Kolkata megacity, the 13th most populous and the 8th largest urban agglomeration in the world with a population of 14.11 million, is also facing waterlogging problem during the monsoon period. The rapid encroachment in the EKW due to unplanned development has caused urban floods and waterlogged condition having various environmental, social and economic impacts (Saha, 1998, Roy Basu et al., 2020). Cursed with a particular bowl-shaped pattern of city surface profile, Kolkata is always prone to flooding. Banerjee (2018) analysed the factors responsible for the waterlogging conditions in Kolkata with special reference to rainfall pattern and suggested some suitable measures to adopt. Urban Water Security is essential in urban planning to manage the water infrastructures and strengthen their water stress resilience and adaptive capacities. Kolkata being a growing megacity in a developing country is facing rising pressures on water–environmental provisions due to the rapid population growth and urbanisation and resultant governance and infrastructural issues. Kolkata was hydrologically endowed with enough fresh water at one point of time; however, the city has been predicted to be one of the water-stressed megacities of India, crippled by poor water quality through contaminations and rapidly shrinking groundwater supplies (Sahu et al., 2013). The challenge is to build resilience in an urban water supply system that is threatened by climatic variability, population growth and ecological degradation (Ray & Shaw, 2016). Heavy pumping of groundwater beneath the urban centres, particularly in regions with low natural topographical gradients, such as deltas and floodplains, can fundamentally alter the hydrological system. Hydrogeological simulation analysis allows evaluation of the impact on past, present and future extractions for the region of Kolkata, in an aquifer that supplies water to over 13 million people (Sahu et al., 2013).

The peripheral areas too have become vulnerable to the problem of waterlogging as the rate of urbanisation is much higher in these areas than that of north and central Kolkata (Sen, 2011). Urban growth in the fringe areas of Kolkata city is mainly unplanned in nature (Majumdar & Chatterjee, 2020). As a consequence, this unplanned urbanisation gives birth to the various adversities like degradation of the drainage system and canals, which further results in frequent waterlogging during the rainy seasons (Mukherjee, 2002). Cities experiencing unprecedented growth often have serious impacts on the local physical environment due to their expansion and increased water demand or disposal. In this issue, Kolkata is an ideal area for research because of its complex physiographic setup, long history, rapid rate of urbanisation and deteriorating environmental health. It is envisaged that this study will highlight the impact of changing physical landscape on the urban environment.

25.2 Objectives and Methodology

Urban geomorphology, a recent but most useful branch of applied geomorphology, is the study of landforms and their related processes, in ways that are beneficial to planning, development and management of existing urban areas or where urban growth is expected. An urban geomorphologist has to play twofold roles: that is, the study through field survey, terrain classification, identification and selection of alternative suitable locations for urban settlements after proper analysis of material composition and nature of geomorphic process; and secondly, study the effects of urban development on natural environment and their remedial measures. This chapter discusses the dynamics of hydro-geomorphic hazards with special reference to the causes and consequences of urban geo-forms and hydro-geomorphic effects in Kolkata. It aims to provide a comprehension analysis on the conditions of drainage, surface water flows and stagnations in the city due to urbanizations beyond the managerial capacities and its consequent hazards in the riverine city of the lower part of the Ganga-Brahmaputra-Meghna Delta.

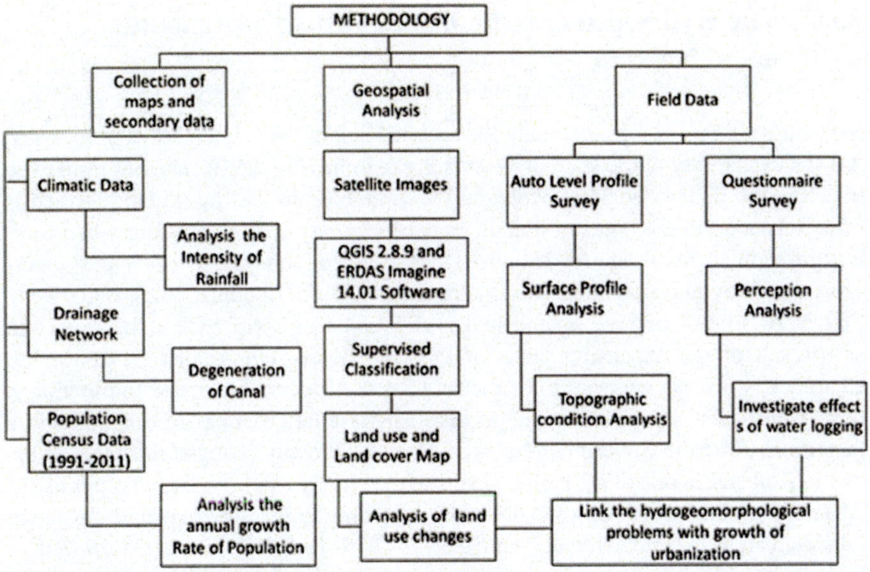

Fig. 25.1 Methodological details of the study

The study is based on systematic analysis of the two most contemporary disastrous impacts of human habitations, that is., urban flooding and the linkages between hydro-geomorphological problems due to rapid growth of urbanisation, for which relevant literatures, government reports and manuals, information and data have been reviewed and analysed. The remote sensing data obtained from the Landsat database of the United States Geological Survey Earth Explorer to detect land use and land cover (LULC) changes in the Kolkata Municipal Corporation (KMC) and Kolkata Metropolitan Area (KMA) have been used and analysed (Fig. 25.1). Landsat data are useful, particularly due to the availability of a long-term digital archive with a spatial resolution of 30 m. Remote sensing imageries of Landsat (Path-Row: 138-44) for the month of December and January 1988 (MSS), 1998 (TM), 2008 (ETM) and 2017(OLI) are used for detection of the changes in LULC. The collected satellite images are further processed by using remote sensing and GIS software. Rainfall data have been collected from India Meteorological Department (IMD). To understand the inherent causes of waterlogging in case study areas and their associated impact on local people, questionnaire-based field surveys, informal interviews and open discussions have been conducted with the authorities of different organisations and local residents. During the field survey, some surface profiles had been taken with Auto Level instrument, to analyse the topographical conditions of the particular area.

25.3 The Hydro-geo-environmental History of Calcutta a.k.a. Kolkata

Kolkata is a linear city grown along the eastern bank of the Hugli (Ganges) river, and it is the capital city of the state of West Bengal, India (Fig. 25.2). The origin may be traced to the deltaic 'soaking ecologies' with its base in the 'liquid land' which is often forgotten in the concrete jungle of contemporary Kolkata, except when forcibly brought to mind, as, for instance, by the recent subsidence along the underground railway in central Kolkata (Bhattacharya, 2018; Dasgupta, 2020; Banerjee & Sikdar, 2020). Calcutta was established as an urban centre way back in the year 1686 as a result of the expansion plans of the British Raj. On August 24 1686, Job Charnock, who was believed to be the founder of Calcutta, first came to the village of Sutanuti as a representative of the British East India Company to establish a factory. In 2001 the Government of West Bengal officially changed the name of the city to Kolkata. In one of the earliest maps of the city by Mark Wood, who published 'Plan of Kolkata' (1784–85), the divisions according to the European and the native inhabited areas is quite evident, later Smart in 1903–10 detailed it out (Mitra, 2011). Kolkata has a long and significant history when it comes to city planning ideas. The location of the city appears to have been originally selected partly because of its defense position and partly because of its favourable trading location. On the east and south, the old settlements of Kolkata sloped away into marshes, swamplands and creeks of the Sundarban forested tracts. Reclamation of the Salt Lake area on the

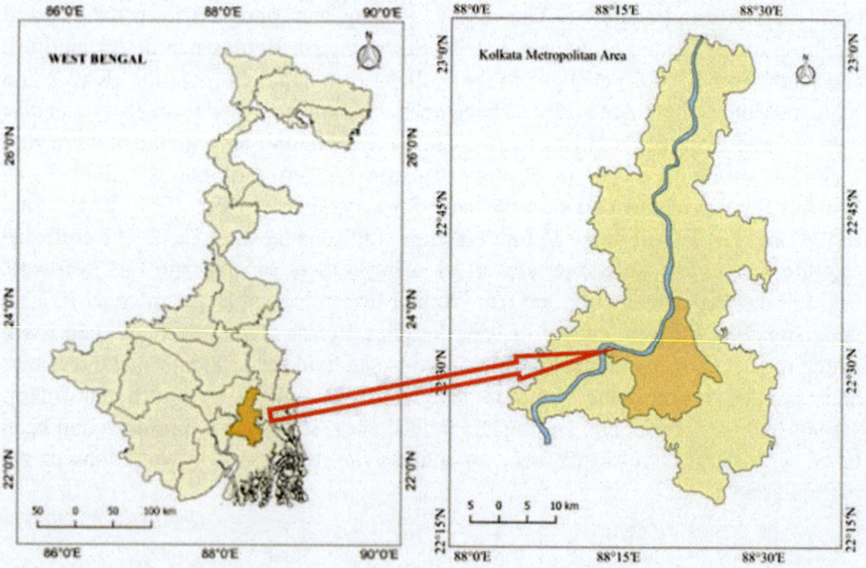

Fig. 25.2 Location of the KMC area. (Source: Prepared by the authors based on maps published by NATMO, Kolkata and the KMC)

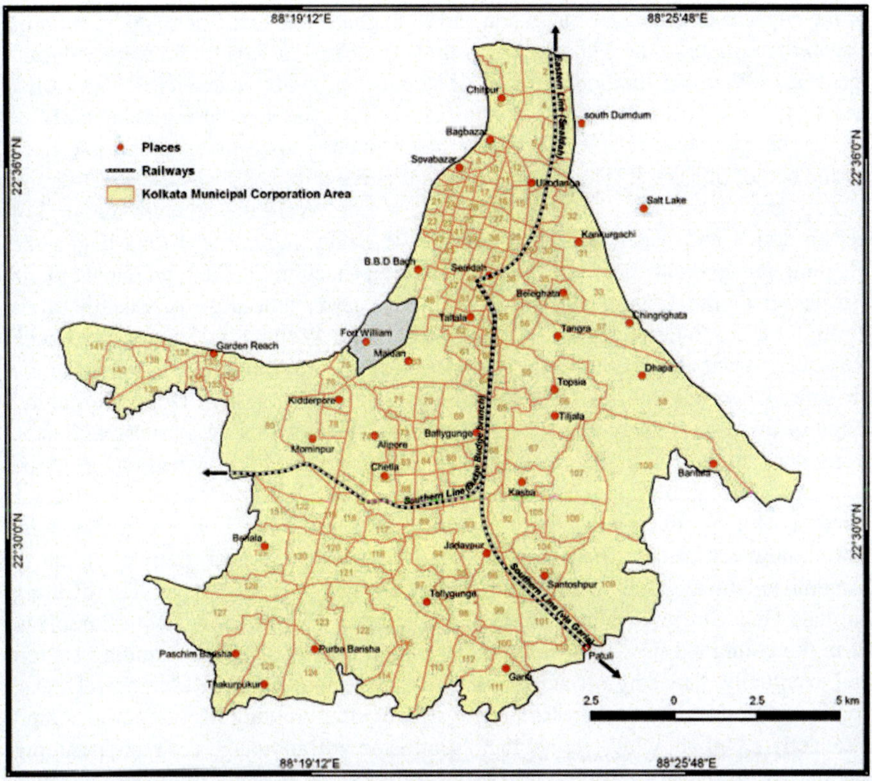

Fig. 25.3 Detailed locations of the places in Kolkata discussed in the study. (Source: Prepared by the authors based on maps published by NATMO, Kolkata and the KMC)

northeastern fringe of the city, however, demonstrated that the spatial expansion was feasible with reclamation projects. Urbanisation processes to the east, south and west, with not only amalgamating wards of 142–144 within KMC limits but also considering similar growth and development plans in the suburban areas as Greater Kolkata with Baranagar to the north, South Dum Dum to the northeast, Joka and Garia to the south and Maheshtala-Garden Reach in the southwest, are rapidly changing the character of the terrain. The present extent of the KMC along with the Wards has been illustrated in details, and the references to the locations in the chapter can be identified from Fig. 25.3.

The primarily north–south alignment of the main roads, form a grid pattern in the old European sector, but elsewhere road planning has a random character due to the difficulty of providing enough river crossings. Khals, nullahs (watercourses) and canals that require bridging also have been important factors in influencing the road pattern. The name, 'Ultadanga', comes from the original name of the place, *UltaDingi*, or a place where small boats which were colloquially called *Dingis* were made or repaired. The Dihi-Panchannagram areas bordered the eastern margins of the city and Ultadanga, Manicktala and Beliaghata were considered to be suburbs beyond the limits of the Maratha Ditch (now the Circular Road) that were incorporated in 1889 within the boundary of KMC (Cotton, 1980; Nair, 1995). The earlier name of Creek Row was Dhinga-bhanga as a ship broke there during the cyclone of 1737. The creek was navigable for large boats. Wellington Square was a tank made on the bed of this creek. The first drainage initiative of the city was the construction of Beliaghata Canal in 1810. Governor General Lord Wellesley stressed the need of public drains and watercourses to control the heavy rainfall and thus began the planning process with his prescriptive 'Minute on Calcutta' in 1803, which led to the setting up of the Lottery Committee in 1817, termed because of raising funds through public lotteries for city development. Both Wellington Square and Creek Row were developed by the Lottery Committee funds. William Clark, proposed the scheme (Town Scheme) in 1859 to serve as a combined system along the natural slope of the terrain away from the River Hugli and towards the eastern wetlands, from west to east (Goode, 1916). Two main intercepting sewers were to be constructed along the Upper Circular Road (now APC Roy Road) from Sova Bazaar Street to Moulali junction, where it was to be joined with the Lower Circular Road sewer (now AJC Bose Road) coming from Zeerut Bridge near Tolly's Nullah to Moulali junction. Other main sewers were planned from the River Hugli along Nimtala Ghat Street, Kolutola Street and Dharmatala Street (now Lenin Sarani) to form the combined flow (Town Outfall) through Palmer's Bridge pumping station and eventually disposing off at Tangra creek, and another from Kidderpore Docks, continued along Judges Court Road (in Alipore) after crossing Mominpur area, and into Tolly's Nullah, while yet another culminated at Ballygange drainage pumping station (Basu et al., 2013). The city now has 9 surface drainage systems, namely Chingrighata Canal, Mominpur Canal, Kolkata Port Trust CPT Canal, Pagaladanga Canal, Kulia Tangra Canal, Ballygunge Drainage Pumping Station Canal, Jora Bridge Canal, Baghajatin Canal and Nonadanga Canal. The locations Magura, Topsia and Chingrihata suggest specialised association of different fish species which might have been caught or traded at these fringe locations of proper Kolkata during its early period of development. The Budge-Budge, Ballygunge and Sealdah railway tracts correspond to the city limits in the first phase of inclusion of the south sub-urban areas with the main city. In the second phase Behala and Jadavpur areas were included to provide better civic administration and amenities to the erstwhile municipality areas which were already heavily congested with unplanned growth due to influx of population in the post-partition of India period. The Kabi Nazrul Islam Sarani or Eastern Metropolitan (E-M) bypass connecting Ballygunge, Kasba, Panchannagram, Beliaghata and Ultadanga facilitated urban growth in the eastern fringe of the city during the post liberal period. This area happened to be the low-lying flood plain of the Hugli river and was drained by the tributaries of the Bidyadhari. Consequent upon the degeneration of the Bidyadhari system, the area was transformed into wetlands which were traditionally used as collection ponds of stormwater, based on which peri-urban agriculture and pisciculture developed. But E-M bypass provided a better connectivity, leading to infrastructural development in the form of arterial roads, housing complexes and other establishments that resulted in reduction of the wetland areas whereby raising the natural elevations and modifying the slope character.

Distribution of the water bodies in the KMC area is highly related to the geomorphic characteristics and history of urbanisation of the city of Kolkata. Previously the elevated land along the eastern levee of the Hugli River was developed for settlements, and therefore the natural water bodies including cut-off channels and creeks were filled up, but a number of ponds were dug for water. The Boat-canal connecting Adi Ganga at Chetla with the Hugli River at Garden Reach runs parallel to the Budge Budge section of the eastern railways. The canal was beneficial for removing the excess water of Alipore and Mominpore area to prevent waterlogging. However, over the years the degeneration of the canal made it dysfunctional for both movements of the boats as well as reduction of waterlogging conditions. A part of Adi Ganga is now a purely tidal channel fed with tidal water ingression that reaches up to Garia from Hugli River beyong which no flow can be observed. The Adi Ganga had degenerated in its southern reach long back, which was cut by William Tolly (1775–76), hence the part from Kudghat to Garia came to be known as Tolly's Nullah. Extension of the Metro Railway from Tollygunge to Garia required erecting of pillars over Tolly's Nullah from the bed of the channel that blocked the in-channel flow and there by facilitating sedimentation both natural and anthropogenic. In the southern suburbs the narrow unmetalled roads were widened and elevated as broad metalled roads which prevented the natural drainage, and in many of the places the homestead levels went below the road surface level, allowing the water to stagnate within the buildings during heavy downpours. Conversion of the numerous ponds for urban renewal into built-up areas in these areas is responsible for the reduction of infiltration and depression storage of the flood water. During the last three decades, thousands of water bodies were filled up. With the installation of tube wells and extension of tap water supply in almost all of KMC areas the importance of surface water bodies had reduced for the people. The contemporary usage of the water bodies is limited to very few people of low-income category. Disregarding the ecology, the prime locations of many of the water bodies have made them vulnerable to degradation as is eventually used for vertical expansion of the city. Case studies conducted in the added areas of the KMC for the last five years show that a number of water bodies, especially the smaller ones, have been seriously neglected to become unfit for human use and ultimately filled up for construction of buildings. Pressure on land and extension of transport facilities beyond the peripheral areas of the city and defective land use policy have favoured conversion of open greenery plots and wetland areas into built-up areas. During the last few decades, however, some voluntary groups of the city people have raised voices of concern for the conservation of wetlands and water bodies of Kolkata (Bose, 2008). But economic and political system is often unscrupulously used to satisfy the greed of land mafia and promoters who have no consideration for the ecological need and aesthetic beauty of the city (Satpati & Haldar, 2020). Hopefully, the citizens are now little more aware of the situation and the importance of conserving the water bodies have been felt, which may improve the situation in the near future; although meanwhile many of the smaller water bodies will be vanished from the map of the KMC.

Over the last 150 years, the city has proliferated in a rapid manner, nearly 100 times in size, and about eight times in terms of population (Haldar and Satpati, 2018). The spatial dimensions of the city of Kolkata (previously Calcutta) increased markedly from 7 sq.km. in 1706 to 95.62 sq.km.in 1955 and 187.33 sq.km in 2010. The South Suburban Municipality was established in 1869 and the Kolkata Municipal Corporation in 1876. It was merged with the Kolkata Municipal Corporation on 4.1.1984, and the last three wards were added in 1.9.2012. As of 2018, the KMC area presently includes about 206.08 square kilometres being divided into 144 wards. It accounts for the 30% the total population of the Kolkata Metropolitan Area (KMA), within only 10% of its area. According to the census data of 2011, the KMC has a population of 44.87 lakh and the density of population stands as 23,995 persons per sq.km. From the study of the census data of 1991, 2001 and 2011, it is found that the concentration of population with respect to the total population has been increasing in south and southeast parts of the city (Fig. 25.4.).

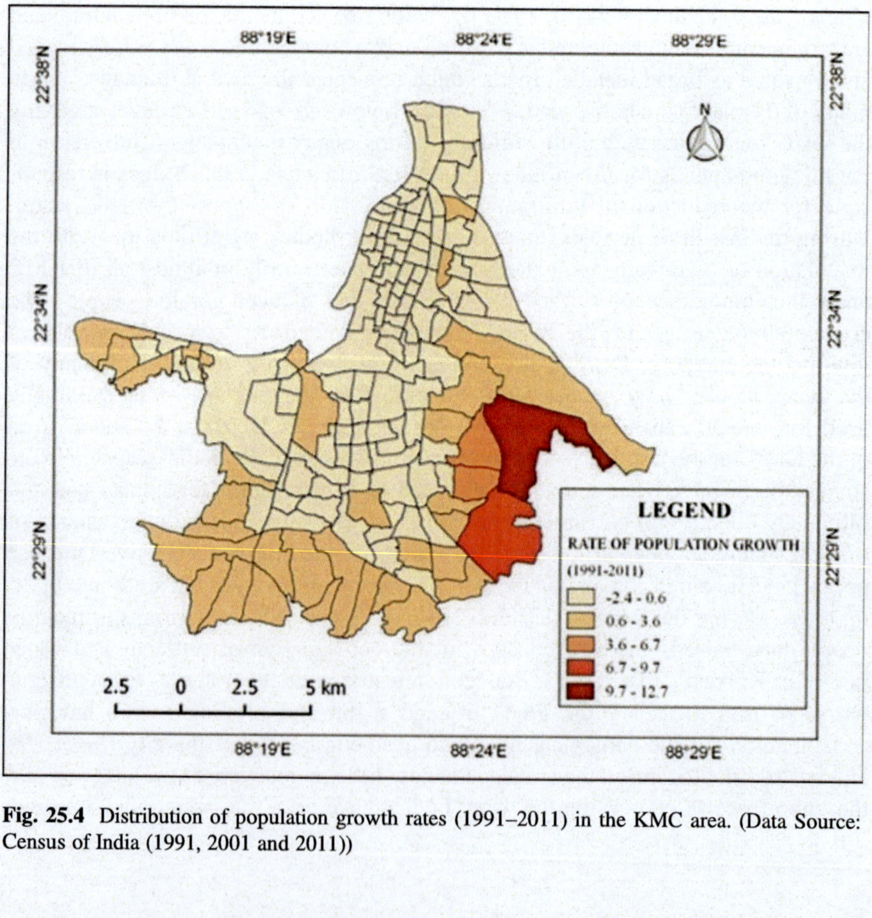

Fig. 25.4 Distribution of population growth rates (1991–2011) in the KMC area. (Data Source: Census of India (1991, 2001 and 2011))

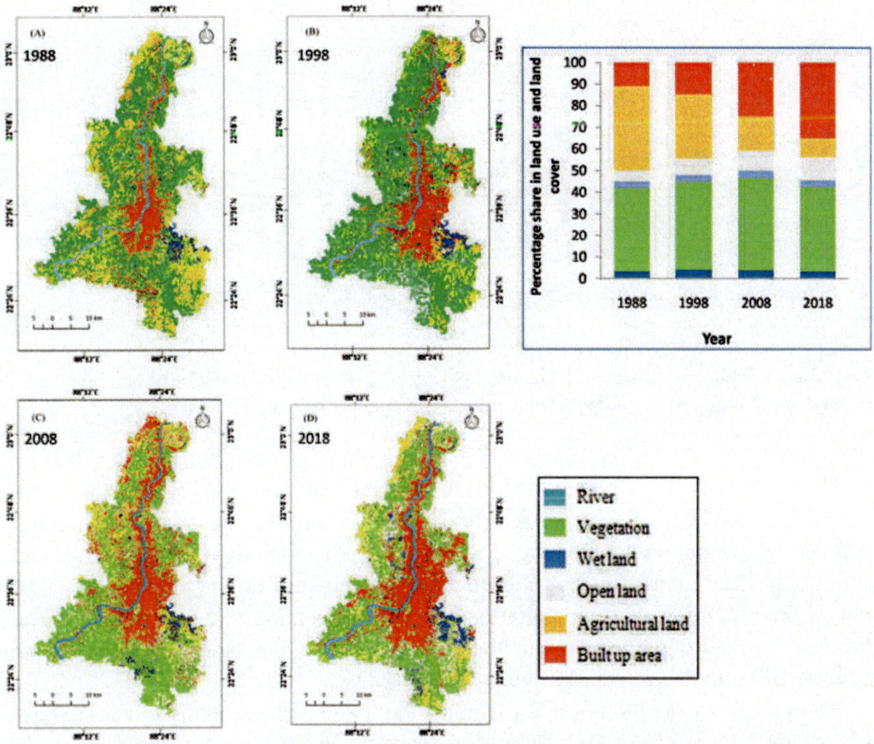

Fig. 25.5 LULC maps of the KMA for the years 1988 (**a**), 1998 (**b**), 2008 (**c**) and 2018 (**d**). (Source: Prepared by the authors based on maps published by NATMO, Kolkata and the KMDA)

A major shift of the residential areas of the inhabitants of the city is observed to be gradually taking place from northern core to the recently expanding edge areas occupying relatively low-lying open plots that were more congenial for absorbing the city's runoff and groundwater recharges. The urbanisation process of the southern and eastern margins too has taken place through reclaiming land from agricultural lands, open spaces and water bodies during the last four decades. The LULC classification of the KMA boundaries shows that in 1988 there were just 11% of the area under built-up coverage which rose to 65% in 2018 (Fig. 25.5). The gradual change in land character of not only Kolkata city but all the surrounding towns and cities in the surrounding area under the Kolkata Metropolitan Development Authorities and beyond can be observed from this map. The drastic fall in agricultural land and sharp rise in built-up confirms the rapid urbanisation policies in the last four decades.

Kolkata falls under the humid tropical climatic regime with the maximum temperature of 24–42 °C (May–June) and the minimum temperature of 8–10 °C (December–January). The temperatures have increased about 0.68 °C over the last century, with increasing trends in annual mean temperature and more pronounced warming in winter and the deferred monsoon season. Rainfall is generally

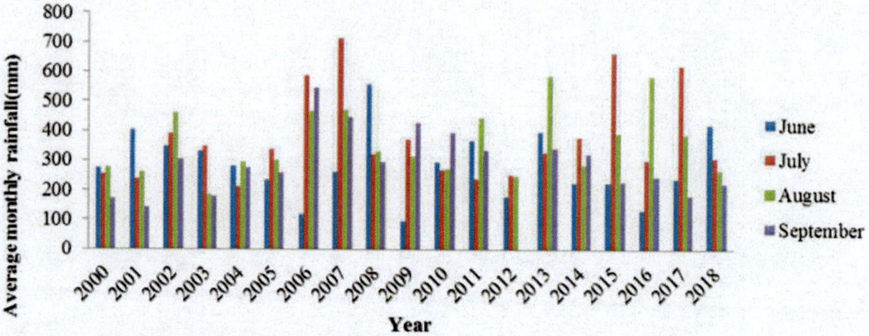

Fig. 25.6 Average monthly rainfall (June to September) in Kolkata (IMD station: Alipur, Kolkata). (Data Source: India Meteorological Department (Government of India), Alipur)

1200–1600 mm/year of which 80% occurs during the monsoons (Fig. 25.6.). The pre- and post-monsoons are characterised by very short downpours while July and August are characterised by long duration, moderate- to low-intensity rains and shows a gradual rising precipitation trend. A marked shift or late torrential rains or few peaked high intensity precipitation is also observed in the recent decades. The problem of waterlogging and urban flooding occurs when the rainfall intensity shoots beyond 100 mm/day on an average.

The city lies over the alluvial plains of the lower deltaic plains of the Ganga-Bhagirathi system where the distributary channels and *khals* (canals) play a major role in land formations. There are records with the Geological Survey of India, which reveals a 762 m thick unconsolidated alluvium below the city, and a borehole sample drilling of tube wells in the Tollygunge area shows a 200 m deep layering of peat, kankar, clay beds and thick sand layers. Over centuries, the Hugli River has deposited large amounts of alluvial silt along its bank, forming a natural levee (Fig. 25.7.). In Kolkata, many areas have a thick deposit of fine silty clay layer with very little permeability for groundwater recharge. This is further responsible for delayed absorption of rainwater into the ground and that resulting in high recharge lag time. Another important consideration is the decreasing pre-monsoon peizometric level (about more than 10 m in the last 50 years); ground water level depth varies from 12–15 m bgl in the developing parts where waterlogging is common and 15.5–19.6 m bgl in the older settled parts of the city (KMC-CGWB, 2007). This report also mentions the fact that areas closer to the River Hugli and southern edges have saline ground water just below the surface clay layer, while in the city core the saline levels are below 160 m. This clay layer below the impervious concrete furthermore hinders surface water infiltration.

Traditionally, the Kolkata city forms a bowl-shaped topography with the high levee along the Hugli banks, low-lying compacted and built-up region lying parallel to this, and after a small rise along the Circular Canal/Road region, gradually slopes to the low-lying east and southeastern edges. Therefore, if all waterlogged areas of the city are superimposed, it can be found that highly waterlogged areas are mainly

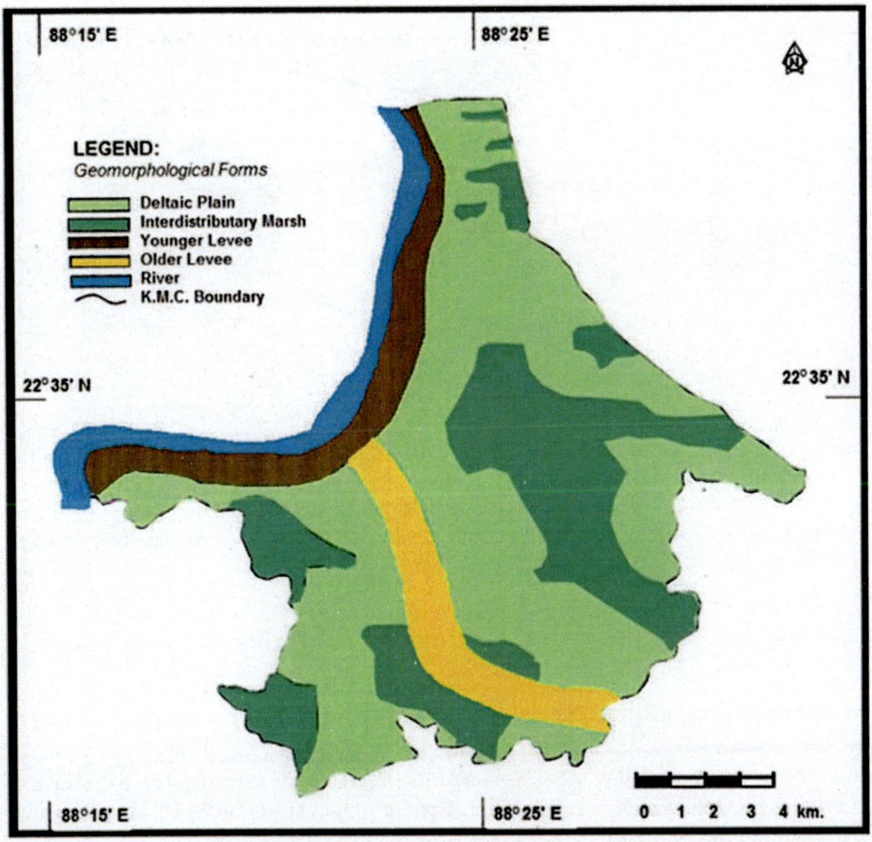

Fig. 25.7 Geomorphology of the KMC area. (Source: Prepared by the authors based on maps published by the KMC)

concentrated in the low lying areas subsumed within the 'high road' networks. Elevation of the KMC area ranges from 3.5 to 9.0 metre above the MSL with an average elevation of 6 metre. The highest elevation of the city is found near the levee region along the banks of the Hugli and the lowest elevation in the eastern margins. Thus, the natural slope of the city is from west (left bank of the Hugli) to east (marshy areas with wetlands). The ground configuration including relief and slope, and transformation of the land use has been creating problems to the drainage system. The unplanned and unregulated urbanisation has led to the formation of sharp rise and fall of the surface level of the city as can be seen in the profiles along roads (Fig. 25.8). The total water requirement of Kolkata was estimated to be around 1000 MLD (megalitres/day) while the per capita demand was 170 LPCD (lit./capita/ day) and the total sewage generated is 734 as per Central Pollution Control Board in 2005 for Kolkata city (CSE, 2005).

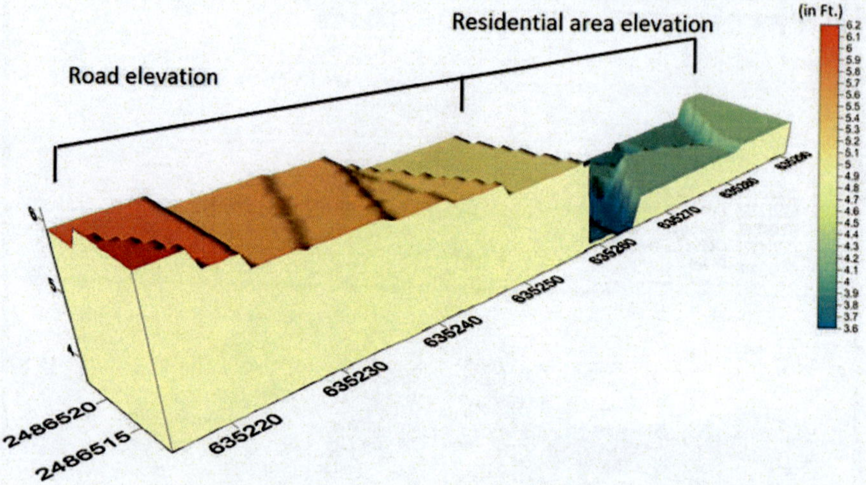

Fig. 25.8 Surface profiles in Nrisingha Dutta Lane, Purba Barisha, Ward No. 123. (Source: Prepared on Surfer 10 software from Auto level Surveys, 2015)

25.4 Urban Flooding and Drainage Congestion

Excess water accumulation due to high-intensity rainfall, at times coupled with high tidal situations, without proper and timely drainage options has flooded the Kolkata city frequently during the monsoon months. Water may stagnate for days in the low-lying pockets which is referred to as waterlogged situation. Both flooding and waterlogging are common urban hazards of the city. The causes of urban flooding and thereafter waterlogging conditions may be attributed to both the anthropogenic and natural factors.

It has been observed by the India Meteorological Department (Government of India) that character of monsoon season has been changing during the recent years when the problem of waterlogging mainly aggravates. It was proposed by the officials at the pumping stations that as per the designed drainage capacity of the municipal area, no water stagnation should occur in Kolkata if the rainfall is within 6 mm per hour yet in the recent analysis it was found that even with 150 mm 24 h rainfall on 29.7.21, extreme waterlogging ranging from 15 cm. to 1m. loomed in numerous pockets of Kolkata. The main part of the sewer network consist of century-old brick sewer of length 180 km. Most of the drainage pumping stations were constructed 50 to 100 years ago (KMC portal, 2021). The designed carrying capacity of the combined sewers did not increase with the rising precipitation intensities and population of the city, so whenever the intensity of rainfall is higher than normal, some parts of the city remain under water for a considerable period of time. The old open sewerage lines are insufficient to discharge the increasing daily wastewater produced by the newly formed residential areas. During monsoon periods, many times the rainfall intensity exceeds the mark of even 15 mm and is beyond the drainage pumping system capacity of the KMC. The main drawback of

Kolkata's drainage system is the combined flow of dry weather and storm weather through the same conduit. In the central part of Kolkata, sewerage network system (Town system) is almost 160 years old, and the southern part of the sewer system (Suburban system) is around 100 years old. The trunk sewers laid along the east–west direction carry the wastewater and storm runoff from the western part of the city to the eastern part through different pumping stations from where water is pumped to the dry weather flow (DWF) channel and storm water flow (SWF) channel for disposal into the Kulti River. The S.W.F (storm water flow) channel of 5100 cusec capacity, has mainly two lead channels meeting at Bantala, namely the Suburban Head Cut Channel for about 7.32 km, receiving drainage from Ballygunge and Chowbhaga pumping station on its way, and the other part is the Town Head Cut Channel of 7.80 km, from the Palmerbazar, Chingrihata and Pagladanga stations as well as from the Dhapa lock pumping station via feeder canal (Chatterjee 2015). Before discharging into the Kulti River the sewage receives partial treatment as it passes through the East Kolkata Wetlands (EKW), the large aquacultural ponds occupying 5000 hectares to the east of the city. The combined drainage, that is, dry weather flow as well as storm weather flow, discharges directly through the canals of Kolkata or flow overland as surface runoff, accumulating in all available depressions if not properly pumped out through the underground pipelines. As a result, Kolkata and its adjoining areas suffer from drainage congestion during the high-intensity rainy days. The peak factor or the ratio of maximum to average flows depends upon contributory population, topography of the site, and hours of water supply. Storm sewers are often not designed for the peak flow of rare occurrence such as once in 10 years or more, but it is necessary to provide sufficient capacity to avoid too frequent flooding of the drainage area (CPHEEO, 1993).

The duration of stagnation of water in Kolkata is about 2–12 h depending on the efficiency of the local authorities in getting portable pumps to drain out the waters. In the recent years, it has been observed that the pumping stations not only have been maintained and operated at their full capacity during the rainfall days, but portable pumps are also being used to link the stagnated water to the nearest major drains. However the channel siltation, rising of the channel beds, reduced flow velocity, settlement encroachment and reclaiming the natural channel areas along the banks have rapidly degenerated the carrying capacity of the drainage system. The capacity of the drains is yet to be increased by a complete overhauling of the pipes or construction of more trunk sewers which is not feasible at times given the extremely high built-up density of the city.

Apart from the technological issues, the geo-environmental causes of urban flooding in Kolkata include the complex anthropogenic alterations on the hydro-geomorphic setting and changes in the land usages, especially the canal systems. The land use conversions, concretisations and rapid urban sprawls have greatly reduced the natural infiltration processes and groundwater recharge as observed from the increasing rates of runoff in the outer areas of Kolkata. Expansions of built-up areas both vertically and horizontally have led to an exodus of population largely influencing the urban supply and discharge rates. Even the encroachment of shanties and illegal housing along drainage courses shortens the width of the canals, thus lowering the carrying capacities as observed from our studies in the Churial canal area at the southern edges of the city (Haldar, 2020). Unplanned growth of settlements has led to a large number of populations occupying vacant lands adjacent to the various drainage channels and natural ponds. Filling up of water bodies and open plots also reduces depression storage, restricts the path available for stormwater flow into the surface drainage outlets and increases runoff flow rate over concrete impervious surfaces. Lastly the luxurious growth of water hyacinth and eutrophication along the canal markedly reduces the flow capacity and velocity.

The intricate canal system of the city, even though had a long history of success, but is almost at the verge of becoming defınct (Fig. 25.9.). Tolly's Nullah, Circular Canal and Kestopur Khal (canal) in the north have cross-flow connections to the Hugli River in the west and Adi Ganga to Bidyadhari River and Kulti River in the southeast. In times of high tide in the Hugli, a backflow occurs in these channels that prevent the drainage outflow; hence, lock gates are installed at the outfall points. This backflow when coupled with high-intensity rainfall also leads to the waterlogging disasters inundating the lower elevation areas including slums, building basements, roads, etc. The canal system of Kolkata was for drawing out liquid wastes, storm flows and sustained estuarine ecology through the East Kolkata

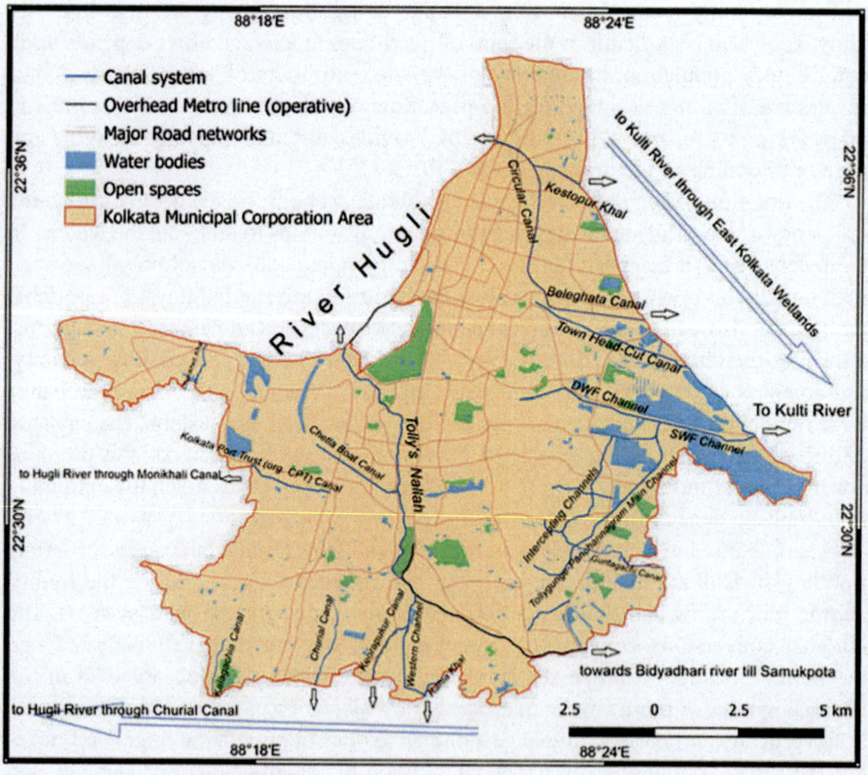

Fig. 25.9 Canal system in the KMC area and direction of the outfall channels. (Source: Prepared by the authors based on maps published by NATMO, Kolkata and the KMC)

Wetlands. However, the present poor state of the channels, through various field studies, was observed that the sedimentation and disposal of solid wastes, especially plastics and slow biodegradable or non-degradable wastes, has clogged major portions of the channel, leading to contamination, deterioration and decay of the urban water bodies, thus aggravating the disaster. These areas have not only become breeding grounds of mosquitoes but lead to odour pollution, ecosystem degradation and, finally, are not a proper visual for a new-age city like Kolkata. The quality of canal water in Churial canal, as measured by KEIP in 2012 was about 7.35 pH, 34 mg/l TSS, 725 mg/l TDS, 5.2 mg/l DO, 82.44 mg/l COD, 160 mg/l BOD, 3500 Colony Forming Unit/100 ml of total coliform which sort up to higher levels during late winters and early summers when rain water mixing is nil; therefore not even meeting the minimum criteria for washing. A February 2016 data revealed that in Churial Canal at Diamond Harbour Road crossing the pH was 7.15, TSS was 40 mg/l, DO 1 mg/l, and COD 71.36 mg/l showing no significant improvement over the years (KEIIP 2012, 2019).

From our studies along Tolly's Nullah in Garia region, it has been found that the depth of the water in the channel in Kudghat is about 5 m, but it is only 2 m below Garia Railway Bridge. The gradient of thalweg is low throughout the area (Roy et al., 2017), and the channel is filled with garbage and sediments from urban debris at many places and the water is murky, brackish and highly polluted. From Tollygunge to Garia, about 200 pillars have been constructed for the Metro Railways along the bed of Tolly's Nullah which has further deteriorated the flow of the water to a large extent. It is also found that this segment of Tolly's Nullah is suffering from the problem of siltation as it is standing at a higher topographic level than the surrounding areas, so it is incapable of draining out monsoon runoff and carrying away any garbage dumped into it. The natural flow is zero between Garia Bazar to Garia Railway Station area. It is also observed that a large number of houses have been constructed along both banks of Tolly's Nullah in the recent years and have gradually reduced the width of the channel. Anthropogenic interferences like slums and squatter encroachments, lack of awareness, environmentally degrading constructions of roads and bridges, liquid and solid waste dumping are increasing the problem of channel maintenance whereby choking the channels (Fig. 25.10.).

Fig. 25.10 Accumulated garbage at Garia Bazar and Metro Rail pillars over Tolly's Nullah, Ward no. 111 & 112. (Source: Photographs captured by the authors, 2019)

25.5 Impacts of Urban Flooding

The problems of urban flooding come with associated hazards which can be classified based on physical environmental, economic and social issues. The physical and environmental problems include damage to road surfaces, damage to roadside houses, dampening and collapse of walls along the waterlogged areas, accumulating sticking muck and rotten solid wastes on the surfaces and massive damages to slums and squatter settlements turning the dwellers temporarily homeless. These wastes when washed into ponds, pollute the water bodies, and causes contamination of drinking water wells or clogging of pipelines due to poor maintenance. In areas where the building structures were originally aligned with the road elevation, it is often observed that over the years the subsequent layering of the road has increased the elevation of the road and surface runoff gets accumulated in the low-lying ground floors of these buildings and stagnates till adequate condition of drying out of the water is attained. The stormwater mixed with the solid waste, domestic waste and silt, further clogs the drainage networks, causing overflow, water stagnation and resultant water and odour pollutions, pests, mosquitoes and other waterborne diseases around that area. Among the social problems, loss of working days, delay and disruption of transport facility, escalation of travelling costs, etc., are most often observed. As per the field survey, almost 96 percent of the inhabitants mentioned that waterlogging greatly hamper daily lives in various ways. The economic impacts of urban flooding have major consequences from the increase in repair and maintenance cost or constructions to non-availability of manpower and other resources due to disruption of transport, communication and power lines (Fig. 25.11.).

From our field surveys in Ward nos. 40, 111, 112, 120, 121 and 123, if there is a continuous high-intensity rainfall of 3–6 h, then the depth of water stagnation may be about 12–20 cm and waterlogging generally takes 4–6 h after rainfall ceases to recede. 40 percent of the respondents stated, it takes at least 1–3 h for rain water to stagnate while it takes a minimum of 2 h to recede. While in Ward nos. 92 and 107, the receding time can be around 30 min to 3 h. Here too the depth varies around 15 cm. Even though a lot of new projects and initiatives have come up to reduce the water stagnation period, more pumping facilities should be undertaken to prevent the waterlogging scenarios.

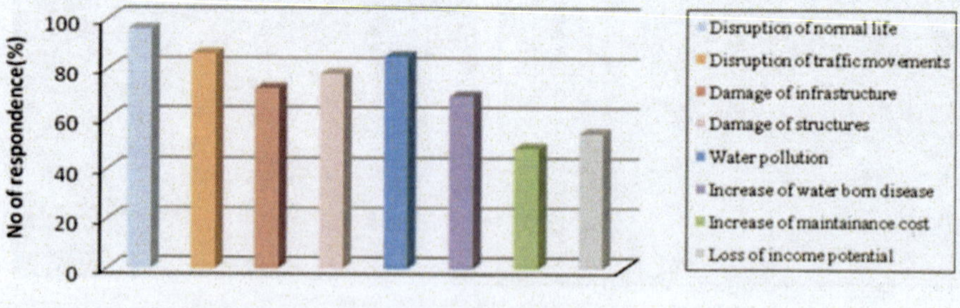

Fig. 25.11 Residents' perceptions survey on problems faced due to waterlogging in selected case study locations in the KMC area. (Source: Primary Survey, 2019)

25.6 Surface Readjustments and Associated Hazards from Urban Renewal

Urban renewal is a programme of land redevelopment in areas of moderate- to high-density urban land use. The process has had a major impact on many urban landscapes. Such projects include improvement of urban infrastructures. But in the past few years, Kolkata has seen a number of ill-adjustments of urban renewal, the most prominent being the over-topping of the road surfaces with the surfacing materials without the traditional 'cut and fill' technique. This technique needs to be applied so as not to raise the road surface level in patches that relatively lowers the adjoining plots and cause major waterlogging problems (Fig. 25.12.).

Open drains along the roads and public thoroughfares are quite common in the upcoming urban fringe areas or usually in the added erstwhile municipalities. The roads are partially covered with asphalt at the centre flanked by pervious soil and uncovered drains to carry both stormwater and sewerage water. Subsequent urban renewal creates widening of roads and construction of pavements by covering the open drains. The usual gradient of the surface is towards either sides of the centreline of the roads, and the surface drainage is designed to be collected as catch pits which are connected to the main drains running below the roads (Fig. 25.13.). At the initial phase the pavements are at a relatively higher level compared to the roads so that even during floods and waterlogging, people can walk freely and safely. But in a tropical environment high-intensity rainfall and waterlogging conditions and continuous movement of heavy vehicles detach the asphalt from the surface, and random pits and potholes are formed on the roads. In the post-monsoon period, the roads are repaired almost every year either as patchwork or complete covering of

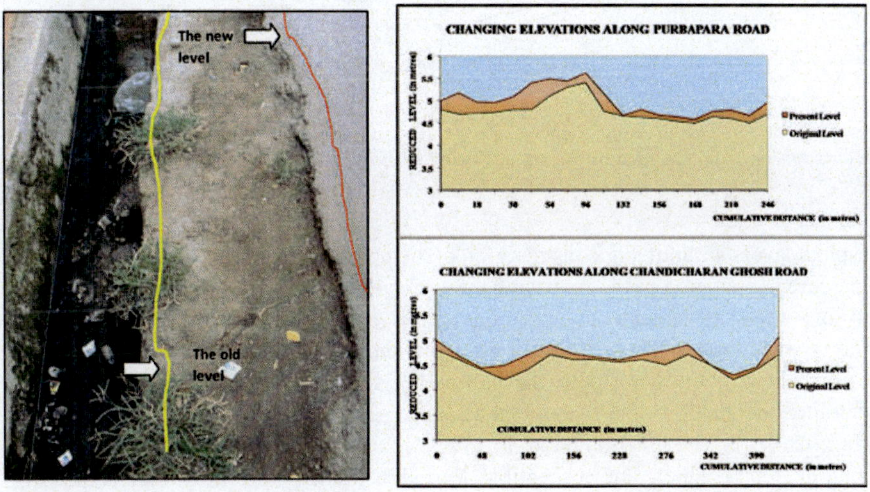

Fig. 25.12 The changes in the profiles of surface level due to overtopping maintenance processes on urban renewal in Behala area of Kolkata, Ward no. 120. (Source: Primary Survey, 2017)

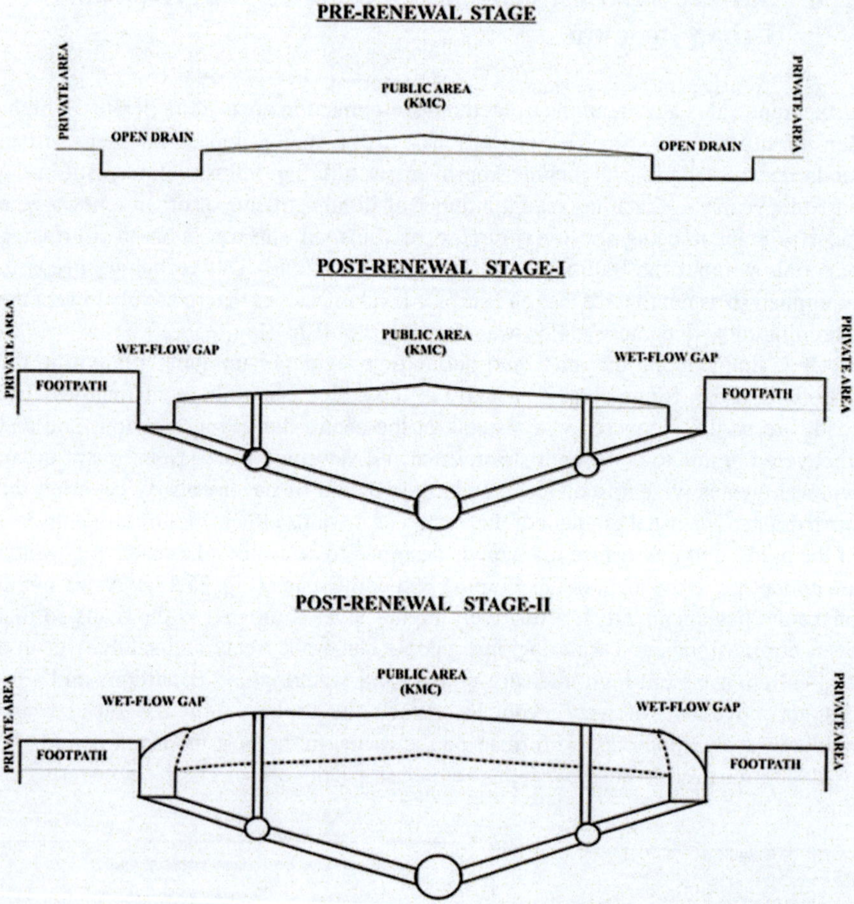

Fig. 25.13 A schematic diagram of the sub-surface drainage network has been conceived and the problem of road level rising higher than the footpath level has been highlighted. (Source: Conceptualised and prepared by the authors based on field observations)

asphalt layers without any consideration of the new levelling or slope. This causes an increase in elevation of the roads and the surface to more than half a metre compared to the adjacent pavements and houses (private areas) constructed during the pre-renewal phase. Thus, in the long run even a moderate-intensity rainfall may cause water overflowing onto the pavements, adjoining constructions and porch/entrance of houses built earlier, causing severe hardships to the residents (Fig. 25.14.). The phenomenon is even worse when there is clogging of the man-hole pits which are susceptible to sedimentation with clay and cementing materials coming from various construction sites.

Fig. 25.14 Adjustments for waterlogging: Relative lowering of the adjoining plots causing surface runoff inflow into the ground floor of the houses, Ward no. 123. (Source: Photographs taken by the authors 2017)

The following profiles (Fig. 25.15.) depict the nature of the road profile of major waterlogged areas along both banks of Tolly's Nullah. As the general elevation of this Garia area is high, accumulated runoff remains stagnant for a considerable period of time causing urban floods. Here, as the data indicate from primary surveys (Table 25.1), the water stagnation depth varies and the problem of waterlogging remains a frequent phenomenon along either side of the road. This is because the road is elevated while the settlement in either side of the road stands at a much lower elevation than the main road. This topographic condition creates the condition favourable to the waterlogging problem along Garia Station Road. The water stagnation rises to knee length (42 cm), the ankle length (8 cm) and estimated a middle length (36 cm). The water depth is highest near Fartabad area, K. B. colony East and Kabi Nazrul metro area eastern side. It can be concluded that areas having high undulations or having no slope are the areas having the highest depths of water stagnation considering that drainage characteristics and wastewater generated are taken as homogeneous throughout.

In an ongoing research this relationship between the changing slope character and its associated impacts is being analysed in details by the authors. That rainfall duration and intensity are not the sole markers in the urban water stagnation and flooding has thus been established from the chapter. The unplanned anthropogenic surface forms and failures collectively termed as geo-forms are one of the most influential factors in aggravating urban hydro-geo-environmental hazards in any built-up regions across the world as is the case in Kolkata city (Haldar & Satpati, 2018). Land subsidence issues too have cropped up in few areas, mainly because of the over-extraction of groundwater. Withdrawal of ground water in KMC area increased progressively from 121.5 ml/day in 1986 to 209.7 mlpd in 1998 and it continued upto 2004. From 2005 restrictions were imposed and KMC started replacing the ground water supply with surface water supply. Such policies and recommendations, mentioned in the last section, can atleast reduce the detrimental impacts and future disasters.

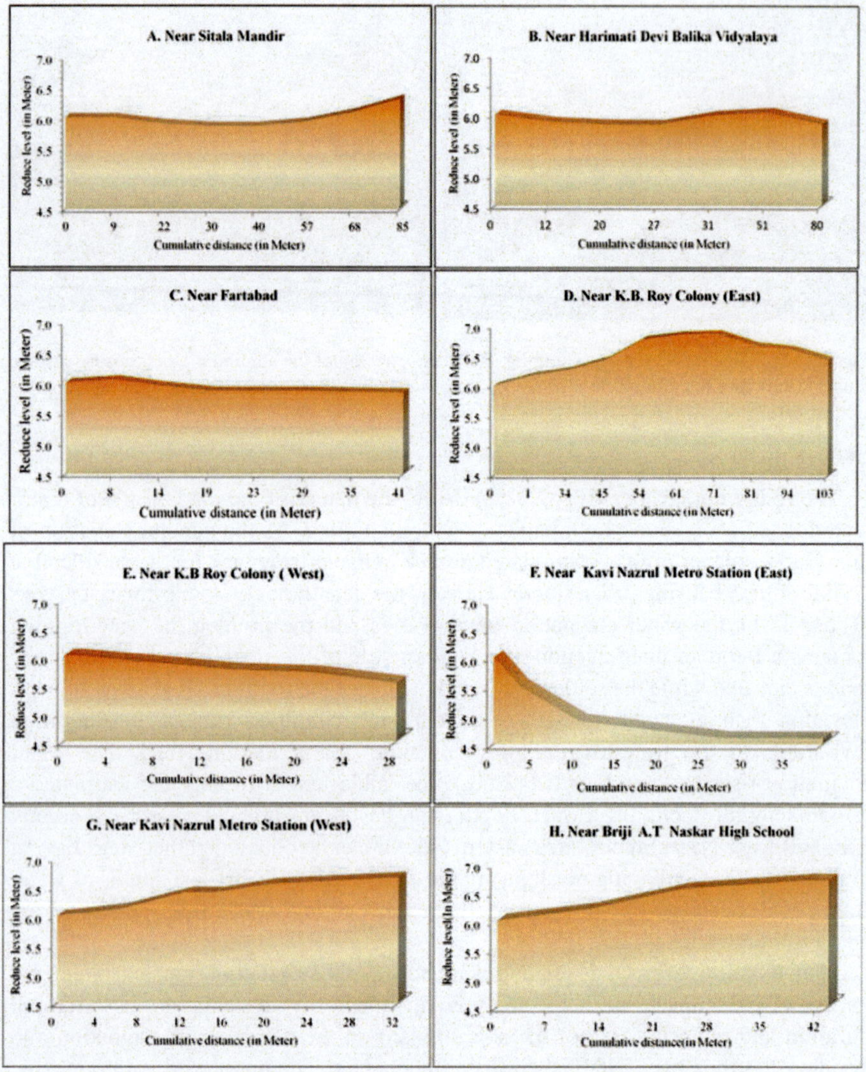

Fig. 25.15 Surface profiles along the Tolly's Nullah. (Source: Primary Survey, 2018)

25.7 Conclusions

The chapter deals with various aspects of hydro-geomorphic hazards of the city of Kolkata within the KMC boundary and proposes to generate awareness on the causes, consequences and probable mitigation measures. Through the study the observations that need to be highlighted are that the lack of timely and effective public infrastructure management, and rapid and uncontrolled expansion of the city in

Table 25.1 Average depth of waterlogging along the bylanes around Tolly's Nullah in Garia area

Code	Area	Water stagnation depth (in cm)
A	Sitala Mandir Area	36
B	Harimati Devi Balika Vidyalaya	36
C	Near Fartabad	42
D	Near K.B. Roy Colony (east)	36
E	K.B Roy Colony (west)	42
F	Kavi Nazrul Metro Station (east)	42
G	Kavi Nazrul Metro Station (west)	6
H	Near Briji A.T Naskar High School	6

Source: Primary Survey, 2018

the last 20 years in terms of population, settlements, built-up zones by engulfing the surrounding fringes that have not only put a resource oriented, economic, social and infrastructural pressure but is a root cause of hydrological, geomorphological and environmental imbalances. The intensity of rainfall has shot up to a significant level in the recent years; the uneven road surfacing techniques have altered the natural flow surface regime in times of high-intensity rainfall; the canals and channels that intersect the city quite adequately have mostly degenerated, and consequently the probability of urban flooding is rising. The availability of technical efficiency is undoubtedly managing the hazard to the best of their ability, but too much dependence on government support and pumping stations alone without managing the ecological aspects will not only be a drain on the government's capital or human resources and capacity, but will gradually make the city landscape unsustainable. In this situation a prior planning on maintaining the natural slope of the city towards the southeast direction is necessary, so that the excess surface water drains into the EKW and the ecological flows in the Ramsar wetland area is preserved. In conclusion, it can be proposed that the problems and their causes, though deep-seated and intermingled with various factors, if timely and accurately identified, with necessary improvement works that are properly executed, adequately operated and maintained, can be easily solved. This shall also open up great opportunities in urban renewal and remodelling of the cityscape to achieve a sustainable city.

25.7.1 *Recommendation Towards a Geo-environmentally Sustainable City*

It has been widely established that the man–land interaction is at its maximum in the urban areas and has prominent concerns for hydro-geomorphic issues. Thus, different approaches need to be dealt with before any Urban Renewal Projects are imposed on the city area. Some of the recommendations proposed may be categorised as: (a) environmental approach, (b) people-centric approach and (c) developmental approach. The environmental proposals include the removal of riparian vegetation from canal banks and creating more open spaces with trees and grasses in between built-up areas to enhance infiltration of surface runoff. Another way may be the de-siltation or uniform dredging from origin to mouth of the canals for proper wastewater flow or finally reallocation of slums and squatters. Among the socio-economic approaches development work must be in accordance with the adjoining environment and should not hamper the existing legal constructions, major connecting thorough-fares and supply of basic amenities. Regular local inspections should be done regarding the civic issues, prevention of urban flooding, spread of diseases, road and transport regulation to prevent accidents. Above all some conservational developmental projects may be the best solution for achieving a sustainable city and combat major hydro-geomorphic and environmental hazards. These infrastructural readjustments have been seen as urban renewal which includes construction and proper maintenance of roads, flyovers, pavements; filling up and restoring all potholes, diggings or hollows as soon as possible, and construction of separate waste disposal lines, underground sewerage system, water treatment plants and pumping station at strategic points.

Almost the entire core area of the city today has underground drainage facility, but most of it is heavily silted up, with some of the major trunk sewers to the extent of siltation is as high as 70%, leading to a significant reduction in the hydraulic capacity of the city's sewerage system. Capacity of this drainage system is to be greatly increased. In the extremely congested areas of north and central Kolkata, the old drainage pumping stations have to be upgraded while new portable pumping systems must be sanctioned for post-disaster measures. Some pumps are too old for which spare parts are no longer available and have to be locally fabricated. This reduces pump efficiency and increases power consumption. Pumping station electrical systems are also old and needs immediate replacement. The efficiency of old electrical motors is low compared to their modern counterparts. Bar screens at the inlet to pumping stations are to be manually cleaned and are difficult to keep clean during the high flow periods of the monsoon. A high proportion of plastic rubbish passes through the bar screens and damages pump impellers. Mechanical racked screens are to be introduced in all pumping station. The hydraulics of the major sewers approaching the major pumping stations is unsatisfactory. The sewers flow in a flooded condition and flow velocities are inadequate for self-cleansing.

Currently there are 73 pumping stations (PSs) under KMC (from just 62 in 2013), 4 pumping station under Irrigation and Waterways Department, 4 pumping stations under KMDA, 1 pumping station under Public Health Engineering (PHE) and 4 pumping stations are under Kolkata Metropolitan Water and Sanitation Authority (KMW & SA) to prevent the water logging problem of Kolkata City. Of these the ones at Bangur, Garden Reach, Hatisure near Tiljala, South Suburban East (Keorapukur) and Baghajatin are generally refered to in reports. The existing capacities of these STPs are 52, 57, 10, 45, and 15 MLD respectively. 80% of the city is drained out through lifting pumping stations with the overall capacity of 2878 cusecs and if all major pumping stations of the city are operated in full efficiency, the S.W.F channel has to accommodate a discharge of about 5100 cusec keeping in mind that for 13 h a day high tide prevails when draining out is difficult (Chatterjee,

2015). Drained out waters generated from core area of the city gets discharged to drainage channel onto the EKW. With the rise in pollutants, BOD and coliform bacteria counts in the discharged water, treatment plants have to be even more efficient and active to manage the waste water flow. Areas of high priority include PSs located in Raja Dinendra Street and its surrounding areas, Bidhan Sarani, Thanthania Kalibari, Muktaram Babu street, Eliot Road, Mirza Galib Street, Market street, Abdul Halim street, Raja Rammohan Sarani (Amherst Street), M.G. Road/ Chittaranjan Avenue and its surrounding areas in between M.G. Road and Kalootala Street, Camac Street, Park Street/AJC Bose road, Gokhale Road, Mominpur–Khidirpur area, Kankurgachi area and Ultadanga HUDCO Junction, among others. For houses over 500 sq.m or 13.5 m height, KMC proposes sewerage lines of more than 600 m diameter while trunk sewers range from 1200–2400 mm. Details on the waterlogged patches and the operational capacities of the pumping stations has been reported by KMC every year in their website to generate awareness among the residents (KMC 2017, KMC 2021).

The two sewerage and drainage Master Plans have been completed for Kolkata – one in 1966 by SMPO and another in 2001 with aid from the World Bank by an association of US consulting companies including Weston International and a second after setting up of the Kolkata Metropolitan Planning Committee (KMPC) in 2001. All of these Master Plans have been studied in reviewing the current state of sewerage and drainage channels, but still upgradation of the drainage networks, keeping in mind the altered slope of the canal basins within the KMC, is yet to be undertaken. The original west to east and north to south slopes have been intercepted by multiple high roads that not only raise the surface but also act as embankments to the natural runoff after a heavy downpour. Saving the EKW plays an integral part in the prevention of urban flooding and water treatment process of Kolkata. Adequate ecological flows in the Kulti River, proper management and efficient control of drainage flows to EKW aquaculture farms are of utmost importance in preventing waterlogging, channel degeneration and formation of murky stagnant waters which are the breeding grounds of mosquitoes.

As Kolkata is an old built-up city it is quite impossible to entirely go through urban renewal. Remedial measures have always included two kinds of measures: (a) structural and (b) non-structural. The first and more pressing structural measure is to clean up the sewer drains and rainwater drainage system regularly and most importantly to make it separate. As increasing pattern of rainfall is inducing the tendency for occurrences of urban flood, so urban roof top farming, creating urban farms or urban forestry can help as measures of reducing rainfall impacts. An innovative technique that has come up in the recent past is the harnessing of the rainwater and creating gardens on the terraces of newly built high-rise buildings to intercept the rains for domestic utilisation and reducing runoff. This is known as the 'Green Roof Technology' and is part of the rainwater harvesting program. Permeable pavements can also be a measure to artificially recharge the groundwater. Moreover, the natural water bodies such as canals, ponds, lakes and wetlands must be conserved and maintained. As far as the non-structural measures are concerned, the citizens should be made aware of the consequences of unplanned structures and road layering, improper disposal of wastes and encroachment through different types of awareness campaigns. An alert and effective order by the local government agencies can raise public awareness for judicious disposal of plastics. Prevention of disposing off the untreated solid and dissolved wastes into the canals and unclogging of surface drains can mitigate the problem of waterlogging and overflowing of open drains. Therefore sludge removal and proper solid water disposal and management must be significantly prioritized. Closure of water bodies by land-filling and dumping, concretisation of parks and open plots, undertaking heavy constructions without proper soil testing and corporation sanctions, excessive exploitation of groundwater through illegal deep tube wells should be immediately stopped. Shanties and illegal hutments encroaching over the canals flowing along the eastern margin of the area should be removed and reinstated in proper localities. Regular clearing of muck, weeds and water hyacinths along with de-silting should be undertaken in the open drains and outflow channels. Municipal help must be called for at regular intervals for the maintenance of the underground sewerage lines. As the fringe areas of the city are most affected during the monsoons, a separate drainage for DWF and SWF is highly preferable.

Finally, to make a quick start and successful progress in resolving any urban problem, a good and healthy relation, cooperation and sharing of equal responsibility among the different departments of the government and the municipal corporation such as the various drainage boards, namely Irrigation Department under the Government of West Bengal, Kolkata Metropolitan Development Authority (KMDA), Kolkata Improvement Trust (KIT), KMC, Kolkata Metropolitan Water and Sanitation Authority (KMW & SA), Kolkata Environmental Improvement Project (KEIP) is necessary. Here in the multidisciplinary effort of city planners, Public Works Department (PWD) engineers, environmentalists, NGOs, local residents, officials and labour force, together with the geographers, hydro-geologists, geo-physicists and specialists in geospatial technology, can come together to formulate holistic reports and proposals for a sustainable city.

References

Banerjee, S. (2018). An analysis of urban flooding scenario in the City of Kolkata, West Bengal. *International Journal of Innovative Knowledge Concepts, 6*(5) ISSN: 2454-2415, DOI: 11.25835/IJIK-24.

Banerjee, S., & Sikdar, P. K. (2020). Land Subsidence due to leakage of aquitard-aquifer pore water in an under-construction tunnel of East-West Metro railway project, Kolkata. *Journal of Geological Society of India, 96*(5). https://doi.org/10.1007/s12594-020-1584-z

Bhattacharya, D. (2018). *Empire and ecology in the Bengal Delta: The making of Calcutta* (p. xvi +241). Cambridge University Press. ISBN: 978-1-108-42574-2; 978-1-108-44334-0. $30.21.

Basu, N., Dey, A. and Ghosh, D. (2013). Kolkata's brick sewer renewal: History, challenges and benefits. ICE Proceedings Civil Engineering. 166. 74-81. 10.1680/cien.12.00016. https://www.icevirtuallibrary.com/doi/pdf/10.1680/cien.12.00016 accessed on 30.01.21

Bose, S. (2008). *Adaptive and integrated management of wastewater and strom water drainage in Kolkata- case study of a Mega City* (pp. 341–355). Springer.

Brandolini, P., Cappadonia, C., Luberti, G. M., et al. (2020). Geomorphology of the Anthropocene in Mediterranean urban areas. *Progress in Physical Geography: Earth and Environment, 44*(4), 461–494. https://doi.org/10.1177/0309133319881108

Brown, A. G., Tooth, S., Bullard, J. E., Thomas, D. S. G., Chiverrell, R. C., Plater, A. J., Murton, J., Thorndycraft, V. R., Tarolli, P., Rose, J., Wainwright, J., Downs, P., & Aalto, R. (2017). The geomorphology of the Anthropocene: Emergence, status and implications. *Earth Surface Processes and Landforms, 42*(1), 71–90. https://doi.org/10.1002/esp.3943

Burak, G., Shuaib, L., Masundire, H., Parnell, S., & Seto, K. C. (2017). Urbanisation in Africa: Challenges and opportunities for conservation. *Environmental Research Letters, 13*(1), 015002. https://doi.org/10.1088/1748-9326/aa94fe

Chatterjee, A. (2015). What will be the impact in the drainage system in Kolkata, if we face similar type of torrential rain that occurred during "Nov-Dec.- 2015" in Chennai? https://medium.com/@anjan.chatterjee/what-will-be-the-impact-in-the-drainage-system-in-kolkata-if-we-face-similar-type-of-torrential-554d52097713. accessed on 30.01.21

Cotton, H. E. A. (1980). *Calcutta Old and New*, first published 1909/reprint 1980, pages 103-4 and 221, General Printers and Publishers Pvt. Ltd.

CPHEEO (1993). Design and Construction of Sewers. Central Public Health and Environment Engineering Organization. http://cpheeo.gov.in/upload/uploadfiles/files/engineering_chapter3.pdf accessed on 30.01.21

CSE (2005). Coastal cities- Kolkata: The Waste-Water Portrait. Centre for Science and Environment http://cdn.cseindia.org/userfiles/kolkata_chapter.pdf accessed on 30.01.21

Dasgupta, S. (2020). Empire and ecology in the Bengal Delta: The making of Calcutta. *Conservation and Society, 18*, 200–201. https://doi.org/10.4103/cs.cs_19_148

Downs, P. W., & Piégay, H. (2019). Catchment-scale cumulative impact of human activities on river channels in the late Anthropocene: Implications, limitations, prospect. *Geomorphology, 338*, 88–104. https://doi.org/10.1016/j.geomorph.2019.03.021

Ebisemiju, F. (1989). Patterns of stream channel response to urbanization in the humid tropics and their implications for urban land use planning: A case study from southwestern Nigeria. *Applied Geography, 9*(4), 273–286. https://doi.org/10.1016/0143-6228(89)90028-3

Eldho, T. I., Zope, P. E., & Kulkarni, A. T. (2018). Urban flood management in coastal regions using numerical simulation and geographic information system. In *Integrating disaster science and management* (pp. 205–219, ISBN 9780128120569). Elsevier. https://doi.org/10.1016/B978-0-12-812056-9.00012-9

Estoque, R., & Murayama, Y. (2015). Intensity and spatial pattern of urban land changes in the megacities of Southeast Asia. *Land Use Policy, 48*, 213–222. https://doi.org/10.1016/j.landusepol.2015.05.017

Fan, P., Chen, J., & John, R. (2016). Urbanization and environmental change during the economic transition on the Mongolian Plateau: Hohhot and Ulaanbaatar. *Environmental Research, 144* (Part B), 96–112. https://doi.org/10.1016/j.envres.2015.09.020

Goode S. W. (1916) Municipal Calcutta: its Institutions in their Growth and Origin. T & A Constable, Edinburgh, UK, reprinted in 2005 by Kolkata Municipal Corporation and MacMillan, New Delhi, India

Gupta, A. K., & Nair, S. S. (2010). Flood risk and context of land-uses: Chennai city case. *Journal of Geography and Regional Planning, 3*(12), 365–372.

Haldar, A. (2020). Canal degeneration and urban flooding in selected areas of Kolkata Municipal Corporation (KMC): A study along the Churial Basin. In S. Gupta (Ed.), *Water Resource and Sustainability*. SAIARD.

Haldar, A., & Satpati, L. N. (2018). Urban Geo-forms: Concepts and Significance in Anthropogeomorphology. *Journal of Indian Geomorphology, 6*. ISSN 2320-073, 108–115.

Haque, S., Amin, S. B., & Karim, S. R. (2018). Disaster Risk and Impact of Urbanization of Megacities in Bangladesh: Evidence from Dhaka and Chittagong. *Journal of Accounting, Finance and Economics, 8*(3) 35–57.

Kalnay, E., & Cai, M. (2003). Impact of urbanization and land-use change on climate. *Nature, 423*, 528–531. https://doi.org/10.1038/nature01675

KEIIP (2012). Kolkata Environmental Improvement Investment Program (Tranche 1): Sewerage and Drainage Subproject. May 2012. Published by KMC, Government of West Bengal for the Asian Development Bank. https://www.adb.org/sites/default/files/project-document/73264/42266-023-ind-iee-02.pdf accessed on 30.01.21

KEIIP (2019). IND: Kolkata Environmental Improvement Investment Program Tranche 3 – Sewerage and Drainage Network. Project Number: 42266-026 February 2019. Published by KMC, Government of West Bengal for the Asian Development Bank. https://www.keiip.in/pdf/42266-026-iee-en_2.pdf accessed on 30.01.21

KMC (2017). Action Plan to Mitigate Flood, Cyclone & Water Logging. Kolkata Municipal Corporation. https://www.kmcgov.in/KMCPortal/downloads/water_logging_09_06_2017.pdf. accessed on 30.01.21

KMC (2021). Action Plan to Mitigate Flood, Cyclone & Water Logging. Kolkata Municipal Corporation. https://www.kmcgov.in/KMCPortal/downloads/Monsoon_Book_2021_07_06_2021.pdf accessed on 30.01.21

KMC (2020). City Disaster Management Plan of Kolkata. Kolkata Municipal Corporation. http://wbdmd.gov.in/writereaddata/uploaded/DP/DPKolkata71707.pdf accessed on 25.07.21

KMC-CGWB (2007). Ground Water Information Booklet Kolkata Municipal Corporation, West Bengal. Prepared by Central Ground Water Board for Kolkata Municipal Corporation. http://cgwb.gov.in/District_Profile/WestBangal/Kolkata%20Municipal%20Corporation.pdf accessed on 30.01.21

KMC portal (2021). Kolkata Municipal Corporation website. https://www.kmcgov.in/KMCPortal/jsp/SewerageAndDrainageServices.html accessed on 30.01.21

Liu, J., Shao, W., Xiang, C., Mei, C., & Li, Z. (2020). Uncertainties of urban flood modeling: Influence of parameters for different underlying surfaces. *Environmental Research, Volume, 182*, 108929., ISSN 0013-9351. https://doi.org/10.1016/j.envres.2019.108929

Majumdar, S., & Chatterjee, U. (2020). Modelling urban growth using Urban growth deterministic model in Kolkata Metropolitan Area: A geo-statistical approach. *Modeling Earth Systems and Environment*. https://doi.org/10.1007/s40808-020-00985-6

Malik, S., Pal, S. C., Sattar, A., Singh, S. K., Das, B., Chakrabortty, R., & Mohammad, P. (2020). Trend of extreme rainfall events using suitable Global Circulation Model to combat the water logging condition in Kolkata Metropolitan Area. *Urban Climate, 32*, 100599., ISSN 2212-0955. https://doi.org/10.1016/j.uclim.2020.100599

Minnig, M., Moeck, C., Radny, D., & Schirmer, M. (2018). Impact of urbanization on groundwater recharge rates in Dübendorf, Switzerland. *Journal of Hydrology, 563*, 1135–1146. https://doi.org/10.1016/j.jhydrol.2017.09.058

Mitra, T. (2011). *Integrating grey zones in the city fabric a study in transformation of housing dynamics in Kolkata India*. Ph.D. thesis, Department of Architecture, Jadavpur University, p. 35. http://hdl.handle.net/10603/157389

Mukherjee, S., Bebermeier, W., & Schütt, B. (2018). An overview of the impacts of land use land cover changes (1980–2014) on urban water security of Kolkata. *Land, 7*(3), 91. https://doi.org/10.3390/land7030091

Mukherjee, S. (2002). Problem of waterlogging in Calcutta. In S. R. Basu (Ed.), *Changing environmental scenario of Indian Subcontinent* (pp. 457–459). ACB Publications.

Nair, P. T. (1995). The growth and development of old Calcutta. In S. Chaudhuri (Ed.), *Calcutta, the Living City* (Vol. I, pp. 14–15). Oxford University Press.

Ning, Y.-F., Dong, W.-Y., & Lin, L.-S. (2017). Analyzing the causes of urban waterlogging and sponge city technology in China. *IOP Conference Series: Earth and Environmental Science, 59*, 012047.

Price Simon, J., Ford Jonathan, R., Cooper Anthony, H., & Catherine, N. (2011). Humans as major geological and geomorphological agents in the Anthropocene: The significance of artificial ground in Great Britain. *Philosophical Transactions of the Royal Society A, 369*, 1056–1084. https://doi.org/10.1098/rsta.2010.0296

Rafiq, F., Ahmed, S., Ahmad, S., & Khan, A. A. (2016). Urban floods in India. *International Journal of Scientific & Engineering Research, 7*(1), 721–734.

Ray, B., & Shaw, R. (2016). 20 – Water Stress in the Megacity of Kolkata, India, and Its Implications for Urban Resilience. In *Urban Disasters and Resilience in Asia* (pp. 317–336, ISBN 9780128021699). Butterworth-Heinemann. https://doi.org/10.1016/B978-0-12-802169-9.00020-3

Roy Basu, A., Bharat, G. K., Chakraborty, P., & Sarkar, S. K. (2020). Adaptive co-management model for the East Kolkata wetlands: A sustainable solution to manage the rapid ecological transformation of a peri-urban landscape. *Science of The Total Environment, 698*, 134203., ISSN 0048-9697. https://doi.org/10.1016/j.scitotenv.2019.134203

Roy, A. K., Mukherjea, A., Talapatro, A. K., Mitra, A. K., Banerjee, A. S., & Sengupta, S. (2017). *Restoration and rejuvenation of Adi Ganga-Tolly Nallah for its socio-economic Impacts on the development of West Bengal.* Calcutta University Geology association.

Rumbach, A. (2017). At the roots of urban disasters: Planning and uneven geographies of risk in Kolkata, India. *Journal of Urban Affairs, 39*(6), 783–799. https://doi.org/10.1080/07352166.2017.1282771

Saha, P. (1998). Canal system and wetlands in the environment of Calcutta and its prospective utilisation. *Geographical Review of India, 60*(3), 265–268.

Sahu, P., Michael, H. A., Voss, C. I., & Sikdar, P. K. (2013). Impacts on groundwater recharge areas of megacity pumping: analysis of potential contamination of Kolkata, India, water supply. *Hydrological Sciences Journal, 58*(6), 1340–1360. https://doi.org/10.1080/02626667.2013.813946

Satpati, L. N., & Haldar, A. (2020). Neoliberal urban sustainability in Old Kolkata, India: Case studies of contested developments. *Regional Science Policy & Practice, 2020*, 1–17. https://doi.org/10.1111/rsp3.12325

Sen, S. (2011). Effect of urban sprawl on human habitation in urban fringe and peri-urban areas in Kolkata Metropolitan Area. *Institution of Town Planners, 8*, 58–66.

Tawhid, K. G. (2004). *Causes and effects of water logging in Dhaka City, Bangladesh.* Department of Land and Water Resource Engineering, KTH.

Youpeng, X., Jintao, X., Jinjia, D., Ying, C., Yixing, Y., Xingqi, Z. (2010). Impacts of urbanization on hydrology in the Yangtze River Delta, China. Water Science and Technology, 62(6):1221–1229. https://doi.org/10.2166/wst.2010.391. PMID: 20861534.

Chapter 26
Monitoring Land Use/Cover Change and Urban Sprawl Using Remote Sensing Data: A Study of Siliguri and Raiganj Urban Agglomerations, India

Bhaswati Roy and Nuruzzaman Kasemi

26.1 Introduction

The study of the spatial outline of urban expansion and urban dynamics is a crucial part of Urban Geography. Urban development is one of the processes of built-up enlargement (Bhatta, 2010; Dadras et al., 2014). There are three different and widely deliberated concepts of urban expansion in the urban and regional planning literature related to the transformation of population, cost-effective performance, and the spatial expansion of built-up areas (Reis et al., 2016). Urban growth is a significant reason for urban sprawl. While cities grow, they enlarge around their peripheral areas. The main dissimilarity between urban expansion and urbanization is that urban growth or expansion refers to a universal increase in either the population size or the land area of an urban area over some time. In contrast, urbanization is the virtual proportion of the country's population living within urban areas. Urban agglomeration (UA) is an extended city or town area that comprises a built-up area of a central place and suburbs connected by continuous urban area. The term (UA) is referred to as the concept of "town group" recognized by the Census of India in 1951 (Verma, 2008). The concept was adopted for 1981, 1991, and 2001 Censuses, and the Census of India maintained the same in the 2011 Census. In general, the term urban agglomeration (UA) is the propensity wherein a census town or the number of less significant towns agglomerate around a significant or major municipality town or city (Shaw, 2005). It is a nonstop urban increase and usually consists of a town or a city and its adjoining outgrowths. It refers to the contiguous built-up area of a town or city. It is a part of urban expansion. Researchers have devised a variety of indexes and models integrated with GIS and remote sensing for

B. Roy (✉) · N. Kasemi
Department of Geography, Raiganj University, Raiganj, West Bengal, India

G. S. Bhunia et al. (eds.), *Anthropogeomorphology*, Geography of the Physical Environment, https://doi.org/10.1007/978-3-030-77572-8_26

quantifying the built-up expansion processes of different cities. Although India is considered a rural country, there is a crystal-clear urban course. The dimension of urban enlargement is quite alarming in India. India's urban population increased drastically from 217 million in 1991 to 377.1 million in 2011, whereas the number of towns has increased from 4689 (1991) to 7935 (2011) (Mukherjee & Singh, 2020). The occurrence of urban sprawl is one of the key effects of land cover change in India. Several definitions for sprawl have been put forward that describe sprawl as a specific form of urban development with dispersed, low-density, auto-dependent, environmentally, and socially impacting qualities (Burchell & Shad, 1999; Galster et al., 2001). Hasse and Lathrop (2003) and Gadakh and Jaybhaye (2015) described urban sprawls as dispersed and inefficient urban expansion. Sudhira et al. (2004) and Ramachandra et al. (2014) identified urban sprawls as uncontrolled and unplanned growth of city boundaries. Karakayaci (2016) and Majumdar and Chatterjee (2020) considered urban sprawls as hinterland among the countryside and urban areas. Even though of all the concerns the concept of sprawl suffers from a crisis of not having any commonly accepted definition. Therefore, rapid growth urban sprawls have major impacts on environmental degradation and socioeconomic condition (Ewing et al., 2002) such as loss of biodiversity (Eko & Ayama, 2013; Forys & Allen, 2005), an increase of greenhouse gas emission (Glaeser & Kahn, 2004; Sung et al., 2013), and enhancing the infrastructural and public service cost (Burchell & Shad, 1999; Liu et al., 2018). It also causes traffic jamming (Ewing et al., 2003). It also has an impact on income inequality (Carruthers & Ulfarsson, 2003). Thus, urban sprawls are an important field for urban planners and policymakers to be concerned about urban areas' actual and absolute development. Presently, most of the scholars emphasize remote sensing and GIS combined with statistical techniques to monitor and analyze urban growths and urban sprawls (Sun et al., 2013; Katyambo & Ngigi, 2017; El Garouani et al., 2017; Shukla & Jain, 2019). Land use/land cover (LULC) change detection (Pal & Ziaul, 2017; Cheruto et al., 2016; Mishra et al., 2019; Reis, 2008; Khan & Jhariya, 2016; Sreenivasulu & Bhaskar, 2010) is necessary to examine and investigate the urban growth. In the recent past, many kinds of researches have been conducted on land use/land cover analysis. Scholars have attempted to depict and identify the changing pattern of land use all over the world. However, scholars are still in need of comprehensive as well as holistic research on LULC. There is increased use of satellite imagery to measure urbanization by quantifying LULC alteration for the whole city region (Moskal et al., 2011). Pauleit et al. (2005) examined the transformation of LULC of 11 selected residential areas of Merseyside, UK, with aerial photographs obtained in 1975 and 2000. Indian scholars have conducted their work on land useland cover change detection, especially by applying remote sensing and GIS technique. Mallupattu and Sreenivasula Reddy (2013) have focused on the LULC change of Triputi, the world-famous pilgrim site of India, using remote sensing and GIS techniques. They showed that non-agricultural land increase gradually compared to agricultural land and the built-up area increased from 5.91 km^2 (1976) to 18.34 km^2 (2013). Rawat and Kumar (2015) have carried out their study on the LULC change of Hawalbagh block Almora district in Uttarakhand based on 20 years of satellite data from 1990 to

2010. They observed that agricultural land (1.52%) and water bodies (0.08%) decreased in this area, whereas built-up area increased 3.55%. In this study, pixel-based comparison has been done to construct change information, thus interpreting the change more efficiently. Mukherjee and Singh (2020) examined the spatio-temporal change in land use along with vegetation cover and its impact on land surface temperature in two cities of India. They observed that urban growth has a major impact on surface temperature increase. Mandal et al. (2019) studied the LULC change along with urban dynamics of mega city Kolkata and its surrounding environment. The result of this analysis showed that highest occurred in the north–north–west (NNW) direction while the growth is least in the east–north–east (ENE) direction. Studies have been conducted in India on changing land use/land cover of cities, but most of them are in major metropolitan cities such as Delhi, Kolkata, and Chennai. Such kinds of researches carried out in smaller towns are comparatively lesser in number. But presently the changing pattern of land use/land cover as well as urban growths is also an alarming issue to the smaller and medium cities as well as urban agglomerations. This study attempted to fill the void by selecting Siliguri, a Class I major town in the Darjeeling district, and Raiganj, a smaller town in the Uttar Dinajpur district (Fig. 26.1).

The main objectives of this study are to investigate the characteristics of changes in the LULC and monitor urban sprawl using remote sensing data. Distinctively the objectives are (i) to explore the LULC changes of two agglomerations between 1990/1991 to 2019, (ii) analyze the spatial, temporal variation of urban growth in this period, (iii) examine the urban sprawls in two agglomerations by using remote sensing data. This assessment is very useful for urban planners and policymakers to improve the urban environment's superiority in the growing towns and cities.

26.2 Study Area

The present study has been carried out in two agglomerations of the North Bengal region with contrasting environmental conditions. River Ganga separates West Bengal into two parts. The northern part of West Bengal, to the north of River Ganga, is traditionally known as North Bengal. There are two divisions in North Bengal, viz., Malda Division (comprising districts of Uttar Dinajpur, Dakshin Dinajpur, and Malda) and Jalpaiguri Division (comprising districts of Alipurduar, Koch Bihar, Darjeeling, Jalpaiguri, and Kalimpong). The percentage of urban population is higher in West Bengal (31.89%) than in the country (31.16%) (Mandal, 2017). As per the 2011 Census, there are 26 urban agglomerations in West Bengal, and each agglomeration has a population above 100,000. In North Bengal, eight agglomerations have a population of 1,925,322. This study brings out the extended pattern of urban growth of Siliguri and Raiganj agglomeration (UA) (Table 26.1), around 3 km from its outer limit of official jurisdiction with remote sensing GIS.

Siliguri (UA), situated on the banks of the Mahananda River, is known as the entrance to northeast India. The geographical location of Siliguri (UA) is

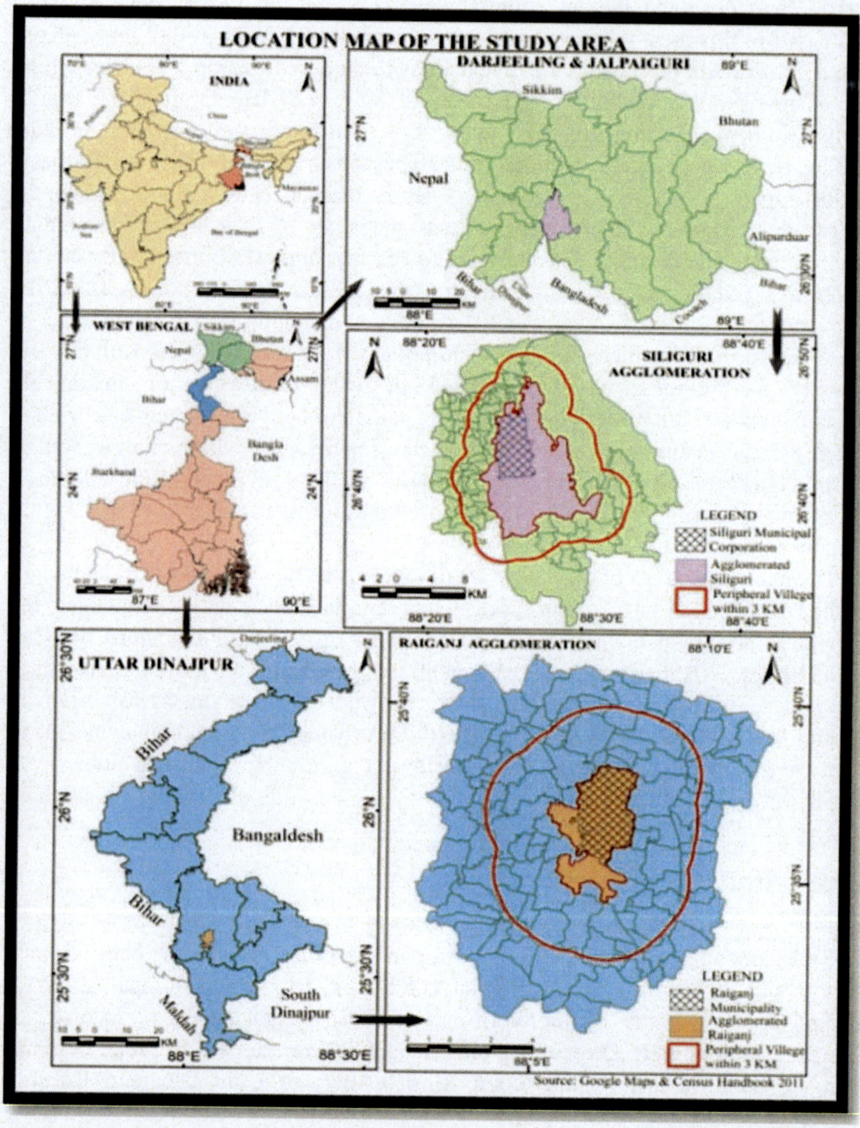

Fig. 26.1 Location map of the study area

26°46′10″N to 26°40′42″N latitude and 88°23′43″E to 88°27′34″E longitude. Due to huge population pressure, the peripheral areas are being converted into urban landuse and have become a noticeable phenomenon in the region. Siliguri (UA) has been experiencing faster urban growth at an alarming rate due to two forces: (i) migratory movements and (ii) deviation in the rate of natural increase. Raiganj UA is located in the southwestern part of Uttar Dinajpur district, and River Kulik

Table 26.1 Basic characteristics of two agglomerations

Name	Name of constituents of UA	District	Population
Siliguri	Siliguri (M Corp.), Kalkut (p) (CT), Dabgram (P) (CT), Binnaguri (CT), Chakiabhita (CT)	Darjeeling	7,01,489
Raiganj	Raiganj (M), Kasba (CT), Nachhratpur Katabari (CT)	Uttar Dinajpur	1,99,758

Source: Census of India 2011

(a tributary of River Nagor) bounds it from northwest to southeast. The town achieved municipal status on 15 August 1951 and is also familiar with Raiganj wildlife sanctuary (commonly known as Kulik Bird Sanctuary), the second biggest bird sanctuary in Asia. The absolute location of this urban agglomeration lies between 88°6′23.812″ E to 88°9′5.932″ E longitude and 25°38′27.102″ N to 25°34′57.153″ N latitude.

26.3 Materials and Methods

Landsat-5 Thematic Mapper (TM) and Landsat-8 Operational Land imager (OLI) at a resolution 30 m of 1990/1991 and 2019 have been used for the study. Four cloud-free satellite images of 139 paths and 42 rows for Raiganj and 139/4142 path/row for the Siliguri region were downloaded from the United States Geological Survey (USGS) website (https://earthexplorer.usgs.gov) to view the land use/cover of the study area. The obtained images were preproceeding with lots of corrections to enhance image quality. The executive maps of Siliguri and Raiganj urban agglomeration and its surrounding peripheral villages have been obtained from the administrative Atlas of West Bengal (Census of India 2011). The executive maps of Siliguri and Raiganj (UA) have been registered in universal Transverse Mercator (UTM) zone 45N and WGS-1984 datum. The rectified map has been used to subset the processed images.

26.3.1 Image Classification and Change Detection

The supervised classification method vis-a-vis hard classification method with maximum likelihood algorithm was used in the present study in ERDAS imagine 14 software. The maximum likelihood algorithm (MLC) is one of the trendiest supervised classification methods based on probability. The fundamental theory presumes that the probability is identical for all classes and that the input bands have regular or normal distributions. The image interpretation key is an essential element for image interpretation or LULC classification (Table 26.2). With the help of image interpretation key and field investigation, the whole study area is

Table 26.2 Land use and land cover class categories for two selected agglomerations

Sl. No.	LULC category	Areas included	Urban agglomeration
1	Built-up area	Residential area, homestead with plantation, industrial area, all paved surface	For Siliguri and Raiganj (UA)
2	Fallow land	Uncultivated land, temporary vacant crop field	For Siliguri and Raiganj (UA)
3	Agricultural land	Cropland, plantation, and other cultivated area	For Siliguri and Raiganj (UA)
4	Tea plantation	Tea gardening area	For Siliguri (UA)
5	Vegetation	Forest, scrubs, grass land, etc.	For Raiganj (UA)
6	Forest	Forest, grass land, other vegetation, etc.	For Siliguri (UA)
7	Water bodies	River, stream, lake, reserves, tank, canal, water logged land	For Siliguri and Raiganj (UA)

Source: Area included by the authors

categorized into six for Siliguri and five for Raiganj, such as the built-up area, water bodies, agricultural land, fallow land, vegetation cover, tea plantation, and forest. The whole classification process was conducted in ERDAS imagine 14 and Arc GIS 10.5 software.

In this chapter, the postclassification change detection method has been used to detect the nature and rate of change, including location. For obtaining the spatial modification in LULC, two-way cross-matrix has been applied. Cross-tabulation study on a pixel-by-pixel source helped to determine the extent of adaptations from an exacting land cover category to other land use or land cover categories and their equivalent area over the time estimated. A conversion or dynamic map of two agglomerations has been produced, which describes the "-from" "-to" changes of different classes.

26.3.2 Accuracy Assessments

As a general principle, the level of accuracy accessible in the classification of remote sensing depends on different factors like the volume, shape, distribution, suitability of training sites, and the frequency of occurrence of particular areas designated to each class, along with the degree to which pixels are mixed, the efficiency and resolution of the sensor, the methods involved in classification, the precision of ground-truthing, etc. (Bhatta et al., 2010). The accuracy assessment method applied for LULC and field survey has been done to compare reality and output from image classification. The producer's accuracy, user's accuracy, and kappa coefficient (K) have been calculated for evaluating the overall accuracy level of the image. The following formula has been used for the accuracy assessment method.

$$\textbf{Producer's accuracy (PA)} = 100\% - \text{error of omission}(\%) \tag{26.1}$$

$$\textbf{User's accuracy (UA)} = 100\% - \text{error of commission}\ (\%) \tag{26.2}$$

$$\textbf{Kappa coefficient (K)} = \frac{\sum \mathrm{a} - \sum \mathrm{ef}}{\mathrm{N} - \sum \mathrm{ef}} \tag{26.3}$$

$$\textbf{Overall accuracy (OA)} = \text{Total number of correct samples} - 100\%\text{total number of samples} \tag{26.4}$$

In this study, for each sequential period, 400 control points were generated to assess the accuracy level. Each point was compared with the features present on the ground using auxiliary GPS survey data, topographic map, and Google Earth, as well as field survey data.

26.3.3 Identification of Urban Sprawls

Urban sprawl is not sporadic, it is a continuous and temporal process, so it is very difficult to measure. Shannon entropy is a commonly used technique for precise measures of urban sprawls. The value of entropy goes ahead from 0 to 1, where 0 signifies the compact allocation of built-up area and values close to 1 indicate the dispersed allocation of built-up area. Thus, higher entropy indicates higher urban sprawl occurrences. In this study, the Shannon entropy value was calculated by using the following formula:

$$Hn = -\sum_{i=1}^{n} P_i \log_e(P_i) \tag{26.5}$$

where P_i is the proportion of the variable in the ith zone, and n is the total number of the zone.

26.4 Results and Discussion

26.4.1 LULC Change and Urban Dynamicity

Land use and land cover (LULC) transformations have long-term implications on the environment at local and global levels. Transformation in land cover reveals changes in soil quality, biotic diversity, agricultural productivity, sedimentation rates, runoff rate, etc., that cannot be well implicit without the proper knowledge of land use/land cover changes that drive this phenomenon. The world's total natural environmental areas, including soil, climate, hydrology, and vegetation, are considered as land. The term land cover generally refers to the arrangement of soil/vegetation system or the biophysical materials (such as forest, water body, cropland, etc.) on the Earth's surface. On the other hand, land use refers to the way in which human societies use this land cover type. For example, the forest can be used for timber production. In short, land use refers to the individual use of land at a particular point in time. Although LULC is considered a key factor in the study of global change, the goals of changing land use vary from developed to developing countries. In developed nations, land use transformation depends on economic reasons, but in the case of developing countries, poverty, the rapid growth of population, and economic status are the main driving forces of changing land use. The effect of rising population and growing socioeconomic requirements on LULC creates uncontrolled and unplanned changes in LULC (Reis, 2008). Land use/land cover changes have a significant impact on global climate changes. The surrounding areas of Siliguri and Raiganj agglomerations have experienced a rapid and substantial change in their LULC configuration. The towns of Siliguri and Raiganj are growing physically every year to accomplish radical and multidimensional activities. The vegetation and agricultural land in both regions has been affected by this fastest expansion of land use that began with economic upbringing.

Siliguri (UA)

At the beginning of the twentieth century, Siliguri was an unknown small village of West Bengal, but at present, it is one of the biggest metropolises of eastern India. The increasing trend of urban growth has been detected in Siliguri (UA) from 1990 to 2019 in the extent of 29 years. Figure 26.2 clearly shows the faster growth of the study area. The urban area had increased from 32.86 km^2 in 1990 to 61.16 km^2 in 2019. As a result of infilling growth and the creation of impervious surfaces, arable land has decreased significantly. There have been massive losses in the fallow land during the last 29 years; it decreased by 3.37%. In recent decades, progress in real estate property has flourished in Siliguri, and property developers have encroached agricultural land and wetlands without considering the associated ecological costs. Agricultural land has been reduced by 5.63% within this period. Besides, many individual households developed in fringe areas. The tea plantation area increased by 2.92%, boosting Siliguri and its surrounding area's economic growth. Table 26.3 and Fig. 26.3 illustrate the result of LULC change within this period.

"From" and "to" changes or computation of the transition matrix are very useful for analyzing the land cover transition's accurate result. In the present study, LULC change detection matrices have been generated from 1990 to 2019 and are shown in Table 26.4. The transition matrices indicate that between 1990 and 2019, 31.95 km^2 of agricultural land has been transformed into a built-up area. During the same period, 36.84 km^2 of agricultural land has been converted into tea plantations. Darjeeling district has more than 100 tea plantations scaling thousands of acres of land, and the principal economy of this area depends on tea production, agriculture, and forestry. For this study, tea plantation is, therefore, considered an important

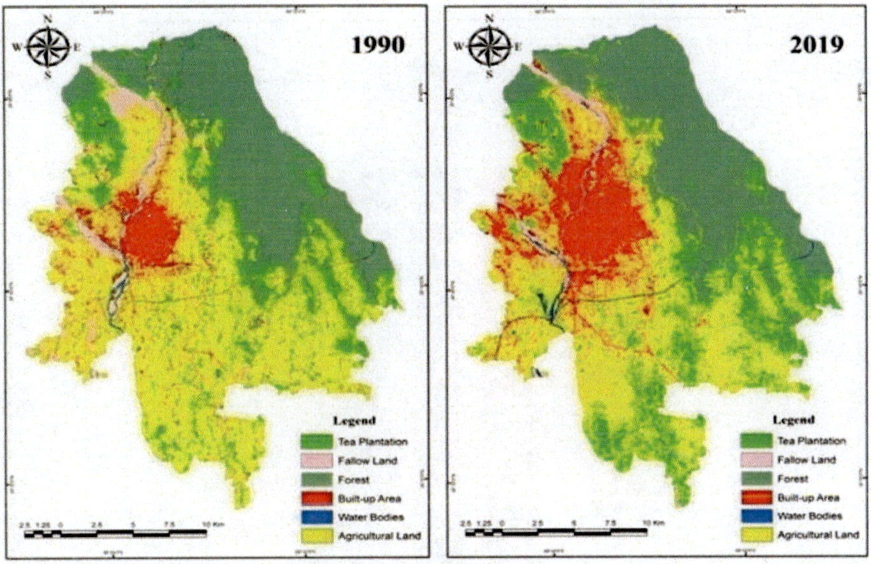

Fig. 26.2 Land use/land cover map of Siliguri (1990–2019) (UA)

Table 26.3 Result of LULC classification for 1990 and 2019

Land use/land cover category	1990		2019		Change 1990–2019		Remarks
	km^2	%	km^2	%	km^2	%	
Built-up area	32.86	6.20	61.16	11.55	28.86	5.45	Increase
Tea plantation	53.38	10.08	68.82	12.99	15.44	2.92	Increase
Forest	146.34	27.63	153.44	28.97	7.10	1.34	Increase
Agricultural land	255.8	48.29	225.97	42.66	−29.83	−5.63	Decrease
Fallow land	30.96	5.85	13.11	2.48	−17.85	−3.37	Decrease
Water bodies	10.36	1.96	7.2	1.36	−3.16	−0.60	Decrease
Total area	529.7	100	529.7	100			

Source: Classified and computed by authors from Landsat-5 and Landsat-8 images

category of land use. 6.04 km^2 areas have been converted from the category of fallow land into built-up land. From the other land use transformation, it is observed that 3.49 km^2 and 1.52 km^2 water bodies have been converted to agricultural land and built-up area. A total of 363.03 km^2 area remains unchanged during this period. The rigorous assessment of the conversion map clarifies that all land use categories have significantly changed from 1990 to 2019. The built-up area increased at the rate of a decrease in agricultural land, fallow land, forest, and water bodies (Figs. 26.4 and 26.5). In contrast, fallow land has decreased and transformed into agricultural land, built-up, forest, tea plantation, and water bodies.

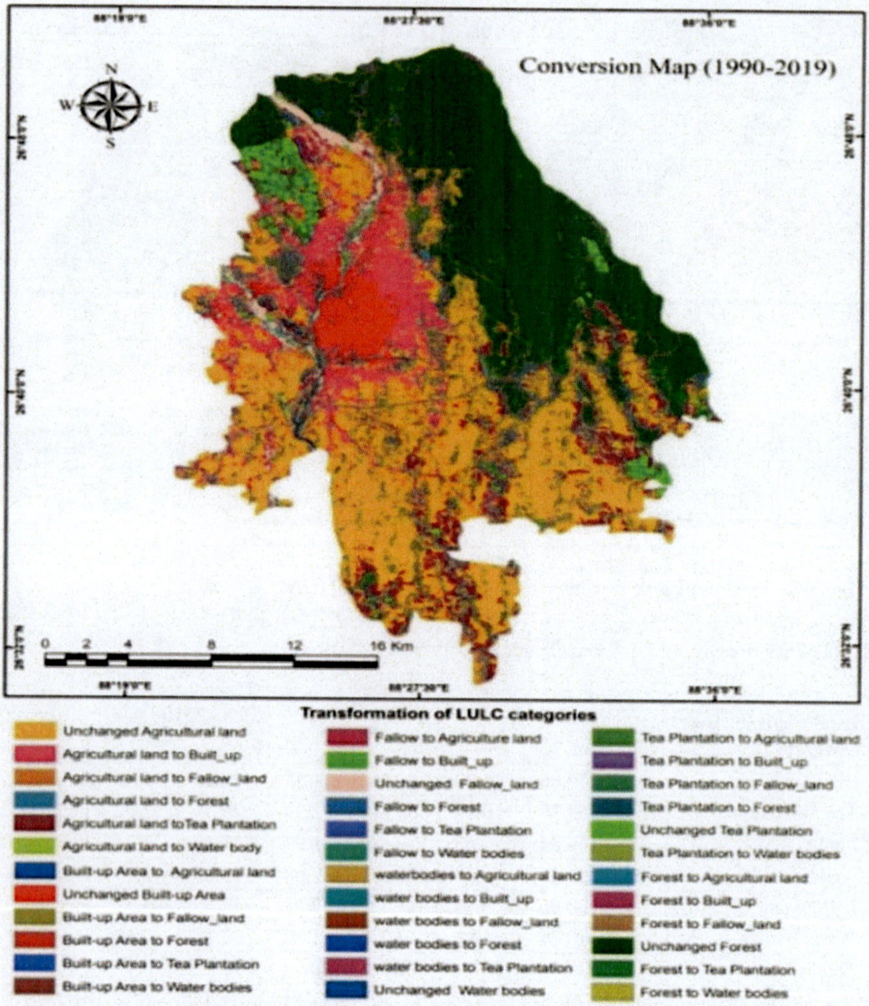

Fig. 26.3 Conversion map of Siliguri urban agglomeration (1990–2019)

Raiganj (UA)

Rapid urbanization in the last few years is the main cause of LULC changes in Raiganj agglomeration. As per Census 2011 (Census of India, 2011), the population of the Raiganj municipality increased from 165,045 in 1991 to 183,682 in 2011.

This growth of the urban population pushed up the high demands for land for residential purposes. Accordingly, the city expanded in different directions without following any comprehensive planning guidelines. Urban features, mainly the built-up area, show significant changes during the study period. It increased by 16.47% within this period. Significant negative changes have been recorded during the past

Table 26.4 Transition matrix for Siliguri (UA)

		LULC time t^2 2019						
	Area in km^2	Built up area	Tea plantation	Forest	Agricultural Land	Fallow land	Water bodies	Total 1990
LULC time t^1 1990	Built up area	19.5	0.64	1.53	9.28	0.7	1.21	32.86
	Tea plantation	1.95	22.06	10.84	18.04	0.37	0.12	53.38
	Forest	0.2	7.06	135.72	3.2	0.06	0.1	146.34
	Agricultural Land	31.95	36.84	3.2	177.16	4.9	1.75	255.8
	Fallow Land	6.04	1.79	0.3	14.8	6.3	1.73	30.96
	Water bodies	1.52	0.43	1.85	3.49	0.78	2.29	10.36
	Total 2019	61.16	68.82	153.44	225.97	13.11	7.2	529.7

Source: Computed by the authors from Landsat-5 and Landsat-8 images

Fig. 26.4 Uttorayon township (Siliguri) developed on the outskirts of thetown

28 years in agricultural land, vegetation cover, and water bodies. Table 26.5 illustrates the land use/land cover changes of Raiganj agglomeration and its surrounding area.

In the year 1991, agricultural land has been found as a dominant class out of five land use classes. It covered 50.61% of the city's area, followed by vegetation, fallow land, water bodies, and built-up area, having 20.52%, 19.31%, 5.28%, and 4.76%, respectively. In 2019, the built-up area increased by 17.65 km^2 (15.95 %) and fallow land increased by 6.7 km^2 (6.03%). The built-up area is considered as the dominant class due to rapid urbanization within this agglomeration boundary. Agricultural land, vegetation, and water bodies have decreased, that is,, −15.47%, −6.50%, and

Fig. 26.5 Residential complex is coming up in vacant land of Shivmandir Siliguri

Table 26.5 Result of LULC classification for 1991 and 2019

Land use/land cover category	1991		2019		Change 1991–2019		Remarks
	km^2	%	km^2	%	km^2	%	
Built-up area	4.76	4.30	22.41	20.25	17.65	15.95	Increase
Vegetation	22.73	20.52	15.54	14.02	−7.19	−6.50	Decrease
Agricultural land	56.08	50.61	38.93	35.13	−17.15	−15.48	Decrease
Fallow land	21.38	19.31	28.08	25.34	6.7	6.03	Increase
Water bodies	5.85	5.28	5.84	5.27	−0.01	−0.01	Decrease
Total	110.80	100	110.80	100			

Source: Computed by the authors from Landsat-5 and Landsat-8 images

−0.01%. Mainly the arable land is encroached by built-up area and fallow land within these 28 years (Fig. 26.6).

Figure 26.7 and Table 26.6 illustrate the conversion of land use change from each category. During the period 1991–2019, there is remarkable transformation observed from agricultural land to the built-up area (Fig. 26.8). About 7.72 km^2 of the agricultural land has been converted into a built-up area. About 8.29 km^2 of vegetation cover has been converted into built-up area, and conversion of fallow land and water bodies into built-up area accounts for 2.8 km^2 and 0.46 km^2, respectively. About 1.44 km^2 of water bodies have been converted into agricultural land, 0.46 km^2 into built-up area, 0.39 km^2 into vegetation, and 1.65 km^2 into fallow land. A total of 46.49 km^2 area remains unchanged during this period.

During the last two decades, both the regions have witnessed speedy population growth and built-up expansion, resulting in higher demand for food, shelter, and

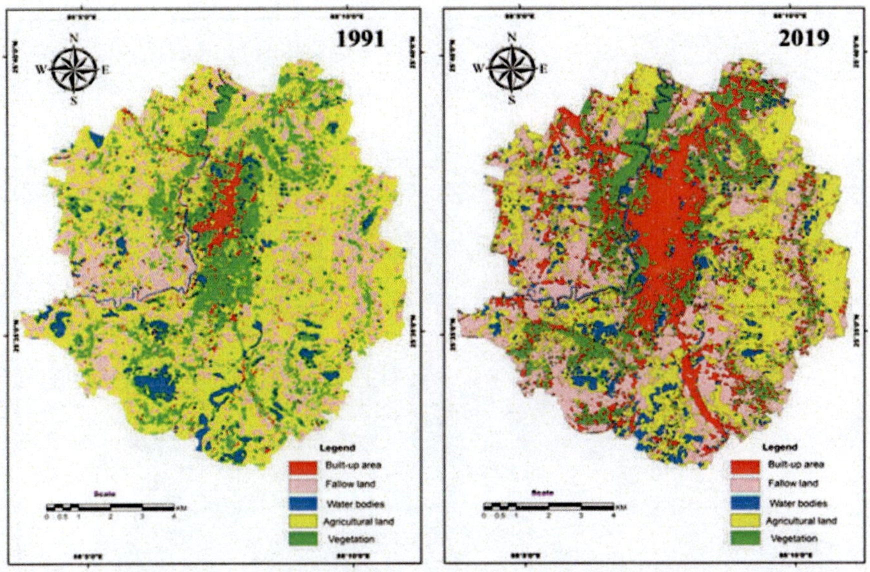

Fig 26.6 Land use/land cover of Raiganj (UA)

other essentials (Figs. 26.4 and 26.9). These have led to an increase in the volume of solid waste in the study area. However, the management and disposal of these wastes have not been done properly, so it has become a serious environmental hazard affecting both humans and nature. At present, the dumping site of Raiganj municipality is located near the River Kulik (Fig. 26.10), which degrades the river health and surrounding environment. The urban agglomeration of Siliguri is one of the largest and thriving phenomena in North Bengal, having great impetus on the regional economic structure. The growth of both the cities of Siliguri and Raiganj has an unprecedented impact on the surrounding areas. Outward urban development with the conversion of cropland and green vegetation is the main problem of the two agglomerations and their periphery at present.

26.4.2 Accuracy Assessment

The error matrix was applied to calculate the accuracy of mapping. The overall accuracy only considers the diagonal elements in the matrix. The kappa coefficient method has been used, which ranges from 0 to 1. Value 1 represents the perfect assessment, and less than 1 indicates a lower ideal agreement. The overall accuracies calculated for Siliguri (UA) in 1990 and 2019 are 89.9 and 91.32, respectively. The kappa coefficients for 1990 and 2019 are 0.89 and 0.95, respectively, which is acceptable for such change analysis (Table 26.7).

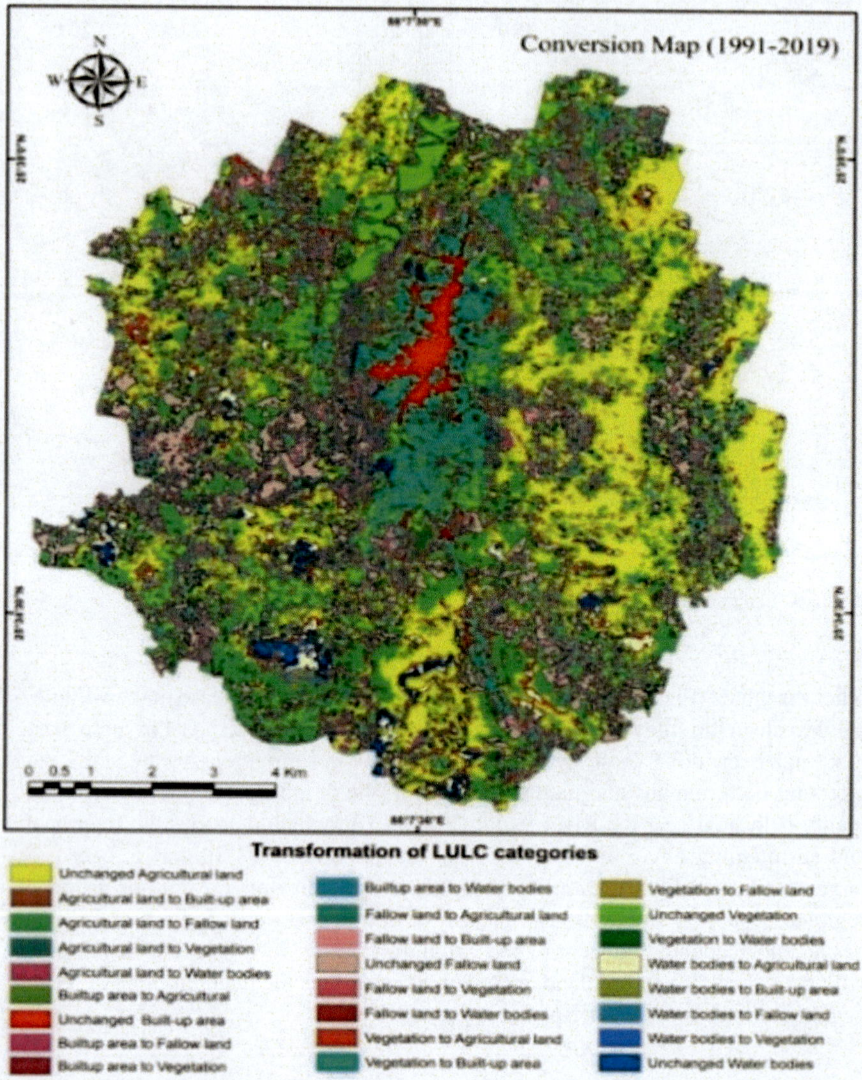

Fig. 26.7 Conversion map of Raiganj urban agglomeration (1991–2019)

In the case of the Raiganj (UA) image postclassification technique, the overall accuracies for 1991 and 2019 are found to be 99.9 and 97.32, respectively (Table 26.8). The kappa coefficient values are 0.99 (1990) and 0.97 (2019), clearly indicating that agreement is very good, and these data have a high accuracy.

Table 26.6 Transition matrix for Raiganj (UA)

LULC time t^1 1991	LULC time t^2 2019						
	Area in km2	Built up area	Vegetation	Agricultural land	Fallow Land	Water bodies	Total 1991
	Built up area	3.14	0.48	0.56	0.18	0.4	4.76
	Vegetation	8.29	7.2	4.53	0.85	1.86	22.73
	Agricultural land	7.72	5.58	25.21	16.17	1.4	56.08
	Fallow Land	2.8	1.89	7.19	9.23	0.27	21.38
	Water bodies	0.46	0.39	1.44	1.65	1.91	5.85
	Total 2019	22.41	15.54	38.93	28.08	5.84	110.80

Source: Computed by the authors from Landsat-5 and Landsat-8 images

Fig. 26.8 Encroachment of agricultural land near Kornojhora for settlement

26.4.3 Urban Sprawls

The growth of urban sprawls in Indian cities is a common concern for policymakers and urban planners. The core area of a city is considered as a central place, and it acts as a CBD is State Bus Stand. There is a mutual interaction between the city core and

Fig. 26.9 Land is acquired for future extension of commercial complex near Siliguri more, Raiganj

Fig. 26.10 Disposal site near Kulik River, Raiganj

Table 26.7 Summary of the error matrices for Siliguri (UA)

	Producer's accuracy		User's accuracy		Overall accuracy		Kappa coefficient	
Data	1990	2019	1990	2019	1990	2019	1990	2019
Built-up area	96.32	98.74	89.79	92.23	89.9	91.32	0.89	0.95
Tea plantation	89.90	92.23	86.97	96.63				
Forest	95.32	99.30	95.23	91.36				
Agricultural land	89.23	97.77	94.32	99.00				
Water bodies	88.52	95.56	88.74	95.54				
Scrubs	97.32	94.74	92.85	98.65				

Source: Based on the error matrix of Landsat-5 (1990) and Landsat-8 (2019)

Table 26.8 Summary of the error matrices for Raiganj (UA)

	Producer's accuracy		User's accuracy		Overall accuracy		Kappa coefficient	
Data	1991	2019	1991	2019	1991	2019	1991	2019
Built-up area	100	99.58	97.56	100	99.9	97.32	0.99	0.97
Vegetation	100	100	99.04	96.83				
Agricultural fallow	99.16	97.37	99.16	99.23				
Water bodies	95.88	95.77	100	9873				
Fallow land	100	95.36	99.04	100				

Source: Based on the error matrix of Landsat 5 (1991) and Landsat-8 (2019)

Table 26.9 Shannon's entropy of Siliguri

Temporal periods	1990	2019	Loge
Shannon's entropy	1.73	1.71	2.08

Source: Computed by the authors from Eq. (26.5)

Table 26.10 Shannon's entropy of Raiganj

Temporal periods	1991	2019	Loge
Shannon's entropy	1.89	2.00	2.08

Source: Computed by the authors from Eq. (26.5)

its peripheral villages or its hinterland. The degree of interaction with central place and hinterland depends upon distance; it is usually very high near the CBD and declines with increasing distance. The nearby village of the urban fringe area gradually developed and merged with the main city. Such a process is related to the form of urban sprawl. The progress of a town is also correlated with urban sprawling. Urban sprawls mainly occur near the urban fringes area along the highway and the edge of the CBD. In the present study, Shannon's entropy model was executed to analyze and measure urban sprawls. The spatial diffusion or absorption of any geographical or physical variables can be successfully measured by the entropy technique (Thomas, 1981). The results of the entropy value of the study area are shown in Tables 26.9 and 26.10. For Siliguri, the highest value was observed in 1990, and it decreased in 2019. Table 26.10 reveals that the degree of

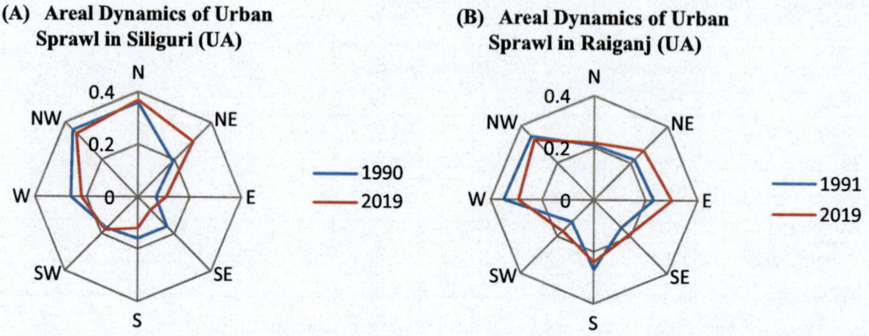

Fig. 26.11 Areal dynamics of urban sprawls in (**a**) Siliguri (UA) and (**b**) Raiganj (UA)

Fig. 26.12 Illegal contruction in the side of Balason River, Matigara, Siliguri

urban sprawl gradually increases in Raiganj (UA). Figure 26.11 describes the direction-wise distribution of urban sprawl.

It shows that sprawl distribution is already overlaid in N, NW, and SW direction in Siliguri (UA) and N, NW, and S direction in Raiganj (UA).

Tables 26.9 and 26.10 also demonstrate that, in general, the value of entropy is greater than 1 in both areas. Thus, it can be inferred that the expansion of both Siliguri and Raiganj is sprawling. Therefore, the general urban growth processes of Siliguri (UA) as well as Raiganj (UA) require a clear urban planning policy. Faster growth in the urban population and the urban economy is the main reason for sprawled development in both areas. Another factor responsible for sprawled growth is that some people prefer to live on the outskirts (Fig. 26.5) of the city. In both the towns of Siliguri and Raiganj, this trend has helped in the increase of outer growth (Fig. 26.12).

26.5 Conclusion

For sustainable urban agglomeration progress, it is essential to assess and monitor the LULC change, urban growth, and sprawls. Ecological impacts due to urban expansion and urban infrastructure crises have been increasing in complexity, producing a severe imbalance among the cities and their surroundings. The need to deal with this complexity in monitoring and measuring the urban dynamics and processes is strongly considered in recent decades. The level and rate of urban expansion depend on combined factors like the region's history, economic and social background, etc. So, it is necessary to look into the progress of the urban expansion of a region. The study reveals significant changes in LULC in both areas. The percentage of built-up growth is very high in the northern direction and in the core of the Raiganj city in 2019, whereas, in Siliguri, the built-up area increases in all the directions from the core. The agricultural land, fallow land, vegetation, and forest areas have constantly been decreasing. In this area, the loss of these categories of land use is mainly because of the increasing built-up area in the form of settlement and commercial areas. This massive urban growth influences production efficiency, economic growth, and urban dwellers' way of living. Due to the increase of infilling growth, the pattern of the urban area has become more compact in the core of Siliguri and Raiganj cities. Shannon entropy helps to understand and restrict the urban expansion and sprawl in concern for sustainable development. The results obtained by the Shannon entropy provide good guidance for modeling urban sprawl processes and understanding the causative factors of urbanization. The Shannon entropy results asserted that the tendency of sprawl in the Raiganj (UA) area is greater than Siliguri. Both the cities are experiencing sprawling, which harms the environment. Therefore, proper planning is the need of the hour for urban sustainability, making Siliguri and Raiganj sustainable cities. Suggestions and inputs of expertise in the urban field are essential at every stage to prepare an accurate plan for community development in these areas. The present study also illustrates the fact that remote sensing data is an important resource for analyzing the changes in spatial phenomenon with better accuracy and low cost. The research shows that the application of remote sensing and GIS can be helpful techniques to monitor the past and present LULC change in small and medium cities also.

References

Bhatta, B. (2010). *Analysis of urban growth and sprawl from remote sensing data, advances in geographic information science*. Springer.

Bhatta, B., Saraswati, S., & Bandyopadhyay, D. (2010). Quantifying the degree-of-freedom, degree-of-sprawl, and degree-of-goodness of urban growth from remote sensing data. *Applied Geography, 30*, 96–111. https://doi.org/10.1016/j.apgeog.2009.08.001

Burchell, R. W., & Shad, N. A. (1999). The evolution of the sprawl debate in the United States. *Hastings Environmental Law Journal, 5*, 25.

Carruthers, J. I., & Ulfarsson, G. F. (2003). Urban sprawl and the cost of public services. *Environment and Planning. B, Planning & Design, 30*, 503–522. https://doi.org/10.1068/b12847

Cheruto, M. C., Kauti, M. K., Kisangau, P. D., & Kariuki, P. (2016). Assessment of land use and land cover change using GIS and remote sensing techniques: A case study of Makueni County, Kenya. *Journal of Remote Sensing and GIS, 05*. https://doi.org/10.4172/2469-4134.1000175

Dadras, M., Shafri, H. Z. M., Ahmad, N., Pradhan, B., & Safarpour, S. (2014). Six decades of urban growth using remote sensing and GIS in the city of Bandar Abbas, Iran. *IOP Conference Series: Earth and Environmental Science, 20*, 012007. https://doi.org/10.1088/1755-1315/20/1/012007

Eko, J., & Ayama, O. R. (2013). Urban Sprawl effects on biodiversity in peripheral agricultural lands in Calabar, Nigeria. *Journal of Environment and Earth Science, 3*, 14.

El Garouani, A., Mulla, D. J., El Garouani, S., & Knight, J. (2017). Analysis of urban growth and sprawl from remote sensing data: Case of Fez, Morocco. *International Journal of Sustainable Built Environment, 6*, 160–169. https://doi.org/10.1016/j.ijsbe.2017.02.003

Ewing, R., Pendall, R., & Chen, D. (2003). Measuring sprawl and its transportation impacts. *Transportation Research Record Journal of the Transportation Research Board, 1831*, 175–183. https://doi.org/10.3141/1831-20

Ewing, R., Pendall, R., & Chen, D. (2002). *Measuring Sprawl and its impact* (p. 42). Smart Growth America.

Forys, E. A., & Allen, C. R. (2005). The impacts of Sprawl on biodiversity: The Ant Fauna of the Lower Florida keys. *Ecology and Society, 10*, art25. https://doi.org/10.5751/ES-01307-100125

Gadakh, B. L., & Jaybhaye, R. G. (2015). Land resource impact indicators of urban sprawl: A case study of Nashik city, Maharashtra. *Journal of Basic Sciences, 7*, 28–33.

Galster, G., Hanson, R., Ratcliffe, M. R., Wolman, H., Coleman, S., & Freihage, J. (2001). Wrestling Sprawl to the Ground: Defining and measuring an elusive concept. *Housing Policy Debate, 12*, 681–717. https://doi.org/10.1080/10511482.2001.9521426

Glaeser, E. L., & Kahn, M. E. (2004). *Sprawl and urban growth, handbook of regional and urban economics*. Elsevier.

Hasse, J. E., & Lathrop, R. G. (2003). Land resource impact indicators of urban sprawl. *Applied Geography, 23*, 159–175. https://doi.org/10.1016/j.apgeog.2003.08.002

Karakayaci, Z. (2016). The concept of urban sprawl and its causes. *Journal of International Social Research, 9*, 815. https://doi.org/10.17719/jisr.20164520658

Katyambo, M. M., & Ngigi, M. M. (2017). GIS, remote sensing, spatial urban growth, Nairobi Metropolitan Region, sustainable development. *American Journal of Geographic Information System, 6*, 19. https://doi.org/10.5923/j.ajgis.20170602.03

Khan, R., & Jhariya, D. C. (2016). Land use land cover change detection using remote sensing and geographic information system in Raipur Municipal Corporation Area, Chhattisgarh. *Scientific Society of Advanced Research and Social Change, 3*, 4.

Liu, Y., Fan, P., Yue, W., & Song, Y. (2018). Impacts of land finance on urban sprawl in China: The case of Chongqing. *Land Use Policy, 72*, 420–432. https://doi.org/10.1016/j.landusepol.2018.01.004

Majumdar, S., & Chatterjee, U. (2020). Modelling urban growth using Urban growth deterministic model in Kolkata Metropolitan Area: A geo-statistical approach. *Modeling Earth Systems and Environment, 2020*. https://doi.org/10.1007/s40808-020-00985-6

Mallupattu, P. K., & Sreenivasula Reddy, J. R. (2013). Analysis of land use/land cover changes using remote sensing data and GIS at an urban area, Tirupati, India. Scientific World Journal, 1–6. https://doi.org/10.1155/2013/268623

Mandal, J., Ghosh, N., & Mukhopadhyay, A. (2019). Urban growth dynamics and changing land-use land-cover of megacity Kolkata and its environs. *Journal of the Indian Society of Remote Sensing, 19*. https://doi.org/10.1007/s12524-019-01020-7

Mandal, S. (2017). Level and trend of urbanization: A study on Purulia District of West Bengal. *Arabian Journal of Geosciences, 22*, 13–20.

Mishra, P. K., Rai, A., & Rai, S. C. (2019). Land use and land cover change detection using geospatial techniques in the Sikkim Himalaya, India. *Egyptian Journal of Remote Sensing and Space Science, S1110982318302035*. https://doi.org/10.1016/j.ejrs.2019.02.001

Moskal, L. M., Styers, D. M., & Halabisky, M. (2011). Monitoring urban tree cover using object-based image analysis and public domain remotely sensed data. *Remote Sensing, 3*, 2243–2262. https://doi.org/10.3390/rs3102243

Mukherjee, F., & Singh, D. (2020). Assessing land use–Land cover change and its impact on land surface temperature using LANDSAT data: A comparison of two urban areas in India. *Earth Systems and Environment*, 23. https://doi.org/10.1007/s41748-020-00155-9

Pal, S., & Ziaul, S. (2017). Detection of land use and land cover change and land surface temperature in English Bazar urban centre. *Egyptian Journal of Remote Sensing and Space Science, 20*, 125–145. https://doi.org/10.1016/j.ejrs.2016.11.003

Pauleit, S., Ennos, R., & Golding, Y. (2005). Modeling the environmental impacts of urban land use and land cover change—A study in Merseyside, UK. *Landscape and Urban Planning, 71*, 295–310. https://doi.org/10.1016/S0169-2046(04)00083-0

Ramachandra, T. V., Bharath, H. A., & Sowmyashree, M. V. (2014). Urban structure in Kolkata: Metrics and modelling through geo-informatics. *Applied Geomatics, 6*(4), 229–244. https://doi.org/10.1007/s12518-014-0135-y

Rawat, J. S., & Kumar, M. (2015). Monitoring land use/cover change using remote sensing and GIS techniques: A case study of Hawalbagh block, district Almora, Uttarakhand, India. *Egyptian Journal of Remote Sensing and Space Science, 18*, 77–84. https://doi.org/10.1016/j.ejrs.2015.02.002

Reis, J. P., Silva, E. A., & Pinho, P. (2016). Spatial metrics to study urban patterns in growing and shrinking cities. *Urban Geography, 37*, 246–271. https://doi.org/10.1080/02723638.2015.1096118

Reis, S. (2008). Analyzing land use/land cover changes using remote sensing and GIS in Rize, North-East Turkey. *Sensors, 8*, 6188–6202. https://doi.org/10.3390/s8106188

Shaw, A. (2005). Peri-Urban Interface of Indian Cities Growth, Governance and Local Initiatives. *Economic and Political Weekly, 40*, 129–136.

Shukla, A., & Jain, K. (2019). Modeling urban growth trajectories and spatiotemporal pattern: A case study of Lucknow City, India. *Journal of the Indian Society of Remote Sensing, 47*, 139–152. https://doi.org/10.1007/s12524-018-0880-1

Sreenivasulu, V., & Bhaskar, P. U. (2010). Change detection in landuse and landcover using remote sensing and GIS techniques. *International Journal of Engineering, Science and Technology, 2*, 5.

Sudhira, H. S., Ramachandra, T. V., & Jagadish, K. S. (2004). Urban sprawl: Metrics, dynamics and modelling using GIS. *International Journal of Applied Earth Observation and Geoinformation, 5*, 29–39. https://doi.org/10.1016/j.jag.2003.08.002

Sun, C., Wu, Z., Lv, Z., Yao, N., & Wei, J. (2013). Quantifying different types of urban growth and the change dynamic in Guangzhou using multi-temporal remote sensing data. *International Journal of Applied Earth Observation and Geoinformation, 21*, 409–417. https://doi.org/10.1016/j.jag.2011.12.012

Sung, C. Y., Yi, Y., & Li, M.-H. (2013). Impervious surface regulation and urban sprawl as its unintended consequence. *Land Use Policy, 32*, 317–323. https://doi.org/10.1016/j.landusepol.2012.10.001

Thomas, R. W. (1981). *Information statistics in geography, concepts and techniques in modern geography*. Geo Abstracts.

Verma, L. N. (2008). *Urban geography*. Rawat Publications.

Chapter 27

Geospatial Mapping of SPM Load Under Urban Industrial Set-up, Durgapur, West Bengal, India, Through Q-GIS Application

Shiboram Banerjee, Debnath Palit, and Arnab Banerjee

Abbreviations

ANOVA	Analysis of variance
AT	Air temperature
CO	Carbon monoxide
CPCB	Central Pollution Control Board
GCP	Ground control points
GPS	Global Positioning System
ISPM	Inhalable suspended particulate matter
NHAI	National Highway Authority of India
NH2	National Highway 2
NOx	Oxides of nitrogen
O_3	Ozone
Q-GIS	Quantum-Geographic Information System
QML	Qt Modelling Language
RH	Relative humidity
RWC	Relative water content
SO_2	Sulfur dioxide
SPM	Suspended particulate matter
SR	Solar radiation

S. Banerjee (✉)
Post Graduate Department of Conservation Biology, Durgapur Government College, Durgapur, West Bengal, India

D. Palit
Department of Botany, Durgapur Government College, Durgapur, West Bengal, India

A. Banerjee
U.T.D, Department of Environmental Science, Sant Gahira Guru Viswavidyalaya, Ambikapur, Chattisgarh, India

G. S. Bhunia et al. (eds.), *Anthropogeomorphology*, Geography of the Physical Environment, https://doi.org/10.1007/978-3-030-77572-8_27

TSP	Total suspended particulate matter
VOC	Volatile organic compounds
WV	Wind velocity

27.1 Introduction

The microscopic solid and semi-solid particles of 0.01–100 μm size ranges suspended in the atmosphere are regarded as particulate matter. Those particles that originate from the natural and anthropogenic sources are considered as primary particulates, and when these primary molecules interact with each other in the atmosphere, they produce other variants recognized as secondary particulates (Offor et al., 2016). When atmospheric gaseous molecule after many physicochemical reactions gets converted into particulate matters, it is known as "secondary aerosol" (Buch, 1972). There are two ways of particulate matter entry into the Earth's environment: one is from Earth's crust and the other is from vehicular exhaust (Verma & Singh, 2006). These particulate matter (PM) pollutions tend to produce major problems in the environment of urban industrial area and even the distant rural areas by traveling through air current (Agrawal, 2005). PMs are harmful in the sense that they penetrate deep into the alveoli of human beings, leading to many respiratory and cardiac problems (Shandilya et al., 2007; Kampa & Castanas, 2008; Mcdonald et al., 2007; Delfino et al., 2011; Ulrich et al., 2012), and in plants particulates are responsible for inhibiting the gaseous exchange by blocking the stomatal door (Heath et al., 2009). Eastern India's most famous Durgapur steel city is located near to the largest coalfield area of Raniganj, and both are known as active industrial sectors. Recently, the air quality has deteriorated drastically due to higher emission of different industries and tremendous traffic load in this urban industrial area. Heterogeneous vegetational community is acting as a buffer to protect the environment from atmospheric pollution, but recently some anthropogenic activities such as deforestation, exotic plantation, monoculture practice, urban expansion, and mismanagement of solid waste have made the situation worse. Furthermore, notorious vehicular movement, fly ash from different industrial dumping ground, and fine soil dust from the surrounding rural areas are also contributing to particulate matter load in atmosphere, especially in the winter season. Moving vehicles on road emit different metals such as power brake of a car releases Cu, Zn is released from tire wear, and fluid leakage and detachment of vehicle are sources of Zn and Fe (Ball et al., 1991). The stability of particulate matters depends on their sizes as it microscopic to ultramicroscopic in size like < 10 μm or < 2.5 μm are more stable in the atmosphere (Shandilya et al., 2007); rather, > 10 or ~100 μm particles precipitate rapidly from the air through wet and dry deposition. These particles not only create adverse impact on surrounding living organisms but also influence the atmospheric chemistry and are regarded as an important air quality indicator (Offor et al., 2016). The compositions of PMs ranged from 24 inorganic elements (like

Na, P, K, Ca, Al, Mg, Si, S, Cl, etc.), elemental carbon, organic carbon to many bacterial and fungal species, and these were detected both in PM_{10} and $PM_{2.5}$ (Wei et al., 2017). $PM_{2.5}$ contains more microbial load than PM_{10}, mainly *Cladosporium, Aspergillus, Penicillium,* and Mesophilic bacteria are found in the particulate matters (Haas et al., 2013; Alghamdi et al., 2014; Cao et al., 2014). The availability of these cluster variants in particulate matters depends on the regional atmospheric status and their spatial distribution. The regional distribution of PMs is dependent primarily on wind direction and the source of air pollution as its intensity is mostly detected in areas adjacent to their origin. Furthermore, humidity condenses the air pollutants, forming different secondary harmful products; in contrast, precipitation cleanses the air and removes all types of dust particles and gaseous pollutants immediately. Solar radiation plays an important role by mediating many chemical reactions and molecular alterations in the atmosphere, producing different secondary by-products of air pollutants. Regional topography is also responsible for the dispersion of air pollutants spatially as the molecular nature of air pollutants is different and their distribution pattern follows the surface morphology of earth. This spatial variation has been influenced by the biochemical and physiological alterations of floral community and also affects the same to animal and human population. All these ultramicroscopic variants are harmful to surrounding biota, and only the green vegetation would be the most effective natural device though they are affected severely (sensitive plants) and moderately or without any manifestations (tolerant plants). More tolerant native varieties plantation in a scientific way at particularly suggested areas would be the first management practice to combat against the air pollution problem. Therefore, in this study the target was to monitor the spatial and temporal variability of suspended particulate matter (SPM) load and formation of a pictograph by funneling the data of SPM through Q-GIS software, which will reflect information about the pollution affected areas at a glimpse. This software-generated modeling clearly revealed the regional variability of SPM load depending upon which different management practices like green belt or green zone development could be formulated for human beings and generation of industrial biodiversity.

Most of the research works in air quality monitoring are emphasized on gross air quality analysis through SOx, NOx, SPM, and RSPM quantification followed by determination of air quality index (AQI) and air pollution index (API). There are lacunae in the field of air pollution monitoring with respect to geospatial mapping of individual air pollutants. We have made an attempt to map the spatial SPM load in an urban industrial set-up of Durgapur Industrial Township.

27.2 Materials and Methods

27.2.1 Selection of Study Sites and Sampling Procedure

Durgapur is the "Ruhr anchal" of West Bengal in India, having million tons of particulate matters from industrial and vehicular origin on a regular basis in its ambient atmosphere. It is located in between the two major river basins Damodar and Ajay at 23.48°N 87.32°E, elevation 213 ft., and was an ideal region for large industrial set-up. Recently, this urban area has become severely polluted through the emission of many large- and small-scale industries like Durgapur Steel Plant (DSP), Durgapur Project Limited (DPL), Alloy Steel Plant (ASP), National Thermal Power Station (NTPC), Philips Carbon (PC), graphite, fertilizer, etc. Furthermore, tremendous vehicular emission and day and night busy NH 2, which traverses the city at middle, augment the air pollution scenario of this urban area. The topography of this region is undulating, and the soil underneath contains different sizes of gravels, making it unfertile for agriculture practice but fertile for growing other native tree species like *Shorea robusta, Mangifera indica, Ficus benghalensis, Ficus religiosa, Azadirachta indica, Anthocephalus cadamba, Dalbergia sissoo, Lagersroemia speciose, Scleichera oleosa, Tamarindus indica, Terminalia arjuna, Tectona grandis,* etc.

The industrial area was divided into nine different study sites depending upon their industrial and traffic pollution load. Some sites were included inside the habitation zone, some were near the industrial/heavy traffic zone, and only one was selected as the forested area apart from the direct impact of air pollution (Table 27.1). All the nine sites were selected for SPM sampling through preliminary field survey before investigation (Fig. 27.1). As this Durgapur industrial area is large (approximately 49 km^2) and there have many big and small industrial sectors along with different crowded traffic spots, therefore all the sites could not be included for sampling. Only a few areas or sites that reflected the regional air pollution load and comparison were selected. This SPM sampling covered the residential areas (S1, S7, S8), major industrial areas (S4, S6), polluted traffic spots (S3, S5), railway station (S9), and pristine forested zone (S2). Thus, we had tried to select the diverse zone for air pollution monitoring program with regard to SPM. The study sites were depicted on the Google Earth Image as ground control point taken from GPS (Etrex-Garmin) (Fig. 27.2). Monitoring of SPM load at nine different experimental sites was performed by Handy Air Sampler (Envirotech APM 821) to study the natural effects of SPM on the plant species. Air temperature, solar radiation, relative humidity, and wind velocity were also recorded by Multimeter along with SPM monitoring. The load of SPM was assessed to characterize the variation in PM load in different selected sites. The sampling was done annually on triplicate basis (mean ± SD, n= 9) for the years 2013, 2014, and 2015. According to the Central Pollution Control Board (CPCB) (2009), the measurement of atmospheric pollutants should be operated over 24 h (for gaseous pollutants 4 h sampling and for particulate matter 8 h sampling of ambient air) two times in a week. In this regard, the sampling was carried out for 8 h from 12:00 noon to 8:00 p.m. in the month of April, August, and December at all the selected study sites for three consecutive years.

27.2.2 SPM Measurement

Crucial spatio-temporal measurement of suspended particulate matter (SPM) revealed the pollution status of an area, and with this scenario-specific management

Table 27.1 Sampling sites and their code names with quick description

Sl. no.	Study sites and their code names	Longitude and latitude	Site description
S1	Netazi Bhawan (NBH)	N 23° 34′ 21.33″	It is inside the township area (habitation + traffic)
		E 87° 17′ 27.85″	
S2	Jemua (JEM)	N 23° 32′ 56.18″	Forested rural area apart from industrial activity
		E 87° 20′ 44.54″	
S3	Durgapur Government College (DGC)	N 23° 32′ 30.26″	It covers the campus of Durgapur Government College area with heavy traffic pollution
		E 87° 19′ 32.00″	
S4	Durgapur Steel Plant (DSP)	N 23° 33′ 37.5″	Severe industrial activity with heavy traffic load of 24 h busy National Highway 2 (NH2)
		E 87° 14′ 59.6″	
S5	City Center (CCR)	N23° 31′ 59.78″	This is the city's only one bus stoppage where all types of traffic activity were observed
		E 87° 18′ 23.49″	
S6	Tamla (TAM)	N 23° 32′ 01.15″	It covers the area where three big industry like DSP, ASP, and DPL are merged
		E 87° 16′ 01.30″	
S7	Parulia (PAR)	N 23° 36′ 12.33″	Slightly outside of the city with moderate traffic load and apart from industrial activity
		E 87° 19′ 48.95″	
S8	Ara (ARA)	N 23° 31′ 01.79″	Residential area, hospital, and moderate to low vehicular load
		E 87° 21′ 37.21″	
S9	Durgapur Station (DST)	N 23° 29′ 59.45″	This covers the city's major railway station, where severe industrial and traffic pollution is noticed
		E 87° 18′ 21.29″	

practice should be formulated on a large industrial area. SPMs were measured by Handy Air Sampler with a special arrangement of filter paper. The ambient air was sucked through this Glass Microfiber filter paper (1.6 μm pore size and 2.5 cm. diameter) and particulate matter having diameter larger than 1.6 μm trapped on the surface of the filter (Fig. 27.3). Therefore, the difference between previous (W_i) and final (W_f) glass fiber filter paper weight was calculated. Flow rate of air through the machine was observed and regulated by flow cytometer and regulator switch,

Fig. 27.1 Nine different study sites of Durgapur industrial area

respectively. A difference between initial (Q_i) and final (Q_f) flow rate was recorded. The concentration of SPM was calculated by considering the two weight differences.

$$\text{Sucked air total volume}\ \left(\mu\text{g/m}^3\right) = \frac{(Qi + Qf) \times T}{2}$$

Where,

Q_i = flow rate, initial (m^3/min),
Q_f = final flow, final (m^3/min),
T = time duration (min)

SPM concentration was calculated by the following formula:

$$\text{Concentration of SPM}\ \left(\mu\text{g/m}^3\right) = \frac{(Wf - Wi) \times 10^6}{V}$$

Where,

V = Total volume of air Sucked
W_f = filter paper weight, final (gram)
W_i = filter paper weight, initial (gram)

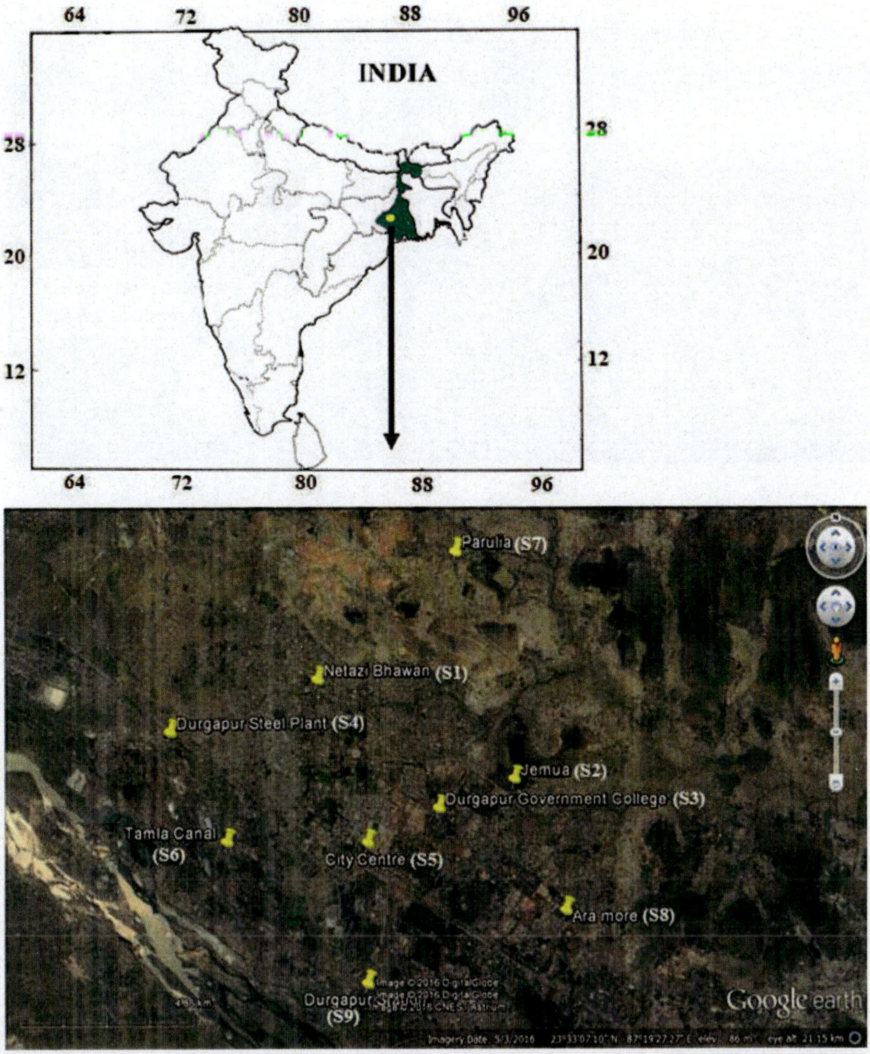

Fig. 27.2 Different point locations of nine study sites shown on Google Earth Image

27.2.3 Spatio-temporal Variation of SPM

In this section, the spatio-temporal variation of SPM in different areas is discussed. The annual average levels of SPM directed by CPCB at industrial, any habitational/ rural, and sensitive areas are 360 μg/m^3, 140 μg/m^3, and 70 μg/m^3, respectively. Similarly, CPCB-recommended RSPM levels on those sites are 120 μg/m^3, 60 μg/ m^3, and 50 μg/m^3 (Table 27.2).

Fig. 27.3 Handy Air Sampler (Envirotech APM 821) with GF/A (diameter 2.5 cm, pore size 1.6 μm) filter kept at the top cleft of black round cap and the nine GF/A filters showing after sampling

Table 27.2 CPCB-recommended SPM and RSPM levels (November 2009)

Particulate matter	Annual and hours basis	Concentration in the surrounding air		
		Industrial sites	Residential, rural, and other sites	Sensitive areas
SPM	Average (annual basis)	360 $\mu g/m^3$	140 $\mu g/m^3$	70 $\mu g/m^3$
	Average (24 h basis)	500 $\mu g/m^3$	200 $\mu g/m^3$	100 $\mu g/m^3$
RSPM	Average (annual basis)	120 $\mu g/m^3$	60 $\mu g/m^3$	50 $\mu g/m^3$
	Average (24 h basis)	150 $\mu g/m^3$	100 $\mu g/m^3$	75 $\mu g/m^3$

27.2.4 Q-GIS Methodology

The nine locations were selected on the basis of pollution intensities with respect to SPM parameters. The latitude and longitude of those nine points were collected with a Global Positioning System (GPS) (Etrex-Garmin). The location data were then downloaded as QML (Qt Modelling Language) file to point out the Ground Control Points (GCP) on the Google Earth Image. The saved Google Earth Image is then georeferenced with the help of those GCPs in Geographic Information System (GIS) platform by using the Quantum-Geographic Information System (Q-GIS) software (Fig. 27.4.). After georeferencing the image, classification was undertaken to point, line, and polygon types by opening the new shapefile layer from create layer option in Q-GIS software. The study sites points were recognized as point location and were saved the same points with their respective name. The lines were saved as different

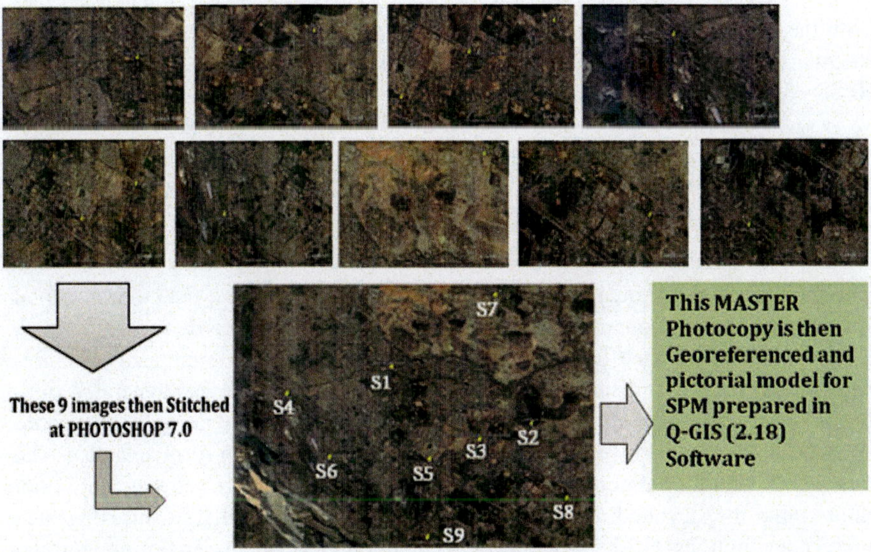

Fig. 27.4 Methodology of Q-GIS map preparation for SPM level on nine different study sites

road names in all over the Durgapur region. The same procedures were followed for the polygon type as these were differentiated into different land uses. Viz., habitation, industry, agriculture, forest, waterbody, eroded and fallow land, etc. The different land uses were colored differently by using style option of polygon property after classifying the land uses and subjected that land uses with respective colors. In the same way, the point locations where the air pollution data were inserted increased the diameter of those points at a constant size (25.00000) after categorizing them by clicking the style option of point property in Q-GIS software. Then, the color was added depending upon the intensity of pollution load, that is, severe air pollution zone gets more intense color and the color becomes lighter than the first in subsequent pollution-affected sites. The digitized image was then used for map composing by using Map Composer tool provided in Q-GIS software. At this stage, different map elements like legends, scale bar, latitude, longitude, and north arrow were incorporated into the map.

27.2.5 Statistical Analysis

Mean values and standard deviation were calculated for all the physical parameters that were replicated thrice. Depending on these data, different spatio-temporal bar graphs were prepared in Microsoft Excel 2007. SPSS 16.0 statistical software was used to perform one-way ANOVA with Tukey's post-hoc analysis (at $p < 0.05$ level) for comparison of means on meteorological parameters and SPM level considering both the dependent variables. Hierarchical cluster analysis of nine different experimental sites was also performed depending on the concentration of SPM level by SPSS 16.0.

27.3 Results and Discussion

The monitoring of air quality is an essential criterion to find out the environmental impact caused by various industrial activities and day-by-day increasing traffic load. The pollutants generated from industrial sectors emitted by different sizes of chimneys and their radii of dispersion are much more than the pollutants generated from the automobile or other civil constructions. Therefore, industrial air pollutants could spread outskirts of the urban area or even reach the surrounding rural area also. The automobile emission deteriorates the ambient air quality severely because it is not diluted into the surrounding atmosphere due to low-level discharge and congested urban constructions. Furthermore, the level of air pollution increased during morning or in evening hours and also in the winter season due to atmospheric inversion. The avenue trees, heterogeneous forested land, and sacred groove trees in the urban area are responsible to ameliorate the surrounding air quality depending on their tolerance level to air pollution. Some plant species were screened on the basis of their ability to absorb, adsorb, detoxify, and tolerate high level of air pollution (Ninave et al., 2001), and are known as tolerant plant species. These types of tree species should be planted around the affected industrial areas after proper scientific monitoring. In contrast, the plant species which are not properly accumulating or detoxifying the air pollutants are regarded as sensitive plants. The sensitivity have been reflected by several biochemical, physiological and morphological anomaly under stressful environmental conditions. Thus, the plants can be used as both passive biomonitors and biomitigators in the urban environment to indicate the atmospheric quality (Rai, 2016) and attenuate the pollution level in a locality (Beckett et al., 1998; Cox, 2003). Ambient air quality data, especially concentration of SPM level in the urban industrial area of Durgapur, were collected from nine different selected study sites to assess its regional variation on a spatio-temporal basis. Some discussion has also been done on the effect of SPM in biota, which control crucial ecological balance in urban industrial area.

27.3.1 Characterization of Meteorological Quality of Urban Atmosphere

In this section, the general meteorological parameters, that is, air temperature (AT), solar radiation (SR), relative humidity (RH), and wind velocity (WV), in all study sites are discussed. The annual averages of all sites for all the parameters were done

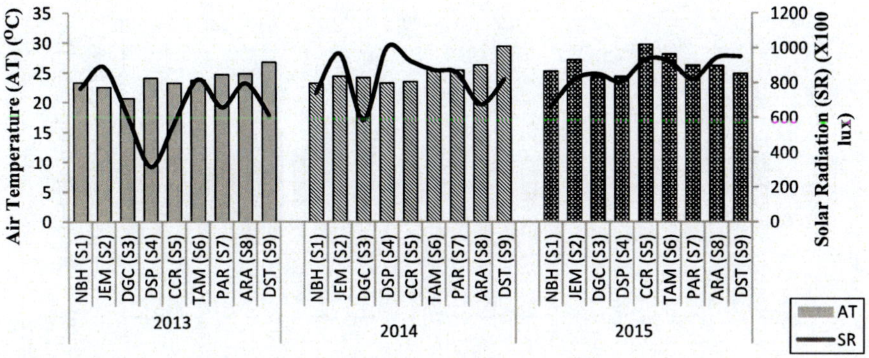

Fig. 27.5 Spatio-temporal variation of air temperature (AT) and solar radiation (SR) in the selected study areas

for overall detection of influential role of meteorological conditions on the distribution of SPM level in the atmosphere. The annual average air temperatures of Durgapur steel city for the three studied years were 23.83 ± 1.68 °C, 25.10 ± 1.96 °C, and 26.33 ± 1.83 °C, which were in increasing trends but reflect no significant temporal variation (Fig. 27.5). The mean maximum and minimum AT were varied from 26.80 °C to 20.70 °C during 2013, 29.50 °C to 23.33 °C in 2014, and 29.80 °C to 24.33 °C during 2015. Air temperature is regarded as the characteristic feature for the formation of various types of secondary pollutants products, and it mediates different types of chemical reactions in the atmosphere. Vegetational community could play a vital role in reducing temperature and hosted million tons of particulate matters.

The annual average of solar radiation was measured, that is, 672.11 ± 173.12 (X100 lux) in 2013, 828.22 ± 138.37 (X100 lux) in 2014, and 858.44 ± 93.91 (X100 lux) in 2015. The mean maximum and minimum SR were varied from 892 (X100 lux) to 317(X100 lux) in the first year, 1004 (X100 lux) to 558 (X100 lux) in the second year, and in the final year it was 953 (X100 lux) to 663 (X100 lux). An increasing trend of SR was noticed in the three studied years but showed no significant temporal variation (Fig. 27.5). Solar radiation of a region depends on the latitude and longitude of the earth, and that time meteorological condition mainly on cloud. Increased level of pollution alters the total biogeochemistry of the region and ultimately on the earth on a broad scale. Air pollution could also influence the cloud formation, rainfall, and finally the intensity of solar radiation.

The mean relative humidity in 2013, 2014, and 2015 was 53.06 ± 5.91%, 58.83 ± 6.13%, and 60.09 ± 4.94%, respectively. The mean maximum and minimum RH in 2013, 2014, and 2015 were 64.80% and 46.33%, 68.96% and 49.56%, and 72.03% and 55.00%, respectively. Year-wise RH has shown an increasing mode with no such statistically significant variation (Fig. 27.6). Increased humidity could aggravate the air pollutants and is responsible for producing different types of secondary aerosol products by mediating various chemical reactions in the atmosphere.

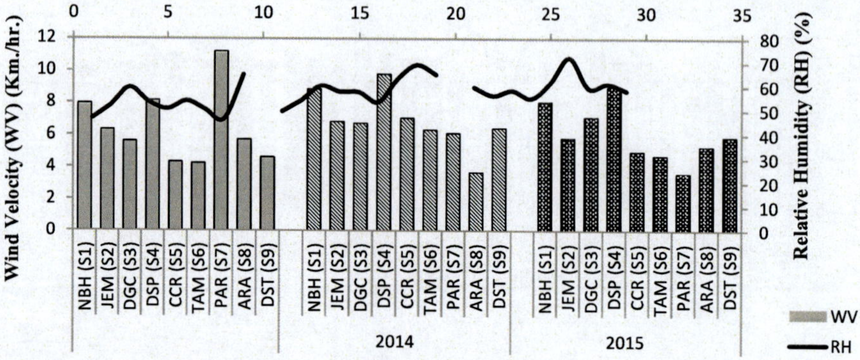

Fig. 27.6 Spatio-temporal variation of relative humidity (RH) and wind velocity (WV) in the selected study areas (average of data obtained in 2013–2015)

The annual average wind velocities of all the study sites for the three subsequent years were 6.44 ± 2.27 km/h, 6.84 ± 1.72 km/h, and 6.03 ± 1.70 km/h. The mean maximum and minimum WV were varied from 11.16 km/h to 4.23 km/h during 2013, 9.76 km/h to 3.66 km/h in 2014, and 8.93 km/h to 3.60 km/h during 2015. WV has shown an opposite trend of decreasing value, which also reflects no such temporal variation (Fig. 27.6). Wind velocity is another meteorological factor that is responsible for distributing the particulate matters (PMs) to the distant place. Wind direction also plays a crucial role for spreading of PMs, and this type of movement depends on the season because seasonal air current carries the pollutants to the proper direction.

One-way ANOVA (at significance level $p < 0.05$) with Tukey's HSD test was performed for observed variation of meteorological parameters (AT, SR, RH, and WV) between the study sites (Table 27.3). The analysis was performed to establish the spatial variation of those parameters under study to reveal the atmospheric status. Table 27.3 depicts that no statistically significant variation was noticed between the study sites when compared with the control S2 (JEM). No significant temporal variation was also recorded for those meteorological parameters (AT, SR, RH, and WV) on the subsequent 3 years (Fig. 27.7).

27.3.2 Suspended Particulate Matter (SPM)

Under air pollutants, oxides of nitrogen (NOx), sulfur dioxide (SO_2), carbon monoxide (CO), surface layer ozone (O_3), volatile organic compounds (VOCs), ammonia (NH_3), etc., are regarded as gaseous air pollutants (Archibald et al., 2017). While ultrafine inhalable (size <10 μm), respirable (<2.5 μm), and coarse particles (10–100 μm) including many heavy metals like Pb, As, Cr, Cd, etc., are specified as particulate matters (Fig. 27.8).

Table 27.3 The result of one-way ANOVA using sites as the explanatory factor for the observed variation of atmospheric parameters (AT, SR, RH, and WV) and SPM

The following sites compared with control site S2 (JEM)								
Sites	NBH (S1)	DGC (S3)	DSP (S4)	CCR (S5)	TAM (S6)	PAR (S7)	ARA (S8)	DST (S9)
Atmospheric parameters	**MD**	**MD**	**MD**	**MD**	**MD**	**MD**	**MD**	**MD**
AT	0.74	1.68	0.79	−0.79	−1.06	−0.73	−1.07	−2.30
SR	169.00	211.00	184.00	83.66	20.33	114.00	87.00	97.66
RH	2.15	−5.17	−1.27	−2.41	−5.66	−3.49	−4.67	−8.48
WV	−1.95	−0.16	−2.63	0.87	1.22	−0.65	1.42	0.68
SPM	187	−245[a]	−1312[a]	−673[a]	−1346[a]	−119	−116	−1090[a]

[a]Indicates significance at $p < 0.05$ level with Tukey's comparisons *MD* mean difference

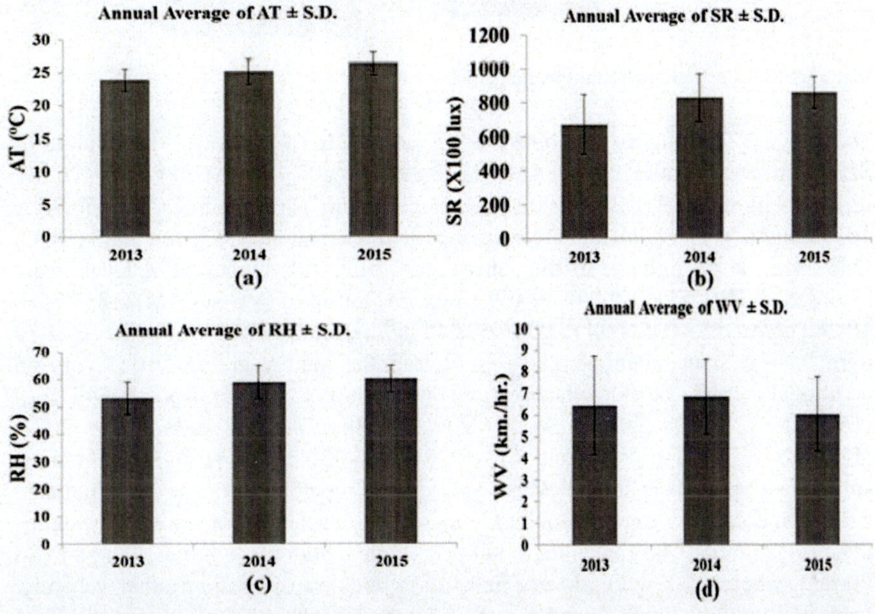

Fig. 27.7 Temporal variations of air temperature (AT), solar radiation (SR), relative humidity (RH), and wind velocity (WV) level are shown graphically between the three consecutive years. [Data given as mean ± S.D. (n= 9)]

In the 3 years of study (2013, 2014, and 2015), the average values of SPM were 538.83 ± 464.14 μg/m^3, 637.01 ± 548.07 μg/m^3, and 717.36 ± 650.77 μg/m^3, respectively (Fig. 27.9). In all the three experimental years, the average values of SPM were found to exceed several folds above the recommended limit that was prescribed by CPCB. The maximum and minimum values were recorded as 1278.89 ± 133.47 μg/m^3 and 64.19 ± 13.40 μg/m^3 in 2013, 1399.68 ± 169.17 μg/m^3 and 70.78 ± 12.16 μg/m^3 in 2014, and in 2015 the measured values of SPM were

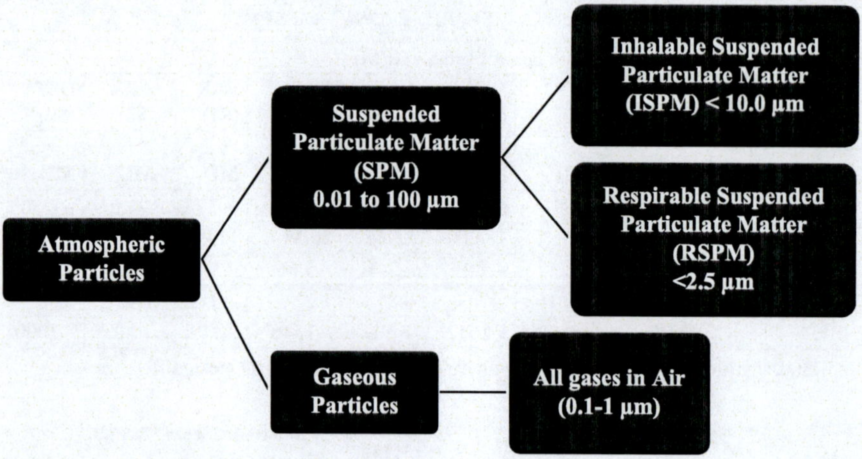

Fig. 27.8 Categorization of atmospheric particles

1659.21 ± 128.24 µg/m^3 and 68.25 ± 7.51 µg/m^3. In 2013, the elevated values of SPM were documented in S6 (TAM) (1278.89 ± 133.47 µg/m^3), S4 (DSP) (1082.57 ± 133.09 µg/m^3), S9 (DSN) (1002.56 ± 112.90 µg/m^3) and site 5 (CCR) (556.76 ± 103.02 µg/m^3). The moderate increase of particulate matters was found in S3 (DGC) (343.29 ± 31.46 µg/m^3). In the subsequent year 2014, higher SPM levels were recorded in S4 (DSP) (1399.68 ± 169.17 µg/m^3) followed by S6 (TAM) (1357.76 ± 181.60 µg/m^3), S9 (DSN) (1181.53 ± 88.45 µg/m^3), and S5 (CCR) (781.01 ± 42.79 µg/m^3). The same phenomenon was noticed the next year, i.e. 2015, with an additional increase of particulate matter concentration in the atmosphere: S4 (DSP) (1659.21 ± 128.24 µg/m^3), S6 (TAM) (1606.88 ± 145.47 µg/m^3), S9 (DSN) (1290.72 ± 75.28 µg/m^3), and S5 (CCR) (886.03 ± 154.41 µg/m^3). In three subsequent years, persistent elevated values were recorded in the severely polluted sites with escalating trends, though the increasing level was not statistically significant (Fig. 27.10). The concentration of particulate matter was alarmingly high in S6 (TAM), where 24 h active heavy industrial goods transportation, other vehicular load, and industrial emissions have their combined influence to increase the SPM level in air. Actually this site was located at the junction between the three big industrial sectors, that is, DSP, ASP, and DPL. Site S4 (DSP) was another sampling site where an increased level was found in the three subsequent years due to tremendous industrial emission along with the reconstruction of NH2, a central government project by the National Highway Authority of India (NHAI) for six-lane roads, the raw material of which mixes tones of particulate matter into the atmosphere. The fly ash was the main packing material for construction of flyover crossing in the NH2, and at this sampling site two flyovers were constructed. So a huge particulate matter emission was found with high recorded values in the subsequent 3 years. The sampling location of S4 (DSP) was just within 250 m from the NH2. It was observed that in control S2 (JEM) the level of particulate pollutants was below the threshold limit and therefore is considered as a low polluted

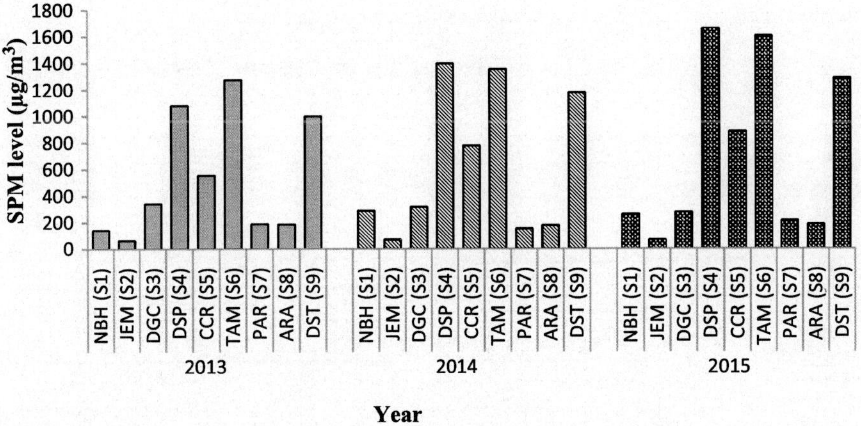

Fig. 27.9 Spatio-temporal variation of the SPM in the selected study areas (average of data obtained in 2013–2015)

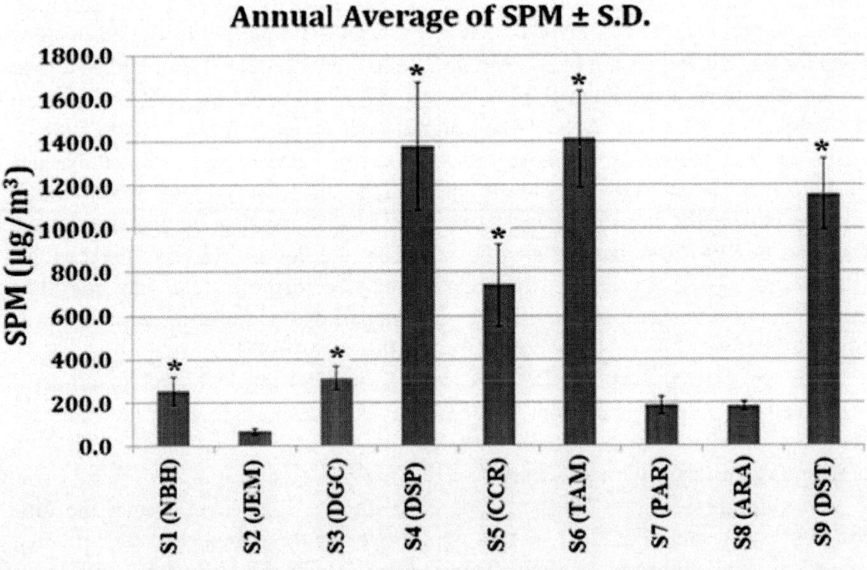

Fig. 27.10 Comparison of SPM level (μg/m^3) between different sites. [Data given as mean ± S.D. (n= 9). Significant differences in SOx level among different study sites from Jemua (S2-control sites) have been indicated by asterisks (*) ($p<0.05$, ANOVA with Tukey's comparisons)]

site. This site is long apart from any type of industrial activity, very low traffic density with its natural and artificial forestry management practice.

A spatial significant variation of SPM concentration was detected for all the sites with respect to the control S2 (JEM), except S7 (PAR) and S8 (ARA). The

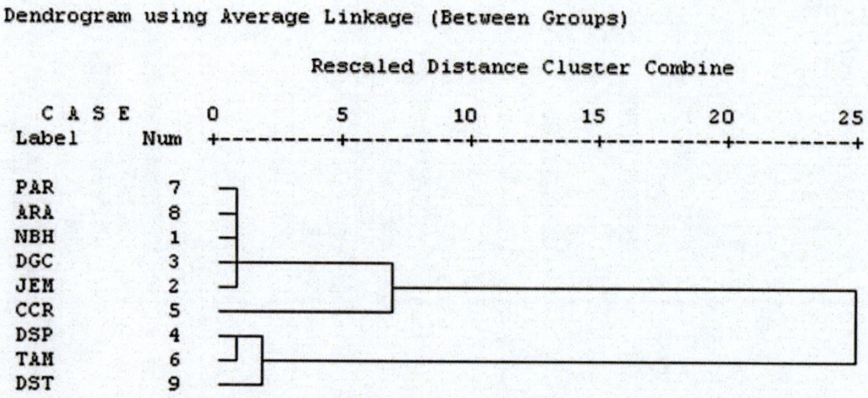

Fig. 27.11 Hierarchical cluster analysis of nine different study sites of Durgapur based on the SPM level. Here, 1 = S1 (NBH), 2 = S1 (JEM), 3 = S1 (DGC), 4 = S1 (DSP), 5 = S1 (CCR), 6 = S1 (TAM), 7 = S1 (PAR), 8 = S1 (ARA), and 9 = S1 (DST)

particulate density was almost the same on three sites for 3 years. The graphical representation in Fig. 27.9 depicts that the +1 and −1 standard deviation of SPM level for S2 (JEM) does not overlap on the Y axis with +1 and −1 standard deviation of the remaining experimental sites, except S7 (PAR) and S8 (ARA) (have no asterisks "*"). One-way ANOVA (at significance level $p < 0.05$) with Tukey's HSD test was performed for observed variation of SPM between the study sites (Table 27.3). The analysis was done to establish the site-wise variation of SPM under study to portray the environmental status. Table 27.3 shows that, among all the sites, the level of SPM has changed significantly from the control S2 (JEM), except S7 (PAR) and S8 (ARA), and a similarity has existed in these three sites for SPM emission, which was not statistically significant. No significant variation was observed between the severely polluted sites, that is, S4 (DSP),

Sites S6 (TAM) and S9 (DST), except S5 (CCR) and S3 (DGC), reflected statistically significant variation. The SPM level was approximately the same for the above-polluted sites in the three experimental years, but S5 (CCR) and S3 (DGC) contained significantly lower values.

A dendrogram was established depending on the SPM level among the nine different study sites (Fig. 27.11). This figure revealed three major clusters; the first one occupied the moderately affected sites like S7 (PAR), S8 (ARA), S1 (NBH), S3 (DGC), and S2 (JEM). In the second cluster, single site S5 (CCR) was isolated, which was linked with S3 (DGC) of the first cluster. Because the values of SPM were slightly similar for those studied sites, therefore a linkage was detected. The third cluster was segregated with three experimental sites, that is, S4 (DSP), S6 (TAM), and S9 (DST), which were spotted as severely affected sites considering the SPM level. These three sites occupied the highest SPM concentrations consecutively for the three studied years 2013, 2014, and 2015.

An air pollution hazard map was prepared considering the atmospheric SPM level on nine different experimental sites (Fig. 27.12); and on the basis of this pictorial

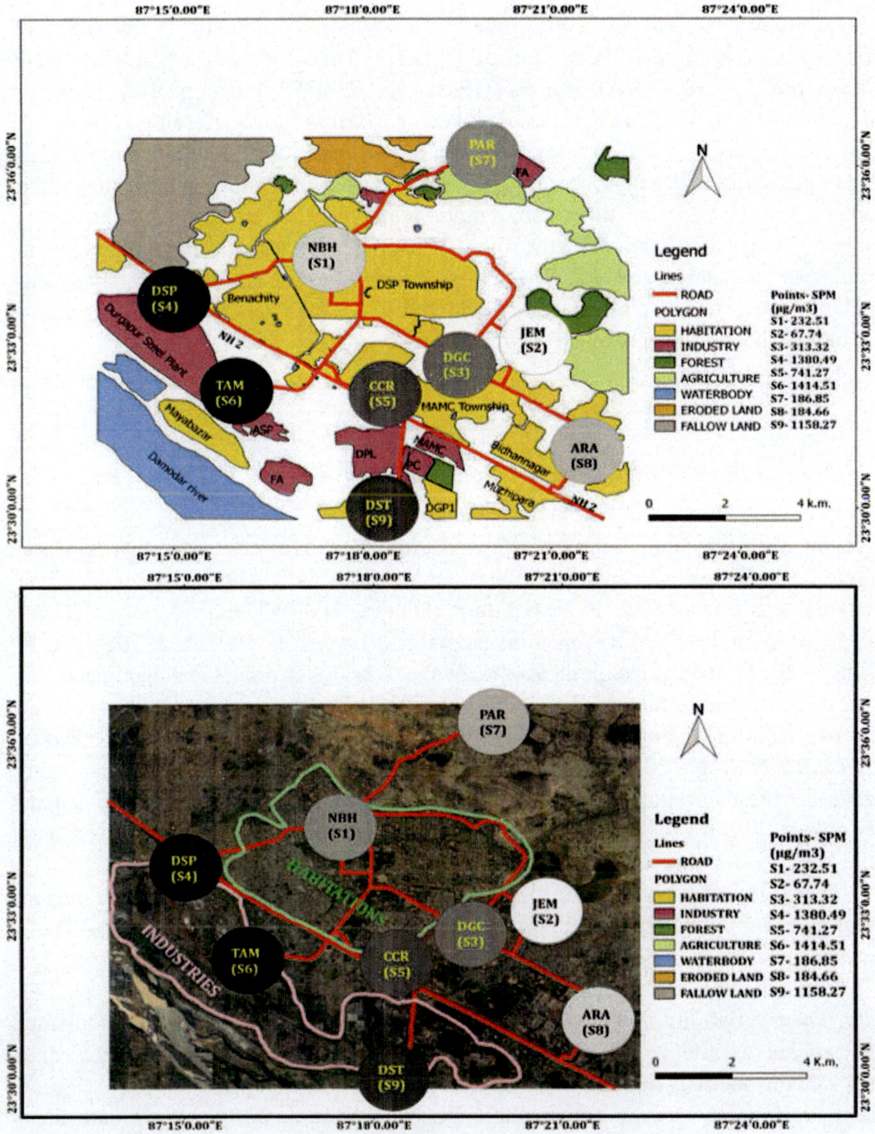

Fig. 27.12 Pictorial model showing the spatial variation of SPM concentration with clear distinction between industrial and habitation zones. Q-GIS map showing habitation, industrial, forest, agriculture, waterbody, eroded land, and fallow land

hazard map, an effective management strategy could be formulated for a sensitive urban industrial area. This pictorial model reflected a distinct color variation between the study sites. Black color was selected for categorization of sites for different levels of SPM concentration in the air. An intense black color was subjected for severely polluted sites like S4 (DSP), S6 (TAM), and S9 (DST); and depending on the concentration of SPM level, moderate black color was automatically fitted for S2 (NBH), S3 (DGC), S5 (CCR), and S7 (PAR). Finally, light black and white were developed for S8 (ARA) and S2 (JEM). As S2 (JEM) occupied white color depending on its SPM level, this site was categorized as the control site. This picture has also been depicted as a clear distinction between habitation and industrial area of this large industrial belt. Depending upon this visual modeling of SPM data, their spatial variation for the three consecutive years and clear zonation would clearly reveal the specified area for green belt development that could ameliorate the air pollution problem effectively. The spatial deterioration of regional air quality considering the SPM level is exactly derived by the Q-GIS model map, which provides a strong suggestion for prospective green belt.

27.3.3 *Impacts of Particulate Matters (PMs) on Vegetation*

Among the biotic community plants are adapted more to the changing environmental conditions. These adaptive capabilities could help the plants to minimize their internal stress caused due to air pollution (Dineva, 2004). The deleterious effect of PMs has been observed by many investigators (Garg et al., 2001; Rai, 2013; Rai & Singh, 2015). The physiological, biochemical, and morphological parameters of vegetation are adversely affected by PMs, and the ultimate impact has been observed on overall biodiversity (Seyyednjad et al., 2011; Younis et al., 2013; Rai & Panda, 2014; Rai & Singh, 2015). The leaf morphology and architecture of terrestrial plants act as a natural acceptor for air pollutants, mainly PMs. It is also amazing that a plant is able to live up to that condition when all the leaves are totally covered with the PMs.

There are some pollution-responsive physiological and biochemical parameters that revealed their spatio-temporal variability in front of air pollution stress. These are pH, relative water content (RWC), ascorbic acid, total chlorophyll (Chl "a" and "b"), total soluble sugar, protein, amino acid (proline, cysteine), carotenoid, phenol, etc. Their variability under air pollution stress segregated the plants under tolerant and sensitive varieties. Micromorphology of plant leaves is another parameter that showed the anomaly of surface architectural pattern under air pollution stress (Prasad & Rao, 1981; Verma & Singh, 2006; Joshi & Swami, 2007) after biochemical changes (Tripathi et al., 2009). Particulate matters enter into the plant leaves through stomata, diminishing their resistance to any biotic or abiotic stress (Shanker et al., 2005) and affecting on overall yield (Reimann et al., 2001). The tolerant are those that could adsorb, detoxify, and translocate the PMs efficiently; on the other hand, the plants that are unable to do that work precisely are known as sensitive to air pollution (Agbaire & Esiefarienrhe, 2009). The component of PMs can alter the pH of plant leaves. Alkaline pH may cause wax deterioration, protein and fat denaturation, direct damage to plant tissue, etc. (Czaja, 1960; Guderian, 1986); on the other hand, acidic pH degrades the chlorophyll molecule (Turk & Wirth, 1975). Sensitive plant species are more prone to be affected with this type of toxic-soluble salt present in PMs (Prajapati & Tripathi, 2008). Relative water content (RWC) is regarded as the body water percentage of plant leaves dependent on the release and uptake of water from the environment through different routes (Jones, 1994). The plants that maintain the high physiological water balance are known as tolerant (Dedio, 1975; Singh & Verma, 2007), and sensitive plants not properly maintain the body water in pollution stress (Masuch et al., 1988). Ascorbic acid is a small molecular natural antioxidant that is responsible for removing the metabolic and pollution-induced free radicals (hydroxyl radical, superoxide, singlet oxygen, H_2O_2, etc.) from the plant's body (Noctor & Foyer, 1998; Blokhina et al., 2003; Kuddus et al., 2011). This strong reducing substance also enhances photosynthetic light reaction (Singh & Verma, 2007), carbon assimilation (Seyyednjad et al., 2011), defense mechanism (Arora et al., 2002), control cell division (Conklin et al., 2000), etc., and these functions are dependent on its concentration. Many experiments have shown that the level of ascorbic acid was higher in resistant tree species, while it becomes lowered in vulnerable plants to air pollution (Prajapati & Tripathi, 2008). Chlorophyll is another parameter present inside of chloroplast grana (Banerjee et al., 2016) that has performed a vital role in metabolism of plants. The degradation of this photosynthetic pigments is caused by the effect of air pollution (Prusty et al., 2005). The plants that kept their chlorophyll content despite under pollution stress are known as tolerant, and the opposite is also true (Banerjee et al., 2019). Air pollution tolerance index (APTI) is the most popular and well-known index that classifies plant species into sensitive, moderately sensitive, and tolerant varieties (Shannigrahi et al., 2004; Kaur & Nagpal, 2017). It is substantiated algebraically through determination of pH, RWC, ascorbic acid, and total chlorophyll (Singh & Rao, 1983). This index provides important information about the influence of atmospheric pollution on physiological and biochemical parameters of floral community (Banerjee et al., 2021).

Sugar is the most important energetic molecule manufactured by all green plants maintaining their physiological and biochemical balance by providing energy through ATP. It is also responsible to remove the air pollutants indirectly, and its level is an indicator that either the plant is under stress or not (Tripathi & Gautam, 2007). Tolerant plants maintain their internal sugar content by properly activating photosynthetic pathway despite pollution attack (Lorenc-Plucinska, 1982), but in sensitive plants soluble sugar is exhausted rapidly due to low photosynthetic rate and higher metabolic ATP production for pollutants removal (Seyyednejad et al., 2009). Protein is another bioorganic molecule and is the main component of all enzymes in plant body responsible for various types of biochemical reactions. The level of protein is dependent on air pollution stress and also on inherent capabilities of each plant species (Iqbal et al., 2010a, b). Tolerant plants can continuously produce proteins for various functions; in contrast, sensitive plants not properly or partially synthesize protein molecule or become denatured under air pollution stress. Amino acids are the main component of all proteins, and their concentrations provide an indication of environmental stress. Its gradual increase means more protein denaturation due to any biotic or abiotic cause (Seyyednejad & Koochak, 2011). Proline is an amino acid that is accumulated more in air pollution-affected plant species to eradicate Reactive Oxygen Species (ROS) (Wang et al., 2009). An increased trend of cysteine was also observed in tolerant plant cells under polluted environments (Tausz et al., 1998; DeKok et al., 2000). Phenol is an aromatic alcohol present in plant body and acts as an antioxidant (Cook & Samman, 1996). Many metabolic reactions and floral maturation have been regulated by phenol (Zobel, 1996).

Along with biochemical and morphological injury in plants, PMs are also affecting the biodiversity of earth (Cape, 2008), especially in urban industrial areas. It has been reported that, after deposition, PMs disturb the soil microbial decomposers, affect rhizospheric bacteria (Grantz et al., 2003), and disintegrate the phyllosphere microbial community. These three types of microbial community would drastically change the nutrient cycling, vegetation dynamics (Fan & Wang, 2000; Grantz et al., 2003), and ecological integrity (Narayan et al., 1994). Genetic diversity, species diversity, and ecological diversity are the assets of the earth, but in recent eras pollution and human population both are bottlenecking those creatures that would move the existence of human beings to the verge of extinction.

27.4 Conclusion

Steel plants, sponge iron factories, power plants, traffic emission, open waste burning, and mining activities are responsible for loading of particulate matters in the atmosphere, and their distributions are regulated by regional meteorological conditions and landscape geometry. This study revealed that the sites near to the industry and heavy traffic area exceeded PMs loads in the surrounding atmosphere recommended by CPCB. A statistically significant variation of SPM level was noticed in all the studied sites with respect to the control site S2 (JEM), except S7 (PAR) and S8 (ARA). No site-wise significant variation of meteorological parameters was detected when compared with the control site. Recently, many experiments have shown that air pollutants causing many deleterious effects on vegetation not only at the origin of PM emission but could also affect agriculture on distant rural areas. Their ultrafine particulate size (<2.5 μm) allows them to clogged many vital biological machineries like stomata of flora deep into the alveoli of fauna to create various biochemical and physiological problem, sometimes the situation becoming lethal. As the floral community is exposed maximum to air pollution, they become more tolerant to atmospheric pollutants in contrast to any faunal species. This proposition suggests that for clean future of human well-being and other biodiversity of planet earth, it would be necessary to plant or conserve the green vegetation by forming green belts not only in the urban industrial areas but also in the rural areas.

References

Agbaire, P. O., & Esiefarienrhe, E. (2009). Air pollution tolerance indices (APTI) of some plants around Otorogun Gas Plant in Delta State, Nigeria. *Journal of Applied Sciences and Environmental Management, 13*(1), 11–14.

Agrawal, M. (2005). Effects of air pollution on agriculture: An issue of national concern. *National Academy Science Letters, 28*, 93–106.

Alghamdi, M. A., Shamy, M., Redal, M. A., Khoder, M., Awad, A. H., & Elserougy, S. (2014). Microorganisms associated particulate matter: A preliminary study. *Science of the Total Environment, 479*, 109–116. https://doi.org/10.1016/j.scitotenv.2014.02.006

Arora, A., Sairam, R. K., & Srivastava, G. C. (2002). Oxidative stress and antioxidative system in plants. *Current Science, 82*, 1227.

Archibald, A. T., Folberth, G., Wade, D. C., & Scott, D. (2017). A world avoided: impacts of changes in anthropogenic emissions on the burden and the effects of air pollutants in Europe and North America. *Faraday Discussions, 200*, 475–500. https://doi.org/10.1039/C7FD00004A

Ball, D., Hamilton, R., & Harrison, R. (1991). The influence of highway related pollutants on environmental quality. In R. Hamilton & R. Harrison (Eds.), *Highway pollution* (pp. 1–47). Elsevier Science.

Banerjee, S., Banerjee, A., & Palit, D. (2016). Changing strategies of biochemical & physiological features of selected plant species on effect of air pollution in eastern Steel City, Durgapur, India. *International Journal of Current Microbiology and Applied Sciences, 5*(9), 733–741.

Banerjee, S., Banerjee, A., Palit, D., & Roy, P. (2019). Assessment of vegetation under air pollution stress in urban industrial area for greenbelt development. *International journal of Environmental Science and Technology, 16*, 5857–5870. https://doi.org/10.1007/s13762-018-1963-9

Banerjee, S., Palit, D., & Banerjee, A. (2021). Variation of tree biochemical and physiological characters under different air pollution stresses. Environmental Science and Pollution Research, 28, 17960–17980. https://doi.org/10.1007/s11356-020-11674-3

Beckett, K. P., Freer-Smith, P. H., & Taylor, G. (1998). Urban woodlands: Their role in reducing the effects of particulate pollution. *Environmental Pollution, 99*, 347–360.

Blokhina, O., Virolinen, E., & Fagerstedt, K. V. (2003). Antioxidants, oxidative damage and oxygen deprivation stress: A review. *Annals of Botany, 91*, 179–194. https://doi.org/10.1093/aob/mcf118

Buch, W. (1972). *Atmospheric pollution*. McGraw Hill.

Cao, C., Jiang, W., Wang, B., Fang, J., Lang, J., Tian, G., et al. (2014). Inhalable microorganisms in Beijing's PM2.5 and PM10 pollutants during a severe smog event. *Environmental Science & Technology, 48*, 1499–1507. https://doi.org/10.1021/es4048472

Cape, J. N. (2008). Interactions of forests with secondary air pollutants: Some challenges for future research. *Environmental Pollution, 155*, 391–397.

Conklin, P. L., et al. (2000). Identification of ascorbic acid-deficient Arabidopsis thaliana mutants. *Genetics, 154*, 847–856.

Cook, N. C., & Samman, S. (1996). Flavonoids- chemistry, metabolism, cardio protective effect, and dietary sources. *Nutritional Biochemistry, 7*, 66–76.

Cox, R. M. (2003). The use of passive sampling to monitor forest exposure to O_3, NO_2 and SO_2: A review and some case studies. *Environmental Pollution, 126*, 301–311.

CPCB. (2009). *Ambient air quality data*. Central Pollution Control Board. http://www.cpcb.nic.in/bulletin/del/2009.html

Czaja, A. T. (1960). Die wirkung von verstaubtem kalk und zement auf pflanzen. *Qualitas Plantarum et Materiae Vegetabiles, 7*, 184–212.

Dedio, W. (1975). Water relations in wheat leaves as screening tests for drought resistance. *Canadian Journal of Plant Science, 55*, 369–378.

DeKok, L. J., Westernman, S., Elisabeth, C., Stuiver, E., & Stulen, I. (2000). Atmospheric H_2S as plant sulfur source: Interaction with pedospheric sulfur nutrition – A case study with *Brassica oleracea* L. In C. Brunold et al. (Eds.), *Sulfur nutrition and sulfur assimilation in higher plants* (pp. 41–56). Paul Haunt.

Delfino, J. R., Staimer, N., & Vaziri, D. N. (2011). Air pollution and circulating biomarkers of oxidative stress. *Air Quality, Atmosphere and Health, 4*, 37–52.

Dineva, S. B. (2004). Comparative studies of the leaf morphology and structure of white ash *Fraxinus americana* L. and London plane tree *Platanus acerifolia* willd growing in polluted area. *Dendrobiology, 52*, 3–8.

Fan, H. B., & Wang, Y. H. (2000). Effects of simulated acid rain on germination, foliar damage, chlorophyll contents and seedling growth of five hardwood species growing in China. *Forest Ecology and Management, 126*, 321–329.

Garg, A., Shukla, P. R., Bhattacharya, S., & Dadhwal, V. K. (2001). Sub-region (district) and sector level SO_2 and NOx emissions for India: Assessment of inventories and mitigation flexibility. *Atmospheric Environment, 35*(4), 703–713.

Grantz, D. A., Garner, J. H. B., & Johnson, D. W. (2003). Ecological effects of particulate matter. *Environment International, 29*, 213–239.

Guderian, R. (1986). Terrestrial ecosystems: particulate deposition. In: Air pollutants and their effects on the terrestrial ecosystem. In A. H. Legge & S. V. Krupa (Eds.), *Advances in environmental science and technology* (Vol. 18, pp. 339–363). Wiley.

Haas, D., Galler, H., Luxner, J., Zarfel, G., Buzina, W., Friedl, H., et al. (2013). The concentrations of cultural microorganisms in relation to particulate matter in urban air. *Atmospheric Environment, 65*, 215–222. https://doi.org/10.1016/j.atmosenv.2012.10.031

Heath, R. L., Lefohn, A. S., & Musselman, R. C. (2009). Temporal processes that contribute to nonlinearity in vegetation responses to ozone exposure and dose. *Atmospheric Environment, 43*, 2919–2928.

Iqbal, M., Jura-Morawiec, J., WŁoch, W., & Mahmooduzzafar, N. (2010a). Foliar characteristics, cambial activity and wood formation in *Azadirachta indica* a. Juss. as affected by coal smoke pollution. *Flora, 205*, 61–71.

Iqbal, M., Mahmooduzzafar, N., Khan, F., & Pervaiz, R. (2010b). Photosynthetic, metabolic and growth responses of *Triumfetta rhomboidea* to coal smoke pollution at different stages of plant ontogeny. *Journal of Plant Interactions, 5*(1), 11–19.

Jones, H. G. (1994). *Plants and microclimate* (2nd ed.). Cambridge University.

Joshi, P. C., & Swami, A. (2007). Physiological responses of some tree species under roadside automobile pollution stress around city of Haridwar, India. *The Environmentalist, 27*, 365–374.

Kampa, M., & Castanas, E. (2008). Human health effects of air pollution. *Environmental Pollution, 151*, 362–367. https://doi.org/10.1016/j.envpol.2007.06.012

Kaur, M., & Nagpal, A. K. (2017). Evaluation of air pollution tolerance index and anticipated performance index of plants and their application in development of green space along the urban areas. *Environmental Science and Pollution Research, 24*, 18881–18895.

Kuddus, M., Kumari, R., & Ramteke, W. P. (2011). Studies on air pollution tolerance of selected plants in Allahabad city. *Indian Journal of Environmental Management, 2*(3), 42–46.

Lorenc-Plucinska, G. (1982). Influence of SO_2 on CO_2 assimilation and carbon metabolism in photosynthetic processes in Scots pine. In *Arboretum Kornickie* (in Polish) (pp. 285–310), rocznik. XXVII.

Masuch, G., Kicinski, H., Kettrup, A., & Boss, K. S. (1988). Single and combined effects of continuous and discontinuous O_3 and SO_2 emission on Norway spruce needle: Histochemical and cytological changes. *International Journal of Environmental Analytical Chemistry, 32*, 213–241.

Mcdonald, A. G., Bealey, W. J., Fowler, D., Dragositsa, U., Skibaa, U., Smitha, R. I., Donovanb, R. G., Brettc, H. E., Hewittd, C. N., & Nemitz, E. (2007). Quantifying the effect of urban tree planting on concentrations and depositions of PM10 in two UK conurbations. *Atmospheric Environment, 41*, 8455–8467.

Narayan, D., Agrawal, M., Pandey, J., & Singh, J. (1994). Changes in vegetation characteristics downwind of an aluminum factory in India. *Annals of Botany, 73*, 557–565.

Ninave, S. Y., Chaudhri, P. R., Gajghate, D. G., & Tarar, J. L. (2001). Foliar biochemical features of plants as indicators of air pollution. *Bulletin of Environmental Contamination and Toxicology, 67*, 133–140. https://doi.org/10.1007/s00128-001-0101-3

Noctor, G., & Foyer, C. H. (1998). Ascorbate and glutathione: Keeping active oxygen under control. *Annual Review of Plant Physiology and Plant Molecular Biology, 49*, 249–279.

Offor, I. F., Adie, G. U., & Ana, G. R. (2016). Review of particulate matter and elemental composition of aerosols at selected locations in Nigeria from 1985–2015. *Journal of Health & Pollution, 6*, 1–18.

Prajapati, S. K., & Tripathi, B. D. (2008). Anticipated performance index of some tree species considered for green belt development in and around an urban area: A case study of Varanasi city, India. *Journal of Environmental Management, 88*, 1343–1349. https://doi.org/10.1016/j.jenvman.2007.07.002

Prasad, B. J., & Rao, D. N. (1981). Growth responses of *Phaseolus aureus* plants to petro- coke pollution. *Journal of Experimental Botany, 32*(6), 1343–1350.

Prusty, B. A. K., Mishra, P. C., & Azeezb, P. A. (2005). Dust accumulation and leaf pigment content invegetation near the national highway at Sambalpur, Orissa, India. *Ecotoxicology and Environmental Safety, 60*, 1193–1204.

Rai, P. K. (2013). Environmental magnetic studies of particulates with special reference to biomagnetic monitoring using roadside plant leaves. *Atmospheric Environment, 72*, 113–129.

Rai, P. K. (2016). Impacts of particulate matter pollution on plants: Implications for environmental biomonitoring. *Ecotoxicology and Environmental Safety, 129*, 120–136.

Rai, P., & Panda, L. S. (2014). Dust capturing potential and air pollution tolerance index (APTI) of some road side tree vegetation in Aizawl, Mizoram, India: An Indo-Burma hot spot region air quality. *Air Quality, Atmosphere and Health, 7*(1), 93–101. https://doi.org/10.1007/s1186

Rai, P. K., & Singh, M. M. (2015). *Lantana camara* invasion in urban forests of an Indo-Burma hotspot region and its ecosustainable management implication through biomonitoring of particulate matter. *Journal of Asia-Pacific Biodiversity, 8*(4), 375.

Reimann, C., Koller, F., Kashuline, G., Nilksvaara, H., & Englmaier, E. (2001). Influence of extreme pollution on the inorganic chemical composition of some plants. *Environmental Pollution, 115*, 239–252.

Seyyednejad, S. M., & Koochak, H. (2011). A study on air pollution-induced biochemical alterations in *Eucalyptus camaldulensis*. *Australian Journal of Basic and Applied Sciences, 5*(3), 601–606.

Seyyednejad, S. M., Niknejad, M., & Yusefi, M. (2009). The effect of air pollution on some morphological and biochemical factors of *Callistemon citrinus* in petrochemical zone in south of Iran. *Asian Journal of Plant Sciences, 8*, 562–565.

Seyyednjad, S. M., Majdian, K., Koochak, H., & Niknejad, M. (2011). Air pollution tolerance indices of some plants around industrial zone in south of Iran. *Asian Journal of Biological Sciences, 4*(3), 300–305.

Shandilya, K. K., Khare, M., & Bhusan, A. (2007). Suspended particulate matter distribution in rural-industrial Satna and in urban-industrial South Delhi. *Environmental Monitoring and Assessment, 128*, 431–445. https://doi.org/10.1007/s10661-006-9337-z

Shanker, A. K., Cervantes, C., Loza-Tavera, H., & Avudainayagam, S. (2005). Chromium toxicity in plants. *Environment International, 31*, 739–753.

Shannigrahi, A. S., Fukushima, T., & Sharma, R. C. (2004). Anticipated air pollution tolerance of some plant species considered for green belt development in and around an industrial/urban area in India: An overview. *International Journal of Environmental Studies, 61*(2), 125–137. https://doi.org/10.1080/0020723032000163137

Singh, S. K., & Rao, D. N. (1983). Evaluation of plants for their tolerance to air pollution. *Proceedings of the Symposium on Air Pollution Control, 1*, 218–224.

Singh, S. N., & Verma, A. (2007). Phytoremediation of air pollutants: A review. In S. N. Singh & R. D. Tripathi (Eds.), *Environmental bioremediation technology 1* (pp. 293–314). Springer.

Tausz, M., Van der Kooij, T. A. W., Muller, M., DeKok, L. J., & Grill, D. (1998). Uptake and metabolism of oxidized and reduced sulfur pollutants by spruce trees. In L. J. DeKok & I. Stulen (Eds.), *Responses of plant metabolism to air pollution and global change* (pp. 457–460). Backhuys Publishers.

Tripathi, A. K., & Gautam, M. (2007). Biochemical parameters of plants as indicators of air pollution. *Journal of Environmental Biology, 28*, 127–132.

Tripathi, A., Tiwari, P. B., & Singh, M. (2009). Assessment of air pollution tolerance index of some trees in Moradabad city. *Indian J Environmental Biology, 30*(4), 545–550.

Turk, R., & Wirth, V. (1975). The pH dependence of SO_2 damage to lichens. *Oecologia, 19*, 285–291.

Ulrich, M. M., Alink, G. M., Kumarathasan, P., Vincent, R., Boere, A. J., & Cassee, F. R. (2012). Health effects and time course of particulate matter on the cardiopulmonary system in rats with lung inflammation. *Journal of Toxicology and Environmental Health, A65*, 1571–1595.

Verma, A., & Singh, S. N. (2006). Biochemical and ultrastructural changes in plant foliage exposed to auto pollution. *Environmental Monitoring and Assessment, 120*, 585–602.

Wang, F., Zeng, B., Sun, Z., & Zhu, C. (2009). Relationship between proline and Hg2þ- induced oxidative stress in a tolerant rice mutant. *Archives of Environmental Contamination and Toxicology, 56*, 723–731.

Wei, X., Lyu, S., Yu, Y., Wang, Z., Liu, H., Pan, D., & Chen, J. (2017). Phylloremediation of air pollutants: Exploiting the potential of plant leaves and leaf-associated microbes. *Frontiers in Plant Science, 8*, 1–23. https://doi.org/10.3389/fpls.2017.013

Younis, U., Bokhari, T. Z., Malik, S. A., Ahmad, S., & Raja, R. (2013). Variations in leaf dust accumulation, foliage and pigment attributes in fruiting plant species exposed to particulate pollution from Multan. *International Journal of Agricultural Science and Research, 3*(3), 1–12.

Zobel, A. M. (1996). Phenolic compounds against in defense air pollution. In M. Yunus & M. Iqbal (Eds.), *Plant response to air pollution* (pp. 241–266). Wiley.

Chapter 28
Urban Expansion Around Kolkata: Evaluating Urbanogenic Interventions in New Town, Rajarhat

Anindya Basu and Adrija Bhattacharjee

28.1 Backdrop

By common parlance, any city having million-plus residents is referred to as a "metropolitan city," while the city having ten million-plus urban agglomeration is termed as a "megacity" (Chatterjee & Chattopadhyay, 2020). The United Nations Human Settlements Programme (UN-Habitat) devised a new jargon "metacity" or "hypercity" referring to large urban conurbations with more than 20 million population (United Nations, 2006).

As per the estimates of UN-Habitat, the urban world population by 2050 will reach 6.6 billion, with 68% of the people living in urban areas (United Nations, 2018a). This increase is really spectacular as in 1950 the share of urban population was only 30% (751 million), while in 2018 it touched 55% (4.2 billion). As anticipated, the degree of urbanization is high in the developed world, but in the present phase the developing countries, especially the Asian ones, have come up in a big way, paving the way for a kind of transition (United Nations, 2018a). Asia has been rightly termed as the "epicenter of the highest urban living" as the average annual rate of change of the urban population here has been as high as 1.47% during 2010–2015 and is expected to rise further (United Nations, 2018a).

It has been estimated that by 2030 there would be 662 cities having a million-plus residential population, out of which 41 cities will be having a population of over 10 million or more (United Nations, 2014). As per UN-Habitat statistics, in Asia, around 43% population resides in million-plus cities, out of which 14% population are residents of megacities (United Nations, 2018a); and following the trend of

A. Basu (✉) · A. Bhattacharjee
Department of Geography, Diamond Harbour Women's University, Sarisha, West Bengal, India

G. S. Bhunia et al. (eds.), *Anthropogeomorphology*, Geography of the Physical Environment, https://doi.org/10.1007/978-3-030-77572-8_28

correlated economic expansion and urban boom, by 2030, 8 out of the 10 largest megacities in the world will be in Asia only, surpassing Europe and America.

India accounted for 11% of the world's urban population in 2011 and is expected to rise by leaps and bounds. Interestingly, over the last century, the urban population increased 11 times (25.8 million in 1901 to 285.4 million in 2011), while the level of urbanization increasing only threefold (10.83% in 1901 to 31.16% in 2011) (Census of India, 2011).

In today's context, the role of humans as a geomorphological agent in reshaping the Earth's surface cannot be ignored anymore. The ever-increasing human population is tirelessly working for the "development" exploiting resources with newer technological know-hows and in turn acts as an effective operational force changing the Earth's surface (Brown, 1970). As the rate of urbanization is growing by leaps and bounds alongside the rapid population explosion, the problems associated with developing urban landscape have also become more prominent.

Kolkata Metropolitan Area, the 13th most populated area in the world (Majumdar & Chatterjee, 2020), faces severe congestion; the area has grown from 144 sq. km in 1970 to 633.2 sq. km in 2010 (Rajashekariah, 2011), resulting in the widespread conversion of wetlands and farmlands giving way to townships like Salt Lake and New Town, Rajarhat.

New Town, Rajarhat, a satellite township to the northeast of the megalopolis of Kolkata, was initiated in 1993 to ease the population burden both in terms of residential and commercial (Kundu, 2016). It was touted as the first "green, eco-friendly, self-sufficient, and smart city" of the state of West Bengal (India Smart City Mission, 2016). As per the master plan, West Bengal Housing Infrastructure Development Corporation (WBHIDCO) has been divided into four action areas having various land uses like residential, commercial, open spaces, waterbodies, etc., along with a Central Business District.

The basic area of concern is the continuous urban expansion and landscape modifications without having much concern about the geological structure of the area and extensive land use and land cover change, leading to the changed socio-economic composition of the region.

There are a handful of papers that have dealt with the issues like rapid population growth in urban areas and resultant unplanned urbanization for several cities worldwide. There are also papers that have mainly discussed about the various anthropogenic factors and investigated their impact on physical environment. But this chapter has tried to identify the physio-structural issues that are affecting the entire fabric of New Town, Kolkata, along with an attempt to figure out the increasing trend of urbanization in the study area and its effects on changing land use and land cover pattern. Evolving vertical land use along with the horizontal pattern has been taken into account to draw an overall scenario of urbanization of New Town before its initiation as a satellite town to present date. Analyzing all the factors, a number of feasible policy measures have been suggested to maintain the sustainability of the township.

28.2 Literature Review

Delving deep into urban expansion and resultant urbanogenic interventions having an idea about several concepts becomes imperative. For that a detailed literature review has been done about systematic urban expansion, nature of urban growth in India, germination of anthropogeomorphology, and role of urbanization as a prominent anthropogenic agent.

28.2.1 Framework Behind Systematic Urban Expansion

Cities are believed to be the real engines of growth and main producers of national wealth (Jacobs, 1984) for which there is a high tendency of population concentration in and around those. World Bank, 2009 highlighted the need for decentralization of a metropolitan region to address the growing inequalities between metropolises and adjacent suburban areas (Hamer, 1994). With the expansion of the metropolitan areas, mega-regions start to emerge, which are a combination of several megacities and act as centers of knowledge economy and play a main role in restructuring urban spaces providing an edge to Asia (Dahiya, 2012a, b). It was forecasted that by 2030 there would be a tripling of urban land area to cater to this growing urban population (Angel et al., 2011; Seto et al., 2012); by 2050, 52% of the global urban population will be from Asian cities. If this urban expansion is unregulated and unmanaged, then it would raise issues for environmental sustainability, urban insufficiencies like lack of resources, civic amenities, etc. (Brueckner & Sridhar, 2012; Libertun de Duren & Compeán, 2015).

To alleviate the pressure of metropolitan downtown region, planned suburban areas were ushered through introduction of satellite cities by Graham Romeyn Taylor in 1915 (Taylor, 1915). Gradually, the dependent, predominantly residential satellite towns (Chen, 2011) evolved to become independent with higher level of functions and the New Town concept came up in a big way in the post–World War II period in Britain where to deal with acute housing shortage, urban settlements were newly built nearby the thriving urban centers with a high degree of self-sufficiency (TCPA, 2014).

Habitat III Conference, Ecuador (2016), indicated that the Global South will come up with a new kind of urbanism, and India, with its unique properties of compact, mixed, and polycentric urban development, will have its own "New Urban Agenda" in line with Sustainable Development Goals (United Nations, 2015, 2018b), which will help in "promoting equitable growth of regions across the urban–rural continuum" through "sustainable urban and territorial planning". For accommodating the growing population in the metropolises of Asia, the rural hinterland often faces haphazard conversion of agricultural plots and fragmented developmental ventures (Tacoli, 2006; Dupont, 2007). This creates a unique challenge for the government and planning agencies to systematically deal with population explosion and land conflicts through sustainable, pragmatic policies.

28.2.2 *Outline of Urban Growth and Expansion: India*

Dutta (2013) categorized the patterns of urban expansion into three: (a) suburban sprawl or peri-urbanization leading to city regions; (b) low-density urbanism through gated community proliferation; and (c) urban compaction – increasing densities within the existing framework. For the Indian cases, the third one having a high density has been the most common one, creating sustainability and livability concerns. Frolking et al. (2013) and Beard et al. (2016) highlighted that in developing countries, more specifically India, the trend of horizontal outgrowth is more prominent than vertical or upward expansion. But over time, the Indian scenario has changed; tall vertical buildings, skyscrapers, gated communities, and to certain extent planned expansions have come into play.

As per the 2011 Census, there are eight megacities or urban agglomerations (UAs) – Mumbai, Delhi, Kolkata, Chennai, Bengaluru, Hyderabad, Ahmedabad, and Pune with a population of more than 5 million. Among these, three are by far frontrunners—Greater Mumbai UA (18.4 million), Delhi UA (16.3 million), and Kolkata UA (14.1 million). For balanced metropolitan growth, emphasis was given on creating smaller satellite towns around metropolises, long back through Fourth Five-Year Plan (1969–1974). The stress on dispersed urban growth continued through governmental plans, and the National Commission on Urbanization (NCU) was established in the late 1980s by the Planning Commission of India (Mehta & Mehta, 1989). Another landmark program, Jawaharlal Nehru National Urban Renewal Mission (JNNURM), was launched in 2005 during the tenure of Tenth Five-Year Plan.

Long back, in post-independent India several satellite towns were set up to accommodate the huge influx of refugee population like Ashoknagar in West Bengal (Sivaramakrishnan, 1977). More systematic satellite township planning was made at a later date like that of Bidhannagar (Salt Lake) near Kolkata (Rao, 1990). Later, in 2008, the Ministry of Urban Development framed strategies for infrastructure development in satellite towns around metropolitan cities through innovative Public–Private Partnership (PPP) models; Rajarhat, Kolkata, was one of the first seven satellite towns selected in the first phase of development (MoUD, 2015a).

Steinert et al. (2009) outlined a framework for "Making Cities Smart and Sustainable" focusing on areas like governance, people, environment, mobility, economy, and living. The Urban and Regional Development Plans Formulation and Implementation (URDPFI) Guidelines 2014 recognized the importance of interrelationship between urban and rural areas for holistic regional development (MoUD, 2015b). In 2015, Smart Cities Mission was launched for sustainable and inclusive development (MoUD, 2015b).

To keep the pace with increasing urban population and development, the Government of India has initiated massive investments for urban areas; in the last five years, INR 48,000 crores (7215 million USD) was sanctioned for developing 100 smart cities (Ministry of Housing and Urban Affairs, 2019). Shyama Prasad Mukherji Rurban Mission (SPMRM) was launched in 2016 to enhance the infrastructure and services in the rurban clusters (Ministry of Rural Development, 2016). The draft National Urban Policy Framework (NUPF) was formulated to set guiding principles of urban management (Ministry of Housing and Urban Affairs, 2018). To handle the rapidly growing megacities with overcrowded core and gradual crowding of the urban fringes, provision of satellite towns or new towns around the mother cities with planned infrastructure and housing have been a feasible option (Short et al., 2007; TCPO, 2007). It has been noted that the population growths in urban peripheries are much higher than that in the city core, which might be due to factors like easy availability of land and better infrastructure (HSMI-HUDCO-NIUA, 2017).

28.2.3 Contextualizing the Role of Humans in Landform Development: Germination of Anthropogeomorphology

There is a huge probability of intensification of this trend as the human-induced changes on the landscape will keep on affecting the natural processes acting (Lóczy & Süto, 2011). Environmental geomorphology handles the interaction between human and geographical environment taking into account the landforms and also the processes involved in the formation of those (Lóránt, 2012).

Gradually, a branch emerged, which primarily investigated the changes in the natural environment induced by human interactions (Bennett et al., 1999). To analyze the dual role of nature and man, the modern geomorphology has developed a new branch anthropogenic geomorphology or anthropogeomorphology, which primarily concentrates on landforms induced by human activities. This branch was first systematically explored by George P. Marsh (1864) through his monumental publications "Physical Geography as modified by human action," after which the Italian geologist Antonio Stoppani (1873) found human action comparable to other landscape-shaping forces and was the first to coin "Anthropozoic era" (Crutzen, 2006), which was followed by Robert L. Sherlock's "Man as a Geological Agent" (1922). The term anthropogeomorphology was devised in this phase (Golomb & Eder, 1964) to denote investigations of the direct human creation of landforms and human modification of natural (biophysical) processes (primarily through land use changes). After World War II, the emphasis on this branch increased in the 1960s and became more popular in the 1970s–1980s when the scientific–technological revolution increased the efficiency of resource extraction of human by manifold. The branch anthropogenic geomorphology is not merely the study of manmade

Table 28.1 Classification of anthropogeomorphic processes

Direct anthropogenic processes		Indirect anthropogenic processes	
Constructive	For example, urban constructions	Acceleration of erosion and sedimentation	For example, tillage
Excavational	For example, mining and quarrying	Subsidence	For example, mining failures
Hydrological	For example, river management techniques	Slope failure	For example, quasi-natural landslides
		Triggering earthquakes	For example, constructing river dams
**Direct anthropogenic processes are further classified into primary and secondary processes based on the nature of the landscape generated.		** Indirect anthropogenic processes are subdivided into qualitative (quasi-natural in character) and quantitative (prominent modification).	

Source: Compiled by authors from Haigh's (1978) and Szabó et al. (2010)

landforms but also investigates the human-induced surface changes involving prediction of probable hazards (Szabó et al., 2010).

Loh and Wackernagel (2004) estimated that nearly one-third of the Earth's surface is affected by direct or indirect anthropogeomorphological activities; the predominant activities being cultivation, plantation, grazing, deforestation, mining, and construction works, varying spatially depending on degree of population pressure, nature of economic activities, and state of technological advancement.

Nir (1983) determined that on the basis of the anthropic erosion rate, agricultural activities are the most substantial landscape-modifying anthropogenic activity, leading to accelerated rates of erosion and sedimentation. Hooke (2000) classified anthropogeomorphic processes into intentional (activities directly modifying the landscape like mining, construction) and unintentional (indirect activities influencing the landscape like agriculture, grazing) ones (Table 28.1). The landforms associated with anthropic activities are classified into three types:

(i) Excavational landforms (E type): These are kinds of negative landforms generated by excavational processes leading to material deficit.
(ii) Accumulation landforms (A type): These are kinds of positive landforms generated by accumulative processes mounting to aggradation.
(iii) Planated landforms (P type): These are formed by planation processes, where reduction of slope is prominent through smoothening processes.

There have been several attempts to quantify and map the impact of human intervention on environment and landforms. Erlich and Erlich (1990) attempted to quantify the potential human environmental impact through an equation – $I=P{\cdot}A{\cdot}T$; where I refers to environmental impact, P refers to population, A refers to per capita affluence while T is technology factor. Nir (1983) devised index of potential anthropic geomorphology on the basis of two parameters – the degree of development (DD) reflecting rate of human impact and the degree of perception (DP) concerning the perception of damage from anthropogeomorphological processes (AGP) taking into account climatic and relief conditions through constants. As per Nir, in urban society the degree of development is highest with the corresponding high rate of anthropogeomorphological processes, and if the urban population has higher exposure to education, then the general public degree of perception about potential hazard would be of considerable level. When the index, that is, IAGP>0.50, the hazard involves considerable damage and corrective measures are urgently required like that in the urban environments. Another similar kind of index, based on anthropic geomorphological sensitivity, was formulated by Rózsa and Novák (2011). The ratio of the anthropogeomorphological transformation (Rag) having six types of variation was calculated by Lóczy and Pirkhoffer (2009) as the ratio of earth movement (tons ha^{-1} yr^{-1}) caused by human activities (Va) and earth movement (tons ha^{-1} yr^{-1}) caused by natural processes (Vn). They opined that in case of human-controlled geo-environment like the built-up areas there is no role of natural processes at all.

28.2.4 Urbanization as a Prominent Anthropogenic Agent

In the Anthropocene, where man is the most active geomorphic agent, altering of the physical landscape has become common. Man-induced geomorphic transformation mirrored by relief and urban development relationship is a prolonged one. Mendöl (1963) mentioned that landscape is one of the pivotal allocating factors for settlement site selection. Human agency is the prime force behind the heavily modified landscapes integrating biophysical and socio-cultural processes. Zalasiewicz et al. (2010) and Harden et al. (2013) reiterated that human agency through settlement establishment and extensive industrial–urban activities became the major factor behind landscape evolution in Anthropocene epoch. Gradually, anthropogeomorphology formed a distinct identity (Brown et al., 2013) through the branch of urban geomorphology. Urban geomorphology grew as an established discipline in several countries (in the Soviet Union: Kotlov, 1961; in the USA: Coates, 1976; in Great Britain: Douglas, 1983). The scale and degree of modification by human agents depend on the nature of socioeconomic dynamics and technological advancement along with the kind of governmental policies in place. However, the interaction between geomorphic processes and urban development is a both-way interactive process (Csima, 2010), where the evaluation of land suitability for urban development along with monitoring of process-response system is done (Cooke, 1977). Changes in slope, land cover and land use, soil character, natural drainage channel geometry, land subsidence, etc., developed due to human modifications for expansion of built-up areas can be included within the purview of urban anthropogeomorphology. Urban surfaces include transportation lines, construction features, bare surfaces, and natural waterbodies.

Szabó et al. (2010) classified anthropogenic processes and landforms on the basis of eight major types of interventions, out of which urbanogenic is a significant one.

Influence of settlement (urban) expansion on the landscape is major, and its impact is known as urbanogenic. In case of most of the urban settlements, that is, urbanogenic interventions, the direct processes involved are formation of satellite towns or new towns (as primary process) and spillover built-up formations (as secondary process). The natural slope is modified through construction works, landfills triggering erosion or mass movements (Szabó, 1993), and creation of artificial hollows through urban engineering like construction of tunnels and underground transportation ways leading to slumps or cave-ins in (Gálos & Kertész, 1997) while the natural drainage lines and waterbodies are affected by built-up structures and runoff conditions are altered. The degree of intervention is more severe for the developing countries (Gupta & Ahmad, 1999). Macro-scale landscaping over a larger area for housing estates, entertainment parks, and industrial blocks require more technical planning to prevent surface erosion and mass movements; often this promotes restoration of disturbed water budget (Chin & Gregory, 2009) and architectural arrangements try to infuse ecological and landscape-aesthetic aspects too (Csima, 2010). Piscitelli (2018) suggested that to realize the potential of urbanity resource identification including the natural ones involving innovative techniques is imperative, along with provision of housing and infrastructure for all creating immense pressure on natural landscape; to maintain a balance between nature and urbanization, the idea of designing new naturescapes was also put forward, adding a new dimension to urbanogenic existences. There has been a widespread concern that in an urbanized world its relationship with adjacent rural stretches is often ignored (Schmid & Brenner, 2011), which hampers the continuum (Zijderveld, 1998).

The anthropogenic features produced in an urban landscape, referred to as urban geoforms, are the resultant impact of surface deformations, which might cause hazards like uneven surfaces, waterlogging, and subsistence (Haldar & Satpati, 2018), and proper identification of these can help in future urban planning and development issues through location-specific risk assessment.

28.3 New Town, Rajarhat: The Case in Point

Kolkata Metropolitan Area, the 13th most populated area in the world, faces severe congestion; the area has grown from 144 sq. km in 1970 to 633.2 sq. km in 2010 (Rajashekariah, 2011), resulting in the widespread conversion of wetlands and farmlands giving way to townships like Salt Lake and New Town, Rajarhat.

New Town, Rajarhat, a satellite township to the northeast of the megalopolis of Kolkata, was initiated in 1993 (Fig. 28.1) to ease the population burden both in terms of residential and commercial (Kundu, 2016). It was touted as the first "green, eco-friendly, self-sufficient, and smart city" of the state of West Bengal (India Smart City Mission, 2016).

As per the master plan, West Bengal Housing Infrastructure Development Corporation (WBHIDCO) has been divided into four action areas having various land uses like residential, commercial, open spaces, waterbodies, etc., along with a

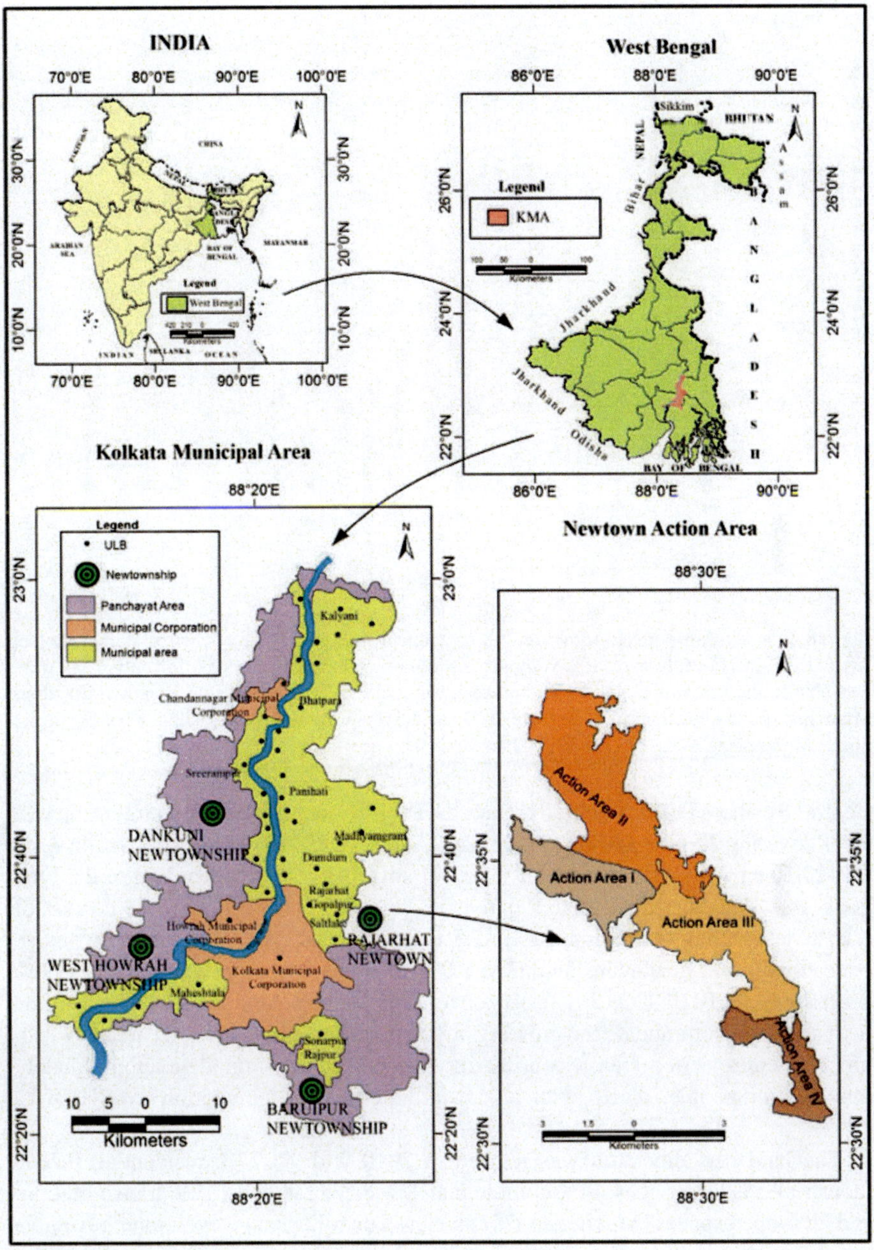

Fig. 28.1 Introducing the study area: New Town, Rajarhat, West Bengal

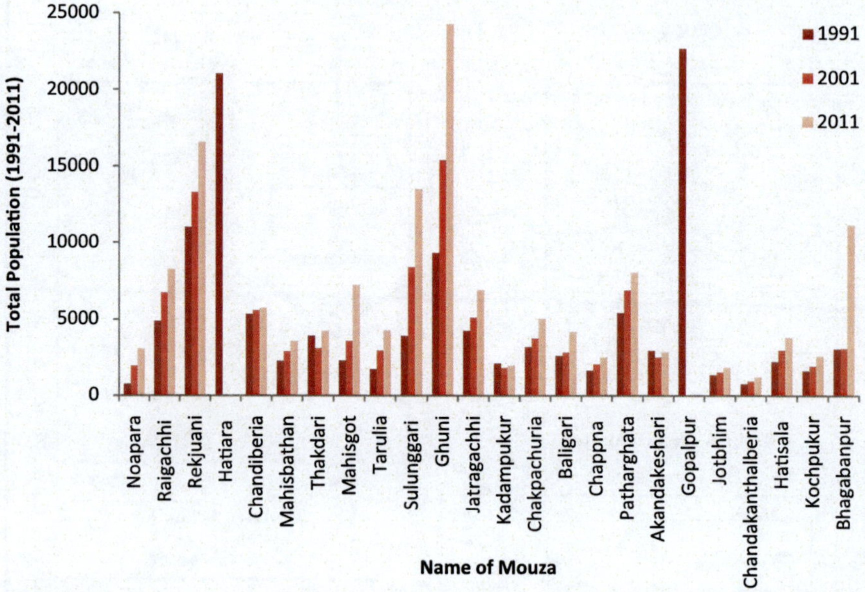

Fig. 28.2 Increasing population across the constituent mouzas of New Town, Rajarhat Project Area (1991–2011) (**In case of two mouzas – Hatiara and Gopalpur – data for only 1991 were available as the first one merged with another town and the latter turned into Rajarhat–Gopalpur Municipality shedding their rural attributes). (Data Source: District Census Handbooks of North and South 24 Parganas, West Bengal 1991–2011)

Central Business District. Initially, Rajarhat Project Area started off with three action areas covering 27.5 sq. km and 24 mouzas (Action Area I – 6.77 sq. km; Action Area II – 13.10 sq. km; Action Area III – 7.8 sq. km) (Fig. 28.2). According to the First Land Use Plan, the then total Rajarhat Project Area was supposed to consist of 30.5% residential, 7.3% industrial, 5.5% commercial area, 1% area for social, educational, and healthcare facilities, 8.8% for transport connectivity, and 47.6% open space (WBHIDCO, 2012). At a later date, Action Area IV was added to the township to accommodate the growing city population, taking the total area to 37 sq. km. The entire action area was primarily designed to accommodate approximately one million people, along with a half-million floating population (WBHIDCL, 1999).

The land use allocation was revised in 2010 with 41.77% residential, 0.23% industrial, 5% IT services, 11% commercial, 8% social facilities, 12% transportation, and 22% open spaces (Mukherjee, 2016). New Town Kolkata Development Authority (NKDA) constituted in 2008 is responsible for rendering civic services and amenities in the entire area comprising 34 mouzas covering Rajarhat Block of North 24 Parganas district, Bhangar II Block of South 24-Parganas and areas under two police stations of Kolkata district.

The notified New Town Action Area previously had a rural feel to it where the residents were engaged with primary sector production activities (Dey et al., 2016);

with the drastic transformation in the landscape from farmlands to self-sufficient township with diverse functions, the natives lost their land and livelihood (Sengupta, 2013). So, the rapid urbanization in the peri-urban area has not only affected environmental quality, but also altered the social composition of the region where now people from upper echelons of the society are able to afford the upscale apartments.

Since inception, New Town has witnessed large-scale destruction of the natural landscape. A complete rural milieu, where fishing and agriculture that were primary activities of the local residents, got completely transformed with the implementation of the township project (Mitra & Banerji, 2018). Due to the fast transformation of the region, both by public and private real estate agencies, there have been severe encroachments of not only arable land but also waterbodies and astonishingly at times with environmental clearance; some of which were even part of RAMSAR site of East Kolkata Wetlands (Majumdar & Kait, 2014). Interestingly, this change is often referred to as the reclamation of waterbodies rather than conversion to tone down the degree unplanned urbanogenic interventions in the name of township development. Motive of recreating waterbodies and urban greeneries once the initial wave of urbanizing is over, but is still long due.

28.4 Objectives

To assess the magnitude of anthropogenic interventions for rampant urbanization in New Town, Rajarhat, the main objectives are formulated:

(i) Theorization of urban expansion and resultant contextualization of urbanogenic activities in modifying landscape
(ii) Identification of physio-structural issues and analysis of increasing urbanizing trend
(iii) Delineation of spatio-temporal change of land use and land cover (LULC), urban clustering over a span of almost 40 years
(iv) Delivery of feasible policy interventions for sustainable urbanization

28.5 Materials and Methods

The study has been conducted to analyze the ongoing urbanogenic interventions in New Town, Rajarhat township. For this, help of both secondary and primary data has been sought with heavy reliance on satellite imageries.

From the available literature, the idea about anthropogeography and its relevance to today's rampant urbanization trend has been established; the published articles and reports have helped in developing an impression about the physical setting (geological, geomorphic) of the study area and in turn in highlighting the resultant vulnerabilities for the developing smart township. Informal, random interaction with the locals was done to get a sound idea about the ground reality and corroborate it with the findings derived from imagery analysis.

As the study area has high concentration of sky scrapers, to get an impression about the vertical expansion of the selected region, Sketch-up v2019 pro has been used to work on the CAD files downloaded from cadmapper.com and the three-dimensional views have been overlaid on their corresponding location in Google Earth Pro. Besides, Rapid Visual Screening (RVS) survey was also done with the help of Google Earth images and corresponding field photographs for the prominent housing projects.

Usage of Landsat imageries (Path 138, Row 44) acquired in the months of January–February for the years 1980, 2000, and 2020 from Landsat 3(MSS), 5 (TM), and 8 (OLI_TIRS) missions having a resolution of about 79, 30, and 30 m, respectively, has been made for implementing few geospatial techniques involving ARC GIS v10.2.2 to bring out the spatio-temporal variations. Attempts have been made to carry out land use–land cover analysis, along with corresponding change detection; calculate urbanization index; compute Shannon's entropy values for detecting degree of urban sprawl; figure out Normalized Difference Built-up Index (NDBI) and detect urban hotspots; and identify the notifiable urban clusters and implement kriging for predicting future urban growth trends. Through the above-mentioned methodological approach, the chapter tries to establish the effect of urbanization on the landscape.

28.6 Seismic Vulnerability and the Superstructure

The Kolkata metropolis, the second largest urban agglomeration of India, houses population of more than 14 million as per Census of India, 2011, giving the city the dubious distinction of one of the most populated parts on the globe (Nath et al., 2014). Due to this extremely high population pressure in Kolkata city proper and a congested Central Business District (Nandy, 2007), the built-up area expanded toward the eastern side encroaching the swampy–marshy lands and leading to the formation of townships like Salt Lake and New Town. Southern Bengal, including Kolkata and its surrounding areas, is a part of Zone V seismic zone as per the revised Indian Standards, comprising only four seismic zones (II, III, IV, and V), having high seismic sensitivity (Bureau of Indian Standards, 2002). The National Disaster Management Authority (2019) formulated Earthquake Disaster Risk Index taking into account hazard (H) of the area, the exposure (E) of local persons to the earthquake hazard, and known vulnerability (V) of the houses in that area and placed Kolkata in the medium risk bracket.

28.6.1 Structural Characteristics

Kolkata overlies the Bengal Basin, a part of Ganga–Brahmaputra river system consisting of fluvio-marine sediment of a pericratonic Tertiary basin (Vaccari et al., 2011). The area is at the margin of a sedimentary basin with a thickness of 7.5 km, placed above a crystalline basement (Murty et al., 2008). The thickness of the sediments gradually increases toward south and east (deepest parts being more than 16 km) (Curray & Moore, 1971; Murphy, 1988). The city is very close to the plate boundary zone (northeast and Indo–Burma ranges), which is one of the most seismically active regions of the world (Mohanty et al., 2013). The tectonic framework of Bengal Basin comprises the western stable shelf region or Indian platform, the central deep basin, and the eastern Chittagong–Tripura fold belt. The narrow Eocene Hinge Zone (EHZ), also referred to as Calcutta–Mymensingh gravity, highly separates the stable shelf region from the central deep basin or geo-synclinal area (Sengupta, 2011; Khandoker, 1989), which has been inactive for long. It demarcates a zone of differential thickening and subsidence rate of the overlying Oligocene and Miocene sections (Salt et al., 1986) and runs right through the middle of the city of Kolkata, cutting across New Town, Rajarhat, and Ranaghat, Nadia, before passing through Bangladesh. Besides, the presence of several other faults, flexures, and lineaments makes it highly vulnerable to earthquake hazards (Ganguly, 1997; Ghatak et al., 2017).

For multicriteria-based seismic hazard microzonation and vulnerability index construction, the factors that Nath et al. (2015) took into account were (a) probabilistic peak ground acceleration at the surface, (b) liquefaction potential index, (c) NEHRP soil site class, (d) sediment class, (e) geomorphology, (f) geology, and (g) groundwater table integrated on a hierarchical framework [analytical hierarchical process (AHP)] with the assignment of appropriate weights and ranks to each theme and feature. Structural Risk Index (SRI) > 0.5 indicates high vulnerability and risk. Along with these structural risk exposure factors, socioeconomic risk exposures were also taken into account, which add to the instability and vulnerability such as population density, nature of land use–land cover, building aspects like typology, age, and height (Ishita & Khandaker, 2010). Besides, often the other parameters like design, shape, building proximity, lateral strength, stiffness, ductility, foundation, material, and its construction practice, etc., are taken into account too (Sarris et al., 2010). McGuire (2004) pointed out that seismic hazard continues to be stable over a considerable period of geological time, but the risk associated with it varies with the increasing vulnerability.

However, out of these several factors were taken into account in building topology, especially building height being an important determining factor and can also be regulated by competent agencies too. It has been noted that the fundamental resonance frequency of structures starts from 2 Hz for low structures (up to four stories) and can be between 0.5 and 1 Hz for tall buildings (for 10–20 stories), so higher structures have the tendency to amplify longer period motions (Kramer, 1996).

28.6.2 *Ground Reality for the Study Area*

The surficial geology is characterized by 10–15 m of silty clay, below which relatively coarser sediments occur made up of either silt or clay or sand having combinations like sandy clay, silty sand, and silty clay (Vaccari et al., 2011; Govindaraju & Bhattacharya, 2012). Relatively, the younger soils like Holocene alluvium amplify the ground motion, leading to liquefaction, large ground deformation, and lateral spreading (Mohanty et al., 2013).

Geomorphologically, this typical deltaic flat land with surface elevation ranging between 5.5 and 9.5 m a.m.s.l is sloping toward the south and east (Roy et al., 2012). The groundwater table depth here (region having deltaic plains, interdistributary marshes, paleochannels, younger levees) is quite high, affecting the stability of the soil column. Youd and Perkins (1978) and Ganapathy and Rajawat (2012) pointed out that the area has affinity to moderate to high susceptibility to liquefaction during strong seismic shaking.

As per the calculated Structural Risk Index (SRI) by Nath et al. (2015), Salt Lake and New Town, Rajarhat, belong to the class 0.75<SRI≤1.0, indicating severe risk condition (refer Table 28.2).

When the building topology factors like material, age, height, etc., are added, then the vulnerability dips to moderate level as the building age is low in New Town area as the township is in its nascent stage and in few pockets the urbanization spree has not reached the peak, leaving open spaces or lower storied buildings with low population density, but with each passing year the age of the buildings would rise and more and more skyscrapers would come up increasing the risk index (Fig. 28.3). As of now, when the socioeconomic risk exposures are also taken into account as per the seismic hazard microzonation map of extended Kolkata, the combined hazard index class for New Town, Rajarhat, is 0:47 <HI≤0.68, indicating a high hazard condition.

Table 28.2 Structural Risk Index (SRI) indicative chart for New Town, Rajarhat

Criteria	Total number of classes	Class assigned (higher the class higher the impact)	Nature of feature
Geomorphology	7 classes	3rd class	Interdistributary marsh
NEHRP site class	4 classes	3rd class	D2 – 240.1–300.0
Sediment class	11 classes	10th class	Silty clay
PGA – peak ground acceleration (g) with 10% probability of exceedance in 50 years at surface	6 classes	6th class	0.242–0.253
Groundwater table [bgl(m)]	8 classes	8th class	7.0–7.7
Liquefaction potential index	4 classes	4th class	> 15
Seismic hazard microzonation	4 classes	4th class	0.68–0.88

Source: Modified by authors from Nath et al. (2015)

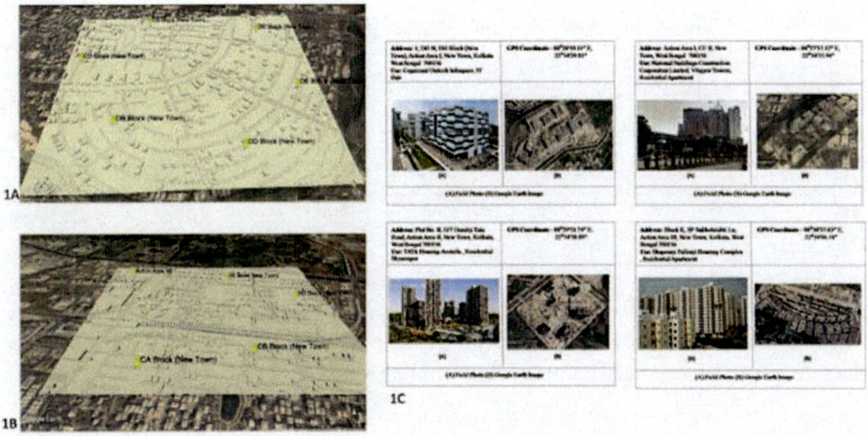

Fig. 28.3 Urban sky scenario in New Town, Rajarhat, West Bengal. Two adjoining portions of Action Area I each of 1 sq. km. (1A: GPS coordinate – 88°28′08.63″ E, 22°34′26.67″ and 1B: GPS coordinate – 88°28′00.09″ E, 22°35′01.44″) have been chosen for 3D mapping with the help of Cadmapper and Sketch-up v2019 and exported KML files have been loaded in Google Earth for visualization. 1C bears the imprint of four prominent housing projects represented through Rapid Visual Screening (RVS) survey

So, with increasing anthropogenic interventions the seismic vulnerability increases. But with the way the New Town is shaping up with huge high-rise complexes both residential and commercial and in some cases by unregulated filling up waterbodies, there remains a spot of bother as to how the human-induced factors will affect the region. Though the governmental agencies involved in shaping up New Town are very careful about the aesthetics and to some extent ecology, they are not too serious about the proactive, holistic, and integrated approach of strengthening disaster mitigation, preparedness, and response (National Disaster Management Authority, 2019). As per a study funded by the Union Ministry of Earth Sciences and conducted by IIT Kharagpur, the greater Kolkata standing on soft alluvium and witnessing a real estate boom might face a severe disaster from impending earthquake and resultant liquefaction. So, there is a need for developing disaster awareness, while going strong with high degree of anthropogenic interventions so that the area can withstand an earthquake beyond 6 in Richter Scale, which many geoscientists fear is in the offing (Thakur, 2015). For ensuring safety and sustainability of the sprawling urban areas, serious risk assessments and resultant urban planning–modeling is imperative, so that the urban fabric when exposed is capable of facing it armed with proper building codes and allied measures (BMTPC, 2006).

The residential projects, mostly in this region, are high-rise apartments with projects having different target beneficiaries comprising HIG, MIG, LIG, and EWS housing (refer Table 28.3).

Both public and private agencies are more interested in providing amenities, accessibility for luxurious households to the affluent section more as reflected by the percentage share of land allocated for MIG and LIG, which is as high as 79% (Bysack, 2004); the ones who wish to project themselves as environmentally aware

Table 28.3 Scenario of differential rental and buying demands of action areas

Action area	Monthly average rent of flat (Rs/1200 sq. ft.)	Rental demand of flats	Flat buying demand	Average sale price of flats (Rs/sq.ft.)	Price change (quarter IV 2014–quarter II 2016)
I	18,300	46%	46%	4900	−3%
II	18,000	17%	19%	4700	−5%
III	19,000	37%	35%	4500	5%

Data Source: Magicbricks (2016)

also harp on open spaces and waterbody availability, but are ignorant about the seismic hazard that looms large. The industrial units, IT hub, and educational hubs in the Central Business District also show similar ignorance.

28.7 Human-Controlled Geo-Environment

In any area, the degree of human impact is highest for the settlement areas as there the concentration and intensity of anthropic activities is the maximum and Kolkata and its surroundings are no exception. In a human-controlled geo-environment like in the cities where the impervious, built-up areas (roads, buildings, etc.) completely modify the natural surface, the natural geomorphological processes get completely mired (Lóczy & Pirkhoffer, 2009).

28.7.1 Physical Attributes and Resultant Influence

New Town, Rajarhat, is a part of alluvial tract of lower Ganga Basin with gentle southeasterly slope along with few local elevation and depressions, with maximum elevation below 9 mts above mean sea level and average being around 3 mts (Saha, 2009) (Fig. 28.4). As per ICAR (2001), the texture of the soil of the region is silty–clayey in nature comprising inceptisol and alfisols types. The township is located within the Bidyadhari Basin area with a gradient toward south–west, leading to the presence of waterbodies (Akher & Chattopadhyay, 2017). The Bagjolakhal is the main drainage artery running through the area along with Krishnapurkhal passing through the southern section, with both having outlet to Kulti Gang through gravity sluices (Roy & Dhar, 1995).

28.7.2 Urban Geoforms and Associated Traits

But the human interventions have affected the physical parameters too. The slopes have been modified through incessant construction works, natural vegetation has

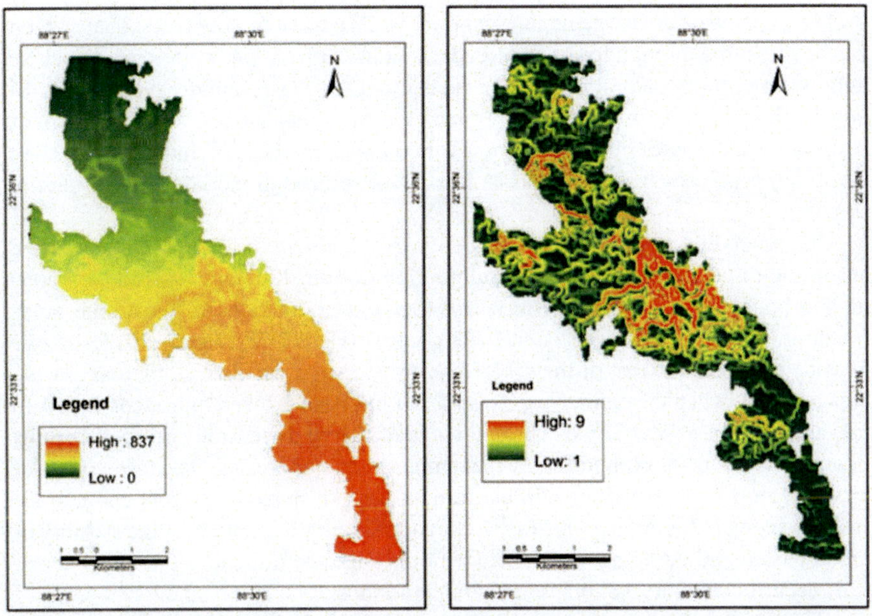

Fig. 28.4 Glimpses of physical attributes of New Town, Rajarhat, West Bengal. Both the maps have been generated from Shuttle Radar Topography Mission (SRTM) DEM, USGS. The first map shows the nature of relief, which is almost flat with highest value being 837 cm or 8.37 mt in the southern section (Action Area IV), and second map depicts the gentle slope of the study area in %, which has the corresponding highest value of 5°, mostly in the central part (Action Area III)

been removed, and marshy wetlands have been filled up in the wake of unplanned urbanization blocking the natural runoff pattern of the region, leading to severe waterlogging. The soil of the area having a high organic peaty content has high moisture-retaining capacity and poor bearing strength and is not suitable for heavy structures without any additional protective measures but is often overlooked in reality.

28.8 Remotely Sensed "Real World"

Though Newtown, Rajarhat, is a lately developed planned city, various pockets have been facing the menace of waterlogging, which even persists for months (Yengkhom, 2011; Chakraborti, 2020), and the main triggering factor behind this is the illegal filling up of waterbodies by real estate developing agencies. Karmakar (2015) reported that even the authorized agency in charge of planning and development of the township, that is, WBHIDCO itself, filled around 33 waterbodies covering an area over 405 sq. m.

The impacts of urbanogenic intervention, in case of New Town, Rajarhat, which can be measured through forms and trends, are raised planation, waterlogged pockets with obstruction of natural runoff. For recreational purpose, another type of anthropogenic intervention, in New Town, Rajarhat, urban green spaces, mostly in form of parks and waterbodies (like the prominent Eco Park), has been included (Roy, 2019) involving processes like planation and excavation through modification of natural slopes.

The relatively permanent urban geoforms that can be easily noticed here according to periodicity are occasional pot holes, rubbish dumps (sporadic); waterlogged pockets (short-term); broken surfaces and undulations in the fringe areas (medium-term); and clogged canals (long-term) (Haldar & Satpati, 2018). If one goes by the spatial extent of the urban geoforms, then it can be classified as per the area covered; when the entire New Town Planning Area is taken into account, it is in mega scale, while New Town Action Area is of macro dimension and smaller units like blocks or neighborhoods represent micro- or nanoscales. The categories of urban geoforms observed are surface formations like raised platforms and surface slope change predominantly due to constructional activities that affect the stability of urban terrain and dumping while surface failures like deformations due to wear and tear, occurrence of cracks, digging, and depressions.

WBHIDCO, which is trying to develop New Town, Rajarhat, as a futuristic Smart City, has prepared a Master Plan keeping infrastructural aspects like roads, drains, sewerage line, water supply lines, and major beautification works within its purview but mostly ignoring control of the physical parameters.

28.8.1 *Preparation of Land Use Land Cover Map*

For LULC, before going into classification processes, preprocessing operations were carried out to do away with radiometric and geometric distortions (Butt et al., 2015). Then using maximum likelihood, supervised classification output with conventional color-coding was generated based on four macro-classes (waterbodies, vegetation, built-up area, and farmland – on the basis of spectral signatures) with the help of training output creation through creation of region of interest (ROIs).

Over the years (span of 40 years), there has been a stark change in the various categories. Over the years, the settlement (built-up area) has increased from 0.8% in 1980 to 7.4% in 2000, with the latest figure being 30.2% in 2020 with the settlement character changing from dispersed rural to concentrated urban with distinct commercial zone. The study area initially had a rural character with high percentage of agricultural land (56.6% in 1980), which came down to 47.0% in 2020. In the wake of urbanization, the vegetated areas also faced a sharp decline of over 10% in the said period, the sparse vegetation remaining in the region can mostly be attributed to the parks constructed mostly in Action Areas I and II. With the increasing trend of real estate development, from the late 1990s, waterbodies were affected severely; from 21.5% in 2000, it came down to 13.7% in 2020 (refer Table 28.4 and Fig. 28.5). So,

Table 28.4 Area and percentages of LULC classes for the period 1980–2020.

Years	LULC classes							
	Waterbody		Vegetation		Agricultural land		Settlement	
	(sq. km.)	(%)	(sq. km.)	(%)	(sq. km.)	(%)	(sq. km.)	(%)
1980	8.4	22.8	7.3	19.8	20.8	56.6	0.3	0.8
2000	7.9	21.5	6.42	17.5	19.6	53.6	2.65	7.4
2020	5.04	13.7	3.33	9.1	17.23	47.0	11.1	30.2

Source: Computed by authors from satellite imageries of Landsat 3, 5, and 8 for 1980, 2000, and 2020, respectively

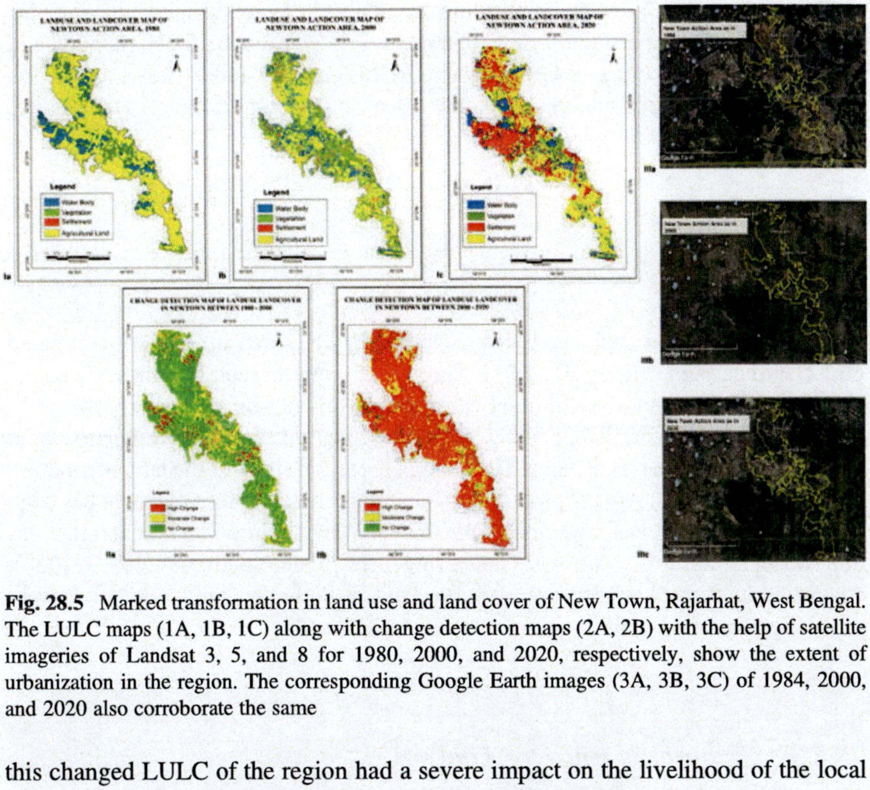

Fig. 28.5 Marked transformation in land use and land cover of New Town, Rajarhat, West Bengal. The LULC maps (1A, 1B, 1C) along with change detection maps (2A, 2B) with the help of satellite imageries of Landsat 3, 5, and 8 for 1980, 2000, and 2020, respectively, show the extent of urbanization in the region. The corresponding Google Earth images (3A, 3B, 3C) of 1984, 2000, and 2020 also corroborate the same

this changed LULC of the region had a severe impact on the livelihood of the local people who from primary producers became mostly laborers.

In order to assess the quality of a classification, it is extremely necessary to evaluate the accuracy of the classification (Reis, 2008; Torahi et al., 2011). It would be noteworthy to mention that an overall accuracy of 88.45% and kappa coefficient of 0.84 were obtained from the 1980 map; the accuracy clocked 92.17% in the 2000 map with a kappa coefficient of 0.89 while for the 2020 map the accuracy was 84.72% and kappa coefficient 0.79.

From the above information, urbanization index can be calculated. Urbanization process is not only dependent on the population increase (Durieux et al., 2008) but is also reliant on the expansion of urbanized areas too (Franco 2016).

Table 28.5 Urbanization indices for the years 1980, 2000, and 2020

	Years		
	1980	2000	2020
Urbanization index (%)	0.81	7.16	30.00

Source: Computed by authors from satellite imageries of Landsat 3, 5, and 8 for 1980, 2000, and 2020, respectively

$$UI = (Ua/Ta) \times 100 \tag{28.1}$$

where UI is the urbanization index, Ua is the urban built-up area in a particular zone, while Ta is the total area of that particular zone.

This index, apart from being an indicator of level of urbanization process, also indicates the status of economic development too (refer Table 28.5). Higher value of UI indicates higher degree of development. In this case, the index shows a meteoric rise of urban settlements in the study area, indicating severe anthropogenic intervention.

28.8.2 *Change Detection*

For change detection analysis, two independently classified images are compared post classification (Petit et al., 2001). Through the matrix table, "from–to" change class is obtained from pixel-to-pixel comparison; this not only detect changes, but identifies the nature of change along with determination of areal extent and spatial pattern of the changes (Fig. 28.5). The change detection study of the land use pattern of New Town Action Area depicts that in the initial phase 1980–2000 there has been a minor change with just initiation of the township beginning. The human intervention was gradually on rise with increasing urbanogenic activities (2000–2020), where the land parcels used as cultivable land, waterbodies, and vegetation were gradually converted to built-up areas (refer Table 28.6).

28.8.3 *Shannon's Entropy Analysis*

Integrating Shannon's entropy with GIS tools, computation has been done to measure the patterns of urban growth and detect the urban sprawl phenomenon (degree of spatial concentration or dispersion of the features over a selected area) in the region (Sarvestani et al., 2011).

$$H = \sum_{i=1}^{n} Pi\,(\log\ Pi)\ [\text{following Yeh\&Li, 2001}] \tag{28.2}$$

Table 28.6 Overall change detection matrix: cross-tabulation of LULC classes between 2000 and 2020 (area in sq.km)

Class	Agricultural land	Settlement	Vegetation	Waterbody	Grand total
Agricultural land	10.44	1.12	2.38	3.38	17.32
Settlement	6.51	0.74	1.44	2.42	11.11
Vegetation	0.76	0.27	1.73	0.50	3.26
Waterbody	2.17	0.33	0.96	1.38	4.84
Total	19.88	2.46	6.51	7.68	36.53

Source: Computed by authors from satellite imageries of Landsat 3, 5, and 8 for 1980, 2000, and 2020, respectively

In this case, H is the value of Shannon's entropy, Pi the ratio of built-up area to total area in a particular zone, and n is the total number of zones into which the study area is divided.

The entire area was divided into three concentric circles of incrementing radii (Zone I: 0–1 km; Zone II: 1–2km; and Zone III: 2–2.5 km) from the center of the city so as to visualize changes at neighborhood levels (Fig.28.6) and detect the nature of urban sprawl (Burgess, 1998).

Computed Shannon's entropy (H) based on the given formulae is:

$$\begin{aligned} H1980 &= -\Sigma ni = 1 \ \text{Pi}(\log \ \text{Pi}) \\ &= -(-0.89794) \\ &= 0.89794 \end{aligned}$$

From Tables 28.7, 28.8, and 28.9, entropy values have been derived; the values of entropy for the years 1980–2020 are closer to the upper limit of Ln(3) = 1.09861, and hence it means that urban growth during the study period has been highly dispersed. The Shannon's entropy values are continuously increasing from 1980 to 2020, which is indicative of increasing sprawl.

It is noticeable that the values start dipping once the sprawl decreases and compaction begins. There is a temporal drop in values of difference in Shannon's entropy, which is an indication that there is a reduction in the levels of sprawl.

28.8.4 Normalized Difference Built-up Index (NDBI)

Spectral indices are used to distinguish between land cover types; these indices are operations between spectral bands that are suitable for extracting information. Normalized Difference Built-up Index (NDBI) is one of them (Zha et al., 2003). The NDBI is applicable for impervious surfaces or built-up areas, which is calculated from remote sensing data using reflectance in the NIR and mid-infrared (MIR) portions of the spectrum. It can be expressed as

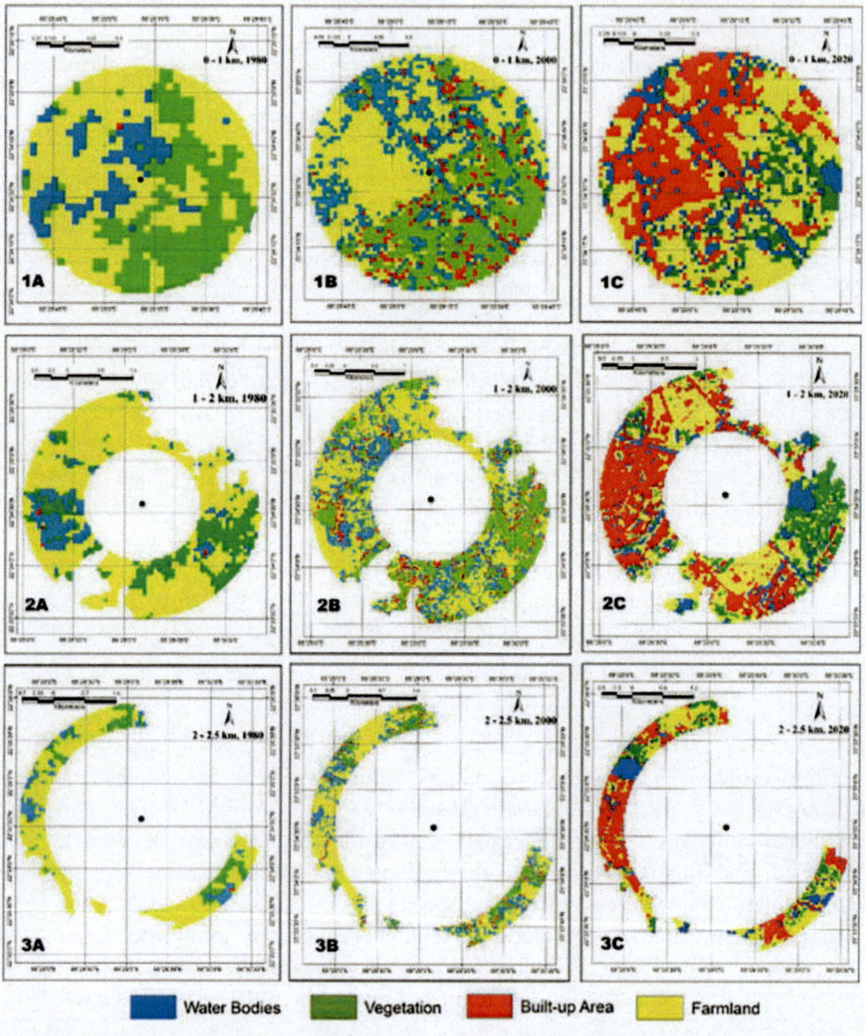

Fig. 28.6 LULC segments for Shannon's entropy analysis of New Town, Rajarhat, West Bengal. Satellite imageries of Landsat 3, 5, and 8 for 1980, 2000, and 2020, respectively, have been taken, and three concentric circles of incrementing radii (Zone I: 0–1 km [1A,1B,1C]; Zone II: 1–2 km [2A,2B,2C]; and Zone III: 2–2.5 km [3A,3B,3C]) from the center of the township have been drawn

$$\mathrm{NDBI} = (\mathrm{SWIR} - \mathrm{NIR})/(\mathrm{SWIR} + \mathrm{NIR}) \tag{28.3}$$

The value lies between (−)1 to (+)1. The negative values tend to indicate the presence of waterbodies while increasing value indicates the presence of more built-up area.

Table 28.7 Urban built-up class area in three zones in 1980, 2000, and 2020

Zones	Years (area in sq m)		
	1980	2000	2020
I	900	230400	1104300
II	5400	501300	2524500
III	2700	180900	1351800
Total area	9000	912600	4980600

Source: Computed by authors from satellite imageries of Landsat 3, 5, and 8 for 1980, 2000, and 2020, respectively

Table 28.8 Pi and Ln(Pi) values for 1980, 2000, and 2020

	1980		2000		2020	
Zones	Pi	Ln(Pi)	Pi	Ln(Pi)	Pi	Ln(Pi)
I	0.1	−2.3026	0.2525	−1.3763	0.2217	−1.5064
II	0.6	−0.5108	0.5493	−0.5991	0.5069	−0.6794
III	0.3	−1.2040	0.1982	−1.6185	0.2714	−1.3042

Source: Computed by authors

Table 28.9 Shannon's entropy in 1980, 2000, and 2020

Year	Shannon's entropy	Difference in Shannon's entropy	Ln(3)
1980	0.89794	–	1.09861
2000	0.99739	0.09945	
2020	1.03232	0.03493	

Source: Computed by authors

From the given map of the study area (Fig.28.7), it becomes evident that the values are higher for Action Areas I and II, where township building was started in the initial phases and was later extended to Action Area III. Since Action Area IV is in the nascent stage of urban development, the impact of urbanogenic activities is the least and thus the value of NDBI is low accordingly. Though Action Area I has the highest built-up area concentration, NDBI value is a tad bit lower due to the presence of vegetation and waterbodies providing moisture content to the soil layer alongside the impervious layer.

28.8.5 Hotspot Analysis

The hotspot analysis (through Getis-Ord Gi* tool) has been used to assesses the presence of high or low values of urban cluster spatially. So, hotspot detection has evolved from the study of point distributions or spatial arrangements of points in a space (CDC, 2010) with the help of z-score, p-value, and confidence level bin (Gi_Bin). The statistically significant spatial clusters of high values or higher

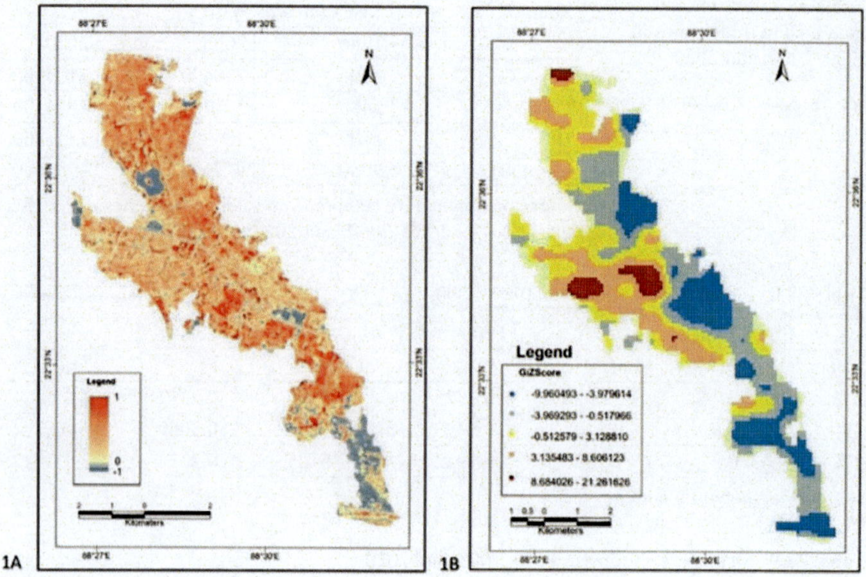

Fig. 28.7 Present built-up scenario in New Town, Rajarhat, West Bengal, using satellite imagery of Landsat 8, 2020. 1A depicts the Normalized Difference Built-up Index (NDBI) while 1B detects the urban hotspots

concentration are termed as hotspots while that of low values or lower concentration are called cold spots; when the Gi* statistic value is near 0, it indicates no spatial clustering.

The hotspot map for New Town, Rajarhat (Fig.28.7), which is a planned township, shows distinct characteristics. In Action Areas I and II, which are comparatively older, the concentration of residential and commercial buildings is more, showing pockets of significant spatial clusters of high values. In Action Areas III and IV, negatives values have been recorded, indicating the presence of cold spots, meaning that the urban scenario has not warmed up much here.

28.8.6 Identifying Urban Patches and Anticipating Trends

Urban growth patterns are broadly categorized into infilling, edge-expanding, and outlying (Wilson et al., 2003). From the LULC map, NDBI map, and hotspot map, patches of settlements of different patterns have been identified (Fig.28.8). Since the efforts of setting up the township started from the late 1990s, patches could be seen from 2000. The urban density varied among the patches; the number of outlying patches was greater in number for 2000 but was dispersed in nature. During 2020,

1A 1B

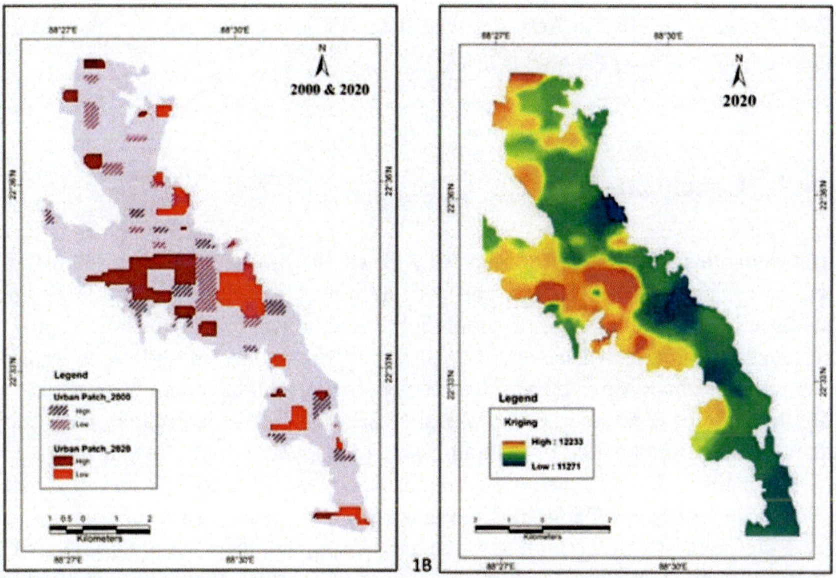

Fig. 28.8 Urbanizing trends then, now, and in New Town, Rajarhat, West Bengal, using satellite imagery of Landsat 5 and 8 of 2000 and 2020. Settlement patches with varying compactness can be identified from 1A while 1B through interpolation and covariance functions tries to identify pockets of future growth

the number of clusters diminished but was of more compact nature as infilling and edge expansion could be identified. Concentration and compactness of the urban patches are the most prominent in the central part in Action Area I.

Now from present scenario, one can try to predict the future trajectory of further urban expansion, with the help of kriging and spatial autocorrelation. Continuous functional surfaces are often generated to represent terrestrial surfaces through the formation of pseudo-three-dimensional surfaces with x, y, and z values (Oliver & Webster, 1990). On the other hand, interpolation procedure is used on such surfaces to predict the values of cells at locations where there is a dearth of sampled points. The principle of spatial autocorrelation determines the degree of interrelationship and tries to figure out a spatial pattern. Kriging is an advanced geostatistical method that predicts values for a geographical area, generating an estimated surface from a scattered set of points with z-values. It is a kind of regression that gives a least square estimate of data (Remy et al., 2009). This method is most effective when there is a known directional bias in the data. Kriging involves a multistep process – exploratory statistical analysis, variogram modeling, usage of covariance functions, and creation of surface. This not only generates an interpolated spatial model but also creates an estimate of uncertainty of each sampled site.

Generally, low values within the output variance of prediction raster indicate a high degree of confidence in the predicted value. So, in New Town the areas having low kriging values like in Action Areas III and IV in the southern section hold more promise for further expansion (Fig.28.8).

28.9 Conclusion

The initiations were taken to build New Town (Rajarhat), a planned township, to reduce the pressure on Kolkata and provide urban facilities like residential apartments at an affordable price, providing job opportunities in IT hubs but keeping the sustainability unaffected. Several unplanned actions in and around the township in the garb of urbanogenic activities have perished the human–nature balanced relationship. Still it is not too late to take preventive as well as alternative measures to maintain the ultimate goal of making New Town a smart city in recent future.

The main findings of the study are that New Town, Rajarhat, which is a part of alluvial tract of lower Ganga Basin, ranks high in terms of structural risk, and with the increase in age of the high skyscrapers the risk will increasingly go up. But the gradual urbanogenic interventions are going on in the study area (1980–2020), resulting in extensive changes in land use and land cover with a sharp decline in vegetation, agricultural land and waterbodies, raised planation, and waterlogged pockets with obstruction of natural runoff. The increasing value of urbanization index indicated that with increasing level of urbanization the township is developing economically too. Considering the continuously increasing value of Shannon's entropy, a highly dispersed pattern of urbanization has been noted for the said time span. The NDBI values for the four action areas proves that urbanogenic activities are highest in Action Areas I and II, followed by Action Area III and are least in Action Area IV as are still in the nascent stage. Hotspot analysis showed that the concentration of built-up environment is high for Action Areas I and II, which have developed comparatively earlier. The degree of compactness of the settlement clusters has also increased over time evolving from the initial dispersed patches.

The way forward for developing a smart new town in the city periphery in this case can be charted as follows:

A large proportion of the township should remain under the control of nonprofit making public body so that the developmental activities would aim at being efficient, self-contained, and socially balanced with affordable housing, educational centers, health facilities, and cultural hubs (Jana & Sarkar, 2018) rather than having only money-making residential complexes.

Without exerting further pressure on the already developed action areas, making them more concretized and clumsier, efforts have to be made to find out new comparatively vacant areas for absorbing future metropolitan growth with residential and commercial facilities through public–private partnership model.

Initiating an extensive landscape program including provision of trees and flowering plants in the parks and playgrounds within and outside the complexes for controlling the urban heat island effects decreases dust and noise pollution apart from adding to the aesthetic appeal of the entire area.

Efficient use of land can be made by planned vertical growth keeping demand–supply gap in mind, but the urban sky growth has to be made (Dhali et al., 2019) by following the building norms set by the New Town Development Authority of restricting building heights within 15 m, keeping the impending seismic vulnerability in mind. Aspects like rainwater harvesting and artificial groundwater recharge need proper boost to promote green living.

As waterbodies have been indiscriminately filled up by the real estate developers, efforts have to be taken to stop further conversions and step up the recreation of artificial waterbodies. These can also be used for commercial fishery and help in revenue generation for WBHIDCO. Eco Urban Village and Eco Park are such milestone projects.

References

Akher, S. K., & Chattopadhyay, S. (2017). Significance of micro-morphology in town planning—A case study of Kolkata new town. *Indian Journal of Regional Science, 49*(1), 113–117.

Angel, S., Parent, J., Civco, D. L., & Blei, A. M. (2011). *Making room for a planet of cities*. Policy focus report. Lincoln Institute of Land Policy, Cambridge. https://www.lincolninst.edu/sites/default/files/pubfiles/making-room-for-a-planet-of-cities-full_0.pdf. Last accessed 7 Sept 2020.

Beard, V. A., Mahendra, A., Westphal, M. I. (2016). *Towards a more equal city: framing the challenges and opportunities*. Working paper. World Resources Institute, Washington, DC. http://www.wri.org/sites/default/files/WRR_Framing_Paper_Final_Nov4.pdf. Last Accessed 4 July 2020.

Bennett, M. R., Doyle, P., & Nichol, D. (1999). Environmental geology: Geology and the human environment. *Engineering Geology, 50*(1), 221–222.

BMTPC. (2006). *Vulnerability atlas of India, first revision, buildings materials and technology promotion council, Ministry of Housing and Urban Poverty Alleviation*. Govt. of India.

Brown AG, Tooth S, Chiverrell RC, Rose J, Thomas DSG, Wainwright J, Bullard JE, Thorndycraft VR, Downs P (2013) The Anthropocene: Is there a geomorphological case? Earth Surface Process and Landform. 38: 431–434. wileyonlinelibrary.com https://doi.org/10.1002/esp.3368. Last accessed 21 Aug 2020.

Brown, E. H. (1970). Man shapes the earth. *The Geographical Journal, 136*(1), 74–85.

Brueckner, J. K., & Sridhar, K. S. (2012). Measuring welfare gains from relaxation of land-use restrictions: The case of India's building-height limits. *Regional Science and Urban Economics, 42*(6), 1061–1067. https://doi.org/10.1016/j.regsciurbeco.2012.08.003

Burgess, P. (1998). *Revisiting "sprawl": Lessons from the past*. Maxine Goodman Levin College of Urban Affairs at Cleveland State University.

Bureau of Indian Standards. (2002). Indian standard criteria for earthquake resistant design of structures, IS 1893: (Part 1). In: *General provisions and buildings* (Fifth Revision), New Delhi.

Butt, A., Shabbir, R., Ahmad, S. S., & Aziz, N. (2015). Land use change mapping and analysis using remote sensing and GIS: A case study of Simly watershed, Islamabad, Pakistan. *The Egyptian Journal of Remote Sensing and Space Science, 18*(2), 251–259.

Bysack, R. (2004). *Housing development in the new towns around Kolkata*. Masters thesis, School of Planning and Architecture.

CDC – Centers for Disease Control and Prevention. (2010). *Introduction to hotspot analysis*. https://www.cdc.gov/dhdsp/maps/GISX/training/module3/files/3_hotspot_analysis_module.PDF. Last accessed 9 September, 2020.

Census of India. (2011) *Rural urban distribution of population*. http://censusindia.gov.in/2011-prov-results/paper2/data_files/india/Rural_Urban_2011.pdf. Last accessed on 28 July 2020.

Chakraborti, S. (2020). Government agencies join hands to resolve Kolkata's waterlogging problem. *Times News Network*. http://timesofindia.indiatimes.com/articleshow/76923897.cms?utm_source=contentofinterest&utm_medium=text&utm_campaign=cppst. Last accessed 9 Sept 2020.

Chatterjee, A., & Chattopadhyay, R. N. (2020). Growth of metropolises and megacities with focus on global south. In *Satellite towns in neo-metropolitan development in India* (pp. 1–28). Springer.

Chen, M. (2011). *From satellite towns to new towns, evolution and transformation of urban spatial structure in Chinese metropolises*. Dissertation, University of Hong Kong, Pokfulam, Hong Kong SAR.

Chin, A., & Gregory, K. (2009). From research to application: Management implications from studies of urban river channel adjustment. *Geography Compass, 3*(1), 297–328.

Coates, D. R. (Ed.). (1976). *Geomorphology and engineering* (p. 360). Dowden, Hutchinson & Ross, Stroudsburg.

Cooke, R. U. (1977). Urban geomorphology. *The Geographical Journal, 142*(1), 59–65. https://doi.org/10.2307/1796025

Crutzen, P. (2006). The anthropocene. In E. Ehlers & T. Krafft (Eds.), *Earth system science in the Anthropocene: Emerging issues and problems* (pp. 13–18). Springer.

Csima, P. (2010). Urban development and anthropogenic geomorphology. In J. Szabó, L. Dávid, & D. Lóczy (Eds.), *Anthropogenic geomorphology: A guide to man-made landforms* (pp. 179–187). Springer.

Curray, J. R., & Moore, D. G. (1971). Growth of the Bengal deep-sea fan and denudation in the Himalayas. *Geological Society of America Bulletin, 82*(3), 563–572.

Dahiya, B. (2012a). 21st century Asian cities: Unique transformation, unprecedented challenges. *Global Asia, 7*(1), 96–104.

Dahiya, B. (2012b). Cities in Asia, 2012: Demographics, economics, poverty, environment and governance. *Cities, 29*(2), S44–S61. https://doi.org/10.1016/j.cities.2012.06.013

Dey, I., Samaddar, R., & Sen, S. K. (2016). Beyond Kolkata: Rajarhat and the dystopia of urban imagination. Routledge. Growth and a century: A micro study of Kolkata, India. *International Research Journal of Social Sciences, 5*(6), 16–23.

Dhali, MK, Chakraborty M, Sahana M (2019) Assessing spatio-temporal growth of urban sub-centre using Shannon's entropy model and principle component analysis: A case from North 24 Parganas, lower Ganga River Basin, India The Egyptian Journal of Remote Sensing and Space Science 22(1): 25-35.

District Census Handbooks of North and South 24 Parganas, West Bengal. Census of India. 1991–2011.

Douglas, I. (1983). *The urban environment* (p. 240). Edward Arnold.

Dupont, V. (2007). Conflicting stakes and governance in peripheries of large indian metropolises—An introduction. *Cities, 24*(2), 89–94. https://doi.org/10.1016/j.cities.2006.11.002

Durieux, L., Lagabrielle, E., & Nelson, A. (2008). A method for monitoring building construction in urban sprawl areas using object-based analysis of spot 5 images and existing GIS data. *ISPRS Journal of Photogrammetry and Remote Sensing, 63*(4), 399–408.

Dutta, V. (2013). Land use dynamics and peri-urban growth characteristics: Reflections on master plan and urban suitability from a sprawling north Indian city. *Environ Urbanization Asia, 3*(2), 277–301. https://doi.org/10.1177/0975425312473226

Erlich, P. R., & Erlich, A. H. (1990). *The population explosion*. Simon and Schuster.

Gálos, M., & Kertész, P. (1997). Mérnökgeológia (Engineering geology). University Notes, M˝uegyetemiKiadó, Budapest. 35.

Ganapathy, G. P., & Rajawat, A. S. (2012). Evaluation of liquefaction potential hazard of Chennai city, India: Using geological and geomorphological characteristics. *Natural Hazards, 64*(2), 1717–1729.

Ganguly, S. (1997). Petroleum geology and exploration history of the Bengal Basin in India and Bangladesh. *Indian Journal of Geology, 69*, 1–25.

Ghatak, C., Nath, S. K., & Devaraj, N. (2017). Earthquake induced deterministic damage and economic loss estimation for Kolkata, India. *Journal of Rehabilitation in Civil Engineering, 5* (2), 1–21.

Golomb, B., & Eder, H. M. (1964). Landforms made by man. *Landscape, 14*(1), 4–7.

Govindaraju, L., & Bhattacharya, S. (2012). Site-specific earthquake response study for hazard assessment in Kolkata city, India. *Natural Hazards, 61*(3), 943–965.

Gupta, A., & Ahmad, R. (1999). Geomorphology and the urban tropics: Building an interface between research and usage. *Geomorphology, 31*(1–4), 133–149.

Franco, S. (2016). *Urbanization and its implications on an emerging mega city*. Doctoral thesis. VIT University. http://hdl.handle.net/10603/151750. Last accessed 17 June 2020.

Frolking, S., Milliman, T., Seto, K. C., & Friedl, M. A. (2013). A global fingerprint of macro-scale changes in urban structure from 1999 to 2009. *Environmental Research Letters, 8*(2), 024004. https://doi.org/10.1088/1748-9326/8/2/024004

Haigh, M. J. (1978). *Evolution of slopes on artificial landforms* (Dept Geol res papers 183). University of Chicago.

Haldar, A., & Satpati, L. N. (2018). Urban geo-forms: Concept and significance in Anthropogeomorphology. *Journal of Indian Geomorphology, 6*, 108–115.

Hamer, M. A. (1994). Economic impacts of third world mega-cities: Is size the issue? In R. Funchs, E. Brennan, J. Chamie, F. Lo, & J. Uitto (Eds.), *Mega city growth and the future* (1st ed., pp. 172–191). United Nations University Press.

Harden, C. P., Chin, A., English, M. R., Fu, R., Galvin, K. A., Gerlak, A. K., McDowell, P. F., McNamara, D. E., Peterson, J. M., Poff, N. L., Rosa, E. A., Solecki, W. D., & Wohl, E. (2013). Understanding human–landscape interactions in the "Anthropocene". *Environmental Management, 53*(1), 4–13.

Hooke, R. L. (2000). On the history of humans as geomorphic agents. *Geology, 28*(9), 843–846.

HSMI-Human Settlement Management Institute HUDCO-Housing and Urban Development Corporation Chair-NIUA-National Institute of Urban Affairs. (2017). *Urban India—Status of demography, economy, social structure, housing and basic infrastructure*. HSMI-HUDCO Chair-NIUA Collaborative Research. National Institute of Urban Affairs, New Delhi. https://www.niua.org/sites/default/files/HUDCO-Consolidated.pdf. Last accessed 5 July 2020

ICAR. (2001). *NBSS & LUP Hal soil series of West Bengal*. NBSS Publ. No. 89, 260.

India Smart City Mission. (2016). *Smart City challenge stage 2, Smart City proposal*. Ministry of Urban Development, Government of India.

Ishita, R. P., & Khandaker, S. (2010). Application of analytical hierarchical process and GIS in earthquake vulnerability assessment: Case study of Ward 37 and 69 in Dhaka City. *Journal of Bangladesh Institute of Planners, 3*, 103–112.

Jacobs, J. (1984). *Cities and the wealth of nations: Principles of economic life*. Random House.

Jana, A., & Sarkar, S. (2018). Disparate housing strategies and practices of public and private enterprises in India: Analysis of middleclass housing and new towns. *Cities, 72*, 339–347.

Karmakar, J. (2015). Encountering the reality of the planning process in peri urban areas of Kolkata: Case study of Rajarhat. *Archives of Applied Science Research, 7*(5), 129–138.

Khandoker, R. A. (1989). Development of major tectonic elements of the Bengal Basin: A plate tectonic appraisal. *Bangladesh Journal of Scientific Research, 7*(2), 221–232.

Kotlov, F. V. (1961). AntropogennyeizmeneniyareliefanaprimeregorodaMoskvy (anthropogenic relief changes on the example of city of Moscow). *Sbornik, 52*, 134–150.

Kramer S L (1996) Geotechnical Earthquake Engineering. Prentice Hall New Jersey, United States.

Kundu, R. (2016). Making sense of place in Rajarhat new town—The village in the urban and the urban in the village. *Economic and Political Weekly, 51*(17), 93–101.

Libertun de Duren, N. R., & Compeán, R. G. (2015). Growing resources for growing cities: Density and the cost of municipal public services in Latin America. *Urban Studies, 53*(14), 3082–3107. https://doi.org/10.1177/0042098015601579

Lóczy, D., & Pirkhoffer, E. (2009). Mapping direct human impact on the topography of Hungary. *ZeitschriftfürGeomorphologie, Supplementary Issues, 53*(2), 215–222.

Lóczy, D., & Süto, L. (2011). Human activity and geomorphology. In K. J. Gregory & A. S. Goudie (Eds.), *The SAGE handbook of geomorphology*. SAGE.

Loh, J., & Wackernagel, M. (2004). Living planet report 2004 World Wide Fund For Nature Gland Switzerland https://vtechworks.lib.vt.edu/bitstream/handle/10919/65909/1051_Loh_Living_planet_report_2004.pdf?sequence=1&isAllowed=y. Last accessed 17 Aug, 2020.

Lóránt, D. (2012). Introduction to anthropogenic geomorphology. In *Studies on environmental and applied geomorphology*. IntechOpen.

Magicbricks. (2016). *Corridor of Growth Rajarhat-Newtown*. http://property.magicbricks.com/microsite/cog/RAJARHAT_NEWTOWN_COG_protected.PDF. Last accessed 25 Aug 2020.

Majumdar, S., & Chatterjee, U. (2020). Modelling urban growth using urban growth deterministic model in Kolkata metropolitan area: A geo-statistical approach. *Model. Earth Syst. Environ., 2020*. https://doi.org/10.1007/s40808-020-00985-6

Majumdar, P., & Kait, K. S. (2014). Dynamics of urban development and wetland management in East Kolkata wetlands. In *Landscape ecology and water management* (pp. 287–305). Springer.

Marsh, G. P. (1864). *Man and Nature*. Reprinted in 1965: The Earth as modified by human action. Belknap Press – Harvard University Press, Cambridge, MA, 560.

McGuire RK (2004) Seismic hazard and risk analysis. Earthquake Engineering Research Institute.

Mendöl T (1963) Általánostelepülésföldrajz. AkadémiaiKiadó.

Mehta, M., & Mehta, D. (1989). Priorities in urban planning and national commission on urbanization. *Economic and Political Weekly, 24*(21), 1178–1184.

Ministry of Housing and Urban Affairs. (2018). *National urban policy framework*. Ministry of Housing and Urban Affairs, New Delhi. https://smartnet.niua.org/sites/default/files/resources/nupf_final.pdf. Last accessed 12 June 2020

Ministry of Housing and Urban Affairs. (2019). *Transforming urban landscape 2014–2019*. Government of India, New Delhi. http://mohua.gov.in/upload/5c7faf00eac57UT%20Book1.pdf. Last accessed 21 July 2020

Ministry of Rural Development. (2016). *Shyama Prasad Mukherji Rurban mission: Integrated cluster action plan (ICAP)*. Ministry of Rural Development, Government of India, New Delhi. http://rurban.gov.in/download/ICAP.pdf. Last accessed 21 July 2020.

MoUD-Ministry of Urban Development. (2015a). *Urban and regional plans formulation and implementation (URDPFI) guidelines, volume I*. Ministry of Urban Development, Government of India. http://mohua.gov.in/upload/uploadfiles/files/URDPFI%20Guidelines%20Vol%20I.pdf. Last accessed 5 Feb 2017

MoUD-Ministry of Urban Development. (2015b). *Smart city mission statement and guidelines*. Ministry of Urban Development, Government of India, New Delhi. https://smartnet.niua.org/sites/default/files/resources/smartcityguidelines.pdf. Last accessed 05 May 2020

Mitra, D., & Banerji, S. (2018). Urbanisation and changing waterscapes: A case study of new town, Kolkata, West Bengal, India. *Applied Geography, 97*, 109–118.

Mohanty, W. K., Verma, A. K., Vaccari, F., & Panza, G. F. (2013). Influence of epicentral distance on local seismic response in Kolkata City, India. *Journal of Earth System Science, 122*(2), 321–338.

Mukherjee, A. (2016). *Villages in the City: Trajectories of Urbanisation in Rajarhat-New Town, Kolkata*. Unpublished MPhil thesis, School of Habitat Studies, Tata Institute of Social Sciences.

Murphy, R. W. (1988). Bangladesh enters the oil era. *Oil & Gas Journal, 86*(9), 76–82.

Murty, A. S. N., Sain, K., & Prasad, B. R. (2008). Velocity structure of the West-Bengal sedimentary basin, India along the Palashi-Kandi profile using a travel-time inversion of wide-angle seismic data and gravity modeling–an update. *Pure and Applied Geophysics, 165* (9–10), 1733–1750.

Nandy, D. R. (2007). Need for seismic microzonation of Kolkata megacity. In: *Proceedings of workshop on microzonation (Vol. 2627) Indian Institute of Science Bangalore*, India https://www.civil.iisc.ernet.in/~microzonation/workshop_files/paper%2015.pdf. Last accessed on 5 July, 2020.

Nath, S. K., Adhikari, M. D., Devaraj, N., & Maiti, S. K. (2015). Seismic vulnerability and risk assessment of Kolkata City, India. *Natural Hazards and Earth System Sciences, 15*(6), 1103.

Nath, S. K., Adhikari, M. D., Maiti, S. K., Devaraj, N., Srivastava, N., & Mohapatra, L. D. (2014). Earthquake scenario in West Bengal with emphasis on seismic hazard microzonation of the city of Kolkata, India. *Natural Hazards and Earth System Sciences, 14*(9), 2549–2575.

National Disaster Management Authority (2019) Earthquake disaster risk index report. Ministry of Home Affairs Government of India.

Nir, D. (Ed.). (1983). *Man, a geomorphological agent: An introduction to anthropic geomorphology*. Keter Publishing House & co-publisher D. Reidel Publishing Company.

Oliver, M. A., & Webster, R. (1990). Kriging: A method of interpolation for geographical information systems. *International Journal of Geographical Information System, 4*(3), 313–332.

Petit, C., Scudder, T., & Lambin, E. (2001). Quantifying processes of land-cover change by remote sensing: Resettlement and rapid land-cover changes in South-Eastern Zambia. *International Journal of Remote Sensing, 22*(17), 3435–3456.

Piscitelli, P. (2018). *8 research paths for the City, Feltrinelli camp 2018, a cura di*. Fondazione Giangiacomo Feltrinelli.

Rajashekariah, K. (2011). *Impact of urbanization on biodiversity: Case studies from India*. WWF-India. https://d2391rlyg4hwoh.cloudfront.net/downloads/impact_of_urbanisation_on_biodiversity.pdf. Last accessed 7 Aug 2020

Rao, P. M. (1990). *Planning for metropolitan cities: A suggestive approach*. Concept Publishing Company.

Reis, S. (2008). Analyzing land use/land cover changes using remote sensing and GIS in Rize, North-East Turkey, Sensors, 8: 6188–6202. 88.

Remy, N., Boucher, A., & Wu, J. (2009). *Applied geostatistics with SGeMS: A user's guide*. Cambridge University Press.

Roy, A. K., & Dhar, N. (1995). Geo environmental, Geo-hydrology, Geotechnical, appraisal of the parts of Calcutta Metropolitan Area for Development of Megacity district. Code-EN V/ER/WSA'95.

Roy, P. K., Banerjee, G., Mazumdar, A., Kar, A., Majumder, A., & Roy, M. B. (2012). A study to ascertain the optimum yield from groundwater source in the eastern part of Kolkata municipal corporation area in West Bengal, India. *European Journal of Sustainable Development, 1*(2), 97–112.

Roy, U. (2019). Transformations of urban green spaces (UGS) in sub-cities: A case of new town of Kolkata and Gurgaon of Delhi. *Proceedings of the Fábos Conference on Landscape and Greenway Planning, 6*(1), 26.

Rózsa, P., & Novák, T. (2011). Mapping anthropic geomorphological sensitivity on a global scale. *ZeitschriftfürGeomorphologie, Supplementary Issues, 55*(1), 109–117.

Saha, S. (2009). *Problem of planning a new town close to Kolkata a case study of Rajarhat new town*. Doctoral thesis, Department of Geography, University of Calcutta. http://hdl.handle.net/10603/164857. Last accessed on 5 July, 2020.

Salt, C. A., Alam, M. M., & Hossain, M. M. (1986). Bengal Basin: current exploration of the hinge zone area of south-western Bangladesh. In: *Proceedings of the South East Asia Petroleum Exploration Society Conf., 20–31 January 1986*, World Trade Centre, Singapore, 65–67.

Sarris, A., Loupasakis, C., Soupios, P., Trigkas, V., & Vallianatos, F. (2010). Earthquake vulnerability and seismic risk assessment of urban areas in high seismic regions: Application to Chania City, Crete Island, Greece. *Natural Hazards, 54*(2), 395–412.

Sarvestani, M. S., Ibrahim, A. L., & Kanaroglo, P. (2011). Three decades of urban growth in the city of shiraz, Iran: A remote sensing and geographic information systems application. *Cities, 28*(4), 320–329.

Schmid, C., & Brenner, N. (2011). Planetary urbanisation. In M. Gandy (Ed.), *Urban constellations* (pp. 10–13). Jovis.

Sengupta, S. (2011). Geological and geophysical studies in western part of Bengal basin, India. *AAPG Bulletin, 50*(5), 1001–1017.

Sengupta, U. (2013). Inclusive development? A state-led land development model in new town, Kolkata. *Environ Plann C Govern Policy, 31*(2), 357–376. https://doi.org/10.1068/c1103. Last accessed 17 Aug 2020.

Seto, K. C., Guneralp, B., & Hutyra, L. R. (2012). Global forecasts of urban expansion to 2030 and direct impacts on biodiversity and carbon pools. *Proc Natl Acad Sci USA, 109*(40), 16083–16088. https://doi.org/10.1073/pnas.1211658109

Sherlock, R. L. (1922). *Man as a geological agent: An account of his action on inanimate nature*. HF & G. Witherby.

Short, J. R., Hanlon, B., & Vicino, T. (2007). The decline of inner suburbs: The new suburban gothic in the United States. *Geography Compass, 1*(3), 641–656.

Sivaramakrishnan, K. C. (1977). *New towns in India—A report on a study of selected new towns in the eastern region* (pp. 1–287). A Homi Bhava Fellowship Award Project with Support from the Indian Institute of Management Calcutta. http://www.cprindia.org/sites/default/files/books/NEW%20TOWNS%20IN%20INDIA_1.pdf. Last accessed 10 July 2020.

Steinert, K., Marom, R., Richard, P., Viega, G., & Witters, L. (2009). Making cities smart and sustainable. In S. Dutta (Ed.), *The global innovation index 2011*. Insead.

Stoppani, A. (1873). Corso di Geologia II. Cap. xxxi. Sec. 1327. Milano.

Szabó, J. (1993). A társadalomhatása a földfelszínre—antropogéngeomorfológia [The impact of society on the Earth's surface—Anthropogenic geomorphology]. Általánostermészetföldrajz—NemzetiTankönyvkiadó [Physical geography] publisher, Budapest, 500–518.

Szabó J, Dávid L, Lóczy D (Eds.) (2010) Anthropogenic geomorphology: a guide to man-made landforms. Springer Science & Business Media.

Tacoli, C. (Ed.). (2006). *The Earthscan reader in rural urban linkages* (Earthscan reader series). Earthscan.

Taylor, G. R. (1915). *Satellite cities: A study of industrial suburbs*. D. Appleton and Company.

TCPA-Town and Country Planning Association. (2014). *New towns and garden cities—Lessons for tomorrow. Stage 1: An introduction to the UKs new town and garden cities* (pp. 1–32). TCPA, London https://www.tcpa.org.uk/Handlers/Download.ashx?IDMF=5bda030e-0b33-42edb4d4-0d4728be4ebd. Last accessed 12 June 2020.

TCPO-Town and Country Planning Organization. (2007). *Evaluation study of DMA towns in National Capital Region (NCR)*. Town and Country Planning Organization, Ministry of Urban Development Government of India. http://tcpomud.gov.in/divisions/mutp/dma/final_dma_report.pdf. Last accessed on 28 July 2020

Thakur, J. (2015, May 14). Kolkata lies right on top of faultline, may face major quake in future. *Hindustan Times*. https://www.hindustantimes.com/india/kolkata-lies-right-on-top-of-faultline-may-face-major-quake-in-future/story-OUcQbgMp2H4I7bN8XV5YmK.html. Last accessed 20 June 2020.

Torahi, A. A., Rai, S. C., & Al, E. T. (2011). Land cover classification and forest change analysis, using satellite imagery – A case study in Dehdez area of Zagros Mountain in Iran. *Journal of Geographic Information System, 3*(1), 1–11.

United Nations. (2006). *Urbanizations: mega & meta cities, new city states. UN-HABITAT: State of World's Cities 2006/7; United Nations, New York, NY, USA* http://mirror.unhabitat.org/documents/media_centre/sowcr2006/SOWCR%202.pdf. Last accessed 27 Aug 2020.

United Nations Department of Economic and Social Affairs, Population Division. (2014). *World urbanization prospects: The 2014 revision*. https://esa.un.org/unpd/wup/publications/files/wup2014-report.pdf. Last accessed 27 Aug 2020.

United Nations. (2015). *Transforming our world: The 2030 agenda for sustainable development*. https://sustainabledevelopment.un.org/content/documents/21252030%20Agenda%20for%20Sustainable%20Development%20web.pdf. Last accessed 27 Aug 2020.

United Nations, Department of Economic and Social Affairs, Population Division. (2018a). *World urbanization prospects: the 2018 revision [key facts]* https://population.un.org/wup/Publications/Files/WUP2018-KeyFacts.pdf. Last accessed 05 July 2020.

United Nations. (2018b). *Sustainable development goal 11: Make cities and human settlements inclusive, safe, resilient and sustainable*. https://sustainabledevelopment.un.org/sdg11. Last Accessed 02 Aug 2020.

Vaccari, F., Walling, M. Y., Mohanty, W. K., Nath, S. K., Verma, A. K., Sengupta, A., & Panza, G. F. (2011). Site-specific modeling of SH and P-SV waves for microzonation study of Kolkata metropolitan city, India. *Pure and Applied Geophysics, 168*(3–4), 479–493.

WBHIDCL-West Bengal Housing Infrastructure Development Corporation Limited. (1999). *New town Kolkata: Project report*. WBHIDCL.

WBHIDCO- West Bengal Housing Infrastructure Development Corporation. (2012). *Landuse and development control plan for new town planning area*. WBHIDCO.

Wilson, H. E., Hurd, J. D., Civco, D. L., Prisloe, M. P., & Arnold, C. (2003). Development of a geospatial model to quantify, describe and map urban growth. *Remote Sensing of Environment, 86*(3), 275–285.

World Bank. (2009). *World development report 2009*. Reshaping Economic Geography. https://openknowledge.worldbank.org/handle/10986/5991. Last accessed 05 July 2020

Yeh, A. G., & Li, X. (2001). Measurement and monitoring of urban sprawl in a rapidly growing region using entropy. *Photogrammetric Engineering & Remote Sensing, 67*(1), 83–90.

Yengkhom, S. (2011). Rajarhat gateway under water. *Times News Network*. http://timesofindia.indiatimes.com/articleshow/9547512.cms?utm_source=contentofinterest&utm_medium=text&utm_campaign=cppst. Last accessed 16 June 2020.

Youd, T. L., & Perkins, D. M. (1978). Mapping liquefaction-induced ground failure potential. *Journal of the Soil Mechanics and Foundations Division, 104*(4), 433–446.

Zalasiewicz, J., Williams, M., Steffen, W., & Crutzen, P. (2010). The new world of the Anthropocene. *Environmental Science and Technology, 44*(7), 2228–2231.

Zha, Y., Gao, J., & Ni, S. (2003). Use of normalized difference built-up index in automatically mapping urban areas from TM imagery. *International Journal of Remote Sensing, 24*, 583–594.

Zijderveld, A. C. (1998). *A theory of urbanity: The economic and civic culture of cities*. Transaction Publishers.

Chapter 29
Seasonal Change Detection of Wetlands Using Remote Sensing and GIS

Aksha Chowdhary and Swapnil Vyas

29.1 Introduction

Wetlands are primarily identified as areas where the land is inundated by shallow water or is saturated by surface or ground water at a frequency and duration every year that water becomes a dominant controlling factor for the environment and the associated plant and animal life (Gopal & Ghosh, 2008). Wetlands occur globally in every climate, and on every continent, except Antarctica. Global estimates of the extent of wetlands vary from 560 million to 1.2 billion ha of wetlands, including marine and freshwater wetlands (Burton & Tiner, 2009). Wetlands constitute *4.7%* of the total geographical area of India (Gond et al., 2000) enriching the economy of the country by utilizing aquatic resources. It is home to 26 Ramsar sites. The physical and biological properties of wetlands provide excellent retention capacity for storm water, thereby delaying floods by impeding the flow of water. Although the Mangrove ecosystem is an exemplary model that controls the damaging possibilities of tides, freshwater wetlands too are an equal match to flood mitigation. Freshwater wetlands, locally known as "beels" in northeastern India, is a significant element both socially and scientifically. They are regularly fed by major tributaries of the Brahmaputra River supplementing with nutrient-rich silt, encouraging growth of hydrophytic vegetation and providing an abundance of fish. Subtropical temperatures and heavy rainfall effect a change in the extent and function of wetlands. These variations have consequences on the local climate, the density and composition of vegetation, chemical parameters, flow of water, and the aquatic and avian diversity. Wetlands play a vital role as landscape elements, providing ecological, biogeochemical, and hydrological benefits as well as socioeconomic, food, and freshwater security. Large-scale alterations to landscapes and the hydrological cycle to meet

A. Chowdhary (✉) · S. Vyas
Department of Geography (Geoinformatics), Savitribai Phule Pune University (SPPU), Pune, Maharashtra, India

G. S. Bhunia et al. (eds.), *Anthropogeomorphology*, Geography of the Physical Environment, https://doi.org/10.1007/978-3-030-77572-8_29

demands of growing increasingly affluent populations have left wetlands impaired. Typically viewed as single units, we conceptualize wetlands as hydrologically and ecologically connected self-organized, dynamic, complex networks. Wetland networks and their many functions are linked to produce persistent self-emergent patterns and landscape resilience (Sanchez et al., 2017).

29.1.1 Application of Space Technology in Monitoring Wetlands

Data acquired by Earth Observation systems cover a wide range of spectral, spatial, and temporal characteristics pertinent to the study of wetlands. This data can be exploited for locating the wetlands and monitoring their seasonal changes based on different parameters over large areas. Two popular approaches, pixel-based and object–based, using parametric and nonparametric algorithms are effectively used in describing wetland structure and habitat.

Pixel-based indices directly employ bands of different wavelengths in the spectrum depending on the sensor used. Indices give important information about the feature under study as each feature responds uniquely to different wavelengths. The Normalized Difference Vegetation Index (NDVI) initially defined for vegetation studies (Townshend & Justice, 1986; Tucker, 1979) is proven useful for detecting water and silt-laden open water from lateritic soils (Caloz & Collet, 1997). The logic behind employing NDVI for water bodies is that near infrared (NIR) band is theoretically well-suited for detecting open water surfaces from optical images due to the strong water absorption in the NIR range (Verdin, 1996). Water bodies appear almost black in optically sensed images. However, because of some complicating water characteristics such as turbidity and/or the presence of aquatic vegetation (seaweed, duckweed, and others), the NIR band alone is not sufficient to properly distinguish open water surfaces. Moreover, NIR, being sensitive to built-up features, often leads to overestimation of water bodies. It is well known that the SWIR is absorbed by water contained in plant cells. To assess water content in a normalized way, the Normalized Difference Water Index (NDWI) (NIR-SWIR)/(NIR+SWIR)) was thus introduced by Gao (1996). This index increases with vegetation water content or from dry soil to free water. The Modified Normalized Difference Water Index (MNDWI) was derived from the NDWI by the use of SWIR instead of NIR (Xu, 2006). To study water characteristics in open water bodies, green band is employed instead of NIR.

Another method of extracting wetlands from an area by utilizing its spectral and geometric characteristics by a computed process is known as feature extraction or object-based image classification. Feature extraction has applications in agriculture, environment, forestry, disaster response, and state and local government mapping.

The applications of object-based image analysis (OBIA) in remote sensing studies of wetlands have been growing over recent decades, addressing tasks from detection and delineation of wetland bodies to comprehensive analyses of within-wetland cover types and their change. Compared to pixel-based approaches, OBIA offers several important benefits to wetland analyses related to smoothing of the local noise, incorporating meaningful nonspectral features for class separation and accounting for landscape hierarchy of wetland ecosystem organization and structure (Dronova, 2015).

29.1.2 Importance of Monitoring Wetlands

From the historic times, human survival is directly or indirectly dependent on wetland ecology and the products derived from it. They are among the world's most productive environments; cradles of biological diversity that provide the water and productivity upon which countless species of plants and animals depend for survival. Wetlands are indispensable for the countless benefits or "ecosystem services" that they provide humanity, ranging from freshwater supply, food and building materials, and biodiversity, to flood control, ground water recharge, and climate change mitigation (Logan et al., 2013). Yet study after study demonstrates that wetland area and quality continue to decline in most regions of the world. As a result, the ecosystem services that wetlands provide to people are compromised. The degradation of wetlands can lead to serious consequences, including increased flooding; ecosystem collapse, deformity, or extinction; decline in water quality; and loss in economic valuation of which loss of fish is conspicuous. Besides, wetlands are also important as a genetic reservoir for various species of plants including rice, which is a staple food for 3/4th of the world's population. The spatial loss of wetlands means broader ramifications to life on earth through loss in food chain links. The quality of water flowing into wetlands may be impaired indirectly by alterations to the water regime or by different point and nonpoint pollution sources. The extraordinary productivity of water in this ecosystem implies that numerous stakeholders as well as domestic users have easy access to and use of wetland resources. The overexploitation of these resources entails intense cropping, overfishing, and developmental and construction activities. The cumulative impacts of these activities threaten biodiversity.

February 2 has been recognized as World Wetland Day to acknowledge the urgency for conservation of this precious resource. In India, the protection of wetlands has received increasing attention in the past few years and the use of Geographic Information System has become popular for wetland monitoring and mapping. We have utilized the prospects of remote sensing to understand the biophysical changes that take place in wetlands. Data obtained from remote sensing saves time and labor, which would otherwise make it impossible to monitor these micro ecosystems. This information can be utilized to study the structural network of wetlands that are affected by seasonal changes, particularly monsoon. Changes in water level and vegetation can further imply changes in the biodiversity and food chain, which becomes pertinent in conservation.

29.2 Materials and Methods

29.2.1 Study Area

Kaziranga National Park spans over districts with a recorded area of 430 sq kms between latitudes 26°30′ N and 26°45′ N, and longitudes 93°08′ E to 93°36′ E within two districts in the Indian state of Assam – the Kaliabor subdivision of Nagaon district and the Bokakhat subdivision of Golaghat district. The annual rainfall recorded in 2018 was >300 mm according to CRIS, India (Barua & Sharma, 1999).

Kaziranga National Park is located on a flat terrain with a gentle almost imperceptible slope from east to west with an elevation of 55–70 m. The park is a part of the Brahmaputra flood plain and thus is rich in alluvial soil resulting from recurrent floods (Fig. 29.1). The area is swampy and waterlogged almost perennially having small and large shallow to deep water bodies spread all over that are locally known as "beels" (Barua & Sharma, 1999). These beels are characteristic of wetlands. The park is also criss-crossed by river channels of which the prominent ones are Mora

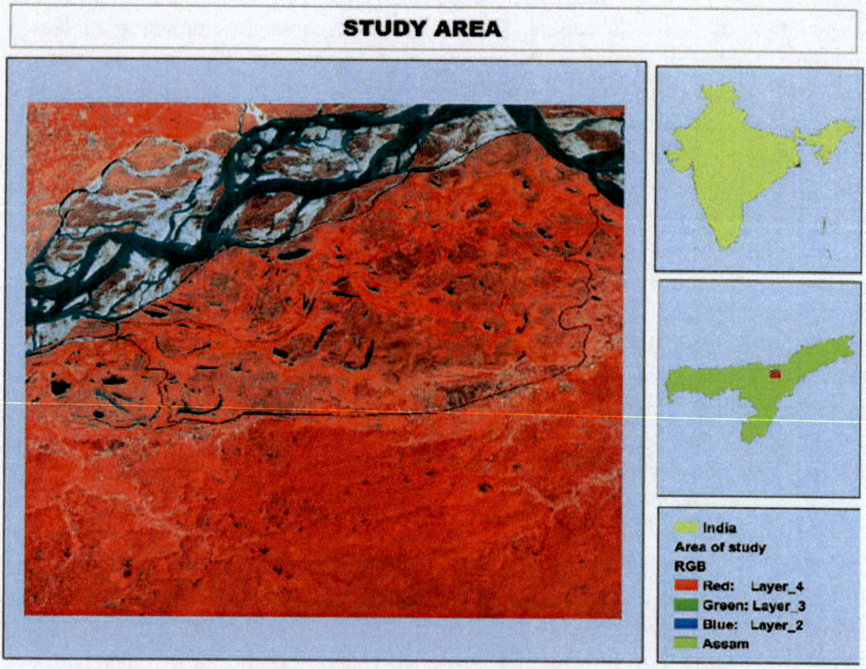

Fig. 29.1 Study area (Sentinel 2a image of Kaziranga dated November 2018)

Diphlu, which runs along the south of the park, and Jiya Diphlu, which runs eastwest in the middle of the park. In the extreme south of the park are Karbi-Anglong hills, which provide shelter to the animals at the time of extreme floods.

Located on the edge of the Eastern Himalaya biodiversity hotspot, the habitat of Kaziranga is such that water bodies and grasslands form a significant part of the park's area. Wetlands cover 7%, grasslands 57% (tall grass 52%; short grass/ marshes 5%), sand cover 7%, and woodland 29% of the total area (Kushwaha, 2000). Out of these ecosystems, wetlands invariably have high value inside the park as it attracts a large number of migratory birds, provides fish to the people living in the fringe villages of the park, is a resting and drinking ground for all the animals adding to its ecotourism value and, most importantly, retains water during seasonal floods.

Two ranges have been focused for this study – Bagori Range and Kohora Range – as they have the highest density of wetlands. A large number of water birds can be observed in these ranges, reiterating the importance of wetlands in this area. The aforementioned important beels are located in these ranges. However, the analysis has been extrapolated to nearby water bodies to understand the interactions between them.

29.2.2 Methodology

Sentinel 2a images for the following dates were obtained for the study:

10 November 2018
3 February 2019
24 April 2019

Pixel-based index calculations were done using the raster calculator in ArcGIS 10.5. Two indices – Normalized Difference Vegetation Index (NDVI) and Modified Normalized Difference Water Index (MNDWI) – were studied for each of the three seasons. Cutoff values for each class were verified from their false color composite (fcc) images, leading to the creation of four classes with distinct signatures: water, wet soil/shallow water/hydrophytic vegetation and vegetation.

Figure 29.2 roughly maps steps followed to obtain the classification values.

Extraction of wetlands using object-based image classification was operated in eCognition Developer 64 using the following steps:

Image segmentation: the **multiresolution segmentation** algorithm breaks up an image based on compactness and shape settings. The image was broken into multiple segments with conditional settings:

Shape = 0.6
Compactness = 0.5
Scale parameter = 75

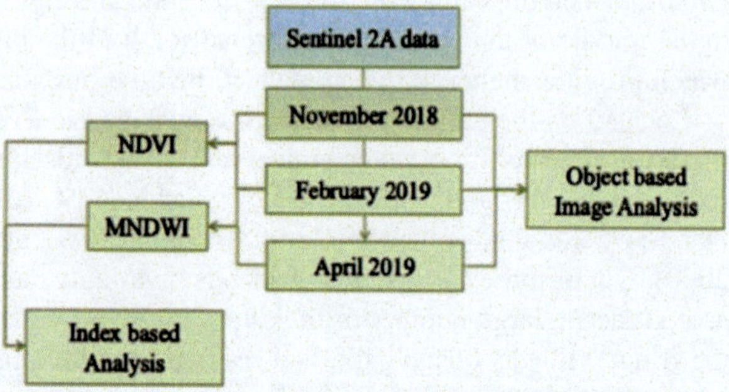

Fig. 29.2 Workflow of the study

A second image segmentation was performed using Spectral Difference algorithm at difference = 20 to aggregate the segments of the same class.

Classification: Both spectral and spatial parameters were monitored and utilized to delineate wetlands among other features. The MNDWI ratio was defined, and a cutoff value >= 0.4 was set as threshold to extract water pixels. Since wetlands are compact water bodies, they can be differentiated from the main river using geometry differences. Thus, all objects that have length >=775 pixels were assigned a new class called "river" and the remaining water bodies were successfully classified as "wetlands."

The wetlands were then exported as a polygon layer for further processing.

Traditional pixel-based image classification assigns a land cover class per pixel. All pixels have the same predefined size and can be distinguished from its neighbors based only on their spectral properties. However, in object-based classification the entire image is segmented by grouping spectrally similar pixels together into vector objects of a predefined shape and compactness. Instead of a per-pixel basis, segmentation automatically digitizes the image.

29.3 Results and Discussion

29.3.1 Classification Based on Indices

Modified Normalized Difference Water Index (MNDWI) The method based on Spectral Water Index, especially *MNDWI*, has a better ability to suppress built-up areas and enhances water bodies than it does in *NDWI*. *MNDWI* uses the 10 m green band (Band 3) and the 20 m short infrared wavelength (Band 11) to compute the index as shown in Eq. (29.1).

$$MNDWI = \frac{\text{Green} - SWIR}{\text{Green} + SWIR} \tag{29.1}$$

A decreasing trend of *MNDWI* values was observed from November to April. Monsoon is the primary reason for higher values of water in November. One can

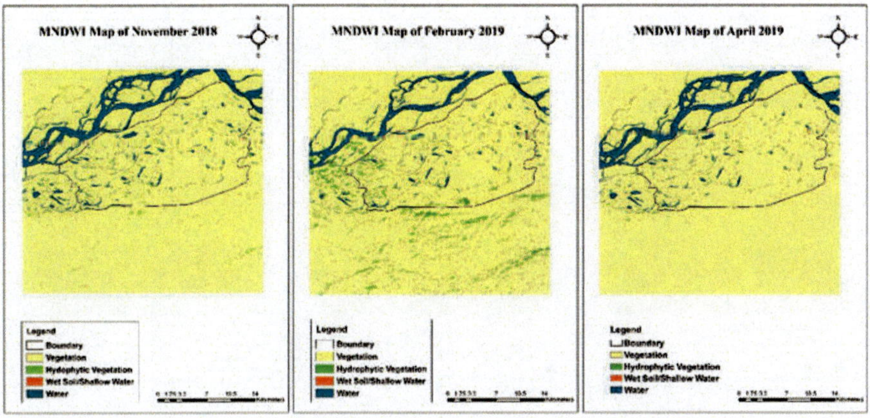

Fig. 29.3 *MNDWI* of wetland extent in November 2018, February 2019, and April 2019

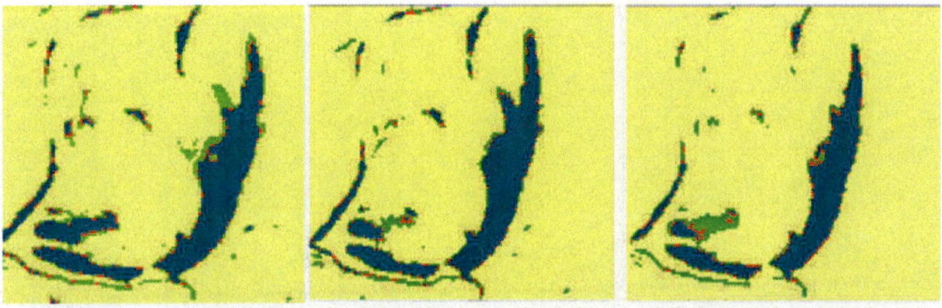

Fig. 29.4 *MNDWI* of regional extent in November 2018, February 2019, and April 2019

also observe the change in perimeter and extent of the wetland over the seasons. Water transitions into shallow water and in some cases to hydrophytic vegetation. Changes in *MNDWI* values over the three timelines, along with the area showing distinct change in index values, are shown in the maps labeled as Fig. 29.3a, and their regional specificity is shown in Fig. 29.4.

*29.3.2 Normalized Differential Vegetation Index (*NDVI*)*

It is the most popular index to delineate vegetation in an area by using the red band and *NIR* band as shown in Eq. (29.2). Vegetation around a wetland, known as hydrophytic vegetation, has an index value in the range of 0.097–0.4 according to classification performed.

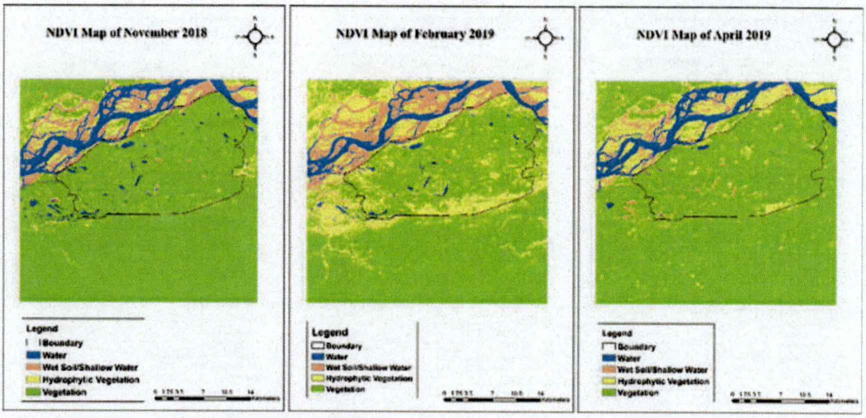

Fig. 29.5 *NDVI* of wetland extent in November 2018, February 2019, and April 2019

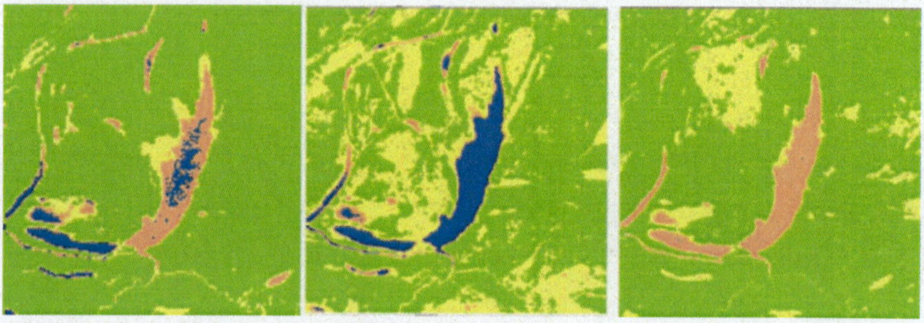

Fig. 29.6 *NDVI* of regional extent in November 2018, February 2019, and April 2019

$$NDVI = \frac{NIR - \text{Red}}{NIR + \text{Red}} \tag{29.2}$$

Water has poor reflectance, whereas vegetation has the highest reflectance in *NIR* band. Since wetlands are characterized by hydrophytic vegetation, hence *NDVI* is an important index to extract wetlands in an area as they will have a positive *NDVI* value.

Seasonal changes lead to change in the *NDVI* values in the wetlands. In the post-monsoon month of November, high *NDVI* values are exhibited, whereas as the dry spell hits in February, the *NDVI* values show a decrease. Dense vegetation is saturated around the streams and water bodies post monsoon. The wetlands are covered by hydrophytic plants like prickly water lily, duckweed, while in February as wetlands become a nesting ground for birds, most of the vegetation is fed on and the water is exposed, leading to lower *NDVI* values.

Changes in *NDVI* values over the three timelines, along with the area showing distinct change in index values, are shown in the maps labeled as Fig. 29.5, and their regional specificity is shown in Fig. 29.6.

29.3.3 Object-Based Image Analysis

The seasonal changes that were detected from object-based image analysis can be understood from the decrease in the number and geometry of wetlands from November to April. Extraction of wetlands using the aforementioned parameters shows a decrease of 16.87% in the number of wetlands from November to February and a further decrease of 35.33% from February to April. A reasonable explanation for this result is that, as the season changes from monsoon to spring, the water-saturated soil dries up and becomes more compact, leading to a decrease in water level and consequently to its *MNDWI* value. Thus, extracting water bodies with an *MNDWI*>=0.4 eliminates the shallow or saturated bodies categorizing them as seasonal wetlands. The total area covered by all the extracted wetlands also shows a decreasing trend, which reiterates the reduction in water retention capacity of the wetland network over the months. Figure 29.7 is a graph showing the structural changes in wetlands.

An intersection of all the wetlands across the three seasons shows that only half the extracted wetlands are permanent, showing variation in water and vegetation magnitudes. To comprehend the structural changes in each of these permanent wetlands, their perimeter was calculated. As opposed to the popular approach of calculating area for change detection, in this study perimeter was chosen as an appropriate measure. This is justified by the fact that wetlands extend or shrink

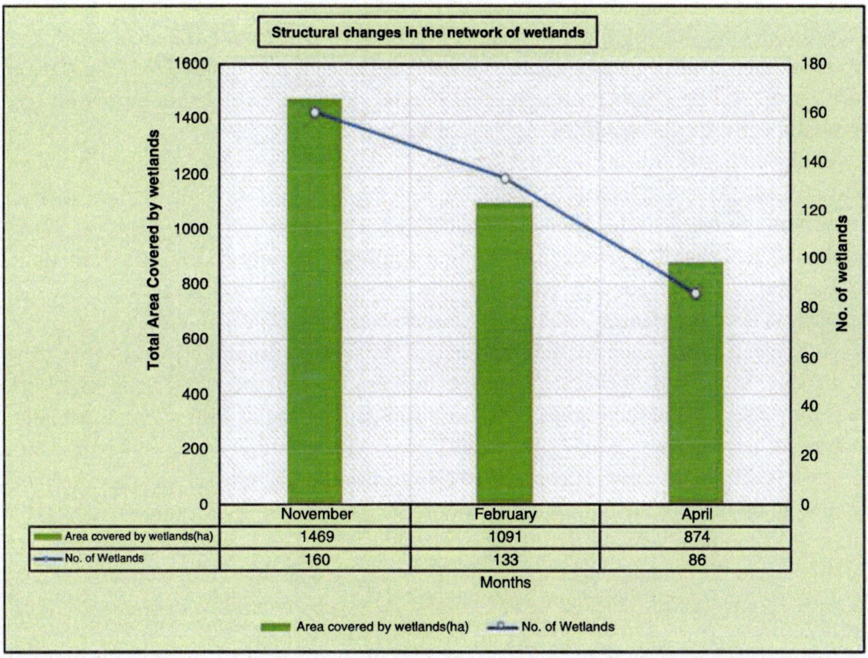

Fig. 29.7 Changes in number and area of wetlands over the seasons

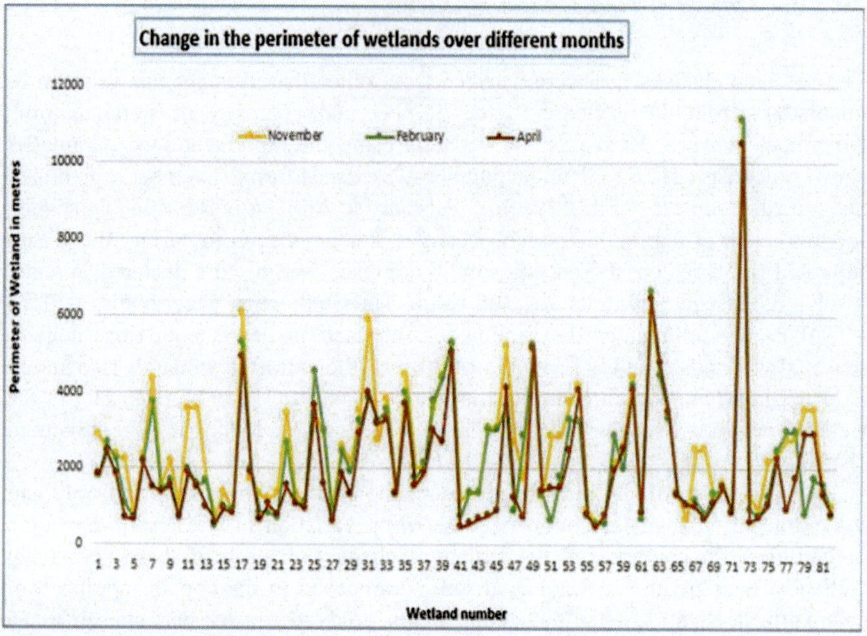

Fig. 29.8 Month-wise change in perimeter of each perennial wetland

according to the flow of water in them. Therefore, by calculating the area, we might overlook their change in shape and geometry and reckon it to have reduced in size or vice versa. A line graph representing the trend of change in perimeter for all the wetlands w.r.t. to the month of November is shown in Fig. 29.8.

Using both spectral and spatial features to study the seasonal change in water bodies is effective and useful in understanding the behavior of wetlands in response to rainfall, flood activity, temperature change, and other physical factors. Since wetlands are dynamic systems exhibiting a colossal diversity of life, it is important to monitor them seasonally and correlate these changes to the activity of the associated flora and fauna.

As expected, the values of *MNDWI* and *NDVI* are dominantly controlled by monsoons. In the post-monsoon month of November, the water levels are high and so is the vegetation in the region. The loose soil gets saturated with water, especially in areas of negative elevation, giving rise to numerous shallow water bodies. As soil moisture starts to dry, many seasonal wetlands disappear apart from lowering water values. Thus in February, the number of wetlands reduces. The hydrophytic vegetation in some patches is replaced by terrestrial plants. For the chosen year, that is, 2019, the satellite image was captured after a seasonal shower in that month, hence in some water bodies, the MNDWI values shoot up or remain unchanged. This observation reinforces the fact that rain is the dominant controlling factor in wetland ecology. Finally, as temperatures soar and atmospheric and soil moisture reduce, the wetland network in the month of April is at its basal level. Many wetlands exhibit a shrinkage in perimeter and area, simultaneously bringing down their water level. At this time, most of the hydrophytic vegetation has been replaced, leading to a biological change in the ecosystem.

Apart from understanding ecological change, seasonal monitoring is important from a management point of view as well. By delineating the perennial wetlands, which in this region are often historic "beels," economic activities like aquaculture and community fishing can be encouraged. Additionally, wetland systems are potent elements of integrated farming and aquaponics. By maintaining the wetland network, the local communities can benefit economically, socially, and aesthetically.

29.4 Conclusion

Conservation of wetlands has to be given utmost importance considering the rate at which these sensitive eco-zones are being degraded. Since wetlands are distributed widely across the country in both accessible as well as inaccessible areas, it is important to use the technology of remote sensing data to map and monitor these areas so that a systematic approach can be devised for its conservation. While mapping wetlands, field data carries an equal weightage as it lists down the geochemistry and water quality of the area without which conservation efforts are futile. It is also necessary to study the land use/land cover of the vicinity so as to build an effective and robust policy framework guiding the conservation ideals.

In this study, wetlands located in Kaziranga and its surrounding areas were extracted with the aid of Sentinel 2a Multispectral Imagery and classified using different parameters. Primarily, three software, both for spatial analysis and image processing, aided the process, viz., ArcGIS 10.5, ERDAS IMAGINE 9, and eCOG NITION 64. An MNDWI lower limit of 0.4 was used as the basis for classification of water bodies, and values from 0.28 to 0.4 represent water pixels mixed with vegetation. The selection of these values yielded best results when compared to the fcc of the respective satellite images in the three time periods. However, field work for validation should be carried out in subsequent studies to achieve greater accuracy. The object-based rule set used for extracting wetlands was determined from previous studies on wetlands in addition to trial and error. The study sets a basis for conducting several other analyses on the structure and function of wetlands. An example of a rewarding study would be to track the annual change in wetlands and their variable seasonal changes every year. Such a database will give us a transparent outlook not only to the situation of existing wetlands, but also to the multiple factors that affect them like temperature, land use land cover, water, monsoon, vegetation, etc.

Results validate the laid-down conditions for classification of wetlands by the National Wetland Inventory Assessment (NWIA).

References

Barua, M., & Sharma, P. (1999). Birds of Kaziranga National Park. *Forktail, 15*, 47–60.

Burton, T. M., & Tiner, R. W. (2009). Ecology of wetlands: Global distribution of wetlands. In *Encyclopedia of inland waters* (pp. 507–515).

Caloz, R., & Collet, C. (1997). Geographic information systems (GIS) and remote sensing in aquatic botany: Methodological aspects. *Aquatic Botany, 58*(3–4), 209–228.

Dronova, I. (2015). Object based image analysis in wetland research: A review. *Remote Sensing, 7* (5), 6380–6413.

Gao, B.-C. (1996). NDWI – A normalized difference water index for remote sensing of vegetation liquid water from space. *Remote Sensing of Environment, 58*(3), 257–266.

Gond, V., Bartholome, E., Ouattara, F., Nonguierma, A., & Bado, L. (2000). Mapping and monitoring small ponds in dryland with the vegetation instrument – Application to West Africa. In *Conference: Vegetation 2000, Volume: Proceedings of the Vegetation 2000 Conference, Belgirate – Italy, Saint G. ED, CNES Toulouse & JRC – ISPRA, 3–6 April 2000.*

Gopal, B., & Ghosh, D. (2008). Natural Wetlands, Jawaharlal National University, New Delhi, India. In *Encyclopedia of ecology* (pp. 2493–2504).

Hoffer, R. M. (1978). Biological and physical considerations in applying computer aided analysis techniques to remote sensor data. In *Remote sensing: The quantitative approach* (pp. 227–289).

Kushwaha, S., Roy, P., Azeem, A., Boruah, P. & Lahan, P. (2000). Land area change and rhino habitat suitability analysis in Kaziranga National Park, Assam. Tigerpaper, 27, 9–17.

Logan, L. H., Karlsson, E. M., Gall, H. E., Park, J., Emery, N., Owens, P., et al. (2013). Freshwater wetlands: Balancing food and water security with resilience of ecological and social systems. In *Climate vulnerability: Understanding and addressing threats to essential resources* (Vol. 2, pp. 105–116). Elsevier Inc.

Sanchez, J. M., Ferguson, T. & Savage, K. (2017, October 2). Hydrological dynamics of two seasonal floodplain wetlands in the South Carolina Piedmont. *Journal of Young Investigators*. Department of Environmental Sciences and Department of Biology, Wofford University, 429 N Church St, Spartanburg, SC 29303, USA.

Townshend, J. R. G., & Justice, C. O. (1986). Analysis of the dynamics of African vegetation using the normalized difference vegetation index. *International Journal of Remote Sensing, 7*(11), 1435–1445.

Tucker, C. J. (1979). Red and photographic infrared linear combinations for monitoring vegetation. *Remote Sensing of Environment*, *8*(2), 127–150. Elsevier.

Verdin, J. P. (1996). Remote sensing of ephemeral water bodies in Western Niger. *International Journal of Remote Sensing, 17*(4), 733–748.

Xu, H. (2006). Modification of normalized difference water index (NDWI) to enhance open water features in remotely sensed imagery. *International Journal of Remote Sensing, 27*(14), 3025–3033.

Chapter 30
Restoration and Conservation of Wetlands: A Geospatial Approach

Adikanda Ojha and Jajnaseni Rout

30.1 Introduction

Water is an important resource for all living beings. Therefore, the use of water and its supply from sources is very important. Wetlands are an ecosystem from mangrove to subarctic peatlands that have affected human. The earliest civilizations were established near the river, lake, and floodplains. The Mesopotamian civilization is authoritatively accepted to have started around 4000–3500 BC between the Euphrates and Tigris River. The other ancestral civilization, Egypt, commenced in the Nile Valley around 3200 BC. This represents the importance of the water and wetlands. The fact that people are in these regions is a reflection of how important it is for biotic diversity. Therefore, wetlands are a very critical ecosystem, and some of them are the most productive habitats. Monitoring is the long-term regular observation and recording of current and altering situations. In the environmental assessment, these data were utilized to evaluate wetlands based on decision-making and planning processes. Consequently, wetland surveys have possessed a multidisciplinary perspective. The fact that the recognition of wetlands supplies many values for people and is an important case for global conservation has led to an increase in research and management activity. Because of urbanization, economic growth, industrialization, and increasing population, more wastes were discharged into nature. Wetlands carry through some beneficial functions in the protection of whole balance of the nature.

Wetlands are ecologically sensitive systems and provide many significant services to the human population. The evaluation of wetlands with a multidisciplinary perspective in the natural sciences and social sciences provides efficient results. This perspective can give an increased understanding of the processes and problems

A. Ojha
GIS Analyst, State Pollution Control Board, Bhubaneswar, Odisha, India

J. Rout (✉)
Guest Faculty, Ravenshaw University, Cuttack, Odisha, India

G. S. Bhunia et al. (eds.), *Anthropogeomorphology*, Geography of the Physical Environment, https://doi.org/10.1007/978-3-030-77572-8_30

associated with such strategies. It is clear that wetlands expose noteworthy economic value (depending on the cost–benefit analysis) and are under severe stress. The reasons for wetland loss and deterioration implicate excessive use, land degradation, urbanization, pollution, climate change, decreased biotic diversity, and invasive species. Since wetlands are complex multifunctional systems, they are likely to be the most beneficial if conserved as integrated ecosystems (within a catchment area) rather than their individual component parts.

Anthropogenic activities (urbanization, water and land uses, land cover changes, industrial activity, pollution, climatic change, etc.) have direct and indirect effects on wetlands. The degradation degree of an ecosystem is depended on temporal variation. Ecosystem recovery level and duration have two main factors. Firstly, anthropogenic pressures can increase or decrease due to the usage grade. Secondly, wetland's carrying capacity is changed due to spatial and temporal variation. For these reasons, positive and negative feedback mechanisms at the wetland are critical control systems. Therefore, the wetland is considered as holistic ecosystem perspective from its basin scale. In the natural ecosystems, feedback control processes are repeated between environmental factors and population growth rates in their carrying capacities. However, when the human population intervenes and extremely uses wetlands, this tolerance is destructed, and ecosystems wander off their homeorhesis. Wetlands in and around city provide significant services such as water supply and climate regulation. However, the value of these ecosystems remains largely unrecognized by policy- and decision-makers.

Odisha is an east coastal state of India and is surrounded with Bay of Bengal in eastern side, Chhatisgarh in western, Andhra Pradesh in southern and Jharkhand, West Bengal, in northern part. It is sacred with many natural resources and is also a riverine state of India. The coastline of the state is 480 km. Many rivers and rivulets are flowing to Bay of Bengal within this state. The major rivers are Mahanadi, which is from Chhatisgarh, Brahmani, which is from Jharkhand, Rushikulya, Baitarani, Subarnarekha, and Vanshadhara. There are several wetlands existing in the state. Chilika Lake and Bhitarkanika Sanctuary are the largest wetlands of Odisha, which has also Ramsar Convention importance. Chilika is the largest brackish water lagoon, and other freshwater lakes are Ansupa, Tampara, and Kanjia. The wetlands are covered with emergent and submergent vegetation, fish, bird, and other fauna. It is the home of a very high biodiversity with many types of common and exceptional species of plants and animals. Presently the biodiversity is threatened by growing anthropogenic pressures, which need to conserve. Many factors are responsible for the degradation of the wetlands: one is the intensive agriculture with heavy use of fertilizers and pesticides, which make the wetland polluted and eutrophic. The other is the over-exploitation of wetland resources. To understand the factors and processes of wetland, it is important to have knowledge about its hydrology and catchment.

The first Ramsar site of Odisha is Chilika, which has also maintained its biodiversity by using management issues. The lagoon was severely threatened by problems like shrinkage of water spread and decrease in depth due to siltation, fall in salinity, macrophyte infestation, eutrophication, and loss of biodiversity when the old mouth was closed due to longshore sediment transport. The weak circulation of lagoon water and poor tidal influx further complicated the conditions, leading to deterioration of the environment. Considering the threats faced by the lake, the National Wetlands, Mangroves and Coral Reefs Committee of the Ministry of Environment and Forests, Government of India, and the state government identified the lake as a priority site for conservation and management. The most effective ameliorative action was the hydrological intervention by opening the new lake mouth and channel to the sea through the barrier beach at Satapada. This improved the spatial and temporal salinity gradients of the lake to maintain the unique characteristics of an estuarine ecosystem. This intervention was undertaken after detailed scientific studies, including three-dimensional mathematical modeling and hydraulics studies on a model prototype, by the Central Water and Power Research Station, Pune, and National Institute of Oceanography, Goa. In September 2000, the desiltation of the channel connecting the lake to the sea and opening of a new mouth to restore the natural flows of water and salinity levels were carried out. These actions resulted in a notable increase in the lake's fish yield and a reduction of freshwater weeds. The new mouth reduced the length of the outflow channel by 18 km. Opening of the new mouth provided a favorable increased salinity regime throughout the lake with less fluctuations and improved water clarity. The other restoration activities by CDA involved catchment management in a participatory micro-watershed management in a whole ecosystem approach. A significant decrease in freshwater weed proliferation has been reported since the opening of the inlet (Patnaik, 2005).

Basavaraj and Nijagunappa (2011) have worked on identification of groundwater potential zone using geoinformatics in Ghataprabha basin, North Karnataka. This chapter presents a systematic planning of groundwater exploitation using modern techniques that is essential for proper utilization and management of this precious but shrinking natural resource. With the advent of powerful and high-speed personal computers, efficient techniques for water management have evolved of which geoinformatics technology includes remote sensing (RS), Geographic Information System (GIS), and Global Positioning System (GPS) are of great significance. Finally, it is concluded that the geoinformatics technology is very efficient and useful for the identification of groundwater potential zones. Kumar et al. (2008) have worked on potential sites for WHS using remote sensing and GIS technology at Bakhar watershed of Mirzapur District, UP, India. Availability of groundwater varies spatially and temporally depending upon the terrain. The scarcity of water affects the environmental and developmental activities of an area. Construction of small water harvesting structures across streams/watersheds is gaining momentum in recent years. Raghu and Reddy (2011) have worked on hydro geomorphological mapping at village level using high-resolution satellite data and impact analysis of check dams in part of Akuledu Vanka watershed, Anantapur district, Andhra Pradesh. The satellite data facilitates to update the extent of built-up area, road, and drainage network. Further, the revenue villages enclosed in the watershed are digitized, mosaicked, and superimposed on hydrogeomorphology map. This helps to give site-specific recommendation on groundwater prospects survey number wise,

Table 30.1 Summary statistics of wetlands in Orissa (area in ha)

Sl. no.	Category	Number of wetlands	Total wetland area	% of wetland area	Open water	
					Post-monsoon area	Pre-monsoon area
1	Inland wetlands – natural	3111	238867	34.57	228701	218773
2	Inland wetlands – man-made	7871	220207	31.87	197205	126476
	Total – inland	**10982**	**459074**	**66.45**	**425906**	**345249**
3	Coastal wetlands – natural	560	143978	20.84	60699	52384
4	Coastal wetlands – man-made	724	21678	3.14	21677	21677
	Total – coastal	**1284**	**165656**	**23.98**	**82376**	**74061**
	Subtotal	**12266**	**624730**	**90.42**	**508282**	**419310**
	Wetlands (<2.25 ha)	66174	66174	9.58	–	–
	Total	**78440**	**690904**	**100**	**508282**	**419310**

that is, for individual farmers. In addition, the impact analysis of check dams constructed in the watershed is also discussed. Studies showed that after construction of check dams the water levels in wells increased, abandoned wells got rejuvenated, and new borewells came up, resulting in increased irrigated area. With the advent of high-resolution satellite data, site-specific recommendations for groundwater exploration can be given at cadastral level for effective management of groundwater resources at smallest possible revenue boundary, that is, in the individual fields of the farmers. With the satellite imageries, the Space Application Centre, Ahmadabad, has been able to identify and delineate the extent of wetlands of Odisha. Area estimates of various wetland categories for Orissa have been carried out using GIS layers of wetland boundary, water spread, aquatic vegetation, and turbidity. In the state of Orissa, 12,266 wetlands have been delineated, with 66,174 small wetlands (< 2.25 ha). The total wetland area is estimated to be 690,904 ha. Inland wetlands dominated the extent of wetlands constituting about 66%. Further, inland natural and man-made wetlands shared approximately similar extents with about 34% and 32% of area under wetlands. Out of 24% of coastal wetlands, the natural accounted for about 20% and the rest 3% is shared by man-made wetlands. The major wetland types are river/stream (223522 ha) comprising about 32% of extent wetlands followed by reservoir/barrage (189972 ha), tank/ponds (29301 ha), lagoon (89023 ha), intertidal mudflats (25514 ha), and mangroves (23395 ha). There are a large number of small wetlands (<2.25 ha) contributing about 10% of the wetlands in Odisha (Table 30.1) (Odisha Wetland Atlas, 2013).

Figure 30.1 shows that the study area is Ansupa, which is a freshwater lake of Odisha. The lake is situated between 20°26′28.43″ to 20°28′34.44″ latitude and 85°35′56.74″ to 85°36′30.01″ longitude of Banki block in Cuttack district. It is

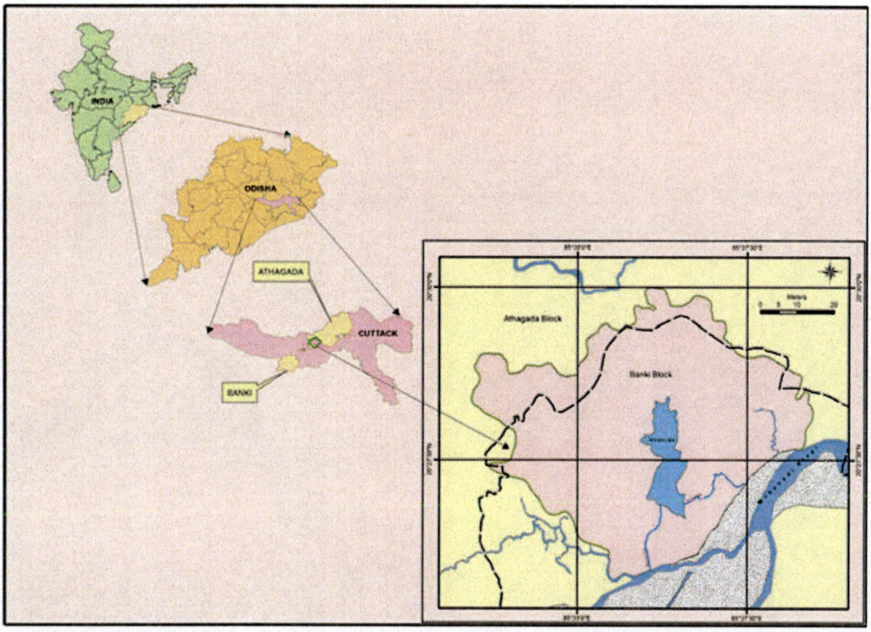

Fig. 30.1 Location map of study area

situated on the left side of the river Mahanadi. Ansupa Lake is about 3 km. Length and breadth vary from 250 m to 500 m. The area of the lake is about 382 acres comprising 152 acres of Kadalibadi mouza, 138 acres of Bishnupur mouza, and 92 acres of Subrnapur mouza. Ansupa Lake is a topographic depression surrounded by small hillocks as Saranda (124 m) on the west, Bishnupur (65 m) on the east, Dhangarh (160 m) on the north, and Betla pahara (105 m) on the northwest. Water from high land areas usually drain into Mahanadi river, but due to the local depressions poor slopes gets collected in the low-lying areas, giving rise to the water bodies.

The lake is connected with Mahanadi in its southern side with the channel known as "Kabula Nala" through which flood water of Mahanadi enters the lake. To the southwest of Ansupa, there is another channel known as "Huluhula nala," which has come from Tigiria tahsil into Mahanadi traversing through which flood water of Mahanadi enters the lake. The lake is bounded by the Saranda hills on its western side, Bishnupur hill on its northeastern side, and by villages like Subarnapur, Malbiharipur, Kadalibadi, and Bishnupur. The main objective of the study is to find out the causes responsible for degrading the wetland and potentialities for the development of the wetland. Geospatial technology is now recognized as an essential tool for observing, evaluating, describing, and making decisions of all natural resources (Fig. 30.2).

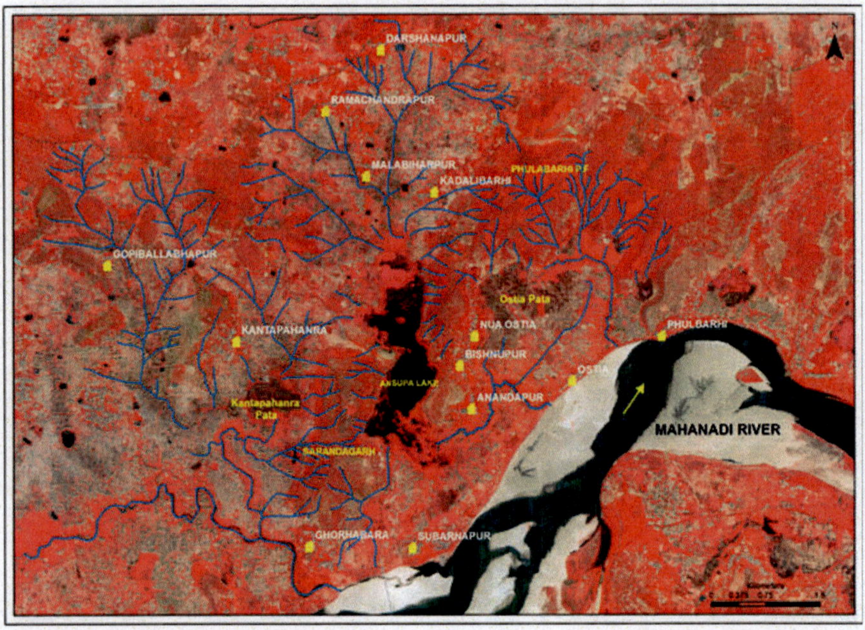

Fig. 30.2 Hydrological map of study area

30.2 Materials and Methods

- Ground truthing sampling points were taken with the help of GPS for accuracy assessment.
- Primary data collection through random sampling method, participatory rural appraisal (PRA) exercise, focused group discussion (FGD), and key informant interview (KII) method.
- Survey of India Toposheet with scale 1: 25,000.
- Satellite image of Cartosat II, Quick Bird and IRS P6 Resources at II LISS IV.

The name of the water body, surrounding settlement/villages, and other local information can be collected from the toposheet. The present scenario of the wetland was prepared from the present satellite imageries. We can study the historical status of the wetland from the time-series satellite imageries. We can estimate the seasonal variation of the water spread area from the satellite imageries. Human encroachments are also identified with the help of satellite imagery with ground truthing. With the availability of spatial technologies, it is possible to characterize, map, and monitor wetlands. Figure 30.3 shows the flow chart of the methodology used for the study (Barua et al., 1998 and Deka & Goswami, 1993).

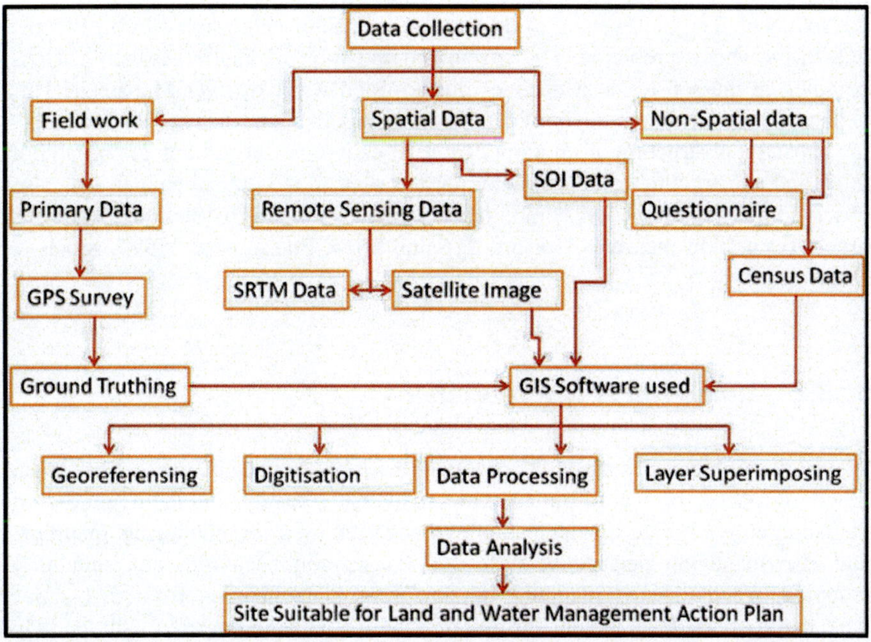

Fig. 30.3 Flow chart of the geomatic application for the research work

30.3 Results and Discussion

30.3.1 Problems

Siltation, shrinking of the water spread area, decreasing fish landing, weed infestation, eutrophication, decline in the biodiversity, and deteriorating water quality are the major problems of the deteriorating wetland. Inadequate exchange of water between lake and river, siltation due to change in the land use/land cover pattern in the catchment, and lack of understanding about values and function of the lake ecosystem are among the stakeholders. Heavy siltation is identified as one of the most detrimental factors for degradation of the lake ecosystem. It leads to a decrease in the depth of the lake alarmingly. The two inlet channels of the lake have been closed, which leads to hydraulic loss and poor flushing actions. It was identified as the main obstacle for the discharge of oxygen-depleted water from the bottom of the lake and sediment during monsoon to the river Mahanadi through outlet channel and favors eutrophication in the lake during the post-monsoon months. Due to the siltation and poor flushing action, the water spread area of the lagoon has been observed to be shrinking alarmingly. The types of weed in Ansupa are emergent, floating, and submerged types. The prolific spread of the invasive species is restricting the feeding as well as the breeding ground of the many economic species of the fishes. It is also causing navigational problems and impending the free flow of the sediment. The biodiversity of the lagoon has also been observed to be affected due to the above problems. The species composition of fish in Ansupa Lake is complex in the sense that it declines due to flood water of river Mahanadi. This decline can be attributed to combinations of factors like over-fishing beyond carrying capacity, obstructing of migratory route of economic species due to choking of inlet, and constructing of new structure at the neck of the outlet channel. It is also observed that the lake ecosystem is tending toward a bog ecosystem and warrants urgent restoration measures (Mishra & Sahu, 2014, Prasad et al., 1997, Singh & Dubey, 1994).

30.3.2 Management of Restoration

The above problems destroyed the beauty of the lake. So, the government has taken initiation for the survival of the lake, its biodiversity, beauty, and the livelihood of the community. The geospatial technology has been used for monitoring, mapping, and also managing perspective. The soil conservation activities site suitability mapping, weed infested mapping, water spread area mapping from time-series data, and tourism development site suitability analysis were prepared with the help of geospatial technology.

Soil Moisture Conservation

Treatment of the catchment area through development of the micro watersheds was taken up by DRDA, Cuttack, through the Junior Soil Conservation Officer, deployed from the Soil Conservation Department, who was declared as the PIA. Four micro watershed projects, that is, (1) Kadalibadi watershed, (2) Ostia watershed, (3) Subarnapur watershed, and (4) Kantapanhara watershed, were selected and different soil and moisture conservation measures along with other livelihood activities were carried out by the PIA in a participatory mode. The lake is bounded by the Saranda hills on its western side, Bishnupur hill on its northeastern side, and by villages like Subarnapur, Malbiharipur, Kadalibadi, and Bishnupur. It is very highly essential to recharge the groundwater by making interventions for water storage system as well as arresting the silt during flow of rainwater from ridge to valley. Soil moisture conservation measures, field bunding, baldhill plantation, avenue plantation, misc plantation, etc., will definitely be the proper intervention for the purpose.

These lands include cultivable and noncultivable wastelands having poor field bunds and lacking of proper disposal of runoff water. The activities are implemented like (1) rainwater management structures; primary graded bund cum recharge trench below the foot hill; (2) field bunding cum continuous contour bunding; (3) in order to improve groundwater recharge and to ensure better growth of plants, staggered trenches; (4) controlling nala bank erosion measures loose boulder structures;

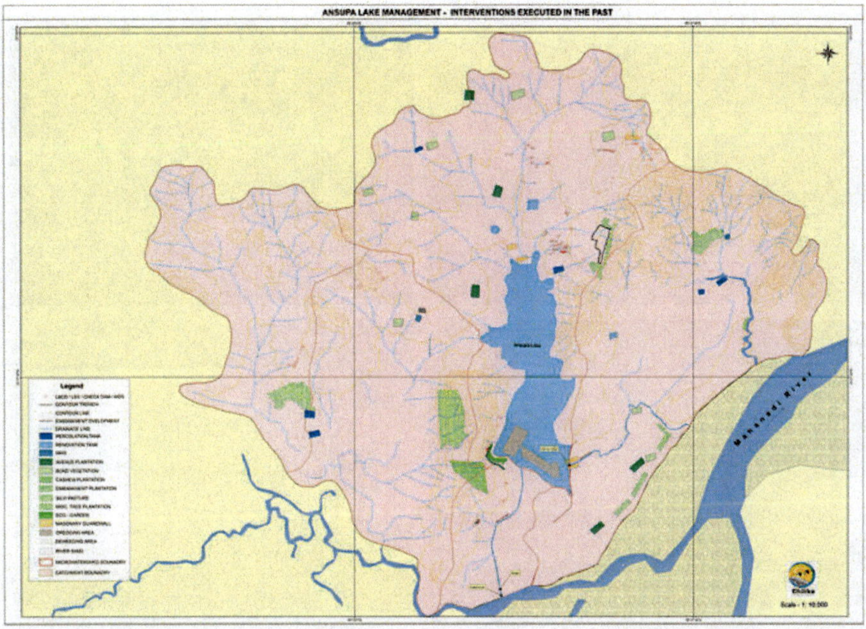

Fig. 30.4 Land and water interventions in the catchment area

(5) bank stabilization by loose boulder structures; (6) gully control measures; (7) masonry drop structures; (8) percolation tanks; and (9) renovation of wells. After percolation of tanks and renovation of the wells, the groundwater level has been increased in the catchment area, which is recognized by the surrounding community (Dilip & Venkatesh 2004; Krishna 1996; Kumar et al., 2014). The surrounding village people contribute that the moisture content has increased in the agricultural lands, thus enhancing the crop production. The water level of Ansupa Lake has risen up to 3 ft because of the above soil conservation measures/ treatments in the catchment area. Figure 30.4 shows the map of the activities that are implemented in the catchment area for the restoration of the lake. The catchment programs are implemented for reducing the siltation from the catchment to the lake. Figure 30.5 shows the photographs of the activities done on the catchment area for the reduction of siltation.

Tourism Development

Ansupa, with its beautiful blue water and surrounding green hill, has been attracting people from all over the country. During the winter season, the lake and its surroundings are flocked by the tourists, especially the picnickers. Considering the tourism potential, picnic spots have been developed by the Tourism Department of Govt. of Odisha and infrastructures like peripheral roads, landscaping, drinking

Fig. 30.5 The photographs of executed activities in the catchment

water facilities, platforms, boats for boating, public toilet, etc., have been developed at the suitable locations. Pallishree mela at Sarandhagarh was also organized by the executive director, ORMAS, to popularize Ansupa. Although there is ample potential for tourism, tourism in the lake has not flourished as expected due to ecological degradation and proliferation of aquatic weeds, and lack of visitor amenities. However, after the deweeding, the tourism sector has developed tremendously. The Chilika Development Authority (CDA) has developed an ecogarden comprising within a 10-acre area and enhanced another 5-acre land for this purpose. There is a children's park, and one rock garden has been developed within the ecogarden, which attracts the school children very much. A bamboo setum was developed comprising 37 species of bamboos. A parking place (0.83 acres) and a waiting place have been developed in the south of the ecogarden. Figure 30.6 shows the photographs of the boating facilities at the lake for the public and children's park, which were made in the ecogarden near the lake.

Fig. 30.6 Boating facility in the lake and children's park in the ecogarden

Activities by Fishery Department

The activities being carried out currently by the fishery department include installation of a pen over an area of 1.35 acres for growing grass carp fingerlings, training of 168 fishermen including SHGs, etc. These activities have been undertaken by the Assistant Director of Fisheries, Cuttack, under Fisheries & ARD Department, Government of Odisha.

Livelihood Support for Women SHGs

About 400 fisherman are living around the lake. Their main occupation is fishing from Ansupa Lake as well as from river Mahanadi. At present, four women SHGs are involved in the management of the ecogarden at Ansupa. They are collecting the entry and boating fees from tourists. These SHGs have established a nursery in the bank of lake by taking group loan from watershed committee. Similarly, 10 more SHG groups from the locality are keen for inclusion in the livelihood generation program. It is proposed for introducing beekeeping, mushroom cultivation, tailoring, and vending kiosks. This would be implemented through the SHGs involving 10 groups comprising 150 individuals from the villages in the catchments (Nyonand et al., 1997, Pande et al., 1998, Pandit et. al., 2007, Thapa, 2001, Yoganand & Gebremedhin, 2006).

Communication Education and Public Awareness (CEPA)

Ansupa Lake is a common property resource. For sustainable management of the lake, the active participation of the local communities is highly essential. The much desired awareness and community participation can be achieved through CEPA in the catchment villages. To step up the wetland education on lake ecology and sustainable management of the lake, CEPA program assumes high priority. To achieve this, Communication, Education and Public Awareness (CEPA) tool needs to be put into practice to generate the desired awareness among the local community about the values and& functions of wetland ecosystem and wise use of its resources to maintain the ecological integrity of the wetland and uninterrupted flow of ecosystem services.

It is proposed to generate awareness about the values and function of the wetland through carefully planned awareness programs. The target group would be resource users, students, and other stakeholders. Different CEPA tools would be used to achieve these objectives. For the purpose, customized signage, brochure, and other education materials would be developed and at the same time curriculum for environmental education for the school children would be developed. The display of signboard to the nearest town Athagarh to the lake side would bring more visibility of the lake and its prominence and message for its conservation and sustainable use. Fixing of signboards in local language and national language at the periphery of the lake would be enhanced in this program (Ganeriwala, 1997, Habtamu, 2011, Kaushik et al., 2007, Krantz, 2001).

The local nongovernment organizations and village-level institutions will be actively involved in the awareness and environmental education programs. Necessary kits and materials, that is, posters, pamphlets, brochures, video clips, documentary on various traditional method of fishing, traditional fishing gears, wise use practices, and other resource use, will be developed. Emphasis would be laid on the capacity building of stakeholders, school children, NGOs, and CBOs. The CEPA activities targeting the student from nearby school would be given priority.

Weed Management

Abundant growth of aquatic macrophytes in the lake plays an obstacle to the growth of fishery resources as well as adversely affects the tourism activities in the lake. Out of 199 ha of water spread area of the lake, 135 ha has been infested by macrophytes. These macrophytes are categorized into three major classes such as Emergent, Submerged, and Floating. The macrophytes, which are in great abundance in the lake, are Pistia stratiotes, Eichhornia crassipes, Monochoria hastata, and Salvinia cucullata. The submerged species are represented by Potamogeton pectinatus, Najas foveolata, Hydrilla verticillata, Ceratophyllum demersum, Utricularia inflexa, and Ottelia alismoides.

The channel joining the river and lake is almost dead, but low water-lying areas are still marked and such places are rich in species content and dominated by amphibious life forms. In general, the tall elements like Typha angustata, Saccharum spontaneum, Scleria terrestris, and Scirpus grossus are found in abundance. During the rainy season Limnophila heterophylla, Setaria pumlla, and Enydra flactuans are commonly met with. Hence to make the lake more tourist-friendly and more productive from the fishery point of view, it is essential to get rid of the profuse growth aquatic weeds through removal, that is,. by way of mechanical removal or manual. In addition to this, reduction of runoff from agriculture field by promoting

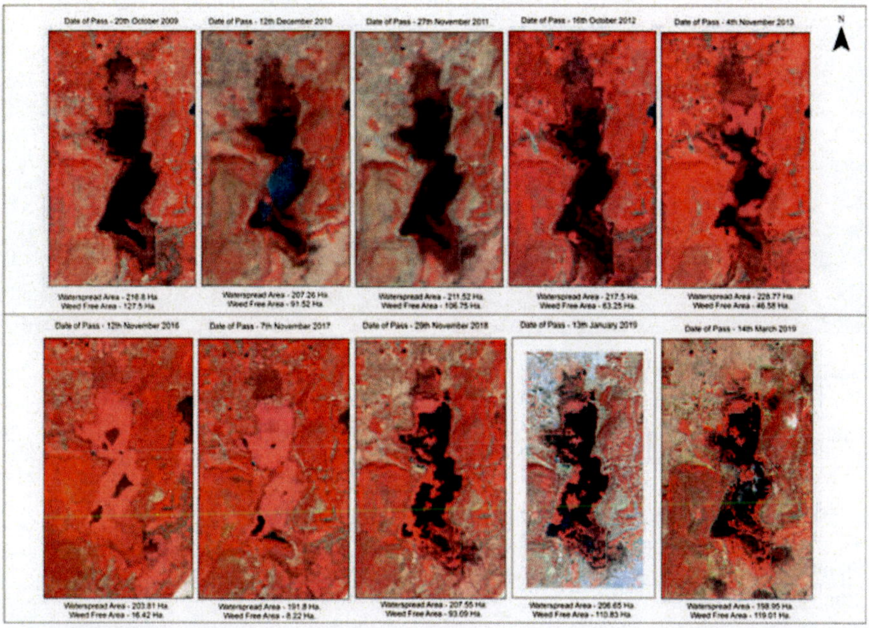

Fig. 30.7 Time-series satellite image of study area

sustainable agriculture needs attention. Figure 30.7 describes the time-series satellite imageries of how the lake turned into dead and after intervention how it is revived (Gajbhiye & Sharma, 2012, Kundu, 2014, Lakshmamma et al., 2011, Tolessa & Rao, 2013).

Deweeding

Removal of aquatic weeds from the lake manually has been tried in the past, and the process is successful only in the shore line, that is, up to a water depth of 1.00 m. The activity has continued through engagement of local labor in the shore line especially during September/October. Growth of aquatic weeds has almost spread all over the lake, and it was eradicated by the local fishermen community. We cannot control the weed growth from the entire lake due to manual process, but we could achieve only 20% of the lake area. However, during December 2018, one Swiss-made Amphibian Weed Harvester was procured by the financial assistance of ICZMP and engaged in deweeding work in the lake. Due to mechanical means, we could be able to clear the weed growth up to 85% of the lake. The satellite imagery also proves the clearance of the weed from the lake area. After the eradication of weeds, the lake biodiversity has increased and rare birds like Blackbellied tern and Indian Skimmer were seen in the lake. The aesthetic value of the lake has increased in various ways, which attract tourists of different age groups. Figure 30.8 shows the manual and mechanical

Fig. 30.8 The deweeding process done by manually and also by machine

Table 30.2 The water spread area and weed-free area of Ansupa Lake

Year	Water spread area (Ha)	Weed free area (Ha)	Weed area (Ha)
20th May 1999	161.79	26.36	135.43
10th November 2005	208.69	103.33	105.36
20th October 2009	216.8	127.5	89.3
12th December 2010	207.26	91.52	115.74
27th November 2011	211.52	106.75	104.77
16th October 2012	217.5	63.25	154.25
4th November 2013	228.77	46.58	182.19
12th November 2016	203.81	16.42	187.39
7th November 2017	191.8	8.22	183.58
29th November 2018	207.55	93.09	114.46
13th January 2019	206.65	110.83	95.82
14th March 2019	198.95	119.01	79.94

procedure of the deweeding timing. It is the most important activity for the lake reviving.

The time-series satellite imagery has been used for visual interpretation of the weed-free area. Table 30.2 shows the area, and Fig. 30.7 also shows that the lake has been covered with emergent weed till the year November 2017. After taking initiative for deweeding the lake by Chilika Development Authority after November 2017, the weed cover area has reduced. In the year November 2017, the weed-free area was 8.22 ha; and after that in the year March 2019, the weed-free area has increased to 119 ha. (Rout & Ojha 2019). Ansupa Lake has international importance because it is the home of many migratory birds. But the lake was silted up to a great extent and is full of water hyacinth and algae, which makes boating inside the lake impossible. Figure 30.7 shows the trend of the weed-free area until March 2019. If the water of the lake was so clear, then ground surface was visible clearly and facilitated the fisherman to locate the fish in the water. Due to heavy erosion from its catchment, the water spread area has reduced (Mishra & Panda 2008, Mishra & Panda 2015). The Soil Conservation Department has taken up plantation of cashew

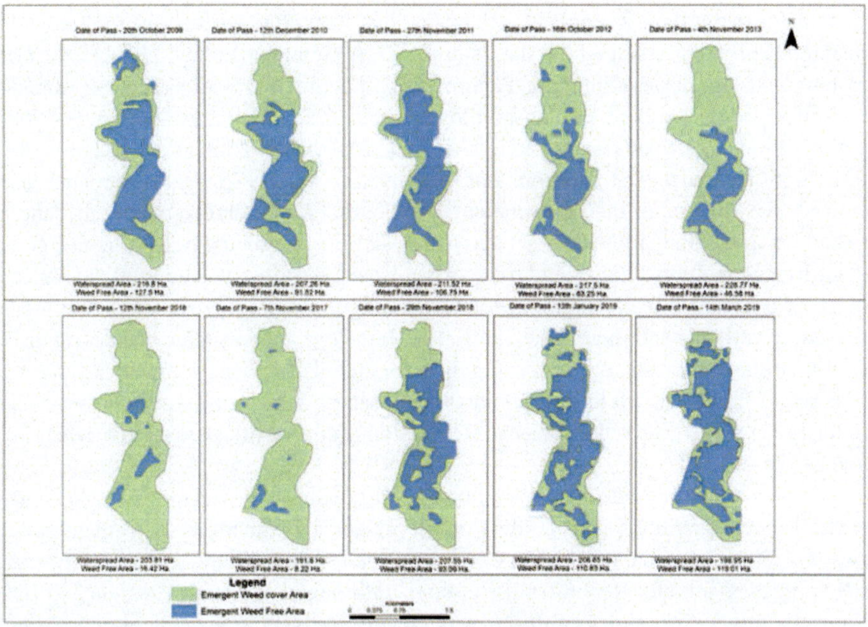

Fig. 30.9 Map showing the weed-free water area after deweeding

around Ansupa, which of course has arrested the silt inflow to some extent to the lake. The Forest Department has developed a reserve forest over an area of 15 acres on the Saranda hill. For improvement of the Ansupa Lake, treatment of the entire catchment of the Ansupa needs to be taken up. The entire runoff from the catchment drains to the Lake Ansupa. Figure 30.9 shows the exact area of water spread and weed infested of the lake, which was prepared by using remote sensing and GIS software (Gopal & Chauhan, 2001).

30.4 Conclusion

Though the lake area has degraded, but the government has taken appropriate measures through the Chilika Development Authority (CDA) to restore the lake ecology and its environment. The government initiation has not only increased the water quality of the lake but also developed the catchment area and ecological status of the environment, which makes it a success story. This increased the greenery of Saranda hill and Bishnupur hill increased the aesthetic value of the lake. Incidentally it may be mentioned here that for the development of Ansupa Lake, a committee named "Ansupa Lake Development Committee" has been established at Subarnapur village, Cuttack, under the leadership of a few eminent people of the locality.

India does not have a separate wetlands policy. Key policy directions for wetland management are contained in the National Environment Policy (2006), which recommends adoption of integrated approaches, specifically river basin management and site-specific prudent use as guiding actions. The regulatory framework for conservation of wetlands is defined by the Wetland (Conservation and Management) 2010 notified under The Environment (Protection) Act, 1986. "Environment," as defined in Section 2 of the Environment (Protection) Act, included water, air, land, and the interrelationship that exists between water, air, land, and human beings and other living creatures, plants, and microorganisms and property. The Wetland Rules of 2012 is the first major national framework for regulating detrimental activities related in wetlands and their catchments. The state governments have been entrusted with the task of identification of wetlands to be included under the ambit of the act. The rules, however, exclude river channels, paddy fields, and coastal wetlands (already covered under the Coastal Regulation Zone notification, most recently updated in 2011).

Efforts have been made in the program to build capacity of wetland managers and create awareness among stakeholders on the values and functions of wetland ecosystems. The ecological restoration of Chilika Lake, which led to the removal of the site from the Monteux Record (a list of sites with negative changes maintained by Ramsar Convention) and significant recovery of biodiversity and livelihoods, was conferred the Ramsar Award in 2002 and recognized as a model initiative. Nongovernment organizations such as Wetlands International South Asia (WISA), World Wide Fund for Nature – India (WWF), Bombay Natural History Society (BNHS), and Salim Ali Center for Ornithology (SACON) implement programs related to wetland management planning, water bird monitoring, community participation, research, capacity building, and awareness generation. The result of the study will be very helpful for administrators, managers, decision–makers, and also for research scholars to take decisions for management purpose. The policies and programs devised by various government agencies need to be effectively monitored. Its successful implementation needs involvement of the local community, which will lead to sustainable development of the ecosystem. Here, the science has used, that is, remote sensing and GIS with GPS for ground verification.

Wetland degradation usually impacts environmental quality and can lead to major changes in the community composition. Therefore, a recent paradigm that alters within wetland science toward integration of social, all environmental, and life sciences is further appealing to the historical linkage between wetland and special kinds of science today. Modern wetland science has become a multidisciplinary, interdisciplinary, and sometimes transdisciplinary study that melds the social with the life sciences to understand wetlands as social–ecological systems. The wetland ecosystems have vital values and functions in the world. Humans (controlling factor), as an ecosystem stakeholder, benefit from this. Rapidly developing technology enables us to better understand the planet we live in. Due to technological development and increasing human population, all ecosystems are inevitably deteriorated by domestic, agricultural, and industrial pollution; climate change; reducing biodiversity; invasive species; and change of land use. Sustainability includes a greater and more explicitly long-term situation and target than environmental quality increment. Sustainable environmental management depends mainly on ecosystem stability, ecological tolerance, and biotic diversity. Sustainable environmental management plans need to be implemented and controlled.

References

Barua, P., Sharma, P., & Goswami, D. C. (1998). A database for wetlands of Assam: A study using remote sensing technique. In *Proceedings NAGI (En Region) National Conference on environment and sustainable development* (pp. 27–34).

Basavaraj, H., & Nijagunappa, R. (2011). Identification of groundwater potential using geoinformatics in Ghataprabha basin, North Karnataka, India. *International Journal of Geomatics and Geosciences, 2*(1), 91–109.

Deka, S. K., & Goswami, D. C. (1993). Water and sediment characteristics of wetlands: A geo-environmental study of Deepar beel, Assam. *Gauhati University Journal Science, XXXII*, 93–105.

Dilip, G. D., & Venkatesh, B. (2004). Site suitability analysis for soil and water conservation structures. *Journal of the Indian Society of Remote Sensing, 32*(4), 399–405.

Gajbhiye, S., & Sharma, S. K. (2012). Land use and land cover change detection of Indra river watershed through remote sensing using multi-temporal satellite data. *International Journal of Geomatics and Geosciences, 3*(1), 89–96.

Ganeriwala, A. K. (1997). Watershed development – A total for rural uplift. *Yojana, 41*(3), 15.

Gopal, B., & Chauhan, M. (2001). South Asian wetlands and their biodiversity: The role of monsoons. In B. Gopal, W. J. Junk, & J. A. Davis (Eds.), *Biodiversity in wetlands: Assessment, functions and conservation* (Vol. 2, pp. 257–275). Backhuys Publishers.

Habtamu, T. (2011). Assessment of sustainable watershed management approach: Case study Lenche Dima, Tsegur Eyesus and Dijjil Watershed. In *Project paper of Cornell University*.

Kaushik, P. K., et al. (2007). Participatory approach to watershed management in India. *Indian Forester, 133*, 1659–1668.

Krantz, L. (2001). The sustainable livelihood approach to poverty reduction- An introduction. In *Swedish International Development Cooperation Agency (Sida)*. Division for Policy and Socio-Economic Analysis.

Krishna, A. P. (1996). Remote sensing approach for watershed base resource management in the Sikkim Himalaya: A case study. *Journal of the Indian Society of Remote Sensing, 24*(2), 69–83.

Kumar, G. M., et al. (2008). Delineation of potential sites for water harvesting structures using remote sensing and GIS. *Journal of Indian Society of Remote Sensing, 36*(4), 323–334.

Kumar, G., et al. (2014). Watershed impact evaluation using remote sensing. *Current Science, 106* (10), 1369–1378.

Kundu, R. (2014). Morphometric and land resource analysis using remote sensing and geographic information system – A case study in Agargaon watershed near Nagpur, Maharashtra. *Neo Geographia, III*(III), 53–65.

Lakshmamma, et al. (2011). Morphometric analysis of Gundal watershed, Gunlupet taluk, Chamarajanagar district, Karnataka, India. *International Journal of Geomatics and Geosciences, 3*(2), 758–775.

Mishra, A. K., & Panda, G. K. (2008). Micro watershed projects of Bolangir district and their role in water resource development. *Eastern Geographer, XIV*(1), 79–86.

Mishra, P., & Panda, G. K. (2015). Geospatial technology based catchment treatment planning – Case study of Sandul irrigation project catchment. *SGAT Bulletin, 16*(1), 59–66.

Mishra, D., & Sahu, N. C. (2014). Response of farmers to climate change in Odisha: An empirical investigation. *International Journal of Environmental Sciences, 4*(5), 786–800.

Nyonand, A. K., et al. (1997). Rehabilitation of degraded ecosystem through watershed management – A success story. *Indian Journal of Soil Conservation, 25*(2), 162–166.

Odisha Wetland Atlas (2013). Ministry of Environment and Forest, Govt.of India, Space Application Center, ISRO, Ahmedabad, March 2013

Pande, V. C., et al. (1998). Farm resource development- A case study of watershed management in semi arid tropics of Gujarat. *Indian Journal of Soil Conservation, 26*(1), 52–56.

Pandit, B. H., et al. (2007). Watershed management and livelihoods: Lessons from Nepal. *Journal of Forest and livelihood, 6*(2), 67–75.

Patnaik, A. K. (2005). Rejuvenation of Chilika lagoon: Restoration of a coastal wetland with community participation. *EPTRI–ENVIS Newsletter, 11*, 2–4.

Prasad, S. N., et al. (1997). Impact of watershed management on runoff water resource development and productivity of arable lands in south eastern Rajasthan. *Indian Journal of Soil Conservation, 25*(1), 68–72.

Raghu, V., & Reddy, K. M. (2011). Hydrogeomorphological mapping at village level using high resolution satellite data and impact analysis of check dams in part of Akuledu Vanka watershed, Anantapur District, Andhra Pradesh. *Indian Journal of Geophysics Union, 15*(1), 1–8.

Report on National Environment Policy, 2006. Govt of India Ministry of Environment of Forest, Approved by the Union Cabinet on 18th May 006

Rout, J., & Ojha, A. (2019). Geospatial technology application on sustainable wetland management & monitoring: A case of Ansupa Lake, Odisha. *Journal of Remote Sensing and GIS, 10*(2), 13–18.

Singh, A., & Dubey, A. (1994). *Geoenvironmental planning of watershed in India*. Chugh Publications.

Thapa, R. S. (2001). *Contribution of watershed management program to the livelihoods of the local people: A case study from Kathekhola sub watershed, Baglung, Nepal.* Master Thesis, Institute of Geodesy, GIS and Land Management.

Tolessa, G. A., & Rao, P. J. (2013). Watershed development prioritization of Tandava river basin, Andhra Pradesh, India – GIS approach. *International Journal of Engineering Science Invention, 2*(2), 12–20.

Yoganand, B., & Gebremedhin, T. G. (2006). Participatory watershed management for sustainable rural livelihoods in India. In *Workshop paper presented at the Regional Research Institute, West Virginia University*.

Chapter 31
Tourism and Tribal Economy: Application of GIS Technology on Sundarbans Region

Ranajit Sardar and Sukla Basu

31.1 Introduction

Tourism is an industry with one of the strongest effects on economy, because it helps in developing other sectors through multiplier effect. It is a composite of activities, facilities, services, and industries that deliver a travel experience, that is, transportation, accommodation, eating and drinking establishments, entertainment, recreation, historical and cultural experiences, destination attractions, shopping, and other services available to travel away from home. In other words, tourism is defined as the sum of phenomenon and relationships arising from the interaction of the tourist and host communities in the process of attracting and hosting this tourist (Fadahunsi, 2011). This definition shows that tourism is interactive in nature. Geography essentially plays a pivotal role in tourism because of the geographical base. It offers a fascinating environment for geographers and GIS experts to analyze the different components of tourism in spatial context. Today's world is profoundly affected by Information Communication Technology (ICT) revolution, which has enabled information and knowledge circulation at unprecedented speed, changing all aspects of life, and economic, political, and sociocultural mosaic of the world. In this information era, it is the ability of a country to use information and communication technology effectively and efficiently that increasingly determines the relevance and competitiveness of a country in the global economy. But in my study area, there was absence of technological information for tourist guidance. No work has been done about the GIS-based tourism application and tribal economy relating to tourism in Indian Sundarbans.

R. Sardar (✉)
Acharya Prafulla Chandra College, New Barrackpore, West Bengal, India

S. Basu
Department of Rural Studies, West Bengal State University, Barasat, West Bengal, India

G. S. Bhunia et al. (eds.), *Anthropogeomorphology*, Geography of the Physical Environment, https://doi.org/10.1007/978-3-030-77572-8_31

Geographical Information System is an integral part of the information system and forms a chain of operation from a survey, collection to storage, analysis, and output of spatial information for supporting decision making (Dockkey, 2002). Sustainable tourism is rightly known as a catalyst for the overall development of the developing part of the world. On account of this, it has attracted the interest of the government, communities, and researchers. These entire stakeholders have reached consensus that tourism needs to be developed in a planned way and Geographical Information System can assist as a decision support system for planning strategies. GIS has diverse applications for a wide range of activities such as environmental planning, property management, infrastructure setting, and emergency event planning, automobile navigation system, urban studies, and market analysis and business demographics (Chen, 2007). Recently, GIS application in tourism sector has rapidly increased: as, for example, park management, facility monitoring, visual resource assessment, and tourist area identification, etc.

31.2 Tourist Potential of the Sundarbans Area

Sundarbans is located in the southern part of West Bengal and Bangladesh. This is known as Sundarbans Biosphere Reserve (SBR) and Sundarbans Tiger Reserve (STR) formed in 1973 under the Project Tiger of Ministry of Environment and Forest. It forms a significant part of the Indian Sundarbans and is one of the original nine tiger reserves in India. Not only that there are various opportunities of economic development like honey collection, fishing, etc. The natural beauty of Sundarbans and scenic beauty attracts people from all over the world. Undoubtedly it is the most attractive scenic spot. As it is a bioreserve forest, there are different kinds of plant and animal species over here. Tiger is the main attractive and ferocious animal and also the national animal. Dears, pigs, and crocodiles are the other animals inhabiting together. Neti Dhopani, Dobanki, Bannie Camp, Khalsi Khali, Haldibari, Pakhiralay, and Sajnakhali, Sudhanyakhali, are important spots to visit in the Sundarbans Tiger Reserve (STR). There are also markets, ATMs, hospital facilities, etc., although it is a remote area from Kolkata, almost 84 km away. As it is a famous tourist spot, the tribal people are recently depending on the tourism sector for sustainable livelihood instead of the primary sector economy. The Department of Information and Cultural Affairs of West Bengal (2015) has launched a scheme (Folk Artist) for the tribal people. They have made groups, and each folk artist group should have a minimum of 11 members and a maximum of up to 15 members. The male and female ratio of these groups is 2:3. In these groups, each member should be a tribal. They are performing their dance at any other governmental program also. They are earning remuneration per month of Rs 3500/- per member for their performance. The Adibasi Kalyan Folk Group (AKFG) is one of the examples of such a folk group. In this regard, women work participation rate is greater than men. These dance groups also promote the tourism in the Sundarbans Region. They are also now earning the money from public tourism sector. It is a great example of economic inclusion of Munda tribe of the Sundarbans Region.

31.3 Use of GIS Application on Tourism Planning

Geographical or spatial data refers to the location on the Earth's surface usually expressed as a grid coordinates or in degrees of latitude and longitude. Most institutions or organizations make use of implicit geographical references as place names, address, and postcode and road number, etc.; implicit spatial references can usually be geocoded into explicit spatial references. Advanced technology, particularly software and hardware, has resulted in the development of system that provides a range of searching, querying, presentation, and analytical function. Travel preferences are often hidden and are not explicitly known when users start to plan their trips, particularly if visiting an unfamiliar place. According to Van and Wessel (1994), there has always been a direct relationship between tourism and cartography. Tourism is concerned with the traveling between close and distance places and maps are the important tools to identify the specific location; therefore, it is more concerned with maps and thematic information than other industries. Maps of the travel route and general information about the area to visit are used in selecting the destination, planning travel, accommodation, etc. Therefore, tourist information system should contain large amounts of detailed up-to-date information about the destination (Brown & Weber, 2012).

The decision to adopt tourism as an agent of development has been largely based on the expectation that tourism can increase foreign exchange earnings, create employment, attract foreign investment, and positively contribute to local economies and national balance of payment (Dye & Shaw, 2006). The success of any tourism business in any destination is determined by tourism planning, tourism development, and research and tourism marketing. Some of the key features of GIS that could be benefiting tourism planning include their ability to manipulate data and spatial attributes (Jovanovic & Njegus, 2008) and provide necessary value-added information (Bahaire & White, 1999). The ease is in allocating resources between what are often conflicting demands (Williams et al., 1996), their adaptability requirements, needs, and data changes over time (Beedasy & Whyatt, 1999), and their ability to identify pattern and spatial relationship (McAdam, 1999). Although the number of GIS applications in tourism and recreation management and planning is increasing, there are still many more potential opportunities (Porter & Tarrant, 2001), but tourism resource inventory is one of the very first and basic applications, and such inventories may be further used for resource management resource allocation and land use planning decisions (Williams et al., 1996). Tourism marketing is another fascinating area wherein geo-demographics and lifestyle analysis can be performed by a GIS, which actually have a significant contribution in the needs of post-modernism tourism marketing (Elliott-White & Finn, 1998). Therefore, managers, entrepreneurs, and other stakeholders responsible for tourism marketing could benefit from GIS in order to locate and analyze the characteristics of potential customers. The synthesizing of environmental, social, and economic parameters in GIS format is easily possible because GIS is an integrating technology capable of working along with other technologies, such as remote sensing, global positioning system, etc., which could further facilitate more tools for sustainable tourism planning and marketing. Furthermore, service-oriented architecture can be applied to remanage the GIS resources and provide a dynamic and reliable service system that could meet the information and service requirements of different users over the Internet instantly (Emhmed & Chellapan, 2012). The Internet development contributed to the growing importance of GIS in various areas, opening new perspectives for people who need to use spatial data (Yang et al., 2004), when making decisions, planning and analyzing the effect of changes, looking for pattern, etc. We may look at maps, tables, charts, lists, graphs, reports, and sometimes it is rather difficult or nearly impossible to pull all these sources of information together and make sense out of them. Hence lays the benefits of GIS and its application.

GIS has the capability to handle several kinds of information that can be related to location or area. Therefore, GIS gives integrated tourism information. The database created in GIS format will answer the following frequently asked question in a user-friendly manner:

- Where is the tourist spot located?
- What is the shortest route direction?
- What is geographical environment?
- What is the best time to visit?
- What are the types of accommodation available?
- Where is the location of ATM services and tourist products available to purchase?

31.3.1 *Research question formulation*

Basically the tribal people usually use to depend on primary activities in the Sundarbans Region like fishing, wood collecting, etc. But now Sundarbans Tourism is developing; therefore, they are focusing their economy on tourism sectors like hotel business, hotel cooking, Jhumur dance, etc. In this regard, two definite research problems arise:

1. What is the extent of tribal involvement in the tourism sector for their economy?
2. What is the state of development of the GIS-enabled inventory for existing tourist infrastructures of the Sundarbans Region?

To look into these questions, multiple methods of researching have been adopted.

31.4 Materials and Methods

The base map of Gosaba Block of Sundarbans Region and Sundarbans Tiger Reserve was generated from Google Earth and District Census Handbook (2011), and *Esri's Arc-GIS 10.1* version was used to support the existing ground-level situation and also for developing the drainage system and subsequently various thematic layers were generated, that is, existing land use, road network, water bodies, hotel or resort, ATM facilities, etc. Intensively field investigations have been done regarding the relationship between tourism and tribal economy selecting four tribal villages (Hamilton Abad, Kamakhyapur, Chotta Molla Khali, Kumirmari) of Gosaba Block because these villages are situated near the Sundarbans Region and also most of the tribal people are concentrated in this block and Kultali Block (2.5%). Focus Group Discussion (FGD) has also been done on different Jhumur dance group like Loko Sanskriti Sampraday and others.

31.5 Results and Discussion

31.5.1 Tribal Economy and Tourism

Economy is one of the major subsystems of any societal system. Comprehension of the structure and dynamics of the economic system is a prerequisite for social work practice in tribal India as most of the challenges to tribal welfare are embedded in them. Moreover, tribal economies in India are at different stages ranging from food gathering to industrial labor types. Thus, economy can be understood as an institutional arrangement that facilitates acquisition, production, and distribution of material means of survival for individual and community life. Tribes develop their economies through a wide range of activities, including agriculture, mining, and tourism. Nowadays tribal nations in the Sundarbans Region pursue their economic development by performing the Jhumur dance for tourists. Tribal economy is embedded in and revolves around the Sundarbans. Not only tribal economy, but also culture and social organizations are interwoven with the forest. Tribals depend on forest for fulfilling their basic needs earlier. Now with increasing situation of the tourism industry, they partly depend on forest for their development of economic infrastructure. The Adibasi Kalyan Folk Group (AKFG) is one of the examples of a folk group led by Ratan Sardar (Card No-S24-1527). As male and female participants in a group are 2:3 accordingly so that women work participation rate is now increasing in the Sundarbans Region. These dance groups also promote the tourism in the Sundarbans Region. They are also now earning money from the public tourism sector. It is a big example of economic inclusion of Munda tribe of the Sundarbans Region. There were seen several Jhumur dance groups, namely Sundarbans Adibasi Lokosankriti Sampraday (Card No-S24-1526) led by Bithika Sardar (Hamiltan Abad), Tusu Mata Sampraday (Card No- S24-1528) by Bablu Sardar

Table 31.1 Relationship seasonal involvements of tribal people in their economy

Chi-square value	DF	P-value
94.180	1	0.000

(Kamakshapur), Sonali sampraday (Card No-S24-1532) by Sarswati Sardar (Chotta Molla Khali), Radhakrishna Sardar (Card No-S24-1523) by Shila Sardar (Kumirmari), Shidhu kanhu Sampraday (Card No-S24-1525) by Nripen Sardar (Kantamari), and Bandana Sampraday (Card No-S24-1535) by Bandana Sardar (Shyamnagar).

FGD revealed that each member of the group is getting remuneration Rs 3500/- per month from the government and earn Rs. 1200/- per performance before a tourist party in any tourist lodge or resort or guest house during the peak season (September–March). Each group performs three or four times every day in the peak season. In this way, every member earned almost Rs 5000/- per month. Earlier they earned money from fishing, honey collection, wood collection, etc. The amount earned is also less and is also risky because it would have to be collected from deep forest. According to Santosh Sardar, a resident of Hamiltan Abad, 20% of total population of the village (i.e., 2817 according to 2011 Census) is engaged in fishing and honey collection in the Sundarbans Tiger Reserve (STR). The present Lahiripur Gram Panchayat Pradhan (Prakriti Ghorami) indicated that the economic benefits of tourism are largely more in Hamiltan Abad and its surroundings. They left this job and are now engaged in different hotels and restaurants for cooking and other works and earn Rs 3500/- per month. There are many hotels (more than 50) and resorts that have been constructed, and different works have opened up for the local people. So, if tourists are not allowed in the Sundarbans Region, not only the tribal but other people also will not be able to earn livelihood. They will be compelled to depend on primary sector, which is very risky. Therefore, tourism definitely plays a vital role to develop their economy. Nowadays tourists are allowed to visit there except for three months during the rainy season, specifically June, July, and August. The tourist season starts from September every year. So, tourism definitely promotes tribal economy of the Sundarbans Region.

This chi-square result (Table 31.1) proves that the involvement of tribal people is basically season wise, because p-value is 0.000. As the p-value is 0, the null hypothesis is true; and if the p-value is 1, the null hypothesis is false. But it is true that all season tourists visited the Sundarbans. Therefore, tribal economy more or less depends on the Sundarbans Region.

31.6 Distribution Pattern of Amenities

Spatial pattern is the manner of arrangement of events in space or simply how spatial features are distributed. A spatial pattern can be described as regular, random, or clustered, which is the product of certain processes at a particular time and space

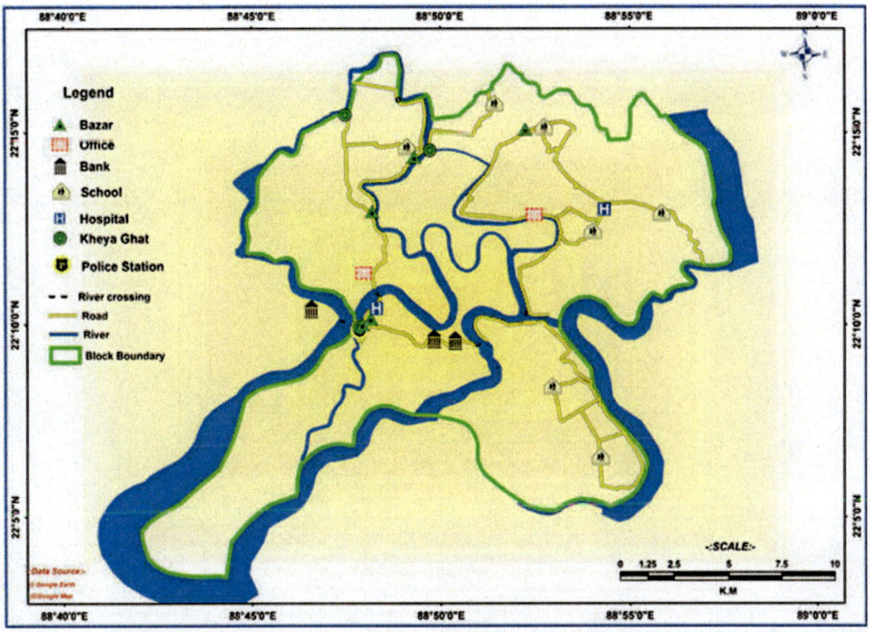

Fig. 31.1 Amenities map of Gosaba. (Source: Map drawn by the first author)

(Li & Zhang, 2007). Visualization is the process of creating and viewing graphical images of data with the aim of increasing human understanding (Hearnshaw & Unwin, 1994). It is based on the premise that humans are able to reason and learn more effectively in a visual setting than when using textual and numerical data (Tufte, 1999). Specific location and accommodation is the basic needs for the pace of tourism industry of any destination. At present, there are 65 hotels or resorts including government and private comprising 6985 rooms and with 13,980 bed capacity. The map (Fig. 31.1) shows that tourists can easily select any hotel on basic location, category, facilities, type of food, tariff, and can finalize the accommodation process through email and phone, which is available for tourist party in lean or peak season. The month of September is the starting season and continues up to March. But there is a huge pressure of flow of tourists during the peak season; room availability then may be short of supply. In this situation, room rent escalates, ranging from Rs 600–1200 per room.

31.6.1 Identification of Preferred Accommodation

Identification of accommodation is the basic need for the tourist party. If there was no availability of accommodation in any specific tourist destination, no tourist will be there as per their choice. Always every visitor prefers guest houses that are in

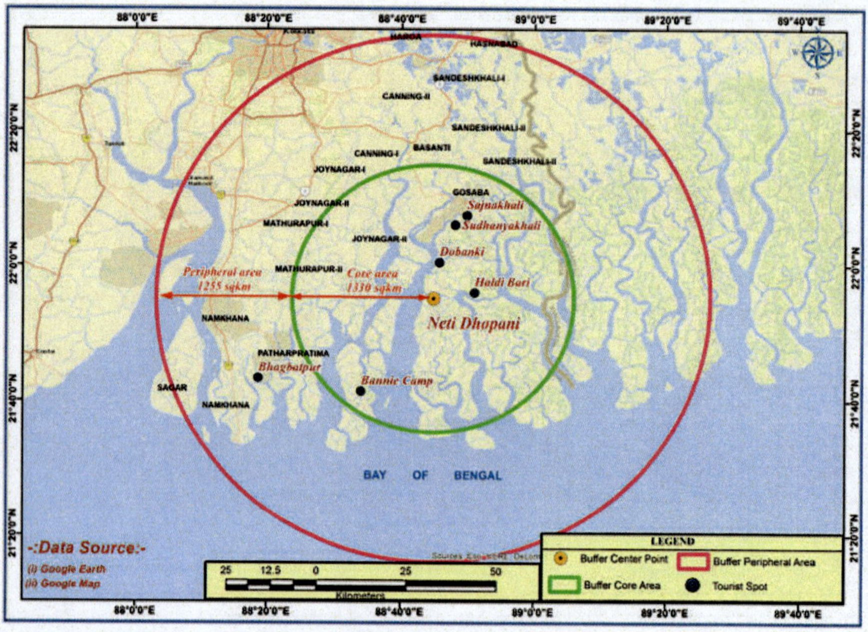

Fig. 31.2 Buffer map of Sundarbans. (Source: Map drawn by first author)

close proximity to the visiting place or spot. Sundarbans being a tourist place, the accommodation is concentrated in a particular area, that is, it is located near the Sajnekhali Sanctuary, so all are lured to seek accommodation in that place. As the Sundarbans is natural beauty and unique geographical location in the world, people try to visit at least one time as per their capability. In a map, one place is located in the peripheral area and the rest of the visiting places are in core area (Fig. 31.2). The buffer map has been created with 1330 km^2 of core area and with 1255 km^2 of periphery area out of 2585 sq km of Indian Sundarbans Tiger Reserve (STR).

31.6.2 Visualization of Tourist Place and Heritage Bacon Bungalow

Bacon Bungalow is popularized as Hamilton Bungalow, which is at present an important part of the Sundarbans Tourist place. On 20th December, 1873 Mr. Tilmen Henkel, magistrate of Jessore, wrote to Governer General Hastings for considering whether the jungle portion of the Sundarbans could be given on lease. For that cleansing the jungle started. Befor 1877, there was no fixed law as such for revenue collection and from this very year a proper revenue collection system was started, although unused land was kept out of revenue jurisdiction. Mr. Gomes, the then commissioner of Sundarbans was directed by the Board of Revenue to chalk out a draft of law, which later came to be known as Large Capitals Rules in 1879 (Pargiter, 1934). According to that law Sir Daniel Hamilton (1903) was clearing the jungle with the help of tribal people for settleing up the new settlement due to establishing a Zamindar system and business. In 1932, Kavi Guru Rabindra Nath Tagore was invited and came to this bungalow and stayed there from 30 to 31 December. That is why this is the important part of the Sundarbans for visitors as a historical place of Gosaba Block in Indian Sundarbans Region (ISR). Neti Dhopani is another mythological place. In mythological view, once upon a time Neta Maa, a Dhopa, used to wash cloths of Gods of Heaven at this place. Now this place is known as "Neti Dhopani." Sudhannya Khali, Sajnakhali, Bannie Camp, Bhagabatpur Crocodile Project, Dobanki, and KhalsiKhali (Fig. 31.2) are also some important visiting places in the STR.

31.6.3 Identification of Route Direction Kolkata to STR

A traveler has limited spatial knowledge about the specific tourist destination and map, and can perform an essential function in the question of spatial information about the travel destination. Web-based G.I.S Route Finder System not only models and processes the real road network to digital format but also provides user with different route finder options to proper management of travel time, safety, reliability, passenger convenience, and mitigating traffic congestion (Singh & Singh, 2012). The map shows a tourist how to reach Gosaba and Kultali Block for visiting STR. There was only one way to visit ISR, and traveling to different places was only through boats because all routes were by water to reach the tourist place. The distance can be easily measured; as the map shows, the distances between Kolkata to Gosaba and Kultali are 85km and 72km (Fig. 31.3), respectively, as two blocks are the gateway of Sundarbans in the southern part of Sundarbans.

31.7 Future Plan for Sundarbans Tourism

Distances from Sealdah to Kultali and Sealdah to Gosaba on Sate High way 1 are 72 km and 85 km, respectively. No railway exists from Jaynagar to Kultali and Canning to Gosaba. But Sealdah to Canning and Sealdah to Jaynagar railway connectivity already exists. Tourists have to journey either via canning or Jaynagar. It is time consuming and expensive also to reach Sundarbans through both routes. If the extension of railway line from Canning to Gosaba and Jaynagar to Kultali Block is made available, then many tourists can reach easily this destination cheaply and comfortably (Fig. 31.4). Extension of railway will be beneficial to tourists in respect of traveling expense, and at the same time local tribal people will be benefited for hotel business and others tourism-related occupations. Therefore, this extension will not only develop the tourism sector of Sundarbans but also upgrade the livelihood

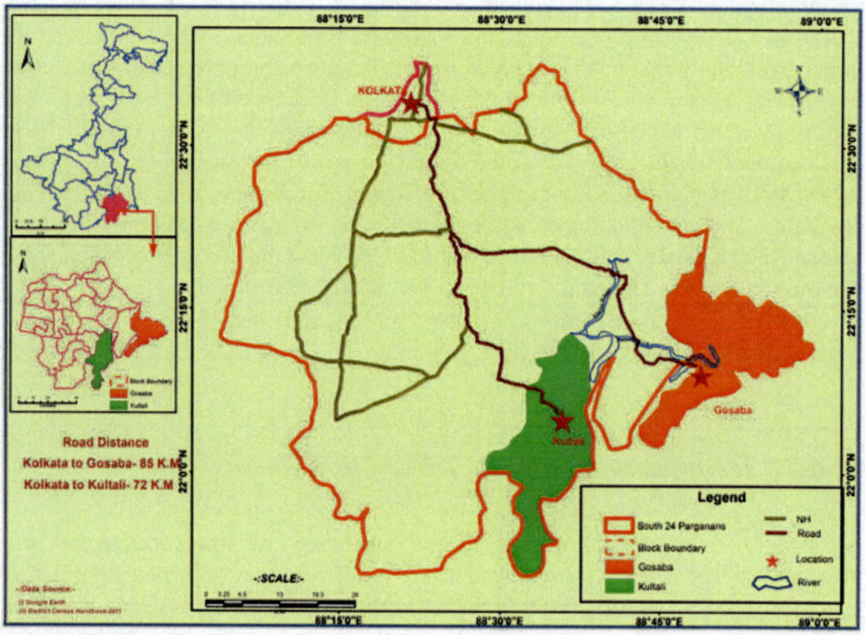

Fig. 31.3 Road connectivity map of Kolkata to Gosaba and Kultali. (Source: Map drawn by the first author)

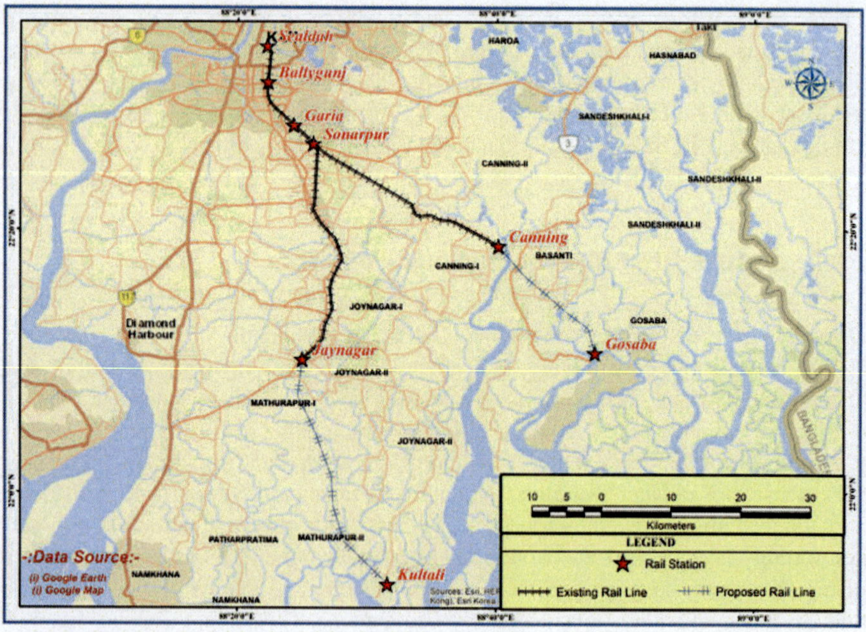

Fig. 31.4 Future plan for train route direction. (Source: Map suggested by the first author)

and socioeconomic condition of the local stakeholder. As this is the only route to enter the Sundarbans, it would be very much effective for the development of Sundarbans Tourism.

31.8 Sustainable Tourism in Sundarbans

One of the major essentials of Sundarbans Tourism (ST) Development and management is active participation of all relevant stakeholders in the tourism development and management process (UNEP and UNWTO, 2005; UNWTO, 2013). With a similar view, other scholars (Gunn, 1994; Ioannides, 1995; Robson & Robson, 1996; Briassoulis, 2002; Byrd, 2007) claim that the success and proper implementation of ST in a destination notably depend on the support from and involvement of stakeholders. A stakeholder (Freeman, 1984, p. 46) is "any group or individual who can affect or is affected by the achievement of the organizations objectives."

There are many stakeholders in the field of ST; however, it is not entirely predetermined and mostly depends on the nature and context of a tourism destination. Figure 31.5 demonstrates the key stakeholders in sustainable tourism of ST, a typical tourism destination source (Swarbrooke, 2001).

The development and/or maintenance of specific sites, infrastructures, and routes in the designated areas of Sundarbans will offer quality ecotourism and recreation experience. In order to take advantage of the increasing tourist flow, the Ecotourism Department of Forest will seek public/private partnerships consistent with the guidelines and principles established by the Government of West Bengal to improve the ecotourism services and facilities. The first necessary step in this regard is to impart communications and hospitality skills to residents and especially to tour operators. This would include training them on similar initiatives that have been successful. In the Sundarbans, people will also need financial assistance so that they can develop environmentally sustainable home stays and restaurants that would meet international standards of comfort and hygiene. For preventing any damage to the ecosystem, the guards in the forest departments of India are to be trained so that they can help tourists touring the area. Boatmen taking the tourist around belong to the economically vulnerable section in Sundarbans. In such a way, it will be possible to maintain the sustainable tourism in the Sundarbans Region.

31.9 Major Findings

- Electricity supply is often interrupted and mobile phones sometimes do not work properly.
- The boats currently used for cruises do not comply with government-mandated regulations and standards in terms of safety and hygiene.

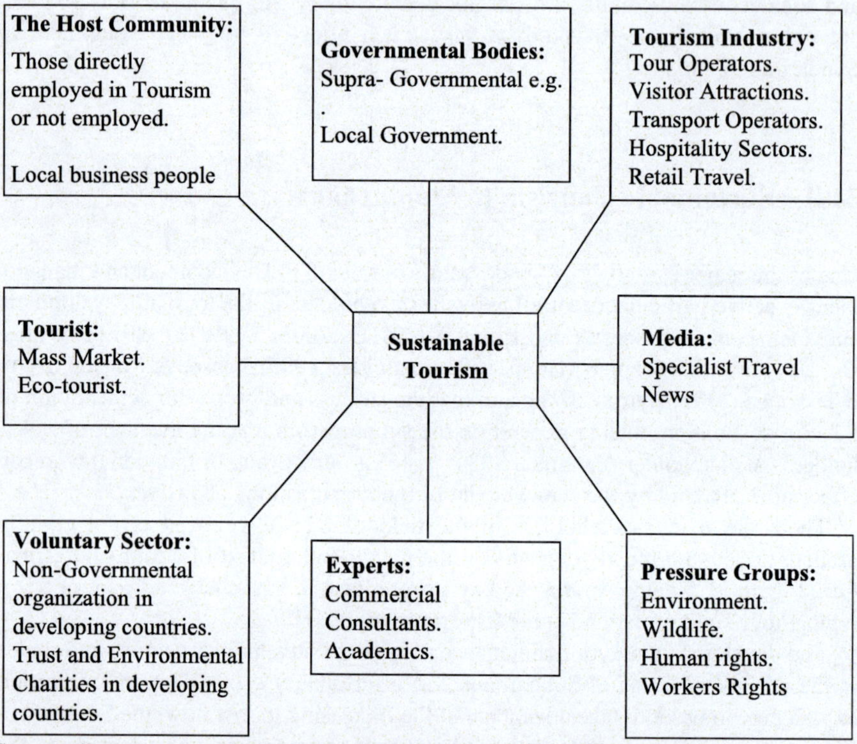

Source: (Swarbrooke, 2001) the key Stakeholders in Sustainable Tourism.

Fig. 31.5 The key stakeholders of sustainable tourism. (Source: Swarbrooke, 2001)

- There are hardly any hotels that would be acceptable to most international tourists.
- Though the number of eating places for tourists is limited, appropriate hygiene standard is always maintained.
- Lack of communication skills is also the cause of worsening situation of tourism sector in the Sundarbans.
- Local tribal people are earning money from tourist party through performing the Jhumur dance.
- Tourism promotes tribal economy of the Sundarbans Region.
- GIS-based tourist information system is increasing in the Sundarbans Region.

31.10 Conclusion

GIS-based ways and methods of analyzing spatial data are needful for the description and analysis of different dimensions of travel patterns. It should be the dynamic and interactive process. The researcher, planner, and administrator should interact and manipulate spatial and attribute of facilities, and natural and cultural features constituting the tourism products of the destination. GIS could immensely help not only to visual aspects of tourism products but can also provide alteration at regular intervals to update the information. GIS-based tourist information system can help tourists in taking decisions and planning for tourist destination. In this study, a small effort has been made for suggesting development of Sundarbans Tourism. If the different tourist maps and other favorable user-friendly maps for tourist are uploaded at the Sundarbans Tourism website, it will be very useful for everyone. Information about Sundarbans through the website is only possible through GIS-based information platform. And at the same time as the study on tribal economy and tourism, it has been found that tribal economy recently depends on tourism of Indian Sundarbans.

References

Bahaire, T. M., & White, E. (1999). The application of geographical information systems (GIS) in sustainable tourism planning: A review. *International Journal of Sustainable Tourism, 7*(2), 20–27.

Beedasy, J., & Whyatt, D. (1999). Diverting the tourists: A spatial decision-support system for tourism planning on a developing Island. *ITC-Journal, 3*(4), 163–174.

Briassoulis, H. (2002). Sustainable tourism and the question of the commons. *Annals of Tourism Research, 29*(4), 1065–1085.

Brown, & Weber. (2012). Using public participation GIS (PPGIS) on the Geo-web to monitor tourism development performances. *Journal of Sustainable Tourism, 3*, 1–20.

Byrd, E. T. (2007). Stakeholders in sustainable tourism development and their roles: Applying stakeholder theory to sustainable tourism development. *Tourism Review, 62*(2), 6–13.

Chen, R. J. C. (2007). Geographic Information Systems (GIS) applications in retail tourism portal and teaching curriculum. *Journal of Retailing and Consumer Services, 14*, 28–295.

Dockkey, K. (2002). *GIS for a policy decision support in national tourism portal*. Korea Culture And Tourism Policy Institute, viewed 28 October 2020, http://www.kctpi.re.kr

Dye, A. S., & Shaw, L. S. A. (2006). GIS based spatial decision support system for tourists of great smoky mountains national park. *Journal of Retailing and Consumer Services, 14*, 269–227.

Elliott-White, M. P., & Finn, M. (1998). Growing in sophistication: The application of GIS in post-modern marketing. *International of Journal of Travel and Tourism Marketing, 7*(1), 65–84.

Emhmed, A., & Chellapan, K. (2012). Modeling a homogenate GIS architecture based on service oriented architecture for the tourism mapping needs. *International Journal of Scientific & Engineering Research, 3*(1), 1–5.

Fadahunsi, J. T. (2011). Application of geographical information system (GIS) technology to tourism management in ile-ife, osun state, Nigeria. *Pacific Journal of Science and Technology, 12*(2), 274–283.

Freeman, R. E. (1984). *Strategic management: A stakeholder approach*. Pitman.

Gunn, C. A. (1994). *Tourism planning: Basic concepts cases* (3rd ed.). Taylor and Francis.

Hearnshaw, H. M., & Unwin, D. (1994). *Visualization in geographical information systems*. Wiley.
Ioannides, D. (1995). A flawed implementation of sustainable tourism; the experience of Akamas, Cyprus. *Tourism Management, 16*(8), 583–592.
Jovanovic, & Njegus. (2008). The application of GIS and its components in tourism. *Yugoslav Journal of Operations Research, 18*(2), 261–272.
Li, F., & Zhang, L. (2007). Comparison of point pattern analysis methods for classifying the spatial distributions of spruce-fir stands in the North-East USA. *Forestry, 80*(3), 337–349.
McAdam, D. (1999). The value and scope of geographical information systems in tourism management. *Journal of Sustainable Tourism, 7*(1), 77–92.
Pargiter, F. E. (1934). *Revenue history o Sundarbans from 1765–1870*. Calcutta, p. 124.
Porter, R., & Tarrant, M. (2001). A case study of environmental justice and federal tourism sites in southern Appalachia: A GIS application. *Journal of Travel Research, 40*, 27–40.
Robson, J., & Robson, I. (1996). From shareholders to stakeholders: Critical issues for tourism marketers. *Tourism Management, 17*(7), 583–540.
Singh, A. K., & Singh, P. (2012). An approach for web based GIS Route finder system. *International Journal of Advanced Research in Computer Science and Software Engineering, 2*(5), 184–189.
Swarbrooke, J. (2001). *Sustainable Tourism Management* (2nd ed.). CAB International.
Tufte, E. R. (1999). *Visual explanations: Images and quantities, evidence and narrative*. Graphics Press.
UNEP and UNWTO. (2005). *Making tourism more sustainable – A guide for policy makers*. UNEP and UNWTO, viewed 07 August, 2020, http://www.unep.fr/shared/publications/pdf/dtix0592xpa-tourismpolicyen.pdf
UNWTO. (2013). *Sustainable tourism for development guidebook* (1st ed.). UNWTO, viewed 07 August, 2020, http://www.provincia.pisa.it/uploads/2013_11_7_09_48_11.pdf
Van, A., & Wessel. (1994). *Travelling by the computer: Application of GIS in tourism and recreation*. Proc, Egis, viewed 07 August, 2020 http://www.odyssey.ursus.maine.edu/gisweb/spatdb/egis/e
Williams, P., et al. (1996). Keeping track of what really counts: Tourism resource inventory systems in British Columbia. In *International case studies in tourism planning, policy & development New York* (pp. 404–421). Wiley.
Yang, C., et al. (2004). Performance –improving techniques in web-based GIS. *International Journal of Geographical Information of Science, 19*, 319–342.

Index

G. S. Bhunia et al. (eds.), *Anthropogeomorphology*, Geography of the Physical Environment, https://doi.org/10.1007/978-3-030-77572-8

D

T

U